Physics and Mathematics of Nonlinear Phenomena

(PMNP 2013)

Journal of Physics: Conference Series Volume 482

Gallipoli, Italy
22-29 June 2013

ISBN: 978-1-63266-709-0
ISSN: 1742-6588

Printed from e-media with permission by:

Curran Associates, Inc.
57 Morehouse Lane
Red Hook, NY 12571

Some format issues inherent in the e-media version may also appear in this print version.

Printed by Curran Associates, Inc. (2014)

For permission requests, please contact the Institute of Physics
at the address below.

Institute of Physics
Dirac House, Temple Back
Bristol BS1 6BE UK

Phone: 44 1 17 929 7481
Fax: 44 1 17 920 0979

techtracking@iop.org

Additional copies of this publication are available from:

Curran Associates, Inc.
57 Morehouse Lane
Red Hook, NY 12571 USA
Phone: 845-758-0400
Fax: 845-758-2634
Email: curran@proceedings.com
Web: www.proceedings.com

TABLE OF CONTENTS

Journal of Physics: Conference Series **482** (2014) 011001 doi:10.1088/1742-6596/482/1/011001

Preface

Modern theory of nonlinear integrable equations is nowdays an important and effective tool of study for numerous nonlinear phenomena in various branches of physics from hydrodynamics and optics to quantum filed theory and gravity. It includes the study of nonlinear partial differential and discrete equations, regular and singular behaviour of their solutions, Hamitonian and bi- Hamitonian structures, their symmetries, associated deformations of algebraic and geometrical structures with applications to various models in physics and mathematics.

The PMNP 2013 conference focused on recent advances and developments in

- Continuous and discrete, classical and quantum integrable systems
- Hamiltonian, critical and geometric structures of nonlinear integrable equations
- Integrable systems in quantum field theory and matrix models
- Models of nonlinear phenomena in physics
- Applications of nonlinear integrable systems in physics

The **Scientific Committee** of the conference was formed by

- Francesco Calogero (University of Rome 'La Sapienza', Italy)
- Boris A. Dubrovin (SISSA, Italy)
- Yuji Kodama (Ohio State University, USA)
- Franco Magri (University of Milan 'Bicocca', Italy)
- Vladimir E. Zakharov (University of Arizona, USA, and Landau Institute for Theoretical Physics, Russia)

The **Organizing Committee**: Boris G. Konopelchenko, Giulio Landolfi, Luigi Martina, *Department of Mathematics and Physics "E. De Giorgi" and the Istituto Nazionale di Fisica Nucleare*, and Raffaele Vitolo, *Department of Mathematics and Physics "E. De Giorgi"*.

The conference was supported by:

- **E.I.N.S.T.E.IN Consortium**
- **MIUR** (Ministero dell'Istruzione, dell'Universita' e della Ricerca; Italy). Project: **P.R.I.N. Geometric and analytic theory of Hamiltonian systems in finite and infinite dimensions**
- **University of Salento** and **Department of Mathematics and Physics "E. De Giorgi"**, Lecce, Italy.

The conference was attended by about 140 participants. We thank all of them for their talks, posters, convival discussions and the papers presented here. All these contributed to the success of the PMNP 2013 conference.

B.G. Konopelchenko, G. Landolfi, L. Martina, R. Vitolo, *Department of Mathematics and Physics "E. De Giorgi"*

Plenary speakers

- Francesco Calogero (University of Rome 'La Sapienza', Italy), Isochronous systems are not rare;
- Boris A. Dubrovin (SISSA, Trieste, Italy), Integrable systems in small dispersion expansion
- Athanassios S. Fokas (University of Cambridge, UK), Explicit formulae for the asymptotic behavior of the unknown boundary values for integrable nonlinear evolution PDEs on the half-line
- Yuji Kodama (Ohio State University, USA), Construction of KP solitons from wave patterns
- Franco Magri (University of Milan 'Bicocca', Italy), Haantjes manifolds
- Andrey K. Pogrebkov (Steklov Mathematical Institute, Moscow, Russia), IST of KPII equation for perturbed multisoliton solutions
- Paolo M. Santini (University of Rome 'La Sapienza', Italy), Overview of results on commuting vector fields and integrable PDEs of hydrodynamic type in multidimensions
- Junkichi Satsuma (Aoyama Gakuin University, Japan), Ultradiscrete analysis on linear and nonlinear systems
- Alexey B. Shabat (Landau Institute of Theoretical Physics, RAN, Moscow, Russia), Symmetric polynomials and conservation laws
- Kanihisa Takasaki (Kyoto University, Japan), Modified melting crystal model and Ablowitz- Ladik hierarchy
- Harry Yeh (Oregon State University, USA), Laboratory Realization of Kodama's KP-Soliton Solutions
- Anton V. Zabrodin (ITEP, Moscow, Russia), Bethe ansatz and Hirota equation in integrable models
- Vladimir E. Zakharov (University of Arizona, USA, and Landau Institute for Theoretical Physics, Russia), Integrable turbulence and nonlinear stage of modulational instability

List of talks

Simonetta Abenda	*Grassmannians and multi-soliton solutions to KP-II an algebraic-geometric approach*
Igor Barashenkov	*PT-symmetry breaking and PT-symmetric solitons in a necklace of coupled optical waveguides*
Julia Bernatska	*SU(3) magnet: finite-gap integration on the lowest genus curve*
Leonid Bogdanov	*Projective differential geometry of multidimensional dispersionless integrable hierarchies.*
Sonia Boscolo	*Nonlinear pulse shaping and soliton dynamics in mode- locked fibre lasers*
Georgy I. Burde	*Alternative Sets of Multisoliton Solutions of Some Integrable Equations*
Diletta Burini	*Nonlocal Spectral Formulation of a Three Fluid System and Numerical Investigation of Traveling Wave Solutions*
Silvana De Lillo	*Traveling Waves in Elastic Rods with Arbitrary Curvature and Torsion*
Vladislav Dubrovskiy	*About linear superpositions of Nizhnik-Veselov-Novikov equation*
Anton Dzhamay	*Discrete Schlesinger Transformations and Difference Painlevé Equations*
Victor A. Eremenko	*Nonlinear interaction of the concentrated waves*
Gregorio Falqui	*Inertia and stratification in incompressible Euler fluids in 2D channels*
Renato Fedele	*Relativistic charged-particle beam motion as thermal matter wave envelope: modulational instability and soliton formation*

Journal of Physics: Conference Series **482** (2014) 011001 doi:10.1088/1742-6596/482/1/011001

Eugene Ferapontov	*On the central quadric ansatz: integrable models and Painleve reductions*
Vladimir Gerdjikov	*Riemann-Hilbert Problems with canonical normalization and families of commuting operators*
Yves Grandati	*Extended Krein-Adler theorem for state-adding Darboux-Baecklund transformations*
Tamara Grava	*Asymptotic expansion in one matrix-Models*
Petr Grinevich	*Discrete SL2 Connections and Self-Adjoint Difference Operators on the Triangulated 2-manifolds.*
John Harnad	*Symmetric polynomials, generalized Jacobi-Trudi identities and Tau functions*
Véronique Hussin	*Grassmannian sigma models and constant curvature solutions*
Rossen Ivanov	*Euler-Poincaré equations for G-strands*
Anton Izosimov	*Bi-Hamiltonian structures and stability*
Leonid Kalyakin	*Stability under white noise perturbation*
Karima Khusnutdinova	*On KdV and KP equations in non-Cartesian geometries*
Arthemy Kiselevv	*On the geometry of the BV-Laplacian*
Lyubomir Kovachev	*A New Class of Localized Solutions of the Linear and Nonlinear Wave Equations*
Joseph Krasil'shchik	*Integrable structures on deformed cotangent coverings to PDEs*
Boris Kruglikov	*Dispersionless integrable PDEs and Einstein-Weyl geometry*
Anjan Kundu	*Novel Hierarchies and Hidden Dimensions in Integrable Field Models: Theory and Application*
Grzegorz Kwiatkowski	*Quantum corrections to soliton solutions and applications*
Sergey Leble	*Pseudopotentials by Moutard transformations and differen- tial geometry*
Jyh-Hao Lee	*NLS with Quantum Potential and the Related Reaction- Diffusion Systems*
Francois Leyvraz Waltz	*A macroscopic system with undamped periodic compressional oscillations*
Sara Lombardo	*Rational solitons of wave resonant interaction models*
Natalia I. Manaenkova	*Applications of Tools of Nonlinear Time Series Analysis to Ionospheric Data*
Kenichi Maruno	*Self-adaptive mesh discrete integrable systems*
Annalisa Marzuoli	*Hamiltonian dynamics of a quantum of space: quadratic symmetry algebras and discrete orthogonal polynomials*
Vladimir Matveev	*Large parametric asymptotic of the multi-rogue waves solutions of the NLS equation and rogue wave solutions of the KP-I equation*
Oleg Mokhov	*Commuting ordinary differential operators of arbitrary rank and genus*
Antonio Moro	*Integrable viscous conservation laws and applications*
Richard Morris	*Symmetries, Conservation Laws and Reductions of a Fourth-order Wave Equation*
Folkert Mueller-Hoissen	*Soliton interactions in a tropical limit*
Vladimir Novikov	*Integrable systems in 2+1-dimensions: deformations of dispersionless limits*
Victor Novokshenov	*Double scaling limit of the generalized Hermite and Okamoto polynomials*
Maria Clara Nucci	*Spectral realization of the Riemann zeros by quantizing H = w(x)(p + lp^2/p): the Lie-Noether symmetry approach*
Oktay Pashaev	*Two Circle Theorem, q-Periodic Functions and Entangled Qubit States*

Maxim Pavlov	*Whitham Equations and Multi-dimensional Euler–Darboux–Poisson System*
Alexander Protogenov	*Topological invariants and homotopy groups of SU(3)*
Orlando Ragnisco	*TBA*
Colin Rogers	*On q-logarithmic integrable Hamiltonian reductions in spatial anisentropic gasdynamics*
Vladimir Roubtsov	*Quadratic Poisson structures on free associative algebra representations and double brackets*
Jan Sanders	*Automorphic Lie Algebras*
Christian Scimiterna	*Classification of C- and S-integrable equations defined on a three or four-points lattice*
Michael Semenov Tian-Shansky	*Poisson Structure of Difference Lax Equations, Exchange Algebras and Difference Galois Theory*
Matteo Sommacal	*Propagating two-dimensional magnetic droplets*
Piergiulio Tempesta	*Applications of Formal Groups to Nonlinear Physics and Number Theory*
Giorgio Tondo	*Generalized Lenard Chains, Separation of Variables and Superintegrability*
Dmytro Uvarov	*Towards proving integrability of AdS 4 x CP^3 superstring equations*
Tihomir Valchev	*On Quadratic Bundles Related to Hermitian Symmetric Spaces*
Pavel Winternitz	*Lie point symmetries of differential difference equations*
Chaozhong Wu	*Tau functions and Virasoro symmetries for Drinfeld-Sokolov hierarchies*
Derchyi Wu	*Integrable twisted hierarchies with D2 symmetries*
Ismet Yurdusen	*Superintegrable systems with spin*
Yair Zarmi	*Particles hidden in soliton dynamics in more than one space dimension*

List of posters

Nora Alexeeva	*Optical solitons in PT-Symmetric nonlinear couplers with gain and loss*
Anita Bagora	*Dust Fluid Tilted Bianchi Type V Cosmological Models With Electromagnetic field*
Roy Choudhury	*Building Integrable NLPDE Hierarchies with Temporally AND Spatially Varying Coefficients*
Fiammetta Conforto	*About the Quasi Steady State Approximation for a Reaction Diffusion System Describing a Chain of Irreversible Chemical Reactions*
Laurent Delisle	*Classical and Susy solutions of the BLMP equation*
Gianluca De Pasquale	*Numerical solution of the (2+1)D NLSE in a semi-strip*
Elizaveta Dubrovina	*Water dissociation in a bipolar membrane*
Valeriy Dvoeglazov	*Notoph-Graviton-Photon Coupling*
Zahida Ehsan	*Excitation of strong radiations through beating of Lasers (EMWs) and its interdisciplinary applications*
Mahmoud M. El-Borai	*Integrated semi groups and the Cauchy problem for some none linear fractional differential equation*
Gianpiero Gervino	*Nonlinear nuclear equation of state and thermodynamical instabilities in warm and dense nuclear matter*
Mariagiovanna Gianfreda	*C operators in PT -symmetric quantum mechanics*
Benoit Huard	*Classification of Integrable Davey-Stewartson Type Equations*
Abdul H. Kara	*Symmetries and conservation laws of differential and difference equations*
Nino Khatiashvili	*On some effective solutions of the Non-Linear Schrodinger Equation*
Svetlana Kordyukova	*Double boundary layer in asymptotics of nonlinear problem*
Andrea Lavagno	*Power-law distributions in protoneutron stars*
Tony Lyons	*Dark soliton and peakon solutions of the Qiao Equation*
Guido Macorini	*Towards the dressing phase in the AdS_3/CFT_2 duality*
Abdul Mannan	*Nonlinear evolution of concentric solitons in shallow water and their envelope soliton counterparts*
Luigi Martina	*Waves in the Skyrme–Faddeev model*
Phetogo Masemola	*On the invariances, conservation laws, and conserved quantities of a transformed pt symmetric coupler*
De-Xin Meng	*Interactions of solitons for a variable-coefficient generalized dispersive water-wave system*
Zoltan Néda	*Chaos on the conveyor belt*
Antonio Palacios	*Nonlinear Dynamics of a Network of Spin Torque Nano-Oscillators*
Carlo Palmisano	*Resonant driving of a quartic oscillator*
Dmitry Pikulin	*Complete bifurcation analysis of DC-DC converters under current mode control*
Sarah Post	*Contractions of quadratic algebras and limits of superintegrable systems*
Chris Radford	*Static Stellar Systems are Spherically Symmetric*
Angela Ricciardello	*Hexacopter flight modeling*
Danilo Riglioni	*Classical and quantum Superintegrable systems from coalgebra symmetry*
Marianna Ruggieri	*Optimal System of one Dimensional Subalgebras for Generalized KdV Equations*
Alexey Samokhin	*On certain classes of symmetries for delay-difference equations*

Yoshinao Shiraki	*Weighting Function for Improving the Convergence Performance of IRWLS-Based All-Pass IIR Filters*
Gabriele Sicuro	*Lie symmetries for fractional partial differential equations*
Maria Paola Speciale	*Solitary wave solutions for a perturbed coupled KdV equations derived from two-layer fuids*
Fatema Tanjia	*Relativisitic charged particle beam dynamics via the generation of wake field*
Nasser-eddine Tatar	*Hopfield neural networks with non-Lipschitz activation functions*
Vladimir Tyuterev	*Symmetry breaking and bifurcations in semi-classical and quantum analyses of molecular vibration modes*
Federica Vitale	*The inverse scattering transform for the defocusing nonlinear Schr"odinger equation with nonzero boundary condition*
Raffaele Vitolo	*On the classification of homogeneous differential-geometric third-order Poisson brackets*
Herbert Wiegel	*Kink-Antikink scattering in phi^4 and phi^6 models*
Ge Yi	*The para-Jacobi polynomials and their applications in some solvable isochronous systems*

List of other participants

Oluwaseun Mattew Adeyemi	Ibadan	NG
Matteo Beccaria	Lecce	IT
Khadija Bilal		DZ
Mahmoud el-Borai	Alexandria	EG
Steven Brett	IOP Publishing	UK
Federica Cataldini	Lecce	IT
Marta Dell'Atti	Lecce	IT
Elisaveta Dubrovina	London	UK
Joel Ekstrand	Trieste	IT
Mohammed Amin Ghezzar	Mostaganem	DZ
Francesco Giglio	Lecce	IT
Davide Guzzetti	Trieste	IT
David Janjgava	Tblisi	GE
Robert Jenkins	Trieste	IT
Gianni Manno	Jena	DE
Dario Merzi	Trieste	IT
C.Radhakrishnan Nair	Trivandrum	IN
Barbara Prinari	Lecce	IT
Rakeshwar Purohit	Udaipur	IN
Simone Quaranta	Lecce	IT
Carmen Ropchan	Edmonton	CA
Kamel Tahri	Tlemcen	DZ
Daniele Valeri	Trieste	IT

Journal of Physics: Conference Series **482** (2014) 011002 doi:10.1088/1742-6596/482/1/011002

Peer review statement

All papers published in this volume of *Journal of Physics: Conference Series* have been peer reviewed through processes administered by the proceedings Editors. Reviews were conducted by expert referees to the professional and scientific standards expected of a proceedings journal published by IOP Publishing.

Symmetric coupling of angular momenta, quadratic algebras and discrete polynomials[1]

V Aquilanti[†], **D Marinelli**[‡] **and A Marzuoli**[§]

[†] Dipartimento di Chimica, Biologia e Biotecnologie, via Elce di Sotto 8, Università di Perugia, 06123 Perugia, I

[‡] Dipartimento di Fisica, Università di Pavia & INFN Sezione di Pavia, via Bassi 6, 27100 Pavia, I

[§] Dipartimento di Matematica 'F Casorati', Università di Pavia, via Ferrata 1 & INFN Sezione di Pavia, via Bassi 6, 27100 Pavia, I

E-mail: [†]`vincenzoaquilanti@yahoo.it`, [‡]`dimitri.marinelli@pv.infn.it`, [§]`annalisa.marzuoli@pv.infn.it`

Abstract. Eigenvalues and eigenfunctions of the volume operator, associated with the symmetric coupling of three $SU(2)$ angular momentum operators, can be analyzed on the basis of a discrete Schrödinger–like equation which provides a semiclassical Hamiltonian picture of the evolution of a 'quantum of space', as shown by the authors in [1]. Emphasis is given here to the formalization in terms of a quadratic symmetry algebra and its automorphism group. This view is related to the Askey scheme, the hierarchical structure which includes all hypergeometric polynomials of one (discrete or continuous) variable. Key tool for this comparative analysis is the duality operation defined on the generators of the quadratic algebra and suitably extended to the various families of overlap functions (generalized recoupling coefficients). These families, recognized as lying at the top level of the Askey scheme, are classified and a few limiting cases are addressed.

1. Introduction and a brief review

In [1] a family of orthogonal polynomials has been introduced based on a three–term recursion relationship which plays the role of a discrete Schrödinger equation describing the action of a 'volume' operator. This operator occurs in the symmetric treatment of the quantum few–body problem as well as in spin–network modeling of a quantum of space, as pointed out originally in [2]. In this section a short introduction to the necessary mathematical background will be given, together with a summary of a few significant results found by the authors in [1]. Improved insights into algebraic and analytical aspects of the subject will be provided in the next sections.

The theory of (re)coupling of eigenstates of three $SU(2)$ angular momentum operators $\mathbf{J}_1$, $\mathbf{J}_2$, $\mathbf{J}_3$ to states of sharp total angular momentum $\mathbf{J}_4$ (with projection J_0 along the quantization axis) is usually carried out in the setting of 'binary couplings' (see [3], Topic 12 and original references therein). Referring to the ordered triple as above, the admissible schemes are $\mathbf{J}_1 + \mathbf{J}_2 = \mathbf{J}_{12}$; $\mathbf{J}_{12} + \mathbf{J}_3 = \mathbf{J}$ and $\mathbf{J}_2 + \mathbf{J}_3 = \mathbf{J}_{23}$; $\mathbf{J}_1 + \mathbf{J}_{23} = \mathbf{J}_4$, respectively. The corresponding

[1] Talk delivered by Annalisa Marzuoli

eigenvectors are denoted

$$|j_{12}>:=|(j_1j_2)j_{12}\,j_3\,j_4m> \text{ and } |j_{23}>:=|j_1(j_2j_3)j_{23}\,j_4m'>, \tag{1}$$

where small js are labelings of the eigenvalues associated with the angular momentum operators (*e.g.* $\mathbf{J}_1^2|j_{12}>=j_1(j_1+1)\,|j_{12}>$, $\mathbf{J}_{12}^2|j_{12}>=j_{12}(j_{12}+1)\,|j_{12}>$) running over $\{0,1/2,1,3/2,\dots\}$, in $\hbar$ units, and m (m') is the eigenvalue of J_0 with $-j_4\le m,m'\le j_4$ in integer steps. Thus the ket vectors above belong to Hilbert spaces representing simultaneous eigenspaces of the two, partially overlapping sets of commuting operators $\mathbf{J}_1^2$, $\mathbf{J}_2^2$, $\mathbf{J}_3^2$, $\mathbf{J}_4^2$, J_0 and $\mathbf{J}_{12}^2$ ($\mathbf{J}_{23}^2$, respectively). The Racah–Wigner $6j$ symbol is the unitary (actually orthogonal by Condon-Shortley convention) transformation relating the two sets (1) according to

$$<j_{23}\,|\,j_{12}>:=(-1)^{\Phi}\left[(2j_{12}+1)\,(2j_{23}+1)\right]^{1/2}\begin{Bmatrix} j_1 & j_2 & j_{12}\\ j_3 & j_4 & j_{23}\end{Bmatrix}, \tag{2}$$

where $\Phi\equiv j_1+j_2+j_3$ and the weights $(2j+1)$ are the dimensions of the spin–j representations of $SU(2)$ which provide the standard normalization of such a 'recoupling coefficient' as encoded in the shorthand notation in the left–hand side. Therefore a basis transform is simply written as $|j_{23}>=\sum_{j'_{12}}<j_{23}\,|\,j'_{12}>\,|j'_{12}>$ while the inverse one is achieved by the transpose $<j_{12}\,|\,j_{23}>$ (all non–null matrix elements obey the selection rule $m=m'$ by Wigner–Eckart theorem). Recall in passing that the $6j$ symbol in (2) encodes naturally the symmetry of an Euclidean tetrahedron, a fact which is at the basis of the huge amount of literature about 'spin–network' models for 3–dimensional discretized quantum gravity and quantum computing flourished in the past two decades (see [4] and references therein).

The treatment of the 'symmetric' coupling scheme for the addition of three $SU(2)$ angular momenta $\mathbf{J}_1$, $\mathbf{J}_2$, $\mathbf{J}_3$ to give $\mathbf{J}_4$ (with projection J_0) is characterized in terms of a 'volume' operator $K=\mathbf{J}_1\cdot\,\mathbf{J}_2\times\mathbf{J}_3$. Unlike what happens with binary coupling schemes, the $\mathbf{J}$s appear now to be all on the same footing, indicating that the volume operator can be thought of as acting democratically on either a composite system of four objects with vanishing total angular momentum (a configuration that can be associated with a not necessarily planar quadrilateral vector diagram $\mathbf{J_1}+\mathbf{J_2}+\mathbf{J_3}+\mathbf{J_4}=0$), or a system of three objects with total angular momentum $\mathbf{J}_4$ (see again [3], Topic 12, last section, and original references therein). The present scheme is characterized by the six commuting Hermitian operators $\mathbf{J}_1^2$, $\mathbf{J}_2^2$, $\mathbf{J}_3^2$, K, $\mathbf{J}^2$ and J_0, so that eigenvectors and eigenvalues of K are given formally (consistently with the notation used in (1)) as

$$|k>:=|(j_1j_2j_3)\,k\,j_4m> \quad\text{with}\quad K\,|k>=\lambda_k\,|k>\,. \tag{3}$$

Eigenvalues and matrix elements of K are naturally found within an imaginary antisymmetric representation based on a three–terms recursion relationship [2], which can be turned into a real, time–independent Schrödinger equation which governs the dynamics of a 'quantum of space' as a function of a discrete variable denoted ℓ (see below). This has been achieved in [1] (which can be referred to also for a complete list of previous and related papers) where the introduction of discrete, potential–like functions highlights the surprising crucial role of 'hidden' symmetries, first discovered by Regge [5] for the $6j$ symbols. The Schrödinger equation is discretized with respect to a lattice variable given by the label of the operator $\mathbf{J}_{12}$ which characterizes the first of the binary schemes in (1) and reads

$$\lambda_k\,\Psi_\ell^{(k)}+\alpha_{\ell+1}\,\Psi_{\ell+1}^{(k)}+\alpha_\ell\,\Psi_{\ell-1}^{(k)}=0 \text{ with } \ell\equiv j_{12}\in\{j_{12}^{\min},\,j_{12}^{\min}+1,\,\dots,\,j_{12}^{\max}\}, \tag{4}$$

where the matrix elements α_ℓ are expressed in terms of geometric quantities, namely

$$\alpha_\ell=F(\ell;j_1+1/2;j_2+1/2)F(\ell;j_3+1/2;j_4+1/2)\,/\,[(2\ell+1)(2\ell-1)]^{1/2}\,. \tag{5}$$

Journal of Physics: Conference Series **482** (2014) 012001 doi:10.1088/1742-6596/482/1/012001

Here $F(A,B,C)) = \frac{1}{4}[(A+B+C)\,(-A+B+C)\,(A-B+C)\,(A+B-C)]^{\frac{1}{2}}$ is the Heron's formula for the area of a triangle with side lengths A, B and C. Thus α_ℓ is proportional to the product of the areas of the two triangles sharing the side of length ℓ and forming a quadrilateral of sides $j_1+\frac{1}{2}$, $j_2+\frac{1}{2}$, $j_3+\frac{1}{2}$ and $j_4+\frac{1}{2}$. Such a parameter quadrilateral, together with its Regge–conjugate (see below), is the guiding tool of the combinatorial and geometric analysis, in the asymptotic limit, of the Hamiltonian dynamics governing both tetrahedral and 'fluttery' quadrilateral configurations, see sections 2 and 3 of [1] for more details.

2. Quadratic symmetry algebras

Following [6, 7], the quantum version of a classical dynamical algebra associated with a pair of 'mutually integrable' dynamical variables calls into play a triple K_1, K_2, K_3 of linear operators acting on a (suitably defined) Hilbert space with $K_{1,2}$ Hermitian and algebraically independent and $K_3 := [K_1, K_2]$ anti–Hermitian. The request that these generators do fulfill the Jacobi identity constrains the fundamental commutation relations to be of the form ($\{\,,\,\}$ is the anticommutator)

$$\begin{aligned} [K_1, K_2] &= K_3 \\ [K_2, K_3] &= 2R\,K_2K_1K_2 + A_1\,\{K_1, K_2\} + A_2\,K_2^2 + C_1\,K_1 + D\,K_2 + G_1 \\ [K_3, K_1] &= 2R\,K_1K_2K_1 + A_1\,K_1^2 + A_2\,\{K_1, K_2\} + C_2\,K_2 + D\,K_1 + G_2\,, \end{aligned} \qquad (6)$$

where $R, A_{1,2}, C_{1,2}, D, G_{1,2}$ are real parameters (the structure constants) and the right–hand sides of the last two relations contain only Hermitian terms. Such a kind of algebraic structures was actually introduced by Sklyanin [8] and they are called 'quadratic' algebras for the obvious reason that the commutators (Poisson brackets in the classical cases) are combinations of quadratic (and linear) terms in each of the generators. Mutual integrability is a sharper requirement with respect to the original formulation, and amounts to look at the symmetry algebra as a dynamical one –where K_1 is a constant of the motion for K_2 taken as the Hamiltonian operator, as well as the other way around. Further improvements in the the study of classical, quantum and q-deformed symmetries along these lines have been provided over the past few decades by a number of authors. Often the admissible structures associated with (2) and listed in the table below [6] are referred to as 'Zhedanov's algebras' in the literature. Note that for completeness the last line includes the two 'standard' Lie algebras on three generators (whose commutation relations are by definition linear).

Classification of quadratic algebras

	R	A_1	A_2	$C\,\&\,D$
AW(3) (*Askey–Wilson*)	*	*	*	*
R(3) (*Racah*)	0	*	*	*
H(3) (*Hahn*)	0	0	*	*
J(3) (*Jacobi*)	0	0	*	0
Lie algebras: $su(2)$, $su(1,1)$	0	0	0	*

The denominations of the algebras, Askey–Wilson, Racah, ..., are strongly reminiscent of the Askey–Wilson scheme of hypergeometric orthogonal polynomials of one (continuous or discrete) variable [9]. This is not accidental: rather, this remark turns out to be crucial in order to recognize the deep connection between algebraic symmetries of (quantum) systems and special function theory in a quite straightforward way. Indeed the 'overlap functions' stemming from the analysis of the eigenvalue problems for the operators K_1, K_2, K_3 which generate the quadratic algebras are, under mild conditions, orthogonal families of Wilson, Racah, Hanh, Jacobi, ..., Hermite polynomials. In what follows an account of a few technical details is given for the case of the Racah algebra $\mathbf{R(3)}$ which corresponds to set $R = 0$ in (2).

Suppose that the Hermitian operators K_1 and K_2 –defined on a separable Hilbert space and possibly depending on a same (finite) set of real parameters– are both ladder operators, namely possess discrete, evenly–spaced spectra, and start considering the eigenvalue problem for K_1

$$K_1\,\psi_p \;=\; \chi_p\,\psi_p\,,\quad p = 0, 1, 2, \dots \text{ with } \chi_{p+1} = \chi_p + 1\,. \tag{7}$$

Then it can be easily shown that the operator K_2 is tridiagonal in this basis

$$K_2\,\psi_p \;=\; \mathfrak{a}_{p+1}\,\psi_{p+1} \;+\; \mathfrak{a}_p\,\psi_{p-1} \;+\; \mathfrak{b}_p\,\psi_p \tag{8}$$

and, similarly, by exchanging the role of K_1 and K_2, one would get

$$K_2\,\phi_s \;=\; \mu_s\,\phi_s,\quad s = 0, 1, 2, \dots \text{ with } \mu_{s+1} = \mu_s + 1 \tag{9}$$

and

$$K_1\,\phi_s \;=\; \mathfrak{c}_{s+1}\,\phi_{s+1} \;+\; \mathfrak{c}_s\,\phi_{s-1} \;+\; \mathfrak{d}_s\,\phi_s\,. \tag{10}$$

The (real) matrix coefficients $\mathfrak{a}, \mathfrak{b}, \mathfrak{c}, \mathfrak{d}$ can be evaluated explicitly in terms of the commutation relations (2) and contain also the parameters which the operators may depend on (such parameters are dropped in the present simplified treatment aimed to point out the overall structural properties). Once chosen suitable normalizations for the two sets of eigenbases (7) and (9), it is possible to introduce two families of overlap functions by resorting to the Dirac braket convention (in which for instance $< x|\psi >$ stands for the eigenfunction $|\psi >$ of a system in the position representation)

$$< \phi_s\,|\psi_p > \equiv < s\,|\psi_p > \equiv < s\,|p > \quad \text{and} \quad < \psi_p\,|\phi_s > \equiv < p\,|\phi_s > \equiv < p\,|s > \tag{11}$$

which are both hypergeometric orthogonal polynomials of one discrete variable (the spectral parameter μ_s and χ_p respectively) to be identified, up to suitable rearrangements of the hidden parameters, with the Racah polynomial on the top of the Askey scheme [9].

In the K_1–eigenbasis the operator K_3 satisfies

$$K_3\,\psi_p \;=\; (\chi_{p+1} - \chi_p)\,\mathfrak{a}_{p+1}\,\psi_{p+1} \;-\; (\chi_p - \chi_{p-1})\,\mathfrak{a}_p\,\psi_{p-1}\,, \tag{12}$$

where eigenvalues χ and matrix elements $\mathfrak{a}$ are iteratively evaluated from (7) and (8). K_3 has a discrete, in general not evenly–spaced spectrum found as a solution of

$$K_3\,\varphi_n \;=\; \nu_n\,\varphi_n,\quad n = 0, 1, 2, \dots\,. \tag{13}$$

It is worth noting that in general the diagonalization of K_3 cannot be carried out analytically, except in a few cases in which at least the lowest eigenvalues turn out to be representable in closed algebraic forms. The associated families of (normalized) overlap functions are denoted

$$< \varphi_n\,|\psi_p > \equiv < n|p > \quad \text{and} \quad < \psi_p\,|\varphi_n > \equiv < p|n > \quad (n = 0, 1, 2, \dots; p = 0, 1, 2, \dots) \tag{14}$$

and can be shown to be orthogonal (on different suitably defined lattices), each depending on one discrete variable, but in principle they might not be included into the Askey scheme.

Similarly, other two families of (normalized) overlap functions associated with the pair K_2, K_3 can be defined by notation consistency as

$$< \varphi_n \, | \phi_s > \equiv < n|s > \quad \text{and} \quad < \phi_s \, | \varphi_n > \equiv < s|n > \quad (n = 0, 1, 2, \ldots ; s = 0, 1, 2, \ldots). \qquad (15)$$

A crucial feature of the Racah algebra $\mathbf{R(3)}$ and associated overlap functions is the duality property. It relies on the following transformation of the generators

$$K_1 \leftrightarrows K_2 \; ; \quad K_3 \to -K_3 \qquad (16)$$

which can be easily shown to represent an automorphism of the Racah algebra $\mathbf{R(3)}$. The notion of duality is extended to (all of) the sets of overlap functions introduced above. More precisely

i) Under the automorphism (16) the discrete, evenly–spaced variables of the two hypergeometric families of overlap functions associated with K_1, K_2 given in (11) and their degrees as polynomials are interchanged. Since in the present case the operator K_3 is not called into play, the stronger property of 'self–duality' of these families holds true: both of them are recognized as Racah polynomials, as already mentioned above.

ii) Referring to the families in (14), under the automorphism (16) the discrete, not evenly–spaced spectral variable ν_n of the first family, which is orthogonal on the evenly–spaced lattice $p = 0, 1, 2, \ldots$, is turned into the second family, where the variable is p and the polynomial degree is given in terms of the labels $n = 0, 1, 2, \ldots$ of the eigenvalues of K_3. A similar property is shared by the families associated with the pair K_2, K_3 given in (15).

More details on the nature of the automorphism group and on the statements about the overlap functions will be reported in the next section when dealing with a specific 'realization' of the Racah algebra.

3. Generalized recoupling theory, Regge symmetry and duality

The realization of the Racah algebra $\mathbf{R(3)}$ within the setting of generalized $SU(2)$ recoupling theory was actually the issue addressed originally in [7] which has inspired further work on quadratic algebras. Combining the definitions and notation of section 2 with those of section 1 it is straightforward to recognize the following correspondence

$$\begin{aligned} K_1 &= \mathbf{J}_{12}^2 \, ; \quad K_2 \; = \; \mathbf{J}_{23}^2 \, ; \\ K_3 &= [\mathbf{J}_{12}^2, \mathbf{J}_{23}^2] \; = \; -4i \, \mathbf{J}_1 \cdot (\mathbf{J}_2 \times \mathbf{J}_3) \; \equiv -4i \, K \end{aligned} \qquad (17)$$

between the abstract ordered set of operators K_1, K_2, K_3 and its realization as $\mathbf{J}_{12}^2$, $\mathbf{J}_{23}^2$, K. The next step would consist in associating eigenvalue equations and three–term recursion relations of the abstract approach with their realizations in generalized quantum (re)coupling theory. Here we do not enter into much details about this matter since the translation of (8) based on the pair K_1, K_2 represents the three–term recursion relation for the $6j$ coefficient in disguise (see *e.g.* [10]). The analysis for the pair K_1, K_3 which gives the abstract three–term relation as written in (12) is examined in details in [2] (and references therein) while its symmetrized counterpart is nothing but the discretized Schrödinger–like equation displayed already in (4) [1].

Focusing on the specific issue regarding the families of solutions of such relationships, one would directly be lead to establish the correspondence

$$\text{overlap functions} \;\longrightarrow\; \text{binary and symmetric recoupling coefficients}\,, \qquad (18)$$

where the arrow stands for the specific realization (17) of $\mathbf{R(3)}$. To achieve this goal in a transparent and consistent way a few more steps are needed, the first one of which consists in establishing suitable notations for all of the recoupling coefficients. The $6j$ symbol in (2) and the functions $\Psi_\ell^{(k)}$ in (4) are thus denoted and defined respectively as

$$< j_{23}\,|\,j_{12} > \equiv < \tilde{\ell}\,|\,\ell > \quad \text{and} \quad \Psi_\ell^{(k)} \; := < \ell\,|\,k > \,. \tag{19}$$

Actually this is not a mere question of notation, since in this way the objects $< \bullet|\circ >$ may reveal their 'double' meaning as *i)* quantum mechanical transition amplitudes, namely the square modulus $|<\bullet|\circ>|^2$ is the probability that a system, prepared in the state $|\circ>$, be measured to be in the state $|\bullet>$; *ii)* eigenfunctions of the operator whose quantum number is in $|\circ>$ in the representation labeled by the eigenvalue of the other operator, namely through the projection onto $<\bullet|$. The latter interpretation will be under focus in what follows and more details about the correspondence (18) can be worked out by introducing explicitly the (so far ignored) parameters of the problem. Upon replacement of the original (ordered) set of labeling of the four angular momenta forming a quadrilateral according to

$$(\,j_1, j_2, j_3, j_4\,) \;\mapsto\; (\,a, b, c, d\,)\,, \tag{20}$$

the functionals are rewritten as

$$< \tilde{\ell}\,|\,\ell > \,(a,b,c,d) \,\propto \begin{Bmatrix} a & b & \ell \\ c & d & \tilde{\ell} \end{Bmatrix} \quad \text{and} \quad \Psi_\ell^{(k)}(a,b,c,d) \, = < \ell\,|\,k > \,(a,b,c,d)\,. \tag{21}$$

Recall that geometrically the first functional is associated with a tetrahedron (ℓ and $\tilde{\ell}$ being a pair of opposite edges) and the second one to a quadrilateral (actually two triangles hinged by one of its diagonal, ℓ or $\tilde{\ell}$) bounding, so to speak, a portion of volume of amount λ_k, the eigenvalue of the volume operator given in (3). In order to select in a convenient way the Hilbert space on which the volume operator acts and all the functionals above can be defined consistently, the role of Regge symmetries, originally introduced for the $6j$ [5], is crucial. Such symmetries in their original formulation are recognized as functional relations on the arguments (namely they cannot be derived by interchanging the $6j$ arguments as happens for the so–called 'classical' or tetrahedral symmetries) and read

$$\begin{Bmatrix} a & b & \ell \\ c & d & \tilde{\ell} \end{Bmatrix} = \begin{Bmatrix} s-a & s-b & \ell \\ s-c & s-d & \tilde{\ell} \end{Bmatrix} := \begin{Bmatrix} a' & b' & \ell \\ c' & d' & \tilde{\ell} \end{Bmatrix}, \tag{22}$$

where $s = (a+b+c+d)/2$ is the semi–perimeter of the parameter quadrilateral and in the last equality the new set (a', b', c', d') is defined. It can be checked that the total number of classical and Regge symmetries is 144, which equals the order of the product permutation group $S_4 \times S_3$.

Denoting by a the smallest value among the eight parameters $(a, b, c, d, a', b', c', d')$, it can be shown that a consistent ordering of the other parameters compatible with all the due inequalities is given by $\{\, a \le b \le d \;;\; d - (b-a) \le c \le d + (b-a)\,\}$. This sort of gauge fixing implies that the whole problem becomes finite–dimensional and workable out for each fixed values of the parameters $(a, b, c, d) \in \mathbb{R}^4$. Moreover: *i)* the tetrahedron $< \tilde{\ell}\,|\,\ell > (a,b,c,d)$ can be chosen as the reference one, calling $< \tilde{\ell}\,|\,\ell > (a', b', c', d')$ its Regge–conjugate; *ii)* the same thing holds for the quadrilateral denoted $< \ell\,|\,k > (a,b,c,d)$ and its conjugate $< \ell\,|\,k > (a', b', c', d')$. More technical details about this specific parametrization and the denomination Regge–'conjugate' (as well as the proof that the volume operators and all quantities in its three–term recursion relation (4) are Regge–invariant) can be found in [11] and [1] respectively.

Coming back to the statement regarding the correspondence (18), the remarks above should have made clear that Regge symmetry is strictly related to the duality property of the Racah algebra discussed at the end of section 2. Note that in [7] it had been already recognized that (classical + Regge) symmetries do have the group structure given by $S_4 \times S_3$, to be identified with the automorphism group of the Racah algebra.

4. Classification of discrete polynomial families

In this section the focus will be on interconnections among the families of discrete orthogonal polynomials in view of the formalization presented in section 2 and summarized there in items **i)** and **ii)**. This analysis –not addressed elsewhere to our knowledge– is just sketched here, leaving aside a number of technical details that can be found in [12]. The various cases, together with the most significant properties of each family, are summarized in the following table.

Finite families of discrete orthogonal polynomials [(a, b, c, d) **fixed**]

#	family	orthogonality on lattice	eigenvalue (related to the variable)	degree related to
I.A	$<\tilde{\ell}\,\vert\,\ell>$	$\sum_{\tilde{\ell}} \overline{<\tilde{\ell}\,\vert\,\ell'>} <\tilde{\ell}\,\vert\,\ell> = \delta_{\ell'\ell}$	$\ell(\ell+1)$	$\tilde{\ell}$
I.B	$<\ell\,\vert\,\tilde{\ell}>$	$\sum_{\ell} \overline{<\ell\,\vert\,\tilde{\ell}'>} <\ell\,\vert\,\tilde{\ell}> = \delta_{\tilde{\ell}'\tilde{\ell}}$	$\tilde{\ell}(\tilde{\ell}+1)$	ℓ
II.A	$<\ell\,\vert\,k>$	$\sum_{\ell} \overline{<\ell\,\vert\,k'>} <\ell\,\vert k> = \delta_{k'k}$	λ_k	ℓ
II.B	$<k\,\vert\ell>$	$\sum_{k} \overline{<k\,\vert\,\ell'>} <k\,\vert\ell> = \delta_{\ell'\ell}$	$\ell(\ell+1)$	k
III.A	$<\tilde{\ell}\,\vert\,k>$	$\sum_{\tilde{\ell}} \overline{<\tilde{\ell}\,\vert\,k'>} <\tilde{\ell}\,\vert k> = \delta_{k'k}$	λ_k	$\tilde{\ell}$
III.B	$<k\,\vert\tilde{\ell}>$	$\sum_{k} \overline{<k\,\vert\,\tilde{\ell}'>} <k\,\vert\tilde{\ell}> = \delta_{\tilde{\ell}'\tilde{\ell}}$	$\tilde{\ell}(\tilde{\ell}+1)$	k

Comparing the notations adopted here –the bar stands for complex conjugation or simply transposition in the real cases– with those of section 2, it is straightforward to recognized that the classes **I**, **II** and **III** are in correspondence with the overlap functions in (11), (14) and (15) (restricted to finite sets by suitable choices of the omitted parameters), respectively.

Looking at the family **IA**, observe that $\overline{<\tilde{\ell}\,\vert\,\ell'>} := <\ell'\,\vert\,\tilde{\ell}> = <\tilde{\ell}\,\vert\,\ell'>$ by the convention chosen for $6j$ symbols in (2) (and similarly for **IB**). Thus 'self–duality' relations for class **I** read either way

$$\sum_{\tilde{\ell}} <\ell'\,\vert\tilde{\ell}><\tilde{\ell}\,\vert\,\ell> = \delta_{\ell'\,\ell} \quad \text{and} \quad \sum_{\ell} <\tilde{\ell}'\,\vert\,\ell><\ell\,\vert\,\tilde{\ell}> = \delta_{\tilde{\ell}'\,\tilde{\ell}}\,, \tag{23}$$

once fulfilled the completeness relations $\Sigma\,\vert\,\tilde{\ell}><\tilde{\ell}\,\vert = \mathbb{I}$ and $\Sigma\,\vert\,\ell><\ell\,\vert = \mathbb{I}$ for the binary coupled eigenbases introduced in (1). Note that the operators associated with class **I** ($\mathbf{J}^2_{12}$ and $\mathbf{J}^2_{23}$) represent a 'Leonard pair' so that the associated overlap functions (recoupling coefficients) are necessarily hypergeometric of Racah type [13]. More generally, in connection with the analysis of the other classes, a stringent result holds true: any *finite* system of orthogonal polynomials whose dual is a finite system of orthogonal polynomials must be (possibly q–deformed) Racah or one of its limiting cases which constitute finite systems (refer to [14] for a modern monograph on hypergeometric polynomials in the Askey–Wilson scheme). Indeed here all of the families are consistently defined, for fixed parameters (a, b, c, d), as finite sets (recall the choice on the ordering discussed in connection with Regge symmetry) but the recognition of classes **II** and **III** as belonging to the Askey scheme is certainly not straightforward. (More precisely, the reduction process to specific hypergeometric functions of type ${}_4F_3$ would require to find out a closed algebraic form for the sets of eigenvalues of the volume operator for given parameters, a task not yet accomplished.)

For what concerns duality within class **II**, a first remark is about the bar operation: $\overline{<\ell\,|\,k>}$ is $< k\,|\ell >$, but the latter, unlike what happens for the $6j$, is not necessarily equal to $<\ell\,|\,k>$ because this property actually depends on the volume operator K being Hermitian (imaginary antisymmetric) [2] or real symmetric (see [1] also for plots of the family of eigenfunctions $<\ell\,|\,k>$). Anyway, both options can be included through a suitable notation into the duality relations

$$\sum_{\ell} < k'\,|\ell >< \ell\,|\,k > = \delta_{k'\,k} \quad \text{and} \quad \sum_{k} < \ell'\,|\,k >< k\,|\,\ell > = \pm\delta_{\ell'\,\ell} \tag{24}$$

according to the choiche of the representation of K. Duality relations in class **III** are similar to (24), with $\tilde{\ell}$ taking the role of ℓ.

To conclude this general overview on duality relationships, a further remarkable property –transversal with respect to the classes– has to be mentioned, namely

$$\sum_{k} < \tilde{\ell}|\,k >< k\,|\,\ell > = = \pm < \tilde{\ell}\,|\,\ell > . \tag{25}$$

Such a 'triangular relation' (and the other ones that can be derived by using the properties of the single classes given above) closely resembles the Racah identity satisfied by three $6j$ symbols and might be used also to explore a formalization of the whole subject within the general scheme of tensor categories.

5. Limiting cases

The issue of asymptotic (semiclassical) limits of angular momentum functions is of continuous interest in many fields, ranging from special function theory [14] to applied quantum mechanics [15]. Here just a few remarks concerning two limiting cases of families **II.A** and **III.B** are sketched.

The reference model of asymptotics is the well–know limit of the $6j$ symbol for three large entries (see [16, 10]), $6j \to 3j$, where the latter is the Wigner symbol, the symmetrized version of a Clebsch–Gordan coefficient. The counterpart of this operation in the Askey scheme is achieved by moving one step downwards from top, namely from ${}_4F_3$ (Racah) to ${}_3F_2$ (Hahn and dual Hahn) hypergeometric families.

A new change of notation is needed which consists in restoring the string (j_1, j_2, j_3, j_4) for the parameters (see (20)) and in writing down as an array the functions in (21) (equivalently, in family **II.A**) according to

$$\Psi_{\ell}^{(k)}(j_1, j_2, j_3, j_4) \;\to\; \begin{Bmatrix} j_1 & j_2 & | & \ell \\ j_3 & j_4 & | & \lambda_k \end{Bmatrix}, \tag{26}$$

where the vertical bars in front of the last column of this symbol indicate that not all of the entries are constrained by standard triangular inequalities, as happens for the $6j$. To address any limit in which (some of) the arguments of the symbols become large –a fact that implies that all of the arguments can be 'running'– a convenient notation is to substitute capital to small letters. Thus the formal limiting process for the symbol in (26) when the arguments of the lower row become large can be displayed as a generalized $3j$ coefficient, denoted $3\mathfrak{j}$, related in turn to a generalized dual Hahn polynomial; schematically

$$\begin{Bmatrix} j_1 & j_2 & | & \ell \\ J_3 & J_4 & | & \Lambda_k \end{Bmatrix} \rightarrowtail \begin{pmatrix} j_1 & j_2 & | & \ell \\ J_4-\Lambda_k & \Lambda_k-J_3 & | & J_3-J_4 \end{pmatrix} \leftrightarrow 3\mathfrak{j} \text{ (dual Hahn family)}\,. \qquad (27)$$

On applying a similar procedure to family **III.B**, and denoting $\tilde{L}$ the previous generic argument $\tilde{\ell}$ (playing the role of j_{23}), the resulting correspondence would read

$$\begin{Bmatrix} j_1 & j_2 & | & \lambda_k \\ J_3 & J_4 & | & \tilde{L} \end{Bmatrix} \rightarrowtail \begin{pmatrix} j_1 & j_2 & | & \lambda_k \\ J_4-\tilde{L} & \tilde{L}-J_3 & | & J_3-J_4 \end{pmatrix} \leftrightarrow 3\mathfrak{j} \text{ (Hahn family)}\,. \qquad (28)$$

A few comments on these results are in order, leaving aside a more careful analysis and most technical details reported in [12]. As already noticed, the symbols in round brackets on the right–hand sides of (27) and (28) are generalized counterparts of $3j$ coefficients, the arguments in the lower row being interpreted as magnetic quantum numbers. They actually share with standard $3j$s a suitable formulation of Regge symmetry [17] and their properties as orthogonal families are inferred from three–term recursion relationships. The latter can in turn be derived as limits of the three–term recursions at the upper level (in particular, the relation for (27) can be quite easily worked out). The motivation for associating dual Hahn and Hahn families respectively is related with the specific lattices these three–term recursion relations are defined on. Thus it is found that the relation for (27) mimics the behavior of the relation of a $3j$ on a quadratic lattice ($\ell(\ell+1)$), so that it is functionally similar to the standard dual Hahn polynomial family. Conversely, the relation for (28) mimics the behavior of the relation of a $3j$ on a linear lattice (given by scaling the quantum number $m \equiv J_3 - J_4$) and thus these functions represent counterparts of the Hahn polynomial family.

6. Outlook

Further developments can be addressed in parallel, from algebraic–analytical and geometric viewpoints. A schematic list of ongoing works (and still open questions) follows:

- improved interpretations of Regge symmetries on the geometric (*scissor–congruent tetrahedra* [18]) and algebraic (*quaternionic reparametrization* [12]) sides;
- convolution rules for overlap functions (specifically, symmetric recoupling coefficients) of Racah algebra;
- composition rules of collections of quadrilaterals able to provide new classes of integrable quantum systems to be associated with extended quantum geometries;
- q–deformed extensions and limiting cases of the dual sets of orthogonal polynomials also in view of applications in quantum chemistry.

In particular, a systematic study of limiting procedures –to be carried out on recurrence relations, on families of polynomials and possibly directly on the defining relations (2) of the underlying quadratic algebras– seems particularly promising also in view of recent analytical and numerical work on strictly related issues [19, 20, 21, 22].

Acknowledgments

D M and A M acknowledge partial support from PRIN 2010-2011 *Geometrical and analytical theories of finite and infinite dimensional Hamiltonian systems.*

References

[1] Aquilanti V, Marinelli D and Marzuoli A 2013 Hamiltonian dynamics of a quantum of space: hidden symmetries and spectrum of the volume operator, and discrete orthogonal polynomials *J. Phys. A: Math. Theor.* **46** 175303

[2] Carbone G, Carfora M and Marzuoli A 2002 Quantum states of elementary three-geometry *Class. Quantum Grav.* **19** 3761

[3] Biedenharn L C and Louck J D 1981 *The Racah–Wigner Algebra in Quantum Theory* (Encyclopedia of Mathematics and its Applications Vol 9) ed G-C Rota (Reading, MA: Addison–Wesley)

[4] Carfora M, Marzuoli A and Rasetti M 2009 Quantum tetrahedra *J. Phys. Chem. A* **113** 15376

[5] Regge T 1959 Symmetry properties of Racah's coefficients *Nuovo Cimento* **11** 116

[6] Granovskii Ya I, Lutzenko I M and Zhedanov A S 1992 Mutual integrability, quadratic algebras, and dynamical symmetry *Ann. Phys.* **217** 1

[7] Granovskii Ya I and Zhedanov A S 1988 Nature of the symmetry group of the 6j-symbol *Sov. Phys. JETP* **67** 1982

[8] Sklyanin E K 1982 Some algebraic structures connected with the Yang–Baxter equation *Funct. Anal.* **16** 263

[9] Askey R 1975 *Ortogonal Polynomials and Special Functions* (Philadelphia, PE: SIAM)

[10] Varshalovich D A, Moskalev A N and Khersonskii V K 1988 *Quantum Theory of Angular Momentum* (Singapore: World Scientific)

[11] Aquilanti V, Haggard H M, Hedeman A, Jeevanjee N, Littlejohn R G and Liang Y 2012 Semiclassical mechanics of the Wigner 6j-symbol *J. Phys. A: Math. Theor.* **45** 065209

[12] Marinelli D 2013 *Single and Collective Dynamics of Discretized Geometries* PhD Thesis (Pavia: University Press)

[13] Leonard D A 1982 Orthogonal polynomials, duality and association schemes *SIAM J. Math. Analys.* **13** 656

[14] Koekoek R, Lesky P A and Swarttouw R F 2010 *Hypergeometric Orthogonal Polynomials and Their q-Analogues* (Heidelberg Dordrecht London New York: Springer)

[15] Ragni M, Bitencourt A C P, Da S Ferreira C, Aquilanti V, Anderson R W and Littlejohn R G 2009 Exact computation and asymptotic approximations of 6j symbols: illustration of their semiclassical limits *Int. J. Quantum Chem.* **110** 731

[16] Ponzano G and Regge T 1968 *Semiclassical limit of Racah coefficients* in *Spectroscopic and Group Theoretical Methods in Physics* ed F Bloch et al (Amsterdam: North–Holland) p 1

[17] Regge T 1958 Symmetry properties of Clebsch–Gordan's coefficients *Nuovo Cimento* **10** 544

[18] Roberts J 1999 Classical 6j-symbol and the tetrahedron *Geom. Topology* **3** 21

[19] Bitencourt A C P, Marzuoli A, Ragni M, Anderson R W and Aquilanti V 2012 Exact and asymptotic computations of elementary spin networks: classification of the quantum–classical boundaries *Lect. Notes Comput. Science* **7333** Part I ed B Murgante et al (Berlin Heidelberg: Springer–Verlag) p 723

[20] Anderson R W, Aquilanti V, Bitencourt A C P, Marinelli D and Ragni M 2013 The screen representation of spin networks: 2D recurrence, eigenvalue equation for 6j-symbols, geometric interpretation and Hamiltonian dynamics *Lect. Notes Comput. Science* **7972** (Berlin Heidelberg: Springer–Verlag) p 46

[21] Ragni M, Littlejohn R G, Bitencourt A C P, Aquilanti V and Anderson R W 2013 The screen representation of spin networks: images of 6j-symbols and semiclassical features *Lect. Notes Comput. Science* **7972** (Berlin Heidelberg: Springer–Verlag) p 60

[22] Calderini D, Coletti C and Aquilanti V 2013 Continuous and discrete algorithms in quantum chemistry: polynomial sets, spin networks and Strumian orbitals *Lect. Notes Comput. Science* **7972** (Berlin Heidelberg: Springer–Verlag) p 32

Higher symmetries of cotangent coverings for Lax-integrable multi-dimensional partial differential equations and Lagrangian deformations

H Baran[1], I S Krasil′shchik[1,2], O I Morozov[3] and P Vojčák[1]

[1] Mathematical Institute, Silesian University in Opava, Na Rybníčku 1, 746 01 Opava, Czech Republic

[2] Independent University of Moscow, B. Vlasevsky 11, 119002 Moscow, Russia

[3] Institute of Mathematics and Statistics, University of Tromsø, Tromsø 9037, Norway

E-mail: `josephkra@gmail.com`

Abstract. We present examples of Lax-integrable multi-dimensional systems of partial differential equations with higher local symmetries. We also consider Lagrangian deformations of these equations and construct variational bivectors on them.

1. Introduction

Different approaches to integrability of partial differential equations (PDEs), [11, 22], are based on their diverse but related properties such as existence infinite hierarchies of (local or nonlocal) symmetries and/or conservation laws, zero-curvature representations (ZCR), bi-Hamiltonian structures, recursion operators, etc.

Much progress was achieved in the study of PDEs with two independent variables. In particular, in a big number of examples it was shown that a PDE integrable in the sense of presence of a ZCR with a non-removable parameter (we call such equations *Lax-integrable*) have infinite hierarchies of higher local symmetries, see, e.g., [11]. In the multidimensional case the situation looks different. As far as we know, no nontrivial examples of multi-dimensional equations with local higher symmetries were found, cf. [20, §6].

In this paper, we present five examples of multi-dimensional systems with higher local symmetries. All these systems are defined as cotangent coverings, [8, 9], for Lax-integrable PDEs with three or four independent variables, namely, for the r-th dispersionless Dym equation, [2, 10, 12, 15, 17],

$$u_{ty} = u_x\, u_{xy} - u_y\, u_{xx}, \qquad v_{ty} = u_x\, v_{xy} - u_y\, v_{xx} + 2\,(u_{xx}\, v_y - u_{xy}\, v_x), \tag{1}$$

the r-th dispersionless KP equation, [2, 5, 10, 16, 17, 18],

$$u_{yy} = u_{tx} + u_y\, u_{xx} - u_x\, u_{xy}, \qquad v_{yy} = v_{tx} + u_y\, v_{xx} - u_x\, v_{xy} + 2\,(u_{xy}\, v_x - u_{xx}\, v_y), \tag{2}$$

the Veronese web equation, [3, 7, 21],

$$u_{xy} = \frac{\alpha\, u_x\, u_{ty} + (1-\alpha)\, u_y\, u_{tx}}{u_t},$$

$$\begin{aligned} v_{xy} &= \frac{\alpha\, u_x\, v_{ty} + (1-\alpha)\, u_y\, v_{tx}}{u_t} \\ &+\frac{1}{u_t^2}\,((\alpha\, u_x\, u_{ty} + (1-\alpha)\, u_y\, u_{tx})\, v_t \\ &+((1-2\,\alpha)\, u_t\, u_{ty} - (1-\alpha)\, u_y\, u_{tt})\, v_x - (\alpha\, u_x\, u_{tt} + (1-2\,\alpha)\, u_t\, u_{tx})\, v_y) \\ &-\frac{v}{u_t^3}\,((1-\alpha)\, u_y\, u_{tt}\, u_{tx} - u_t\, u_{tx}\, u_{ty} + \alpha\, u_x\, u_{tt}\, u_{ty}) \end{aligned} \tag{3}$$

with $\alpha \neq 0, 1$, the universal hierarchy equation, [10, 17],

$$u_{yy} = u_y\, u_{tx} - u_x\, u_{ty}, \qquad v_{yy} = u_y\, v_{tx} - u_x\, v_{ty} + 2\,(u_{ty}\, v_x - u_{tx}\, v_y), \tag{4}$$

and for the 4-dimensional analogue of (1), [14],

$$u_{ty} = u_z\, u_{xy} - u_y\, u_{xz}, \qquad v_{ty} = u_z\, v_{xy} - u_y\, v_{xz} + 2\,(u_{xz}\, v_y - u_{xy}\, v_z), \tag{5}$$

introduced in [10]. We show that these systems have local symmetries of the third order. Equations (1), (2) and (3) belong to the families of PDEs

$$u_{ty} = u_x\, u_{xy} + \varkappa u_y\, u_{xx}, \tag{6}$$

$$u_{yy} = u_{tx} + \left(\tfrac{1}{2}\,(\varkappa + 1)\, u_x^2 + u_y\right)\, u_{xx} + \varkappa\, u_x\, u_{xy}, \tag{7}$$

$$u_{xy} = \frac{\alpha\, u_x\, u_{ty} + \beta\, u_y\, u_{tx}}{u_t} \tag{8}$$

and are featured as the only representatives of their families whose ZCRs have a non-removable parameter. We make an observation that all the symmetries of the cotangent coverings of Equations (6) with $\varkappa \neq -1$, (7) with $\varkappa \neq -1$, and (8) with $\beta \neq 1-\alpha$ up to order 5 are the point symmetries. This leads to a conjecture that the cotangent coverings of Equations (6), (7), (8) have local higher symmetries whenever these equations are Lax-integrable. If this claim turns out to be correct, it will be a valuable refinement of the observations of [20, §6].

In the last section, we consider *Lagrangian deformations* of cotangent coverings to some of the equations considered above and construct variational bivectors, see [8], on these deformations.

2. Basics

We briefly recall here the basic definitions from the geometry of PDEs, including the definition of the cotangent covering, see [8, 9], and its Lagrangian deformations. Let $\pi\colon \mathbb{R}^n \times \mathbb{R}^m \to \mathbb{R}^n$ be the trivial bundle with the coordinates $(x^1, \ldots, x^n)$ in $\mathbb{R}^n$ and $(u^1, \ldots, u^m)$ in $\mathbb{R}^m$ and $J^\infty(\pi)$ be the space of its infinite jets. Local coordinates on $J^\infty(\pi)$ are $(x^i, u^\alpha, u^\alpha_I)$, where $I = (i_1, \ldots, i_k)$ is a multi-index, and for every local section f of π and the corresponding infinite jet $j_\infty(f)\colon \mathbb{R}^n \to J^\infty(\pi)$ the coordinate u^α_I is defined by $u^\alpha_I|_{j_\infty(f)} = \dfrac{\partial^{\#I} f^\alpha}{\partial x^I} = \dfrac{\partial^{i_1+\cdots+i_n} f^\alpha}{(\partial x^1)^{i_1}\ldots(\partial x^n)^{i_n}}$. We put $u^\alpha = u^\alpha_{(0,\ldots,0)}$.

The vector fields

$$D_{x^k} = \frac{\partial}{\partial x^k} + \sum_{\#I \geq 0}\sum_{\alpha=1}^{m} u^\alpha_{I+1_k}\,\frac{\partial}{\partial u^\alpha_I}, \qquad k = 1, \ldots, n,$$

$(i_1, \ldots, i_k, \ldots, i_n) + 1_k = (i_1, \ldots, i_k + 1, \ldots, i_n)$, are called *total derivatives*. They commute everywhere on $J^\infty(\pi)$.

The *evolutionary derivation* associated to an arbitrary smooth function $\varphi\colon J^\infty(\pi) \to \mathbb{R}^m$ is the vector field

$$\mathbf{E}_\varphi = \sum_{\#I\geq 0}\sum_{\alpha=1}^{m} D_I(\varphi^\alpha)\,\frac{\partial}{\partial u_I^\alpha} \tag{9}$$

with $D_I = D_{(i_1,\dots i_n)} = D_{x^1}^{i_1}\circ\cdots\circ D_{x^n}^{i_n}$.

A system of PDEs $F_a(x^i, u_I^\alpha) = 0$, $\#I \leq s$, $a = 1,\dots,r$, of order $s \geq 1$ with $r \geq 1$ defines the submanifold $\mathcal{E} = \{(x^i, u_I^\alpha) \in J^\infty(\pi) \mid D_K(F_a(x^i, u_I^\alpha)) = 0,\ \#K \geq 0\}$ in $J^\infty(\pi)$.

A function $\varphi\colon \mathcal{E} \to \mathbb{R}^m$ is called a *(generator of an infinitesimal) symmetry* of $\mathcal{E}$ when $\mathbf{E}_\varphi(F) = 0$ on $\mathcal{E}$. The symmetry φ is a solution to the *defining system*

$$\ell_{\mathcal{E}}(\varphi) = 0, \tag{10}$$

where $\ell_{\mathcal{E}} = \ell_F|_{\mathcal{E}}$ with the matrix differential operator

$$\ell_F = \left(\sum_{\#I\geq 0} \frac{\partial F_a}{\partial u_I^\alpha}\, D_I\right).$$

Solutions to (10) constitute the Lie algebra with respect to the *Jacobi bracket* $\{\varphi, \psi\} = \mathbf{E}_\varphi(\psi) - \mathbf{E}_\psi(\varphi)$ denoted by $\mathbf{sym}(\mathcal{E})$. The subalgebra of *contact symmetries* of $\mathcal{E}$ is $\mathbf{sym}_0(\mathcal{E}) = \mathbf{sym}(\mathcal{E}) \cap C^\infty(J^1(\pi), \mathbb{R}^m)$. Symmetries from $\mathbf{sym}(\mathcal{E}) \setminus C^\infty(J^1(\pi), \mathbb{R}^m)$ are said to be *higher* ones.

Dually, a *cosymmetry* is a function $\psi\colon \mathcal{E} \to \mathbb{R}^r$ that satisfies the equation

$$\ell^*_{\mathcal{E}}(\psi) = 0,$$

where $\ell^*_{\mathcal{E}}$ denotes the formally adjoint operator

$$\ell^*_F = \left(\sum_{\#I\geq 0} (-1)^{\#I} D_I \circ \frac{\partial F_a}{\partial u_I^\alpha}\right)^T$$

restricted to $\mathcal{E}$.

A differential operator $\Delta\colon C^\infty(\mathcal{E}, \mathbb{R}^r) \to C^\infty(\mathcal{E}, \mathbb{R}^m)$ is called a *variational bivector* if it takes symmetries of the equation $\mathcal{E}$ to its cosymmetries and satisfies $(\ell_{\mathcal{E}} \circ \Delta)^* = \ell_{\mathcal{E}} \circ \Delta$. Any such an operator defines a skew-symmetric bracket $\{\omega_1, \omega_2\}_\Delta = L_{\Delta(\delta\omega_1)}(\omega_2)$ on the space of conservation laws, where $\delta = \delta^{0,n-1}$ is the differential in the term E_1 of the Vinogradov $\mathcal{C}$-spectral sequence (see [19]) and L denotes the Lie derivative. If the variational Schouten bracket $[\![\Delta, \Delta]\!]$ vanishes then Δ is a Hamiltonian operator, see [8].

Denote $\mathcal{W} = \mathbb{R}^l$, $l \leq \infty$, with coordinates $w^0, \dots, w^s, \dots$. Locally, a *differential covering* over $\mathcal{E}$ is a trivial bundle $\tau\colon \tilde{\mathcal{E}} = \mathcal{E} \times \mathcal{W} \to \mathcal{E}$ equipped with the *extended total derivatives*

$$\tilde{D}_{x^k} = D_{x^k} + \sum_{s=0}^{l} T_k^s(x^i, u_I^\alpha, w^j)\,\frac{\partial}{\partial w^s} \tag{11}$$

such that $[\tilde{D}_{x^i}, \tilde{D}_{x^j}] = 0$. This yields the system of the *covering equations*

$$w^s_{x^k} = T_k^s(x^i, u_I^\alpha, w^j). \tag{12}$$

This over-determined system of PDEs is compatible whenever $(x^i, u_I^\alpha) \in \mathcal{E}$.

The *cotangent covering* is defined as follows. Let, as before, $\mathcal{E} \subset J^\infty(\pi)$ be given by the relations $F_1 = \cdots = F_r = 0$, where $F_a = F(x^i, u_I^j)$. Consider the bundle $\pi' \colon \mathbb{R}^r \times \mathbb{R}^m \times \mathbb{R}^n \to \mathbb{R}^n$ with coordinates v^α in $\mathbb{R}^r$. Then $\mathcal{T}^*\mathcal{E} \subset J^\infty(\pi')$ is the equation obtained from $\mathcal{E}$ by adding the relations $\ell_\mathcal{E}^*(v^\alpha) = 0$. The natural projection $\tau^* \colon \mathcal{T}^*\mathcal{E} \to \mathcal{E}$ is an infinite-dimensional covering. Note that $\mathcal{T}^*\mathcal{E}$ is always an Euler-Lagrange equation with the Lagrangian density $\mathcal{L} = (F_1 v^1 + \cdots + F_r v^r)\, dx^1 \wedge \cdots \wedge dx^n$. Consequently, the spaces of symmetries $\mathbf{sym}(\mathcal{E})$ and cosymmetries $\mathbf{cosym}(\mathcal{E})$ coincide for $\mathcal{T}^*\mathcal{E}$.

3. Higher symmetries of cotangent coverings

In this section we present the results of computation of the third order symmetries for Systems (1)–(5). The computations were held in the JETS software, [1, 6]. We use the notation $u^1 = u$, $u^2 = v$, $x^1 = t$, $x^2 = x$, $x^3 = y$, and $x^4 = z$ in the formulas of the previous section.

Proposition 1. *The cotangent coverings of the Lax-integrable cases of the r-th dispersionsless Dym equation, r-th modified dispersionsless KP equation, the Veronese web equation, the universal hierarchy equation, and the 4-dimensional r-th dispersionsless Dym equation have the following symmetries up to the third order:*

1) *System* (1):

$$
\begin{aligned}
\varphi_0(A_0) &= \left(A_0\, u_t + A_{0,t}\,(x\, u_x - u) + \tfrac{1}{2}\, A_{0,tt}\, x^2, A_0\, v_t + A_{0,t}\,(x\, v_x + 2\, v)\right), \\
\varphi_1(A_1) &= (A_1\, u_x + A_{1,t}\, x, A_1\, v_x), \\
\varphi_2(A_2) &= (A_2, 0), \\
\varphi_3(A_3) &= \left(0, A_3\,(2\, u_t - 3\, u_x^2) + A_{3,t}\,(2\, x\, u_x + u) - \tfrac{1}{2}\, A_{3,tt}\, x^2\right), \\
\varphi_4(A_4) &= (0, 2\, A_4\, u_x - A_{4,t}\, x), \\
\varphi_5(A_5) &= (0, A_5), \\
\varphi_6(B_0) &= (B_0\, u_y, B_0\, v_y), \\
\varphi_7(B_1) &= \left(0, B_1\, u_y^{-2}\right), \\
\varphi_8 &= (x\, u_x - 2\, u, x\, v_x), \\
\varphi_9 &= (0, v), \\
\varphi_{10} &= \left(0, u_{ttt} - 3\, u_x\,(u_{ttx} - u_x\, u_{txx} - u_{tx}\, u_{xx}) - u_x^3\, u_{xxx} - \tfrac{3}{2}\,(u_{tx}^2 + u_x^2\, u_{xx}^2)\right), \\
\varphi_{11} &= \left(0, u_{ttx} - 2\, u_x\, u_{txx} + u_x^2\, u_{xxx} - u_{xx}\,(u_{tx} - u_x\, u_{xx})\right), \\
\varphi_{12} &= \left(0, u_{txx} - u_x\, u_{xxx} - \tfrac{1}{2}\, u_{xx}^2\right), \\
\varphi_{13} &= (0, u_{xxx}), \\
\varphi_{14} &= \left(0, (2\, u_y\, u_{xxy} - u_{xy}^2)\, u_y^{-2}\right), \\
\varphi_{15} &= \left(0, (u_y\, u_{xyy} - u_{xy}\, u_{yy})\, u_y^{-3}\right), \\
\varphi_{16} &= \left(0, (2\, u_y\, u_{yyy} - 3\, u_{yy}^2)\, u_y^{-4}\right),
\end{aligned}
$$

where $A_i = A_i(t)$, $B_i = B_i(y)$ are arbitrary functions of their arguments;

2) *System* (2):

$$
\begin{aligned}
\varphi_0(A_0) &= \big(A_0\, u_t + (x\, u_x + y\, u_y - u)\, A_{0,t} + (\tfrac{1}{2}\, u_x\, y^2 - x\, y)\, A_{0,tt} - \tfrac{1}{6}\, y^3\, A_{0,ttt}, \\
&\qquad A_0\, v_t + (x\, v_x + y\, v_y + 2\, v)\, A_{0,t} + \tfrac{1}{2}\, y^2\, v_x\, A_{0,tt}\big), \\
\varphi_1(A_1) &= \left(A_1\, u_y + A_{1,t}\,(y\, u_x - x) - \tfrac{1}{2}\, A_{1,tt}\, y^2, A_1\, v_y + A_{1,t}\, y\, v_x\right), \\
\varphi_2(A_2) &= (A_2\, u_x - A_{2,t}\, y, A_2\, v_x), \\
\varphi_3(A_3) &= (A_3, 0),
\end{aligned}
$$

$$\begin{aligned}
\varphi_4(A_4) &= \big(0, 2A_4\,(u_t + 2\,u_x^3 + 3\,u_x\,u_y) + A_{4,t}\,(3\,y\,u_x^2 + 2\,x\,u_x + 2\,y\,u_y + u) \\
&\qquad + A_{4,tt}\,y\,(y\,u_x + x) + \tfrac{1}{6}\,A_{4,ttt}\,y^3\big)\,, \\
\varphi_5(A_5) &= \big(0, A_5\,(2\,u_y + 3\,u_x^2) + A_{5,t}\,(2\,y\,u_x + x) + \tfrac{1}{2}\,A_{5,tt}\,y^2\big)\,, \\
\varphi_6(A_6) &= (0, 2\,A_6\,u_x + A_{6,t}\,y)\,, \\
\varphi_7(A_7) &= (0, A_7)\,, \\
\varphi_8 &= \big(2\,x\,u_x + y\,u_y - 3\,u, 2\,x\,v_x + y\,v_y + 4\,v + 2\,t\,u_t + u_x\,(2\,x + 3\,y\,u_x + 4\,t\,u_x^2) \\
&\qquad + 2\,u_y\,(3\,t\,u_x + y) + u\big)\,, \\
\varphi_9 &= (y\,u_x + 2\,t\,u_y, y\,v_x + 2\,t\,v_y)\,, \\
\varphi_{10} &= (0, v)\,, \\
\varphi_{11} &= (0, u_{xxx})\,, \\
\varphi_{12} &= \big(0, u_{xxy} + u_x\,u_{xxx} + \tfrac{1}{2}\,u_{xx}^2\big)\,, \\
\varphi_{13} &= \big(0, u_{txx} + u_{xxx}\,(u_x^2 + u_y) + u_x\,u_{xxy} + u_{xx}\,(u_{xy} + u_x\,u_{xx})\big)\,, \\
\varphi_{14} &= \big(0, 2\,u_{txy} + 4\,u_x\,u_{txx} + 2\,u_x\,u_{xxx}\,(u_x^2 + 2\,u_y) + 2\,u_{xxy}\,(u_x^2 + u_y) + u_{xx}^2\,(3\,u_x^2 + 2\,u_y) \\
&\qquad + u_{xy}^2 + 2\,u_{xx}\,(u_{tx} + 2\,u_x\,u_{xy})\big)\,, \\
\varphi_{15} &= \big(0, u_{ttx} + u_{txx}\,(3\,u_x^2 + 2\,u_y) + 2\,u_x\,u_{txy} + u_{xxx}\,(u_x^4 + 3\,u_y\,u_x^2 + u_y^2) + u_x\,u_{xy}^2 \\
&\qquad + u_x\,u_{xxy}\,(u_x^2 + 2\,u_y) + u_{tx}\,(u_{xy} + 3\,u_x\,u_{xx}) + u_x\,u_{xx}^2\,(2\,u_x^2 + 3\,u_y) \\
&\qquad + u_{xx}\,(u_{ty} + u_{xy}\,(3\,u_x^2 + 2\,u_y))\big)\,, \\
\varphi_{16} &= \big(0, 2\,u_{tty} + 6\,u_x\,u_{ttx} + 4\,u_x\,u_{txx}\,(2\,u_x^2 + 3\,u_y) + 2\,u_{txy}\,(2\,u_y + 3\,u_x^2) \\
&\qquad + 2\,u_x\,u_{xxx}\,(u_x^4 + 4\,u_x^2\,u_y + 3\,u_y^2) + 2\,u_{xxy}\,(u_x^4 + 3\,u_x^2\,u_y + u_y^2) + 3\,u_{tx}^2 \\
&\qquad + 6\,u_{tx}\,(u_x\,u_{xy} + u_{xx}\,(u_y + 2\,u_x^2)) + u_{xx}^2\,(5\,u_x^4 + 12\,u_x^2\,u_y + 3\,u_y^2) \\
&\qquad + u_{xy}^2\,(3\,u_x^2 + 2\,u_y) + 2\,u_{ty}\,(u_{xy} + 3\,u_x\,u_{xx}) + 4\,u_x\,u_{xx}\,u_{xy}\,(2\,u_x^2 + 3\,u_y)\big)\,, \\
\varphi_{17} &= \big(0, u_{ttt} + 3\,(2\,u_x^2 + u_y)\,u_{ttx} + 3\,u_x\,u_{tty} + (5\,u_x^4 + 12\,u_x^2\,u_y + 3\,u_y^2)\,u_{txx} \\
&\qquad + (5\,u_y\,u_x^4 + 6\,u_y^2\,u_x^2 + u_x^6 + u_y^3)\,u_{xxx} + u_x\,(u_y + u_x^2)\,(3\,u_y + u_x^2)\,u_{xxy} \\
&\qquad + 12\,u_x\,u_y\,u_{tx}\,u_{xx} + 10\,u_x^3\,u_{tx}\,u_{xx} + 6\,u_x\,u_{tx}^2 + 2\,u_x\,(2\,u_x^2 + 3\,u_y)\,u_{txy} \\
&\qquad + 5\,u_x^4\,u_{xx}\,u_{xy} + 12\,u_x^2\,u_y\,u_{xx}\,u_{xy} + 3\,u_y\,u_{tx}\,u_{xy} + 6\,u_x^2\,u_{tx}\,u_{xy} + 3\,u_x\,u_y\,u_{xy}^2 \\
&\qquad + 3\,u_y^2\,u_{xx}\,u_{xy} + 3\,u_x^5\,u_{xx}^2 + 6\,u_x\,u_y^2\,u_{xx}^2 + 10\,u_y\,u_x^3\,u_{xx}^2 \\
&\qquad + 3\,(u_{tx} + 2\,u_{xx}\,u_x^2 + u_y\,u_{xx} + u_x\,u_{xy})\,u_{ty} + 2\,u_x^3\,u_{xy}^2\big)\,,
\end{aligned}$$

where $A_i = A_i(t)$ are arbitrary functions of t;

3) *System* (3):

$$\begin{aligned}
\varphi_1(P_1) &= (P_1\,u_t, P_1\,v_t + P_{1,t}\,v)\,, \\
\varphi_2(Q_1) &= (Q_1\,u_x, Q_1\,v_x)\,, \\
\varphi_3(R_1) &= (R_1\,u_y, R_1\,v_y)\,, \\
\varphi_4(S_1) &= (S_1, -S_{1,u}\,v)\,, \\
\varphi_5 &= (u, 0)\,, \\
\varphi_6(P_2) &= \big(0, P_2\,u_t^{-1}\big)\,, \\
\varphi_7(Q_2) &= \big(0, Q_2\,u_t\,u_x^{-2}\big)\,, \\
\varphi_8(R_2) &= \big(0, R_2\,u_t\,u_y^{-2}\big)\,, \\
\varphi_9(S_2) &= (0, S_2\,u_t)\,, \\
\varphi_{10} &= \big(0, (2\,u_t\,u_x\,u_{ttx} - 2\,u_x\,u_{tt}\,u_{tx} - u_t\,u_{tx}^2)\,u_t^{-2}\,u_x^{-2}\big)\,,
\end{aligned}$$

$$\begin{aligned}
\varphi_{11} &= \left(0, (2\,u_t\,u_y\,u_{tty} - 2\,u_y\,u_{tt}\,u_{ty} - u_t\,u_{ty}^2)\,u_t^{-2}\,u_y^{-2}\right),\\
\varphi_{12} &= \left(0, (2\,u_t\,u_x\,u_{txx} - u_x\,u_{tx}^2 - 2\,u_t\,u_{tx}\,u_{xx})\,u_t^{-1}\,u_x^{-3}\right),\\
\varphi_{13} &= \left(0, (2\,u_t\,u_y\,u_{tyy} - u_y\,u_{ty}^2 - 2\,u_t\,u_{ty}\,u_{yy})\,u_t^{-1}\,u_y^{-3}\right),\\
\varphi_{14} &= \left(0, (2\,u_t\,u_{ttt} - 3\,u_{tt}^2)\,u_t^{-3}\right),\\
\varphi_{15} &= \left(0, u_t\,(2\,u_x\,u_{xxx} - 3\,u_{xx}^2)\,u_x^{-4}\right),\\
\varphi_{16} &= \left(0, u_t\,(2\,u_y\,u_{yyy} - 3\,u_{yy}^2)\,u_y^{-4}\right),
\end{aligned}$$

where $P_i = P_i(t)$, $Q_i = Q_i(x)$, $R_i = R_i(y)$, $S_i = S_i(u)$ *are arbitrary functions of their arguments;*
4) *System* (4):

$$\begin{aligned}
\varphi_0(A_0) &= (A_0\,u_t - A_{0,t}\,u, A_0\,v_t + 2\,A_{0,t}\,v),\\
\varphi_1(A_1) &= (A_1, 0),\\
\varphi_2(A_2) &= (0, 2\,A_2\,u_t + A_{2,t}\,u),\\
\varphi_3(A_3) &= (0, A_3),\\
\varphi_4(B_0) &= (B_0\,u_x + B_{0,x}\,y\,u_y, B_0\,v_x + B_{0,x}\,y\,v_y),\\
\varphi_5(B_1) &= (B_1\,u_y, B_1\,v_y),\\
\varphi_6(B_2) &= \left(0, (2\,B_2\,u_x - B_{2,x}\,y\,u_y)\,u_y^{-3}\right),\\
\varphi_7(B_3) &= \left(0, B_3\,u_y^{-2}\right),\\
\varphi_8 &= (y\,u_y + u, y\,v_y),\\
\varphi_9 &= (0, v),\\
\varphi_{10} &= (0, u_{ttt}),\\
\varphi_{11} &= \left(0, (u_y^2\,u_{ttx} - u_x\,u_y\,u_{tty} + u_x\,u_{ty}^2 - u_y\,u_{tx}\,u_{ty})\,u_y^{-3}\right),\\
\varphi_{12} &= \left(0, (2\,u_y\,u_{tty} - u_{ty}^2)\,u_y^{-2}\right),\\
\varphi_{13} &= \left(0, (2\,u_x^2\,u_y^2\,(u_x\,u_{ttx} - 3\,u_{txy}) - 2\,u_x^4\,u_y\,u_{tty} - 2\,u_y^3\,(u_{xxy} - 2\,u_x\,u_{txx})\right.\\
&\quad +3\,u_x\,u_y^2\,(u_x\,u_{tx}^2 - 2\,u_{ty}\,u_{xx} - 2\,u_{tx}\,u_{xy}) - 4\,u_x^2\,u_y\,u_{ty}\,(2\,u_x\,u_{tx} - 3\,u_{xy})\\
&\quad \left.+u_y^2\,(3\,u_{xy}^2 + 2\,u_y\,u_{tx}\,u_{xx}) + 5\,u_x^4\,u_{ty}^2)\,u_y^{-6}\right),\\
\varphi_{14} &= \left(0, (2\,u_y^4\,u_{xxx} + 2\,u_x^4\,u_y\,(u_y\,u_{ttx} - 5\,u_{tx}\,u_{ty}) - 4\,u_x^3\,u_y^2\,(2\,u_{txy} - u_{tx}^2)\right.\\
&\quad +6\,u_x^2\,u_y^2\,(u_y u_{txx} - 2\,u_{ty}\,u_{xx} - 2\,u_{tx}\,u_{xy}) - 2\,u_x^5\,(u_y\,u_{tty} - 3\,u_{ty}^2)\\
&\quad \left.-6\,u_y^3\,(u_x\,u_{xxy} + u_{xx}\,u_{xy} - u_x\,u_{tx}\,u_{xx}) + 4\,u_x\,u_y\,u_{xy}\,(3\,u_y\,u_{xy} + 5\,u_x^2\,u_{ty}))\,u_y^{-7}\right),\\
\varphi_{15} &= \left(0, (u_x^2\,(2\,u_y\,u_{tty} - 3\,u_{ty}^2) + u_y^2\,(2\,u_{txy} - 2\,u_x\,u_{ttx} - u_{tx}^2)\right.\\
&\quad \left.-2\,u_y\,u_{ty}\,(u_{xy} - 2\,u_x\,u_{tx}))\,u_y^{-4}\right),\\
\varphi_{16} &= \left(0, (u_x^3\,(u_y\,u_{tty} - 2\,u_{ty}^2) + u_x\,u_y^2\,(2\,u_{txy} - u_{tx}^2) - u_y^2\,(u_x^2\,u_{ttx} - u_{ty}\,u_{xx} - u_{tx}\,u_{xy})\right.\\
&\quad \left.-u_y^3\,u_{txx} + 3\,u_x\,u_y\,(u_x\,u_{tx}\,u_{ty} - u_{ty}\,u_{xy}))\,u_y^{-5}\right),
\end{aligned}$$

where $A_i = A_i(t)$, $B_i = B_i(x)$ *are arbitrary functions of their arguments;*
5) *System* (5):

$$\begin{aligned}
\varphi_0(A_0) &= (A_0\,u_x - A_{0,x}\,u + A_{0,t}\,z, A_0\,v_x + 2\,A_{0,x}\,v),\\
\varphi_1(A_1) &= (A_1, 0),\\
\varphi_2(A_2) &= (0, 2\,A_2\,u_x + A_{2,x}\,u - A_{2,t}\,z),\\
\varphi_3(A_3) &= (0, A_3),\\
\varphi_4(B_0) &= (B_0\,u_y, B_0\,v_y),
\end{aligned}$$

$$
\begin{aligned}
\varphi_5(B_1) &= \left(0, B_1\, u_y^{-2}\right),\\
\varphi_6 &= \left(t\, u_t + u, t\, v_t\right),\\
\varphi_7 &= \left(z\, u_z - u, z\, v_z\right),\\
\varphi_8 &= \left(u_z, v_z\right),\\
\varphi_9 &= \left(0, v\right),\\
\varphi_{10} &= \left(0, u_{xxx}\right),\\
\varphi_{11} &= \left(0, (2\, u_y\, u_{xxy} - u_{xy}^2)\, u_y^{-2}\right),\\
\varphi_{12} &= \left(0, (u_y\, u_{xyy} - u_{xy}\, u_{yy})\, u_y^{-3}\right),\\
\varphi_{13} &= \left(0, (2\, t\, u_y^2\, u_{xxx} - z\,(2\, u_y\, u_{xxy} - u_{xy}^2))\, u_y^{-2}\right),\\
\varphi_{14} &= \left(0, (t\, u_y\,(2\, u_y\, u_{xxy} - u_{xy}^2) - 2\, z\,(u_y\, u_{xyy} - u_{xy}\, u_{yy}))\, u_y^{-3}\right),\\
\varphi_{15} &= \left(0, (z^2\,(u_y\, u_{xyy} - u_{xy}\, u_{yy}) - t\, z\, u_y\,(2\, u_y\, u_{xxy} - u_{xy}^2) + t^2\, u_y^3\, u_{xxx})\, u_y^{-3}\right),\\
\varphi_{16} &= \left(0, (2\, u_y\, u_{yyy} - 3\, u_{yy}^2)\, u_y^{-4}\right),\\
\varphi_{17} &= \left(0, (z\,(2\, u_y\, u_{yyy} - 3\, u_{yy}^2) - 2\, t\, u_y\,(u_y\, u_{xyy} - u_{xy}\, u_{yy}))\, u_y^{-4}\right),\\
\varphi_{18} &= \left(0, (z^2\,(2\, u_y\, u_{yyy} - 3\, u_{yy}^2) - 4\, t\, z\, u_y\,(u_y\, u_{xyy} - u_{xy}\, u_{yy})\right.\\
&\qquad \left. + t^2\, u_y^2\,(2\, u_y\, u_{xxy} - u_{xy}^2))\, u_y^{-4}\right),\\
\varphi_{19} &= \left(0, (z^3\,(2\, u_y\, u_{yyy} - 3\, u_{yy}^2) - 6\, t\, z^2\, u_y\,(u_y\, u_{xyy} - u_{xy}\, u_{yy})\right.\\
&\qquad \left. + 3\, t^2\, z\, u_y^2\,(2\, u_y\, u_{xxy} - u_{xy}^2) - 2\, t^3\, u_y^4\, u_{xxx})\, u_y^{-4}\right),
\end{aligned}
$$

where $A_i = A_i(t, x)$, $B_i = B_i(y, z)$ *are arbitrary functions of their arguments.*

All the functions above are assumed to be smooth and the second subscript (as in, say, $A_{3,t}$*) denotes the corresponding derivative.*

Thus, we have the following higher symmetries:

For Systems (1), (3), (4):	$\varphi_{10}, \ldots, \varphi_{16}$,
For System (2):	$\varphi_{11}, \ldots, \varphi_{17}$,
For System (5):	$\varphi_{10}, \ldots, \varphi_{19}$.

4. Bivectors on the deformed cotangent covering

Let $\mathcal{E}$ be an equation and $\tau^*\colon \mathcal{T}^*\mathcal{E} \to \mathcal{E}$ be its tangent covering. As it was mentioned in Section 2, $\mathcal{T}^*\mathcal{E}$ is an Euler-Lagrange equation. Let $\mathcal{T}^*\mathcal{E}$ be given by $F(x, u) = 0$, $\ell^*_{\mathcal{E}}(x, u, v) = 0$. Then we say that the system

$$
\tilde{\mathcal{E}}\colon \qquad \ell^*_{\mathcal{E}}(x, u, v) + G(x, u, v) = 0, \quad F(x, u) + H(x, u, v) = 0 \tag{13}
$$

is a *Lagrangian deformation* (LD) if $\tilde{\mathcal{E}}$ is also an Euler-Lagrange equation. We know at least two meaningful examples of this construction: (i) The system

$$
u_{ty} = u_{xy}u_x - u_{xx}u_y, \qquad v_{ty} = 2(u_{xx}v_y - u_{xy}v_x) + u_xv_{xy} - u_yv_{xx} - 2(u_{xx}u_y + 2u_{xy}u_x),
$$

see [15], is an LD of (1); (ii) The Dunajski equation, [4],

$$
\Theta_{wx} + \Theta_{zy} + \Theta_{xx}\Theta_{yy} - \Theta_{xy}^2 = f, \qquad f_{xw} + f_{yz} + \Theta_{yy}f_{xx} + \Theta_{xx}f_{yy} - 2\Theta_{xy}f_{xy} = 0
$$

is an LD of the 2nd Heavenly Equation.

The identical map from $\mathbf{cosym}(\tilde{\mathcal{E}})$ to $\mathbf{sym}(\tilde{\mathcal{E}})$ is always a bivector on any LD $\tilde{\mathcal{E}}$ and we pose the following problem: *Given an equation* $\mathcal{E}$, *can* $\mathcal{T}^*\mathcal{E}$ *be deformed in such a way that the resulting* LD $\tilde{\mathcal{E}}$ *will admit nontrivial bivectors?*

We studied LDs for Equations (1), (4), and (5) and obtained the following result:

Proposition 2. *Equations* (1), (4), *and* (5) *possess two-parameter families of* LDs *with* $H = 0$ *and* $G = \delta L_{\varkappa_1,\varkappa_2}/\delta u$ *and three-dimensional spaces of bivectors* $\Delta = \alpha_0 \operatorname{id} + \alpha_1\Delta^1 + \alpha_2\Delta^2$, $\Delta^i = (\Delta^i_{jk})$, $i, j, k = 1, 2$, *where*

For Equation (1):

$$\begin{aligned}
L_{\varkappa_1,\varkappa_2} &= (\varkappa_1 u_y + \varkappa_2)u_x^2,\\
\Delta^1_{11} &= u_y D_x^{-1} \circ u_y^{-2} \circ D_y,\\
\Delta^1_{12} &= 0,\\
\Delta^1_{21} &= (v_y - \varkappa_2)D_x^{-1} \circ u_y^{-2} \circ D_y + u_y^{-2} D_x^{-1} \circ (v_y - \varkappa_2) \circ D_y,\\
\Delta^1_{22} &= u_y^{-2} D_x^{-1} \circ u_y \circ D_y,\\
\Delta^2_{11} &= \frac{1}{2}u_x + \frac{1}{2}D_x^{-1} \circ (D_t - 2u_x D_x),\\
\Delta^2_{12} &= 0,\\
\Delta^2_{21} &= v + \varkappa_1 u D_x - \frac{1}{2}D_x^{-1} \circ (2\varkappa_1 u D_x + v_{xx}) - \varkappa_1 D_x^{-1} \circ (D_t - 2u_x D_x),\\
\Delta^2_{22} &= \frac{1}{2}u D_x - u_x + \frac{1}{2}D_x^{-1} \circ (D_t - u D_x^2).
\end{aligned}$$

For Equation (4):

$$\begin{aligned}
L_{\varkappa_1,\varkappa_2} &= u_t(\varkappa_1 u_y + \varkappa_2 u_x),\\
\Delta^1_{11} &= u D_t^{-1} \circ u_y^{-2} \circ D_y,\\
\Delta^1_{12} &= 0,\\
\Delta^1_{21} &= (v_y + \varkappa_2)D_t^{-1} \circ u_y^{-2} \circ D_y + u_y^{-2} D_t^{-1} \circ (v_y + \varkappa_2) \circ D_y,\\
\Delta^1_{22} &= u_y^{-2} D_t^{-1} \circ u_y \circ D_y,\\
\Delta^2_{11} &= u D_t - D_y^{-1} \circ (u_{yt} + u D_y D_t),\\
\Delta^2_{12} &= 0,\\
\Delta^2_{21} &= -2v D_t - D_y^{-1} \circ (v_{yt} + 2\varkappa_2 D_t - 2v D_y D_t),\\
\Delta^2_{22} &= u D_t + D_y^{-1} \circ (u D_y D_t - 2u_{yt}).
\end{aligned}$$

For Equation (5):

$$\begin{aligned}
L_{\varkappa_1,\varkappa_2} &= u_x(\varkappa_1 u_z + \varkappa_2 u_y),\\
\Delta^1_{11} &= -u D_x + D_z^{-1} \circ (u_{xz} + D_t + u D_x D_z),\\
\Delta^1_{12} &= 0,\\
\Delta^1_{21} &= -D_z^{-1} \circ (v_{xz} + 2(v_z + \varkappa_2)D_x),\\
\Delta^1_{22} &= -u D_x + 2u_x + D_z^{-1} \circ (u_{xz} + D_t + u D_x D_z),
\end{aligned}$$

$$\begin{aligned}
\Delta^2_{11} &= u_y D_x^{-1} \circ u_y^{-2} \circ D_y,\\
\Delta^2_{12} &= 0,\\
\Delta^2_{21} &= -u_y^{-2} D_x^{-1} \circ (v_y + \varkappa_1) \circ D_y - (v_y + \varkappa_1) D_x^{-1} \circ u_y^{-2} \circ D_y,\\
\Delta^2_{22} &= u_y^{-2} D_x^{-1} \circ u_y D_y.
\end{aligned}$$

5. Concluding remarks

Remark 1. The identical bivector in Equation (1), after transforming the system to the evolutionary form, corresponds to the first Hamiltonian operator from [15], while the bivector Δ^1 corresponds to the second one.

Remark 2. It can be shown that the LDs for Equations (4) and (5) presented above are in a sense trivial, because the deformed equations are transformed to the initial cotangent space by a point transformation. As for Equation (1), there exists a covering $\tau\colon \tilde{\tilde{\mathcal{E}}} \to \tilde{\mathcal{E}}$ (that corresponds to a differential substitution) such that $\tilde{\tilde{\mathcal{E}}}$ is equivalent to $\mathcal{T}^*\mathcal{E}$.

Remark 3. As it was mentioned above, all computations were done using the JETS (MAPLE 15), [1, 6], software. We used the IBM Blade cluster consisting of 3 working machines, each of them quad-core Intel Xeon: one E5460, 2.66 GHz, 56 GB RAM two E5430, 2.66 GHz, 16 GB RAM. The classification tree contained 3446 branches, 15 of them were abandoned due to lack of memory. The results presented are due to successful processing of 11 branches only. Other branches are to be processed in future. We also plan to consider LDs of a more general nature.

Remark 4. Actually, presenting results of Proposition 2 in the operator form is not complete and is due to the common tradition only. It is more adequate to deal with them as with Bäcklund transformations between the tangent (see [8]) and cotangent spaces of the equation at hand (cf. [14]), but the operator form seems to be "more visual".

Acknowledgments
The work of H. Baran and P. Vojčák was supported by RVO funding for IČ47813059. The work of H. Baran was also supported by Czech Science Foundation (GAČR) project P201/11/0356. I. Krasil′shchik was supported by the European Social Fund under the project CZ.1.07/2.3.00/20.0002.

References

[1] Baran H, Marvan M Jets. A software for differential calculus on jet spaces and diffieties. `http://jets.math.slu.cz`
[2] Błaszak M 2002 Classical R-matrices on Poisson algebras and related dispersionless systems *Phys. Lett.* A **297** 191–5
[3] Burovskiy P A, Ferapontov E V, Tsarev S P 2010 Second order quasilinear PDEs and conformal structures on in projective space *Int. J. Math.* **21** 799–841
[4] Dunajski M 2002 Anti-self-dual four-manifolds with a parallel real spinor *Proc. Roy. Soc. Lond.* **A 458** 1205–1222
[5] Dunajski M 2004 A class of Einstein-Weil spaces associated to an integrable system of hydrodynamic type *J. Geom. Phys.* **51** 126–37
[6] Marvan M 2009 Sufficient set of integrability conditions of an orthonomic system *Foundations of Comp. Math.* **9** 651–74
[7] Marvan M, Sergyeyev A 2012 Recursion operators for dispersionless integrable systems in any dimensions *Inverse Problems* **28** 025011
[8] Krasil′shchik J, Verbovetsky A 2011 Geometry of jet spaces and integrable systems *J. Geom. Phys.* **61** 1633–74
[9] Krasil′shchik I S, Verbovetsky A M, Vitolo R 2012 A unified approach to computation of integrable structures *Acta Appl. Math.* **120** 199–218

[10] Martínez Alonso L, Shabat A B 2004 Hydrodynamic reductions and solutions of a universal hierarchy *Theor. Math. Phys.* **140** 1073–85

[11] Mikhailov A V (Ed) 2009 *Integrability* Lecture Notes in Physics **767** (Berlin: Springer)

[12] Morozov O I 2009 Contact integrable extensions of symmetry pseudo-groups and coverings of (2+1) dispersionless integrable equations *J. Geom. Phys.* **59** 1461–75

[13] Morozov O I 2013 A Recursion Operator for the Universal Hierarchy Equation via Cartan's Method of Equivalence, Central Euro. J. Math., accepted, arXiv:1205.5748 [nlin.SI]

[14] Morozov O I 2013 A Four-Dimensional Generalization of the Integrable rdDym Equation *Preprint* arXiv:1309.4993 [nlin.SI]

[15] Ovsienko V 2010 Bi-Hamiltonian nature of the equation $u_{tx} = u_{xy}\,u_y - u_{yy}\,u_x$ *Adv. Pure Appl. Math.* **1** 7–17

[16] Ovsienko V, Roger C 2007 Looped cotangent Virasoro algebra and non-linear integrable systems in dimension $2+1$ *Comm. Math. Phys.* **273** 357–88

[17] Pavlov M V 2003 Integrable hydrodynamic chains *J. Math. Phys.* **44** 4134–56

[18] Pavlov M V 2006 The Kupershmidt hydrodynamics chains and lattices 2006 *Intern. Math. Research Notes* **2006** 46987

[19] Vinogradov A M 1984 The $\mathcal{C}$-spectral sequence, Lagrangian formalism, and conservation laws. I. The linear theory. II. The nonlinear theory *J. Math. Anal. Appl.* **100** 1–129.

[20] Vinogradov A M 1989 Symmetries and conservation laws of partial differential equations: basic notions and results *Acta Appl. Math.* **15** 3–21

[21] Zakharevich I 2000 Nonlinear wave equation, nonlinear Riemann problem, and the twistor transform of Veronese webs *Preprint* math-ph/0006001

[22] Zakharov V E (Ed) 1992 *What is integrability?* Springer series in Nonlinear Dynamics (Berlin: Springer)

Towards the Dressing Phase in the AdS_3/CFT_2 Duality

M Beccaria[1] **and G Macorini**[1]

[1] Dipartimento di Matematica e Fisica "Ennio De Giorgi", Universita' del Salento & INFN, Via Arnesano, 73100 Lecce, Italy

E-mail: `Matteo.Beccaria@le.infn.it`, `macorini@nbi.ku.dk`

Abstract. In this paper we describe some recent progresses in the study of the leading quantum correction at strong coupling of the dressing phase appearing in the Bethe Ansatz for string theory on the $AdS_3 \times S^3 \times T^4$ background. Using the $SU(2)$ rigid circular string as guiding example, we find that the phase is different than in the $AdS_{5,4}$ cases. We discuss in detail the determination of the phase using both a Word-Sheet approach and the Algebraic Curve formalism.

1. Introduction

The discovery of the integrability in the context of the AdS/CFT correspondence opened the possibility for spectacular developments in the past years: in particular the spectral problem for the most studied example of the duality - $\mathcal{N} = 4$ SYM - II B string theory on $AdS_5 \times S^5$ - has been completely solved in terms of a system of Thermodinamic Bethe Equations. Many other interesting objects in the theory have been investigated exploiting the integrability of the model, ranging from correlation functions, Wilson loops, scattering amplitudes etc. The situation is similar for the second main example of integrable theories in the AdS/CFT duality: ABJM - II A string theory on $AdS_4 \times CP^3$ (for a review see [1]).

In both these cases the first step in the solution of the spectral problem has been its reformulation in terms of a spin chain description, mapping the calculation of the anomalous dimensions for (long) gauge invariant operators in the solutions of some sets of Bethe Equations (BE).

It is well known that other backgrounds with an AdS_n factor allow for integrable string theory, at least at the classical level (see for example [2]): recently the backgrounds $AdS_3 \times S^3 \times S^3 \times S^1$ and $AdS_3 \times S^3 \times T^4$ received much attention, and a first conjecture for the the Bethe Equations describing the spectrum of these models has been proposed in [3], followed by the second proposal of [4]. As it happens in the previous $AdS_{5,4}$ cases, these BE contains a dressing factor, which is not fixed by the symmetry of the model.

The main goal of this paper is to report some recent results [5, 6] in the study of the dressing phase appearing in the aforementioned set of BE for the $AdS_3 \times S^3 \times T^4$ background: technically this backgound can be thought as a special case of the more general $AdS_3 \times S^3 \times S^3 \times S^1$: setting the radius of AdS_3 to 1, then the radii of the two 3-spheres can be parametrized as $R_+^2 =$

α^{-1}, $R_-^2 = (1-\alpha)^{-1}$, and the $AdS_3\times S^3\times T^4$ model with $R_2=\infty$ (after recompactification) corresponds to $\alpha=1$; strings on $AdS_3\times S^3\times T^4$ and $AdS_3\times S^3\times S^3\times S^1$ are described respectively by the GS superstring action on the supercosets $PSU(1,1|2)\times PSU(1,1|2)/SU(1,1)\times SU(2)$ and $D(2,1;\alpha)\times D(2,1;\alpha)/SU(1,1)\times SU(2)\times SU(2)$: It is important to note that, contrary to the $AdS_{5,4}$ case, the supercoset description of the background, which is the core of the integrabilty of the model, is only partial for the AdS_3 cases, the S^1 and T^4 factors being missing. Due to the difficulty to describe the complete background in the integrability setup, and the fact that the $\alpha\to 1$ limit is not smooth, it is simpler to consider separately the two cases; in this paper we concentrate on the $AdS_3\times S^3\times T^4$ case only.

In more detail the aim of the paper is to investigate the next to leading correction at strong coupling of the dressing phase: the phase for the scattering of two magnons with momenta p_j and p_k can be written in general as [7, 8]

$$\vartheta(p_j,p_k) = 2\sum_{\substack{r=2\\ s\geq r+1\\ r+s\ \mathrm{odd}}}^{\infty} c_{r,s}(\lambda)\left(\frac{\lambda}{16\,\pi^2}\right)^{\frac{r+s-1}{2}}\Big[q_r(p_j)\,q_s(p_k)-q_s(p_j)\,q_r(p_k)\Big]\,. \tag{1}$$

where $q_n(p)$ is the elementary magnon n-th charge. The strong coupling expansion of the coefficient functions $c_{r,s}(\lambda)$ is

$$c_{r,s}(\lambda) = c_{r,s}^{(0)} + \frac{1}{\sqrt{\lambda}}\,c_{r,s}^{(1)} + \ldots, \tag{2}$$

where the leading term $c_{r,s}^{(0)} = \delta_{r+1,s}$ is fixed by the matching with the classical finite-gap equations, and it is the same AFS contribution [9] as in the $AdS_{5,4}$ cases.

Our strategy to compute the $c_{r,s}^{(1)}$ relies on the study of a simple solution, the $SU(2)$ circular spinning string as a guiding example. While the 1-loop correction to the dressing phase and the $c_{r,s}^{(1)}$ coefficients can be computed, in principle in a more general way, without refering to any particular solution, in the Algebraic Curve (AC) framework, following the approach of [10], we will need anyway an independent calculation of the 1-loop energy, done for the particular solution of the rigid circular string using the World-Sheet (WS) method, which is independent of the integrability of the model: we need this particular example both to understand regularization issues and to check the independence of the result from the missing modes in the AC, realtive to the T^4 component of the background.

The paper is organised as follows: in section 2 we briefly review the AC setup for the $AdS_3\times S^3\times T^4$ model, and present the calculation of the coefficients applying the AC method naively. Section 3 is devoted to the same calculation in the WS approach. In section 4 we discuss then the origin of the mismatch between the two results for the particular case of the circular string and show that the difference is due to regularization issues in the sum over the modes. The Section 5 briefly describes the second example of the folded string solution, and finally we discuss open problems.

2. Algebraic Curve determination of the $c_{r,s}^{(1)}$

Introducing the function of the spectral parameter $\widehat{\alpha}(x)$ and the resolvents G, H defined in terms of the Bethe roots $x_{a,k}$ as

$$\widehat{\alpha}(x) = \tfrac{4\pi}{\sqrt{\lambda}}\tfrac{x^2}{x^2-1}\,,$$

$$G_a(x) = \textstyle\sum_{k=1}^{K_a}\frac{\widehat{\alpha}(x_{a,k})}{x-x_{a,k}}, \qquad H_a(x) = \sum_{k=1}^{K_a}\frac{\widehat{\alpha}(x)}{x-x_{a,k}}, \qquad \overline{G}(x) = G(1/x),\quad \overline{H}(x) = H(1/x). \tag{3}$$

the finite-gap equations can be written as:

$$2\pi n_1 = -H_2 - \overline{H}_{\overline{2}} + \frac{G_2(0) + x\,G_2'(0)}{x^2-1} - \frac{G_{\overline{2}}(0) + x\,G_{\overline{2}}'(0)}{x^2-1}, \tag{4}$$

$$2\pi n_2 + \frac{4\pi\mathcal{J}x}{x^2-1} = 2H_2 - H_1 - H_3 + 2\overline{H}_{\overline{2}} - \overline{H}_{\overline{1}} - \overline{H}_{\overline{3}} + 2\,\frac{G_{\overline{2}}(0) - G_2(0)}{x^2-1}, \tag{5}$$

$$2\pi n_3 = -H_2 - \overline{H}_{\overline{2}} + \frac{G_2(0) + x\,G_2'(0)}{x^2-1} - \frac{G_{\overline{2}}(0) + x\,G_{\overline{2}}'(0)}{x^2-1}, \tag{6}$$

$$2\pi n_{\overline{1}} = -H_{\overline{2}} - \overline{H}_2 - \frac{G_2(0) + x\,G_2'(0)}{x^2-1} + \frac{G_{\overline{2}}(0) + x\,G_{\overline{2}}'(0)}{x^2-1}, \tag{7}$$

$$2\pi n_{\overline{2}} - \frac{4\pi\mathcal{J}x}{x^2-1} = 2H_{\overline{2}} - H_{\overline{1}} - H_{\overline{3}} + 2\overline{H}_2 - \overline{H}_1 - \overline{H}_3 - 2\,\frac{G_{\overline{2}}(0) - G_2(0)}{x^2-1}, \tag{8}$$

$$2\pi n_{\overline{3}} = -H_{\overline{2}} - \overline{H}_2 - \frac{G_2(0) + x\,G_2'(0)}{x^2-1} + \frac{G_{\overline{2}}(0) + x\,G_{\overline{2}}'(0)}{x^2-1}. \tag{9}$$

Following on the work of [11], we define the quasimomenta in terms of the resolvents as ($\mathcal{J}$ is the angular momentum of the string):

$$p_1 = -p_4 = -\frac{1}{2}H_1 - \frac{1}{2}\overline{H}_{\overline{1}} - \frac{1}{2}H_3 - \frac{1}{2}\overline{H}_{\overline{3}} - \frac{2\pi\mathcal{J}x}{x^2-1} + \frac{x}{x^2-1}[G_2'(0) - G_{\overline{2}}'(0)], \tag{10}$$

$$p_2 = -p_3 = H_2 + \overline{H}_{\overline{2}} - \frac{1}{2}H_1 - \frac{1}{2}\overline{H}_{\overline{1}} - \frac{1}{2}H_3 - \frac{1}{2}\overline{H}_{\overline{3}} - \frac{2\pi\mathcal{J}x}{x^2-1}, \tag{11}$$

$$p_{\overline{1}} = -p_{\overline{4}} = -\frac{1}{2}H_{\overline{1}} - \frac{1}{2}\overline{H}_1 - \frac{1}{2}H_{\overline{3}} - \frac{1}{2}\overline{H}_3 + \frac{2\pi\mathcal{J}x}{x^2-1} + \frac{x}{x^2-1}[G_{\overline{2}}'(0) - G_2'(0)], \tag{12}$$

$$p_{\overline{2}} = -p_{\overline{3}} = H_{\overline{2}} + \overline{H}_2 - \frac{1}{2}H_{\overline{1}} - \frac{1}{2}\overline{H}_1 - \frac{1}{2}H_{\overline{3}} - \frac{1}{2}\overline{H}_3 + \frac{2\pi\mathcal{J}x}{x^2-1}. \tag{13}$$

Up to winding contributions, we have

$$p_{1,2,3,4}(x) = p_{\overline{1},\overline{2},\overline{3},\overline{4}}(1/x)\,. \tag{14}$$

The above finite-gap equations are obtained with $p_i - p_j = 2\pi n_{ij}$ where

$$(i,j) = (1,2),(2,3),(3,4), \qquad (\overline{1},\overline{2}),(\overline{2},\overline{3}),(\overline{3},\overline{4}). \tag{15}$$

The eight functions $p_i(x)$ and their branch cuts define the classical algebraic curve for the $AdS_3 \times S^3 \times T^4$ background: Note that here the algebraic curve is a connected sum of two pieces interchanged by the $x \to 1/x$ transformation, while in the $AdS_5 \times S^5$ case the curve is a single connected invariant piece. In the recent paper [4] a second, different, set of BE has been derived, using the bootstrap method: It is worth to note that these two versions of the BE, while they are inequivalent at the quantum level, share the same finite-gap limit, and so they lead to equivalent classical algebraic curves.

2.1. Semiclassical one-loop dressing factor

In the semiclassical quantisation of the AC, the first correction to the phase is encoded in a set of potentials V_I correcting the classical quasimomenta that characterize the curve: these potentials compute the effect of the quantum fluctuations around the classical solution; for each quasi-momentum p_I the correction V_I is obtained summing over all the fluctuations connecting the sheets of the AC. The total phase corrections to the Bethe equations are obtained by evaluating

$\mathcal{V} = V_I - V_J$.For the middle node 2 equation (the other BA equations are not corrected) $\mathcal{V}(x) \equiv V_2(x) - V_3(x)$

$$\mathcal{V}(x) = \int_{-1}^{1} \frac{dy}{2\pi} \left[\left(G_2(y) + \overline{G}_{\overline{2}}(y) \right)' \frac{\widehat{\alpha}(x)}{x-y} + \left(\overline{G}_2(y) + G_{\overline{2}}(y) \right)' \frac{\widehat{\alpha}(1/x)}{1/x - y} \right]. \tag{16}$$

If we excite only the node 2, it reduces to

$$\mathcal{V}_2(x) = \int_{-1}^{1} \frac{dy}{2\pi} \left[G_2'(y) \frac{\widehat{\alpha}(x)}{x-y} + \overline{G}_2'(y) \frac{\widehat{\alpha}(1/x)}{1/x-y} \right], \tag{17}$$

where the notation is $\int_{-1}^{1} = \frac{1}{2}\int_{C^+} + \frac{1}{2}\int_{C^-}$ and the half circumferences $C^{\pm}$ (and their orientation) are defined in the caption of figure 4 of [10].

At this point we can evaluate $\mathcal{V}_2(x)$ for large x: Using the relation between the resolvent G_2 and the conserved charges Q_n,

$$G_2(y) = -\sum_{n=0}^{\infty} Q_{n+1}\, y^n, \tag{18}$$

the resulting function of y is not singular anywhere on the circle $|y| = 1$ and the integration is trivial. The result is

$$\mathcal{V}_2(x) = \frac{\widehat{\alpha}(x)}{2\pi} \sum_{r=2}^{\infty} \sum_{s=1}^{\infty} \widehat{c}_{r,s} \frac{Q_r}{x^s}, \tag{19}$$

$$\widehat{c}_{r,s} = -4\left(1 - \frac{1}{2}\delta_{s,1}\right) \frac{1-(-1)^{r+s}}{2} \frac{r-1}{r+s-2}. \tag{20}$$

Here $\widehat{c}_{r,s}$ are the (naive) prediction of the algebraic curve method for the values of the $c_{r,s}^{(1)}$ coefficients which parametrize the phase. A serious problem of this result is that the coefficients $\widehat{c}_{r,s}$ are not antisymmetric, which is an important consistency requirement.

If we now apply the AC method to the particular case of the $SU(2)$ circular string, we get for the 1-loop contribution of the dressing phase to the energy as:

$$\delta E_1^{AC} = \frac{1}{\sqrt{m^2+\mathcal{J}^2}} \left(m^2 + \mathcal{J}^2 \log \frac{\mathcal{J}^2}{m^2+\mathcal{J}^2} \right) - \frac{m^2\left(2\mathcal{J}\left(\mathcal{J}-\sqrt{\mathcal{J}^2+m^2}\right)+m^2\right)}{2(\mathcal{J}^2+m^2)^{3/2}}. \tag{21}$$

In the next sections we show that the part that breaks the antisymmetry in the coefficients also induces a mismatch with the WS calculation for the non-analytic term in the one-loop energy. This disagreement turns out to be due to a regularization ambiguity in the sum over fluctuation frequencies. Once this regularization problem is fixed, the algebraic curve approach result agrees with the string theory result, the the antisymmetry of $c_{r,s}$ is recovered, and the result can be consistently interpreted as a phase.

3. One-loop correction to the energy for the $SU(2)$ string: WS approach

The 1-loop correction to the energy in the WS approach is obtained as usual as a sum over the fluctuation frequences around the classical solution: at the classical level the solution is identical than that of the AdS_5 case, and the frequences in the AdS_3 background can be easily obtained by truncating a certain subset of the fluctuations fields and adjusting the number of independent fluctuations, i.e. removing two bosonic frequencies that correspond to fluctuations in the transverse directions of S^5, and halve the AdS and fermionic contributions. There are also four additional bosonic and four fermionic massless modes coming from the T^4 part of the background, but their contributions cancel among themselves. The result for the one-loop correction to the energy is given by:

$$E_1 = \sum e(n) \ , \tag{22}$$

$$e(n) = \sqrt{1 + \frac{(n+\sqrt{n^2-4m^2})^2}{4(\mathcal{J}^2+m^2)}} + \sqrt{1+\frac{n^2}{\mathcal{J}^2+m^2}} - 2\sqrt{1+\frac{n^2-m^2}{\mathcal{J}^2+m^2}} \ . \tag{23}$$

It is straightforward to check that this sum is UV finite. The dressing contribution to E_1 is obtained isolating the non-analytic part in the sum, following the same method as in [7]: expanding $e(n)$ at large $\mathcal{J}$ and separating out the convergent (i.e. regular) and divergent (i.e. singular) parts, we can write $e(n) = e^{\text{sum}}_{\text{reg}}(n) + e^{\text{sum}}_{\text{sing}}(n)$. To deal with the singular part we define $e^{\text{int}}(x) = e(\mathcal{J}x)$ and expand it for large $\mathcal{J}$ at fixed x, getting $e^{\text{int}}(x) = e^{\text{int}}_{\text{reg}}(x) + e^{\text{int}}_{\text{sing}}(x)$ where $e^{\text{int}}_{\text{sing}}$ is the part whose integral is divergent at $x = 0$. The regular part in one regime is in fact equal to the singular part in the other regime.

$$e^{\text{int}}_{\text{sing}}(x) = e^{\text{sum}}_{\text{reg}}(\mathcal{J}x), \qquad e^{\text{sum}}_{\text{sing}}(n) = e^{\text{int}}_{\text{reg}}(n/\mathcal{J}), \tag{24}$$

so that finally we get:

$$E_1 = E_1^{\text{analytic}} + E_1^{\text{non-analytic}} \ , \tag{25}$$

$$E_1^{\text{analytic}} = \sum e^{\text{sum}}_{\text{reg}}(n), \qquad E_1^{\text{non-analytic}} \equiv \delta E_1 = \int_{-\infty}^{\infty} \mathcal{J} dx e^{\text{int}}_{\text{reg}}(x) \ . \tag{26}$$

leading to:

$$\delta E_1^{AdS_3} = \frac{1}{\sqrt{\mathcal{J}^2+m^2}}\left(m^2 + \mathcal{J}^2 \log\frac{\mathcal{J}^2}{m^2+\mathcal{J}^2}\right) , \tag{27}$$

Clearly there is a mismatch with the previous AC results, the difference being the term:

$$\Delta E_1 = -\frac{m^2\left(2\mathcal{J}\left(\mathcal{J}-\sqrt{\mathcal{J}^2+m^2}\right)+m^2\right)}{2\left(\mathcal{J}^2+m^2\right)^{3/2}} \tag{28}$$

4. Origin of the mismatch

Both problems, the non-antisymmetry of the coefficients and the mismatch ΔE_1, can be traced back to a regularization issue in the sum over the frequencies of the quantum fluctuations: in the WS approach the natural cut-off is a common mode number N, while in the AC approach the cut-off is a common radius for the contour integral in the spectral plane defining the potentials; this difference is translated in a reordering of the terms in the sum over the frequences to compute the 1-lopp energy $E_1 \sim \sum_{n\in Z} (\omega_n^B - \omega_n^F)$, and is the origin of the ΔE_1 term. We can repeat the calculation of $\mathcal{V}$ enforcing the regularization of the AC curve to reproduce the WS result for the circular string, solving both problems at once. At the level of the potential $\mathcal{V}(x)$ the choice of the new regularization corresponds to an integration by parts, i.e. we start with the potential correcting the Bethe equation

$$\mathcal{V}(x) = \int_{-1}^{1}\frac{dy}{2\pi}\left[\left(G_2(y)+\overline{G}_{\overline{2}}(y)\right)'\frac{\widehat{\alpha}(x)}{x-y} + \left(\overline{G}_2(y)+G_{\overline{2}}(y)\right)'\frac{\widehat{\alpha}(1/x)}{1/x-y}\right]. \tag{29}$$

and integrating by parts we define the potential $\widehat{\mathcal{V}}(x)$, correspondin to the new regularization:

$$\widehat{\mathcal{V}}(x) = \int_{-1}^{1} \frac{dy}{2\pi}\Big[\Big(G_2(y) + \overline{G}_{\overline{2}}(y)\Big)' \frac{\widehat{\alpha}(x)}{x-y} - \Big(\overline{G}_2(y) + G_{\overline{2}}(y)\Big)\Big(\frac{\widehat{\alpha}(1/x)}{1/x-y}\Big)'\Big]. \tag{30}$$

The large x expansion of $\widehat{\mathcal{V}}(x)$ is

$$\widehat{\mathcal{V}}(x) = \frac{\widehat{\alpha}(x)}{2\pi} \sum_{r=1}^{\infty}\sum_{s=1}^{\infty} \frac{c_{r,s}^{(1)}\, Q_r - \bar{c}_{r,s}^{(1)}\, \overline{Q}_r}{x^s}, \tag{31}$$

where the expansion coefficients are now *antisymmetric*:

$$c_{r,s}^{(1)} = 2\,\frac{1-(-1)^{r+s}}{2}\,\frac{s-r}{r+s-2}, \qquad \bar{c}_{r,s}^{(1)} = -2\,\frac{1-(-1)^{r+s}}{2}\,\frac{r+s-2}{s-r}. \tag{32}$$

These coefficients (only $c_{r,s}^{(1)}$ is actually contributing) now lead precisely to the string theory expression $\delta E_1^{AdS_3}$ for the non-analytic term in the circular $SU(2)$ string case.

Using the new regularization, we can now compute the scattering phases between magnons, following [10]: We identify the dressing phase contribution in the BE as:

$$e^{i\,\widehat{\mathcal{V}}(x_{2,i})} = \prod_{k\neq j}^{K_2} \sigma^2(x_{2,j}, x_{2,k}) \prod_{k}^{K_{\overline{2}}} \sigma^{-2}(x_{2,j}, x_{\overline{2},k}) = \prod_{k\neq j}^{K_2} e^{i\,\vartheta(x_{2,j},x_{2,k})} \prod_{k}^{K_{\overline{2}}} e^{-i\,\widetilde{\vartheta}(x_{2,j},x_{\overline{2},k})}. \tag{33}$$

Using the discrete definition of the function G and integrating over y we obtain

$$\vartheta(x,y) = -\frac{\widehat{\alpha}(x)\,\widehat{\alpha}(y)}{2\pi\,(x-y)^2}\Big[2\,\log\Big(\frac{x+1}{x-1}\frac{y-1}{y+1}\Big) + 2\,\frac{(x-y)\,(x^2+y^2-2)}{(x^2-1)(y^2-1)}\Big], \tag{34}$$

$$\widetilde{\vartheta}(x,y) = -\frac{\widehat{\alpha}(x)\,\widehat{\alpha}(y)}{2\pi\,(1-x\,y)^2}\Big[2\,\log\Big(\frac{x+1}{x-1}\frac{y-1}{y+1}\Big) - 2\,\frac{(x-y)\,(x^2y^2-1)}{(x^2-1)\,(y^2-1)}\Big]. \tag{35}$$

5. Folded string and open problems

The choice of regularization used in the previous section allows to separate the original potential $\mathcal{V} = \mathcal{V}_{phase} + \delta\mathcal{V}$, where only the first part can be consistently interpreted as a phase, while the second is understood as a regularization effect. Nevertheless the prescription is based on the agreement between the AC and WS approaches in the particular case of the $SU(2)$ curcular string. If we consider as a second example the $SL(2)$ folded string solution, we can repeat the same steps: computing the 1-loop dressing contribution to the energy for the folded string, the AC and WS give different result, and the mismatch is again due to the different regularizations. While we can remove the discrepancy and get a consistent result on the string theory side, if we compare the string prediction with the BE, using the phase derived in the previous section we still have a discrepancy.

$$\text{WS} \equiv \text{AC-reg. mismatch}: \quad E_1^{\text{dressing}} = \frac{\coth^{-1}\Big(\sqrt{\mathcal{J}^2+1}\Big)}{2\mathcal{J}^3\sqrt{\mathcal{J}^2+1}}\,\mathcal{S}^2 + \mathcal{O}(\mathcal{S}^3),$$

$$\text{BE with } c_{r,s} \text{ coeff.}: \quad E_1^{\text{dressing}} = \Big[\frac{\coth^{-1}\Big(\sqrt{\mathcal{J}^2+1}\Big)}{2\mathcal{J}^3\sqrt{\mathcal{J}^2+1}} + \frac{1}{2\,\mathcal{J}^4\,\sqrt{\mathcal{J}^2+1}}\Big]\,\mathcal{S}^2 + \mathcal{O}(\mathcal{S}^3).$$

where the first line is the string theory result for a folded string with semiclassical spin $\mathcal{S}$ and angular momentum $\mathcal{J}$, the second line is the result obtained from the BE equations, assuming the coefficients $c_{r,s}$ for the dressing phase.

This disagreement calls for a deeper understanding of the role of $\delta\mathcal{V}$; while the regularization/antisymmetrization prescription works perfectly in the $SU(2)$ sector, the comparison between string theory and BE predictions for more general cases is still problematic.

Another open issue is related to the fact that the magnon phases in eq.(35) don't satisfy the crossing relations found in [4]: to get crossing respecting phases the coefficients $c_{r,s}$ with $r = 1$ should be modified with an additional factor 1/2 [12, 13]. The relative phases for the magnon scattering can be found starting from the original potential in eq.(16), and making the resulting phase antisymmetric by hand $\theta^{AC,asym}(x,y) = 1/2[\theta^{AC}(x,y) - \theta^{AC}(y,x)]$. But this phase, while crossing symmetric, gives a contribution to the 1-loop energy not in agreement with the string theory WS prediction, even in the case of the circular string solution.

References

[1] N. Beisert, C. Ahn, L. F. Alday, Z. Bajnok, J. M. Drummond, L. Freyhult, N. Gromov and R. A. Janik *et al.*, Lett. Math. Phys. **99** (2012) 3 [arXiv:1012.3982 [hep-th]].
[2] K. Zarembo, *Algebraic Curves for Integrable String Backgrounds*, arXiv:1005.1342.
[3] A. Babichenko, J. Stefanski, B., and K. Zarembo, *Integrability and the AdS(3)/CFT(2) correspondence*, *JHEP* **1003** (2010) 058
[4] R. Borsato, O. Ohlsson Sax, A. Sfondrini, B. Stefanski, Jr. and A. Torrielli, *The all-loop integrable spin-chain for strings on $AdS_3 \times S^3 \times T^4$: the massive sector* arXiv:1303.5995 [hep-th].
[5] M. Beccaria, F. Levkovich-Maslyuk, G. Macorini and A. A. Tseytlin, *Quantum corrections to spinning superstrings in $AdS_3xS^3xM^4$: determining the dressing phase* JHEP **1304** (2013) 006
[6] M. Beccaria and G. Macorini, *Quantum corrections to short folded superstring in $AdS_3xS^3xM^4$* JHEP **1303** (2013) 040
[7] N. Beisert and A. A. Tseytlin, *On quantum corrections to spinning strings and Bethe equations*, *Phys.Lett.* **B629** (2005) 102–110,
[8] N. Beisert and T. Klose, "Long-range gl(n) integrable spin chains and plane-wave matrix theory," J. Stat. Mech. **0607**, P07006 (2006) [hep-th/0510124].
[9] G. Arutyunov, S. Frolov, and M. Staudacher, *Bethe ansatz for quantum strings*, *JHEP* **0410** (2004) 016
[10] N. Gromov and P. Vieira, *Constructing the AdS/CFT dressing factor*, *Nucl.Phys.* **B790** (2008) 72–88,
[11] N. Beisert, V. Kazakov, K. Sakai, and K. Zarembo, "The Algebraic curve of classical superstrings on $AdS_5 \times S^5$", Commun. Math. Phys. **263** (2006) 659–710 [hep-th/0502226].
[12] R. Borsato, O. O. Sax, A. Sfondrini, B. Stefanski and A. Torrielli, *Dressing phases of AdS3/CFT2* arXiv:1306.2512 [hep-th].
[13] M. C. Abbott, *The AdS3 x S3 x S3 x S1 Hernandez-Lopez Phases: a Semiclassical Derivation* arXiv:1306.5106 [hep-th].

Physics and Mathematics of Nonlinear Phenomena 2013 (PMNP2013) IOP Publishing
Journal of Physics: Conference Series **482** (2014) 012004 doi:10.1088/1742-6596/482/1/012004

SU(3) magnet: finite-gap integration on the lowest genus curve

Julia Bernatska, Petro Holod

National University of 'Kyiv Mohyla Academy', G. Skovorody St., 2, 04655 Kyiv, UA

E-mail: BernatskaJM@ukma.kiev.ua, Holod@ukma.kiev.ua

Abstract. We consider the integrable system of isotropic SU(3) Landau-Lifshits equation as a Hamiltonian system on a coadjoint orbit of the SU(3) loop group. We connect the mentioned equation with an isotropic SU(3) magnet because it describes the mean fields of magnetic and quadrupole moments in a spin-1 lattice. For the system of isotropic SU(3) Landau-Lifshits equation we perform separation of variables in Sklyanin's manner, and integrate in the lowest finite gap where the spectral curve is elliptic.

1. Introduction

The interest to integrable continuous spin chains starts from [1], where the isotropic Landau-Lifshitz equation was considered as a spin equation of motion, soliton and periodic solutions were obtained by the method of traveling wave. Besides numerical investigations of soliton-soliton scattering and stability of soliton solutions in [2, 3], the inverse scattering method was applied to the continuous Heisenberg spin chain in [4]. All the mentioned papers deal with SU(2) spin chain, and the Landau-Lifshitz equation serves as an equation of motion. A generalization to SU(n) Heisenberg spin chain (Hamiltonian with bilinear exchange only) derived from the vector nonlinear Schödinger equation was presented in [5]. In [6] the integrable SU(3) spin chain with the most general Hamiltonian was introduced as a system with the Lax-pair representation, and shown to be gauge equivalent to a set of nonlinear Schödinger-type equations. Separation of variables for the classical integrable SL(3) magnetic chain was developed in [7]. In [8] the isotropic SU(3) Landau-Lifshitz equation together with two- and three-component nonlinear Schödinger-type equations were constructed by the orbit method. In [9] the isotropic SU(3) Landau-Lifshitz equation is derived as a continuum limit of the quantum spin-1 lattice system; this and a more general equations were constructed by means of the orbit method. The latter generalization of the Landau-Lifshitz equation, obtained over a generic coadjoint orbit of SU(3), is similar to the equation of motion for the SU(3) spin chain with the most general Hamiltonian from [6].

In the present paper we apply Sklyanin's separation of variables to the isotropic SU(3) Landau-Lifshitz equation over a generic orbit. In contrast to the classical integrable SL(3) magnetic chain from [7], whose phase space is endowed with a quadratic Poisson bracket and the monodromy matrix serves as an L-operator, we consider the system whose dynamic variables describe the mean fields of magnetic and quadrupole moments and phase space is endowed with a Lie-Poisson bracket. Nevertheless, we obtain similar expressions for variables of separation. Then we proceed to finite-gap integration of the isotropic SU(3) Landau-Lifshitz equation.

According to [10] a half of variables of separation serve as poles of the Baker-Akhiezer function. For the system over a generic orbit the number of these poles exceeds the genus of the spectral curve by two, so integration leads to inversion of an extended Abel-Jacobi map by means of generalized θ-functions [11, 12]. The similar situation is considered in [13]. We focus on the simplest one-gap case with an elliptic spectral curve, and express Abelian integrals in terms of the Weierstrass elliptic function.

The structure of the paper is the following. In Section 2 we describe quantum and classical models of SU(3) spin-1 chain. The quantum spin operators are represented over the 3-dimensional space of irreducible representation of the group SU(2). We construct quadrupole operators from the tensor operators of weight 2. These operators allow to linearize the system with the bilinear-biquadratic exchange Hamiltonian. By the mean field approximation we take the continuum limit and obtain a dynamical system coinciding with the one constructed over a degenerate coadjoint orbit of SU(3). At the same time another dynamical system is constructed over a generic coadjoint orbit of SU(3). For the latter system in Section 3 we develop the scheme of separation of variables in Sklyanin's manner, and prove that variables of separation are quasi-canonical. We solve the problem of computing dynamic variables from the variables of separation. Then we perform integration for the one-gap case in terms of the Weierstrass elliptic function.

2. Quantum and classical models of SU(3) magnet

2.1. Quantum spin-1 lattice

Suppose we have a lattice of atoms of spin 1. Every atom is modeled by a tuple of spin and quadrupole operators. In contrast to the well-known Heisenberg model considering two spin projections and representing spin operators by the Pauli matrices, we take into account all three projections of the spin momentum and represent spin operators over the 3-dimensional space:

$$\hat{S}_x = \frac{1}{\sqrt{2}}\begin{pmatrix} 0 & 0 & 1 \\ 0 & 0 & 1 \\ 1 & 1 & 0 \end{pmatrix}, \quad \hat{S}_y = \frac{1}{\sqrt{2}}\begin{pmatrix} 0 & 0 & -\mathrm{i} \\ 0 & 0 & \mathrm{i} \\ \mathrm{i} & -\mathrm{i} & 0 \end{pmatrix}, \quad \hat{S}_z = \begin{pmatrix} 1 & 0 & 0 \\ 0 & -1 & 0 \\ 0 & 0 & 0 \end{pmatrix}.$$

The corresponding matrix algebra $\mathrm{Mat}_{3\times 3}$ is not exhausted by these three operators, so we add six operators more. We choose the following operators and call them quadrupole:

$$\hat{P}_{xy} = \hat{S}_x\hat{S}_y + \hat{S}_y\hat{S}_x, \quad \hat{P}_{xz} = \hat{S}_x\hat{S}_z + \hat{S}_z\hat{S}_x, \quad \hat{P}_{yz} = \hat{S}_y\hat{S}_z + \hat{S}_z\hat{S}_y,$$
$$\hat{P}_{xx} = \hat{S}_x^2, \quad \hat{P}_{yy} = \hat{S}_y^2, \quad \hat{P}_{zz} = \hat{S}_z^2.$$

Introducing the quadrupole operators we have an opportunity to take into account the second order of exchange interaction and use the bilinear-biquadratic exchange Hamiltonian, easily reduced to a bilinear form:

$$\hat{\mathcal{H}} = -\textstyle\sum_{n,\delta}\Big\{J\big(\hat{\boldsymbol{S}}_n, \hat{\boldsymbol{S}}_{n+\delta}\big) + K\big(\hat{\boldsymbol{S}}_n, \hat{\boldsymbol{S}}_{n+\delta}\big)^2\Big\}$$
$$= -(J - \tfrac{1}{2}K)\textstyle\sum_{n,\delta}\big(\hat{\boldsymbol{S}}_n, \hat{\boldsymbol{S}}_{n+\delta}\big) - \tfrac{1}{2}K\textstyle\sum_{n,\delta}\big(\hat{\boldsymbol{P}}_n, \hat{\boldsymbol{P}}_{n+\delta}\big) - \tfrac{4}{3}KN,$$

where J and K are interaction constants, n runs over all N sites of the lattice and δ runs over the nearest-neighbour sites. We use the notation: $\hat{\boldsymbol{S}}_n = \big(\hat{S}_x^{(n)}, \hat{S}_y^{(n)}, \hat{S}_z^{(n)}\big)$, $\hat{\boldsymbol{P}}_n = \big(\hat{P}_{xy}^{(n)}, \hat{P}_{xz}^{(n)}, \hat{P}_{yz}^{(n)}, \hat{P}_{xx}^{(n)} - \hat{P}_{yy}^{(n)}, \sqrt{3}(\hat{P}_{zz}^{(n)} - \tfrac{2}{3}\mathbb{I})\big)$, where every entry $\hat{O}^{(n)}$ is a tensor product over all sites of $N-1$ identity operators $\mathbb{I}$ and the operator $\hat{O}$ at the nth position.

Next, we apply the mean field approximation under the broken symmetry stimulated by an external magnetic field. This guarantees nonzero values of the quasiaverage fields corresponding to the spin and quadrupole operators:

$$\mu_1(x_n) = \langle\hat{S}_x^{(n)}\rangle, \quad \mu_2(x_n) = \langle\hat{S}_y^{(n)}\rangle, \quad \mu_3(x_n) = \langle\hat{S}_z^{(n)}\rangle,$$

$$\tau_{12}(x_n) = \langle \hat{P}_{xy}^{(n)} \rangle, \quad \tau_{13}(x_n) = \langle \hat{P}_{xz}^{(n)} \rangle, \quad \tau_{23}(x_n) = \langle \hat{P}_{yz}^{(n)} \rangle,$$
$$\tau_{11}(x_n) = \langle \hat{P}_{xx}^{(n)} \rangle - \tfrac{2}{3}, \quad \tau_{22}(x_n) = \langle \hat{P}_{yy}^{(n)} \rangle - \tfrac{2}{3}, \quad \tau_{33}(x_n) = \langle \hat{P}_{zz}^{(n)} \rangle - \tfrac{2}{3}.$$

We arrange the mean fields into the magnetic moment vector-field $\boldsymbol{\mu} = (\mu_1, \mu_2, \mu_3)$, and the quadrupole moment tensor-field $\hat{\tau} = (\tau_{ij})$, which is traceless. In the mean field Hamiltonian the operators of neighbors in number z are replaced by the corresponding mean fields:

$$\begin{aligned}\hat{\mathcal{H}}_{\mathrm{MF}} &= -(J - \tfrac{1}{2}K)z \sum_n \big(\hat{\boldsymbol{S}}_n, \langle \hat{\boldsymbol{S}}_n \rangle\big) - \tfrac{1}{2}Kz \sum_n \big(\hat{\boldsymbol{P}}_n, \langle \hat{\boldsymbol{P}}_n \rangle\big) - \tfrac{4}{3}KNz = \\ &= -(J - \tfrac{1}{2}K)z \sum_n \big(\mu_1 \hat{S}_x^{(n)} + \mu_2 \hat{S}_y^{(n)} + \mu_3 \hat{S}_z^{(n)}\big) - \tfrac{1}{2}Kz \sum_n \big(\tau_{12}\hat{P}_{xy}^{(n)} + \\ &+ \tau_{13}\hat{P}_{xz}^{(n)} + \tau_{23}\hat{P}_{yz}^{(n)} + (\tau_{11} - \tau_{22})\big(\hat{P}_{xx}^{(n)} - \hat{P}_{yy}^{(n)}\big) + 3\tau_{33}\big(\hat{P}_{zz}^{(n)} - \tfrac{2}{3}\mathbb{I}\big)\big) - \tfrac{4}{3}KNz.\end{aligned}$$

In what follows we consider the case of $K = J$, when the Hamiltonian is uniform in the spin and quadrupole operators:

$$\hat{\mathcal{H}}_{\mathrm{MF}} = \tfrac{1}{2}Jz \sum_n \sum_a \mu_a \hat{S}_a^{(n)} - \tfrac{4}{3}JNz.$$

We call this case isotropic SU(3) because the operators $\{\hat{S}_x^{(n)}, \hat{S}_y^{(n)}, \hat{S}_z^{(n)}, \hat{P}_{xy}^{(n)}, \hat{P}_{xz}^{(n)}, \hat{P}_{yz}^{(n)}, \hat{P}_{xx}^{(n)} - \hat{P}_{yy}^{(n)}, \sqrt{3}(\hat{P}_{zz}^{(n)} - \tfrac{2}{3}\mathbb{I})\}$ form a basis of the algebra $\mathfrak{isu}(3)$; we denote them all together by $\{\hat{S}_a^{(n)}\}_{a=1}^8$, and the corresponding mean fields by $\{\mu_a\}_{a=1}^8$, which are functions in the spatial coordinate x and the time t. For the mean field model we write Heisenberg's equations of motion

$$\mathrm{i}\hbar \frac{d\hat{S}_a^{(n)}}{dt} = [\hat{S}_a^{(n)}, \hat{\mathcal{H}}_{\mathrm{MF}}],$$

and take the continuum limit. Neglecting correlations between fluctuations of the operator quantum fields

$$\langle \hat{S}_a^{(n)} \hat{S}_b^{(n)} \rangle \approx \langle \hat{S}_a^{(n)} \rangle \langle \hat{S}_b^{(n)} \rangle = \mu_a(x_n, t)\mu_b(x_n, t)$$

we obtain the following equation:

$$\hbar \frac{\partial \mu_a}{\partial t} = J\ell V_0\, C_{abc}\mu_b \frac{\partial^2 \mu_c}{\partial x^2}, \tag{1}$$

where ℓ is the lattice distance, V_0 is an infinitesimal volume, C_{abc} are structure constants of the algebra $\mathfrak{isu}(3)$ with respect to the basis mentioned above. It is easy to see that the equation is Hamiltonian and write the corresponding effective Hamiltonian:

$$\hbar \frac{\partial \mu_a}{\partial t} = \{\hat{\mathcal{H}}_{\mathrm{eff}}, \mu_a\} = V_0 C_{abc}\mu_b \frac{\delta \hat{\mathcal{H}}_{\mathrm{eff}}}{\delta \mu_c}, \qquad \hat{\mathcal{H}}_{\mathrm{eff}} = \frac{J\ell}{2} \int \sum_a \left(\frac{\partial \mu_a}{\partial x}\right)^2 dx.$$

In the case of d-dimensional isotropic lattice the effective Hamiltonian has the form

$$\hat{\mathcal{H}}_{\mathrm{eff}} = \frac{J}{2\ell^{d-2}} \int \sum_a \left\langle \frac{\partial \mu_a}{\partial \boldsymbol{x}}, \frac{\partial \mu_a}{\partial \boldsymbol{x}} \right\rangle d\boldsymbol{x}.$$

For more details see [9].

2.2. Dynamical system of isotropic SU(3) magnet

Here we construct a dynamical system on a coadjoint orbit in the dual space $\widetilde{\mathfrak{g}}^*$ to the loop algebra $\widetilde{\mathfrak{g}} = \mathfrak{isu}(3) \times \mathcal{P}(\lambda, \lambda^{-1})$ with respect to the bilinear form $\langle \hat{A}(\lambda), \hat{B}(\lambda) \rangle = \mathrm{res}_{\lambda=0}\, \lambda^{-N}\, \mathrm{Tr}\, \hat{A}(\lambda)\hat{B}(\lambda)$. In $\mathfrak{isu}(3)$ we introduce the same basis: $\{\hat{S}_x, \hat{S}_y, \hat{S}_z, \hat{P}_{xy}, \hat{P}_{xz}, \hat{P}_{yz}, \hat{P}_{xx} - \hat{P}_{yy}, \sqrt{3}(\hat{P}_{zz} - \frac{2}{3}\mathbb{I})\}$ and denote its elements by $\{\hat{S}_a\}$. The loop algebra $\widetilde{\mathfrak{g}}$ is homogeneous and consists of the eigenspaces $\mathfrak{g}_m$, $m \in \mathbb{Z}$ spanned over the basis elements $\{\lambda^m \hat{S}_a\}_{a=1}^8$. Evidently, $\widetilde{\mathfrak{g}}^*$ coincides with $\widetilde{\mathfrak{g}}$.

According to the Kostant-Adler scheme [14] $\widetilde{\mathfrak{g}}$ is decomposed into two subalgebras: $\widetilde{\mathfrak{g}}_+$ and $\widetilde{\mathfrak{g}}_-$ including the eigenspaces of nonnegative and negative degrees, respectively. Consider the space $\mathcal{M} \subset \widetilde{\mathfrak{g}}^*$ of elements

$$\hat{L}(\lambda) = \frac{\lambda^N}{2}\begin{pmatrix} m+q & 0 & 0 \\ 0 & -m+q & 0 \\ 0 & 0 & -2q \end{pmatrix} + \frac{1}{2}\begin{pmatrix} \mu_3(\lambda) + \tau_{33}(\lambda) & \tau_{11}(\lambda) - \tau_{22}(\lambda) - 2\mathrm{i}\tau_{12}(\lambda) & \frac{1}{\sqrt{2}}\left(\mu_1(\lambda) - 2\mathrm{i}\tau_{23}(\lambda) - \mathrm{i}\mu_2(\lambda) + 2\tau_{13}(\lambda)\right) \\ \tau_{11}(\lambda) - \tau_{22}(\lambda) + 2\mathrm{i}\tau_{12}(\lambda) & -\mu_3(\lambda) + \tau_{33}(\lambda) & \frac{1}{\sqrt{2}}\left(\mu_1(\lambda) - 2\mathrm{i}\tau_{23}(\lambda) + \mathrm{i}\mu_2(\lambda) - 2\tau_{13}(\lambda)\right) \\ \frac{1}{\sqrt{2}}\left(\mu_1(\lambda) + 2\mathrm{i}\tau_{23}(\lambda) + \mathrm{i}\mu_2(\lambda) + 2\tau_{13}(\lambda)\right) & \frac{1}{\sqrt{2}}\left(\mu_1(\lambda) + 2\mathrm{i}\tau_{23}(\lambda) - \mathrm{i}\mu_2(\lambda) - 2\tau_{13}(\lambda)\right) & -2\tau_{33}(\lambda) \end{pmatrix} \quad (2)$$

with constant m and q. We denote the entries of $\hat{L}$ by L_{ij}, namely: $L_{ij}(\lambda) = \sum_{m=0}^{N-1} L_{ij}^{(m)} \lambda^m$ and use the set $\{L_{ij}^{(m)}\}$, $m = 0, 1, \ldots, N-1$ as dynamic variables in $\mathcal{M}$. Obviously, $\mathcal{M} \subset (\widetilde{\mathfrak{g}}_+)^*$ with respect to the bilinear form, and the factor-algebra $\widetilde{\mathfrak{g}}_+ / \sum_{m \geqslant N} \mathfrak{g}_m$ acts effectively on $\mathcal{M}$. We call $\mathcal{M}$ the N-gap phase space. In $\mathcal{M}$ we introduce the following Lie-Poisson bracket

$$\{f_1, f_2\} = \sum_{m,n=0}^{N-1} \sum_{a,b=1}^{\mathrm{rank}\,\mathfrak{g}} P_{ab}^{mn}(N-1) \frac{\partial f_1}{\partial L_a^{(m)}} \frac{\partial f_2}{\partial L_b^{(n)}}, \quad (3)$$

$$P_{ab}^{mn}(N-1) = \langle \hat{L}(\lambda), [\hat{S}_a^{-m+N-1}, \hat{S}_b^{-n+N-1}] \rangle, \qquad f_1, f_2 \in \mathcal{C}(\mathcal{M}),$$

giving the Poisson structure

$$\{L_{ij}^{(m)}, L_{kl}^{(n)}\} = L_{kj}^{(m+n+1-N)} \delta_{il} - L_{il}^{(m+n+1-N)} \delta_{kj}.$$

Then we introduce the invariant functions $h_0, h_1, \ldots, h_{2N-1}, f_0, f_1, \ldots, f_{3N-1}$ generated by

$$I_2(\lambda) \equiv \tfrac{1}{2}\, \mathrm{Tr}\, \hat{L}^2(\lambda) = \sum_{k=0}^{2N} h_k \lambda^k, \quad I_3(\lambda) \equiv \tfrac{1}{3}\, \mathrm{Tr}\, \hat{L}^3(\lambda) = \sum_{k=0}^{3N} f_k \lambda^k.$$

All these functions are in involution with respect to the Lie-Poisson bracket (3), so they are integrals of motion. Moreover, $h_0, h_1, \ldots, h_{N-1}, f_0, f_1, \ldots, f_{N-1}$ are functionally independent and annihilate the bracket, they define an orbit by means of the orbit equations

$$h_0 = c_0, \quad h_1 = c_1, \quad \ldots, \quad h_{N-1} = c_{N-1}, \quad f_0 = b_0, \quad f_1 = b_1, \quad \ldots, \quad f_{N-1} = b_{N-1}. \quad (4)$$

Other $3N$ integrals of motion serve as Hamiltonians for the dynamical system.

We recall the orbit structure of SU(3). It has two types of coadjoint orbits: generic $\mathcal{O}_{\mathrm{gen}}^{\mathrm{SU}(3)}$ of dimension 6, and degenerate $\mathcal{O}_{\mathrm{deg}}^{\mathrm{SU}(3)}$ of dimension 4:

$$\mathcal{O}_{\mathrm{gen}}^{\mathrm{SU}(3)} = \frac{\mathrm{SU}(3)}{\mathrm{U}(1) \times \mathrm{U}(1)}, \qquad \mathcal{O}_{\mathrm{deg}}^{\mathrm{SU}(3)} = \frac{\mathrm{SU}(3)}{\mathrm{SU}(2) \times \mathrm{U}(1)}.$$

Every eigenspace $\mathfrak{g}_m \subset \mathcal{M}$, $m = 0, 1, \ldots, N-1$, is a foliation into such orbits under the coadjoint action of SU(3). An initial condition defines what orbit is chosen in every $\mathfrak{g}_m$, and we call their direct product an orbit of the dynamical system. So the phase space $\mathcal{M}$ is a foliation into the leaves which are direct products of N samples of coadjoint orbits of SU(3). The system of isotropic SU(3) magnet lives in $\mathfrak{g}_0$, the evolution in all other eigenspaces is defined by $\{L^{(0)}_{ij}\}$ and their derivatives with respect to x.

To proceed we introduce the spectral curve $\mathcal{R}$: $P(w,\lambda) \equiv \det\big(\hat{L}(\lambda) - w\,\mathbb{I}\big) = 0$. In the generic case it is trigonal of genus $3N-2$:

$$w^3 - w\, I_2(\lambda) - I_3(\lambda) = 0, \tag{5}$$

the orbit equations are written explicitly (4). In the degenerate case the spectral curve is decomposed as follows:

$$\big[w \mp 2\big(\tfrac{1}{3}\, I_2(\lambda)\big)^{1/2}\big]\big[w \pm \big(\tfrac{1}{3}\, I_2(\lambda)\big)^{1/2}\big]^2 = 0$$

because of the relation $\big(\frac{1}{3} I_2(\lambda)\big)^3 = \big(-\frac{1}{2} I_3(\lambda)\big)^2$, and is reduced to a hyperelliptic one. The orbit equations are not so obvious as in the generic case.

The dynamical system in question is constructed from the Hamiltonian flows of h_N and h_{N+1}:

$$\frac{L^{(m)}_a}{\partial x} = \{L^{(m)}_a, h_N\} \qquad\qquad \frac{L(\lambda)}{\partial x} = [\nabla h_N, L(\lambda)]$$
$$\text{or}$$
$$\frac{L^{(m)}_a}{\partial t} = \{L^{(m)}_a, h_{N+1}\} \qquad\qquad \frac{L(\lambda)}{\partial t} = [\nabla h_{N+1}, L(\lambda)],$$

where ∇ denotes the matrix gradient

$$\nabla \mathcal{H} = \sum_{m=0}^{N-1} \sum_{a=1}^{\operatorname{rank}\mathfrak{g}} \frac{\partial \mathcal{H}}{\partial L^{(m)}_a} \hat{Z}_a^{N-m-1}, \qquad L^{(m)}_a = \langle \hat{L}(\lambda), \hat{Z}_a^{N-m-1}\rangle.$$

Commutativity of the Hamiltonian flows implies the compatibility condition

$$\frac{\partial \nabla h_N}{\partial t} - \frac{\partial \nabla h_{N+1}}{\partial x} + [\nabla h_N, \nabla h_{N+1}] = 0 \quad \Rightarrow \quad \frac{\partial L^{(0)}_a}{\partial t} = \frac{\partial L^{(1)}_a}{\partial x}, \tag{6}$$

which is the zero-curvature equation. We obtain different equations of motion over two types of orbits. Over a degenerate orbit the equation (6) leads to

$$\frac{\partial \mu_a}{\partial t} = \tfrac{2}{3h_0} C_{abc}\mu_b \frac{\partial^2 \mu_c}{\partial x^2} + \tfrac{h_1}{2h_0} \frac{\partial \mu_a}{\partial x}$$

with $\mu_a(x,t) \equiv L^{(0)}_a(x,t)$. The reader can recognize this equation as (1), obtained for the mean fields when correlations between fluctuations are neglected. Over a generic orbit the zero-curvature equation (6) acquires the more complicate form

$$\frac{\partial \mu_a}{\partial t} = \tfrac{2}{4c_0^3 - 27b_0^2} C_{abc}\bigg(c_0^2 \mu_b \frac{\partial^2 \mu_c}{\partial x^2} - \tfrac{27}{2} b_0 \Big(\mu_b \frac{\partial^2 \eta_c}{\partial x^2} + \eta_b \frac{\partial^2 \mu_c}{\partial x^2}\Big) + \tfrac{27}{4} c_0 \eta_b \frac{\partial^2 \eta_c}{\partial x^2}\bigg) +$$
$$+ \tfrac{2c_1c_0^2 - 9b_1b_0}{4c_0^3 - 27b_0^2} \frac{\partial \mu_a}{\partial x} + \tfrac{9}{2} \tfrac{2b_1c_0 - 3c_1b_0}{4c_0^3 - 27b_0^2} \frac{\partial \eta_a}{\partial x}. \tag{7}$$

The latter also serves as equation of motion with the effective Hamiltonian ($c_1 = b_1 = 0$)

$$\hat{\mathcal{H}}_{\text{eff}} = \tfrac{1}{4c_0^3 - 27b_0^2} \int \sum_a \left(c_0^2 \Big(\frac{\partial \mu_a}{\partial x}\Big)^2 - 27 b_0 \frac{\partial \mu_a}{\partial x}\frac{\partial \eta_a}{\partial x} + \tfrac{27}{4} c_0 \Big(\frac{\partial \eta_a}{\partial x}\Big)^2 \right) dx,$$

where $\eta_a = d_{abc}\mu_b\mu_c$ with the symmetric tensor $d_{abc} = \operatorname{Tr}(\hat{S}_a\hat{S}_b\hat{S}_c + \hat{S}_b\hat{S}_a\hat{S}_c)$. This dynamical system was firstly presented in [6], and constructed by means of the orbit method in [9].

Journal of Physics: Conference Series **482** (2014) 012004 doi:10.1088/1742-6596/482/1/012004

3. Integration of the system over a generic orbit

3.1. Separation of variables

We start the process of solving (7) from separation of variables. In [10] Sklyanin declared

Conjecture. *Suppose the orbit* $\mathcal{O}^{SU(3)}_{gen}$ *has the coordinates* $\{L^{(m)}_{11}, L^{(m)}_{12}, L^{(m)}_{21}, L^{(m)}_{22}, L^{(m)}_{31}, L^{(m)}_{32} : m=0,\ldots,N-1\}$ *as above. Then the new coordinates* $\{(\lambda_k, w_k) : k=1,\ldots,3N\}$, *defined by the formulas*

$$\mathcal{B}(\lambda_k) = 0, \qquad w_k = \mathcal{A}(\lambda_k),$$

where

$$\mathcal{B}(\lambda) = \big[L_{11}(\lambda) - L_{22}(\lambda)\big]L_{31}(\lambda)L_{32}(\lambda) - L_{12}(\lambda)L^2_{31}(\lambda) + L_{21}(\lambda)L^2_{32}(\lambda),$$
$$\mathcal{A}(\lambda_k) = L_{11}(\lambda_k) - \frac{L_{12}(\lambda_k)L_{31}(\lambda_k)}{L_{32}(\lambda_k)} = L_{22}(\lambda_k) - \frac{L_{21}(\lambda_k)L_{32}(\lambda_k)}{L_{31}(\lambda_k)}$$

serve as variables of separation.

The polynomial $\mathcal{B}$ of degree $3N$ gives a half of variables of separation, we denote them by $\{\lambda_k\}$. The function $\mathcal{A}$ gives the other half of separation variables, denoted by $\{w_k\}$. The same expressions for $\mathcal{B}$ and $\mathcal{A}$ were used by Sklyanin for the SL(3) magnetic chain model with a quadratic Poisson bracket.

Separation of variables was developed in many papers, among them [10, 13, 15, 16]. Using their ideas it is easy to prove the following statement for our system.

Proposition. *The variables of separation have the following properties:*

(1) *a pair* (w_k, λ_k) *is a point of the spectral curve* (5)*;*

(2) *a pair* (w_k, λ_k) *is quasi-canonically conjugate with respect to the Lie-Poisson bracket* (3)*:*

$$\{\lambda_k, \lambda_l\} = 0, \qquad \{\lambda_k, w_l\} = -\lambda_k^N \delta_{kl}, \qquad \{w_k, w_l\} = 0;$$

(3) *the corresponding Liouville 1-form is* $\Omega = -\sum\limits_k \lambda_k^{-N} w_k\, d\lambda_k$.

We realize this separation of variables for the system of isotropic SU(3) magnet over a generic orbit. In other words, we solve the problem how to express dynamic variables in terms of the variables of separation. We start from the dynamic variables, which are coefficients of entries of the L-matrix (2), and construct a sequence of maps leading to variables of separation. These maps are evidently invertible.

First of all we need to eliminate some dynamic variables from the set $\{L^{(m)}_{ij},\, L^{(m)}_{11} + L^{(m)}_{22} + L^{(m)}_{33} = 0 : m=0,\ldots,N-1\}$ in order to parameterize the orbit. There are $8N$ dynamic variables on a $6N$-dimensional orbit, so we need to eliminate $2N$ variables: one can choose the variables corresponding to a pair of commuting nilpotent elements of the Cartan-Weyl basis, that is one of the following sets: $\{L^{(m)}_{13}, L^{(m)}_{23}\}$, $\{L^{(m)}_{12}, L^{(m)}_{32}\}$, $\{L^{(m)}_{21}, L^{(m)}_{31}\}$, $\{L^{(m)}_{12}, L^{(m)}_{13}\}$, $\{L^{(m)}_{21}, L^{(m)}_{23}\}$, $\{L^{(m)}_{31}, L^{(m)}_{32}\}$. The orbit equations (4) are linear in any of these sets of dynamic variables.

For the eliminated set $\{L^{(m)}_{13}, L^{(m)}_{23}\}$ the orbit equations are written as follows:

$$\begin{pmatrix} c_0 \\ b_0 \\ c_1 \\ b_1 \\ \vdots \\ c_{N-1} \\ b_{N-1} \end{pmatrix} = \begin{pmatrix} L^{(0)}_{31} & L^{(0)}_{32} & 0 & 0 & \ldots & 0 & 0 \\ \mathcal{K}^{(0)}_{31} & \mathcal{K}^{(0)}_{32} & 0 & 0 & \ldots & 0 & 0 \\ L^{(1)}_{31} & L^{(1)}_{32} & L^{(0)}_{31} & L^{(0)}_{32} & \ldots & 0 & 0 \\ \mathcal{K}^{(1)}_{31} & \mathcal{K}^{(1)}_{32} & \mathcal{K}^{(0)}_{31} & \mathcal{K}^{(0)}_{32} & \ldots & 0 & 0 \\ \vdots & \vdots & \vdots & \vdots & \ddots & \vdots & \vdots \\ L^{(N-1)}_{31} & L^{(N-1)}_{32} & L^{(N-2)}_{31} & L^{(N-2)}_{32} & \ldots & L^{(0)}_{31} & L^{(0)}_{32} \\ \mathcal{K}^{(N-1)}_{31} & \mathcal{K}^{(N-1)}_{32} & \mathcal{K}^{(N-2)}_{31} & \mathcal{K}^{(N-2)}_{32} & \ldots & \mathcal{K}^{(0)}_{31} & \mathcal{K}^{(0)}_{32} \end{pmatrix} \begin{pmatrix} L^{(0)}_{13} \\ L^{(0)}_{23} \\ L^{(1)}_{13} \\ L^{(1)}_{23} \\ \vdots \\ L^{(N-1)}_{13} \\ L^{(N-1)}_{23} \end{pmatrix} + \begin{pmatrix} H_0 \\ F_0 \\ H_1 \\ F_1 \\ \vdots \\ H_{N-1} \\ F_{N-1} \end{pmatrix},$$

$$H_j = -\mathcal{K}_{33}^{(j)} + \sum_{m+n=j} L_{33}^{(m)} L_{33}^{(n)}, \qquad F_j = \sum_{m+n=j} L_{33}^{(m)} \mathcal{K}_{33}^{(n)},$$

where $\mathcal{K}_{ij}$ are cofactors of L_{ji}. Solving this linear problem we express the variables $\{L_{13}^{(m)},\, L_{23}^{(m)}\}$ in terms of other dynamic variables, then we eliminate them from the integrals of motion h_N, $h_{N+1}, \ldots, h_{2N-1}, f_N, f_{N+1}, \ldots, f_{3N-1}$, serving as Hamiltonians:

$$\begin{pmatrix} h_N \\ f_N \\ h_{N+1} \\ f_{N+1} \\ \vdots \\ h_{2N-1} \\ f_{2N-1} \\ f_{2N} \\ f_{2N+1} \\ \vdots \\ f_{3N-1} \end{pmatrix} = \begin{pmatrix} L_{31}^{(N)} & L_{32}^{(N)} & L_{31}^{(N-1)} & L_{32}^{(N-1)} & \ldots & L_{31}^{(1)} & L_{32}^{(1)} \\ \mathcal{K}_{31}^{(N)} & \mathcal{K}_{32}^{(N)} & \mathcal{K}_{31}^{(N-1)} & \mathcal{K}_{32}^{(N-1)} & \ldots & \mathcal{K}_{31}^{(1)} & \mathcal{K}_{32}^{(1)} \\ 0 & 0 & L_{31}^{(N)} & L_{32}^{(N)} & \ldots & L_{31}^{(2)} & L_{32}^{(2)} \\ \mathcal{K}_{31}^{(N+1)} & \mathcal{K}_{32}^{(N+1)} & \mathcal{K}_{31}^{(N)} & \mathcal{K}_{32}^{(N)} & \ldots & \mathcal{K}_{31}^{(2)} & \mathcal{K}_{32}^{(2)} \\ \vdots & \vdots & \vdots & \vdots & \ddots & \vdots & \vdots \\ 0 & 0 & 0 & 0 & \ldots & L_{31}^{(N)} & L_{32}^{(N)} \\ \mathcal{K}_{31}^{(2N-1)} & \mathcal{K}_{32}^{(2N-1)} & \mathcal{K}_{31}^{(2N-2)} & \mathcal{K}_{32}^{(2N-2)} & \ldots & \mathcal{K}_{31}^{(N)} & \mathcal{K}_{32}^{(N)} \\ \mathcal{K}_{31}^{(2N)} & \mathcal{K}_{32}^{(2N)} & \mathcal{K}_{31}^{(2N-1)} & \mathcal{K}_{32}^{(2N-1)} & \ldots & \mathcal{K}_{31}^{(N+1)} & \mathcal{K}_{32}^{(N+1)} \\ 0 & 0 & \mathcal{K}_{31}^{(2N)} & \mathcal{K}_{32}^{(2N)} & \ldots & \mathcal{K}_{31}^{(N+2)} & \mathcal{K}_{32}^{(N+2)} \\ \vdots & \vdots & \vdots & \vdots & \ddots & \vdots & \vdots \\ 0 & 0 & 0 & 0 & \ldots & \mathcal{K}_{31}^{(2N)} & \mathcal{K}_{32}^{(2N)} \end{pmatrix} \begin{pmatrix} L_{13}^{(0)} \\ L_{23}^{(0)} \\ L_{13}^{(1)} \\ L_{23}^{(1)} \\ \vdots \\ L_{13}^{(N-1)} \\ L_{23}^{(N-1)} \end{pmatrix} + \begin{pmatrix} H_N \\ F_N \\ H_{N+1} \\ F_{N+1} \\ \vdots \\ H_{2N-1} \\ F_{2N-1} \\ F_{2N} \\ F_{2N+1} \\ \vdots \\ F_{3N-1} \end{pmatrix}. \tag{8}$$

On the other hand the Hamiltonians $h_N, h_{N+1}, \ldots, h_{2N-1}, f_N, f_{N+1}, \ldots, f_{3N-1}$ can be computed from the spectral curve equation written for $3N$ points (w_k, λ_k):

$$\begin{aligned} w_k^3 = w_k\Big(c_0 + c_1\lambda_k + \cdots + c_{N-1}\lambda_k^{N-1} + h_N\lambda_k^N + \cdots + h_{2N-1}\lambda_k^{2N-1} + c_{2N}\lambda^{2N}\Big) + \\ + \Big(b_0 + b_1\lambda_k + \cdots + b_{N-1}\lambda_k^{N-1} + f_N\lambda_k^N + \cdots + f_{3N-1}\lambda_k^{3N-1} + b_{3N}\lambda_k^{3N}\Big), \end{aligned} \tag{9}$$

where $c_{2N} = \frac{1}{4}(m^2 + 3q^2)$, $b_{3N} = \frac{1}{4}q(m^2 - q^2)$. Equating expressions (8) for the Hamiltonians to the obtained from (9) we get the equations

$$L_{31}(\lambda_k)w_k + \mathcal{K}_{31}(\lambda_k) = 0, \qquad L_{32}(\lambda_k)w_k + \mathcal{K}_{32}(\lambda_k) = 0. \tag{10}$$

giving the function $\mathcal{A}$. The consistent equation with respect to $\{w_k\}$ gives the polynomial $\mathcal{B}$.

The equations (10) are linear in the set $\{L_{31}^{(0)}, L_{32}^{(0)}, \ldots, L_{31}^{(N-1)}, L_{32}^{(N-1)}, \mathcal{K}_{31}^{(0)}, \mathcal{K}_{32}^{(0)}, \ldots, \mathcal{K}_{31}^{(2N-1)}, \mathcal{K}_{32}^{(2N-1)}\}$, that is

$$\widehat{W}\boldsymbol{L}_1 = -L_{31}^{(N)}\boldsymbol{w} - \mathcal{K}_{31}^{(2N)}\boldsymbol{\lambda}, \qquad \widehat{W}\boldsymbol{L}_2 = -L_{32}^{(N)}\boldsymbol{w} - \mathcal{K}_{32}^{(2N)}\boldsymbol{\lambda},$$

$$\widehat{W} = \begin{bmatrix} \lambda_1^{N-1}w_1 & \ldots & \lambda_1 w_1 & w_1 & \lambda_1^{2N-1} & \ldots & \lambda_1 & 1 \\ \lambda_2^{N-1}w_2 & \ldots & \lambda_2 w_2 & w_2 & \lambda_2^{2N-1} & \ldots & \lambda_2 & 1 \\ \vdots & \ddots & \vdots & \vdots & \vdots & \ddots & \vdots & \vdots \\ \lambda_{3N}^{N-1}w_{3N} & \ldots & \lambda_{3N}w_{3N} & w_{3N} & \lambda_{3N}^{2N-1} & \ldots & \lambda_{3N} & 1 \end{bmatrix}, \quad \boldsymbol{w} = \begin{pmatrix} \lambda_1^N w_1 \\ \lambda_2^N w_2 \\ \vdots \\ \lambda_{3N}^N w_{3N} \end{pmatrix}, \quad \boldsymbol{\lambda} = \begin{pmatrix} \lambda_1^{2N} \\ \lambda_2^{2N} \\ \vdots \\ \lambda_{3N}^{2N} \end{pmatrix},$$

$$\boldsymbol{L}_1^t = \Big[L_{31}^{(N-1)}, \;\ldots,\; L_{31}^{(1)},\; L_{31}^{(0)},\; \mathcal{K}_{31}^{(2N-1)}, \;\ldots,\; \mathcal{K}_{31}^{(1)},\; \mathcal{K}_{31}^{(0)}\Big],$$

$$\boldsymbol{L}_2^t = \Big[L_{32}^{(N-1)}, \;\ldots,\; L_{32}^{(1)},\; L_{32}^{(0)},\; \mathcal{K}_{32}^{(2N-1)}, \;\ldots,\; \mathcal{K}_{32}^{(1)},\; \mathcal{K}_{32}^{(0)}\Big],$$

and can be easily solved for this set of dynamic variables. The next step is obtaining $\{L_{11}^{(m)}$, $L_{12}^{(m)}, L_{21}^{(m)}, L_{22}^{(m)} : m = 1, \ldots, N-1\}$ from the expressions for $\{\mathcal{K}_{31}^{(m)}, \mathcal{K}_{32}^{(m)} : m = 1, \ldots, 2N-1\}$, which is also a linear problem:

$$\begin{pmatrix} \mathcal{K}_{31}^{(0)} \\ \mathcal{K}_{31}^{(1)} \\ \vdots \\ \mathcal{K}_{31}^{(N-1)} \\ \mathcal{K}_{31}^{(N)} \\ \mathcal{K}_{31}^{(N+1)} \\ \vdots \\ \mathcal{K}_{31}^{(2N-1)} \end{pmatrix} = \begin{pmatrix} L_{32}^{(0)} & 0 & \dots & 0 & -L_{31}^{(0)} & 0 & \dots & 0 \\ L_{32}^{(1)} & L_{32}^{(0)} & \dots & 0 & -L_{31}^{(1)} & -L_{31}^{(0)} & \dots & 0 \\ \vdots & \vdots & \ddots & \vdots & \vdots & \vdots & \ddots & \vdots \\ L_{32}^{(N-1)} & L_{32}^{(N-2)} & \dots & L_{32}^{(0)} & -L_{31}^{(N-1)} & -L_{31}^{(N-2)} & \dots & -L_{31}^{(0)} \\ L_{32}^{(N)} & L_{32}^{(N-1)} & \dots & L_{32}^{(1)} & -L_{31}^{(N)} & -L_{31}^{(N-1)} & \dots & -L_{31}^{(1)} \\ 0 & L_{32}^{(N)} & \dots & L_{32}^{(2)} & 0 & -L_{31}^{(N)} & \dots & -L_{31}^{(2)} \\ \vdots & \vdots & \ddots & \vdots & \vdots & \vdots & \ddots & \vdots \\ 0 & 0 & \dots & L_{32}^{(N)} & 0 & 0 & \dots & -L_{31}^{(N)} \end{pmatrix} \begin{pmatrix} L_{21}^{(0)} \\ L_{21}^{(1)} \\ \vdots \\ L_{21}^{(N-1)} \\ L_{22}^{(0)} \\ L_{22}^{(1)} \\ \vdots \\ L_{22}^{(N-1)} \end{pmatrix},$$

$$\begin{pmatrix} \mathcal{K}_{32}^{(0)} \\ \mathcal{K}_{32}^{(1)} \\ \vdots \\ \mathcal{K}_{32}^{(N-1)} \\ \mathcal{K}_{32}^{(N)} \\ \mathcal{K}_{32}^{(N+1)} \\ \vdots \\ \mathcal{K}_{32}^{(2N-1)} \end{pmatrix} = \begin{pmatrix} -L_{32}^{(0)} & 0 & \dots & 0 & L_{31}^{(0)} & 0 & \dots & 0 \\ -L_{32}^{(1)} & -L_{32}^{(0)} & \dots & 0 & L_{31}^{(1)} & L_{31}^{(0)} & \dots & 0 \\ \vdots & \vdots & \ddots & \vdots & \vdots & \vdots & \ddots & \vdots \\ -L_{32}^{(N-1)} & -L_{32}^{(N-2)} & \dots & -L_{32}^{(0)} & L_{31}^{(N-1)} & L_{31}^{(N-2)} & \dots & L_{31}^{(0)} \\ -L_{32}^{(N)} & -L_{32}^{(N-1)} & \dots & -L_{32}^{(1)} & L_{31}^{(N)} & L_{31}^{(N-1)} & \dots & L_{31}^{(1)} \\ 0 & -L_{32}^{(N)} & \dots & -L_{32}^{(2)} & 0 & L_{31}^{(N)} & \dots & L_{31}^{(2)} \\ \vdots & \vdots & \ddots & \vdots & \vdots & \vdots & \ddots & \vdots \\ 0 & 0 & \dots & -L_{32}^{(N)} & 0 & 0 & \dots & L_{31}^{(N)} \end{pmatrix} \begin{pmatrix} L_{11}^{(0)} \\ L_{11}^{(1)} \\ \vdots \\ L_{11}^{(N-1)} \\ L_{12}^{(0)} \\ L_{12}^{(1)} \\ \vdots \\ L_{12}^{(N-1)} \end{pmatrix}.$$

The rest of variables, eliminated at the first stage, we compute from the orbit equations.

Note. The proposed scheme fails in application to the L-matrix (2), because degrees of the polynomials $\mathcal{B}$, $\mathcal{K}_{31}$, $\mathcal{K}_{32}$ are less than the maximal. We avoid this failure if rotate the L-matrix by means of similar transformation with the matrix

$$\hat{R} = \begin{pmatrix} \frac{1}{\sqrt{2}} & \frac{1}{2} & \frac{1}{2} \\ -\frac{1}{\sqrt{2}} & \frac{1}{2} & \frac{1}{2} \\ 0 & -\frac{1}{\sqrt{2}} & \frac{1}{\sqrt{2}} \end{pmatrix}, \qquad \hat{R}^{-1}\hat{L}(\lambda)\hat{R} = \frac{\lambda^N}{2} \begin{pmatrix} q & \frac{m}{\sqrt{2}} & \frac{m}{\sqrt{2}} \\ \frac{m}{\sqrt{2}} & -\frac{q}{2} & \frac{3q}{2} \\ \frac{m}{\sqrt{2}} & \frac{3q}{2} & -\frac{q}{2} \end{pmatrix} + \cdots$$

3.2. Finite-gap integration

As mentioned above the Hamiltonians h_N, h_{N+1} give rise to a stationary and an evolutionary flows, resulting in the equations

$$\frac{d\lambda_k}{dx} = \frac{\big[L_{32}^{(0)} L_{31}(\lambda_k) - L_{31}^{(0)} L_{32}(\lambda_k)\big] \frac{\partial}{\partial w_k} P(w_k, \lambda_k)}{\mathcal{B}_{3N} \lambda_k \prod_j' (\lambda_k - \lambda_j)},$$

$$\frac{d\lambda_k}{dt} = \frac{\big[\big(L_{32}^{(0)} + L_{32}^{(1)}\lambda_k\big) L_{31}(\lambda_k) - \big(L_{31}^{(0)} + L_{31}^{(1)}\lambda_k\big) L_{32}(\lambda_k)\big] \frac{\partial}{\partial w_k} P(w_k, \lambda_k)}{\mathcal{B}_{3N} \lambda_k^2 \prod_j' (\lambda_k - \lambda_j)}$$

for $3N$ points (w_k, λ_k) on the spectral curve $\mathcal{R}$ of genus $3N-2$. We introduce the basis of holomorphic Abelian differentials on the curve:

$$\frac{\lambda^n \, d\lambda}{\frac{\partial}{\partial w} P(w, \lambda)}, \quad n = 0, 1, \dots, 2N-2, \qquad \frac{w\lambda^n \, d\lambda}{\frac{\partial}{\partial w} P(w, \lambda)}, \quad n = 0, 1, \dots, N-2,$$

where zeros of $\frac{\partial}{\partial w} P(w, \lambda) = 3w^2 - I_2(\lambda)$ belong to the set of zeros of the discriminant Q of the trigonal curve (5): $Q(\lambda) = \big(\frac{1}{3} I_2(\lambda)\big)^3 - \big(\frac{1}{2} I_3(\lambda)\big)^2$, giving branch points to the curve. Also we need two Abelian differentials of the third kind:

$$\frac{\lambda^{2N-1} \, d\lambda}{\frac{\partial}{\partial w} P(w, \lambda)}, \qquad \frac{w\lambda^{N-1} \, d\lambda}{\frac{\partial}{\partial w} P(w, \lambda)}.$$

In general a finite gap integration of similar systems was done in [13]. In more details integration in terms of generalized θ-functions giving inversion of an extended Abel-Jacobi map can be found in [11, 12].

Here we consider the one-gap system having only the stationary flow, resulting in the following equations for the variables $\{\lambda_k\}$:

$$\frac{\partial \lambda_k}{\partial x} = -\frac{\big[3w_1^2 - I_2(\lambda_k)\big](\lambda_i - \lambda_j)}{\det \widehat{W}},$$

where $(k\,ij)$ is a cyclic permutation of (123). The map $(\lambda_1, \lambda_2, \lambda_3) \mapsto (\varphi_1, \varphi_2, \varphi_3)$ is nonsingular if $\det \widehat{W} \neq 0$, and leads to the Abelian integrals

$$\varphi_1 = \int_{\lambda_0}^{\lambda_1} \frac{d\lambda}{3w^2 - I_2(\lambda)} + \int_{\lambda_0}^{\lambda_2} \frac{d\lambda}{3w^2 - I_2(\lambda)} + \int_{\lambda_0}^{\lambda_3} \frac{d\lambda}{3w^2 - I_2(\lambda)},$$
$$\varphi_2 = \int_{\lambda_0}^{\lambda_1} \frac{\lambda d\lambda}{3w^2 - I_2(\lambda)} + \int_{\lambda_0}^{\lambda_2} \frac{\lambda d\lambda}{3w^2 - I_2(\lambda)} + \int_{\lambda_0}^{\lambda_3} \frac{\lambda d\lambda}{3w^2 - I_2(\lambda)},$$
$$\varphi_3 = \int_{\lambda_0}^{\lambda_1} \frac{w\, d\lambda}{3w^2 - I_2(\lambda)} + \int_{\lambda_0}^{\lambda_2} \frac{w\, d\lambda}{3w^2 - I_2(\lambda)} + \int_{\lambda_0}^{\lambda_3} \frac{w\, d\lambda}{3w^2 - I_2(\lambda)};$$
$$\frac{d\varphi_1}{dx} = 0, \quad \frac{d\varphi_2}{dx} = 0, \quad \frac{d\varphi_3}{dx} = -1 \quad \Rightarrow \quad \varphi_1 = C_1, \quad \varphi_2 = C_2, \quad \varphi_3 = C_3 - x.$$

The first differential is holomorphic, and two others are meromorphic of the third kind.

The spectral curve of the one-gap system is elliptic. We reduce it to the Weierstrass form by rotating about an inflection point $(\lambda_{\rm f}, w_{\rm f})$, which is a solution of the equation

$$\big[3w_{\rm f}^2 + I_2(\lambda_{\rm f})\big]\big[\tfrac{3}{2}I_2''(\lambda_{\rm f})I_3(\lambda_{\rm f}) + \tfrac{1}{2}I_3''(\lambda_{\rm f})I_2(\lambda_{\rm f}) - I_2'(\lambda_{\rm f})I_3'(\lambda_{\rm f})\big] +$$
$$+ w_{\rm f}\big[2I_2''(\lambda_{\rm f})I_3^2(\lambda_{\rm f}) + \tfrac{9}{2}I_3''(\lambda_{\rm f})I_3(\lambda_{\rm f}) - I_2(\lambda_{\rm f})[I_2'(\lambda_{\rm f})]^2 - 3[I_3'(\lambda_{\rm f})]^2\big] = 0.$$

The rotation by means of a linear transformation together with a linear-fractional transformation realizes the map $(w, \lambda) \mapsto (z, \xi)$:

$$\begin{pmatrix} \lambda - \lambda_{\rm f} \\ w - w_{\rm f} \end{pmatrix} = \begin{pmatrix} 3w_{\rm f}^2 - I_2(\lambda_{\rm f}) & -\frac{1}{2}\left[w_{\rm f} I_2'(\lambda_{\rm f}) + I_3'(\lambda_{\rm f})\right]^{-1} \\ w_{\rm f} I_2'(\lambda_{\rm f}) + I_3'(\lambda_{\rm f}) & \frac{1}{2}\left[3w_{\rm f}^2 - I_2(\lambda_{\rm f})\right]^{-1} \end{pmatrix} \frac{2}{z - d + b(\xi - a)} \begin{pmatrix} -(\xi - a) \\ c \end{pmatrix}, \tag{11}$$

where

$$a = \tfrac{1}{12}\Bigg[\frac{\big(6w_{\rm f}[w_{\rm f} I_2'(\lambda_{\rm f}) + I_3'(\lambda_{\rm f})] - I_2'(\lambda_{\rm f})[3w_{\rm f}^2 - I_2(\lambda_{\rm f})]\big)^2}{[3w_{\rm f}^2 - I_2(\lambda_{\rm f})]^2} -$$
$$- 4\frac{\big(3[w_{\rm f}I_2'(\lambda_{\rm f}) + I_3'(\lambda_{\rm f})]^3 + h_2[3w_{\rm f}^2 - I_2(\lambda_{\rm f})]^2[w_{\rm f}I_2'(\lambda_{\rm f}) + I_3'(\lambda_{\rm f})] + 3f_3[3w_{\rm f}^2 - I_2(\lambda_{\rm f})]^3\big)I_2'(\lambda_{\rm f})}{2[3w_{\rm f}^2 - I_2(\lambda_{\rm f})][w_{\rm f}I_2'(\lambda_{\rm f}) + I_3'(\lambda_{\rm f})]}\Bigg],$$
$$b = \frac{6w_{\rm f}[w_{\rm f} I_2'(\lambda_{\rm f}) + I_3'(\lambda_{\rm f})] - I_2'(\lambda_{\rm f})[3w_{\rm f}^2 - I_2(\lambda_{\rm f})]}{[3w_{\rm f}^2 - I_2(\lambda_{\rm f})]},$$
$$c = [w_{\rm f}I_2'(\lambda_{\rm f}) + I_3'(\lambda_{\rm f})]^3 - h_2[3w_{\rm f}^2 - I_2(\lambda_{\rm f})]^2[w_{\rm f}I_2'(\lambda_{\rm f}) + I_3'(\lambda_{\rm f})] - f_3[3w_{\rm f}^2 - I_2(\lambda_{\rm f})]^3,$$
$$d = \frac{\big([w_{\rm f}I_2'(\lambda_{\rm f}) + I_3'(\lambda_{\rm f})]^3 - h_2[3w_{\rm f}^2 - I_2(\lambda_{\rm f})]^2[w_{\rm f}I_2'(\lambda_{\rm f}) + I_3'(\lambda_{\rm f})] - f_3[3w_{\rm f}^2 - I_2(\lambda_{\rm f})]^3\big)I_2'(\lambda_{\rm f})}{2[3w_{\rm f}^2 - I_2(\lambda_{\rm f})][w_{\rm f}I_2'(\lambda_{\rm f}) + I_3'(\lambda_{\rm f})]}.$$

The map (11) reduces the spectral curve (5) to the Weierstrass form

$$z^2 = 4\xi^3 - g_2\xi - g_3.$$

Then the parametrization $\xi = \wp(u)$, $z = \wp'(u)$ makes a uniformization of the curve. In the new variables the Abelian integrals are reduced to the form

$$\varphi_1 = \sum_{k=1}^{3} \int_{u_0}^{u_k} du, \qquad \varphi_2 = \sum_{k=1}^{3} \int_{u_0}^{u_k} \left(\lambda_{\mathrm{f}} - \frac{2[\wp(u) - a]}{\wp'(u) - d + b[\wp(u) - a]}\right) du,$$

$$\varphi_3 = \sum_{k=1}^{3} \int_{u_0}^{u_k} \left(w_{\mathrm{f}} + \frac{2c}{\wp'(u) - d + b[\wp(u) - a]}\right) du.$$

Acknowledgments
This work is supported by the International Charitable Fund for Renaissance of Kyiv-Mohyla Academy. We thank the organizing committee of the conference 'Physics and Mathematics of Nonlinear Phenomena 2013' for the substantial financial support and warm hospitality.

References

[1] Nakamura K and Sasada T 1974 *Phys. Lett. A* **48** 321–322
[2] Lakshmanan M, Ruijgrok Th W and Thompson C J 1976 *Physica A* **84** 577–590
[3] Tjon J and Wright J 1977 *Phys. Rev. B* **15** 3470–3476
[4] Takhtajan L A 1977 *Phys. Lett. A* **64** 235–237
[5] Orfanidis S J 1980 *Phys. Lett. A* **75** 304–306
[6] Sasaki R and Ruijgrok Th W 1982 *Physica A* **113** 388–400
[7] Sklyanin E K 1992 *Commun. Math. Phys.* **150** 181–191
[8] Holod P, Kisilevich A and Kondratyuk S *Symmetry in Nonlinear Mathematical Physics* **2** 343–352
[9] Bernatska J and Holod P 2009 *J. Phys. A: Math. Theor.* **42** 075401
[10] Sklyanin E K 1995 *Progr. Theor. Phys. Suppl.* **118** 35–60
[11] Fedorov Yu 1999 *Acta Appl. Math.* **55** 251–301
[12] Braden H W and Fedorov Yu N 2008 *J. Geom. Phys.* **58** 1346–1354
[13] Adams M R, Harnad J and Hurtubise J 1993 *Commun. Math. Phys.* **155** 385–413
[14] Adler M and van Moerbeke P 1980 *Adv. Math.* **38** 318–379
[15] Scott D R D 1994 *J. Math. Phys.* **35** 5831–5843
[16] Gekhtman M I 1995 *Commun. Math. Phys.* **167** 593–605

Projective differential geometry of multidimensional dispersionless integrable hierarchies

L.V. Bogdanov
L.D. Landau ITP RAS, Moscow, Russia

B.G. Konopelchenko
Department of Mathematics and Physics "Ennio De Giorgi", University of Salento and INFN, Lecce, Italy

In the memory of S.V. Manakov

Abstract. We introduce a general setting for multidimensional dispersionless integrable hierarchy in terms of differential m-form Ω_m with the coefficients satisfying the Plücker relations, which is gauge-invariantly closed and its gauge-invariant coordinates (ratios of coefficients) are (locally) holomorphic with respect to one of the variables (the spectral variable). We demonstrate that this form defines a hierarchy of dispersionless integrable equations in terms of commuting vector fields locally holomorphic in the spectral variable. The equations of the hierarchy are given by the gauge-invariant closedness equations.

1. Introduction

In this work we develop further the ideas of the work [1] and introduce a general setting for multidimensional dispersionless integrable hierarchy in terms of some differential m-form Ω_m, in the spirit of construction of universal Whitham hierarchy given in [2], which corresponds to the case of Ω_2 with a Hamiltonian reduction. The coefficients of this form should satisfy the Plücker relations (we call it a Plücker form), it should be gauge-invariantly closed and its gauge-invariant coordinates (ratios of coefficients) should be (locally) holomorphic with respect to one of the variables (the spectral variable). We demonstrate that this form defines a hierarchy of dispersionless integrable equations in terms of commuting vector fields locally holomorphic in the spectral variable. The equations of the hierarchy are given by the gauge-invariant closedness equations.

First we consider the correspondence between Plücker forms and distributions and demonstrate that involutivity of the distribution is equivalent to the gauge-invariant closedness equations for Plücker form.

Then we introduce a spectral variable, suggesting that gauge-independent coordinates of the form (ratios of coefficients) are holomorphic with respect to one of the variables and consider some simple examples of integrable systems arising from gauge-invariant closedness equations.

We demonstrate that nonlinear vector Riemann-Hilbert problem (or $\bar{\partial}$-problem) is a natural tool to construct gauge-invariantly closed Plücker forms holomorphic in the complex plane, and show how to obtain polynomial Ω_m and the corresponding multidimensional dispersionless hierarchy.

2. Integrable distributions and closed Plücker forms

Let us consider a domain with a set of local coordinates $\mathbf{x} = (x_0, x_1, \dots, x_N)$. Distribution is a K-dimensional subspace of the tangent space $\Delta_x \subset T_x$, depending smoothly on $\mathbf{x}$ (there exists a basis of smooth vector fields). Involutive distribution is defined by the relation $[\Delta, \Delta] \subset \Delta$, where the standard commutator of vector fields is used. According to Frobenius theorem, the distribution is integrable (corresponds to a foliation) $\Leftrightarrow$ the distribution is in involution.

There are several dual formulations of Frobenius theorem in terms of differential forms, where the subspace in cotangent space dual to the distribution (codistribution) is considered. Here we will consider decomposable differential $m = (N + 1 - K)$ forms (we will call them Plücker forms) which are in one-to-one correspondence (up to a gauge) with the codistribution, and will formulate the property of these forms which is equivalent to the involutivity of original K-dimensional distribution.

Let us define a Plücker form as an m-form

$$\Omega_m = \sum_{0 \leqslant i_0 \leq \cdots \leqslant i_{m-1} \leqslant N} \pi_{i_0\, i_1\, \dots\, i_{m-1}}(x) dx_{i_0} \wedge dx_{i_1} \wedge \cdots \wedge dx_{i_{m-1}}$$

with coefficients satisfying Plücker relations (see e.g. [3])

$$\sum_{l=0}^{m} (-1)^l \pi_{i_0 \dots i_{m-2}\, j_l} \pi_{j_0 \dots \check{j}_l \dots j_m} = 0,$$

where indices i_p and j_p range over all possible values from zero to N and the notation $\check{j}_p$ means the omission of the index. Due to Plücker relations, the form Ω_m is decomposable

$$\Omega_m = \omega_0 \wedge \cdots \wedge \omega_{m-1},$$

defines a vector subspace in cotangent space and a distribution as a dual object. It is also easy to construct a Plücker form for a given distribution, it is defined up to a gauge. We will call the Plücker forms which differ only by a gauge (multiplication by some function) equivalent.

Having the correspondence between Plücker forms and distributions, it is natural to ask a question what property of Plücker form corresponds to the involutivity of the distribution. To have a geometrical meaning, this property should be gauge-invariant. The answer to this question can be found using the gauge-invariant closedness conditions for Plücker forms introduced in [1].

Let us consider standard closedness equations

$$\left[\frac{\partial \pi_{i_0\, i_1\, \dots i_{m-1}}}{\partial x_{i_m}} \right] = 0, \tag{1}$$

where the bracket $[\dots]$ means antisymmetrization over all indices. We consider a pair of equations with the choice of indices $(0, 1, \dots, m-1, q)$, $(0, 1, \dots, m-1, r)$, $q, r \in \{1, \dots, K = N - m + 1\}$, $q \neq r$, introducing the basic set of gauge independent affine coordinates

$$a_{qk} = (-1)^k J^{-1} \pi_{0 \dots k-1\, k+1 \dots m-1\, m-1+q},$$
$$J = \pi_{0\, 1 \dots m-1}$$

Journal of Physics: Conference Series **482** (2014) 012005 doi:10.1088/1742-6596/482/1/012005

where $k \in \{0, \ldots, m-1\}$, $q \in \{1, \ldots, K = N - m + 1\}$. All the affine coordinates are expressed through the basic set using Plücker relations, and the closedness equations take the form

$$\frac{\partial J}{\partial x_{q+(m-1)}} + \sum_{l=0}^{m-1} \frac{\partial (J a_{ql})}{\partial x_l} = 0, \qquad \frac{\partial J}{\partial x_{r+(m-1)}} + \sum_{l=0}^{m-1} \frac{\partial (J a_{rl})}{\partial x_l} = 0, \tag{2}$$

$$\frac{\partial a_{qk}}{\partial x_{r+(m-1)}} - \frac{\partial a_{rk}}{\partial x_{q+(m-1)}} + \sum_{l=0}^{m-1} \left(a_{rl} \frac{\partial a_{qk}}{\partial x_l} - a_{ql} \frac{\partial a_{rk}}{\partial x_l} \right) = 0. \tag{3}$$

Subsystem (3), invariant under the gauge transformations mentioned above, is the gauge invariant part of the system (1) for Plücker form [1]; we will call it gauge-invariant closedness equations for Plücker form. Equations (2) can be viewed as the equations for the gauge variable J which transforms as $J \to \rho J$ under the gauge transformations. It is a straightforward check that equations (2) are compatible due to the subsystem (3). We note that the system (2), (3) can be rewritten in the form

$$D_q \ln J + \sum_{n=0}^{m-1} \frac{\partial a_{qn}}{\partial x_n} = 0, \quad D_r \ln J + \sum_{n=0}^{m-1} \frac{\partial a_{rn}}{\partial x_n} = 0, \tag{4}$$

$$D_r a_{qk} - D_q a_{rk} = 0, \quad k = 0, \ldots, m-1 \tag{5}$$

where D_q and D_r, $q, r \in \{1, \ldots, K\}$, are vector fields

$$D_q = \frac{\partial}{\partial x_{q+(m-1)}} + \sum_{n=0}^{m-1} a_{qn} \frac{\partial}{\partial x_n}, \quad D_r = \frac{\partial}{\partial x_{r+(m-1)}} + \sum_{n=0}^{m-1} a_{rn} \frac{\partial}{\partial x_n}. \tag{6}$$

Equations of the type (4) were considered in [4], [5] as equations for the Jacobian of solutions of linear equations defined by vector fields. So the subsystem (3) is equivalent to the commutativity $[D_q, D_r] = 0$ of vector fields D_q and D_r.

Considering the set of gauge-invariant closedness equations (3), we come to the following conclusions:

Proposition 1 *The set of gauge-invariant closedness equations (3) is equivalent to the existence of a gauge, in which the Plücker form is closed in the standard sense.*

The gauge variable J corresponding to closed form is defined by linear equations (2).

Proposition 2 *The Plücker form is gauge-invariantly closed $\Leftrightarrow$ the corresponding distribution is in involution*

The Proposition 2 can be easily proved by purely geometrical means, using the Frobenius theorem and some well-known properties of differential forms. It is known that closed decomposable differential form has a decomposition in terms of exact 1-forms (see e.g. [6, 7]). The fact that the Plücker form is decomposable implies that (gauge invariantly) closed Plücker form possesses a decomposition into exact 1-forms (up to a gauge).

Proposition 3 *Gauge-invariantly closed Plücker form can be represented as*

$$\Omega_m = g df_0 \wedge \cdots \wedge df_{m-1}, \tag{7}$$

where g, $f_0, \ldots, f_{m-1}$ are some functions.

Examples of closedness equations
Let us consider some simple examples of closedness equations (2), (3) for Plücker forms, for more detail see [1]. We will use the representation

$$\Omega_m = J\tilde{\Omega}_m = J(dx_0 \wedge \cdots \wedge dx_{m-1} + \dots), \tag{8}$$

where $J = \pi_{01\dots m-1}$ is a gauge-dependent variable, and the Plücker form $\tilde{\Omega}_m$ is gauge-independent, it may be considered as the initial form in affine gauge.

The simplest case corresponds to $m = 1$, let us take $N = 2$:

$$\Omega_1 = J(dx_0 - a_{10}dx_1 - a_{20}dx_2).$$

In this case we have no Plücker relations, and the closedness equations are

$$\frac{\partial J}{\partial x_1} + \frac{\partial (Ja_{10})}{\partial x_0} = 0, \quad \frac{\partial J}{\partial x_2} + \frac{\partial (Ja_{20})}{\partial x_0} = 0, \tag{9}$$

$$\frac{\partial a_{10}}{\partial x_2} - \frac{\partial a_{20}}{\partial x_1} + a_{20}\frac{\partial a_{10}}{\partial x_0} - a_{10}\frac{\partial a_{20}}{\partial x_0} = 0. \tag{10}$$

For the case $m = 2$, $N = 3$ we have

$$\begin{aligned}\Omega_2 \;=\;& J(dx_0 \wedge dx_1 - a_{11}dx_0 \wedge dx_2 - a_{21}dx_0 \wedge dx_3 + a_{10}dx_1 \wedge dx_2 + \\ &+a_{20}dx_1 \wedge dx_3 - (a_{11}a_{20} - a_{10}a_{21})dx_2 \wedge dx_3),\end{aligned} \tag{11}$$

where the Plücker relations are taken into account in the last term. The closedness equations read

$$\frac{\partial J}{\partial x_2} + \sum_{m=0}^{1} \frac{\partial (Ja_{1m})}{\partial x_m} = 0, \quad \frac{\partial J}{\partial x_3} + \sum_{m=0}^{1} \frac{\partial (Ja_{2m})}{\partial x_m} = 0, \tag{12}$$

$$\frac{\partial a_{1k}}{\partial x_3} - \frac{\partial a_{2k}}{\partial x_2} + \sum_{l=0}^{1}\left(a_{2l}\frac{\partial a_{1k}}{\partial x_l} - a_{1l}\frac{\partial a_{2k}}{\partial x_l}\right) = 0, \qquad k = 0, 1. \tag{13}$$

3. The spectral variable. Dispersionless integrable systems

To introduce dispersionless integrable systems, we need the solutions of gauge-invariant closedness equations (3) holomorphic with respect to one variable (the spectral variable). Let us consider (gauge invariantly) closed Plücker form Ω_m with affine coordinates (ratios of coefficients) holomorphic with respect to $\lambda = x_0$ in some complex domain. This form defines a hierarchy of dispersionless integrable equations in terms of commuting vector fields locally holomorphic in λ. The equations of the hierarchy are given by the gauge-invariant closedness equations.

More specifically, we consider the forms meromorphic in the complex plane (in the affine gauge).

This setting for $m = 2$ can be reduced to Whitham hierarchy (Krichever [2]), for $m = 3$ to heavenly equation hierarchy (Takasaki [8], [9]) and connected Dunajski equation hierarchy [10], [11].

Important reductions are volume (or area) conservation corresponding to closedness in the affine gauge ($J = 1$) and hyper CR (Cauchy-Riemann) reduction $\Omega_m \wedge d\lambda = 0$.

Most known examples correspond to the case when there exists polynomial (or Laurent polynomial) set of affine coordinates (affine gauge). Below we will restrict ourselves to the polynomial case. Examples corresponding to Laurent polynomials were considered in [1], [12].

A closely related geometric picture of dispersionless integrable systems in terms of coisotropic deformations was introduced in [13], [14], and connection between this picture and the setting of the present work was discussed in [1].

Polynomial Ω_1 The case $m = 1$ is non-generic, in this case there is no hierarchy of commuting systems, and it is not clear how to solve it in general. The equations corresponding to this case were considered in the framework of inverse scattering method with variable spectral parameter [15]. The gauge-invariant closedness condition for the form

$$\Omega_1 = J(d\lambda + u dx - (1 + v\lambda + \lambda^2)dy),$$

which is given by equation (10), leads to the Liouville equation

$$\varphi_{xy} = e^{\varphi}, \tag{14}$$

where $u = \frac{1}{2}e^{\varphi}$, the Lax pair for this equation reads

$$\begin{aligned}\partial_x \psi &= u \partial_\lambda \psi \\ \partial_y \psi &= -(1 + v\lambda + \lambda^2)\partial_\lambda \psi.\end{aligned}$$

The case of third order polynomial

$$\Omega_1 = J(d\lambda + u dx - (1 + w\lambda + w'\lambda^2 + \lambda^3)dz)$$

leads to 'higher Liouville equation' introduced in [15],

$$\varphi_{xxz} - \varphi_{xz}\varphi_x = \tfrac{3}{2}e^{\varphi}, \tag{15}$$

where $w = \varphi_z$, $w' = e^{-\varphi}\varphi_{xz}$, the Lax pair for this equation is

$$\begin{aligned}\partial_x \psi &= u \partial_\lambda \psi \\ \partial_z \psi &= -(1 + w\lambda + w'\lambda^2 + \lambda^3)\partial_\lambda \psi.\end{aligned}$$

We should emphasize that equations (14), (15) are not commuting and they do not belong to a hierarchy. It is also possible to introduce 'higher Liouville equations' of arbitrary order.

Polynomial Ω_2 For the case $m = 2$, $N = 3$ we consider the form (11) with polynomial coefficients

$$\begin{aligned}a_{10} &= u_0(x), \quad a_{11} = u_1(x) + \lambda, \\ a_{20} &= v_0(x) + \lambda v_1(x), \quad a_{21} = v_2(x) + \lambda v_3(x) + \lambda^2.\end{aligned}$$

Considering gauge-invariant closedness equations (13) and denoting $x = x_1$, $y = x_2$, $t = x_3$ (see [1] for more detail), one gets the Manakov-Santini system [16], [17]

$$\begin{aligned}u_{xt} + u_{yy} + (uu_x)_x + v_x u_{xy} - v_y u_{xx} &= 0, \\ v_{xt} + v_{yy} + u v_{xx} + v_x v_{xy} - v_y v_{xx} &= 0.\end{aligned} \tag{16}$$

Considering more variables and higher order polynomials, it is possible to introduce the form Ω_2 corresponding to the Manakov-Santini hierarchy [21].

Reductions There are two important classes of reductions of dispersionless systems (hierarchies) defined in terms of the form Ω_m.

(i) The form Ω_m is closed in standard sense in affine gauge (J=1). In general, the closedness in the affine gauge leads to representation of the hierarchy in terms of volume-preserving (divergence-free) vector fields. In this case the equations of gauge-invariant closedness (3) are complemented by equations, implied by linear subsystem (2) for $J = 1$,

$$\sum_{l=0}^{m-1} \frac{\partial a_{ql}}{\partial x_l} = 0, \quad q \in \{1, \ldots, K\}.$$

For the Manakov-Santini system this reduction leads to the condition $v = 0$, defining the dKP equation.

(ii) Reduction $\Omega_m \wedge d\lambda = 0$. In this case it is possible to consider $\Omega_m = d\lambda \wedge \Omega'_{m-1}$ with Ω'_{m-1} not containing $d\lambda$ and gauge-invariantly closed, and λ plays a role of a parameter. Vector fields do not contain a derivative over spectral variable. In the case of Manakov-Santini system (16) this reduction leeds to the condition $u = 0$ and the equation

$$v_{xt} + v_{yy} + v_x v_{xy} - v_y v_{xx} = 0$$

considered in [18], [19], [20]. For general Ω_m this reduction defines the hyper CR (Cauchy-Riemann) hierarchies.

4. Nonlinear Riemann-Hilbert problem

Let us consider a form

$$\Omega_m = d\Psi^0 \wedge d\Psi^1 \wedge \ldots d\Psi^{m-1} = J\tilde{\Omega}_m,$$

where Ψ^k are some functions (series in λ), J is some coefficient of the form in coordinates $\mathbf{x}$, $\tilde{\Omega}_m$ is a gauge-invariant (affine) factor. It is easy to see that Ω_m is a closed Plücker form, and the only thing we need to construct a solution of some dispersionless integrable system is to provide definite analytic properties of the affine factor.

Question *How to provide some simple analytic properties of the affine factor? What kind of functions Ψ^k correspond to a polynomial affine factor?*

It is easy to see that $\tilde{\Omega}_m$ is invariant under diffeomorphism

$$(\Psi^0, \Psi^1, \ldots, \Psi^{m-1}) \to \mathbf{F}(\Psi^0, \Psi^1, \ldots, \Psi^{m-1})$$

Let the functions Ψ^k be holomorphic (meromorphic) inside and outside the unit circle (or some curve in the complex plane), having a discontinuity on it. If they satisfy a nonlinear vector Riemann-Hilbert problem on the unit circle

$$(\Psi^0, \Psi^1, \ldots, \Psi^{m-1})_{\text{in}} = \mathbf{F}(\Psi^0, \Psi^1, \ldots, \Psi^{m-1})_{\text{out}}, \tag{17}$$

then the affine factor $\tilde{\Omega}_m$ is holomorhic (meromorphic) in the complex plane.

Thus nonlinear vector Riemann-Hilbert problem gives a tool to construct closed Plücker forms with holomorphic (meromorphic) affine factor, generating commuting vector fields with holomorphic (meromorhic) coefficients.

Equivalently, it is possible to use a nonlinear vector $\bar{\partial}$-problem in some domain of the complex plane

$$\bar{\partial}\Psi^k = F^k(\Psi^0, \Psi^1, \ldots, \Psi^{m-1}), \quad 0 \leqslant k \leqslant m-1.$$

It provides the analiticity of affine factor $\tilde{\Omega}_m$ in the domain.

General hierarchy for the polynomial case
We will demonstrate how for a special choice of structure of functions Ψ^k, using nonlocal Riemann-Hilbert problem, it is possible to obtain polynomial affine factor and the corresponding hierarchy. We consider the formal series

$$\Psi^0 = \lambda + \sum_{n=1}^{\infty} \Psi^0_n(\mathbf{t}^1, \dots, \mathbf{t}^{m-1})\lambda^{-n}, \tag{18}$$

$$\Psi^k = \sum_{n=0}^{\infty} t^k_n (\Psi^0)^n + \sum_{n=1}^{\infty} \Psi^k_n(\mathbf{t}^1, \dots, \mathbf{t}^{m-1})(\Psi^0)^{-n}, \tag{19}$$

where $1 \leqslant k \leqslant m-1$, depending on $m-1$ infinite sequences of independent variables $\mathbf{t}^k = (t^k_0, \dots, t^k_n, \dots)$, $t^k_0 = x_k$, $\lambda = x_0$.

Let us consider the Riemann-Hilbert problem (17) on the unit circle, where the functions Ψ^k are analytic inside and outside the circle and in the neighborhood of infinity are given by the series of the form (18), (19), Then the form $\tilde{\Omega}_m$ is analytic in the complex plane, morover, due to the structure of the series it is polynomial. Thus it corresponds to dispersionless hierarchy for the polynomial case. Equations of the hierarchy are generated by the relation (see [11], [21])

$$\left(J^{-1} d\Psi^0 \wedge d\Psi^1 \wedge \dots d\Psi^{m-1}\right)_- = 0$$

where $(\cdots)_-$ denotes the projection on the part of $(\cdots)$ with negative powers in λ and $J = \pi_{0\,1\,\dots\,m-1} = \det(\partial_l \Psi^k)_{k,l=0,\dots,m-1}$. Generating relation represents analyticity condition for affine factor of the closed Plücker form, which can be provided by the Riemann-Hilbert problem (17).

Using the Jacobian matrix

$$(\mathrm{Jac}_0) = \left(\frac{D(\Psi^0, \dots, \Psi^{m-1})}{D(x_0, \dots, x_{m-1})}\right), \quad \det(\mathrm{Jac}_0) = J,$$

it is possible to write Lax-Sato equation of the hierarchy inthe form

$$\partial^k_n \mathbf{\Psi} = \sum_{i=0}^{m-1} \left((\mathrm{Jac}_0)^{-1})_{ik} (\Psi^0)^n)\right)_+ \partial_i \mathbf{\Psi}, \quad 1 \leqslant k \leqslant m-1, \tag{20}$$

where $1 \leqslant n < \infty$, $\mathbf{\Psi} = (\Psi^0, \dots, \Psi^{m-1})$. First flows of the hierarchy read

$$\partial^k_1 \mathbf{\Psi} = (\lambda \partial_k - \sum_{p=1}^{m-1} (\partial_k u_p)\partial_p - (\partial_k u_0)\partial_\lambda)\mathbf{\Psi}, \quad 1 \leqslant k \leqslant m-1, \tag{21}$$

where $u_0 = \Psi^0_1$, $u_k = \Psi^k_1$, $1 \leqslant k \leqslant m-1$.

A compatibility condition for any pair of linear equations (e.g., with ∂^k_1 and ∂^q_1, $k \neq q$) implies closed nonlinear N-dimensional system of PDEs for the set of functions u_k, u_0, which can be written in the form

$$\begin{aligned} &\partial^k_1 \partial_q \hat{u} - \partial^q_1 \partial_k \hat{u} + [\partial_k \hat{u}, \partial_q \hat{u}] = (\partial_k u_0)\partial_q - (\partial_q u_0)\partial_k, \\ &\partial^k_1 \partial_q u_0 - \partial^q_1 \partial_k u_0 + (\partial_k \hat{u})\partial_q u_0 - (\partial_q \hat{u})\partial_k u_0 = 0, \end{aligned} \tag{22}$$

where $\hat{u}$ is a vector field, $\hat{u} = \sum_{p=1}^{m-1} u_p \partial_p$. For $m = 3$ this system after volume-preservation reduction corresponds to the Dunajski system (generalizing heavenly equation).

Acknowledgments
The research of LVB was partially supported by the President of Russia grant 6170.2012.2 (scientific schools), the research of BGK was partially supported by the PRIN 2010/2011 grant 2010JJ4KBA.003.

References
[1] L. V. Bogdanov and B. G. Konopelchenko, Grassmannians Gr(N-1,N+1), closed differential N-1-forms and N-dimensional integrable systems, 2013 J. Phys. A: Math. Theor. 46 085201
[2] I. M. Krichever, The τ-function of the universal Whitham hierarchy, matrix models and topological field theories, Comm. Pure Appl. Math. 47, 437-475 (1994).
[3] W. V. D. Hodge and D. Pedoe, Methods of algebraic geometry, vols. 1,2, Cambridge Univeristy Press, Cambridge, 1994.
[4] L. V. Bogdanov, On a class of reductions of the Manakov-Santini hierarchy connected with the interpolating system, J Phys. A: Math. Theor. **43** (2010) 115206 (11pp)
[5] L. V. Bogdanov, Interpolating differential reductions of multidimensional integrable hierarchies, Theoretical and Mathematical Physics, 167(3): 705–713 (2011)
[6] S. Sternberg, Lectures on differential geometry, Prentice Hall, Inc. Englewood Cliffs, N. J., 1964.
[7] G. de Rham, Varietes differentiables. Formes, courants, formes harmoniquies, Hermann et Ciè, Paris, 1955.
[8] K. Takasaki, An infinite number of hidden variables in hyper-Kähler metrics, J. Math. Phys. **30**(7), 1515–1521 (1989)
[9] K. Takasaki, Symmetries of hyper-Kähler (or Poisson gauge field) hierarchy, J. Math. Phys. **31**(8), 1877–1888 (1989)
[10] M. Dunajski, Anti-self-dual four-manifolds with a parallel real spinor, Proc. Roy. Soc. Lond. A **458**, 1205 (2002)
[11] L. V. Bogdanov, V. S. Dryuma and S. V. Manakov, Dunajski generalization of the second heavenly equation: dressing method and the hierarchy, *J Phys. A: Math. Theor.* **40** (2007) 14383–14393.
[12] L. V. Bogdanov, Non-Hamiltonian generalizations of the dispersionless 2DTL hierarchy, J. Phys. A: Math. Theor. 43 (2010) 434008
[13] B. Konopelchenko and F. Magri, Coisotropic deformations of associative algebras and dispersionless integrable hierarchies, Commun. Math. Phys., **274**, 627-658 (2007).
[14] B. Konopelchenko and G. Ortenzi, Coisotropic deformations of algebraic varieties and integrable systems, J. Phys. A: Math. Theor., **42**, 415207, (2009).
[15] S. P. Burtsev, V. E. Zakharov, A. V. Mikhailov, Inverse scattering method with variable spectral parameter, Teoret. Mat. Fiz., 70:3 (1987), 323–341
[16] S. V. Manakov and P. M. Santini, The Cauchy problem on the plane for the dispersionless Kadomtsev-Petviashvili equation, JETP Lett. **83** (2006) 462-6.
[17] S. V. Manakov and P. M. Santini, A hierarchy of integrable PDEs in 2+1 dimensions associated with 2-dimensional vector fields, Theor. Math. Phys. **152** (2007) 1004–1011.
[18] L. Martínez Alonso and A. B. Shabat, Energy-dependent potentials revisited: a universal hierarchy of hydrodynamic type, Phys. Lett. A **300** (2002) 58–64
[19] M.V. Pavlov, Integrable hydrodynamic chains, J. Math. Phys. **44**(9) (2003) 4134–4156
[20] L. Martínez Alonso and A. B. Shabat, Hydrodynamic reductions and solutions of a universal hierarchy, Theor. Math. Phys. **140** (2004) 1073–1085
[21] L.V. Bogdanov, A class of multidimensional integrable hierarchies and their reductions, Theoretical and Mathematical Physics **160**(1) (2009) 887–893

On variational formulations and conservation laws for incompressible 2D Euler fluids

R Camassa[1], **G Falqui**[2], **G Ortenzi**[2], **and M Pedroni**[3]

[1]Carolina Center for Interdisciplinary Applied Mathematics, Department of Mathematics, University of North Carolina, Chapel Hill, NC 27599, USA
[2]Dipartimento di Matematica e Applicazioni, Università di Milano-Bicocca, Milano, Italy
[3]Dipartimento di Ingegneria, Università di Bergamo, Dalmine (BG), Italy

E-mail: camassa@amath.unc.edu, gregorio.falqui@unimib.it, giovanni.ortenzi@unimib.it, marco.pedroni@unibg.it

Abstract. With the aim of presenting a unified viewpoint for the variational and Hamiltonian formalism of two-dimensional incompressible stratified Euler equations, we revisit some of the formulations currently discussed in the literature and examine their mutual relations. We concentrate on the example of two-layered systems and its one-dimensional reduction, and use it to illustrate general consequences of density stratification on conservation laws which have been partially overlooked until now. In particular, we focus on the conservation of horizontal momentum for stratified ideal fluid motion under gravity between rigid lids.

1. Introduction

We consider the Euler equations for an ideal incompressible and inhomogeneous fluid, subject to gravity, in two dimensions:

$$\mathbf{v}_t + \mathbf{v}\cdot\nabla\mathbf{v} = -\frac{\nabla p}{\rho} - g\mathbf{k}, \qquad \nabla\cdot\mathbf{v} = 0, \qquad \rho_t + \nabla\cdot(\rho\mathbf{v}) = 0. \tag{1}$$

Here $\mathbf{v} = (u, w)$ is the velocity field, ρ and p are the density and pressure fields, respectively, g is the constant gravity acceleration, $\mathbf{k}$ is the unit vertical (upward) vector, and all physical variables depend on spatial coordinates (x, z) and time t. Besides their well known theoretical interest, this set of equations can be viewed as governing the motion of real fluids with sufficient accuracy whenever viscosity, compressibility and diffusivity effects can be neglected within the time scales of interest in the time evolution. These motion equations need to be supplemented with suitable boundary conditions. For instance, in the case of a fluid rigidly confined by horizontal "plates" of infinite length, located at $z = 0$ and $z = h$, the boundary conditions are those of no flux at infinity and across rigid boundaries

$$\mathbf{v}(x, z, t) \to 0 \ \text{ for } |x| \to \infty, \quad w(x, 0, t) = w(x, h, t) = 0, \tag{2}$$

and hydrostatic equilibrium at the far ends of the channel: $\dfrac{\partial p}{\partial z} = -g\rho$ for $|x| \to \infty$.

The variational formulation of the Euler equations for an ideal fluid has been the subject of several research efforts, as it presents some non-standard challenges with respect to classical field theory formalism such as that of the Maxwell equations, as first pointed out by Lin [9] (see also, e.g., the book [1], the review articles [10], [11], and references cited therein). We revisit this subject with the aim of pointing out some implications of the variational formalism, in the context of stratified fluids, that may have escaped some of the attention devoted to this subject. Our analysis stems from that of Benjamin [2], about the so-called Boussinesq model of 2D Euler fluids, which allows for an initial density distribution depending nontrivially on spatial coordinates, and assumes all subsequent changes to be incompressible. In [2] the conservation laws of such a system were determined, and, as a sort of side remark, it was pointed out that, for motion in a strip bounded by two rigid horizontal lids, total horizontal momentum is not conserved (despite the horizontal translation invariance of the system). Its time-variation is proportional to the pressure imbalance between the far ends of the channel, which however arises not as an imposed boundary condition but as a subtle consequence of the interplay between the fluid's incompressibility, stratification and inertia.

Recently, we have isolated and studied this phenomenon, substantiating Benjamin's observation with analytical and numerical results both for one-dimensional long-wave models and for the full two-dimensional Euler equations. In particular, in [3] we obtained some results on this pressure imbalance in the case of two-layer fluids. Along the same lines, a specific class of initial conditions — leading to momentum evolution for two-layer systems — was studied in [4] by means of an asymptotic expansions in the small-density-variation limit. The opposite case whereby the upper fluid has a very small density was the starting point of the analysis in [5]. This case suggested a topological selection mechanisms for conserved quantities, that turned out to be valid also in the case of continuous stratifications. We also briefly addressed the Hamiltonian aspects of these results, both in the sense of Benjamin and in that of Zakharov-Kuznetsov (see, for a review, [14]).

The focus of the present work is to give a more systematic account of the variational set-up of the Boussinesq model. In particular, after reviewing different Hamiltonian formulations of the problem, we point out their mutual relations, with a view towards our favorite test case of fluid motion in an infinite horizontal channel (which leads to lack of horizontal momentum invariance). The layout of the paper is as follows: Section 2 is devoted to Benjamin's approach, while Section 3 deals with Clebsch variables, both in the Zakharov-Kuznetsov representation and in that of Seliger and Whitham [12]. Lack of momentum conservation and its relation with the horizontal invariance of the system — and with pressure imbalance — are discussed in Section 4. In the final section we also quickly present a novel Dirac reduction whose outcome is the Hamiltonian structure of the dispersionless limit of two-layer systems, and discuss its relation with the 2D Hamiltonian picture.

2. Physical variables: the Benjamin's picture

Benjamin [2] proposed and discussed a set-up for the Hamiltonian formulation of the Boussinesq model. His results can be summarized as follows. The basic variables are the density ρ together with kind of "momentum vorticity" σ defined by

$$\sigma = \nabla \times (\rho \boldsymbol{v}) = (\rho w)_x - (\rho u)_z . \tag{3}$$

The equations of motion for these two fields, ensuing from (1), are

$$\begin{aligned} &\rho_t + u\rho_x + w\rho_z = 0 \\ &\sigma_t + u\sigma_x + w\sigma_z + \rho_x\big(gz - \tfrac{1}{2}(u^2+w^2)\big)_z + \tfrac{1}{2}\rho_z\big(u^2+w^2\big)_x = 0 \,. \end{aligned} \tag{4}$$

They can be written in the form

$$\rho_t = -\left[\rho, \frac{\delta H}{\delta \sigma}\right], \qquad \sigma_t = -\left[\rho, \frac{\delta H}{\delta \rho}\right] - \left[\sigma, \frac{\delta H}{\delta \sigma}\right], \tag{5}$$

where, by definition, $[A, B] := A_x B_z - A_z B_x$, and $H = \int_{\mathcal{D}} \rho \left(\frac{1}{2}|\mathbf{v}|^2 + gz \right) \mathrm{d}x\, \mathrm{d}z$, $\mathcal{D}$ being the fluid domain. In this formalism, the Hamiltonian H is written in terms of the stream function ψ, which is related to Benjamin's variables (ρ, σ) via

$$\sigma = (\rho w)_x - (\rho u)_z = -(\rho\psi_x)_x - (\rho\psi_z)_z = -\rho\nabla^2\psi - \nabla\rho \cdot \nabla\psi. \tag{6}$$

More precisely, once ρ and σ are given, ψ is the unique solution of (6) vanishing on the plates, so that H turns out to be a functional of ρ and σ only. As shown by Benjamin, equations (5) are actually a Hamiltonian system with respect to a non-canonical Hamiltonian structure. This means that equations (4) can be written as

$$\rho_t = \{\rho, H\}, \qquad \sigma_t = \{\sigma, H\}$$

for the Poisson brackets defined by the Hamiltonian operator

$$J_B = -\begin{pmatrix} 0 & \rho_x\partial_z - \rho_z\partial_x \\ \rho_x\partial_z - \rho_z\partial_x & \sigma_x\partial_z - \sigma_z\partial_x \end{pmatrix}. \tag{7}$$

Indeed, the variational differential of the Hamiltonian H is $(\delta_\rho H\ ,\ \delta_\sigma H) = \big(g\, z - |\nabla\psi|^2/2\ ,\ \psi\big)$, so that equations (4) follow by applying the Hamiltonian operator (7).

We close this section by with two remarks. The first concerns the structure of the Poisson tensor (7), and provides a proof, somewhat alternative to the direct one given in the appendix of [2], of the Jacobi identities for (7). We notice that, being linear in the field variables, this is actually a Lie-Poisson structure. The related Lie algebra can be first identified with the vector space of pairs of functions $(\rho(x, z), \sigma(x, z))$ equipped with the Lie bracket

$$[(\rho_1, \sigma_1), (\rho_2, \sigma_2)] = (\nabla\rho_1 \times \nabla\sigma_2 - \nabla\rho_2 \times \nabla\sigma_1, \nabla\sigma_1 \times \nabla\sigma_2). \tag{8}$$

This is the semidirect product between functions ρ and solenoidal vector fields $X^0_\sigma = (-\sigma_z, \sigma_x)$. Indeed, the RHS of (8) is expressed in terms of the natural action of vector fields on functions, $X^0_{\sigma_1}(\rho_2) - X^0_{\sigma_2}(\rho_1)$, with commutator $[X^0_{\sigma_1}, X^0_{\sigma_2}] = X^0_{\nabla\sigma_1\times\nabla\sigma_2}$. Second, as observed in [5], the previous construction heavily relies on the assumption that ρ (and σ) be constant along the plates (or, in general, on the boundary of the fluid domain). Otherwise, formula (7) does not even give rise to a skew-symmetric bracket.

3. Clebsch variables: the Zakharov-Kuznetsov picture

The Euler system (1) admits a variational formulation. As well known (see e.g., [12]), in the Euler representation the variation of the "physical" Lagrangian does not give rise to equations (1). The way out is to use the components of the Euler equations which are non-dynamical in the velocity as constraints in the Lagrangian. In the picture set forth by Zakharov and Kuznetsov (see, for a review, [14]) the action can be written as the integral of the difference between kinetic and potential energy, plus terms with Lagrange multipliers for constraints,

$$A = \int_{t_0}^{t_1} \left(\int_{\mathcal{D}} \mathcal{L}\, \mathrm{d}x\, \mathrm{d}z \right) \mathrm{d}t\,,$$

where

$$\mathcal{L} = \frac{\rho}{2}|\boldsymbol{v}|^2 - \rho g z + \Phi\nabla\cdot\boldsymbol{v} + \lambda\left(\rho_t + \nabla\cdot(\boldsymbol{v}\rho)\right). \tag{9}$$

By varying the action with respect to all the fields in $\mathcal{L}$, we get

$$\begin{aligned} &\frac{\delta A}{\delta \Phi} = 0 \Rightarrow \nabla \cdot \boldsymbol{v} = 0\,, && \frac{\delta A}{\delta \lambda} = 0 \Rightarrow \rho_t + \nabla \cdot (\rho \boldsymbol{v}) = 0\,, \\ &\frac{\delta A}{\delta \rho} = 0 \Rightarrow \lambda_t + \boldsymbol{v} \cdot \nabla \lambda - \frac{1}{2}|\boldsymbol{v}|^2 + g\,z = 0, && \frac{\delta A}{\delta \boldsymbol{v}} = 0 \Rightarrow \rho \boldsymbol{v} - \nabla \Phi - \rho \nabla \lambda = 0. \end{aligned} \tag{10}$$

These equations need to be augmented by the appropriate boundary conditions. Next, it is interesting to show how solutions of equations (10) are related to solutions of the original Euler equations (1). Taking the gradient of the third equation in the system (10), and the convective derivative $D/Dt \equiv \partial/\partial t + \boldsymbol{v} \cdot \nabla$ of the fourth one, yields $\frac{D(\rho \boldsymbol{v})}{Dt} = \nabla \left(\frac{D\Phi}{Dt} \right) - \rho g\, \boldsymbol{k}$, whence Euler equations follow, once we enforce the relation

$$\nabla \left(\frac{D\Phi}{Dt} \right) = -\nabla p\ . \tag{11}$$

It is important to notice that only a specific class of solutions of the Euler equations can be obtained by using this Clebsch representation. Indeed, solving (10) for the velocity gives

$$\boldsymbol{v} = \frac{\nabla \Phi}{\rho} + \nabla \lambda \Rightarrow \boldsymbol{\omega} = \nabla \times \boldsymbol{v} = \frac{1}{\rho^2} \nabla \Phi \times \nabla \rho\,. \tag{12}$$

Thus, the Clebsch representation (10) is compatible only with those fluid motions whose vorticity vanishes in the regions where the density ρ is constant. In particular, for a two-layer fluid, the vorticity may be non-zero only along the interface between the two homogeneous layers.

3.1. The Hamiltonian picture

With this proviso in mind, let us now discuss how, by using these Clebsch variables, one can obtain a canonical Hamiltonian formulation in which ρ and λ are conjugate variables. The Lagrangian density (9), thought of as a function of the three field-variables (ρ, Φ, λ) and their "generalized" velocities $(\rho_t, \Phi_t, \lambda_t)$, can be also written (up to an integration by parts) as

$$\mathcal{L}' = \lambda \rho_t + \rho \left(\frac{|\boldsymbol{v}|^2}{2} - g\,z \right) - \boldsymbol{v} \cdot (\nabla \Phi + \rho \nabla \lambda)\,. \tag{13}$$

Since here $\boldsymbol{v}$ is to be considered a shorthand notation for its expression (12) in terms of (ρ, Φ, λ), we obtain, by taking the (partial) Legendre transformation w.r.t. the variable ρ, the Routhian density

$$\mathcal{R} = \frac{\partial \mathcal{L}'}{\partial \rho_t} \rho_t - \mathcal{L}' = \frac{1}{2\rho} |\nabla \Phi + \rho \nabla \lambda|^2 + \rho\, g\, z. \tag{14}$$

According to the Routhian formalism, the variational equations (10) are — as it can be easily verified — written as

$$\begin{aligned} \rho_t &= \frac{\delta R}{\delta \lambda} = -\nabla \cdot (\nabla \Phi + \rho \nabla \lambda) \\ \lambda_t &= -\frac{\delta R}{\delta \rho} = -\left(\frac{1}{\rho} \nabla \Phi + \nabla \lambda \right) \cdot \nabla \lambda - gz + \frac{1}{2} |\frac{1}{\rho} \nabla \Phi + \nabla \lambda|^2 \\ 0 &= \frac{\delta R}{\delta \Phi} = \nabla^2 \lambda + \nabla \cdot \frac{\nabla \Phi}{\rho} \end{aligned} \tag{15}$$

the last equation being just the zero-divergence equation $\nabla \cdot \boldsymbol{v} = 0$. (We will always use calligraphic letters for the densities and roman letters for the corresponding functionals.)

To obtain a fully fledged Hamiltonian formalism, we can proceed as follows. We regard the third of (15) as a constraint allowing the variable Φ to be determined by the canonical variables λ and ρ. Indeed, the incompressibility equation $\nabla \cdot \boldsymbol{v} = 0$ reads as $\nabla \cdot ((\nabla\Phi)/\rho + \nabla\lambda) = 0$ that is, an elliptic problem for Φ the solution of which — supplemented by suitable boundary conditions — yields $\Phi = \Phi(\lambda, \rho)$.

Substituting this in $\mathcal{R}$ yields the Hamiltonian density $\mathcal{H} = \mathcal{H}(\lambda, \rho)$. The two conjugated variables λ, ρ still evolve according to the system

$$\begin{aligned} \rho_t &= \frac{\delta H}{\delta \lambda} = -\nabla \cdot \Big(\nabla\Phi(\lambda,\rho) + \rho\nabla\lambda\Big), \\ \lambda_t &= -\frac{\delta H}{\delta \rho} = -\Big(\frac{1}{\rho}\nabla\Phi(\rho,\lambda) + \nabla\lambda\Big) \cdot \nabla\lambda - gz + \frac{1}{2}\left|\frac{1}{\rho}\nabla\Phi(\rho,\lambda) + \nabla\lambda\right|^2 \end{aligned} \tag{16}$$

with, now,

$$H = \int_{\mathcal{D}} \Big(\frac{1}{2\rho}|\nabla\Phi(\rho,\lambda) + \rho\nabla\lambda|^2 + g\rho\, z\Big)\, \mathrm{d}x\, \mathrm{d}z\,. \tag{17}$$

We can summarize this three-step procedure in the following diagram:

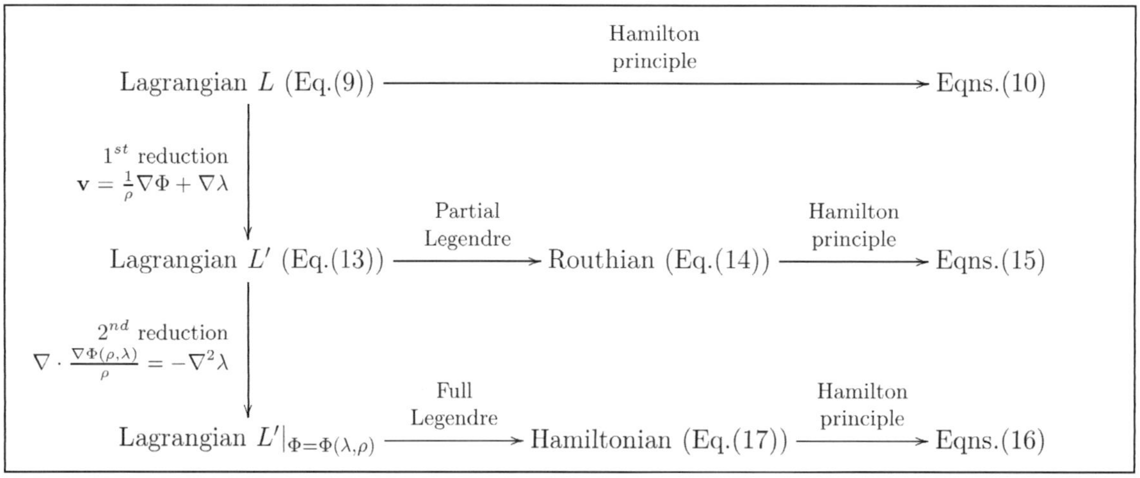

A straightforward computation, based on the fact that the Jacobian matrix of the "coordinate transformation"

$$(\rho, \lambda) \mapsto (\rho, \sigma) = (\rho, \rho_x\lambda_z - \rho_z\lambda_x) \tag{18}$$

can be represented by the matrix of differential operators

$$\begin{pmatrix} 1 & 0 \\ \lambda_z\partial_x - \lambda_x\partial_z & \rho_x\partial_x - \rho_z\partial_x \end{pmatrix},$$

shows how the Lie–Poisson structure (7) can be formally obtained from the canonical bracket.

3.2. Clebsch variables: the Seliger-Whitham picture

A variational Lagrangian formalism for the Euler fluids under consideration, allowing for arbitrary values of the vorticity, was set forth by Seliger and Whitham [12]. Its main features are the following. Rather than (as we did in the previous section) fixing one of the Clebsch variable to be the density ρ, one can consider (in the case of a general 3D Euler fluid) a full set of Clebsch variables $(\vartheta, \eta, \alpha, \beta)$ and write the constrained Lagrangian density as

$$\mathcal{L}_{SW} = \mathcal{T} + \rho g z - \rho\epsilon(\rho, S) + \vartheta\frac{D}{Dt}(\rho v) + \eta\frac{D}{Dt}(\rho S) + \beta\frac{D}{Dt}(\rho\alpha).$$

Here, $\rho\epsilon$ is the internal energy (which can be set to a constant in the incompressible case), S is the entropy density, and α, according to Lin's recipe [9], is one of the fluid's initial Lagrangian coordinate, while ϑ, η, β are Lagrange multipliers.

Among other properties of such general Clebsch formulation, Seliger and Whitham show that, by using the equations of motion (that is, in field-theoretical parlance, *on-shell*), the Lagrangian density is nothing but the pressure field p. Also (see §6 of [12]), for the case of (2D) incompressible evolutions, the number of Clebsch variables can be reduced to three, with the variational equations taking the form:

$$\boldsymbol{v} = \nabla\chi + \alpha\nabla\mu, \qquad \frac{D\alpha}{Dt} = 0, \qquad \frac{D\mu}{Dt} = p\frac{\rho'(\alpha)}{\rho^2(\alpha)}. \tag{19}$$

Here, the density ρ is a function of the sole variable α, while μ and χ are suitable functions of the original Clebsch variables $(\vartheta, \eta, \alpha, \beta)$ and of the entropy density S. More importantly, the Lagrangian density — i.e., the pressure — is expressed through χ, α, μ as

$$\mathcal{L}_{SW} = p = -\rho(\alpha)\left(\chi_t + \alpha\mu_t + \frac{1}{2}|\boldsymbol{v}|^2 + g\,z\right). \tag{20}$$

Let us compare the Seliger & Whitham with the Zakharov & Kuznetsov approaches (hereafter referred to as SW and ZK, respectively) in the Lagrangian set-up (in the case $\rho'(\alpha) \neq 0$). We re-define the Clebsch variable of Section 3 as

$$\varphi = \Phi + \lambda\rho. \tag{21}$$

Taking the convective derivative of both sides of the first relation we get

$$\frac{D\varphi}{Dt} = \frac{D\Phi}{Dt} + \lambda\frac{D\rho}{Dt} + \rho\frac{D\lambda}{Dt} = \frac{D\Phi}{Dt} + \lambda(\rho_t + \nabla\cdot(\rho\boldsymbol{v})) - \lambda\rho\nabla\boldsymbol{v} + \rho\left(\frac{|\boldsymbol{v}|^2}{2} - g\,z\right), \tag{22}$$

where we have used the third of equations (10). Hence, by taking into account that thanks to (11) we can identify (up to constants) the pressure p with $-\frac{D\Phi}{Dt}$, and we set $\varphi - \lambda\rho = \Phi$, we get

$$\mathcal{L} - \varphi_t - \nabla\cdot(\varphi\boldsymbol{v}) = p. \tag{23}$$

In analogy with the SW case, we obtain that the (on shell) ZK Lagrangian density differs from the pressure by terms that do not affect the action.

4. Pressure imbalances and momentum non conservation: the case of the channel

When the two-dimensional domain, as in our example case, is the infinite strip $\mathcal{S} = \mathbb{R}\times[0,h]$, translation along the x-axis is a symmetry of the system, hence by Noether's (first) theorem we expect to obtain a conservation law for the Euler equations.

Let us first examine the problem from the Hamiltonian viewpoint of ZK, in which the density ρ and the Clebsch variable λ are canonically conjugate variables. With respect to the canonical brackets, the functional generating translations along x is

$$I_{ZK} = \int_{\mathcal{S}} \rho\lambda_x\,\mathrm{d}x\,\mathrm{d}z\,. \tag{24}$$

As can be easily verified, this quantity is a constant of the motion.

Benjamin's formalism [2] is explicity taylored to symmetries. The generator of translations along the horizontal directions (the "impulse") is

$$I_B = \int_{\mathcal{D}} z\sigma(x,z)\,\mathrm{d}x\,\mathrm{d}z\,. \tag{25}$$

Under the transformation (18) it agrees, up to divergences, with the generator (24). In particular, for motion between two rigid horizontal lids, I_B can be written as the sum of a bulk and a boundary term (called B_6 by Benjamin),

$$I_B = \int_S \rho\, u \, \mathrm{d}x \, \mathrm{d}z - B_6,$$

where the first term is the ordinary total horizontal momentum, while the second term B_6 equals $\int_{\mathbb{R}} [z\rho u]_{z=0}^{z=h} \, \mathrm{d}x$. The Euler equation for the horizontal momentum

$$(\rho u)_t = -\frac{1}{2}\rho_x(u^2+w^2) - (\frac{1}{2}\rho(u^2+w^2)+p)_x + w\,\sigma \;, \tag{26}$$

in the case of a horizontally constant density at the top lid (where $w=0$) yields, as noticed by Benjamin,

$$\frac{dB_6}{dt}\left(=\frac{d}{dt}\int_S \rho u \, \mathrm{d}x \, \mathrm{d}z\right) = -h\,(p(+\infty,h)-p(-\infty,h))\,. \tag{27}$$

Hence, total horizontal momentum does not necessarily coincide with the constant of the motion I_B; it can evolve in time due to asymptotic pressure imbalances (see [4] for further details).

To discuss this issue in the Lagrangian formalisms of Section 2, let us recall that for a Lagrangian system in two spatial dimensions with N fields $(\varphi_1, \varphi_2, \ldots, \varphi_N)$, the expression for the conservation law associated with horizontal translations is

$$\sum_{\alpha=1}^{N}\left(\frac{\partial}{\partial t}\Big(\frac{\partial L}{\partial \varphi_{\alpha,t}}\varphi_{\alpha,x}\Big) + \frac{\partial}{\partial x}\Big(\frac{\partial L}{\partial \varphi_{\alpha,x}}\varphi_{\alpha,x} - L\Big) + \frac{\partial}{\partial z}\Big(\frac{\partial L}{\partial \varphi_{\alpha,z}}\varphi_{\alpha,x}\Big)\right) = 0, \tag{28}$$

so that for the Lagrangian (9) we get

$$\frac{\partial}{\partial t}(\lambda\rho_x) + \nabla\cdot\boldsymbol{J} = 0, \quad \boldsymbol{J} = (\Phi u_x - u(\lambda\rho)_x - \frac{1}{2}\rho|\mathbf{v}|^2 + \rho g z, \Phi w_x - w(\lambda\rho)_x). \tag{29}$$

This identifies the conserved density of the (Lagrangian) ZK formalism with $\lambda\rho_x$, which is consistent with the outcome of the Hamiltonian formalism (24).

A corresponding result can be obtained in the Seliger-Whitham formalism. Consider the Lagrangian density $\mathcal{L}_{SW}$ of equation (20) and apply the general formula (29); we get

$$\frac{\partial J_0}{\partial t} + \frac{\partial J_{(x)}}{\partial x} + \frac{\partial J_{(z)}}{\partial z} = 0, \qquad \text{where } J_0 = -\rho\chi_x - \rho\alpha\mu_x = -\rho u \text{ by Eq.(19)}.$$

Since $\mathcal{L}_{SW}$ is independent of spatial derivatives, $(J_{(x)}, J_{(z)}) = (-p, 0)$. Thus,

$$\frac{d}{dt}\int_S \rho u \, \mathrm{d}x \, \mathrm{d}z = -\int_0^h (p(+\infty,z) - p(-\infty,z))\, \mathrm{d}z, \tag{30}$$

which reduces to (27) under the assumption that the pressure be hydrostatic at the far ends of the channel.

5. Two-layer fluids in a channel: the dispersionless case

In this final section we shall briefly discuss the Hamiltonian structures that can be used to describe a sharply stratified fluid constrained within an infinite channel. The set-up we consider (see figure 1) is as follows. The upper (resp. lower) fluid has thickness $\eta_1(x,t)$ (resp. $\eta_2(x,t)$), constant density ρ_1 (resp. ρ_2), and velocity field (u_1, w_1) (resp. (u_2, w_2)). The total height of the channel is $h = \eta_1(x,t) + \eta_2(x,t)$, while $P(x,t) = p(x, \eta_2(x,t), t)$ is the interfacial pressure.

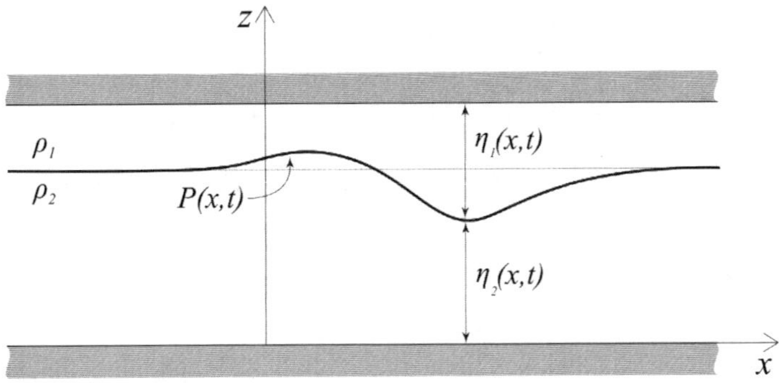

Figure 1: A two-layer system. Asymptotically the interface has the same height at $+\infty$ and at $-\infty$. The asymptotic velocities are zero.

By putting

$$\overline{f}_1(x,t) = \frac{1}{\eta_1(x,t)} \int_{\eta_2(x,t)}^{h} f_1(x,t,z)\,\mathrm{d}z, \qquad \overline{f}_2(x,t) = \frac{1}{\eta_2(x,t)} \int_0^{\eta_2(x,t)} f_2(x,t,z)\,\mathrm{d}z,$$

and vertically averaging the Euler equations in the channel (see [13, 6, 7]), we obtain

$$\begin{aligned} &\eta_{j\,t} + \left(\eta_j \overline{u}_j\right)_x = 0, \\ &\rho_j(\eta_j\overline{u}_j)_t + \rho_j\left(\eta_j\overline{u_j u_j}\right)_x = -(\eta_j\overline{p}_j)_x + (-1)^j(\eta_2)_x P\,, \qquad j = 1,2\,. \end{aligned}$$

Disregarding terms of order greater than zero in the parameter ϵ (the ratio between vertical and horizontal length scales), the previous set of equations becomes, in terms of the momenta $m_j = \rho_j\eta_j\overline{u}_j$,

$$\begin{aligned} \eta_{j_t} &= -(m_j/\rho_j)_x \\ m_{j_t} &= -\left(m_j^2/(\rho_j\,\eta_j)\right)_x - (-1)^j g\rho_j\eta_j\eta_{j_x} - \eta_j P_x. \end{aligned} \tag{31}$$

These equations are actually a system of Hamiltonian equations in one spatial dimension with Hamiltonian density

$$H_{dl} = \int_{-\infty}^{+\infty} \left(\frac{1}{2}\left(\frac{{m_1}^2}{\rho_1\eta_1} + \frac{{m_2}^2}{\rho_2\,\eta_2} \right) + (\eta_1+\eta_2-h)P + \frac{g}{2}\left(\rho_2\eta_2^2 - \rho_1\eta_1^2\right) \right) \mathrm{d}x \tag{32}$$

and the product Lie-Poisson Hamiltonian operator in the variables $(m_1, \eta_1, m_2, \eta_2)$

$$J := -\begin{pmatrix} J_1 & 0 \\ 0 & J_2 \end{pmatrix}, \quad J_k := \begin{pmatrix} m_k\partial_x + \partial_x m_k & \eta_k\partial_x \\ \partial_x\eta_k & 0 \end{pmatrix}, \quad k = 1,2. \tag{33}$$

Equations (31) are the dispersionless limit of the set of equations obtained in [7]. We will refer to them as *dispersionless* strongly nonlinear system.

Note that, in the Hamiltonian (32), the interfacial pressure P (the only quantity that couples the two set of equations (31)) plays the role of Lagrangian multiplier enforcing the constraint (a primary constraint, in Dirac's terminology [8])

$$\phi_1 \equiv \eta_1 + \eta_2 - h = 0. \tag{34}$$

The time evolution of this primary constraint yields $m_{1x}/\rho_1 + m_{2x}/\rho_2 = 0$, which, thanks to the null-velocity (and hence momentum) conditions at the far ends of the channel, yields

$$\phi_2 \equiv \frac{m_1}{\rho_1} + \frac{m_2}{\rho_2} = 0. \tag{35}$$

It turns out that these two constraints actually form a complete consistent set. Indeed, on the constrained "surface" defined by $\phi_i = 0$, for $i = 1, 2$, we have $\phi_{1t} = -\phi_{2x} = 0$; by evaluating $(\frac{m_1}{\rho_1} + \frac{m_2}{\rho_2})_t$ for $x \to \infty$, we see that no further dynamical constraints arise.

These two constraints are actually of "second class," so that they allow the definition of Dirac reduced brackets. To check this, we need only to compute the matrix of Poisson brackets $C_{ij} := \{\phi_i, \phi_j\}$. Since

$$\{\phi_1, \phi_1\} = 0, \quad \{\phi_1, \phi_2\} = -\partial_x\left(\frac{\eta_1}{\rho_1} + \frac{\eta_2}{\rho_2}\right), \quad \{\phi_2, \phi_2\} = -\left(\frac{m_1}{\rho_1^2} + \frac{m_2}{\rho_2^2}\right)\partial_x - \partial_x\left(\frac{m_1}{\rho_1^2} + \frac{m_2}{\rho_2^2}\right),$$

the matrix C can proven to be invertible (as a matrix with elements in the ring of pseudo-differental operators). A direct computation shows that, by choosing on the constrained "surface" the coordinates

$$\mu = \left(1 + \frac{\eta_1/\rho_1}{(h-\eta_1)/\rho_2}\right) m_1, \qquad \zeta = \eta_1 - h = -\eta_2, \tag{36}$$

the reduced Dirac Poisson tensor acquires the local Lie-Poisson form

$$J^D = -\begin{pmatrix} \mu\partial_x + \partial_x\mu & \zeta\partial_x \\ \partial_x\zeta & 0 \end{pmatrix}, \tag{37}$$

while the (reduced) Hamiltonian density is

$$\mathcal{H}^D(\mu, \zeta) = -\frac{\zeta\mu^2}{2(h+\zeta)(\rho_2(h+\zeta) - \rho_1\zeta)} + \frac{g}{2}\left(\rho_2\zeta^2 - \rho_2(h+\zeta)^2\right). \tag{38}$$

Let us also remark that the generator of the x-translation for the tensor (37) is the coordinate μ, which does not coincide with the total "horizontal" momentum $m_1 + m_2$. Hence, at the dispersionless level in the layer averaged formalism, we again obtain that the total horizontal momentum is not the conserved quantity related by Noether's theorem with translational symmetry.

Let us finally show that, in the long-wave regime, the Poisson bracket from the Dirac reduction is actually related with that of Benjamin (7), discussed in Section 2. To this end, let us write the physical variables of the two-layer fluid with the aid of the Heaviside θ-function as

$$\begin{aligned} &\rho = \rho_2 + (\rho_1 - \rho_2)\theta(z - \eta_2(x,t)) \\ &u(x,z,t) = u_2(x,z,t) + (u_1(x,z,t) - u_2(x,z,t))\theta(z - \eta_2(x,t)) \\ &w(x,z,t) = w_2(x,z,t) + (w_1(x,z,t) - w_2(x,z,t))\theta(z - \eta_2(x,t)). \end{aligned} \tag{39}$$

Assuming that the motion is irrotational in the bulk of each of the layers (which is consistent with the long wave approximation of the strongly nonlinear model), we get at first

$$\sigma = -\big(\rho_1 u_1(x,z,t) - \rho_2 u_2(x,z,t) + \eta_{2x}(\rho_1 w_1(x,z,t) - \rho_2 w_2(x,z,t))\big)\delta(z - \eta_2(x,t)). \tag{40}$$

The first order approximation in the expansion in $\epsilon = h/L \simeq w_i/u_i$ is, by replacing u_i with the layer mean velocities $\bar{u}_i(x,t)$,

$$\sigma = -\big(\rho_1\bar{u}_1(x,t) - \rho_2\bar{u}_2(x,t)\big)\delta(z - \eta(x,t)). \tag{41}$$

Integrating over the fluid layer gives the layer-mean

$$\langle\sigma\rangle = \frac{1}{h}\int_0^h \sigma(x,z)\,\mathrm{d}z = -\frac{1}{h}\big(\rho_1\bar{u}_1(x,t) - \rho_2\bar{u}_2(x,t)\big) = \frac{1}{h}\left(\frac{m_2}{\eta_2} - \frac{m_1}{\eta_1}\right). \tag{42}$$

By using the constraints (34, 35) it turns out that $\langle\sigma\rangle = -\mu/(h\zeta)$, so that the pair $(\zeta, \langle\sigma\rangle)$ are a system of coordinates for the spatial 1-D system of the averaged dispersionless equations obtained by the reduced Hamiltonian (38). In these coordinates the Poisson tensor is turned into the constant tensor

$$J^D = \frac{1}{h}\begin{pmatrix} 0 & \partial_x \\ \partial_x & 0 \end{pmatrix}, \tag{43}$$

and the equations of motion are written as the conservation laws

$$\zeta_t = -\left(\frac{h\zeta(h+\zeta)\langle\sigma\rangle}{\rho_2(h+\zeta)-\rho_1\zeta}\right)_x, \quad \langle\sigma\rangle_t = -\left(\frac{h\langle\sigma\rangle^2(\rho_2(h+\zeta)^2-\rho_1\zeta^2)}{2(\rho_2(h+\zeta)-\rho_1\zeta)^2} - \frac{g}{h}(\rho_2-\rho_1)\zeta\right)_x.$$

Let us look at the relations (the first one is obtained by integrating the first of (39) along z)

$$\zeta(x) = \frac{1}{\rho_1-\rho_2}\int_0^h (\rho(x,z)-\rho_1)\,\mathrm{d}z\,, \qquad \langle\sigma\rangle(x) = \frac{1}{h}\int_0^h \sigma(x,z)\,\mathrm{d}z \tag{44}$$

as defining a map from the manifold M^2 spanned by Benjamin's variables $(\rho(x,z), \sigma(x,z))$ to the "reduced" manifold M^1 of the strongly nonlinear dispersionless equations, spanned, as we have just seen, by the coordinates $(\zeta(x), \langle\sigma\rangle(x))$.

By using standard techniques of Hamiltonian reduction we can now be shown that the Poisson tensor J^D is the Hamiltonian reduction (under the map (44)) of Benjamin's Poisson tensor (7).

Indeed, let $(\alpha_\zeta(x), \alpha_{\langle\sigma\rangle}(x))$ be a covector on M^1; the pair $\left(\frac{1}{\rho_1-\rho_2}\alpha_\zeta(x),\ \frac{1}{h}\alpha_{\langle\sigma\rangle}(x)\right)$ is its pull-back along the map (44) to M^2; applying the tensor (7) to this pull-back we get the vector

$$(\dot\rho, \dot\sigma) = \left(\frac{\rho_1-\rho_2}{h}\delta(z-\eta_2)\partial_x(\alpha_{\langle\sigma\rangle}),\ \delta(z-\eta_2)\partial_x(\alpha_\zeta) + \frac{1}{h}(\rho_2\overline{u}_2 - \rho_1\overline{u}_1)\delta'(z-\eta_2)\partial_x(\alpha_{\langle\sigma\rangle})\right).$$

Pushing this vector to M^1 via (44), since $\int_0^h \delta'(z-\eta_2)\,\mathrm{d}z = 0$, yields the vector

$$(\dot\zeta, \dot{\langle\sigma\rangle}) = \left(\frac{1}{h}\partial_x(\alpha_{\langle\sigma\rangle})\ \frac{1}{h}\partial_x(\alpha_\zeta)\right),$$

which is nothing but the image under the Hamiltonian operator (43) of the covector $(\alpha_\zeta(x), \alpha_{\langle\sigma\rangle}(x))$.

References

[1] Arnold V I and Khesin B A *Topological Methods in Hydrodynamics* Springer (1998)
[2] Benjamin T B 1986 On the Boussinesq model for two-dimensional wave motions in heterogeneous fluids *J. Fluid Mech.* **165** 445–474
[3] Camassa R, Chen S, Falqui G, Ortenzi G and Pedroni M 2012 An inertia 'paradox' for incompressible stratified Euler fluids *J. Fluid Mech.* **695** 330–340
[4] Camassa R, Chen S, Falqui G, Ortenzi G and Pedroni M 2013 Effects of inertia and stratification in incompressible ideal fluids: pressure imbalances by rigid confinement *J. Fluid Mech.* **726** 404–438
[5] Camassa R, Chen S, Falqui G, Ortenzi G and Pedroni M 2013 Topological selection in stratified fluids: an example from air-water systems *submitted*
[6] Camassa R and Levermore C D 1997 Layer-mean quantities, local conservation laws, and vorticity *Phys. Rev. Lett.* **78** 650–653
[7] Choi W and Camassa R 1999 Fully nonlinear internal waves in a two-fluid system *J. Fluid Mech.* **396** 1–36
[8] Dirac P A M *Lectures on quantum mechanics* Dover Publications (2001)
[9] Lin C C *Hydrodynamics of Liquid Helium II* Proc. Int. School of Physics E. Fermi, New York, Academic Press (1963) Reprinted in 1987 *Selected papers of C C Lin*, Singapore, World Scientific.
[10] Morrison P J 1998 Hamiltonian description of the ideal fluid *Reviews of Modern Physics* **70** 467–521
[11] Salmon R 1998 Hamiltonian Fluid mechanics *Ann. Rev. Fluid. Mech.* **20** 225–256
[12] Seliger R L and Whitham G B 1968 Variational Principles in Continuum Mechanics *Proc. Roy. Soc. London A Math. Phys. Sci.* **305** 1–25
[13] Wu T Y 1981 Long waves in ocean and coastal waters *J. Eng. Mech.* **107** 501–522
[14] Zakharov V E and Kuznetsov E A 1997 Hamiltonian formalism for nonlinear waves *Phys. Usp.* **40** 1087–1116

KP web-solitons from wave patterns: an inverse problem

Sarbarish Chakravarty[1] and Yuji Kodama[2]

[1]Department of Mathematics, University of Colorado, Colorado Springs, CO 80933
[2] Department of Mathematics, Ohio State University, Columbus, OH 43210

E-mail: [1] `schakrav@uccs.edu`, [2] `kodama.1@osu.edu`

Abstract. Nonlinear interactions among small amplitude, long wavelength, obliquely propagating waves on the surface of shallow water often generate web-like patterns. In this article, we discuss how line-soliton solutions of the Kadomtsev-Petviashvili (KP) equation can approximate such web-pattern in shallow water wave. We describe an "inverse problem" which maps a certain set of measurable data from the solitary waves in the given pattern to the parameters required to construct an exact KP soliton that describes the non-stationary dynamics of the pattern. We illustrate the inverse problem using explicit examples of shallow water wave pattern.

1. Introduction

Web-like wave patterns as shown in Figure 1 below formed on the surface of shallow water are often observed at a beach with nearly flat bottom. These patterns are non-stationary in general, and are formed due to the interaction of solitary waves coming onto the shore. On a smaller scale, such wave patterns can be artificially created in table-top experiments [13] (see also movies available at the website http://www.needs-conferences.net/2009/), as well as in more accurate water tank experiments in the laboratory [15, 19]. The purpose of this note is to provide a mathematical model describing the dynamics of such wave patterns based on the Kadomtsev-Petviashvili (KP) equation [10]

$$(4u_t + 6uu_x + u_{xxx})_x + 3u_{yy} = 0, \tag{1.1}$$

where $u = u(x,y,t)$ represents the (normalized) wave amplitude. The KP equation describes the

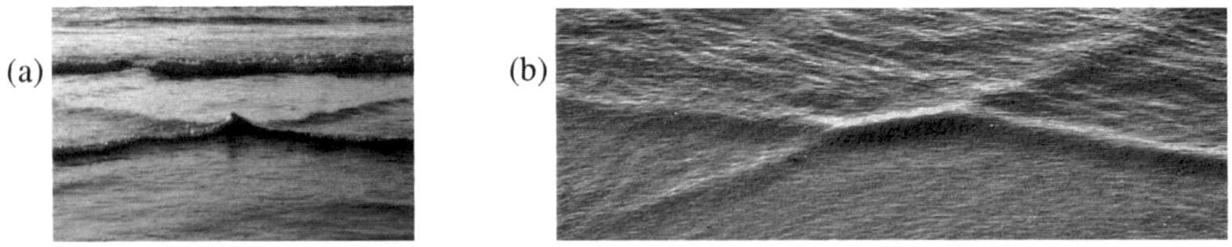

Figure 1. Wave patterns on a beach. Photographs by M. J. Ablowitz and D. E. Baldwin [2]

propagation of small amplitude, long wavelength, uni-directional waves with small transverse variation (i.e., quasi-two-dimensional waves). An important family of solitary wave solutions of the KP equation are the line-soliton solutions which are globally non-singular, and do not decay along distinct lines in the xy-plane. Such solutions consist of arbitrary number of asymptotic line solitons for $|y| \gg 0$ and complex web-like interaction patterns of line solitons in the finite region of the xy-plane. Because of such interaction patterns, these solutions are sometimes called the *web-solitons*. Several simple yet exact web-soliton solutions including the Y-shape soliton and several others have been experimentally demonstrated in recent laboratory wave tank experiments by Harry Yeh's group at Oregon State University [15, 19]. The general line-soliton solutions have been studied extensively by the authors who gave a complete classification of these solutions using geometric and combinatorial techniques [5, 6, 12].

In this article, we consider simple examples of shallow water wave patterns such as in Figure 1 which are assumed to be formed under the same physical conditions where the KP equation is valid. We then formulate the *inverse problem* which determines the information required to construct the exact line-soliton solution of the KP equation, simply referred to as the *KP line-solitons*, from the wave pattern data consisting of the amplitude and slopes of the solitary waves, and the locations of those waves on the xy-plane for fixed t.

Resonant interaction among solitary waves plays an important role in the formation of surface wave patterns that can be approximated by the KP line-solitons. Resonant interactions for the KP solitons in the form of a Y-shape wave-form were first investigated by Miles [16]. It turns out that such Y-shape wave-form is an exact solution of the KP equation, and is referred to as Y-soliton. Subsequently, more general type of resonant and partially-resonant line-soliton solutions have been discovered (see e.g., [4, 11, 3]). In this paper, we demonstrate using explicit examples of shallow water wave patterns that the resonant line-soliton solutions of the KP equation can be used to approximate such patterns and their dynamics fairly well.

Recent interest in the study of the KP line-soliton solutions has generated a wealth of data in the form of photographs and video recordings of wave patterns on flat beaches (see e.g., [18, 1, 2]. There are several works in recent literature on the modeling of shallow water wave patterns by the KP line-solitons (see e.g., [1, 17, 18]). Most of these works are qualitative studies and compares a pattern with a KP soliton at a fixed instant of time $t = t_0$. In contrast, our algorithm makes explicit use of the measurements from a given wave pattern to obtain the analytical form of the KP line-soliton solution which then describes the *non-stationary* dynamics of the pattern for both $t \geq t_0$ and $t \leq t_0$. It should be noted that the analysis presented here is only a leading order approximation since it is limited to the comparison between the wave patterns and the exact KP theory. In other words, we do not consider higher order corrections to the KP equation due to large amplitude, finite angle (departure from quasi-two-dimensionality), uneven bottom, or any such physical perturbations.

The paper is organized as follows: In Section 2, we provide some basic background for the one line-soliton and the resonant Y-soliton solutions of the KP equation. Then in Section 3 we describe an algorithmic method to construct an exact KP line-soliton solution from a given wave pattern, and illustrate our algorithm using explicit examples of actual shallow water wave patterns. Finally, we conclude with some remarks in Section 4.

2. The KP line-soliton solutions

This section contains background information on the KP line-soliton solutions and brief descriptions of some particular solutions (see [5, 6, 12] for the details).

The solutions of the KP equation (1.1) are given in terms of the τ-function $\tau(x,y,t)$ as

$$u(x,y,t) = 2(\ln \tau)_{xx}\,. \tag{2.1}$$

For line-solitons the τ-function consists of (i) M distinct real parameters ordered as $k_1 < k_2 < \cdots < k_M$, and (ii) an $N \times M$ real matrix A of full rank with $N < M$. The τ-function is explicitly given by

$$\tau(x,y,t) = \sum_I \Delta_I(A)\, E_I(x,y,t)\,, \tag{2.2}$$

where the sum is over all (ordered) N-element subsets I of the set $\{1,\ldots,M\}$. The coefficient $\Delta_I(A)$ is the $N\times N$ minor of the matrix A with the column set I, and $E_I(x,y,t) := K_I \exp\Theta_I(x,y,t)$ where

$$K_I = \prod_{l>m}(k_{i_l}-k_{i_m}), \qquad \Theta_I(x,y,t) = \sum_{m=1}^{N}(k_{i_m}x + k_{i_m}^2 y - k_{i_m}^3 t). \tag{2.3}$$

The soliton solution $u(x,y,t)$ is regular if and only if $\Delta_I(A)\geq 0$ for all the k-element subsets I [14]. In this case, the matrix A is called a *totally non-negative* matrix.

Remark 2.1 *In [7], it was shown that the solution (2.1) can be also expressed using the τ-function in the Grammian form, i.e. $u = 2(\ln\tau_h)_{xx}$ with*

$$\tau_h(x,y,t) = \det(I+CF), \qquad F_{ij} = \int^{x} f_i g_j\, dx$$

where C is an $N\times(M-N)$ constant matrix, and F is an $(M-N)\times N$ matrix for $f_i = e^{\phi(p_i)}$ and $g_j = e^{-\phi(q_j)}$ with $\phi(k) = kx + k^2 y - k^3 t$. Those parameters $q_1,q_2,\ldots,q_N$, $p_1,p_2,\ldots,p_{M-N}$, and the coefficient matrix C are uniquely determined by k_j's and the matrix A in (2.2).

In [5, 6], we showed that the general soliton solution given by equations (2.1) and (2.2) consists of N and $(M-N)$ asymptotic line-solitons as $y\gg 0$ and as $y\ll 0$, respectively. Each asymptotic soliton is uniquely parametrized by a pair of distinct k-parameters $\{k_i,k_j\}$ for $i<j$, and we let $[i,j]$ denote the index pair for this soliton (see section 2.1 for further explanation). The index pair $[i,j]$ is uniquely characterized by a map π such that $\pi(i)=j$ if $[i,j]$ labels an asymptotic soliton for $y\gg 0$, and $\pi(j)=i$ if $[i,j]$ labels an asymptotic soliton for $y\ll 0$. The map π turns out to be fixed-point free permutation of the index set $\{1,\ldots,M\}$ known as *derangement* which is conveniently represented by a chord diagram as shown in examples below. Thus the soliton solution generated by the τ-function (2.2) is represented by the *chord diagram* associated to the derangement π.

Next we briefly describe the one line-soliton and the Y-solitons which form the building blocks of a web-soliton solution of the KP equation.

2.1. One line-soliton solution

Asymptotic analysis of the τ-function in (2.2) reveals that the solution $u(x,y,t)$ is exponentially vanishing in regions of the xy-plane where a single exponential E_I is dominant over all other exponentials in (2.2), while it is localized along certain lines denoted by an index pair $[i,j]$ where a pair of dominant exponentials E_I, E_J are in balance. In the neighborhood of this line $[i,j]$, the τ-function can be approximated by

$$\tau \approx \Delta_I(A)E_I + \Delta_J(A)E_J,$$

where $I = I_0\cup\{i\}$ and $J = I_0\cup\{j\}$ with $|I_0| = N-1$. The solution $u(x,y,t)$ along this line approximates a single line-soliton solution,

$$u(x,y,t) \approx A_{[i,j]}\,\mathrm{sech}^2\,\Theta_{[i,j]}(x,y,t),$$

and will be referred to as the $[i,j]$-soliton. Here the phase $\Theta_{[i,j]}$ is given by

$$\begin{aligned}\Theta_{[i,j]} &:= \tfrac{1}{2}(\Theta_I-\Theta_J) = \tfrac{1}{2}(k_i-k_j)\left(x+\tan\Psi_{[i,j]}y - C_{[i,j]}t + x_{[i,j]}\right),\\ x_{[i,j]} &= \frac{1}{k_i-k_j}\ln\frac{\Delta_I(A)K_I}{\Delta_J(A)K_J},\end{aligned} \tag{2.4}$$

and K_I, Θ_I are as in (2.3). The soliton amplitude $A_{[i,j]}$, slope $\Psi_{[i,j]}$ and velocity in the positive x-direction $C_{[i,j]}$, are defined respectively by

$$A_{[i,j]} = \frac{1}{2}(k_i-k_j)^2, \quad \tan\Psi_{[i,j]} = k_i+k_j, \quad C_{[i,j]} = k_i^2+k_j^2+k_ik_j > 0. \tag{2.5}$$

A contour plot of this line-soliton solution is shown in Figure 2. Note that the angle $\Psi_{[i,j]}$ is measured counterclockwise from the y-axis, and $-\frac{\pi}{2} < \Psi_{[i,j]} < \frac{\pi}{2}$.

The line $[i,j]$ forms the boundary between the regions of dominant exponentials E_I and E_J. The parameters k_i and k_j are exchanged during the transition from the region of one dominant exponential to the other by crossing the $[i,j]$-soliton. Hence the line $[i,j]$ can be viewed as representing a permutation of the index set $\{i,j\}$. The chord diagram in Figure 2 depicting the exchange $k_i \leftrightarrow k_j$ represents the permutation $\pi = \begin{pmatrix} i & j \\ j & i \end{pmatrix}$ of the index set $\{i,j\}$.

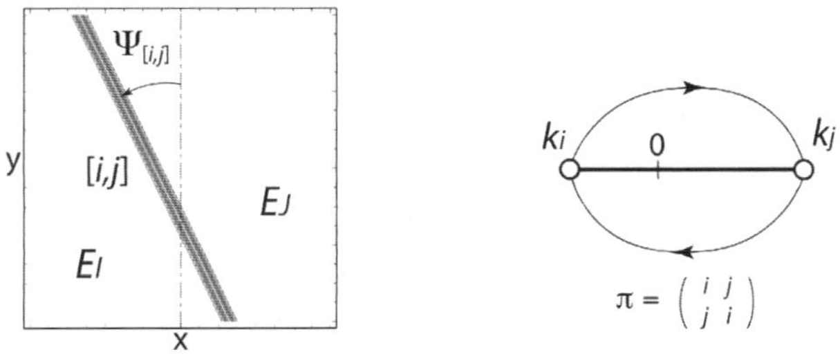

Figure 2. Contour plot and the chord diagram representation of the $[i,j]$-soliton

2.2. Y-soliton solutions

The Y-soliton is formed due to the resonant interaction of three line solitons labeled $[i,j]$, $[j,l]$ and $[i,l]$ with $i < j < l$. If we represent each line soliton as a traveling wave $u = \Phi(\mathbf{K}_{[a,b]} \cdot \mathbf{r} - \Omega_{[a,b]}t)$ with $\mathbf{r} = (x,y)$, and $\mathbf{K}_{[a,b]} = \frac{1}{2}(k_b - k_a, k_b^2 - k_a^2)$, $\Omega_{[a,b]} = \frac{1}{2}(k_b^3 - k_a^3)$ being the wave-vector and the frequency of the wave, then the soliton triplet satisfy the resonant conditions

$$\mathbf{K}_{[i,j]} + \mathbf{K}_{[j,l]} = \mathbf{K}_{[i,l]}, \qquad \Omega_{[i,j]} + \Omega_{[j,l]} = \Omega_{[i,l]}.$$

Near the trivalent vertex, the τ-function has the form

$$\tau \approx \Delta_I(A)E_I + \Delta_J(A)E_J + \Delta_L(A)E_L.$$

There are two cases for the index sets $\{I,J,L\}$ leading to the contour plots in Figures 3.

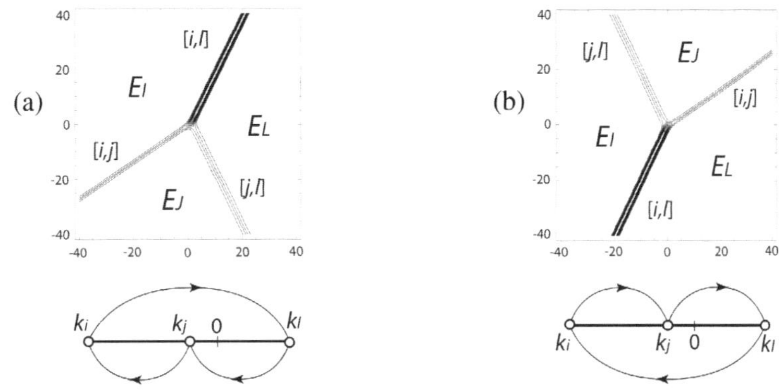

Figure 3. Contour plots and chord diagrams of the Y-solitons. Resonant soliton triplet $[i,j],[j,l]$ and $[i,l]$. The horizontal line in each chord diagram gives the values of the k-parameters $k_i < k_j < 0 < k_l$.

(a) The index sets are given by $I = I_0 \cup \{i\}, J = I_0 \cup \{j\}$ and $L = I_0 \cup \{l\}$ where I_0 is the common set of indices with $|I_0| = N-1$. In this case the $[i,l]$-soliton is above the trivalent vertex while the solitons $[i,j]$ and $[j,l]$ appear below it to form a λ-shape as shown in Figure 3(a). Following the transitions of the dominant exponentials clockwise in Figure 3(a), one recovers the permutation $\pi = \begin{pmatrix} i & j & l \\ l & i & j \end{pmatrix}$ given by the three line solitons $[i,j]$, $[j,l]$, $[i,l]$. The chord diagram corresponding to this permutation is shown below the contour plot.

(b) The index sets are $I = J_0 \cup \{i,j\}, J = J_0 \cup \{i,l\}$ and $L = J_0 \cup \{j,l\}$ with $|J_0| = N-2$. The solution in this case is related to that of Case (a) by an inversion $(x,y,t) \to (-x,-y,-t)$ so that the solitons $[i,j]$ and $[j,l]$ appear above the trivalent vertex and the $[i,l]$-soliton appears below it as shown in Figure 3(b). The line solitons for this Y shape represent the permutation $\pi = \begin{pmatrix} i & j & l \\ j & l & i \end{pmatrix}$ as shown by the chord diagram below the contour plot.

Due to the ordering $k_i < k_j < k_l$, the $[i,l]$-soliton has the largest amplitude $A_{[i,l]} = \frac{1}{2}(k_l^2 - k_i^2)$, and the $[j,l]$-soliton has the largest slope $\tan\Psi_{[j,l]} = k_j + k_l$, while the $[i,j]$-soliton has the smallest slope. Note that all these information can be gathered directly from the chord diagrams presented in Figure 3, which is one of the main reasons we use the chord diagrams in our analysis. One should also note that the k-parameters in the Y-soliton are uniquely determined only from the slopes of the three solitons. Recall that the slopes of the solitons $[i,j]$, $[j,l]$, $[i,l]$ are respectively given by

$$\tan\Psi_{[i,j]} = k_i + k_j, \quad \tan\Psi_{[j,l]} = k_j + k_l, \quad \tan\Psi_{[i,l]} = k_i + k_l\,.$$

Then one can compute the values of the k-parameters as

$$\begin{aligned} k_i &= \tfrac{1}{2}\left(\tan\Psi_{[i,l]} + \tan\Psi_{[i,j]} - \tan\Psi_{[j,l]}\right), \\ k_j &= \tan\Psi_{[i,j]} - k_i, \qquad k_l = \tan\Psi_{[i,l]} - k_i \end{aligned} \tag{2.6}$$

Remark 2.2 *The KP equation also admits solutions where two line solitons interact to form an X-vertex. In this case, the interaction is* non-resonant *and the wave pattern is* stationary. *An important feature of X-soliton solution is the existence of a constant phase shift which appears as a shift of the crest-line of each line-soliton at the interaction region, and is determined by the amplitudes and slopes of the solitary waves forming X-shape pattern [6]. In a real physical situation, the phase shift is typically at most twice the soliton wavelength, and large phase shifts several times the soliton wavelength are highly improbable (see [8] for more details). A long stem observed in a real wave pattern does not correspond to an X-soliton, rather it corresponds an entirely different non-stationary, resonant solution of the KP equation. For example, the stationary wave pattern in Figure 1(a) shows an X-type interaction, while the wave pattern in Figure 1(b) is not of X-type. The long stem appearing in the middle of Figure 1(b) is an intermediate solitary wave which interact resonantly with other two solitons at the trivalent vertices.*

A general line-soliton solution of the KP equation can be regarded as a collection of one-solitons and Y-solitons which fit together to form a web-structure with X- and Y-vertices in the interaction region. A combinatorial description of this solution is obtained by gluing the local chord diagrams corresponding to those components to form the complete chord diagram for the derangement π associated with the general soliton solution. We will implement this technique of "gluing" of chords in the next section in order to reconstruct a KP soliton solution from a *real* wave pattern.

3. The Inverse problem

In this section we discuss our main result which is the problem of constructing an exact soliton solution of the KP equation from a given wave pattern. As examples, we consider the surface wave patterns on shallow water observed on flat ocean beaches. The construction procedure consists of the following steps:

(i) Set the xy-coordinates, so that all solitary waves in the pattern are traveling almost in the positive x-direction (towards the shore).
(ii) Using (2.5) and (2.6), find the k-parameters by measuring the amplitudes and slopes of the solitary waves in the pattern.
(iii) Define the derangement π and its chord-diagram from the k-parameters obtained in the step (ii).
(iv) Determine the form of the matrix A from the derangement π using combinatorial techniques [14].
(v) Using (2.4), find the matrix elements of A from the minors $\Delta_I(A)$ by measuring the locations of the solitary waves in the pattern.

For a simple illustration of our method, in this note, we consider wave patterns with only two trivalent vertices consisting of two Y-type (one Y-shape and other λ-shape) waves which have a common intermediate solitary wave. Such patterns are observed fairly regularly in flat ocean beaches. We distinguish the Y-type waves by assigning a white trivalent vertex for the λ-shape wave and a black trivalent vertex for the Y-shape wave. There are 5 distinct cases for such wave pattern as shown in Figure 4. The resonant vertices in each of these patterns are labeled A (upper vertex) and B (lower

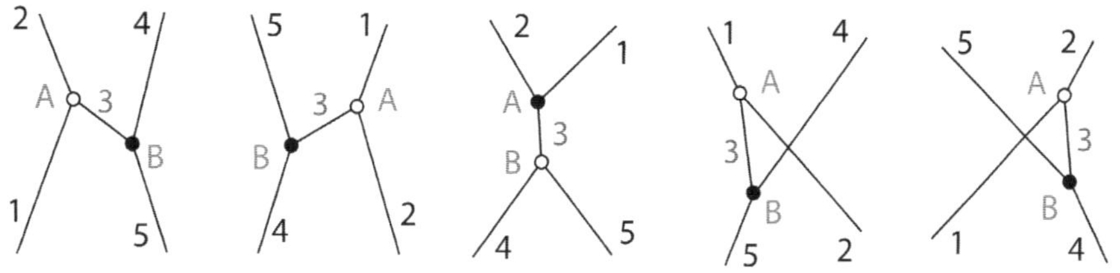

Figure 4. Five distinct wave patterns obtained by gluing two Y-shape waves.

vertex). The solitary waves incident at the vertex A are numbered as 1, 2 and 3, while the solitary waves numbered as 3, 4 and 5 are incident at the vertex B. Thus each pattern consists of two Y-type waves glued together by the common solitary wave 3 joining vertices A and B. We first identify each solitary wave with an $[i, j]$-soliton for some parameters $k_i < k_j$ to be determined. If we denote the parameters $\{k_a, k_b, k_c\}$ at vertex A, and $\{k_{a'}, k_{b'}, k_{c'}\}$ at vertex B, then each one-soliton can be parametrized as follows:

Soliton 1	Soliton 2	Soliton 3	Soliton 4	Soliton 5
$\{k_a, k_b\}$	$\{k_a, k_c\}$	$\{k_b, k_c\}, \{k_{b'}, k_{c'}\}$	$\{k_{a'}, k_{b'}\}$	$\{k_{a'}, k_{c'}\}$

Note that there are two parametrizations (with respect to the vertices A and B) for the intermediate soliton 3. Let $\tan\Psi_i$, $i = 1, 2, \ldots, 5$ denote the slope of each one-soliton. In particular, the slope of soliton 3 is given as $\tan\Psi_3 = k_b + k_c = k_{b'} + k_{c'}$. Then the k-values can be solved uniquely from the slope measurements at the vertices A and B for each Y-soliton as given by (2.6), i.e.

$$\begin{cases} k_a = \frac{1}{2}(\tan\Psi_1 + \tan\Psi_2 - \tan\Psi_3), \\ k_b = \tan\Psi_1 - k_a, \quad k_c = \tan\Psi_2 - k_a, \end{cases}$$
$$\begin{cases} k_{a'} = \frac{1}{2}(\tan\Psi_4 + \tan\Psi_5 - \tan\Psi_3), \\ k_{b'} = \tan\Psi_4 - k_{a'}, \quad k_{c'} = \tan\Psi_5 - k_{a'} \end{cases} \tag{3.1}$$

Without loss of generality, the two sets of k-parameters for soliton 3 can be ordered as $k_b < k_c$ and $k_{b'} < k_{c'}$. If the given wave pattern is an exact KP soliton, then one should have $k_b = k_{b'}$ and $k_c = k_{c'}$. But from equations (3.1) we have

$$k_b - k_{b'} = -(k_c - k_{c'}) = \tfrac{1}{2}[(\tan\Psi_1 - \tan\Psi_2) - (\tan\Psi_4 - \tan\Psi_5)],$$

which in general, has a non-zero value due to the deviation from the exact KP theory which is only a leading order approximation to a real wave pattern. To approximate the pattern by a KP soliton, we take the average of the two sets of k-values obtained from the measurements at the vertices A and B. That is, we define the k-parameters of soliton 3 as

$$\tilde{k}_b = \tfrac{1}{2}(k_b + k_{b'}), \qquad \tilde{k}_c = \tfrac{1}{2}(k_c + k_{c'}),$$

which preserves the slope of soliton 3 since $\tan\Psi_3 = \tilde{k}_b + \tilde{k}_c$. Then we have the following re-parametrization of the line solitons:

Soliton 1	Soliton 2	Soliton 3	Soliton 4	Soliton 5
$\{k_a, \tilde{k}_b\}$	$\{k_a, \tilde{k}_c\}$	$\{\tilde{k}_b, \tilde{k}_c\}$	$\{k_{a'}, \tilde{k}_b\}$	$\{k_{a'}, \tilde{k}_c\}$

In what follows, we provide a simple, step-by-step algorithm to reconstruct the exact KP soliton solution from a given wave pattern with two trivalent resonant vertices.

3.1. Step-by-step construction of the KP line-soliton solution

For specificity, we use a real example of the wave pattern in Figure 5 which corresponds to the first pattern in Figure 4.

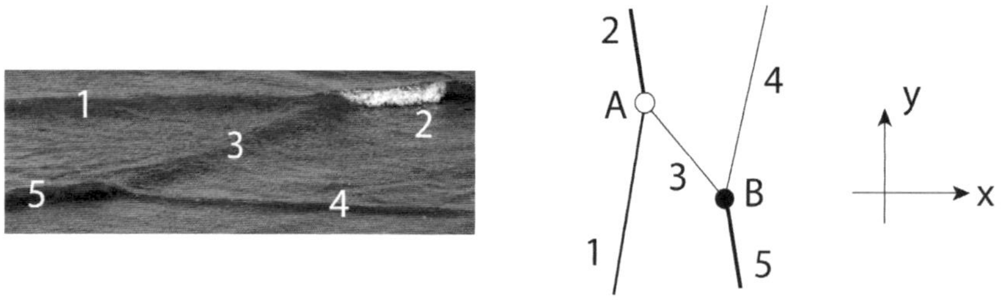

Figure 5. A real example of the wave pattern corresponding to the first pattern in Figure 4. Photograph of the wave pattern taken by D. E. Baldwin [2]

Step 1: The k-parameters We first set the xy-coordinates, so that the wave is propagating along almost x-direction, and choose the origin as the vertex A in Figure 5. Then we trace the wave crests as a graph, and measure the angles of all the line solitons with respect to the positive y-axis. For this example, we estimate the angles for the line solitons to be

$$\Psi_1 = -15^\circ, \quad \Psi_2 = 3^\circ, \quad \Psi_3 = 25^\circ, \quad \Psi_4 = -13^\circ, \quad \Psi_5 = 5^\circ.$$

Next from (3.1), one obtains $k_a = -0.341, k_b = 0.073$ and $k_c = 0.393$ at the vertex A, and $k_{a'} = -0.305, k_{b'} = 0.074$ and $k_{c'} = 0.392$ at the vertex B. After averaging to get $\tilde{k}_b, \tilde{k}_c$ and then arranging the k-parameters in increasing order gives

$$(k_a, k_{a'}, \tilde{k}_b, \tilde{k}_c) = (k_1, k_2, k_3, k_4) = (-0.341, -0.305, 0.0735, 0.3925).$$

Note that the slope data at the resonant vertices is sufficient to obtain the k-parameters; the amplitude data for the solitons, which is difficult to measure from a photograph, is not necessary in this case. The amplitude of each one-soliton can be calculated from (2.5) once the k-values have been found.

From this set of k-parameters, the procedure to find the derangement π for the data may be illustrated as a process of gluing the chord diagrams of the two Y-solitons with vertices A and B through the common chord as illustrated in Figure 6. At the vertex A, one has a Λ-soliton of the type formed by line solitons labeled 1,2 and 3, and the corresponding chord diagram is indicated by the top diagram in the middle of Figure 6. The dots in this diagram denote the k-parameters $k_a < k_b < k_c$. Similarly, the bottom chord diagram corresponds to the Y-soliton at the vertex B formed by solitons 3,4 and 5 with the dots marking the k-parameters $k_{a'} < k_{b'} < k_{c'}$. We then superpose the top and bottom diagrams and erase the common chord for the intermediate soliton 3 and recover the complete chord diagram of the asymptotic solitons shown by the rightmost diagram in Figure 6. In this diagram the k-parameters are marked (by the 4 dots) as $k_a = k_1, k_{a'} = k_2$ and the average values $\tilde{k}_b = k_3, \tilde{k}_c = k_4$ with $k_1 < k_2 < k_3 < k_4$. This diagram corresponds to the derangement $\pi = (4312)$ in the one-line notation of permutation. The top chords in this diagram are associated to the asymptotic solitons 2 and 4 for $y \gg 0$ which are identified as $[1,4]$- and $[2,3]$-soliton. Similarly, the bottom chords in this diagram are associated to the $[1,3]$- and $[2,4]$-asymptotic line solitons for $y \ll 0$. This is summarized in Figure 6.

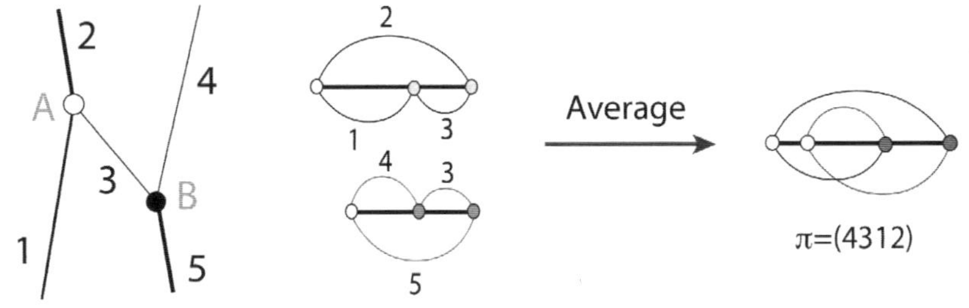

Figure 6. Chord gluing algorithm for the wave pattern in Figure 5

Step 2: Form of the A-matrix The totally non negative matrix A in the τ-function (2.2) can be derived from the derangement π obtained in Step 1 by employing combinatorial methods developed in a recent paper by Kodama and Williams [14], we refer the interested readers to that article and omit the details in this short note. Alternatively, the interested reader may also consult the work of Chakravarty and Kodama [5, 6] where in particular, the A-matrices for those patterns in Figure 4 were explicitly given in reduced row echelon form for the corresponding derangements π. For the current example with $\pi = (4312)$, the A-matrix is given by

$$A = \begin{pmatrix} 1 & 0 & -b & -c \\ 0 & 1 & a & 0 \end{pmatrix}$$

with three real parameters $\{a, b, c\}$. Note that A is a totally nonnegative matrix, i.e., $\Delta_{i,j}(A) \geq 0$ for all $1 \leq i < j \leq 4$, if all $a, b, c > 0$. Then the resulting exact solution of KP is non-singular for all x, y and t.

Step 3: Evaluation of the A-matrix elements The last step is to explicitly find the entries of the A-matrix. These are related to the locations of certain line-solitons in the given wave pattern as explained below. Figure 7 shows the asymptotic solitons labeled by their index pairs $[i, j]$ corresponding to the wave pattern of our example. The index set $I = \{i, j\}$ indicates each region in the xy-plane where the exponential E_I is dominant. The corresponding minor $\Delta_I(A)$ of the matrix A with column set I is given in terms of the parameters a, b and c. The line corresponding to the $[i, j]$-soliton is defined by $\Theta_{[i,j]} = 0$. The expression (2.4) for $\Theta_{[i,j]}$ contains the constant $x_{[i,j]}$ which involves the ratios of $\Delta_I(A)$ and $\Delta_J(A)$ where I and J label the adjacent regions separated by the line $[i, j]$.

We pick a point (x_0, y_0) on the line corresponding to the $[1,4]$-soliton in Figure 7. Setting the time $t = 0$, we have the equation of this line,

$$x + (k_1 + k_4)y + \frac{1}{k_1 - k_4} \ln \frac{\Delta_{1,2}(A)(k_2 - k_1)}{\Delta_{2,4}(A)(k_4 - k_2)} = 0\,.$$

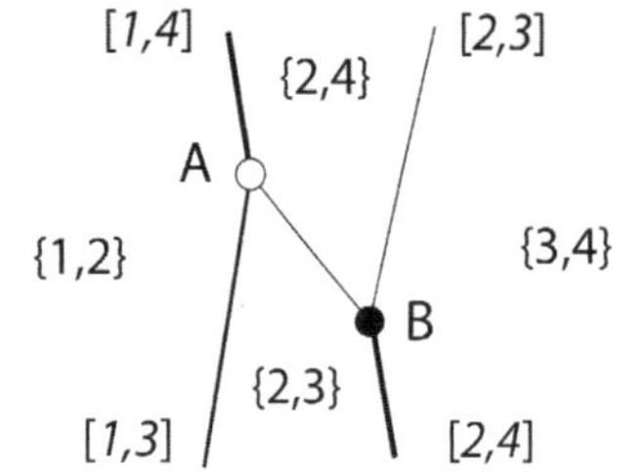

$$\Delta_{1,2} = 1\,, \quad \Delta_{2,3} = b$$

$$\Delta_{2,4} = c\,, \qquad \Delta_{3,4} = ac$$

Figure 7. Dominant exponentials and the minors of the matrix A for the wave pattern in Figure 5

Using the facts that $\Delta_{1,2}(A) = 1$ and the point (x_0, y_0) is on the line, one obtains

$$\Delta_{2,4}(A) = c = \frac{k_2 - k_1}{k_4 - k_2} \exp\left(-(k_4 - k_1)x_0 - (k_4^2 - k_1^2)y_0\right) .$$

Similarly, selecting a point (x_0, y_0) on line corresponding to the $[1,3]$-soliton, yields

$$\Delta_{2,3}(A) = b = \frac{k_2 - k_1}{k_3 - k_2} \exp\left(-(k_3 - k_1)x_0 - (k_3^2 - k_1^2)y_0\right) ,$$

and finally, by choosing a point on the $[2,3]$-line, one can get $a = \Delta_{3,4}(A)/\Delta_{2,4}(A)$. In this way, we find the parameters a, b, c (hence the A-matrix).

For this example, we chose the origin as the vertex A common to lines corresponding to the $[1,3]$- and $[1,4]$-solitons, and estimate the location of the vertex B as $(50, -107.2)$ which lie on the $[2,3]$- and $[2,4]$-soliton lines. Taking (x_0, y_0) as those points, we obtain the values $a = 0.110 \times 10^{-11}$, $b = 0.095$, $c = 0.052$ which gives the exact A-matrix in Step 2. From the A-matrix and the k-parameters found in Step 1, one can construct the τ-function and the soliton solution using (2.1). Plots of the exact solution, $u = 2(\ln\tau)_{xx}$, are shown in Figure 8. The middle plot at $t = 0$ corresponds to the photograph of the wave pattern in Figure 5, the left and right plots are for $t < 0$ and $t > 0$, respectively. Note that the wave pattern is non-stationary, and the stem (soliton 3) is an intermediate soliton rather than a stationary phase shift.

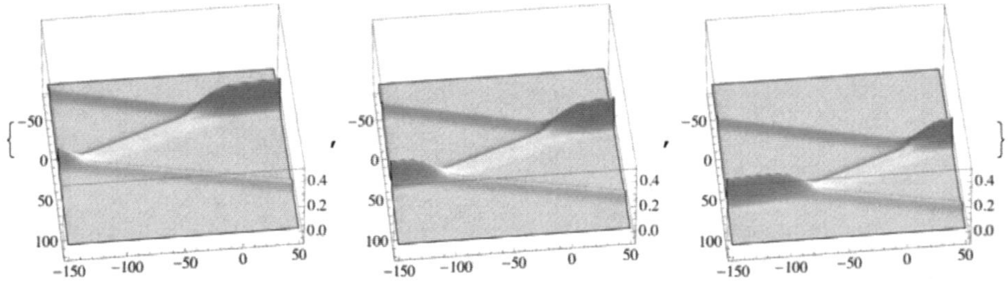

Figure 8. Evolution of the KP soliton constructed from the wave pattern in Figure 5.

One should note that the τ-function for the solution obtained here contains *five* exponential terms, i.e.

$$\tau = \sum_{1 \le i < j \le 4} \Delta_{i,j}(A) E_{i,j} \qquad \text{with} \quad \Delta_{i,j}(A) \neq 0 \ \text{except} \ \Delta_{1,4}(A) = 0,$$

while the τ-function for an X-soliton has only four exponential terms. The main difference is that the X-soliton solutions form a stationary pattern whereas the solution corresponding to our example above is non-stationary. The Grammian form (used by some authors) for the solution in Figure 8 is obtained from Remark 2.1 if one chooses $\{k_1,k_2,k_3,k_4\} = \{q_1,q_2,p_1,p_2\}$ and $C = D_1^{-1}JD_2$ where

$$J = \begin{pmatrix} -b & -c \\ a & 0 \end{pmatrix}, \quad D_1 = \begin{pmatrix} q_1-q_2 & 0 \\ 0 & q_2-q_1 \end{pmatrix},$$
$$D_2 = \begin{pmatrix} (p_1-q_1)(p_1-q_2) & 0 \\ 0 & (p_2-q_1)(p_2-q_2) \end{pmatrix}.$$

3.2. Example: Wave patterns in Lake Peipsi, Estonia

To further illustrate our algorithm we present another example where we consider a succession of photographs of an evolving wave pattern and compare the dynamics with the time evolution of the exact KP line-soliton solution reconstructed by the algorithm described in Section 3.1.

The top panel in Figure 9 shows a snapshot of a wave pattern observed on a beach of Lake Peipsi, Estonia. This photograph is part of a video footage showing the dynamics of the pattern over time. We apply our step-by-step algorithm to this snapshot chosen to be at $t = 0$ to construct the exact soliton solution $u(x,y,0)$, then compare $u(x,y,t)$ for $t \neq 0$ with other snapshots taken from the video footage. The middle picture of the top panel in Figure 9 shows the graph traced from the wave crests in the pattern. This corresponds to the fourth pattern in Figure 4. From the graph, the angles of the line-solitons are estimated to be $\Psi_1 = 14°$, $\Psi_2 = 37°$, $\Psi_3 = 0°$, $\Psi_4 = -42°$, $\Psi_5 = -20°$. Following Step 1 in Section 3.1, we obtain the k-parameters as $(k_1,k_2,k_3,k_4) = (-0.632,-0.268,0.268,0.517)$. Then we have the derangement $\pi = (2413)$ as shown in the top right panel of Figure 9. Here solitons 2 and 5 correspond to the $[3,4]$- and $[1,3]$-solitons respectively, for $y \ll 0$, whereas solitons 1 and 4 correspond to $[2,4]$- and $[1,2]$-solitons for $y \gg 0$. Finally, we construct the exact soliton solution following Steps 2 and 3 in Section 3.1. The bottom two panels of Figure 9 compares the time evolutions of the exact soliton solution and the actual wave pattern. The last snapshot in the middle panel corresponds to $t = 0$ from which the exact soliton solution was constructed, the previous snapshots in this panel correspond to $t < 0$. Also note that the leftmost snapshot corresponds to the wave pattern of the third case in Figure 4 indicating that the evolving solution can coincide with more than one configuration shown in Figure 4.

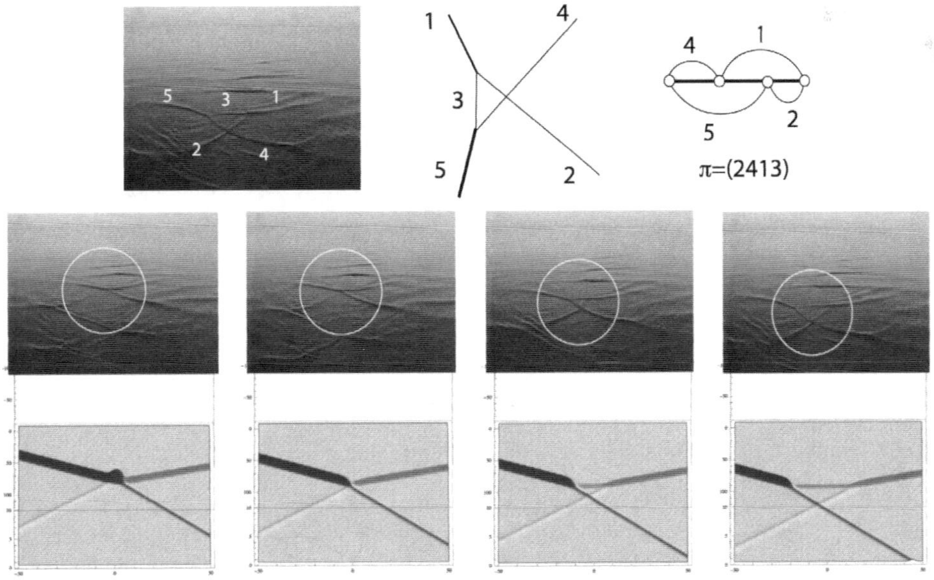

Figure 9. Waves in Lake Peipsi, Estonia. Snapshots from a video courtesy Ira Didenkulova.

4. Conclusion

In this note we provided an algorithm to construct an exact solution of the KP equation approximating a given non-stationary pattern of small amplitude, long wavelength and primarily unidirectional shallow water waves. In this "inverse problem" one can directly read off the parameters for the exact soliton solution by measuring the angles and locations of the solitary waves in the given pattern with respect to a fixed reference frame. We have illustrated our method by applying it to photographs and videos of real wave patterns and have shown that the constructed exact solutions and their time evolution are in good agreement with the dynamics of these non-stationary patterns. Our method is only a leading order approximation to the pattern as we do not consider higher order corrections to the KP equation to take into account effects such as large amplitude, obliqueness of interaction, non constant water depth at the beach etc.

Acknowledgements

One of the authors (YK) would like to thank the organizers for giving him the opportunity to deliver an invited talk on the topics of this paper and financial support during the conference PMNP 2013 at Gallipoli. The authors are grateful to Mark Ablowitz, Douglas Baldwin and Ira Didenkulova for the photographs and the video used in this article. This research was supported by NSF grants DMS-1108694 (SC) and DMS-1108813 (YK).

References

[1] Ablowitz M J and Baldwin D E 2012 Nonlinear shallow ocean-wave soliton interactions on flat beaches *Phys. Rev. E* **86** 036305
[2] Ablowitz M J and Baldwin D E, Photographs and videos at http://www.markablowitz.com/line-solitons and http://www.douglasbaldwin.com/nl-waves.html.
[3] Biondini G and Chakravarty S 2006 Soliton solutions of the Kadomtsev-Petviashvili II equation *J. Math. Phys.* **47** 033514
[4] Biondini G and Kodama Y 2003 On a family of solutions of the Kadomtsev-Petviashvili equation which also satisfy the Toda lattice hierarchy *J. Phys. A: Math. Gen.* **36** 10519-10536
[5] Chakravarty S and Kodama Y 2008 Classification of the line-solitons of KPII *J. Phys. A: Math. Theor.*, **41** 275209
[6] Chakravarty S and Kodama Y 2009 Soliton Solutions of the KP Equation and Application to Shallow Water Waves *Stud. Appl. Math.* **123** 83-151
[7] Chakravarty S, Lewkow T and Maruno K-I 2010, On the construction of KP line-solitons and their interactions *Applicable Analysis* **89** 529–545.
[8] Chakravarty S and Kodama Y 2013 Construction of KP solitons from wave patterns (arXiv:1309.2689).
[9] Hirota R 2004 *The Direct Method in Soliton Theory* (Cambridge:Cambridge University Press)
[10] Kadomtsev B B and Petviashvili V I 1970 On the stability of solitary waves in weakly dispersive media *Sov. Phys. Dokl.* **15** 539-541
[11] Kodama Y 2004 Young diagrams and N-soliton solutions of the KP equation *J. Phys. A: Math. Gen.* **37** 11169-90
[12] Kodama Y 2010 KP solitons in shallow water *J. Phys. A: Math. Theor.* **43** 434004
[13] Kodama Y, Oikawa M and Tsuji H 2009 Soliton solutions of the KP equation with V-shape initial waves *J. Phys. A:Math. Theor.* **42** 312001
[14] Kodama Y and Williams L 2013 The Deodhar decomposition of the Grassmannian and the regularity of KP solitons *Adv. Math.* **244** 979-1032
[15] Li W, Yeh H and Kodama Y 2011 On the Mach reflection of a solitary wave: revisited *J. Fluid Mech.* **672** 326-357
[16] Miles J W 1977 Resonantly interacting solitary waves, *J. Fluid Mech.* **79** 171-179
[17] Peterson P 2002 Reconstruction of multi-soliton interactions using crest data for (2+1)-dimensional KdV type equations, *Physica D* **171** 221-235
[18] Soomere T and Engelbrecht J 2006 Weakly two-dimensional interaction of solitons in shallow water *Eur. J. Mech. B Fluids* **25** 636-648
[19] Yeh H 2013 Laboratory Realization of Kodama's KP-Solitons, Lecture delivered at the NSF/CBMS Regional Conference in the Mathematical Sciences "Solitons in Two-Dimensional Water Waves and Applications to Tsunami", UTPA, May 20–24, 2013 (Lecture slides are available at http://faculty.utpa.edu/kmaruno/nsfcbms-tsunami.html).

Journal of Physics: Conference Series **482** (2014) 012008 doi:10.1088/1742-6596/482/1/012008

About the Quasi Steady State Approximation for a Reaction Diffusion System Describing a Chain of Irreversible Chemical Reactions

F Conforto[1] and L Desvillettes[2]

[1] Dept. of Mathematics and Computer Science, University of Messina
V.le F. Stagno d'Alcontres, 31, 98166 Messina, Italy
[2] CMLA, ENS Cachan, CNRS
61 Av. du Pdt. Wilson, 94235 Cachan Cedex, France

E-mail: [1] fiammetta.conforto@unime.it, [2] desville@cmla.ens-cachan.fr

Abstract. We present here a rigorous proof of the quasi steady state approximation in the context of a system of reaction-diffusion equations coming out of a typical chain of irreversible chemical reactions.

1. Introduction

The quasi steady state approximation (or QSSA) is a standard procedure in the study of chemical reactions kinetics allowing to eliminate from the equations the evolution of the species having a short time of existence w.r.t. the typical time of reaction of the chain. It consists in neglecting the speed of variation of the unstable species (that is, to replace it by 0). One can find in [15] a detailed description of the QSSA and the assumptions underlying its validity.

This approximation has been studied extensively at the mathematical level in the case when the spatial structure is not taken into account (that is, the chemistry is modeled by ODEs satisfied by the global concentrations of the species), see for instance [15, 2, 8, 14] and the references therein.

Our goal is to study the rigorous validity of the QSSA when the evolution of the species is described by reaction diffusion PDEs (each species having its own diffusion). In order to do so, we intend to use methods based on Lyapounov functionals and duality lemmas. Such a study was performed in [1] and in [5] in the case of a typical system of reversible equations. We wish in this paper to study the possible extensions of those methods in the context of a typical chain of irreversible reactions, where the entropy used in [1] is not available anymore. The analysis that we propose is in fact closer to that of [5] (the results that we provide could probably at least partly be recovered by following the arguments of [5], we propose however slightly different arguments).

Note that the QSSA has also been investigated in [16] in cases when there is a bounded invariant region for the unknowns. Finally the fast reaction asymptotics is another interesting problem for reaction-diffusion PDEs, which shares common features with the QSSA. It has been studied for example in [9, 6] for irreversible reactions, and in [3, 4] for reversible ones.

Here, we are interested in a typical chain of chemical reactions, such as

$$\left\{\begin{array}{ccc} A+B & \longrightarrow & C+D, \\ D+E & \longrightarrow & A+C, \\ C+E & \longrightarrow & A, \end{array}\right. \tag{1}$$

or its simpler version

$$\left\{\begin{array}{ccc} A+B & \longrightarrow & C+D, \\ D+E & \longrightarrow & A+C, \\ C & \longrightarrow & A. \end{array}\right. \tag{2}$$

We also introduce the corresponding systems of reaction diffusion for unknowns $a := a(t,x) \geq 0$, $b := b(t,x) \geq 0$, $c := c(t,x) \geq 0$, $d := d(t,x) \geq 0$ and $e := e(t,x) \geq 0$, which represent the concentrations at time $t \in \mathbb{R}_+$ and point $x \in \Omega$ (Ω being a bounded smooth open subset of $\mathbb{R}^N$, $N \in \mathbb{N}^*$) of the species A, B, C, D and E.

For the chain (1), it writes

$$\partial_t a - d_1\, \Delta_x a = -k_1\, a\, b + k_2\, d\, e + k_3\, c\, e, \tag{3}$$
$$\partial_t b - d_2\, \Delta_x b = -k_1\, a\, b, \tag{4}$$
$$\partial_t c - d_3\, \Delta_x c = k_1\, a\, b + k_2\, d\, e - k_3\, c\, e, \tag{5}$$
$$\partial_t d - d_4\, \Delta_x d = k_1\, a\, b - k_2\, d\, e, \tag{6}$$
$$\partial_t e - d_5\, \Delta_x e = -k_2\, d\, e - k_3\, c\, e, \tag{7}$$

together with Neumann boundary conditions (for all $x \in \partial\Omega$)

$$\nabla_x a \cdot n(x) = 0, \quad \nabla_x b \cdot n(x) = 0, \quad \nabla_x c \cdot n(x) = 0, \quad \nabla_x d \cdot n(x) = 0, \quad \nabla_x e \cdot n(x) = 0, \tag{8}$$

and initial data

$$a(0,\cdot) = a_{in}, \quad b(0,\cdot) = b_{in}, \quad c(0,\cdot) = c_{in}, \quad d(0,\cdot) = d_{in}, \quad e(0,\cdot) = e_{in}. \tag{9}$$

The parameters $d_1, .., d_5$ are the diffusion rates of the species $A, .., E$, and $k_1, .., k_3$ are the rates of the three reactions in the chain.

In the simpler chain (2), we get

$$\partial_t a - d_1\, \Delta_x a = -k_1\, a\, b + k_2\, d\, e + k_3\, c, \tag{10}$$
$$\partial_t b - d_2\, \Delta_x b = -k_1\, a\, b, \tag{11}$$
$$\partial_t c - d_3\, \Delta_x c = k_1\, a\, b + k_2\, d\, e - k_3\, c, \tag{12}$$
$$\partial_t d - d_4\, \Delta_x d = k_1\, a\, b - k_2\, d\, e, \tag{13}$$
$$\partial_t e - d_5\, \Delta_x e = -k_2\, d\, e - k_3\, c, \tag{14}$$

still with boundary and initial conditions (8), (9).

We now suppose that C is unstable (w.r.t. other species). This amounts to assume that $k_3 >> k_1, k_2$, so that we take the notation $k_3 = 1/\varepsilon$. The QSSA predicts that (in the limit $\varepsilon \to 0$) the concentration c should be of order ε, provided that $c_{in}(x) = 0$ for all $x \in \Omega$ (so that no initial layer appears). In order to state a rigorous result of convergence, we introduce explicitly the dependence of the concentrations w.r.t. ε, and get for the chain (1):

$$\partial_t a_\varepsilon - d_1\,\Delta_x a_\varepsilon = -k_1\,a_\varepsilon\,b_\varepsilon + k_2\,d_\varepsilon\,e_\varepsilon + \frac{k_3}{\varepsilon}\,c_\varepsilon\,e_\varepsilon, \tag{15}$$

$$\partial_t b_\varepsilon - d_2\,\Delta_x b_\varepsilon = -k_1\,a_\varepsilon\,b_\varepsilon, \tag{16}$$

$$\partial_t c_\varepsilon - d_3\,\Delta_x c_\varepsilon = k_1\,a_\varepsilon\,b_\varepsilon + k_2\,d_\varepsilon\,e_\varepsilon - \frac{k_3}{\varepsilon}\,c_\varepsilon\,e_\varepsilon, \tag{17}$$

$$\partial_t d_\varepsilon - d_4\,\Delta_x d_\varepsilon = k_1\,a_\varepsilon\,b_\varepsilon - k_2\,d_\varepsilon\,e_\varepsilon, \tag{18}$$

$$\partial_t e_\varepsilon - d_5\,\Delta_x e_\varepsilon = -k_2\,d_\varepsilon\,e_\varepsilon - \frac{k_3}{\varepsilon}\,c_\varepsilon\,e_\varepsilon, \tag{19}$$

together with Neumann boundary conditions (for all $x \in \partial\Omega$)

$$\nabla_x a_\varepsilon \cdot n(x) = 0, \quad \nabla_x b_\varepsilon \cdot n(x) = 0, \quad \nabla_x c_\varepsilon \cdot n(x) = 0, \quad \nabla_x d_\varepsilon \cdot n(x) = 0, \quad \nabla_x e_\varepsilon \cdot n(x) = 0, \tag{20}$$

and initial data

$$a_\varepsilon(0,\cdot) = a_{in}, \quad b_\varepsilon(0,\cdot) = b_{in}, \quad c_\varepsilon(0,\cdot) = 0, \quad d_\varepsilon(0,\cdot) = d_{in}, \quad e_\varepsilon(0,\cdot) = e_{in}. \tag{21}$$

In the case of the simpler chain (2), eq. (15) – (19) are replaced by

$$\partial_t a_\varepsilon - d_1\,\Delta_x a_\varepsilon = -k_1\,a_\varepsilon\,b_\varepsilon + k_2\,d_\varepsilon\,e_\varepsilon + \frac{k_3}{\varepsilon}\,c_\varepsilon, \tag{22}$$

$$\partial_t b_\varepsilon - d_2\,\Delta_x b_\varepsilon = -k_1\,a_\varepsilon\,b_\varepsilon, \tag{23}$$

$$\partial_t c_\varepsilon - d_3\,\Delta_x c_\varepsilon = k_1\,a_\varepsilon\,b_\varepsilon + k_2\,d_\varepsilon\,e_\varepsilon - \frac{k_3}{\varepsilon}\,c_\varepsilon, \tag{24}$$

$$\partial_t d_\varepsilon - d_4\,\Delta_x d_\varepsilon = k_1\,a_\varepsilon\,b_\varepsilon - k_2\,d_\varepsilon\,e_\varepsilon, \tag{25}$$

$$\partial_t e_\varepsilon - d_5\,\Delta_x e_\varepsilon = -k_2\,d_\varepsilon\,e_\varepsilon - \frac{k_3}{\varepsilon}\,c_\varepsilon, \tag{26}$$

while the initial and boundary conditions (20) and (21) are left unchanged.

We start by stating a Theorem which shows that the QSSA is indeed rigorously justified in the case of the simplified system (22) – (26).

Theorem 1: Let Ω be a smooth (C^2), bounded and connected open subset of $\mathbb{R}^N$ ($N \in \mathbb{N}^*$), let $d_1, .., d_5 > 0$, $k_1, .., k_3 > 0$, and let a_{in},..,e_{in} be initial data in $C^2(\overline{\Omega}, \mathbb{R}_+)$, compatible with the Neumann boundary condition (20). Then, for any $\varepsilon > 0$, there exists a unique global strong ($C^2(\mathbb{R}_+ \times \overline{\Omega}; \mathbb{R}_+)$) solution $(a_\varepsilon, .., e_\varepsilon)$ to the system (22) – (26) with initial and boundary conditions (20) and (21). Moreover, this solution $(a_\varepsilon, .., e_\varepsilon)$ converges in $\cap_{p\in[1,+\infty[} L^p_{loc}(\mathbb{R}_+ \times \Omega)$ strong towards $(a, b, 0, d, e)$, where (a, b, d, e) is the unique global strong ($C^2(\mathbb{R}_+ \times \overline{\Omega}; \mathbb{R}_+)$) solution of the following limit system:

$$\partial_t a - d_1 \Delta_x a = 2\,k_2\,d\,e, \tag{27}$$

$$\partial_t b - d_2\,\Delta_x b = -k_1\,a\,b, \tag{28}$$

$$\partial_t d - d_4\,\Delta_x d = k_1\,a\,b - k_2\,d\,e, \tag{29}$$

$$\partial_t e - d_5 \Delta_x e = -2\,k_2\,d\,e - k_1\,a\,b, \tag{30}$$

together with initial and boundary conditions (20) and (21) [except those for c].

As expected in the QSSA framework, c_ε converges to 0 (that is, the unstable species has a very small concentration). Moreover, the limit system is closed thanks to the convergence towards 0 of the variation of c_ε, that is $k_1\,a_\varepsilon\,b_\varepsilon + k_2\,d_\varepsilon\,e_\varepsilon - \frac{k_3}{\varepsilon}\,c_\varepsilon \to 0$.

In the case of the full system, the QSSA can only be partially proven, in the sense that instead of getting $c = 0$ in the limit, we only get $c\, e = 0$.

We end up in fact with the following Theorem:

Theorem 2: Let Ω be a smooth (C^2), bounded and connected open subset of $\mathbb{R}^N$ ($N \in \mathbb{N}^*$), let $d_1, .., d_5 > 0$, $k_1, .., k_3 > 0$, and let a_{in},..,e_{in} be initial data in $C^2(\overline{\Omega}, \mathbb{R}_+)$, compatible with the Neumann boundary condition (20). Then, for any $\varepsilon > 0$, there exists a unique global strong ($C^2(\mathbb{R}_+ \times \overline{\Omega}; \mathbb{R}_+)$) solution $(a_\varepsilon, .., e_\varepsilon)$ to the system (15) – (19) with initial and boundary conditions (20) and (21). Moreover, this solution $(a_\varepsilon, .., e_\varepsilon)$ converges up to extraction of a subsequence in $\cap_{p\in[1,+\infty[} L^p_{loc}(\mathbb{R}_+ \times \Omega)$ strong towards (a, b, c, d, e), where (a, b, c, d, e) is a global weak solution of the following limit system:

$$\partial_t(a + c) - \Delta_x(d_1\, a + d_3\, c) = 2\, k_2\, d\, e, \tag{31}$$

$$\partial_t b - d_2\, \Delta_x b = -k_1\, a\, b, \tag{32}$$

$$c\, e = 0, \tag{33}$$

$$\partial_t d - d_4\, \Delta_x d = k_1\, a\, b - k_2\, d\, e, \tag{34}$$

$$\partial_t(e - c) - \Delta_x(d_5\, e - d_3\, c) = -2\, k_2\, d\, e - k_1\, a\, b. \tag{35}$$

together with initial and boundary conditions (20) and (21) [except those for c].

Note that if we knew that $c = 0$ in the above system, we would be able to conclude that the convergence of the whole sequence holds, since there would be uniqueness of a (weak) solution of the system. This identity seems however difficult to prove without extra assumptions. As can be seen, the QSSA is not completely proven in this case (more precisely, it is proven only in the regions where $e \neq 0$: no information about those regions is available unfortunately).

Note also that in the above Theorems, we made no effort to give optimal conditions on the initial data (part of the conclusions would hold even in if the initial data were only continuous, or even belonged to some well chosen L^p space. Also the convergences (of a_ε towards a, etc.) are shown to hold a.e., but part of them in fact hold in (well chosen) Sobolev spaces.

Finally we did not try here to study the case when the unstable species has a non zero initial datum. In that case, an initial layer should appear, which needs a specific treatment (cf. [5]).

The rest of this paper is devoted to the Proof of Theorem 1 (section 2) and Theorem 2 (section 3).

2. Proof of the Theorem for the simplified model

We begin here the

Proof of Theorem 1: Let us first briefly recall how (for any $\varepsilon > 0$), existence and uniqueness of a strong (smooth, nonnegative) solution to system (22) - (26) can be obtained. First, local (in time) solutions are proven to exist thanks to an iterative method like in [10].

Then, this solution is shown to be global (in time) thanks to a continuation process using a *priori* estimates. We recall here the necessary estimates. First, thanks to the maximum principle, it is clear that (for all $t \geq 0$ such that the solution exists)

$$0 \leq e_\varepsilon(t, x) \leq ||e_{in}||_{L^\infty(\Omega)}, \qquad 0 \leq b_\varepsilon(t, x) \leq ||b_{in}||_{L^\infty(\Omega)}. \tag{36}$$

Observing that

$$\partial_t(a_\varepsilon + c_\varepsilon + 2\, e_\varepsilon) - \Delta_x(d_1\, a_\varepsilon + d_3\, c_\varepsilon + 2\, d_5\, e_\varepsilon) = -2\, \frac{k_3}{\varepsilon}\, c_\varepsilon\, e_\varepsilon \leq 0,$$

and

$$\partial_t(b_\varepsilon + d_\varepsilon) - \Delta_x(d_2\, b_\varepsilon + d_4\, d_\varepsilon) = -k_2\, d_\varepsilon\, e_\varepsilon \le 0,$$

we see using the duality method (cf. [12], [11]) that (for any $T > 0$ such that the solution exists)

$$||a_\varepsilon||_{L^2([0,T]\times\Omega)} \le C_T, \qquad ||c_\varepsilon||_{L^2([0,T]\times\Omega)} \le C_T, \qquad ||d_\varepsilon||_{L^2([0,T]\times\Omega)} \le C_T,$$

where $C_T > 0$ does not depend on ε (in fact it only depends on the L^2 bounds of the initial data a_{in}, b_{in}, d_{in}, e_{in}, and of the the diffusion rates $d_1, .., d_5 > 0$). It would even be possible to replace L^2 in the estimates above by $L^{2+\delta}$ for some $\delta > 0$ using the arguments of [7], but we shall show in the sequel that an even better estimate holds.

Indeed, a more careful examination of the duality method shows that (cf. [13])

$$||d_\varepsilon||_{L^p([0,T]\times\Omega)} \le C_T \tag{37}$$

for all $p \neq \infty$. Also, as a consequence

$$||d_\varepsilon\, e_\varepsilon||_{L^p([0,T]\times\Omega)} \le C_T. \tag{38}$$

Then, writing

$$\partial_t(a_\varepsilon + e_\varepsilon) - \Delta_x(d_1\, a_\varepsilon + d_5\, e_\varepsilon) = -k_1\, a_\varepsilon\, b_\varepsilon \le 0,$$

and using again the duality arguments of [13], we end up with

$$||a_\varepsilon||_{L^p([0,T]\times\Omega)} \le C_T \tag{39}$$

for all $p \neq \infty$.

Finally, observing that

$$\partial_t(a_\varepsilon + c_\varepsilon) - \Delta_x(d_1\, a_\varepsilon + d_3\, c_\varepsilon) = 2\, k_2\, d_\varepsilon\, e_\varepsilon,$$

and using estimate (38) together with (one last time) the duality arguments of [13], we see that

$$||c_\varepsilon||_{L^p([0,T]\times\Omega)} \le C_T \tag{40}$$

for all $p \neq \infty$. Note that all those estimates are independent on $\varepsilon > 0$.

As a consequence of the previous estimates (the second inequality having already been stated in (38)),

$$||a_\varepsilon\, b_\varepsilon||_{L^p([0,T]\times\Omega)} \le C_T, \qquad ||d_\varepsilon\, e_\varepsilon||_{L^p([0,T]\times\Omega)} \le C_T, \tag{41}$$

and

$$||c_\varepsilon/\varepsilon||_{L^p([0,T]\times\Omega)} \le C_T/\varepsilon, \tag{42}$$

so that thanks to the properties of the heat equation, and eq. (22) - (26),

$$||a_\varepsilon, b_\varepsilon, c_\varepsilon, d_\varepsilon, e_\varepsilon||_{L^\infty([0,T]\times\Omega)} \le C_{T,\varepsilon}. \tag{43}$$

According to [10], existence (for a given $\varepsilon > 0$) of a global (locally bounded) solution to system (22) - (26) holds. Then, the smoothness of this solution is a direct consequence of bootstraps arguments using the heat kernel properties, and uniqueness also holds (cf. also [10]).

We now turn to the proof of the QSSA. Note that estimates (43) cannot be used since the constant appearing there is not ε-independent, whereas estimates (36) – (40) are available since they are ε-independent.

We observe that integrating eq. (26) over $[0,T]\times\Omega$,

$$\int_\Omega e_\varepsilon(T,x)\,dx + \frac{k_3}{\varepsilon}\int_0^T\int_\Omega c_\varepsilon(t,x)\,dxdt \le \int_\Omega e_{in}(x)\,dx.$$

In other words,

$$||c_\varepsilon/\varepsilon||_{L^1([0,T]\times\Omega)} \le C_T. \tag{44}$$

which, together with estimate (40), entails the convergence towards 0 of c_ε in $\cap_{p\in[1,+\infty[}L^p_{loc}(\mathbb{R}_+\times\Omega)$ strong.

Then, thanks to estimates (36), (37) and (39), we see that up to the extraction of a subsequence, $a_\varepsilon \rightharpoonup a$, $d_\varepsilon \rightharpoonup d$ in $\cap_{p\in[1,+\infty[}L^p_{loc}(\mathbb{R}_+\times\Omega)$ weak (where a, d are some elements of $\cap_{p\in[1,+\infty[}L^p_{loc}(\mathbb{R}_+\times\Omega)$), and $b_\varepsilon \rightharpoonup b$, $e_\varepsilon \rightharpoonup e$ in $L^\infty(\mathbb{R}_+\times\Omega)$ weak (where b, e are some elements of $L^\infty(\mathbb{R}_+\times\Omega)$).

Using estimate (44) together with (41) and eq. (22) - (26), and remembering the compactness properties of the heat equation, we see that the convergences above also hold for a.e. t, x, and therefore in $\cap_{p\in[1,+\infty[}L^p_{loc}(\mathbb{R}_+\times\Omega)$ strong.

As a consequence, the sequence $a_\varepsilon\, b_\varepsilon$ converges (up to extraction of a subsequence) towards $a\,b$ in $\cap_{p\in[1,+\infty[}L^p_{loc}(\mathbb{R}_+\times\Omega)$ and the sequence $d_\varepsilon\, e_\varepsilon$ converges (up to extraction of a subsequence) towards $d\,e$ in $\cap_{p\in[1,+\infty[}L^p_{loc}(\mathbb{R}_+\times\Omega)$.

We can therefore pass to the limit in the sense of distributions (more precisely, with test functions taken in $C^\infty_c(\mathbb{R}_+\times\overline{\Omega})$)) in eq. (23) and (25) and get eq. (28) and (29) in the sense of distributions (more precisely, those equations are satisfied in the weak sense including the boundary condition (8) [for b and d] and the initial data (9) [for b and d]; this means once again that the test functions are taken in $C^\infty_c(\mathbb{R}_+\times\overline{\Omega})$).

We then write the result of the sum of eq. (22) and eq. (24) on one hand, and the result of the difference between eq. (26) and eq. (24) on the other hand:

$$\partial_t(a_\varepsilon + c_\varepsilon) - \Delta_x(d_1\,a_\varepsilon + d_3\,c_\varepsilon) = 2\,k_2\,d_\varepsilon\,e_\varepsilon, \tag{45}$$

$$\partial_t(e_\varepsilon - c_\varepsilon) - \Delta_x(d_5\,e_\varepsilon - d_3\,c_\varepsilon) = -2\,k_2\,d_\varepsilon\,e_\varepsilon - k_1\,a_\varepsilon\,b_\varepsilon. \tag{46}$$

We can also pass to the limit in this equation in the sense of distributions (once again, more precisely, with test functions taken in $C^\infty_c(\mathbb{R}_+\times\overline{\Omega})$), and get eq. (27) and (30) in the sense of distributions (more precisely, those equations are satisfied in the weak sense including the boundary condition (8) [for a and e] and the initial data (9) [for a and e]; this means once again that the test functions are taken in $C^\infty_c(\mathbb{R}_+\times\overline{\Omega})$).

In this way, we end up with a weak solution of the limit system (27) – (30) together with initial and boundary conditions (20) and (21) [except those for c]. According to the method described in [10] for example, we know that this solution is the unique strong (smooth, nonnegative) solution of the system, so that the whole family $a_\varepsilon, b_\varepsilon, c_\varepsilon, d_\varepsilon, e_\varepsilon$ (and not just a subsequence) converges towards $a, b, 0, d, e$.

This concludes the Proof of Theorem 1. □

3. Proof of the Theorem for the complete model

We begin here the

Proof of Theorem 2: We observe that estimates (36) – (41) still hold for the (local in time) solution of system (15) - (19), the proof being identical to the one of the previous section.

We now replace estimate (42) by

$$||c_\varepsilon\, e_\varepsilon/\varepsilon||_{L^p([0,T]\times\Omega)} \le C_T/\varepsilon, \tag{47}$$

so that thanks to the properties of the heat equation, and eq. (15) - (19), estimates (43) hold. Using then the results of [10], existence (for a given $\varepsilon > 0$) of a global in time (locally bounded) solution to system (15) - (19) holds. We recall that the smoothness of this solution is a direct consequence of bootstraps arguments using the heat kernel properties, and that uniqueness also holds (cf. also [10]).

We now investigate the validity of the QSSA for this chain reaction. We integrate eq. (19) over $[0,T]\times\Omega$, and end up with

$$\int_\Omega e_\varepsilon(T,x)\,dx + \frac{k_3}{\varepsilon}\int_0^T\int_\Omega c_\varepsilon(t,x)\,e_\varepsilon(t,x)\,dxdt \le \int_\Omega e_{in}(x)\,dx.$$

Then,

$$||c_\varepsilon\, e_\varepsilon/\varepsilon||_{L^1([0,T]\times\Omega)} \le C_T, \tag{48}$$

which, together with estimates (36) and (40), entails the convergence towards 0 of $c_\varepsilon\, e_\varepsilon$ in $\cap_{p\in[1,+\infty[} L^p_{loc}(\mathbb{R}_+\times\Omega)$ strong.

Using as in the previous section estimates (36) – (39), but also (40), we see that up to the extraction of a subsequence, $a_\varepsilon \rightharpoonup a$, $c_\varepsilon \rightharpoonup c$, $d_\varepsilon \rightharpoonup d$ in $\cap_{p\in[1,+\infty[} L^p_{loc}(\mathbb{R}_+\times\Omega)$ weak (where a, c, d are some elements of $\cap_{p\in[1,+\infty[} L^p_{loc}(\mathbb{R}_+\times\Omega)$), and $b_\varepsilon \rightharpoonup b$, $e_\varepsilon \rightharpoonup e$ in $L^\infty(\mathbb{R}_+\times\Omega)$ weak (where b, e are some elements of $L^\infty(\mathbb{R}_+\times\Omega)$).

Using now estimate (48) together with (41) and eq. (15) - (19), and remembering the compactness properties of the heat equation, we see that the convergences above also hold for a.e. t, x, and therefore in $\cap_{p\in[1,+\infty[} L^p_{loc}(\mathbb{R}_+\times\Omega)$ strong.

As in the previous section, we therefore see that the sequence $a_\varepsilon\, b_\varepsilon$ converges (up to extraction of a subsequence) towards $a\,b$ in $\cap_{p\in[1,+\infty[} L^p_{loc}(\mathbb{R}_+\times\Omega)$ and that the sequence $d_\varepsilon\, e_\varepsilon$ converges (up to extraction of a subsequence) towards $d\,e$ in $\cap_{p\in[1,+\infty[} L^p_{loc}(\mathbb{R}_+\times\Omega)$.

We now pass to the limit in the system (15) - (19). Using estimate (48) and Fatou's lemma, we first see that $c\,e = 0$, so that (33) holds.

We can also pass to the limit in the sense of distributions (more precisely, with test functions taken in $C_c^\infty(\mathbb{R}_+\times\overline{\Omega})$)) in eq. (16) and (18) and get eq. (32) and (34) in the sense of distributions (more precisely, those equations are satisfied in the weak sense including the boundary condition (8) [for b and d] and the initial data (9) [for b and d]; this means once again that the test functions are taken in $C_c^\infty(\mathbb{R}_+\times\overline{\Omega})$).

We then write the result of the sum of eq. (15) and eq. (17) on one hand, and the result of the difference between eq. (19) and eq. (17) on the other hand:

$$\partial_t(a_\varepsilon + c_\varepsilon) - \Delta_x(d_1\, a_\varepsilon + d_3\, c_\varepsilon) = 2\,k_2\, d_\varepsilon\, e_\varepsilon, \tag{49}$$

$$\partial_t(e_\varepsilon - c_\varepsilon) - \Delta_x(d_5\, e_\varepsilon - d_3\, c_\varepsilon) = -2\,k_2\, d_\varepsilon\, e_\varepsilon - k_1\, a_\varepsilon\, b_\varepsilon. \tag{50}$$

We can also pass to the limit in these equations in the sense of distributions (once again, more precisely, with test functions taken in $C_c^\infty(\mathbb{R}_+\times\overline{\Omega})$), and get eq. (31) and (35) in the sense of distributions (more precisely, those equations are satisfied in the weak sense including the boundary condition (8) [for a and e] and the initial data (9) [for a and e]; this means once again that the test functions are taken in $C_c^\infty(\mathbb{R}_+\times\overline{\Omega})$).

In this way, we end up with a weak solution of the limit system (31) – (35) together with initial and boundary conditions (20) and (21) [except those for c].

Note however that we have no information on the smoothness of the solutions to system (31) – (35) since it is not written under the standard parabolic form. We also do not know if uniqueness holds for this system, so that we cannot prove that the whole family $a_\varepsilon, b_\varepsilon, c_\varepsilon, d_\varepsilon, e_\varepsilon$ converges towards a, b, c, d, e.

This concludes the Proof of Theorem 2. □

Acknowledgments

This work was performed in the frame of the activities sponsored by GNFM–INdAM, and by ENS of Cachan and University of Messina.

References

[1] Bisi M, Conforto F and Desvillettes L 2007 *Bull. Inst. Acad. Sin. (N.S.)* **2** (4) 823
[2] Blouza A, Coquel F and Hamel F 2000 *Combust. Theory Modelling* **4** 339
[3] Bothe D 2003 *J. Differential Equations* **193** 27
[4] Bothe D and Hilhorst D 2003 *J. Math. Anal. Appl.* **286** 125
[5] Bothe D and Pierre M 2010 *J. Math. Anal. Appl.* **368** (1) 120
[6] Bothe D Pierre M 2012 *Discrete Contin. Dyn. Syst. Ser. S* **5** (1), 49
[7] Canizo J A, Desvillettes L and Fellner K 2013 Preprint n. 2013-07 of CMLA, ENS Cachan, accepted for publication in Communications in Partial Differential Equations
[8] Come G M 1977 *J. Phys. Chem.* **81** 2560
[9] Hilhorst D, Van Der Hout R and Peletier L A 2000 *Nonlinear Anal.* **41** 803
[10] Desvillettes L 2007 *Rivista di Matematica dell'Università di Parma.* **7** (7) 81
[11] Martin R H and Pierre M1992 *Math. Sci. and Engrg.* **185** 363
[12] Pierre M and Schmitt D 2000 *SIAM Rev.* **42** (1) 93
[13] Quittner P and Souplet P 2007 *Superlinear parabolic problems. Blow-up, global existence and steady states* (Berlin: Birkhauser Advanced Texts)
[14] Ren Z and Pope L A 2004 *Combustion and Flame* **137** 251
[15] Segel L A and Slemrod M 1989 *SIAM Review* **31** 446
[16] Yannacopoulos A et al 1996 *Phys. Letters A* **223** 82

Integrable multi-phase thermodynamic systems and Tsallis' composition rule

G De Nittis [a], **P Lorenzoni** [b] **and A Moro** [c]

[a] Department Mathematik, Universität Erlangen-Nürnberg, Germany
[b] Dipartimento di Matematica e Applicazioni, Università di Milano-Bicocca, Italy
[c] Department of Mathematics and Information Sciences, Northumbria University, UK

E-mail: antonio.moro@northumbria.ac.uk

Abstract. We derive a class of equations of state for a multi-phase thermodynamic system associated with a finite set of order parameters that satisfy an integrable system of hydrodynamic type. As particular examples, we discuss one-phase systems such as the van der Waals gas and the effective molecular field model. The case of $N-$phase systems is also discussed in detail in connection with entropies depending on the order parameter according to Tsallis' composition rule.

1. Introduction

The mathematical description of a macroscopic physical system in thermodynamic equilibrium requires a suitable number of state functions together with their conjugated thermodynamic variables and a set of order parameters. The order parameters bring information about the properties of possible phase transitions occurring within the system.

In the present paper, we are interested in the description of a general thermodynamic system in equilibrium described by a Gibbs function Φ. Hence, the first law of thermodynamics reads as (see e.g. [10])

$$d\Phi(\tau^0,\dots,\tau^M) = \sum_{i=0}^{M} \Lambda_i(\tau^0,\dots,\tau^M)\, d\tau^i + dF \tag{1}$$

where Λ^i and τ^i are, respectively, a set of state functions and thermodynamic conjugated variables and

$$F = \sum_{i=0}^{M} f_i(\tau^i).$$

In particular, we have

$$\tau^0 = T, \qquad S = -\Lambda_0 - f_0(T),$$

where T is the temperature of the system and S is the entropy.

The particular form of the set of functions of a single variable $f_i(\tau^i)$ depends of the specific physical nature of the system. The differential balance relation (1) implies the following closure conditions

$$\frac{\partial \Lambda_i}{\partial \tau^j} = \frac{\partial \Lambda_j}{\partial \tau^i}, \tag{2}$$

Journal of Physics: Conference Series **482** (2014) 012009 doi:10.1088/1742-6596/482/1/012009

known as Maxwell's relations.

Let us now assume that the present thermodynamic system can be viewed as a composite system in thermodynamic equilibrium described by the set of order parameters $\theta^k = \theta^k(\tau^0, \dots, \tau^M)$, with $k = 1, \dots, N$. We also assume that the state functions depend on the thermodynamic variables $\tau^0, \tau^1, \dots, \tau^N$ through the order parameters as follows

$$\Lambda_i = \Lambda_i(\theta^1, \dots, \theta^N). \tag{3}$$

We will look for a class of solutions to Maxwell's relations (2) such that the order parameters satisfy a system of integrable equations of hydrodynamic type. In this case, we will show that the order parameters satisfy the system of N equations of state of the form

$$T - \sum_{i=1}^{M} \left(\frac{\partial S}{\partial \theta^k}\right)^{-1} \frac{\partial \Lambda_i}{\partial \theta^k} \tau^i = \lambda^k(\theta^1, \dots, \theta^N), \qquad k = 1, \dots, N. \tag{4}$$

A particular class of such integrable hydrodynamic equations for the order parameters is associated with Tsallis' type entropy [15, 16] that in the simple case of a two-phase system reads as

$$S_q^{(2)} = \theta^1 + \theta^2 + q(\theta^1\theta^2). \tag{5}$$

In this case, θ^1 and θ^2 are interpreted as the entropies of the component systems. Remarkably, Tsallis' composition rule (5) (see [15, 16] and also e.g. [1, 2]) allows to construct recursively a set of state functions for a class of the multi-phase system possessing the multi-component equation of state (4). Let us note that the parameter q introduced above and used throughout the paper corresponds to Tsallis' *entropy index* $1 - q$.

2. Integrable quasilinear systems in Riemann invariants

Let us consider a diagonal systems of PDEs of hydrodynamic type of the following form

$$\frac{\partial \theta^k}{\partial \tau} = \mu^k(\theta^1, \dots, \theta^N) \frac{\partial \theta^k}{\partial T}, \qquad k = 1, \dots, N. \tag{6}$$

The system (6) is said to be *integrable* if the characteristic velocities μ^k satisfy the system of equations [17]

$$\partial_s \left(\frac{\partial_l \mu^k}{\mu^l - \mu^k}\right) = \partial_l \left(\frac{\partial_s \mu^k}{\mu^s - \mu^k}\right), \qquad s \neq l, s \neq k, l \neq k \tag{7}$$

where $\partial_i = \frac{\partial}{\partial \theta^i}$. The equations (7) are the compatility conditions for the linear systems of equations

$$\frac{\partial_l \lambda^k}{\lambda^l - \lambda^k} = \frac{\partial_l \mu^k}{\mu^l - \mu^k}, \qquad l \neq k \tag{8}$$

and

$$(\mu^i - \mu^j)\, \partial_i \partial_j S = \partial_i \mu^j\, \partial_j S - \partial_j \mu^i\, d_i S. \tag{9}$$

The functions λ^i are the characteristic speeds of a symmetry defined by

$$\frac{\partial \theta^k}{\partial \tau'} = \lambda^k(\theta^1, \dots, \theta^N) \frac{\partial \theta^k}{\partial T}, \qquad k = 1, \dots, N,$$

such that $\theta^i_{\tau\tau'} = \theta^i_{\tau'\tau}$ and S is a density of a conservation laws, that is

$$\frac{\partial S}{\partial \tau} = \frac{\partial \Lambda}{\partial T}$$

for a suitable *current* $K(\theta^1, \ldots, \theta^N)$. Hence, the integrability of the system (6) is equivalent to the linearisability via (8) or (9). In virtue of Tsarev' theorem on the generalised hodograph method [17], the general solution of (6) is implicitly defined by a system of algebraic equations

$$\lambda^k(\theta^1, \ldots, \theta^N) = T + \mu^k(\theta^1, \ldots, \theta^N)\,\tau, \qquad k = 1, ..., N \tag{10}$$

involving the general solution of the *linear* system (8). Due to (7), this depends on N arbitrary functions of a single variable.

Although solving the system (8) may be, in general, highly nontrivial, there exists special classes of systems for which the general solution is found by quadratures. Such is the case of *weakly non linear* or *linearly degenerate* systems that are characterised by the condition

$$\partial_k\, \mu^k = 0, \qquad k = 1, ..., N. \tag{11}$$

Importantly, it was conjectured in [13] that smooth initial data for weakly non linear systems do not break in finite time.

One can prove that any linearly degenerate system (6) admits $N - 1$ independent weakly nonlinear symmetries [7]:

$$\begin{aligned}
&\theta^k_{\tau^0} = \theta^k_T, && (\mu^k_{(0)} = 1) && \\
&\theta^k_{\tau^1} = \mu^k_{(1)}(\theta)\theta^k_T, && (\mu^k_{(1)} = \mu^k,\ \ \tau^1 = \tau) && \\
&\theta^k_{\tau^j} = \mu^k_{(j)}(\theta)\theta^k_T, && \left(\partial_k(\mu^k_{(j)}) = 0\right) && j = 2, ..., N-1
\end{aligned}$$

for each $k = 1, ..., N$. For instance, in the case of the system

$$\theta^k_{\tau^1} = \left(\sum_{i=1}^N \theta^i - \theta^k\right)\theta^k_T \tag{12}$$

the linearly degenerate symmetries are given by the formula

$$\mu^k_{(j)}(\theta) = Res_{s=0}\, \frac{1}{s^{N-j}} \frac{(s+\theta^1)\cdots(s+\theta^N)}{(s+\theta^k)}, \qquad j = 2, ..., N-1.$$

The general solution of the system (8) can be written in terms of these special symmetries as

$$\lambda^k(\theta) = \sum_{j=0}^{N-1} c_j(\theta)\mu^k_{(j)}(\theta), \qquad c_j(\theta) = \sum_{i=1}^N \int \frac{W^{(j+1,i)}}{W^{(N,i)}}\varphi_i(\theta^i)d\theta^i \tag{13}$$

where $W^{(j,i)}$ are the cofactors (i, j) of the matrix

$$W = \begin{pmatrix} \mu^1_{(0)} & \mu^2_{(0)} & \cdots & \mu^N_{(0)} \\ \mu^1_{(1)} & \mu^2_{(1)} & \cdots & \mu^N_{(1)} \\ \cdots & \cdots & \cdots & \cdots \\ \mu^1_{(N-1)} & \mu^2_{(N-1)} & \cdots & \mu^N_{(N-1)} \end{pmatrix} = \begin{pmatrix} 1 & 1 & \cdots & 1 \\ \mu^1 & \mu^2 & \cdots & \mu^N \\ \cdots & \cdots & \cdots & \cdots \\ \mu^1_{(N-1)} & \mu^2_{(N-1)} & \cdots & \mu^N_{(N-1)} \end{pmatrix},$$

and $\varphi_1(\theta^1)$,...,$\varphi_N(\theta^N)$ are arbitrary functions of one variable. More details can be found in the papers [7, 8, 4].

3. Multi-phase equations of state

Let us consider the set of state functions of the form (3) depending on a set of order parameters $\theta^k = \theta^k(\tau^0, \ldots, \tau^M)$, with $k = 1, \ldots, N$ and satisfying a set of equations of hydrodynamic type of the form

$$\frac{\partial \theta^k}{\partial \tau^i} = \mu^k_{(i)}(\theta^1, \ldots, \theta^N)\,\frac{\partial \theta^k}{\partial T}, \qquad i = 0, 1, \ldots, M. \tag{14}$$

In particular we have $\mu^k_0 = 1$, for $k = 1, \ldots, N$, and the *characteristic speeds* $\mu^k_{(i)}$ satisfy the condition (8). Under the assumptions (3) and (14) the Maxwell relations (2) read as follows

$$\sum_{k=1}^{N}\left(\frac{\partial \Lambda_i}{\partial \theta^k}\mu^k_{(j)} - \frac{\partial \Lambda_j}{\partial \theta^k}\mu^k_{(i)}\right)\frac{\partial \theta^k}{\partial T} = 0, \qquad k = 1, \ldots, N. \tag{15}$$

Assuming that the functions $\partial\theta^k/\partial T$ are all independent, the equation above implies the following relation between the characteristic speeds $\mu^k_{(i)}$ and the thermodynamic state functions

$$\frac{\partial \Lambda_i}{\partial \theta^k}\mu^k_{(j)} - \frac{\partial \Lambda_j}{\partial \theta^k}\mu^k_{(i)} = 0, \qquad k = 1, \ldots, N, \tag{16}$$

which can be equivalently written when $i = 0$ as follows

$$\mu^k_{(i)}(\theta^1, \ldots, \theta^N) = -\left(\frac{\partial S}{\partial \theta^k}\right)^{-1}\frac{\partial \Lambda_i}{\partial \theta^k}, \qquad k = 1, \ldots, N, \quad i = 1, \ldots, M. \tag{17}$$

We observe that the formula (17) provides the equilibrium thermodynamic interpretation of a classical result due to Lax [11] that relates the characteristic speeds to the entropy and the conjugated conserved flows in the theory of hyperbolic systems of PDEs. Applying the Tsarev theorem, if the system of hydrodynamic type (14) is integrable, then the general solution $\theta^k = \theta^k(\tau^1, \ldots, \tau^M)$, is given (locally) by the following implicit formula

$$T + \sum_{i=1}^{M}\mu^k_{(i)}(\theta^1, \ldots, \theta^N)\,\tau^i = \lambda^k(\theta^1, \ldots, \theta^N) \qquad k = 1, \ldots, N \tag{18}$$

where λ^k is the general solution of (8). Hence, the order parameters θ^k are fully determined as functions of the temperature T and the set of thermodynamic variables $T, \tau^1, \ldots, \tau^M$ via the *multi-phase equation of state* (18).

4. One-phase systems

Let us consider a macroscopic physical system characterised by its Gibbs potential Φ, entropy S, temperature T, and the function of state Λ associated with the conjugate variable τ. The first law of thermodynamics in differential form reads as follows

$$d\Phi(T,\tau) = -S(T,\tau)dT + \Lambda(T,\tau)d\tau. \tag{19}$$

Let us now assume that the entropy function is of the form

$$S = S\left(\Lambda(T,\tau), T\right) \tag{20}$$

characterised by an implicit dependence on the variable τ via the state function $\Lambda(P,T)$. Following [6], in the case of separable entropy functions of the form

$$S = \tilde{S}\left(\Lambda\right) + f(T), \tag{21}$$

the first law of thermodynamics is equivalent to the Riemann-Hopf type equation

$$\frac{\partial \Lambda}{\partial \tau} + \alpha(\Lambda)\frac{\partial \Lambda}{\partial T} = 0 \tag{22}$$

where $\alpha(\Lambda) := \tilde{S}'(\Lambda)^{-1}$. We emphasise that important physical examples such as the van der Waals model and the magnetic molecular field satisfy the separability condition (21) that results from the factorisation of the partition function at the equilibrium[1](see also Remark 3.1 in Ref. [6]). The general solution to the equation (22) is obtained via the classical characteristic method and given by the implicit formula

$$T - \alpha(\Lambda)\tau - f(\Lambda) = 0, \tag{23}$$

where $f(\Lambda)$ is an arbitrary function of its argument. The relation (23) provides a family of equations of state associated with the first law (19). Following [6], a particular equation of state can be specified, provided a suitable finite number of isothermal/isobaric curves is known. If, for instance, both functions $\alpha(\Lambda)$ and $f(\Lambda)$ are unknown, they can be obtained by solving the following linear system

$$T_1 - \alpha(\Lambda)\tau_1(\Lambda) - f(\Lambda) = 0,$$
$$T_2 - \alpha(\Lambda)\tau_2(\Lambda) - f(\Lambda) = 0,$$

where the graphs of functions $\tau_1(\Lambda)$ and $\tau_2(\Lambda)$ represent any two particular isothermal curves respectively at temperatures T_1 and T_2. Functions $\tau_1(\Lambda)$ and $\tau_2(\Lambda)$ can be obtained, for example, via interpolation of experimental data. If the specific form of the entropy, and consequently the function $\alpha(\Lambda)$, is known, the above procedure applies just to a single isothermal curve and it is equivalent to the solution of the PDE (22) with a particular initial datum. We also mention that the equation (22) includes the inviscid Burgers equation which has been observed to be relevant in the description of symmetry breaking in mean field spin models [9, 3].

We now consider two classical examples of equations of state that belong to the family (23).

Example 1: van der Waals model
Let us consider a gas of van der Waals type (see e.g.[5]) and identify the state function Λ with the volume V of the gas and the conjugate variable with the pressure P. The classical van der Waals equation of state for a real gas corresponds to the particular choice

$$\alpha(V) = \frac{V - nb}{nR} \qquad f(V) = \frac{na}{V} - \frac{n^2 b}{RV^2} \tag{24}$$

where n is the number of moles and a and b are constant associated, respectively, with the mean field interaction and the volume of the gas particles.

Example 2: effective molecular field model
Let us consider a magnetic system at temperature T, with magnetisation M and subject to the external magnetic field H. Let us identify

$$\tau \equiv H \qquad \text{and} \qquad M(T, H) \equiv \Lambda(T, H).$$

[1] We thank Adriano Barra for pointing out the statistical mechanical interpretation of the separability condition (21) for van der Waals and magnetic models.

Journal of Physics: Conference Series **482** (2014) 012009 doi:10.1088/1742-6596/482/1/012009

The equation of state for the effective molecular field model is (see [14] pag.84)

$$M = M_0 B_s \left(\frac{\bar{\mu} s}{\kappa T} (H + \lambda M) \right) \tag{25}$$

where: M_0 is the magnetisation at zero absolute temperature in absence of external magnetic field, $\bar{\mu} = g\mu_B$ being g the Landé factor and μ_B the Bohr magneton, κ is the Boltzmann constant, s the spin and λ the interacting constant. The function

$$B_s(y) = \frac{2s+1}{2s} \coth\left(\frac{2s+1}{2s} y \right) - \frac{1}{2s} \coth\left(\frac{y}{2s} \right)$$

is named Brillouin function. For simplicity let us consider the case of spin $s = 1/2$. The equation of state (25) can be re-casted in the *hodograph* form (23)

$$H - \frac{2\kappa}{\bar{\mu}} \text{arctanh}\left(\frac{M}{M_0} \right) T + \lambda M = 0. \tag{26}$$

The case of non-interacting magnetic system corresponds to the particular choice $\lambda = 0$.

5. N-phase systems and Tsallis' composition rule

5.1. Tsallis-type entropies

The Tsallis composition rule for the entropy of the $N-$phase thermodynamic system described by the set of order parameters $\theta^1, \dots, \theta^N$, interpreted as entropies of the component systems, reads as follows

$$S_q^{(N)} = q^{N-1} \left(\prod_{k=1}^{N} \left(\frac{1}{q} + \theta^k \right) - \frac{1}{q^N} \right). \tag{27}$$

We observe that the limit $q \to 0$ corresponds to the case of an additive entropy, i.e. $S_0^{(N)} = \lim_{q\to 0} S_q^{(N)} = \sum_{k=1}^{N} \theta^k$. In the special case $N = 1$, the order parameter θ^1 is just the entropy function of the single phase system under consideration, i.e. $S \equiv S_0^{(1)} = \theta^1$. In the case of a two-phase system $N = 2$, the Tsallis' composition rule takes the form (5). By iteration from the two-component formula (5), the case of a three-phase system can be readily dealt with. Let us assume, for instance, that given a certain two-phase system (θ^1, θ^2), the subsystem associated to the phase θ^1 can be further decomposed in the two phases, say θ^1 and θ^3. Then, replacing $\theta^1 \to \theta^1 + \theta^3 + q(\theta^1\theta^3)$ in Equation (5), the composition formula

$$S_q^{(3)} = \theta^1 + \theta^2 + \theta^3 + q(\theta^1\theta^2 + \theta^2\theta^3 + \theta^1\theta^3) + q^2(\theta^1\theta^2\theta^3). \tag{28}$$

follows for the three-phase system. By iterating further the procedure for an $N-$phase system we obtain the general formula (27).

5.2. State functions and Tsallis' composition rule

We would like to characterise the class of thermodynamic systems that admit the set of state functions given by all Tsallis' type entropies (27) for arbitrary value of the parameter q. According to (17), let us introduce the characteristic speeds for a given state function Λ associated with the additive entropy $S_0^{(N)}$

$$\mu^k = -\frac{\partial_k \Lambda}{\partial_k S_0^{(N)}} = -\partial_k \Lambda.$$

It is evident from (8) and (9) that integrable quasilinear systems of PDEs are characterised by the following quantities

$$\Gamma^i_{ij} = \frac{\partial_j \mu^i}{\mu^j - \mu^i} = \frac{\partial_{ij}\Lambda}{\partial_j\Lambda - \partial_i\Lambda} = -\Gamma^j_{ji}. \tag{29}$$

where μ^i are the characteristic velocities of the system or (due to (8)) of any of its symmetries. As a consequence, Γ^i_{ij} do not depend on the particular choice of the state function Λ. Hence, all functions of state are obtained as solutions of the following Euler-Poisson-Darboux type equation

$$\partial_{ij}\Lambda = \Gamma^i_{ij}\left(\partial_j\Lambda - \partial_i\Lambda\right). \tag{30}$$

In particular we have the following

Proposition 1 *All Tsallis' type entropies $S_q^{(N)}$ defined in (27) satisfy the equation (30) for any value of the parameter q with*

$$\Gamma^i_{ij} = \frac{1}{\theta_i - \theta_j}.$$

Proof. The proof straightforwardly follows from a direct calculation. Let us consider the hydrodynamic type system associated with Tsallis' type entropy

$$\frac{\partial\theta^k}{\partial\tau_q} = \mu^k_{(q)}(\theta^1, \dots, \theta^{k-1}, \theta^{k+1}, \dots, \theta^N)\frac{\partial\theta^k}{\partial T}, \qquad k = 1, \dots, N \tag{31}$$

where

$$\mu^k_{(q)} = \frac{\partial S_q^{(N)}}{\partial\theta^k} = q^{N-2}\left(\prod_{j\neq k=1}^{N}\left(\frac{1}{q} + \theta^j\right)\right) \tag{32}$$

are the characteristic speeds for the state function $\Lambda_q = -S_q^{(N)}$ (with conjugate variable τ_q). The system (31) is an example of a linearly degenerate system as the characteristic speed of the $k-th$ equation is independent on the $k-th$ order parameter. ■

The entropy $S_q^{(N)}$ generates a hydrodynamic flow in the space of thermodynamic variable and the conjugate variable τ_q can be interpreted as the *$q-$temperature* associated with it. The $q-$temperature τ_q clearly reduces to the standard temperature T when $q \to 0$. Although this system admits infinitely many functions of state, a particular interesting explicit class of state functions is associated to the weakly nonlinear symmetries (13).

Acknowledgements
The authors are grateful to A. Barra, I. Bonamassa, M. Leo and P. Tempesta for useful discussions and references on Tsallis type entropy and non-extensive thermodynamics. This research has been partially supported by the ERC grant FroM-PDE; INDAM-GNFM Grant, Progetti Giovani 2011, prot. 50; Humboldt Foundation; the Italian MIUR Research Project *Teorie geometriche e analitiche dei sistemi Hamiltoniani in dimensioni finite e infinite.*

References

[1] Abe S 2001 General pseudoadditivity of composable entropy prescribed by the existence of equilibrium *Phys. Rev.* E, **63** 061105
[2] Abe S 2002 Macroscopic thermodynamics based on composable nonextensive entropies, *Phys.* A **305** 62–68
[3] Barra A, Del Ferraro G and Tantari D 2013 Mean field spin glasses treated with PDE techniques Eur. Phys. J. B 86-332
[4] Blaszak M and Ma W -X 2003 Separable Hamiltonian equations on Riemann manifolds and related integrable hydrodynamic systems J. Geom. Phys. **47**(1) 21-42

[5] Callen H B 1985 *Thermodynamics and an Introduction to Thermostatistics* (Singapore: Wiley)
[6] De Nittis G and Moro A 2012 Thermodynamic phase transitions and shock singularities *Proc. R. Soc.* A **468** 2139 701-719
[7] Ferapontov E V 1991 Integration of weakly nonlinear hydrodynamic systems in Riemann invariants *Phys, Lett.* A **158** 112-118
[8] Ferapontov EV and Fordy A P 1997 Separable Hamiltonians and integrable systems of hydrodynamic type *J. Geom. Phys.* **21** no. 2 169–182
[9] Genovese G and Barra A 2009 A mechanical approach to mean field spin models, *J. Math. Phys.* **50** 053303 1-16
[10] Landau L D, Lifshitz E M and Pitaevskii L P 1980. *Course on Theoretical Physics, vol. 5: Statistical Physics* (Oxford: Butterworth - Heinemann)
[11] Lax P D 1973 *Hyperbolic systems of conservation laws and the mathematical theory of shock waves* CBMS-NSF Regional Conference Series in Applied Mathematics vol 11 (Philadelphia: SIAM)
[12] Lax P D 1975 Periodic solutions of the KdV equation. *Comm. Pure Appl. Math.* **28** (1975) 141 -188
[13] Majda A 1984 *Compressible fluid flow and systems of conservation laws in several space variables* Applied Mathematical Sciences vol 53 (New York: Springer)
[14] Stanley H E 1971 *Introduction to phase transitions and critical phenomena* (New York: Oxford University Press)
[15] Tsallis C 1988 Possible generalization of Boltzmann-Gibbs statistics, Journal of Statistical Physics **52** (1-2) 479-487
[16] Tsallis C 2009, Nonadditive entropy and nonextensive statistical mechanics - An overview after 20 years *Brazilian J. of Phys.* **39** (2A) 337–356
[17] Tsarev S P 1990 Geometry of Hamiltonian systems of hydrodynamic type. Generalized hodograph method, Izvestija AN USSR Math. **54** 1048-1068

KdV soliton interactions: a tropical view

A Dimakis[1] **and F Müller-Hoissen**[2]

[1] Dept. of Financial and Management Engineering, Univ. of the Aegean, 82100 Chios, Greece
[2] Max-Planck-Institute for Dynamics and Self-Organization, 37077 Göttingen, Germany

E-mail: dimakis@aegean.gr, folkert.mueller-hoissen@ds.mpg.de

Abstract. Via a "tropical limit" (Maslov dequantization), Korteweg-deVries (KdV) solitons correspond to piecewise linear graphs in two-dimensional space-time. We explore this limit.

1. Introduction

The "tropical limit" (Maslov dequantization) of soliton solutions of the (scalar, real) KdV equation

$$4\,u_t = u_{xxx} + 6\,u\,u_x\,,$$

where a subscript indicates a partial derivative with respect to an independent variable, describes them as piecewise linear graphs in two-dimensional space-time. The corresponding exploration in this work is based on techniques developed in [1, 2], but our presentation will be fairly self-contained.

The most striking property of KdV solitons is surely the well-known fact that, after an interaction, they regain their "identity" (amplitude, width and speed) and only experience a shift in space. But what really happens during the interaction of solitons? After the ground-breaking work of Kruskal and Zabusky [3], this question has been addressed again and again [4–21]. Some authors argued that solitons always pass through one another (see, e.g., [12, 15]). Others suggested that solitons exchange their identities during the interaction (see, e.g., [5, 6]). Based on a certain decomposition of the 2-soliton solution, some proposed an intermediate wave that transfers energy between the two solitons [16, 20, 21]. There have been several attempts to "individualize" the a priori asymptotically (i.e., for $t \ll 0$) defined solitons also during an interaction. Solitons do not really behave like classical particles, however. As localized disturbances of a continuous medium, they possess a wavelike nature. Instead of speaking of the same incoming ($t \ll 0$) and outgoing ($t \gg 0$) soliton, it is more adequate to speak of "instances" of a certain soliton state. Once solitons start to interact, they loose their individuality. This is also what the tropical limit shows: two KdV solitons (which can be regarded as parallel KP line solitons, see [1], Example 4.5) interact by exchanging a "virtual soliton", see Section 3.4. The present work explores more generally KdV soliton interactions in the tropical limit. The analogy with a quantum scattering theory (also see [18–20]) is striking. But here a kind of second quantization is not necessary since the KdV equation (and any other evolution equation possessing solitons) is already a many-"particle" model. Solitons are the asymptotically free particles, and their interaction can be understood as an exchange of virtual particles. We should stress, however, that we are *not* attempting to interprete KdV as a quantum theory.

Section 2 expresses the KdV soliton solutions in a form convenient for our purposes. Section 3 then deals with the tropical limit.

2. KdV solitons

Let us recall (see, e.g., [22]) that the M-soliton solution of the KdV equation is given by $u = 2(\log \tau_H)_{xx}$ with the Hirota τ-function

$$\tau_H = \sum_{\mu_1,\ldots,\mu_M=0,1} \exp\Big(\sum_{j=1}^{M} \mu_j \eta_j + \sum_{1\le j<k\le M} b_{jk}\,\mu_j \mu_k\Big)\,,$$

where $\eta_j = 2p_j\,(x + p_j^2\,t + \tilde{c}_j)$ and $b_{jk} = \log[(p_k - p_j)^2/(p_k + p_j)^2]$. This involves arbitrary real constants $\tilde{c}_j$ and $0 < p_1 < p_2 < \cdots < p_M$.

Proposition 2.1. *An equivalent expression for the M-soliton KdV solution is*

$$\tau = \sum_{A\in\{-1,1\}^M} e^{\Theta_A} \qquad \textit{where}$$

$$\Theta_A = \sum_{j=1}^{M} \alpha_j\,\theta_j + \delta_A\,, \qquad \theta_j = p_j\,(x + p_j^2\,t + c_j)\,, \qquad \delta_A = \log\Delta_A\,, \qquad \Delta_A = |\Delta(\alpha_1 p_1,\ldots,\alpha_M p_M)|\,,$$

with $A = (\alpha_1,\ldots,\alpha_M)$, $\alpha_j \in \{\pm 1\}$, and Δ is a Vandermonde determinant.

Proof. With τ we associate the equivalent (no change in u) tau function

$$\tau' = \frac{1}{\Delta_{(-1,\ldots,-1)}}\, e^{\theta_1+\cdots+\theta_M}\,\tau = \sum_{\alpha_1,\ldots,\alpha_M=\pm 1} \frac{\Delta_{(\alpha_1,\ldots,\alpha_M)}}{\Delta_{(-1,\ldots,-1)}}\, e^{(\alpha_1+1)\theta_1+\cdots+(\alpha_M+1)\theta_M}\,.$$

With the help of

$$\Delta_{(\alpha_1,\ldots,\alpha_M)} = \prod_{1\le j<k\le M} (p_k - \alpha_j\alpha_k\,p_j) = \prod_{1\le j<k\le M} (p_k - p_j)^{(1+\alpha_j\alpha_k)/2}(p_k + p_j)^{(1-\alpha_j\alpha_k)/2}\,,$$

we find that $\tau' = \tau_H$ if we set

$$\mu_j := (\alpha_j + 1)/2 \in \{0,1\}\,, \qquad \tilde{c}_j := c_j + \frac{1}{2p_j}\log\Big|\prod_{k\neq j}\frac{p_k + p_j}{p_k - p_j}\Big|\,. \qquad \square$$

For an M-soliton solution, there are 2^M different phases Θ_A. For $M > 2$ we extend the variables x and t to $N \ge M$ (KdV hierarchy) variables $t^{(k)}$, $k = 1,\ldots,N$, where $x = t^{(1)}$, $t = t^{(2)}$. Via translations of these variables, we achieve that $c_j = 0$, $j = 1,\ldots,N$, so that

$$\theta_j = \sum_{k=1}^{N} p_j^{2k-1}\,t^{(k)}\,. \tag{2.1}$$

Remark 2.2. After having arranged (2.1) by a shift of the KdV hierarchy variables, an M-soliton solution is invariant under reflection of all independent variables, i.e., $t^{(k)} \mapsto -t^{(k)}$, $k = 1,\ldots,N$.

Example 2.3. For the 2-soliton solution, setting $N = 2$, we obtain

$$u(x,t) = 8(p_2^2 - p_1^2)\frac{p_1^2 \sinh^2\theta_2 + p_2^2\cosh^2\theta_1}{[(p_2-p_1)\cosh(\theta_1+\theta_2) + (p_2+p_1)\cosh(\theta_1-\theta_2)]^2} .$$

Since we set $c_1 = c_2 = 0$, we have $u(x,t) = u(-x,-t)$, so that the solution is adapted to an obvious symmetry of the KdV equation. In particular, $u(x,0)$ is symmetric about $x = 0$ (also see Theorem 1 in [21], where the expressions for the symmetry event are more complicated due to a different form of the KdV equation and a different parametrization of the 2-soliton solution). We find $u_{xx}(0,0) = -4(p_2-p_1)(p_2+p_1)(p_2^2 - 3p_1^2)$. Hence $u(x,0)$ has a minimum at $x = 0$ if $p_2/p_1 < \sqrt{3}$ and a maximum if $p_2/p_1 > \sqrt{3}$ [4, 10, 21], and we find $u(0,0) = 2(p_2^2 - p_1^2)$.

3. Tropical limit of KdV solitons

Let $\mathcal{U}_B$ be the region in $\mathbb{R}^2$ where the phase Θ_B dominates all others, i.e.,

$$\mathcal{U}_B = \left\{(x,t) \in \mathbb{R}^2 \mid \max\{\Theta_A(x,t) \mid A \in \{-1,1\}^M\} = \Theta_B(x,t)\right\} ,$$

considered at fixed values of the higher KdV hierarchy variables. $\mathcal{U}_B$ is connected, it may be empty for some B. In a non-empty set $\mathcal{U}_B$ and sufficiently far away from the boundary, the approximation

$$\log\tau \simeq \max\{\Theta_A \mid A \in \{-1,1\}^M\}$$

is valid, so that u vanishes (since Θ_B is linear in x). As a consequence, u is localized along the boundary lines of non-empty dominating phase regions. The *tropical limit of the KdV soliton solution* is the piecewise linear planar graph consisting of such boundary lines, and the amplitude u on these lines. The structure of this graph is determined by the intersections of the dominating phase regions. In the following, we write $\mathcal{U}_{AB} := \mathcal{U}_A \cap \mathcal{U}_B$.

Remark 3.1. A precise formulation of what we call the tropical limit of KdV solitons is obtained by using the Maslov dequantization formula (see, e.g., [23])

$$\lim_{\epsilon\to 0} \epsilon \sum_{A\in\{-1,1\}^M} e^{\Theta_A/\epsilon} = \max\{\Theta_A \mid A \in \{-1,1\}^M\} , \tag{3.1}$$

which replaces the operation of addition (of exponentials) by the maximum function (applied to the phases). This is a familar step in "tropicalization". It is usually accompanied by also replacing multiplication by addition (the "tropical product").

Remark 3.2. If the logarithmic terms δ_A were negligible, then we would have $\log\tau \simeq \max_{A\in\{-1,1\}^M}\{|\sum_{j=1}^M \alpha_j\theta_j|\} = \sum_{j=1}^M |\theta_j|$, and the tropical limit of the KdV soliton would be given by the *superimposition* of the space-time lines corresponding to the constituent single solitons. (3.1) shows that this simplified limit corresponds to introducing "slow variables" via $x \mapsto x/\epsilon$ and $t \mapsto t/\epsilon$ in a soliton solution. This simply maps the latter to the corresponding solution of the ϵ-dependent KdV equation $u_t - (3/2)\, u\, u_x = \epsilon^2\, u_{xxx}$, which formally approaches its "dispersionless" (or "quasiclassical") limit, the inviscid Burgers (or Hopf, or Riemann) equation $u_t - (3/2)\, u\, u_x = 0$, as $\epsilon \to 0$. The ϵ-dependent soliton solution does *not* tend to a solution of the inviscid Burgers equation, however. In the limit, KdV solitons "disappear" in the sense that their support becomes a set of measure zero in space-time, while u retains a finite value. The associated initial data become infinitely steep, so there is no corresponding local solution of the inviscid Burgers equation via the method of characteristics. It is also not adequate to think of the (simplified) tropical limit as a kind of non-smooth solution of the inviscid Burgers equation (e.g., similar to those obtained via "front tracking" [24]). This is so because irrespective how small ϵ is, the corresponding ϵ-dependent solution necessarily retains dispersion since it remains solitonic. We have to conclude that the tropical limit cannot be regarded as a dispersionless limit.

Introducing

$$p_A := \sum_{j=1}^{M} \alpha_j\, p_j\,, \quad q_A := \sum_{j=1}^{M} \alpha_j\, p_j^3\,, \quad c_A := \delta_A + \sum_{k=3}^{N}\sum_{j=1}^{M} \alpha_j\, p_j^{2k-1}\, t^{(k)}\,,$$

where $A = (\alpha_1, \dots, \alpha_M)$, we have

$$\Theta_A - \Theta_B = (p_A - p_B)\, x + (q_A - q_B)\, t + c_A - c_B\,. \tag{3.2}$$

There is a line at which two phases Θ_A and Θ_B "meet", i.e., where $\Theta_A = \Theta_B$. It is given by

$$x = x_{AB}(t) := -\frac{q_A - q_B}{p_A - p_B}\, t - \frac{c_A - c_B}{p_A - p_B}\,, \tag{3.3}$$

assuming that $p_A \neq p_B$, and (3.2) can be written as

$$\Theta_A - \Theta_B = (p_A - p_B)(x - x_{AB}(t))\,. \tag{3.4}$$

If $(x,t) \notin \mathcal{U}_B$, there is a phase Θ_A such that $\Theta_A(x,t) > \Theta_B(x,t)$. In this case we say that the phase Θ_B is *non-visible* at the event (x,t). An immediate consequence of the last identity is

$$p_A > p_B \quad \Longrightarrow \quad \begin{array}{c} \Theta_B(x,t) \\ \Theta_A(x,t) \end{array} \quad \text{is non-visible for} \quad \begin{array}{l} x > x_{AB}(t) \\ x < x_{AB}(t)\,. \end{array}$$

Near an event on the boundary line (3.3) that is not a higher order coincidence of phases, we have $\tau \simeq e^{\Theta_A} + e^{\Theta_B}$ and thus $u \simeq 2(p_A - p_B)^2 e^{\Theta_A + \Theta_B}/(e^{\Theta_A} + e^{\Theta_B})^2$. Using $\Theta_A = \Theta_B$, this becomes $u \simeq \frac{1}{2}(p_A - p_B)^2$ (also see Appendix D in [1]).

Remark 3.3. Since the p's are positive and distinct, for $M = 2$ we have $p_A \neq p_B$ if $A \neq B$. For $M > 2$, the latter condition holds if $p_1, \dots, p_M$ are linearly independent over $\{\pm 1\}$.

3.1. x-asymptotics

For fixed t and suffiently large x (we write $x \gg 0$), (3.4) implies $\Theta_{(1,\dots,1)} > \Theta_A$ for all $A \in \{-1,1\}^M$, $A \neq (1,\dots,1)$. Furthermore, for $x \ll 0$ we have $\Theta_{(-1,\dots,-1)} > \Theta_A$ for all $A \in \{-1,1\}^M$, $A \neq (-1,\dots,-1)$. Hence $\mathcal{U}_{(1,\dots,1)}$ and $\mathcal{U}_{(-1,\dots,-1)}$ are dominating phase regions for $x \gg 0$, respectively $x \ll 0$.

3.2. Triple phase coincidences

At a triple phase coincidence three different phases satisfy $\Theta_A = \Theta_B = \Theta_C$. If

$$p_{ABC} := p_A(q_B - q_C) + p_B(q_C - q_A) + p_C(q_A - q_B) \neq 0\,,$$

then a corresponding event occurs at the time

$$t_{ABC} := -p_{ABC}^{-1}\Big(p_A(c_B - c_C) + p_B(c_C - c_A) + p_C(c_A - c_B)\Big)\,.$$

Its x-coordinate is given by $x_{ABC} := x_{AB}(t_{ABC})$. We find that

$$x_{AC}(t) - x_{AB}(t) = -\frac{p_{ABC}}{(p_B - p_A)(p_C - p_A)}\,(t - t_{ABC})\,,$$

and

$$\Theta_A - \Theta_C = (p_A - p_C)(x - x_{AB}(t)) - \frac{p_{ABC}}{p_B - p_A}(t - t_{ABC}),$$

so that

$$\Theta_A - \Theta_C = \frac{p_{ABC}}{p_A - p_B}(t - t_{ABC}) \qquad \text{on the line} \quad x = x_{AB}(t)\,. \tag{3.5}$$

As a consequence, the half-line

$$\{x = x_{AB}(t)\,|\,t \gtrless t_{ABC}\} \quad \text{is non-visible if} \quad \frac{p_{ABC}}{p_A - p_B} \lesseqgtr 0\,.$$

At a triple phase event, $\tau \simeq e^{\Theta_A} + e^{\Theta_B} + e^{\Theta_C}$, hence $u \simeq \frac{4}{9}(p_A^2 + p_B^2 + p_C^2 - p_A p_B - p_A p_C - p_B p_C)$.

3.3. Appearances of single solitons and t-asymptotics

For $A = (\alpha_1, \ldots, \alpha_M)$, let $A_{(k)} := (\alpha_1, \ldots, \alpha_{k-1}, -\alpha_k, \alpha_{k+1}, \ldots, \alpha_M)$. On $\mathcal{U}_{AA_{(k)}}$ we obtain the single soliton expression $\tau \simeq e_{-k} + e_k$, up to a factor that drops out in the expression for u. According to (3.3), the boundary line between the two phases Θ_A and $\Theta_{A_{(k)}}$ is given by

$$x = x_{AA_{(k)}} = -\frac{1}{2\alpha_k p_k}(\delta_A - \delta_{A_{(k)}}) - p_k^2\, t - \sum_{j=3}^{N} p_k^{2j-2}\, t^{(j)}\,.$$

Hence, a soliton (i.e., a visible part of such a line), moves from right to left along the x-axis. Its "height" is $u \simeq 2p_k^2$. Furthermore,

$$x_{AA_{(k)}} - x_{BB_{(k)}} = -\frac{1}{2\alpha_k p_k}(\delta_A - \delta_{A_{(k)}}) + \frac{1}{2\beta_k p_k}(\delta_B - \delta_{B_{(k)}}) \tag{3.6}$$

is constant. For $B \neq A, A_{(k)}$, the lines given by $x = x_{AA_{(k)}}$ and $x = x_{BB_{(k)}}$ are thus parallel. There are 2^{M-1} such lines (for fixed k), since this is the number of different pairs $A, A_{(k)}$. It is natural to interprete the visible segments as appearances of the kth soliton.

Now we show that the kth soliton is visible for large $|t|$. We consider (3.5) on the line $x = x_{AA_{(k)}}(t)$, with

$$\begin{aligned} \frac{p_{AA_{(k)}C}}{p_A - p_{A_{(k)}}} &= p_k^2\,(p_C - p_A) - (q_C - q_A) = \sum_{j \neq k} (\gamma_j - \alpha_j)\, p_j\, (p_k^2 - p_j^2)\,, \\ &= \sum_{j<k} (\gamma_j - \alpha_j)\, p_j\, (p_k^2 - p_j^2) + \sum_{j>k} (\alpha_j - \gamma_j)\, p_j\, (p_j^2 - p_k^2)\,. \end{aligned}$$

This is different from zero if $C = (\gamma_1, \ldots, \gamma_M)$ is different from A and $A_{(k)}$. (3.5) implies

$$\begin{aligned} A = (1, \ldots, 1, \alpha_k, -1, \ldots, -1) \quad &\Longrightarrow \quad p_{AA_{(k)}C}/(p_A - p_{A_{(k)}}) < 0 \quad \forall C \in \{-1, 1\}^M \\ &\Longrightarrow \quad \{x = x_{AA_{(k)}}(t)\,|\,t \ll 0\} \quad \text{is visible} \\ A = (-1, \ldots, -1, \alpha_k, 1, \ldots, 1) \quad &\Longrightarrow \quad p_{AA_{(k)}C}/(p_A - p_{A_{(k)}}) > 0 \quad \forall C \in \{-1, 1\}^M \\ &\Longrightarrow \quad \{x = x_{AA_{(k)}}(t)\,|\,t \gg 0\} \quad \text{is visible} \end{aligned}$$

(using $0 < p_1 < p_2 < \cdots < p_M$). Moreover, for any other $A \in \{-1,1\}^M$ there is a $C \in \{-1,1\}^M$ such that, according to (3.5), $\Theta_C > \Theta_A$ for $t \ll 0$, respectively $t \gg 0$, so that no further boundary lines are visible for $|t| \gg 0$. (3.6) implies that

$$x_{(-1,\ldots,-1)(1,-1,\ldots,-1)} < x_{(1,-1,\ldots,-1)(1,1,-1,\ldots,-1)} < \cdots < x_{(1,\ldots,1,-1)(1,\ldots,1)} \qquad \text{for } t \ll 0\,,$$

while we have

$$x_{(-1,\ldots,-1)(-1,\ldots,-1,1)} < x_{(-1,\ldots,-1,1)(-1,\ldots,-1,1,1)} < \cdots < x_{(-1,1,\ldots,1)(1,\ldots,1)} \qquad \text{for } t \gg 0\,.$$

We conclude that, as time proceeds from $-\infty$ to $+\infty$, the solitons reappear in reversed order. Recalling the results in Section 3.1, we thus proved the asymptotic structure displayed in Fig. 1. In particular, this implies that a phase Θ_A, where A is *not* of the form $(-1,\ldots,-1,)$, $(1,\ldots,1)$, $(1,1\ldots,1,-1,-1,\ldots,-1)$ or $(-1,\ldots,-1,1,\ldots,1)$, can only be visible in a *bounded* region of the xt-plane. Moreover, passing in the asymptotic region $|t| \gg 0$ from left to right along the x-axis, from one phase Θ_A to the next, say $\Theta_{A_{(k)}}$, requires $\alpha_k = -1$. Other transitions can therefore only occur in case of virtual solitons (which are bounded in space-time).

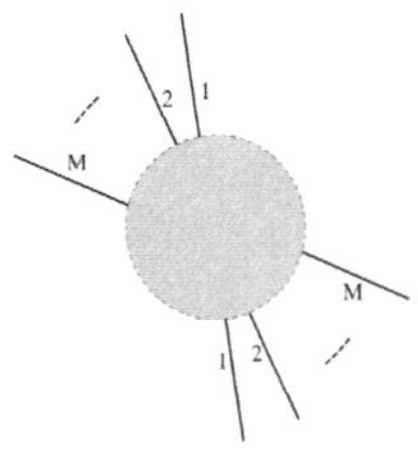

Figure 1. Asymptotic structure of an M-soliton KdV solution in the xt-plane. Time flows from bottom to top. A number k refers to the kth soliton.

3.4. Two solitons

We achieve some simplification of expressions in the following by using the notation $\bar{1} := -1$, and writing, e.g., $\bar{1}\ldots\bar{1}1\ldots1$ instead of $(-1,\ldots,-1,1,\ldots,1)$. The tropical approximation of the 2-soliton solution is given by $\log\tau \simeq \max\{\Theta_{\bar{1}\bar{1}}, \Theta_{1\bar{1}}, \Theta_{\bar{1}1}, \Theta_{11}\}$ with

$$\Theta_{\bar{1}\bar{1}} = -\theta_1 - \theta_2 + \log(p_2 - p_1)\,, \quad \Theta_{1\bar{1}} = \theta_1 - \theta_2 + \log(p_1 + p_2)\,,$$
$$\Theta_{\bar{1}1} = -\theta_1 + \theta_2 + \log(p_1 + p_2)\,, \quad \Theta_{11} = \theta_1 + \theta_2 + \log(p_2 - p_1)\,,$$

and $\theta_j = p_j\,(x + p_j^2 t)$, $0 < p_1 < p_2$. Here we set $N = 2$. We recall that Θ_{11} dominates (all other phases) for $x \gg 0$, whereas $\Theta_{\bar{1}\bar{1}}$ dominates for $x \ll 0$. There are six boundary lines. Those corresponding to asymptotic solitons are

$$\begin{aligned}
\text{1st soliton:} \quad & x = x_{\bar{1}1,1\bar{1}}(t) = -p_1^2 t - \frac{\ell}{2p_1} && t \ll 0 \text{ branch}\,,\\
& x = x_{\bar{1}1,11}(t) = -p_1^2 t + \frac{\ell}{2p_1} && t \gg 0 \text{ branch}\,,\\
& \text{shift:} \quad x_{\bar{1}1,11}(t) - x_{\bar{1}\bar{1},1\bar{1}}(t) = \frac{\ell}{2p_1} && \text{where} \quad \ell := \log\frac{p_2+p_1}{p_2-p_1} > 0\\
\text{2nd soliton:} \quad & x = x_{1\bar{1},11}(t) = -p_2^2 t + \frac{\ell}{2p_2} && t \ll 0 \text{ branch}\,,\\
& x = x_{\bar{1}\bar{1},\bar{1}1}(t) = -p_2^2 t - \frac{\ell}{2p_2} && t \gg 0 \text{ branch}\,,\\
& \text{shift:} \quad x_{\bar{1}\bar{1},\bar{1}1}(t) - x_{1\bar{1},11}(t) = -\frac{\ell}{2p_1}\,.
\end{aligned}$$

The constant shifts exactly correspond to the well-known (asymptotically determined) *phase shifts*, which are the only witnesses of an interaction of KdV solitons. Further boundary lines:

$$x = x_{\bar{1}1,11}(t) = -\frac{1}{p_1+p_2}(p_1^3+p_2^3)\,t = -(p_1^2 - p_1p_2 + p_2^2)\,t\,,$$
$$x = x_{1\bar{1},\bar{1}1}(t) = -(p_1^2 + p_1p_2 + p_2^2)\,t\,.$$

Triple phase coincidences occur at the times

$$t_{\bar{1}\bar{1},1\bar{1},\bar{1}1} = \frac{\ell}{2p_1p_2(p_2+p_1)}\,, \qquad t_{1\bar{1},\bar{1}1,11} = -\frac{\ell}{2p_1p_2(p_2+p_1)}\,,$$
$$t_{\bar{1}\bar{1},\bar{1}1,11} = -\frac{\ell}{2p_1p_2(p_2-p_1)}\,, \qquad t_{\bar{1}\bar{1},1\bar{1},11} = \frac{\ell}{2p_1p_2(p_2-p_1)}\,.$$

These times are ordered as follows: $t_{\bar{1}\bar{1},\bar{1}1,11} < t_{1\bar{1},\bar{1}1,11} < 0 < t_{\bar{1}\bar{1},1\bar{1},\bar{1}1} < t_{\bar{1}\bar{1},1\bar{1},11}$. Since

$$\begin{aligned} \Theta_{11} - \Theta_{1\bar{1}} &= 2\theta_2 - \ell \\ &= -2\,\ell < 0 \qquad \text{on} \quad x = x_{\bar{1}\bar{1},\bar{1}1}(t)\,, \\ \Theta_{11} - \Theta_{\bar{1}1} &= 2\theta_1 - \ell \\ &= -2\,\ell < 0 \qquad \text{on} \quad x = x_{\bar{1}\bar{1},1\bar{1}}(t)\,, \end{aligned}$$

the triple phase events at $t_{\bar{1}\bar{1},\bar{1}1,11}$ and $t_{\bar{1}\bar{1},1\bar{1},11}$ are non-visible. At the remaining two triple phase events $(x_{1\bar{1},\bar{1}1,11}, t_{1\bar{1},\bar{1}1,11})$ and $(x_{\bar{1}\bar{1},1\bar{1},\bar{1}1}, t_{\bar{1}\bar{1},1\bar{1},\bar{1}1})$, we have $\Theta_{11} - \Theta_{\bar{1}\bar{1}} = 2\ell > 0$, respectively $\Theta_{\bar{1}\bar{1}} - \Theta_{11} = 2\ell > 0$, so that these events are visible. Moreover, we have

$$x_{1\bar{1},\bar{1}1,11} - x_{\bar{1}\bar{1},1\bar{1},\bar{1}1} = \frac{p_1^2 + p_1p_2 + p_2^2}{p_1p_2(p_1+p_2)}\,\ell > 0\,.$$

Furthermore,

$$\begin{aligned} \Theta_{11} - \Theta_{1\bar{1}} &= 2p_1p_2(p_2-p_1)\,t - \ell \\ \Theta_{11} - \Theta_{\bar{1}1} &= -2p_1p_2(p_2-p_1)\,t - \ell \end{aligned} \qquad \text{on} \qquad x = x_{\bar{1}1,11}(t)\,,$$

implies $\Theta_{11} < \Theta_{1\bar{1}}$ for $t < 0$ and $\Theta_{11} < \Theta_{\bar{1}1}$ for $t > 0$ on the boundary line $x = x_{\bar{1}1,11}(t)$, so that the whole line is non-visible. We thus arrive at the situation described in Fig. 2. Two solitons interact by exchanging a "virtual soliton".

3.5. Three solitons

Setting $N = 3$, we now have $\theta_j = p_j\,(x + p_j^2 t + p_j^4 s)$, $j = 1,2,3$, with $s := t^{(3)}$. Recall that $0 < p_1 < p_2 < p_3$. The tropical approximation of the 3-soliton solution is given by $\log\tau \simeq \max\{\Theta_0, \Theta_1, \ldots, \Theta_7\}$, where $\Theta_0 := \Theta_{\bar{1}\bar{1}\bar{1}}$, $\Theta_1 := \Theta_{1\bar{1}\bar{1}}$, $\Theta_2 := \Theta_{\bar{1}1\bar{1}}$, $\Theta_3 := \Theta_{11\bar{1}}$, $\Theta_4 := \Theta_{\bar{1}\bar{1}1}$, $\Theta_5 := \Theta_{1\bar{1}1}$, $\Theta_6 := \Theta_{\bar{1}11}$, $\Theta_7 := \Theta_{111}$. Only the Θ_2- and Θ_5-regions are *bounded* in the xt-plane. In the following, we present results derived with the help of computer algebra. It turns out that there are no visible coincidences of more than four of these phases. Among the $\binom{8}{4} = 70$ a priori possible coincidences of four phases, the following can be ruled out: $\{0,1,2,3\}$, $\{0,1,4,5\}$, $\{0,1,6,7\}$, $\{0,2,4,6\}$, $\{0,2,5,7\}$, $\{0,3,4,7\}$, $\{1,2,5,6\}$, $\{1,3,4,6\}$, $\{1,3,5,7\}$, $\{2,3,4,5\}$, $\{2,3,6,7\}$, $\{4,5,6,7\}$, where a number i represents the phase Θ_i. Of the remaining 58 4-phase coincidences, most are non-visible (for all values of the parameters). They can only occur if $s = s_{ijkl}$, where s_{ijkl} is completely determined in terms of the parameters p_m:

$$s_{ijkl} \;=\; \frac{a_{ijkl}}{2p_2(p_2^2-p_1^2)(p_3^2-p_2^2)} + \frac{b_{ijkl}}{2p_1(p_2^2-p_1^2)(p_3^2-p_1^2)} + \frac{c_{ijkl}}{2\,p_3(p_3^2-p_2^2)(p_3^2-p_1^2)}\,,$$

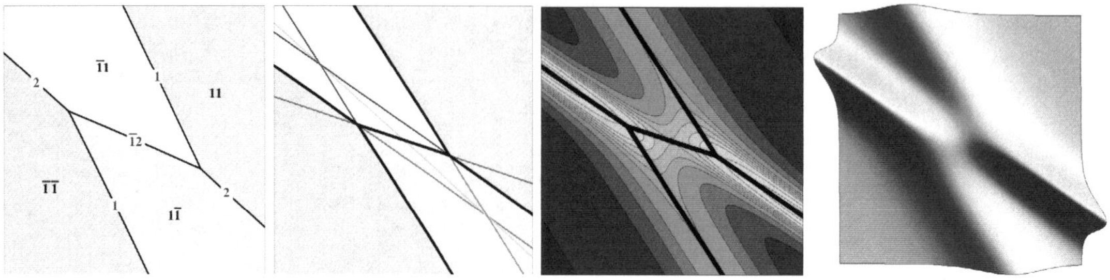

Figure 2. To the left is a space-time plot of the tropical limit graph of a 2-soliton KdV solution ($p_1 = 1, p_2 = 1.5$). The x-axis is horizontal, time proceeds from bottom to top. The colored regions are dominating phase regions. At $t = t_{\bar{1}1,\bar{1}1,11}$ soliton **2** splits into an instance of soliton **1** and a "virtual soliton" $\bar{\mathbf{1}}\mathbf{2}$, formally assigning to it the interpretation of a bound state composed of an antisoliton $\bar{\mathbf{1}}$ and a soliton **2**. At $t = t_{\bar{1}\bar{1},1\bar{1},\bar{1}1}$ it merges with soliton **1** to create a new instance of soliton **2**. The second plot shows all boundary lines between pairs of phases. The whole green line, passing through the middle point, is non-visible. The third plot displays the tropical limit superimposed on a contour plot of the KdV solution. The last is a plot of the KdV solution $u(x,t)$ over the xt-plane.

where

$$\begin{aligned} a_{0124} &= \ell_{12} + \ell_{23} & b_{0124} &= -\ell_{12} - \ell_{13} & c_{0124} &= -\ell_{13} - \ell_{23} \\ a_{1235} &= \ell_{23} - \ell_{12} & b_{1235} &= \ell_{12} - \ell_{13} & c_{1235} &= \ell_{13} - \ell_{23} \\ a_{1245} &= -2\ell_{13} + \ell_{12} + \ell_{23} & b_{1245} &= \ell_{13} - \ell_{12} & c_{1245} &= \ell_{13} - \ell_{23} \end{aligned}$$

with $\ell_{ij} := \log[(p_j + p_i)/(p_j - p_i)]$, and we find

$$s_{2356} = -s_{1245}\,, \qquad s_{2456} = -s_{1235}\,, \qquad s_{3567} = -s_{0124}\,.$$

A 4-phase coincidence described by $\{i,j,k,l\}$ can only be visible if for each subset of three indices the corresponding 3-phase coincidence is visible. Table 1 lists all (under the stated conditions) visible 3-phase coincidences. Fig. 3 provides some more information about the 3-soliton case and Fig. 4 shows an example.

Except for some relations implied by Table 1, like $s_{0124} < s_{1245}$, the order of the values s_{ijkl} depends in a more complicated way on the parameters p_j. In particular, we can draw the following conclusions.

- Only the 4-phase coincidences corresponding to $\{0,1,2,4\}$, $\{1,2,3,5\}$, $\{1,2,4,5\}$, $\{2,3,5,6\}$, $\{2,4,5,6\}$, $\{3,5,6,7\}$ are visible, for certain values of s.
- Θ_2 is non-visible if either $s > s_{1245}$ and $p_3 < \sqrt{p_1^2 + p_1p_2 + p_2^2}$, or $s > s_{1235}$ and $p_3 > \sqrt{p_1^2 + p_1p_2 + p_2^2}$ (since then all 3-phase coincidences with t_{ijk}, $2 \in \{i,j,k\}$, are non-visible). In particular, Θ_2 is non-visible if $s \gg 0$.
- Θ_5 is non-visible if either $s < s_{2356}$ and $p_3 < \sqrt{p_1^2 + p_1p_2 + p_2^2}$, or $s < s_{2456}$ and $p_3 > \sqrt{p_1^2 + p_1p_2 + p_2^2}$. In particular, Θ_5 is non-visible if $s \ll 0$.

Although some general results can be obtained about arbitrarily high order phase coincidences, there is hardly a chance to achieve a classification, for arbitrary M, of all possible evolutions of M-soliton KdV solutions or, equivalently, parallel M-soliton KP-II solutions, in a similar way as for the case of tree-shaped KP line soliton solutions [1, 2]. The simple

Table 1. Results about 3-phase coincidences for 3-soliton solutions.

3-phase coincidence time	visibility condition	
t_{012}, t_{024}	$s < s_{0124}$	
t_{014}	$s_{0124} < s$	
t_{124}	$s_{0124} < s < s_{1245}$	
t_{145}	$s_{1245} < s$	
t_{245}	$s_{2456} < s < s_{1245}$	
t_{125}	$s_{1235} < s < s_{1245}$	if $p_3 < \sqrt{p_1^2 + p_1 p_2 + p_2^2}$
	$s_{1245} < s < s_{1235}$	if $p_3 > \sqrt{p_1^2 + p_1 p_2 + p_2^2}$
t_{123}	$s < s_{1235}$	
t_{135}	$s_{1235} < s$	
t_{235}	$s_{2356} < s < s_{1235}$	
t_{236}	$s < s_{2356}$	
t_{256}	$s_{2356} < s < s_{2456}$	if $p_3 < \sqrt{p_1^2 + p_1 p_2 + p_2^2}$
	$s_{2456} < s < s_{2356}$	if $p_3 > \sqrt{p_1^2 + p_1 p_2 + p_2^2}$
t_{356}	$s_{2356} < s < s_{3567}$	
t_{357}, t_{367}, t_{567}	$s_{3567} < s$	
t_{456}	$s_{2456} < s$	
t_{246}	$s < s_{2456}$	

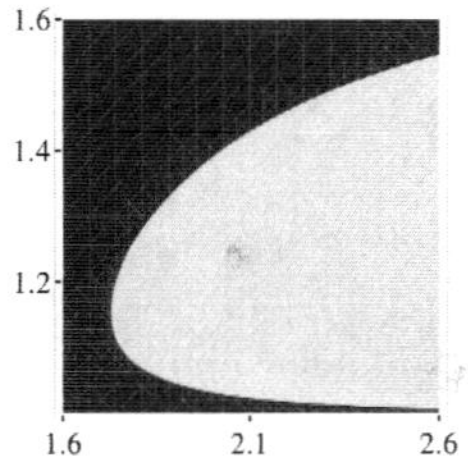

Figure 3. According to Remark 2.2, $u(x,t,s)$ has the property $u(x,0,0) = u(-x,0,0)$. Hence $u(x,0,0)$ has an extremum at $x = 0$, determined by the sign of $\frac{1}{4} u_{xx}(0,0,0) = -(p_1^4 + 3p_2^4 + p_3^4) + 4(p_1^2 p_2^2 - p_1^2 p_3^2 + p_2^2 p_3^2)$. The plot displays for p_3/p_2 (vertical axis) versus p_2/p_1 (horizontal axis) the boundary between the regions where $u(x,0,0)$ has a minimum (light region), respectively a maximum (dark region).

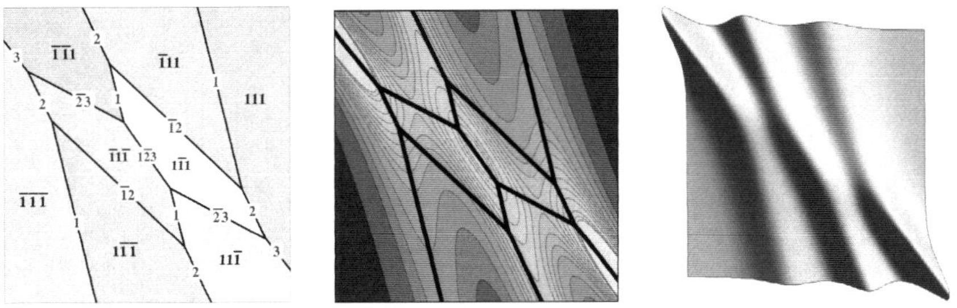

Figure 4. The first plot displays the tropical limit of a 3-soliton KdV solution for $s = 0$ (and $p_1 = 0.5$, $p_2 = 0.7$, $p_3 = 0.9$). Here all 2^3 phases are visible. The three phase regions extending to the bottom ($t \ll 0$) are given, from left to right, by $\Theta_{\bar{1}\bar{1}\bar{1}}, \Theta_{1\bar{1}\bar{1}}, \Theta_{11\bar{1}}$, those extending to the top ($t \gg 0$) by $\Theta_{\bar{1}\bar{1}1}, \Theta_{\bar{1}11}, \Theta_{111}$. In the middle we have two bounded phase regions where $\Theta_{\bar{1}1\bar{1}}$, respectively $\Theta_{1\bar{1}1}$, dominate. The second picture shows the superimposition of the tropical limit on a contour plot of the KdV solution. The third is a plot of the KdV solution $u(x,t,s)$ at $s = 0$.

combinatorics (higher Tamari orders), underlying the latter case, has no counterpart in case of more general KP line soliton solutions (also see, e.g., [25] for an analysis of the general KP case).

F M-H is grateful to Boris Konopelchenko for a very stimulating discussion about tropical and dispersionless limits.

References

[1] Dimakis A and Müller-Hoissen F 2011 KP line solitons and Tamari lattices *J. Phys. A: Math. Theor.* **44** 025203

[2] Dimakis A and Müller-Hoissen F 2012 KP solitons, higher Bruhat and Tamari orders *Associahedra, Tamari Lattices and Related Structures* (Progress in Mathematics vol 299) ed Müller-Hoissen F, Pallo J and Stasheff J (Basel: Birkhäuser) pp 391–423

[3] Zabusky N and Kruskal M 1965 Interaction of solitons in an collisionless plasma and the recurrence of initial states *Phys. Rev. Lett.* **15** 240–243

[4] Lax P 1968 Integrals of nonlinear equations of evolution and solitary waves *Comm. Pure Appl. Math.* **21** 467–490

[5] Bowtell G and Stuart A 1980 Particle motions and the dynamics of Korteweg-de Vries solitons *Progr. Theor. Phys.* **64** 68–80

[6] Bowtell G and Stuart A 1983 A particle representation of the Korteweg-de Vries solitons *J. Math. Phys.* **24** 969–981

[7] Yoneyama T 1984 The Korteweg-de Vries two-soliton solution as interacting two single solitons *Progr. Theor. Phys.* **71** 843–846

[8] Yoneyama T 1984 Interacting Korteweg-de Vries equations and attractive soliton interaction *Progr. Theor. Phys.* **72** 1081–1088

[9] Caenepeel S and Malfliet W 1985 Internal structure of the two-soliton of the KdV equation *Wave Motion* **7** 299–305

[10] Malfliet, W and Van der Velde, L 1985 A study of two interacting KdV-solitons *Lett. Nuovo Cim.* **42** 179–183

[11] Moloney T and Hodnett P 1986 Solution interactions (for the Korteweg-deVries equation): a new perspective *J. Phys. A: Math. Gen.* **19** L1129–L1135

[12] Hodnett P and Moloney T 1989 On the structure during interaction of the two-soliton solution of the KdV equation *SIAM J. Appl. Math.* **49** 1174–1187

[13] Moloney T and Hodnett P 1991 Note on the structure of a three-soliton solution of the Korteweg-Devries equation *SIAM J. Appl. Math.* **51** 940–947

[14] Fuchssteiner B 1987 Solitons in interaction *Progr. Theor. Phys.* **78** 1022–1050

[15] LeVeque R 1987 On the interaction of nearly equal solitons in the KdV equation *SIAM J. Appl. Math.* **47** 254–262

[16] Bryan A and Stuart A 1992 On the dynamics of soliton interactions for the Korteweg-de Vries equation *Chaos Solitons & Fractals* **2** 487–491

[17] Campbell F and Parkes J 1997 The internal structure of the two-soliton solution to nonlinear evolution equations of a certain class *Solitons and Coherent Structures in Physics and Biology, SOLPHYS '97 Proceedings* ed Christiansen P and Sørensen M (Lyngby, Denmark: Technical University of Denmark)

[18] Nguyễn H 2003 Decay of KdV solitons *SIAM J. Appl. Math.* **63** 874–888

[19] Nguyễn H 2004 Soliton collisions and ghost particle radiation *J. Nonl. Math. Phys.* **11** 180–198

[20] Benes N, Kasman A and Young K 2006 On decompositions of the KdV 2-soliton *J. Nonlinear Sci.* **16** 179–200

[21] Bona J and Weissler F 2009 Pole dynamics of interacting solitons and blowup of complex-valued solutions of KdV *Nonlinearity* **22** 311–349

[22] Hirota R 2004 *The Direct Method in Soliton Theory* (Cambridge Tracts in Mathematics vol 155) (Cambridge: Cambridge University Press)

[23] Litvinov G 2010 Tropical mathematics, idempotent analysis, classical mechanics and geometry *Preprint* arXiv:1005.1247

[24] Holden H and Risebro N 2011 *Front Tracking for Hyperbolic Conservation Laws* (Berlin: Springer)

[25] Kodama Y and Williams L 2011 KP solitons and total positivity for the Grassmannian *Preprint* arXiv:1106.0023

About linear superpositions of special exact solutions of Nizhnik-Veselov-Novikov equation

V. G. Dubrovsky[1] and A. V. Topovsky[2]

Novosibirsk State Technical University, Karl Marx prosp. 20, Novosibirsk 630092, Russia

E-mail: [1]dubrovsky@ngs.ru, [2]topovsky.av@gmail.com

Abstract. General scheme for calculations via Zakharov and Manakov $\bar{\partial}$-dressing method of exact solutions, nonstationary and stationary, of Nizhnik-Veselov-Novikov (NVN) equation in the forms of simple nonlinear and linear superpositions of arbitrary number N of exact special solutions $u^{(n)}, n = 1, \ldots, N$ is presented. Simple nonlinear superpositions are given up to a constant by the sums of solutions $u^{(n)}$ and calculated by $\bar{\partial}$-dressing of the first auxiliary linear problem with nonzero asymptotic values of potential at infinity. It is remarkable that in the zero limit of asymptotic values of potential simple nonlinear superpositions convert in to linear ones in the form of the sums of special solutions $u^{(n)}$. It is shown that the sums $u = u^{(k_1)} + \ldots + u^{(k_m)}, 1 \le k_1 < k_2 < \ldots < k_m \le N$ of arbitrary subsets of these N solutions are also exact solutions of NVN equation. The obtained results are illustrated in detail by hyperbolic version of NVN equation, i. e. by NVN-II equation. The presented exact solutions include as superpositions of special line solitons and also superpositions of plane wave type singular periodic solutions. PACS numbers: 02.30.Ik, 02.30.Jr, 02.30.Zz, 05.45.Yv

1. Introduction

This paper is dedicated to the memory of famous russian scientist **Sergey Manakov**. The discovery of $\bar{\partial}$-dressing method which we use here for constructions of exact solutions of NVN equation is one of the most great achievements of Vladimir Zakharov and **Sergey Manakov** in developing of IST.

Among different (2+1)-dimensional integrable nonlinear equations [1, 2, 3, 4, 5, 6, 7, 8, 9] the famous Nizhnik-Veselov-Novikov (NVN) equation is well known and has long history. This equation discovered in the papers of Nizhnik (1980) [10] and Veselov, Novikov (1984) [11] has the following form:

$$u_t + \kappa_1 u_{\xi\xi\xi} + \kappa_2 u_{\eta\eta\eta} + 3\kappa_1(u\partial_\eta^{-1}u_\xi)_\xi + 3\kappa_2(u\partial_\xi^{-1}u_\eta)_\eta = 0 \qquad (1)$$

where $u(\xi, \eta, t)$ is scalar function, κ_1, κ_2 are some constants; $\xi = x + \sigma y$, $\eta = x - \sigma y$, $\sigma^2 = \pm 1$; ∂_ξ^{-1} and ∂_η^{-1} are operators inverse to ∂_ξ and ∂_η, $\partial_\eta^{-1}\partial_\eta = \partial_\xi^{-1}\partial_\xi = 1$. In elliptic case $\sigma = i$ and for constants $\kappa_1 = \overline{\kappa}_2 = \kappa$ equation (1) is known as NVN-I equation or simply as Veselov-Novikov (VN) equation; in hyperbolic case $\sigma = 1$ and for real constants κ_1, κ_2 equation (1) is known as NVN-II equation; below, in calculations valid for both types of equations, we will use common notation - NVN-equation.

NVN equation (1) can be represented as compatibility condition in the form of **Manakov's** triad [12]:

$$[L_1, L_2] = BL_1, \quad B = 3(\kappa_1\partial_\eta^{-1}u_{\xi\xi} + \kappa_2\partial_\xi^{-1}u_{\eta\eta}) \qquad (2)$$

of two linear auxiliary problems

$$L_1\psi = (\partial_{\xi\eta}^2 + u)\psi = 0, \qquad (3)$$

$$L_2\psi = (\partial_t + \kappa_1\partial_\xi^3 + \kappa_2\partial_\eta^3 + 3\kappa_1(\partial_\eta^{-1}u_\xi)\partial_\xi + 3\kappa_2(\partial_\xi^{-1}u_\eta)\partial_\eta)\psi = 0. \tag{4}$$

The invention by **Manakov** [12] such kind of triad representation was the prominent step for exact integration of some (2+1)-dimensional integrable equations.

Several classes of exact solutions of NVN equation (1) have been constructed in last three decades (1980 – 2010) via different methods [10, 11, 13, 14, 15, 18, 16, 17, 19, 20, 21, 22, 23, 24], see also the books [3, 4]. These solutions include finite-zone type solutions (Veselov, Novikov (1984)) [11]; rationally localized solutions or lumps (Grinevich, **Manakov** (1986, 1988)) [13, 14], (Grinevich, Novikov R. (1988)) [15], (Grinevich (1999,2000)) [16, 17], (Dubrovsky, Formusatik, (2001,2003)) [20, 21]; particular examples of solutions with functional parameters (Nizhnik (1980) [10], Matveev and Salle (1991) [18], Dubrovsky, Topovsky and Basalaev (2010,2011) [22, 23]); multi-line soliton solutions (Dubrovsky, Topovsky and Basalaev (2010,2011)) [22, 24] and so on.

Recently in the paper [23] general class of exact solutions with functional parameters and constant asymptotic values $-\epsilon$ at infinity

$$u(\xi,\eta,t) = \tilde{u}(\xi,\eta,t) - \epsilon, \quad \tilde{u}(\xi,\eta,t)|_{\xi^2+\eta^2\to\infty} \to 0 \tag{5}$$

of NVN equation (1) via $\overline{\partial}$-dressing method of Zakharov and **Manakov** [6, 7, 8, 9] has been calculated and subclass of multi-line soliton solutions has been presented [23].

In another paper [24] (see also [22]) it was established that for some special solutions $u^{(1)}$ and $u^{(2)}$ with zero asymptotic values $-\epsilon = 0$ at infinity, for special linear (plane) solitons or for special plane wave type singular periodic solutions, their sum $u^{(1)} + u^{(2)}$ is also exact solution of NVN equation.

In present paper this result [22, 24] is generalized to the case of linear superpositions of arbitrary number N of special line solitons (or special plane wave type singular periodic solutions) $u^{(n)}$, $n = 1, \ldots, N$ in such a way, that the sums of arbitrary subsets of these solutions

$$u = u^{(k_1)} + \ldots + u^{(k_m)}, \quad 1 \leqslant k_1 < k_2 < \ldots < k_m \leqslant N \tag{6}$$

are also exact solutions of NVN equation (1).

For convenience here some useful formulas of $\overline{\partial}$-dressing method for NVN equation (1) [4, 20, 21, 23, 22, 24] are presented. Central object of this method is the scalar wave function χ

$$\chi(\lambda;\xi,\eta,t) = e^{-F(\lambda;\xi,\eta,t)}\psi(\xi,\eta,t), \quad F(\lambda;\xi,\eta,t) = i[\lambda\xi - \frac{\epsilon}{\lambda}\eta + (\kappa_1\lambda^3 - \kappa_2\frac{\epsilon^3}{\lambda^3})t] \tag{7}$$

which satisfies to corresponding $\overline{\partial}$-problem or equivalently to following singular integral equation:

$$\chi(\lambda) = 1 + \iint\limits_C \frac{d\lambda' \wedge d\overline{\lambda'}}{2\pi i(\lambda' - \lambda)} \iint\limits_C \chi(\mu,\overline{\mu})R(\mu,\overline{\mu};\lambda',\overline{\lambda'})d\mu \wedge d\overline{\mu}, \tag{8}$$

here canonical normalization $\chi \to \chi_\infty = 1$, as $\lambda \to \infty$, of wave function is assumed and the kernel R is given by the formula [4, 23, 22, 24]

$$R(\mu,\overline{\mu};\lambda,\bar{\lambda};\xi,\eta,t) = R_0(\mu,\overline{\mu};\lambda,\bar{\lambda})e^{F(\mu;\xi,\eta,t)-F(\lambda;\xi,\eta,t)}. \tag{9}$$

Solutions $u(\xi,\eta,t)$ of NVN equation are expressed via reconstruction formulas

$$u = -\epsilon - i\chi_{-1\eta} = -\epsilon + i\epsilon\chi_{1\xi} \tag{10}$$

through the coefficients χ_{-1} and/or χ_1 of Taylor's series expansions: $\chi = \chi_0 + \chi_1\lambda + \chi_2\lambda^2 + \ldots$ and $\chi = \chi_\infty + \frac{\chi_{-1}}{\lambda} + \frac{\chi_{-2}}{\lambda^2} + \ldots$ in the neighborhoods of points $\lambda = 0$ and $\lambda = \infty$ of complex plane $\mathbb{C}$.

In constructing of exact solutions u of NVN equation (1) two conditions must be satisfied [4, 20, 21, 23, 22, 24]: the condition of potentiality of operator L_1, or the absence in the first auxiliary linear problem (3) of the terms with first derivatives $u_1\partial_\xi$ and $u_2\partial_\eta$, and the condition of reality $\bar{u} = u$ of solutions.

The potentiality condition of operator L_1 in (3), or equivalently in terms of wave function χ, the condition $\chi_0 = 1$, imposes severe restrictions on the kernel R_0 of $\overline{\partial}$-problem. These restrictions were successfully satisfied for broad classes of exact solutions of NVN equation (1), such as lumps [20, 21], solutions with functional parameters, multi-line solitons and plane wave type singular periodic solutions [23, 22, 24].

The condition of reality of solutions $\bar{u} = u$ leads to another following restrictions on the kernel R_0 [23, 22, 24]:

$$R_0(\mu, \overline{\mu}; \lambda, \overline{\lambda}) = \frac{\epsilon^3}{|\mu|^2|\lambda|^2\overline{\mu}\lambda} \overline{R_0\left(-\frac{\epsilon}{\overline{\lambda}}, -\frac{\epsilon}{\lambda}, -\frac{\epsilon}{\overline{\mu}} - \frac{\epsilon}{\mu}\right)} \tag{11}$$

for NVN-I equation and

$$R_0(\mu, \overline{\mu}; \lambda, \overline{\lambda}) = \overline{R_0(-\overline{\mu}, -\mu; -\overline{\lambda}, -\lambda)} \tag{12}$$

for NVN-II equation; these restrictions are obtained in the limit of "weak" fields and lead to broad classes of exact solutions, such as lumps [20, 21], solutions with functional parameters [23] and multi-line soliton solutions [23, 22, 24].

But one can do not use the limit of weak fields and impose the reality condition $u = \overline{u}$ directly to calculated complex solutions of NVN equation satisfying only to potentiality condition. This approach makes it possible to receive besides multi line soliton solutions also plane wave type singular periodic solutions [22, 24].

2. Simple nonlinear superpositions of complex solutions of NVN equation

The choice of delta-functional kernel

$$R_0(\mu, \bar{\mu}; \lambda, \bar{\lambda}) = \pi \sum_n A_n \delta(\mu - \mathrm{M}_n)\delta(\lambda - \Lambda_n) \tag{13}$$

with constant coefficients A_n and discrete spectral parameters M_n, Λ_n leads to simple determinant formula [24, 22]

$$u = -\epsilon + \frac{\partial^2}{\partial\xi\partial\eta} \ln \det A, \quad A_{lk} = \delta_{lk} + \frac{2iA_k}{\mathrm{M}_l - \Lambda_k} e^{F(\mathrm{M}_l) - F(\Lambda_k)} \tag{14}$$

for exact multi-line soliton and plane wave type singular periodic solutions of NVN equation. The main problem in using this last formula is satisfaction to reality and potentiality conditions.

It was shown in [22, 24] that the choice of kernel R_0 in the form

$$R_0(\mu, \bar{\mu}; \lambda, \bar{\lambda}) = \pi \sum_{m=1}^{N} \left[a_m \lambda_m \delta(\mu - \mu_m)\delta(\lambda - \lambda_m) + a_m \mu_m \delta(\mu + \lambda_m)\delta(\lambda + \mu_m)\right] \tag{15}$$

of N paired terms with discrete spectral parameters (μ_m, λ_m) allows to satisfy the potentiality condition $\chi_0 = 1$.

In the simplest cases $N = 1, 2$ of one or two paired terms in (15) one obtains due to (14) the following expressions for $\det A$ [22, 24]:

$$N = 1 : \det A = \left(1 + s_1 e^{\Delta F(\mu_1, \lambda_1)}\right)^2, \tag{16}$$

$$N = 2 : \det A = \left(1 + s_1 e^{\Delta F(\mu_1, \lambda_1)} + s_n e^{\Delta F(\mu_n, \lambda_n)} + w e^{\Delta F(\mu_1, \lambda_1) + \Delta F(\mu_n, \lambda_n)}\right)^2. \tag{17}$$

Here, in (15) – (17) a_n, μ_n, λ_n $(n = 1, \ldots, N)$ are some complex constants; μ_n and λ_n also known as discrete spectral parameters which give spectral characterization for corresponding exact solutions. The quantities s_n, w and $\Delta F(\mu_n, \lambda_n)$ are given by the formulas:

$$s_n := i a_n \frac{\mu_n + \lambda_n}{\mu_n - \lambda_n}; \quad \Delta F(\mu_n, \lambda_n) := F(\mu_n) - F(\lambda_n), \tag{18}$$

$$w := -s_1 s_n \cdot \frac{(\lambda_1 - \lambda_n)(\lambda_n + \mu_1)(\mu_1 - \mu_n)(\lambda_1 + \mu_n)}{(\lambda_1 + \lambda_n)(\lambda_n - \mu_1)(\mu_1 + \mu_n)(\lambda_1 - \mu_n)}. \tag{19}$$

The expression for $\det A$ in the case $N = 2$ (17) is generated by two pairs of terms in (15) with discrete spectral variables (μ_1, λ_1) and (μ_n, λ_n).

The formula for generally complex solution corresponding to one arbitrary pair (μ_n, λ_n), $(n = 1, ..., N)$ of discrete spectral variables due to (14), (16) and (18) has the form:

$$u^{(n)}(\xi, \eta, t) = -\epsilon + \tilde{u}^{(n)}(\xi, \eta, t) = -\epsilon - \epsilon \frac{2s_n(\mu_n - \lambda_n)^2}{\mu_n \lambda_n} \frac{e^{\Delta F(\mu_n, \lambda_n)}}{(1 + s_n e^{\Delta F(\mu_n, \lambda_n)})^2}. \tag{20}$$

It is remarkable that for the case $N = 2$ of two paired terms in (15) with spectral variables $(\mu_1, \lambda_1), (\mu_n, \lambda_n), (n = 2, \ldots, N)$ the choice $w = s_1 s_n$ in (17) i. e.

$$\frac{(\lambda_1 - \lambda_n)(\lambda_n + \mu_1)(\mu_1 - \mu_n)(\lambda_1 + \mu_n)}{(\lambda_1 + \lambda_n)(\lambda_n - \mu_1)(\mu_1 + \mu_n)(\lambda_1 - \mu_n)} = -1, \tag{21}$$

which is equivalent to relation

$$(\lambda_1 - \mu_1)(\lambda_n - \mu_n)(\lambda_1 \mu_1 + \lambda_n \mu_n) = 0, \quad n \neq 1, \tag{22}$$

the expression for $\det A$ (17) greatly simplifies

$$\det A = \left(1 + s_1 e^{\Delta F(\mu_1, \lambda_1)}\right)^2 \left(1 + s_n e^{\Delta F(\mu_n, \lambda_n)}\right)^2 \tag{23}$$

and via (14) leads to very simple formula for complex solution of NVN equation

$$u(\xi, \eta, t) = -\epsilon - 2\epsilon \sum_{m=1,n} \frac{s_m(\mu_m - \lambda_m)^2}{\mu_m \lambda_m} \frac{e^{\Delta F(\mu_m, \lambda_m)}}{(1 + s_m e^{\Delta F(\mu_m, \lambda_m)})^2}. \tag{24}$$

This solution modulo ϵ is the sum $u = \epsilon + u^{(1)} + u^{(n)}$ of two solutions $u^{(1)}$ and $u^{(n)}$ (20).

The solutions $\mu_1 = \lambda_1$ and $\mu_n = \lambda_n$ of (22) correspond to lumps (rationally decreasing at infinity exact solutions u [20, 21]) and will not be considered here. Below it is assumed that $\mu_n \neq \lambda_n$ for all terms $n = 1, \ldots, N$ with discrete spectral variables μ_n, λ_n in (15). Under this requirement the relations (21), (22) reduce to more simple one:

$$\lambda_n \mu_n + \lambda_1 \mu_1 = 0, \quad n = 2, \ldots, N. \tag{25}$$

The relations (25) with arbitrary $n \neq 1$ guarantee that $u(\xi, \eta, t) = -\epsilon + \tilde{u}^{(1)} + \tilde{u}^{(n)}$, as nonlinear superpositions (24) of two solutions $u^{(1)}(\xi, \eta, t) = -\epsilon + \tilde{u}^{(1)}$ and $u^{(n)}(\xi, \eta, t) = -\epsilon + \tilde{u}^{(n)}$, $(n = 2, ..., N)$ of type (20), are exact solutions of NVN equation. This result can be extended to arbitrary number $N > 2$ of paired terms in the sum (15). More exactly the following lemma was proved.

Lemma. The nonlinear superposition

$$u(\xi, \eta, t) = -\epsilon + \tilde{u}^{(1)}(\xi, \eta, t) + \sum_{m=2}^{N} \tilde{u}^{(m)}(\xi, \eta) = -\epsilon - 2\epsilon \frac{s_1(\mu_1 - \lambda_1)^2}{\mu_1 \lambda_1} \frac{e^{\Delta F(\mu_1, \lambda_1)}}{(1 + s_1 e^{\Delta F(\mu_1, \lambda_1)})^2} -$$

$$-2\epsilon \sum_{m=2}^{N} \frac{s_m(\mu_m - \lambda_m)^2}{\mu_m \lambda_m} \frac{e^{\Delta F(\mu_m, \lambda_m)}}{(1 + s_m e^{\Delta F(\mu_m, \lambda_m)})^2} = (N - 1)\epsilon + u^{(1)}(\xi, \eta, t) + \sum_{m=2}^{N} u^{(m)}(\xi, \eta) \tag{26}$$

of arbitrary number $N > 2$ of complex solutions: nonstationary solution $u^{(1)}(\xi, \eta, t) = -\epsilon + \tilde{u}^{(1)}(\xi, \eta, t)$ of the type (20) with nonstationary phase

$$\varphi_1(\xi, \eta, t) := \Delta F(\mu_1, \lambda_1)\big|_{\kappa_1 \lambda_1^3 = \kappa_2 \frac{\epsilon^3}{\mu_1^3}} = i\left[(\mu_1 - \lambda_1)\xi - \left(\frac{\epsilon}{\mu_1} - \frac{\epsilon}{\lambda_1}\right)\eta - 2\left(\kappa_1 \lambda_1^3 - \kappa_2 \frac{\epsilon^3}{\lambda_1^3}\right)t\right], \tag{27}$$

and of $N-1$ stationary solutions $u^{(m)}(\xi,\eta) = -\epsilon + \tilde{u}^{(m)}(\xi,\eta)$, $m = 2,\ldots,N$ of the type (20) with stationary phases

$$\varphi_m(\xi,\eta) := \Delta F(\mu_m,\lambda_m)|_{\kappa_1\lambda_1^3=\kappa_2\frac{\epsilon^3}{\mu_1^3}} = i\left[(\mu_m-\lambda_m)\xi - \left(\frac{\epsilon}{\mu_m}-\frac{\epsilon}{\lambda_m}\right)\eta\right], \quad (m=2,\ldots,N) \quad (28)$$

is also exact solution of NVN equation, when conditions (25) ($\lambda_n\mu_n+\lambda_1\mu_1=0,\quad n=2,\ldots,N.$) on parameters λ_n, μ_n, λ_1, μ_1 and additional restriction

$$\kappa_1\lambda_1^3 - \kappa_2\frac{\epsilon^3}{\mu_1^3} = 0 \quad (29)$$

on the parameters λ_1 and μ_1 are imposed.

Proof. Note at first that according to definitions (7), (18) time-dependent part of phase φ_1 in (27) arises due to restriction (29) on parameters λ_1, μ_1; time-dependent parts of phases φ_m in (28) are absent due to both restrictions (25), (29) on parameters λ_n, μ_n, λ_1, μ_1.

The NVN equation (1) can be represented in the form:

$$\mathrm{N}(u) = \mathrm{L}(u) + \mathrm{Bil}(u,u) = 0 \quad (30)$$

of the sum of linear $\mathrm{L}(u)$ and nonlinear $\mathrm{N}(u)$ parts

$$\mathrm{L}(u) = u_t + \kappa_1 u_{\xi\xi\xi} + \kappa_2 u_{\eta\eta\eta};\quad \mathrm{Bil}(u,u) = 3\kappa_1(u\partial_\eta^{-1}u_\xi)_\xi + 3\kappa_2(u\partial_\xi^{-1}u_\eta)_\eta \quad (31)$$

with bilinear form

$$\mathrm{Bil}(u,v) = 3\kappa_1(u\partial_\eta^{-1}v_\xi)_\xi + 3\kappa_2(u\partial_\xi^{-1}v_\eta)_\eta. \quad (32)$$

The proof of the fact that (26) is solution of NVN equation can be done by direct substitution (26) into (30) and regrouping arising terms:

$$\mathrm{N}\left(-\epsilon + \sum_{k=1}^{N}\tilde{u}^{(k)}\right) = \sum_{k=1}^{N}\mathrm{N}(u^{(k)}) + \sum_{k\neq j}\mathrm{Bil}\left(\tilde{u}^{(k)},\tilde{u}^{(j)}\right). \quad (33)$$

Here due to (26) $u^{(1)}(\xi,\eta,t) = -\epsilon+\tilde{u}^{(1)}(\xi,\eta,t)$ and $u^{(k)}(\xi,\eta) = -\epsilon+\tilde{u}^{(k)}(\xi,\eta)$ are correspondingly nonstationary and stationary ($k=2,\ldots,N$) solutions of NVN equation with phases (27) and (28).

Using formula (20), taking into account (27) and (28) and the relations between derivatives of phases

$$\frac{\partial\varphi_k}{\partial\eta} = \frac{\epsilon}{\lambda_k\mu_k}\frac{\partial\varphi_k}{\partial\xi},\quad k=1,\ldots,N \quad (34)$$

one obtains due to (25)

$$\mathrm{Bil}(\tilde{u}^{(1)},\tilde{u}^{(k)}) = \frac{3\kappa_1}{\epsilon}(\lambda_k\mu_k+\lambda_1\mu_1)(\tilde{u}^{(1)}\tilde{u}^{(k)})_\xi + 3\kappa_2\epsilon\left(\frac{1}{\lambda_k\mu_k}+\frac{1}{\lambda_1\mu_1}\right)(\tilde{u}^{(1)}\tilde{u}^{(k)})_\eta = 0, \quad (35)$$

here $k=1,\ldots,N$. Quite analogously one calculates for $k\neq m$ ($k,m=2,\ldots,N$)

$$\mathrm{Bil}(\tilde{u}^{(k)},\tilde{u}^{(m)}) = 3\left(\frac{\kappa_2\epsilon^2}{\lambda_k^2\mu_k^2}+\frac{\kappa_2\epsilon^2}{\lambda_k\mu_k\lambda_m\mu_m}+\frac{\kappa_1\lambda_k\mu_k}{\epsilon}+\frac{\kappa_1\lambda_m\mu_m}{\epsilon}\right)(\tilde{u}^{(m)}\tilde{u}_\xi^{(k)}) +$$

$$3\left(\frac{\kappa_2\epsilon^2}{\lambda_m^2\mu_m^2}+\frac{\kappa_2\epsilon^2}{\lambda_k\mu_k\lambda_m\mu_m}+\frac{\kappa_1\lambda_k\mu_k}{\epsilon}+\frac{\kappa_1\lambda_m\mu_m}{\epsilon}\right)(\tilde{u}_\xi^{(m)}\tilde{u}^{(k)}) = 6\left(\frac{\kappa_2\epsilon^2}{\lambda_1^2\mu_1^2}-\frac{\kappa_1\lambda_1\mu_1}{\epsilon}\right)(\tilde{u}^{(k)}\tilde{u}^{(m)})_\xi = 0 \quad (36)$$

here in derivation of (35) and (36) the role of constraints (25) and (29) on parameters (λ_1,μ_1) and (λ_k,μ_k), $k=2,\ldots,N$ appears clearly.

Journal of Physics: Conference Series **482** (2014) 012011 doi:10.1088/1742-6596/482/1/012011

Due to (33), (35) and (36) one derives the required result of **Lemma** for nonlinear superposition (26).

Similar to the arguments used in proof of basic **Lemma** of present section one can prove also that every subsum of arbitrary terms $1 \leqslant i < i+1 < \ldots < j-1 < j \leqslant N$ of sum (26)

$$u = -\epsilon + \sum_{n=i}^{j} \tilde{u}^{(n)} \tag{37}$$

under conditions (25) and (29) is exact solution of NVN equation. Simple nonlinear superpositions given by (37) of complex solutions $u^{(n)}$ of NVN equation due to (26) and (27), (28) can be divided on two classes: the class of nonstationary solutions with $i \geqslant 1$ and class of stationary solutions with $i \geqslant 2$.

In the limit $u(\xi, \eta, t)|_{\xi^2+\eta^2 \to \infty} = -\epsilon \to 0$ of zero asymptotic values for potential of the first auxiliary problem (3) nonlinear superposition (26) converts to linear one, in the form of the sum of exact solutions. Underline that basic **Lemma** of present section is proved in general position and can be easily specialized, by corresponding choices of parameters, to calculations of real solutions for both cases of $\bar{\partial}$-dressing for nonzero and zero asymptotic values of potential.

In recent paper [25] the case of $\bar{\partial}$-dressing for zero asymptotic values for potential ($\bar{\partial}$-dressing on zero energy level) by special limiting procedure is treated in detail and corresponding exact real solutions of elliptic version of Nizhnik-Veselov-Novikov (NVN-I or Veselov-Novikov) equation in the form of sums of arbitrary number of simple line solitons and in the form of sums of simple plane wave type singular periodic solutions with specially chosen parameters are calculated.

In following sections 3 and 4 analogous results for hyperbolic version of Nizhnik-Veselov-Novikov (or for NVN-II equation) are presented and corresponding exact solutions are calculated. The constructed exact solutions of NVN-II equation are also the exact potentials of one-dimensional Klein-Gordon equation and/or exact potentials of perturbed string equation.

3. Nonlinear and linear superpositions of line soliton solutions for hyperbolic version of the Nizhnik-Veselov-Novikov equation

For construction of real multi-line solitons via (14) besides potentiality condition satisfied by the kernel R_0 of the type (15) the reality condition $u = \overline{u}$ for solutions u must be fulfilled. This can be done choosing appropriately complex constants a_n and complex discrete spectral parameters (μ_n, λ_n) in (15) – (37) by several ways [22, 24]. For example, by imposing reality condition $u = \bar{u}$ on complex solutions (20), (24), (26) and (37) with additional assumption of real phases $\Delta F(\mu_n, \lambda_n) = \overline{\Delta F(\mu_n, \lambda_n)}$ (18) we have calculated real multi-line soliton solutions.

It was shown in the papers [22, 24] that to such real solutions u leads the following choice of parameters

$$a_n = -\overline{a}_n := -ia_{n0}, \quad \mu_n = -\overline{\mu}_n := i\mu_{n0}, \quad \lambda_n = -\overline{\lambda}_n := i\lambda_{n0}, \quad n = 1, \ldots, N. \tag{38}$$

with real constants a_{n0}, μ_{n0}, λ_{n0}. Due to (38) and under additional assumption of positive values of real constants s_n given by (18)

$$s_n = a_{n0}\frac{\mu_{n0} + \lambda_{n0}}{\mu_{n0} - \lambda_{n0}} = e^{\phi_{0n}} > 0, \quad n = 1, \ldots, N, \tag{39}$$

the solution (20) corresponding to one arbitrary pair (μ_n, λ_n), $(n = 1, \ldots, N)$ of discrete spectral variables takes the form of real nonsingular one-line soliton solution:

$$u^{(n)}(x, y, t) = -\epsilon + \tilde{u}^{(n)}(x, y, t) = -\epsilon - \frac{\epsilon(\lambda_{n0} - \mu_{n0})^2}{2\lambda_{n0}\mu_{n0}} \frac{1}{\cosh^2 \frac{\varphi_n(x,y,t) + \phi_{0n}}{2}} \tag{40}$$

with real phases

$$\varphi_n(x, y, t) := \Delta F(\mu_n, \lambda_n) = (\lambda_{n0} - \mu_{n0})\sqrt{2 + \frac{2\epsilon^2}{\mu_{n0}^2 \lambda_{n0}^2}}(\vec{N}_n\vec{r} - V_n t) \tag{41}$$

where $\vec{r} = (x, y)$ and $\vec{N}_n$ unit vectors of normals to lines of constant values of phases $\tilde{\varphi}_n(x, y, t)$, V_n are corresponding velocities of one-line solitons

$$\vec{N}_n = \frac{\left(1 - \frac{\epsilon}{\mu_{n0}\lambda_{n0}}, 1 + \frac{\epsilon}{\mu_{n0}\lambda_{n0}}\right)}{\sqrt{2 + \frac{2\epsilon^2}{\mu_{n0}^2\lambda_{n0}^2}}}, \quad V_n = \frac{\left(\lambda_{n0}^3 - \mu_{n0}^3\right)\left(\kappa_1 - \kappa_2 \frac{\epsilon^3}{\lambda_{n0}^3\mu_{n0}^3}\right)}{(\lambda_{n0} - \mu_{n0})\sqrt{2 + \frac{2\epsilon^2}{\mu_{n0}^2\lambda_{n0}^2}}}, \quad n = 1, \dots, N. \tag{42}$$

For the cases of nonlinear superpositions (24) ($N = 2$), (26) and (37) ($N \geqslant 2$) of exact solutions of the type (40) the conditions (25) for discrete spectral parameters (μ_n, λ_n), $(n > 1)$ due to (38) lead to following relation

$$\mu_{10}\lambda_{10} + \mu_{n0}\lambda_{n0} = 0, \quad n = 2, \dots, N. \tag{43}$$

Nonsingular two-line soliton solution characterized by two pairs of discrete spectral variables (μ_1, λ_1) and (μ_2, λ_2) due to (24), (38), (39) and (43) takes the form

$$u(x, y, t) = -\epsilon + \sum_{n=1}^{2} \tilde{u}^{(n)}(x, y, t) = -\epsilon - \sum_{n=1}^{2} \frac{\epsilon(\lambda_{n0} - \mu_{n0})^2}{2\lambda_{n0}\mu_{n0}} \frac{1}{\cosh^2 \frac{\varphi_n(x,y,t)+\phi_{0n}}{2}}, \tag{44}$$

where $u^{(n)}(x, y, t) = -\epsilon + \tilde{u}^{(n)}(x, y, t)$, $(n = 1, 2)$ are one-line soliton solutions of the type (40), the phases $\varphi_n(x, y, t), (n = 1, 2)$ under conditions (43) are given by (41).

Due to expressions for vectors of normals (42) and condition (43) it is evident that solitons $u^{(1)}$ and $u^{(2)}$ of superposition (44) move in the plane (x, y) perpendicularly to each other.

One of two one-line solitons $u^{(1)}$ or $u^{(2)}$ (not both) of superposition (44) due to (42) and (43) can be "stopped", i. e. by special choice of spectral parameter λ_1 one can take ones of velocities $V_1 = 0$ or $V_2 = 0$ equal to zero. For example one can choose $V_2 = 0$, this achieves by the use of (42) and (43) for λ_1 satisfying to condition

$$\lambda_{10}\mu_{10} = -\epsilon \sqrt[3]{\frac{\kappa_2}{\kappa_1}}, \tag{45}$$

The results of second section for simple nonlinear superpositions of $N > 2$ complex solutions $u^{(n)}$, $(n = 2, \dots, N)$ can be easily specialized to the case of simple nonlinear superpositions of real one-line solitons (40), such superpositions are given by (26) with parameters a_n, (μ_n, λ_n) and s_n satisfying to (38), (39), the conditions (25) or (43) and (45) also must be fulfilled. Basic **Lemma** of previous section works fully also for real multi-line soliton solutions because basic relations (34), (35) and (36) of **Lemma** remain valid for special choice of parameters (μ_1, λ_1), (μ_n, λ_n) given by the formulas (39), (43) and (45) of present section. The condition (29) due to (38) leads to relation (45), so the simple nonlinear superposition (26) takes in considered case the following real form

$$u(x, y, t) = -\epsilon + \tilde{u}^{(1)}(x, y, t) + \sum_{n=2}^{N} \tilde{u}^{(n)}(x, y) = -\epsilon - \frac{\epsilon(\lambda_{10} - \mu_{10})^2}{2\lambda_{10}\mu_{10}} \frac{1}{\cosh^2 \frac{\varphi_1(x,y,t)+\phi_{01}}{2}} - \tag{46}$$
$$- \sum_{n=2}^{N} \frac{\epsilon(\lambda_{n0} - \mu_{n0})^2}{2\lambda_{n0}\mu_{n0}} \frac{1}{\cosh^2 \frac{\varphi_n(x,y)+\phi_{0n}}{2}},$$

with phases φ_n (27), (28) given due to (38), (39) and (43), (45) by expressions

$$\varphi_1(x, y, t) = (\lambda_{10} - \mu_{10})\sqrt{2 + \frac{2\epsilon^2}{\mu_{10}^2\lambda_{10}^2}}(\vec{N}_1\vec{r} - V_1 t), \quad \varphi_n(x, y) = (\lambda_{n0} - \mu_{n0})\sqrt{2 + \frac{2\epsilon^2}{\mu_{10}^2\lambda_{10}^2}}(\vec{N}_n\vec{r}), \tag{47}$$

In formulas (47) $n = 2, \dots, N$, $\vec{r} = (x, y)$; the unit vectors of normals $\vec{N}_1$, $\vec{N}_2$ and velocity V_1 are given by following formulas

$$\vec{N}_n = \frac{\left(1 - \frac{\epsilon}{\mu_{n0}\lambda_{n0}}, 1 + \frac{\epsilon}{\mu_{n0}\lambda_{n0}}\right)}{\sqrt{2 + \frac{2\epsilon^2}{\mu_{10}^2\lambda_{10}^2}}}, \quad V_1 = \frac{2\kappa_1\left(\lambda_{10}^3 - \mu_{10}^3\right)}{(\lambda_{10} - \mu_{10})\sqrt{2 + \frac{2\epsilon^2}{\mu_{10}^2\lambda_{10}^2}}}. \tag{48}$$

One-line soliton $u^{(1)}(x,y,t) = -\epsilon + \tilde{u}^{(1)}(x,y,t)$ of nonlinear superposition (46) due to (47) and (48) moves in the plane (x,y) perpendicularly to others stationary solitons $u^{(n)}(x,y) = -\epsilon + \tilde{u}^{(n)}(x,y)$, $(n = 2,\ldots,N)$ of this superposition with parallel lines of constant values of phases $\varphi_n(x,y)$. Evidently particular case of (44) with $V_2 = 0$ due to (45) coincides with two-line soliton (for $N = 2$) nonlinear superposition (46).

It was shown in the papers [24, 22] that the limiting procedure for calculation of exact solutions u of NVN equation with zero asymptotic values at infinity $u|_{\xi^2+\eta^2\to\infty} = -\epsilon \to 0$ can be defined by the following way: it is supposed at first that $\epsilon := c_n\mu_{n0}$ with some real constants c_n, and due to (43) one have

$$\mu_{n0} \to 0, \qquad \frac{\mu_{n0}}{\mu_{m0}} = \frac{c_m}{c_n} \neq 0, \quad \lambda_{n0} = -\frac{\lambda_{10}c_n}{c_1}. \tag{49}$$

It is assumed that under procedure (49) the relations (43) remain to be valid.

In the limit (49) two-line soliton solution (44) converts to linear superposition

$$u(x,y,t) = u^{(1)}_{\epsilon=0} + u^{(2)}_{\epsilon=0} = -\sum_{n=1}^{2} \frac{c_n\lambda_{n0}}{2\cosh^2\frac{\tilde{\varphi}_n(x,y,t)+\phi_{0n}}{2}}, \tag{50}$$

of two one-line solitons $u^{(1)}_{\epsilon=0}$ and $u^{(2)}_{\epsilon=0}$

$$u^{(n)}_{\epsilon=0}(x,y,t) = \frac{c_n\lambda_{n0}}{2\cosh^2\frac{\tilde{\varphi}_n(x,y,t)+\phi_{0n}}{2}}, \qquad n = 1,2, \tag{51}$$

with phases $\tilde{\varphi}_n(x,y,t)$ given due to (41) and (49) by formulas

$$\tilde{\varphi}_n(x,y,t) = \frac{\lambda_{n0}}{|\lambda_{10}|}\sqrt{2\lambda_{10}^2 + 2c_1^2}(\vec{N}_n\vec{r} - V_n t), \tag{52}$$

Here $\vec{r} = (x,y)$, $\vec{N}_n$ are unit vectors of normals to lines of constant values of phases $\tilde{\varphi}_n(x,y,t)$; V_n are corresponding velocities of one-line solitons

$$\vec{N}_n = \frac{|\lambda_{10}|\left(1 - \frac{c_n}{\lambda_{n0}}, 1 + \frac{c_n}{\lambda_{n0}}\right)}{\sqrt{2\lambda_{10}^2 + 2c_1^2}}, \qquad V_n = \frac{|\lambda_{10}|\left(\kappa_1\lambda_{n0}^3 - \kappa_2 c_n^3\right)}{\lambda_{n0}\sqrt{2\lambda_{10}^2 + 2c_1^2}}. \tag{53}$$

derived by the use of (42), (43) and (49). By special choice of spectral parameter λ_1 one of two one-line solitons $u^{(1)}_{\epsilon=0}$ or $u^{(2)}_{\epsilon=0}$ (not both) in linear superposition (50) can be "stopped".

Basic **Lemma** of previous section works fully also for considered limit because basic relations (34), (35) and (36) of **Lemma** remain valid for special choice of parameters (μ_1,λ_1), (μ_n,λ_n) given by the formulas (38), (39), (43) and (45) and the rules (49) of zero limit $\epsilon \to 0$. In the limit $\epsilon \to 0$ following to the rules (49), under requirement that relations (43) and (45) remain to be valid, we obtain from (46) linear superposition of N one-line solitons

$$u = u^{(1)}_{\epsilon=0}(x,y,t) + \sum_{n=2}^{N} u^{(n)}_{\epsilon=0}(x,y) = -\frac{c_1\lambda_{10}}{2\cosh^2\frac{\tilde{\varphi}_1(x,y,t)+\phi_{01}}{2}} - \sum_{n=2}^{N}\frac{c_n\lambda_{n0}}{2\cosh^2\frac{\tilde{\varphi}_n(x,y)+\phi_{0n}}{2}}, \tag{54}$$

here the phases $\tilde{\varphi}_n$ given by expressions

$$\tilde{\varphi}_1(x,y,t) = \frac{\lambda_{10}}{|\lambda_{10}|}\sqrt{2\lambda_{10}^2 + 2c_1^2}(\vec{N}_1\vec{r} - V_1 t), \quad \tilde{\varphi}_n(x,y) = \frac{\lambda_{n0}}{|\lambda_{10}|}\sqrt{2\lambda_{10}^2 + 2c_1^2}(\vec{N}_n\vec{r}). \tag{55}$$

are obtained from phases φ_n (47) in limit $\epsilon \to 0$ (49). The first one-line soliton $u^{(1)}_{\epsilon=0}(x,y,t)$ from linear superposition (54) moves with velocity $V_1 = \frac{2\kappa_1|\lambda_{10}|\lambda_{10}^2}{\sqrt{2\lambda_{10}^2+2c_1^2}}$ in the plane (x,y) perpendicularly to other $(N-1)$ stationary one-line solitons $u^{(n)}_{\epsilon=0}(x,y)$.

Following to ideas used in proof of basic **Lemma** of second section one can show that the subsums of arbitrary numbers of solitons $u^{(n)}_{\epsilon=0}$, $(n = 1,\ldots,N)$ from (54) are also solutions of NVN equation. So the set of such solutions can be divided in two subsets: subset of nonstationary linear superpositions (with the first moving line soliton $u^{(1)}_{\epsilon=0}(x,y,t)$ in the sum) of line solitons and subset of stationary linear superpositions (without moving line soliton $u^{(1)}_{\epsilon=0}(x,y,t)$ in the sum) of stationary line solitons.

Journal of Physics: Conference Series **482** (2014) 012011 doi:10.1088/1742-6596/482/1/012011

4. Nonlinear and linear superpositions of plane wave type periodic solutions for hyperbolic version of the Nizhnik-Veselov-Novikov equation

Multi-line soliton solutions are calculated in preceding section by imposing reality condition $u = \bar{u}$ on complex solutions (20), (24), (26) and (37) with additional assumption of real phases $\Delta F(\mu_n, \lambda_n) = \overline{\Delta F(\mu_n, \lambda_n)}$ (18). In contrast the application of reality condition $u = \bar{u}$ with assumption of pure imaginary phases $\Delta F(\mu_n, \lambda_n) = -\overline{\Delta F(\mu_n, \lambda_n)}$ (18) leads to plane wave type periodic solutions and their's superpositions.

Plane wave type solutions can be obtained by this way for example by the following choice of parameters a_n, (μ_n, λ_n) and s_n in (15) – (37) [24, 22]:

$$a_n = \left|\frac{\lambda_{n0} - \mu_{n0}}{\lambda_{n0} + \mu_{n0}}\right| e^{i \arg a_n}, \quad \mu_n = \overline{\mu}_n := \mu_{n0}, \quad \lambda_n = \overline{\lambda}_n := \lambda_{n0}, \quad s_n = -ie^{i \arg a_n} \mathrm{sign}\left(\frac{\lambda_n - \mu_n}{\lambda_n + \mu_n}\right) \tag{56}$$

where $n = 1, \ldots, N$. Simple plane wave type periodic solution corresponding to one arbitrary pair of spectral variables (μ_n, λ_n), $(n = 1, \ldots, N)$, due to (20) and (56) takes the form

$$u^{(n)}(x, y, t) = -\epsilon + \tilde{u}^{(n)}(x, y, t) = -\epsilon - \frac{\epsilon(\lambda_{n0} - \mu_{n0})^2}{2\lambda_{n0}\mu_{n0}} \frac{1}{\cos^2\left(\frac{\varphi_n(x,y,t) + \arg a_n}{2} \mp \frac{\pi}{4}\right)} \tag{57}$$

where sign "$-$" correspond to case of $\frac{\lambda_{n0}-\mu_{n0}}{\lambda_{n0}+\mu_{n0}} > 0$ and sign "+" – case of $\frac{\lambda_{n0}-\mu_{n0}}{\lambda_{n0}+\mu_{n0}} < 0$. The real phases $\varphi_n(x, y, t) := -i\Delta F(\mu_n, \lambda_n)$ in (57) due to (18) and (56) are given by expressions

$$\varphi_n(x, y, t) = (\lambda_{n0} - \mu_{n0})\sqrt{2 + \frac{2\epsilon^2}{\mu_{n0}^2 \lambda_{n0}^2}}(\vec{N}_n \vec{r} - V_n t), \quad n = 1, \ldots, N. \tag{58}$$

In (57), (58) $\vec{r} = (x, y)$; $\vec{N}_n$ are unit vectors of normals to lines of constant values of phases $\varphi_n(x, y, t)$ and velocities V_n of periodic solutions are given by expressions:

$$\vec{N}_n = \frac{\left(1 + \frac{\epsilon}{\mu_{n0}\lambda_{n0}}, 1 - \frac{\epsilon}{\mu_{n0}\lambda_{n0}}\right)}{\sqrt{2 + \frac{2\epsilon^2}{\mu_{n0}^2\lambda_{n0}^2}}}, \quad V_n = -\frac{\left(\lambda_{n0}^3 - \mu_{n0}^3\right)\left(\kappa_1 + \kappa_2 \frac{\epsilon^3}{\lambda_{n0}^3 \mu_{n0}^3}\right)}{(\lambda_{n0} - \mu_{n0})\sqrt{2 + \frac{2\epsilon^2}{\mu_{n0}^2\lambda_{n0}^2}}}, \quad n = 1, \ldots, N. \tag{59}$$

Using general formulas (24) and (26) we also construct nonlinear superpositions of simple wave type periodic solutions of the type (57). The conditions (25) for discrete spectral parameters (μ_n, λ_n), $(n > 1)$ in nonlinear superpositions (24) and (26) due to (56) in considered case lead to following conditions

$$\mu_{10}\lambda_{10} + \mu_{n0}\lambda_{n0} = 0, \quad n = 2, \ldots, N. \tag{60}$$

Nonlinear superposition (24) of two simple plane wave type periodic solutions of the type (57) due to (56) and (60) takes the form

$$u(x, y, t) = -\epsilon + \sum_{n=1}^{2} \tilde{u}^{(n)}(x, y, t) = -\epsilon - \sum_{n=1}^{2} \frac{\epsilon(\lambda_{n0} - \mu_{n0})^2}{2\lambda_{n0}\mu_{n0}} \frac{1}{\cos^2\left(\frac{\varphi_n(x,y,t) + \arg a_n}{2} \mp \frac{\pi}{4}\right)}, \tag{61}$$

where sign "$-$" correspond to case of $\frac{\lambda_{n0}-\mu_{n0}}{\lambda_{n0}+\mu_{n0}} > 0$ and sign "+" – case of $\frac{\lambda_{n0}-\mu_{n0}}{\lambda_{n0}+\mu_{n0}} < 0$. The phases $\varphi_n(x, y, t)$ in solution (61) are given by (61) under conditions (60). Due to expressions for vectors of normals (59) and to parametrization (60) it is evident that lines of constant values of phases $\varphi_n(x, y, t)$, $(n = 1, 2)$ for solutions $u^{(1)} = -\epsilon + \tilde{u}^{(1)}$ and $u^{(2)} = -\epsilon + \tilde{u}^{(2)}$ of superposition (61) move perpendicularly to each other.

One of two simple plane wave type periodic solutions $u^{(1)}$ or $u^{(2)}$ (not both) of superposition (61) due to (59) and (60) can be made stationary, i. e. by special choice of spectral parameter λ_1 one can take ones of velocities $V_1 = 0$ or $V_2 = 0$ equal to zero. For example one can choose $V_2 = 0$, this achieves due to (59) and (60) for λ_1 satisfying to condition

$$\lambda_{10}\mu_{10} = -\epsilon \sqrt[3]{\frac{\kappa_2}{\kappa_1}}, \tag{62}$$

The results of second section for simple nonlinear superpositions of $N > 2$ complex solutions $u^{(n)}$, $(n = 2, \ldots, N)$ can be easily specialized to the case of simple nonlinear superpositions of real plane wave type periodic solutions (57) with parameters a_n, (μ_n, λ_n) and s_n satisfying to (56), the conditions (25) or (60) and (62) also must be fulfilled. Basic **Lemma** of second section works fully also for real solutions of considered plane wave periodic type because basic relations (34), (35) and (36) of **Lemma** remain valid for special choice of parameters (μ_1, λ_1), (μ_n, λ_n) given by the formulas (56), (60) and (62) of present section. The condition (29) due to (56) transforms into (62), so the solution (26) takes the form

$$u(x,y,t) = -\epsilon - \frac{\epsilon(\lambda_{10}-\mu_{10})^2}{2\lambda_{10}\mu_{10}} \frac{1}{\cos^2\left(\frac{\varphi_1(x,y,t)+\arg a_1}{2} \mp \frac{\pi}{4}\right)} - \sum_{n=2}^{N} \frac{\epsilon(\lambda_{n0}-\mu_{n0})^2}{2\lambda_{n0}\mu_{n0}} \frac{1}{\cos^2\left(\frac{\varphi_n(x,y)+\arg a_n}{2} \mp \frac{\pi}{4}\right)}, \tag{63}$$

where sign "$-$" correspond to case of $\frac{\lambda_{n0}-\mu_{n0}}{\lambda_{n0}+\mu_{n0}} > 0$ and sign "$+$" – case of $\frac{\lambda_{n0}-\mu_{n0}}{\lambda_{n0}+\mu_{n0}} < 0$. The phases φ_n (27), (28) in superposition (63) due to (56) and (60), (62) are given by expressions

$$\varphi_1(x,y,t) = (\lambda_{10}-\mu_{10})\sqrt{2+\frac{2\epsilon^2}{\mu_{10}^2\lambda_{10}^2}}(\vec{N}_1\vec{r} - V_1 t), \quad \varphi_n(x,y) = (\lambda_{n0}-\mu_{n0})\sqrt{2+\frac{2\epsilon^2}{\mu_{10}^2\lambda_{10}^2}}(\vec{N}_n\vec{r}), \tag{64}$$

where $n = 2, \ldots, N$, $\vec{r} = (x, y)$; unit vectors $\vec{N}_n$ and velocity V_1 are given by following formulas

$$\vec{N}_n = \frac{\left(1+\frac{\epsilon}{\mu_{n0}\lambda_{n0}}, 1-\frac{\epsilon}{\mu_{n0}\lambda_{n0}}\right)}{\sqrt{2+\frac{2\epsilon^2}{\mu_{10}^2\lambda_{10}^2}}}, \quad V_1 = \frac{-2\kappa_1\left(\lambda_{10}^3-\mu_{10}^3\right)}{(\lambda_{10}-\mu_{10})\sqrt{2+\frac{2\epsilon^2}{\mu_{10}^2\lambda_{10}^2}}}. \tag{65}$$

The lines of constant values of phase $\varphi_1(x,y,t)$ of simple periodic solution $u^{(1)}(x,y,t) = -\epsilon + \tilde{u}^{(1)}(x,y,t)$ of superposition (63) move in plane (x, y) perpendicularly to parallel lines of constant phases $\varphi_n(x,y)$ of others stationary periodic solutions $u^{(n)}(x,y,t) = -\epsilon + \tilde{u}^{(n)}(x,y,t)$, $(n = 2, \ldots, N)$ of this superposition. Evidently particular case of (61) with $V_2 = 0$ coincides due to (62) with the case $N = 2$ of linear superposition (63).

It was shown in the papers [24, 22] that the limiting procedure of calculation of exact solutions u of NVN equation with zero values of parameter $\epsilon = 0$ ($\bar{\partial}$-dressing for solutions with zero asymptotic values at infinity) can be defined by the following way

$$\epsilon = c_n\mu_{n0} \to 0, \quad \frac{\mu_{n0}}{\mu_{m0}} = \frac{c_m}{c_n} \neq 0, \quad \lambda_{n0} = -\frac{\lambda_{10}c_n}{c_1}. \tag{66}$$

Such procedure is applicable also in considered case of plane wave type solutions and their's superpositions. It is assumed that under procedure (66) the relations (60) remain to be valid.

In the limit (66) nonlinear superposition (61) of two plane wave type periodic solutions converts to linear superposition

$$u(x,y,t) = u^{(1)}_{\epsilon=0} + u^{(2)}_{\epsilon=0} = -\sum_{n=1}^{2} \frac{c_n\lambda_{n0}}{2\cos^2\left(\frac{\tilde{\varphi}_n(x,y,t)+\arg a_n}{2} - \frac{\pi}{4}\right)}, \tag{67}$$

of two periodic solitons $u^{(1)}_{\epsilon=0}$ and $u^{(2)}_{\epsilon=0}$

$$u^{(n)}_{\epsilon=0}(x,y,t) = -\frac{c_n\lambda_{n0}}{2\cos^2\left(\frac{\tilde{\varphi}_n(x,y,t)+\arg a_n}{2} - \frac{\pi}{4}\right)}, \quad n = 1, 2, \tag{68}$$

the phases $\tilde{\varphi}_n(x,y,t)$ in (67), (68) are given due to (58) and (66) by formulas

$$\varphi_n(x,y,t) = \frac{\lambda_{n0}}{|\lambda_{10}|}\sqrt{2\lambda_{10}^2 + 2c_1^2}(\vec{N}_n\vec{r} - V_n t), \tag{69}$$

here $\vec{r} = (x, y)$. Unit vectors of normals to lines of constant values of phases $\tilde{\varphi}_n(x, y, t)$ – $\vec{N}_n$ and velocities V_n of simple plane wave type periodic solutions (68) are derived by the use of (59), (60) and (66) and are given by formulas

$$\vec{N}_n = \frac{|\lambda_{10}|\left(1 + \frac{c_n}{\lambda_{n0}}, 1 - \frac{c_n}{\lambda_{n0}}\right)}{\sqrt{2\lambda_{10}^2 + 2c_1^2}}, \quad V_n = -\frac{|\lambda_{10}|\left(\kappa_1\lambda_{n0}^3 + \kappa_2 c_n^3\right)}{\lambda_{n0}\sqrt{2\lambda_{10}^2 + 2c_1^2}}. \tag{70}$$

By special choice of spectral parameter λ_1 one of two of these periodic solutions, $u^{(1)}_{\epsilon=0}$ or $u^{(2)}_{\epsilon=0}$ (not both), in linear superposition (67) can be made stationary.

Basic **Lemma** of second section works fully also for limit of zero ϵ of considered in present section real plane wave periodic solutions because basic relations (34), (35) and (36) of **Lemma** remain valid for special choice of parameters (μ_1, λ_1), (μ_n, λ_n) given by the formulas (56), (60), (62) and the rules (66) of zero limit $\epsilon \to 0$. In the limit $\epsilon \to 0$ ($\bar{\partial}$-dressing for solutions with zero asymptotic values at infinity) following to the rules (66), with assumption that relations (60) and (62) remain to be valid, we obtain from (63) linear superposition of N simple plane wave type periodic solutions

$$u = u^{(1)}_{\epsilon=0}(x, y, t) + \sum_{n=2}^{N} u^{(n)}_{\epsilon=0}(x, y) = -\frac{c_1\lambda_{10}}{2\cos^2\left(\frac{\tilde{\varphi}_1(x,y,t) + \arg a_1}{2} - \frac{\pi}{4}\right)} - \sum_{n=2}^{N} \frac{c_n\lambda_{n0}}{2\cos^2\left(\frac{\tilde{\varphi}_n(x,y) + \arg a_n}{2} - \frac{\pi}{4}\right)}, \tag{71}$$

here phases $\tilde{\varphi}_n$ are obtained from phases φ_n (64) by setting $\epsilon = 0$ in accordance with (66), these phases have the forms:

$$\tilde{\varphi}_1(x, y, t) = \frac{\lambda_{10}}{|\lambda_{10}|}\sqrt{2\lambda_{10}^2 + 2c_1^2}(\vec{N}_1\vec{r} - V_1 t), \quad \tilde{\varphi}_n(x, y) = \frac{\lambda_{n0}}{|\lambda_{10}|}\sqrt{2\lambda_{10}^2 + 2c_1^2}(\vec{N}_n\vec{r}), \quad n = 2, \ldots, N. \tag{72}$$

The lines of constant values of phase $\varphi_1(x, y, t)$ of the first periodic solution $u^{(1)}_{\epsilon=0}$ from linear superposition (71) move with velocity $V_1 = -\frac{2\kappa_1|\lambda_{10}|^3}{\sqrt{2\lambda_{10}^2 + 2c_1^2}}$ perpendicularly to lines of constant phases $\varphi_n(x, y)$ of $(N - 1)$ other stationary periodic solutions $u^{(n)}_{\epsilon=0}$, $(n = 2, \ldots, N)$ of this superposition.

Following to ideas used in proof of basic **Lemma** of second section one can show that the subsums of arbitrary numbers of solutions $u^{(n)}_{\epsilon=0}$, $(n = 1, \ldots, N)$ from (71) are also solutions of NVN-II equation. So the set of constructed in present section solutions can be divided in two subsets: subset of nonstationary linear superpositions (with the first moving line soliton $u^{(1)}_{\epsilon=0}(x, y, t)$ in the sum) and subset of stationary linear superpositions (without moving line soliton $u^{(1)}_{\epsilon=0}(x, y, t)$ in the sum).

Acknowledgments
This research work is supported by scientific Grant of fundamental researches of Novosibirsk State Technical University, in the framework of Program of Strategic Development of NSTU.

References

[1] Novikov S P, Zakharov V E, Manakov S V and Pitaevsky L V 1984 *Soliton Theory: the inverse scattering method* (New York: Plenum Press)
[2] Ablowitz M J and Clarkson P A 1991 *Solitons, nonlinear evolution equations and inverse scattering. London Mathematical Society Lecture Notes Series* vol 149 (Cambridge: Cambridge Univ. Press.)
[3] Konopelchenko B G 1992 *Introduction to multidimensional integrable equations: the inverse spectral transform in 2+1-dimensions* (New York - London: Plenum Press)
[4] Konopelchenko B G 1993 *Solitons in Multidimensions: Inverse Spectral Transform Method* (Singapore: World Scientific)
[5] Manakov S V 1981 The inverse scattering transform for the time-dependent Schrödinger equation and Kadomtsev-Petviashvili equation *Physica D* **3** 420-427
[6] Zakharov V E and Manakov S V 1985 Construction of higher-dimensional nonlinear integrable systems and of their solutions *Funct. Anal. Pril.* **19(2)** 89-101

[7] Zakharov V E 1988 *Commutating operators and nonlocal $\bar{\partial}$-problem Proc. of Int. Workshop Plasma Theory and Nonlinear and Turbulent Processes in Physics* vol I ed V G Bar'yakhtar and V M Chernousenko et al (Kiev: Naukova Dumka) p 152

[8] Bogdanov L V and Manakov S V 1988 The nonlocal $\bar{\partial}$-problem and (2+1)-dimensional soliton equations *J.Phys.A Math.Gen.* **21** L537-L544

[9] Fokas A S and Zakharov V E 1992 The dressing method for nonlocal Riemann-Hilbert problems *Journal of Nonlinear Sciences* **2(1)** 109-134

[10] Nizhnik L P 1980 Integration of multidimensional nonlinear equations by the inverse problem method *Soviet Physics Doklady* **25** 706

[11] Veselov A P and Novikov S P 1984 Finite-zone, two-dimensional, potential Schrödinger operators. Explicit formulas and evolution equations *Sov. Math., Dokl* **30** 588-591

[12] Manakov S V 1976 The method of the inverse scattering problem and two-dimensional evolution equations *Uspehi Mat. Nauk (in Russian)* **31:5(191)** 245246

[13] Grinevich P G and Manakov S V 1986 Inverse scattering problem for the two-dimensional Schrödinger operator, the $\bar{\partial}$-method and nonlinear equations *Funct. Anal. Appl.* **20(2)** 94-103

[14] Grinevich P G and Manakov S V 1988 *Inverse scattering problem for the two-dimensional Schrödinger operator at fixed negative energy and generalized analytic functions Proc. of Int. Workshop Plasma Theory and Nonlinear and Turbulent Processes in Physics (April 1987, Kyiv)* vol 1, ed V G Bar'yakhtar and V.M.Chernousenko et al (Singapore: World Scietific) pp 58-85

[15] Grinevich P G and Novikov S P 1988 Two-dimensional "inverse scattering problem" for negative energies and generalized-analytic functions. I. Energies lower than ground state *Funct. Anal. Appl.* **22(1)** 19-27

[16] Grinevich P G 1999 *Transformation of Dispersion for the Two-Dimensional Schrödinger Operator for one Energy and Connected with it Integrable Equations of Mathematical Physics, Doctorate Thesis* Chernogolovka (in Russian)

[17] Grinevich P G 2000 Scattering tranformation at fixed non-zero energy for the two-dimensional Schrödinger operator with potential decaying at infinity *Russian Math. Surveys* **55(6)** 1015-1083

[18] Matveev V B and Salle M A 1991 *Darbu Transformations and Solitons, in Springer series in Nonlinear Dynamics* (Berlin, Heidelberg: Springer)

[19] Athorne C and Nimmo J J C 1991 On the Moutard transformation for the integrable partial differential equations *Inverse Problems* **7** 809-826

[20] Dubrovsky V G and Formusatik I B 2001 The construction of exact rational solutions with constant asymptotic values at infinity of two-dimensional VN integrable nonlinear evolution equations via $\bar{\partial}$-dressing method *J.Phys.A.: Math. and Gen.* **34A** 1837-1851

[21] Dubrovsky V G and Formusatik I B 2003 New lumps of Veselov-Novikov equation and new exact rational potentials of two-dimensional Schrödinger equation via $\bar{\partial}$-dressing method *Phys. Lett.* **313(1-2)** 68-76

[22] Dubrovsky V G, Topovsky A V and Basalaev M Yu 2010 New exact solutions with constant asymptotic values at infinity of the NVN integrable nonlinear evolution equation via $\bar{\partial}$-dressing method, *http://arxiv.org/abs/0912.2155v2* pp. 143.

[23] Dubrovsky V G, Topovsky A V and Basalaev M Yu 2010 New exact solutions with functional parameters of the Nizhnik-Veselov-Novikov equation with constant asymptotic values at infinity *Theoret. and Math. Phys.* **165(2)** 1470 - 1489

[24] Dubrovsky V G, Topovsky A V and Basalaev M Yu 2011 New exact solutions of two-dimensional integrable equations using the $\bar{\partial}$-dressing method *Theoret. and Math. Phys.* **167(3)** 725 - 739

[25] Dubrovsky V G, Topovsky A V 2013 About simple nonlinear and linear superpositions of special exact solutions of Veselov-Novikov equation *Journal of Math. Phys.* **54(3)** 033509-033509-13.

Nonlinear interaction of the concentrated waves

V Eremenko [1] **and N Manaenkova** [2]

[1] Leading Research Scientist, Institute of Terrestrial Magnetism, Ionosphere and Radio Wave Propagation Russian Academy of Science (IZMIRAN), Moscow, Troitsk, 142190, Russia

[2] Senior Research Scientist, Institute of Terrestrial Magnetism, Ionosphere and Radio Wave Propagation Russian Academy of Science (IZMIRAN), Moscow, Troitsk, 142190, Russia

eremenko@izmiran.ru, nat_man@mail.ru

Abstract. The nonlinear problem of the wave propagation is considered. In addition to Kerr nonlinearity the question of the existence of concentrated solutions is analyzed for the threshold and saturable nonlinearity. It is shown that both in the case of threshold nonlinearity, and in the case of saturable nonlinearity solitary waves - concentrated solutions of the corresponding wave equations exist. For the nonlocal nonlinearity, it is taken into account that the diffusion process transforms the interaction of the electromagnetic field with the environment. This phenomenon is described by the system of differential equations including the equation for the perturbation of the dielectric permittivity. The mathematical problem is reduced to the eigenvalue problem for nonlinear integro-differential equation of Hartree type. The computational procedure is constructed.

1. Introduction

Modern radio communications allow so high increasing of radiated signal intensity that it generates modification of medium of propagation and the problem of determining the wave field becomes nonlinear. Theoretical estimates of the effects of ionospheric plasma heating by powerful radiation began to appear long time ago [1, 2]. Experimental confirmation of the interaction of powerful shortwave radiation with ionospheric plasma at oblique propagation [3, 4] made more active the development of the theory of this interaction [5, 6].

As a rule, the theoretical researches of the nonlinear propagation of powerful wave beams are restricted by models of local nonlinearity. The most commonly used model is the Kerr nonlinearity, in which the nonlinear perturbation of the dielectric permittivity is proportional to the second degree of the modulus of the wave amplitude. This approach allowed us to describe the basic phenomena arising at the nonlinear interaction of radiation with the environment. However such approach has obvious restrictions. Actually, there is no nonlinear effects when the intensity of the wave field is not enough powerful. As soon as the wave amplitude exceeds certain threshold value, so called "disruption" of the environment takes place and the nonlinear dependence of the dielectric permittivity upon the wave field amplitude arises.

Moreover the nonlinear dependence becomes more complex with further growth of the intensity. The upper limit of growth of the dielectric permittivity, namely saturation, is observed for most of the real materials. In this case it is reasonable to use a model medium with saturable nonlinearity.

However, this description of the nonlinear interaction of radiation with the medium is local. This approach is possible only in the case of negligible heat conductivity - when the size of the wave beam is much more than the characteristic scale of the thermal diffusion. Otherwise, you must take into account the spreading of the electron density perturbation from the area of the nonlinear heating. Nonlinear effect becomes nonlocal and, in this case, it is necessary to consider the solution of the diffusion equation for the dielectric permittivity and the solution of the wave equation jointly, as a system.

2. Analysis

Let us consider a typical picture of the ray paths in the near-Earth space (Fig. 1). In the area of rays focusing, signal intensity is increased and, consequently, existence of the nonlinear interaction "wave – ionosphere" is possible.

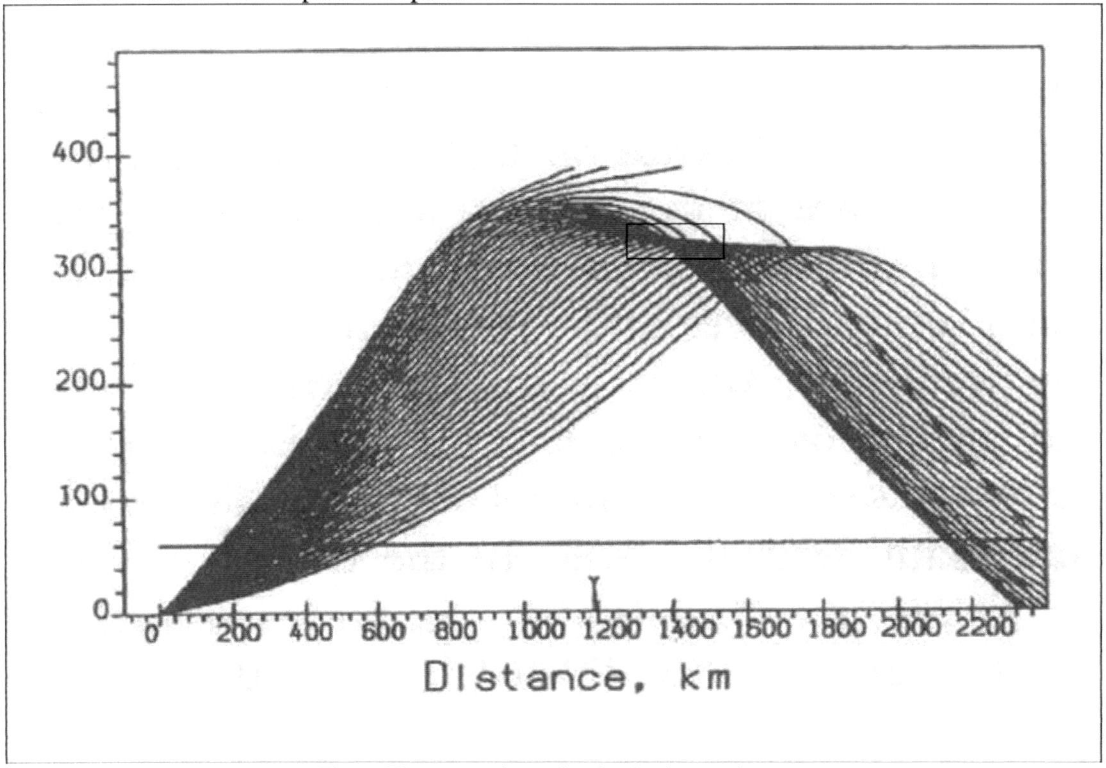

Fig.1. The ray trajectories in the ionosphere.

In order to describe the wave field in this small area we use the Helmholtz equation for the wave amplitude V.

$$\Delta V + k^2 \cdot \varepsilon \cdot V = 0\ ,$$

where k is the wave number and ε is the dielectric permittivity.

For the small wave intensity the dielectric permittivity depends only on the coordinates. Therefore, in order to determine the distribution of the wave field in the space, it is enough to solve the linear problem with the appropriate boundary conditions. But, for the high wave intensity, the dielectric permittivity becomes dependent on the wave amplitude, and it is necessary to solve the nonlinear problem for the description of the wave propagation.

We will consider the propagation of the narrow shortwave beams. Thus we will construct the Helmholtz equation solution concentrated in the small vicinity of the ray trajectory. In this vicinity, we introduce the orthogonal coordinate system: ξ is the length of the trajectory arch; η is the distance along the orthogonal direction to the ray (Fig. 2).

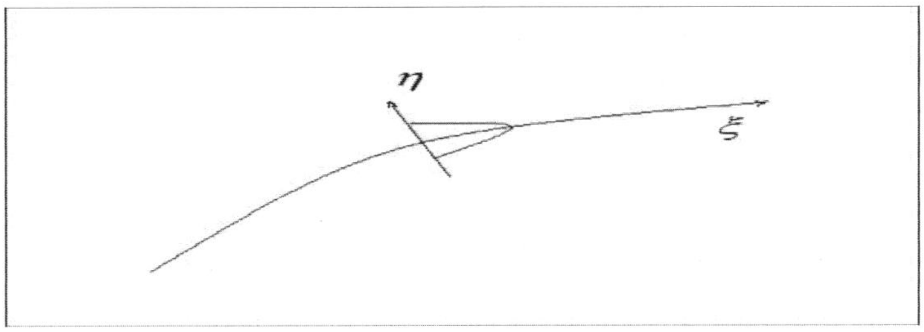

Fig.2. The wave beam and quasi - ray coordinates.

We represent the complex-valued function V in the terms of $V = u \cdot \exp(ik\psi)$, where u and ψ - are real functions. Instead of Helmholtz equation, it is possible to write the equations:

$$\Delta u + k^2\left[\varepsilon - (\nabla\psi)^2\right]u = 0, \qquad div\left(u^2\nabla\psi\right) = 0$$

In this approach, the derivatives along the trajectory are essentially less then the derivatives across the ray direction. Therefore, we can write in the main approximation:

$$\frac{\partial^2 u}{\partial\eta^2} + k^2\left[\varepsilon - \left(\frac{d\psi}{d\xi}\right)^2\right]u = 0, \qquad \frac{\partial\psi}{\partial\eta} = 0$$

The electric field heats the environment and creates the perturbation of the dielectric permittivity. Thus ε consists of the two parts: $\varepsilon = \varepsilon_r + \varepsilon_n$, ε_r being the regular part independent from wave field and ε_n being the perturbation caused by the electric field.

In a small area (Fig. 1) we can assume that the ε_r is almost constant. Thus for the slightly inhomogeneous medium we can write the main approximation and we get the typical problem of nonlinear wave propagation.

$$\frac{d^2u}{d\eta^2} = \lambda^2 u - \varepsilon_n\left(u^2\right)u, \quad \text{were} \quad \lambda = \frac{d\psi}{d\xi} \quad \text{is the phase velocity}.$$

This equation has the first integral.

$$\left(\frac{du}{d\eta}\right)^2 - \lambda^2 u^2 + F\left(u^2\right) = E, \quad E - const, \quad F(u^2) = \int_0^{u^2}\varepsilon_n(x)dx$$

This equation admits the existence of concentrated solutions if $E = 0$, provided that the equation $F(x) - \lambda^2 x = 0$ has solutions $x = 0$ and $x = x_0 > 0$. Without loss of generality, we choose the center of the solution $\eta = 0$. Then $u = \sqrt{x_0}$, when $\eta = 0$, and $u = 0$ at the infinity. The amplitude maximum of the wave beam is $u_0 = \sqrt{x_0}$, and we can write the formal solution.

$$y = \pm \int_{u_0}^{u} \frac{dx}{\sqrt{\lambda^2 x^2 - F\left(x^2\right)}} \tag{1}$$

2.1 Kerr nonlinearity

In particular, if the correction to the nonlinear dielectric permittivity has the form $\varepsilon_n = \alpha |u|^2$, then $F(u^2) = \frac{\alpha}{2}(u^2)^2$, and the equation has a simple root (Fig. 3).

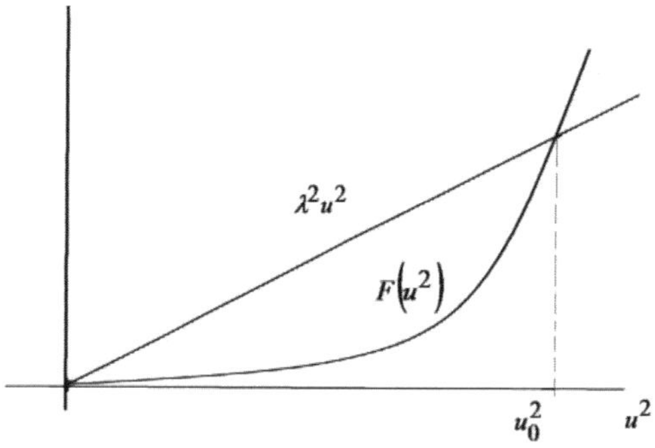

Fig.3. The Ker nonlinearity.

Thus, solution has the form of the simple soliton with the phase velocity $\lambda = \sqrt{\frac{\alpha}{2} u_0^2}$.

It is a well-known solution. This approach allowed us to describe basic phenomena arising at the nonlinear interaction of radiation with the environment.

2.2 Threshold nonlinearity

However as mentioned above, such approach has obvious restrictions. Actually, there is no nonlinear effects when the intensity of the wave field is not enough powerful. As soon as the wave amplitude exceeds certain threshold value, there is a "disruption" of the environment and the nonlinear dependence of the dielectric permittivity upon the wave field amplitude arises. In this case, we can write the model of the threshold nonlinearity. The nonlinear perturbation of the dielectric permittivity for this situation can be represent by the formula. $\varepsilon_n(|u|^2) = \alpha |u|^2 \cdot \theta\left(|u|^2 - A^2\right)$, $\theta(x)$ - is the Heaviside theta function, A is the threshold value.

In this case $F(|u|^2) = \int_0^{|u|^2} \varepsilon_n(t)dt = \frac{\alpha}{2}\theta\left(|u|^2 - A^2\right) * (|u|^4 - A^4)$, and the expression (1) results in the elliptic integral. Nevertheless it is obvious (Fig. 4) that in this case, we have also the concentrated solution with the phase velocity $\lambda = \sqrt{\frac{\alpha}{2}\left(u_0^2 - \frac{A^4}{u_0^2}\right)}$.

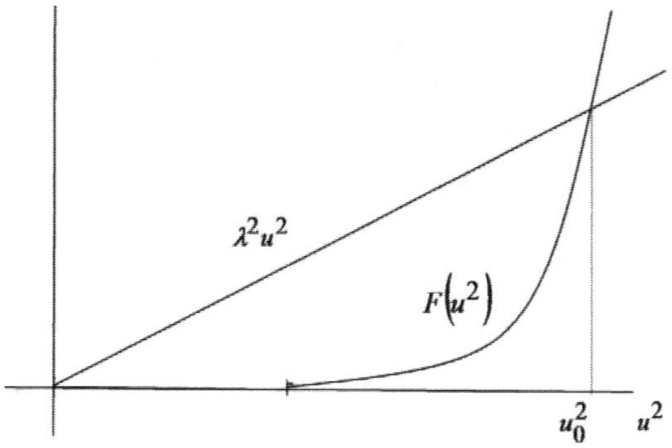

Fig.4. The threshold nonlinearity

The concentrated solution with the threshold nonlinearity is very close to the usual soliton, but it is somewhat narrower in the center of the beam and it has a "long" tails. The interaction between these beams differs from the interaction of the soliton collisions.

2.3 Saturable nonlinearity

The nonlinear dependence becomes more complex with magnification of the wave power. The upper limit, of the dielectric permittivity is observed for most of real materials. In this case it is reasonable to use a model medium with saturable nonlinearity, in which the dependence of the dielectric permittivity on the wave intensity is described by fractional-linear function [7-9].

The perturbation of the dielectric permittivity $\varepsilon_n = \dfrac{\alpha|u|^2}{1+\beta|u|^2}$.

In this case $F(|u|^2) = \int\limits_0^{|u|^2} \varepsilon_N(t)dt = \dfrac{\alpha|u|^2}{\beta} - \dfrac{\alpha}{\beta^2}\ln(1+\beta|u|^2)$

We can see from Fig. 5 that it is also possible the existence of the concentrated solutions.

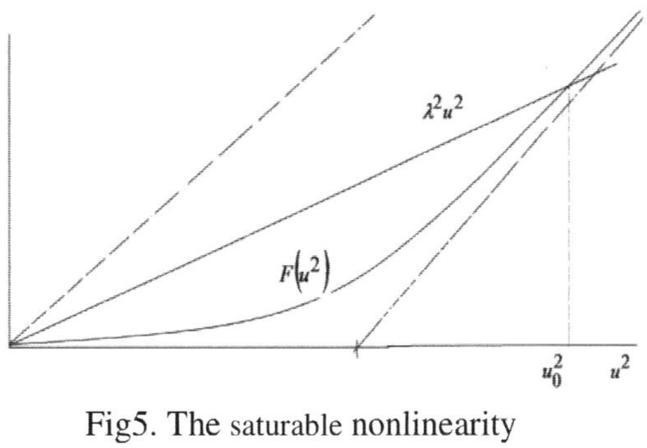

Fig5. The saturable nonlinearity

The phase velocity has the limit $\lambda = \sqrt{\dfrac{\alpha}{\beta}}$, when the wave intensity tends to the infinity. In this case, the environment becomes linear. However, at any finite amplitude, the concentrated solution exists.

2.4 Nonlocal nonlinearity

The previous descriptions of the nonlinear problems were local. That approaches are possible only in the case of negligible heat conductivity, when the size of the wave beam is much more than the characteristic scale of the thermal diffusion. Otherwise, it is necessary to take into account the spreading of the electron density perturbation from the area of the nonlinear heating. Nonlinear effect becomes nonlocal. In this case for the ε_n, we can write the diffusion equation.

$$L^2 \cdot \frac{d^2\varepsilon_n}{d\eta^2} - \varepsilon_n + \alpha \cdot |u|^2 = 0$$

where L is the scale of the diffusion process, α is the nonlinear coefficient. We consider the small region near the ray and we can hold that L is constant in this area. The diffusion equation for ε_n can be integrated in the frameworks of this assumption. Thus, we have got the simple expression for the nonlinear perturbation ε_n:

$$\varepsilon_n(\eta) = \frac{\alpha}{2L}\int_{-\infty}^{\infty} |u(t)|^2 \cdot \exp\left(-\frac{|\eta - t|}{L}\right) dt$$

Now, it is easy to see that we can reduce the problem of the nonlinear interaction in view of diffusion to the one nonlinear integro-differential equation of Hartree type. It is convenient to introduce the fundamental solution of the diffusion equation

$$G(x) = \frac{1}{2L}\exp\left(-\frac{|x|}{L}\right)$$

And now instead of (1), we write the main equation.

$$\frac{d^2u}{d\eta^2} + u \cdot \int_{-\infty}^{\infty} u^2(t) \cdot G(y-t)dt = \lambda^2 \cdot u$$

It is easy to show that when L tends to zero, $G(x)$ tends to the Dirac delta function and we have the well known equation with the cubic nonlinearity. The classical soliton is the solution of this equation $u(\eta) = u_0 \cdot ch^{-1}(p\eta)$ with $p = \sqrt{\frac{\alpha u_0^2}{2}}$. When $L \to \infty$, the problem becomes linear and not interesting for us.

Now, we can see that the wave amplitude u is described by the ordinary differential equation, because it depends on the longitudinal variable ξ just as on parameter. It is obvious that we have got the mathematical eigenvalue problem for the nonlinear integro-differential equation.

We cannot hope to solve this problem by analytical methods. Therefore we constructed the computational approach which is based on the iterative algorithm. Any concentrated solution, for example a soliton $\tilde{v}(y) = ch^{-1}(y)$, is used as the first approximation. Then, we determine the perturbation of the dielectric permittivity, that is we calculate the convolution

$$\tilde{v}(y) * G(y) = \int_{-\infty}^{\infty} \tilde{v}(t) G(y-t) dt$$

This procedure uses the fast Fourier transformation. After that, we solve the linear problem for the eigenvalue p and the eigen function v for the second order differential operator with conditions: $v \to 0$ when $y \to \pm\infty$. This problem is solved by means of the iterations of p from interval $(0,1)$ using the standard Runge-Kutta method. The solution v of this problem is used now as the second approximation. This procedure is continued until $\|v-\tilde{v}\| \geq \delta$, δ is the advance determined precession of the calculations. We used $\delta \approx 10^{-4} \div 10^{-5}$. This method is very quickly converging and it does not depend on the form of initial approximation.

The computational solution allow us to assert that the concentrated solution exist for any value of the diffusion scale. The concentrated solution has not any singularities for any value of $L \neq \infty$ and it looks like the soliton but more wide. The dependence of the wave amplitude v from the cross variable y is represented in the Fig.6 (solid line) for the value $L=10$. The classical soliton $(L=0)$ is represented here by dotted line. Naturally, the diffusion process enlarges the soliton.

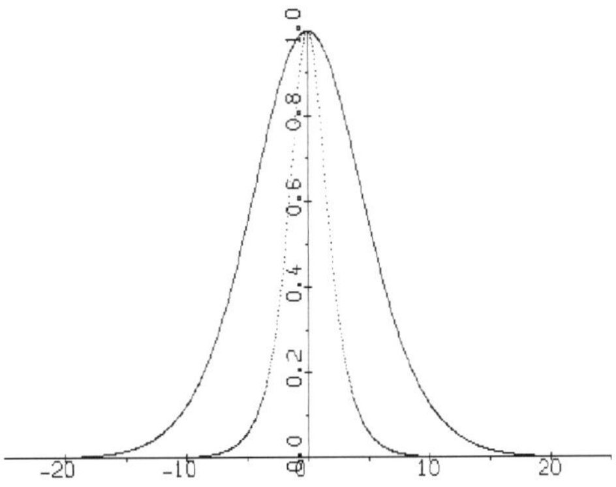

Fig.6. The dependence of wave amplitude versus across coordinate.

The dependence of eigenvalue p versus diffusion scale is represented in the Fig.7. We can see that the solution width, which is determined by the value of p, is very slowly growing with the growth of parameter L.

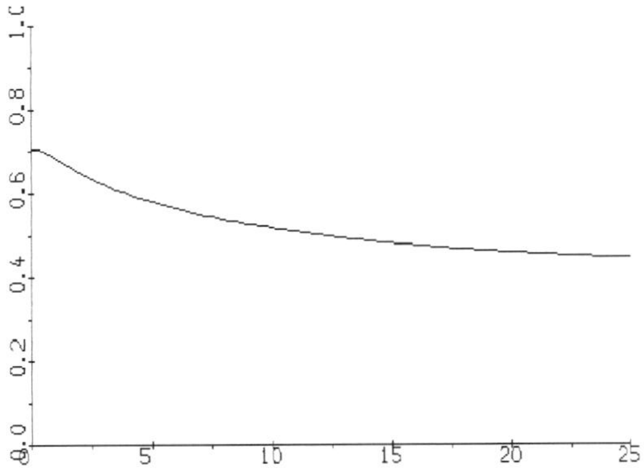

Fig.7. The dependence of eigen value versus diffusion scale.

4. Conclusions

We see that both in the case of threshold nonlinearity, and in the case of saturable nonlinearity solitary waves - concentrated solutions of the corresponding wave equations exist. If it is not possible to write down analytical expressions for these solutions, they are under construction by means of the computer modeling quite successfully. Although these waves are very similar to the one-soliton solutions, their interaction is slightly more complicated, than for the standard solitons. Nevertheless, it is possible to see that their behavior is similar to solitons conduct.

In most cases the problem of the saturable nonlinearity can be reduced to finding the solutions of nonlinear integro - differential equation like the Hartree equation. The computational procedure for the construction of concentrated solutions of this equation, though cumbersome, but converges quickly, and in fact does not depend on the form of the initial approximation. But it is necessary that the initial approximation was concentrated. The resulting numerical solution allows us to state that concentrated solution exists for nonlocal nonlinearity at any scale values of thermal conductivity. This solution has no peculiarities and looks like a wide soliton. The width of the concentrated solution slowly increases with the scale of thermal conductivity, this fact is quite believable.

5. References

[1] Ginzburg V and Gurevich A 1960 Nonlinear phenomena in plasma occurring in an alternating electromagnetic field *Successes of physical sciences* **70** pp 201-246.

[2] Farley D 1963 Artificial heating of the electrons in the F-region of the ionosphere *Journal of Geophysical Research* **68** pp 401-413

[3] Bochkarev G et al 1979 Interaction decameter radio waves at frequencies near the F2 MUF on oblique propagation *Geomagnetism and Aeronomy* **19** pp 830-833

[4] Bochkarev G Eremenko V et al 1982 Nonlinear Interaction of Decameter Radio Waves at close Frequencies on oblique propagation *Journal of Atmos. and Terr.Phys* **44** N 12

[5] Bochkarev G Eremenko V et al 1980 Modeling the impact of the powerful waves on the ionosphere at oblique incidence *Geomagnetism and Aeronomy* **20** pp 848-853

[6] Molotkov I and Cherkashin Y 1994 Some Nonlinear Effects Wave Beams Propagation in the Ionosphere *Journal of Atmos. and Terr.Phys.*56 N 11

[7] Gotz Sand and Herrmann J 1991 Soliton propagation in materials with sutarable nonlinearity *J.Opt.Soc. Am. B* **8** N 11

[8] Eremenko V and Molotkov I 1998 Features of behavior of wave beams in media with saturable nonlinearity *Radiotechnology and eletronika* **43** N 7

[9] Molotkov I Eremenko V Anderson D and Lisak M , 2000 Stationary High Intensity Wave Beams in Media with Saturable Nonlinearity *Physica Scripta* **61** pp 472-475

[10] Eremenko V and Cherkashin Y 1999 Nonlocal nonlinear interaction of high-frequency waves with the ionosphere *Math. Colleges and universities. Radiophysics* **42** pp 700-703

Self-modulation of a relativistic charged-particle beam as *thermal matter wave* envelope

R Fedele[1,4], D Jovanović[2,4], S De Nicola[3,4], A Mannan[4,5] and F Tanjia[5]

[1] Dipartimento di Fisica, Università di Napoli "Federico II", Napoli, Italy
[2] Institute of Physics, University of Belgrade, Belgrade, Serbia
[3] CNR-SPIN Napoli, Italy
[4] Dipartimento di Matematica e Fisica, Seconda Università degli Studi di Napoli, Sede di Caserta, Italy
[5] INFN Sezione di Napoli, Italy

E-mail: renato.fedele@na.infn.it

Abstract. The self-modulation, resulting from its interaction with the surrounding medium, of a relativistic charged-particle beam traveling through an overdense plasma, is investigated theoretically. The description of the transverse nonlinear and collective beam dynamics of an electron (or positron) beam in a plasma-based accelerator is provided in terms of a *thermal matter wave envelope* propagation. This is done using the quantum-like description provided by the *thermal wave model*. It is shown that the charged-particle beam dynamics is governed by a Zakharov-type system of equations, comprising a nonlinear Schrödinger equation that is governing the spatiotemporal evolution of the thermal matter wave envelope and a Poisson-like equation for the wake potential that is generated by the bunch itself.

1. Introduction

Since its early development, the study of the charged particle motion in electric and magnetic fields has played a special role in the construction of the analogy between optics and mechanics. The problems of charged particle motion, such as those occurring in electrostatic and magnetostatic lenses (dipole, quadrupole and other multipoles), magnetic mirrors, magnetic bottles, mass spectrometers and many others [1], have been used as classic examples to show the correspondence between the geometrical optics and classical mechanics. The phenomenological wealth exhibited in this context has generated a separate discipline known as *electron optics*. Remarkably, within the context of electron optics, important scientific and technological developments in electron microscopy [1, 2, 3, 4, 5, 6] and in particle accelerators [7], respectively, have been registered.

Thanks to those developments, electron optics has been expanded mainly in two directions that characterized the analogy between optics and mechanics in two different ways, also at the level of the wave description.

One of them concerned the behavior of the single particle, or the systems of relatively few particles, in which the extension from the *geometrical* to the *wave-like* description coincided with the development of the quantum-mechanical description of electron optics, such as in the contexts in which the quantum behavior of the single particle prevails over the collective behavior, and the concept of temperature does play a role. Valuable examples are encountered in the recent studies of vortex states (angular orbital momentum effects and spin) with integer or fractional topological charge that are associated with the single charged-particle matter wave in the electron microscope [8]; or in studies of the transport of relativistic charged-particle beams traveling in a high-energy accelerating machines based on the quantum mechanical behavior of the single-particle [9, 10, 11].

The other branch of the electron optics is concerned, instead, with the situations in which the behavior of a beam, constituted by an extremely large number of charged particles, is affected by the electromagnetic interactions that are established within such a system, while the effects of the temperature cannot be ignored due to the large number of particles. Therefore, in general, the behavior of such a system is expected to be collective and affected by the thermal spreading among the particles (thermal regimes). In this regime, electron optics has received a strong development within both conventional and non-conventional accelerator physics [7, 12]. Hereafter, we refer to this branch to as *electron optics in thermal regime* (EOTR). A peculiar aspect of the paraxial charged-particle beam propagation, as described by the electron optics in thermal regime, is the mixing of the *electron rays* (charged-particle trajectories) that chracterizes the motion of *non laminar beams* [7]. In fact, due to the thermal spreading among the particles, the electron rays in a beam, that is traveling *in vacuo* with a relativistic speed, deviate with respect to the propagation direction with small and random slopes, whose r.m.s. deviation $\sim v_T/c \ll 1$ (v_T and c are the transverse thermal velocity and the light speed, respectively).

In this paper, we deepen the description of the spatiotemporal evolution of a relativistic non laminar charged particle beam. This is done within the context of the quantum-like formalism provided by the thermal wave model (TWM) that attributes a complex function to the beam as a whole, called the beam wave function (BWF). Its squared modulus is proportional to the beam density. In this way, TWM attributes to the beam a sort of a quantum-like matter wave, whose dispersion properties are related to the thermal spreading of the beam particles (*thermal matter waves*). As an example, we describe the self-modulation of a long non laminar electron (or positron) beam in a unmagnetized overdense plasma as the modulation of a thermal matter wave envelope, whose spatiotemporal evolution is provided by a pair of Zakharov-like equations. The latter take into account the plasma wake field (PWF) that is excited by the beam and experienced as a feedback by the beam itself (self-interaction).

2. Wave envelope representation of a charged particle beam

The thermal spreading introduces an uncertainty in the electron ray positions in the transverse plane at any longitudinal position. Then, the picture that comes from the envelope of the above electron ray mixing resembles the one that is exhibited by the paraxial ray diffraction in the light rays of an electromagnetic radiation beam. It is well known that from this comparison one can conclude that the statistical behavior of the charged particles in a paraxial beams, which is a fully classical process, *simulates* the paraxial diffraction exhibited by the electromagnetic radiation beams. Its experimental evidence is manifested in all the processes that are relevant to EOTR [7] and is supported by theoretical kinetic descriptions (Boltzmann/Vlasov equation). To deepen this aspect, let us consider a charged-particle beam travelling in vacuo along the z-axis with speed βc ($\beta \approx 1$). The assumption of paraxial beam implies that *the slopes of the electron rays are very small with respect to* z . Consequently, if at each longitudinal location z the pair $(x(z), y(z))$ represents the transverse location of an electron ray (z plays the role of a clock coordinate), the paraxial approximation can be expressed by the conditons $p_x \equiv dx/dz \ll 1$, and $p_y \equiv dy/dz \ll 1$. Note that (x, p_x) and (y, p_y) constitute two pairs of canonical conjugate variables [1, 7] and the beam transverse motion, affected by the thermal spreading, should be in principle described statistically by the first- and the second-order moments of the classical phase-space distribution function $\rho(x, y, p_x, p_y, z)$ [13]. This function represents the probability density of finding an electron ray at the transverse phase space location (x, y, p_x, p_y) and time

$t = z/\beta c$.

3. Is the thermal spreading of a non laminar beam a diffusion process?

One is naturally lead to consider the transverse beam thermal spreading as a diffusion process in the real space. For the sake of simplicity, let us now restrict our analysis to the (1+1)D case where, for instance, the transverse motion takes place along the x-axis only (z being still the time-like variable). Then $\rho = \rho(x, p, z)$, where we have replaced p_x with $p = dx/dz$. We introduce the second-order moments of the latter as follows: $\sigma_x(z) \equiv \langle (x - \langle x \rangle)^2 \rangle^{1/2}$ (r.m.s. dispersion in x electron ray position), $\sigma_p(z) \equiv \langle (p - \langle p \rangle)^2 \rangle^{1/2}$ (r.m.s. dispersion in p electron ray slope), and $\sigma_{xp} \equiv \langle (x - \langle x \rangle)(p - \langle p \rangle) \rangle^{1/2}$ (electron ray correlation term), where $< ... >$ stands for the phase space average. Then the diffusion coefficient can be defined as [14]: $\epsilon(z) \equiv 2\left[\sigma_x^2(z)\sigma_p^2(z) - \sigma_{xp}^2(z)\right]^{1/2}$, which is called *r.m.s. transverse emittance of the beam* [7]. Since the beam propagates in vacuo, it is well known that ϵ is the invariant of motion [7] and $\sigma_x(z)$ satisfies the following envelope equation (Ermakov-Pinney equation) :

$$\frac{d^2\sigma_x}{dz^2} - \frac{\epsilon^2}{4\sigma_x^3} = 0 \ . \tag{1}$$

Equation (1), is in full agreement with the experimental observations concerning with the beam propagation in vacuo (e.g., in the final stage of a linear collider) [15, 16]. Since we are trying to describe the thermal spreading of the beam as a diffusion process, it is natural (but not obvious!) to assume the existence of a real function f(x,z), connected somehow with the phase space distribution function satisfying the following diffusion equation [17]:

$$\frac{\partial f}{\partial z} = \frac{\epsilon}{2}\frac{\partial^2 f}{\partial x^2} . \tag{2}$$

Assuming that initially ($z = z_0$) the beam has a Gaussian transverse shape, a Gaussian non-stationary normalized solution of Eq. (2) can be found:

$$f(x, z) = \frac{1}{\sqrt{2\pi}\sigma_f(z)} \exp\left[\frac{x^2}{2\sigma_f^2(z)}\right] \ , \tag{3}$$

where the rms $\sigma_f(z)$ satisfies to the following equation

$$\frac{d^2\sigma_f}{dz^2} + \frac{\epsilon^2}{4\sigma_f^3} = 0 \ . \tag{4}$$

Note that (4) does not formally coincide with (1). Thus, taking for Eqs. (1) and (4) the same initial conditions, namely: $\sigma_f(z_0) = \sigma_x(z_0) \equiv \sigma_0$, $(d\sigma_f/dz)_{z=z_0} = (d\sigma/dz)_{z=z_0} \equiv \sigma_0'$, they do not give the same solution. In fact, the envelope equation (1) gives the following solution: $\sigma_x(z) = \sqrt{2\mathcal{E}(z - z_0)^2 + \sigma_0^2}$, where $2\mathcal{E} \equiv \epsilon^2/4\sigma_{j0}^2 + (\sigma_{j0}')^2 > 0$, whilst the envelope equation (4) gives $\sigma_f(z) = \sqrt{\sigma_0^2 + \epsilon(z - z_0)}$. These solutions differ each other substantially. In particular, the former cannot reduce to zero for any values of z, whilst the latter vanishes for a finite value of this variable. In other words, the former preserves the non laminar beam from the focusing in a single point, in full agreement with the experimental evidence, whilst the latter does not preserve such a circumstance. We can conclude that Eq. (4) does not describe the envelope transverse non laminar beam motion.

4. The Schrödinger-like equation for non laminar beams

Since Eq. (4) is a consequence of the parabolic diffusion equation (2), the latter cannot be taken to describe the beam transport in a vacuum in the presence of finite emittance, as well. It turns out that we need to find an evolution equation which recovers Eq. (1) and takes also into account the information related to the slopes of the particle trajectories. To this end, we observe that (4) transforms into (1) by means of the formal substitution:

$$\epsilon \rightarrow i\epsilon \tag{5}$$

where i is the imaginary unity. Correspondingly, (2) transforms into the following Schrödinger-like equation for the free motion:

$$i\frac{\partial \Psi}{\partial z} = -\frac{\epsilon}{2}\frac{\partial^2 \Psi}{\partial x^2}, \tag{6}$$

where now, instead of f, we have the function Ψ which in principle may be complex. Taking into account also the other transverse component, y, of the beam motion, Eq. (6) can be easily generalized to the following (2+1)D Schrödinger-like equation

$$i\frac{\partial \Psi}{\partial z} = -\frac{\epsilon}{2}\nabla_\perp^2 \Psi, \tag{7}$$

where we have assumed, for simplicity, the same emittance for each transverse direction. Here $\nabla_\perp = (\partial/\partial x, \partial/\partial y)$ denotes the transverse gradient.

Let us represent Ψ in the following form:

$$\Psi(x,y,z) = \sqrt{n(x,y,z)}\exp\left[\frac{i}{\epsilon}\theta(x,y,z)\right] \quad , \tag{8}$$

with $\theta(x,y,z)$ being a real function and $n(x,y,z)$ a positive and real function, satisfying the following condition: $\int_{-\infty}^{\infty} |\Psi(x,y,z)|^2\, dx\, dy = \int_{-\infty}^{\infty} n(x,y,z)\, dx\, dy = N$ (N being the total number of particles of the beam. It is immediately clear, following the language of quantum mechanics, that we have now two suitable information: the transverse probability density of the beam particles, i.e. $n(x,y,z) = |\Psi(x,y,z)|^2$, and the transverse current velocity, i.e. $\mathbf{V}(x,y,z) = \nabla_\perp \theta(x,y,z)$.

It is easy to see that Eq. (7) admits the following Gaussian solution:

$$\Psi(x,y,z) = \frac{1}{\sqrt{2\pi\sigma_x(z)\sigma_y(z)}}\exp\left[-\frac{x^2}{2\sigma_x^2(z)} - \frac{y^2}{2\sigma_y^2(z)}\right]\exp\left[\frac{i}{\epsilon}\theta(x,y,z)\right], \tag{9}$$

where σ_j $(j = x, y)$ satisfies Eq. (1) and θ, is given by:

$$\theta(x,y,z) = \frac{x^2}{2R_x(z)} + \frac{y^2}{2R_y(z)} + \phi_x(z) + \phi_y(z)$$

with

$$\frac{1}{R_j(z)} = \frac{1}{\sigma_j(z)}\frac{d\sigma_x(z)}{dz} \quad \text{and} \quad \frac{d\phi_j(z)}{dz} = -\frac{\epsilon}{4\sigma_j(z)}.$$

Note that, like in the electromagnetic beam optics, $R_x(z)$ and $R_y(z)$ describe the local curvature of the wave front associated with the eikonal $\theta(x,y,z)$ and they are directly connected with the slope of the particle trajectories. Since the gradient of $\theta(x,y,z)$ defines the current velocity $\mathbf{V}(x,y,z)$ and, on the other hand, is directly connected with the particle trajectory slopes, we can also conclude that the transverse free motion of the beam is governed by a Schrödinger-like equation for a complex function, called beam wave function (BWF) which carries two

information pertinent to the fluid picture: the beam density, proportional to its squared modulus ($n(x, y, z) = |\Psi(x, y, z)|^2$), and the beam current velocity, corresponding to the gradient of its phase ($\mathbf{V}(x, y, z) = \nabla_\perp \theta(x, y, z)$).
Eq. (7) is the starting point of the TWM [18], which has been formulated almost 22 years ago. In its early formulation, TWM has extended Eq (7) to the case of an arbitrary potential energy, say U, such that:

$$i\frac{\partial \Psi}{\partial z} = -\frac{\epsilon}{2}\nabla_\perp^2 \Psi + U\Psi . \qquad (10)$$

With the use of TWM, a number of linear and nonlinear problems in both conventional and plasma-based particle acceleration were successfully described [19, 20, 21, 22, 23, 24, 25]. In particular, TWM has been applied to the Gaussian particle-beam optics and dynamics for a quadrupole-like device [18], to luminosity estimates in final focusing stages of linear colliders in the presence of small aberrations [26], whilst the TWM predictions have been compared with tracking-code simulations and a fair agreement has been demonstrated (the analysis has been carried out in both configuration space and phase space [23, 27, 28, 29]). Remarkably, a self consistent theory of the interaction between a relativistic electron (positron) beam and a cold unmagnetized plasma has been also developed [19]. Recently, the TWM description of the self-consistent beam-plasma interaction has been also developed in strongly magnetized plasmas, where the collective vortex beam states (orbital angular momentum states) has been predicted [30, 31]. The approach was also successfully applied to the longitudinal beam dynamics. For instance, it was useful to predict soliton-like states of the charged-particle beams or to provide a *wave key of reading* for the nonlinear and collective effects exhibited by the coherent instabilities in high-energy accelerating machines, showing that it can be formulated as the deterministic or the statistical approach to modulational instability, where a Landau-type damping plays a basic role [22, 32, 33, 34].

5. Self-modulation of *a thermal matter wave envelope* in the strongly nonlocal regime

We want to apply the above description to the case of relativistic non laminar electron (or positron) beam which is travelling through an overdense plasma, of unperturbed density n_0, with spot size $\sigma_\perp$. It is also assumed that the beam is sufficiently long, i.e., its longitudinal extent, say σ_z, is much greater than a plasma length, say λ_p. The electron (or positron) beam is launched into the plasma in the presence of a preformed 3D large amplitude electrostatic plasma wave traveling along the z-axis at the phase velocity βc, where $\beta \approx 1$. The beam is launched along the z-axis with velocity βc, as well. We assume that the electric field associated with the plasma wave has the form

$$\mathbf{E}_p(r, \xi) = \mathbf{E}_{\perp 0}(r)\cos k_p \xi + \hat{z}E_{z\,0}(r)\sin k_p \xi , \qquad (11)$$

where r is the radial cylindrical coordinate, $\xi = z - \beta ct$ and $k_p = 2\pi/\lambda_p$ is the plasma wavenumber. here, $E_{\perp 0}(r)$ and $E_{z\,0}(r)$ account for the radial profile of the the transverse and the longitudinal components, respectively, of the electric field $\mathbf{E}_p$. To obey the electrostatic condition $\nabla \times \mathbf{E}_p = 0$, $E_{\perp 0}(r)$ and $E_{z\,0}(r)$ must be related each other. In particular, if we assume that $E_{z\,0}$ has the following Gaussian radial profile, viz.,

$$E_{z\,0}(r) = A_0 \exp\left(-r^2/r_\perp^2\right) , \qquad (12)$$

where A_0 and $r_\perp$ are the constant amplitude and the constant width, respectively, then we easily get

$$E_{\perp 0}(r) = -\frac{1}{k_p}\frac{dE_{z\,0}(r)}{dr} = \frac{2r/r_\perp}{k_p r_\perp} E_{z\,0}(r) . \qquad (13)$$

Since the particle beam is assumed much longer than a plasma wavelength, there are two effects experienced by the beam particles that are manifested at time scales much longer than $2\pi/c\,k_p = 2\pi/\omega_p$ (ω_p being the electron's plasma frequency). They are the *ponderomotive effect* produced by the plasma wave and the *PWF excitation*. For the time being, let us analyze them separately.

5.1. Ponderomotive effect of the plasma wave

Due to the expression of the electric field given by (11), on the time scales much longer than $T_p = 2\pi/\omega_p$, each particle of the beam experiences a second-order nonlinear force, say f_{NL}, called ponderomotive force [35], viz.,

$$\mathbf{f}_{NL}(r) = -\frac{1}{2}\frac{e^2}{m\gamma_0\omega_p^2}\nabla_\perp\langle\mathbf{E}_p^2(r,\xi)\rangle = \frac{1}{4}\frac{e^2}{m\gamma_0\omega_p^2}\nabla_\perp\left[\mathbf{E}_{\perp 0}^2(r) + E_{z\,0}^2(r)\right] \equiv -\nabla_\perp\Omega_{NL}(r) \tag{14}$$

where $\langle ... \rangle$ denotes the average over a plasma period T_p; m, e and γ_0 are rest mass, electric charge and initial longitudinal relativistic gamma factor of the electron, respectively. It is easy to see that the dimensionless ponderomotive potential energy $U_{ln}(r) \equiv \Omega_{NL}(r)/m\gamma_0 c^2$ can be cast in the form

$$U_{NL}(r) = \left(\frac{e\,A_0}{2\,m\,\gamma_0\,\omega_p\,c}\right)^2\left[1 + \left(\frac{2r/r_\perp}{k_p\,r_\perp}\right)^2\right]\exp\left(-2r^2/r_\perp^2\right). \tag{15}$$

5.2. PWF excitation in the long beam limit

According to the PWF theory [36, 37] for long beams in overdense regime, the plasma provides an *adiabatic shielding* of the beam. In fact, the perturbation of the plasma charge density (cylindrical coodinates) $en_1(r,\varphi,\xi)$ that is originated by the insertion of the long beam in the plasma tends to compensates the local beam charge density $q\rho_b(r,\varphi,\xi)$, i.e., $en_1(r,\varphi,\xi) \approx q\rho_b(r,\varphi,\xi)$. For cylindrically symmetric beams, $\rho_b(r,\varphi,\xi) = \rho_b(r,\xi)$ becomes the source of the *plasma wake potential energy*, i.e., $U_w(r,\xi)$, through the following Poisson-like equation [36, 37]

$$\left(\nabla_\perp^2 - k_p^2\right)U_w = k_p^2\frac{\rho_b}{n_0\gamma_0}. \tag{16}$$

In turn, the PWF, i.e., $\mathbf{W} \propto \nabla U_w$ acts on the particles of the beam itself (self-interaction). The simultaneous effect of U_{NL} and U_w characterizes the transverse dynamics of the long beam. Note that, thanks to the long beam limit, no longitudinal dynamics is practically involved in our analysis. In fact, due to this assumption: (i) the effects of the plasma wave come into play only through the ponderomotive potential that manifests on the time scales longer than a plasma period (therefore the ponderomotive force has a transverse component, only); (ii) the longitudinal component of the plasma wake field, i.e., $W_z \propto \partial U_w/\partial\xi$, is much smaller than the transverse one, i.e., $\mathbf{W}_\perp \propto \nabla_\perp U_w$. In these conditions, the longitudinal beam dynamics can be ignored, so that our analysis becomes 2D.

Within the context of TWM [18], the latter is governed by Eq (10), where $U = U_{NL} + U_w$, i.e., the following 2D Schrödinger-like equation (here given in cylindrical coodinates)

$$i\epsilon\frac{\partial\psi}{\partial\xi} = -\frac{\epsilon^2}{2}\left[\frac{1}{r}\frac{\partial}{\partial r}\left(r\frac{\partial\psi}{\partial r}\right) + \frac{1}{r^2}\frac{\partial^2\psi}{\partial\varphi^2}\right] + (U_{NL} + U_w)\,\psi\,. \tag{17}$$

Here, we cast the normalization of ψ in such a way that $\rho_b(r,\varphi,\xi) = (N/\sigma_z)\,|\psi(r,\varphi,\xi)|^2$ and $\int_0^{2\pi} d\varphi\int_0^\infty r\,dr\,|\psi(r,\varphi,\xi)|^2 = 1$. Note that Eq. (17) describes the feedback that the beam produces on itself through U_w. On the basis of the above interpretation of ψ, Eqs. (16) and (17) can be easily cast, in cylindrical symmetry, as

$$\frac{1}{r}\frac{\partial}{\partial r}\left(r\frac{\partial U_w}{\partial r}\right) - k_p^2 U_w = \frac{k_p^2 N}{n_0\gamma_0\sigma_z}\left|\psi_m\right|^2 . \tag{18}$$

$$i\epsilon\frac{\partial\psi_m}{\partial\xi} = -\frac{\epsilon^2}{2}\frac{1}{r}\frac{\partial}{\partial r}\left(r\frac{\partial\psi_m}{\partial r}\right) + \left(\frac{\epsilon^2 m^2}{2r^2} + U_{NL} + U_w\right)\psi_m , \tag{19}$$

where we have cast the BWF as $\psi(r,\varphi,\xi) = \psi_m(r,\xi)\, e^{im\varphi}$, with m arbitrary integer. Pair of Eqs. (18) and (19) describes the self-consistent transverse spatiotemporal evolution on time scales longer than a plasma period of a charged-particle beam while experiencing its PWF interaction in the long beam limit in the presence of a pre-formed 3D large amplitude plasma wave. It was earlier derived by Fedele and Shukla in 1992 for the case of $m = 0$, and $U_{NL} = 0$ [19]. It can be regarded as a sort of Zakharov pair of equations.

5.3. Strongly nonlocal regime

We confine here our attention to the case of a charged particle beam with a spot size, say $\sigma_\perp$, much smaller than the width, $r_\perp$, of the radial profile of the preformed plasma wave (strongly focussed beam), i.e., $\sigma_\perp \ll r_\perp$. Here, $\sigma_\perp$ is defined as the rms of the positions of the beam particles, i.e., $\sigma_\perp(\xi) = \int_0^{2\pi} d\varphi \int_0^\infty r^2 \left|\psi(r,\varphi,\xi)\right|^2 r\, dr$. Then, if we require that $k_p r_\perp \approx 2\pi$ (i.e., $r_\perp \approx \lambda_p$), we can assume that $k_p\sigma_\perp \ll 1$. Then, Eq. (18) becomes (strongly nonlocal regime)

$$\frac{1}{r}\frac{\partial}{\partial r}\left(r\frac{\partial U_w}{\partial r}\right) = \frac{k_p^2 N}{n_0\gamma_0\sigma_z}\left|\psi\right|^2 . \tag{20}$$

On the other hand, due to the above conditions the beam particles are confined in a regione where $r \ll r_\perp$. Therefore, in Eq. (19) it is enough to expand $U_{NL}(r)$ in powers of $r/r_\perp$ up to the second degree. So, apart from an additive constant, we obtain

$$U_{NL}(r) \approx \frac{1}{2}K_{NL}r^2, \;\; K_{NL} = \left(\frac{e\,A_0}{m\,\gamma_0\,\omega_p\,c\,r_\perp}\right)^2\left(\frac{2}{k_p^2 r_\perp^2} - 1\right). \tag{21}$$

Therefore, K_{NL} is positive (negative) for $r_\perp < r_\perp^*$ ($r_\perp > r_\perp^*$) or vanishes for $r_\perp = r_\perp^*$, where the threshold $r_\perp^* = \lambda_p/2\pi \approx 0.226\,\lambda_p$. According to the method for aberrationless solution presented in [38, 39], but confining our analysis to the case of $m = 0$, pair of equations (19) and (20) allows to get the following Schrödinger equation

$$i\epsilon\frac{\partial\Psi}{\partial\xi} = -\frac{\epsilon^2}{2}\frac{1}{r}\frac{\partial}{\partial r}\left(r\frac{\partial\Psi}{\partial r}\right) + \frac{1}{2}\overline{K}(\xi)r^2\,\Psi , \tag{22}$$

where

$$\overline{K}(\xi) = K_{NL} + \frac{\eta}{\sigma_\perp^2(\xi)} , \quad \eta = \frac{k_p^2 N}{2\pi\,\sigma_z\, n_0\gamma_0} . \tag{23}$$

and $\psi_0(r,\xi)$ has been renamed as Ψ. Note that the constant η is a positive quantity. Here, $\sigma_\perp$ obeys to the following Ermakov-Pinney type equation for $\sigma_\perp(\xi)$ (i.e.,), viz.,

$$\frac{d^2\sigma_\perp}{d\xi^2} + K_{NL}\,\sigma_\perp + \frac{\eta}{\sigma_\perp} - \frac{\epsilon^2}{\sigma_\perp^3} = 0 , \tag{24}$$

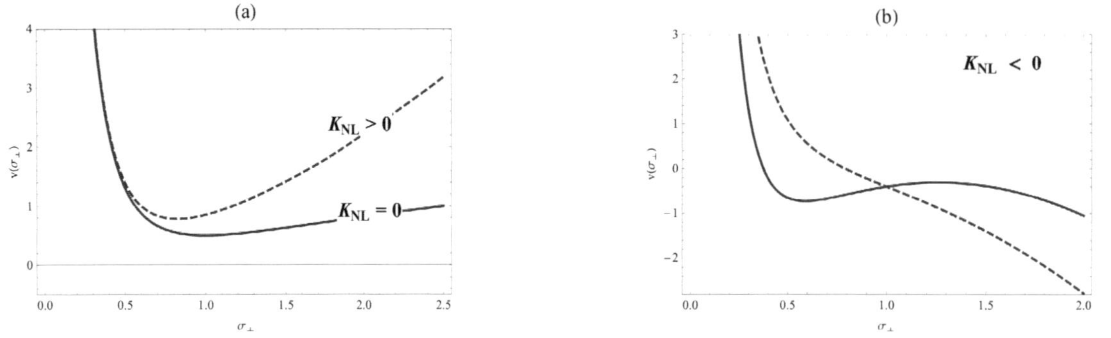

Figure 1. (a) Qualitative plots (arbitrary units) of $V(\sigma_\perp)$ for $K_{NL} \geq 0$ keeping constant both η and ϵ. $V(\sigma_\perp)$ is a trapping potential for any value of the total energy $\mathcal{E}$. (b) Qualitative plots (arbitrary units) $V(\sigma_\perp)$ for a fixed negative value of K_{NL} and ϵ. The two different plot represent two different classes of potentials: the solid plot corresponds to the ones that are trapping up to a certain threshold of the total energy $\mathcal{E}$, whilst the dashed plot is non trapping for any value of $\mathcal{E}$.

Equation (24) is also usually referred to as *Sacherer's envelope equation* [7]. Provided that $\overline{K}$ is positive, Eq. (22) admits bounded states. A complete set of normalized eigenfunctions of Eq. (22) is given by the Laguerre-Gauss modes, viz.,

$$\Psi_n(r,\xi) = \frac{1}{\sqrt{\pi}\sigma_\perp} \exp\left(-\frac{r^2}{2\sigma_\perp^2} + \frac{ir^2}{2\epsilon\rho} + i\phi_n\right) L_n\left(\frac{r^2}{\sigma_\perp^2}\right), \tag{25}$$

where n is an arbitrary integer, L_n are the simple Laguerre polynomials, and $\rho = \rho(\xi)$ and $\phi_n = \phi_n(\xi)$ satisfy the following differential equations

$$\frac{1}{\rho} = \frac{1}{\sigma_\perp}\frac{d\sigma_\perp}{d\xi}, \qquad -\frac{\sigma_\perp^2}{2\epsilon}\frac{d\phi_n}{d\xi} = n + \frac{1}{2}. \tag{26}$$

From Eq. (23), on can easily see that the positivity of $\overline{K}$ implies

$$\frac{\eta}{\sigma_\perp^2} > -K_{NL}. \tag{27}$$

Note that this condition fixes an inequality which depends on the time-like variable ξ. In principle, it can be violated for suitable variation of $\sigma_\perp(\xi)$. To predict the behavior of the beam, let us refer to the Sagdeev pseudo potential [35] that it is given by the first integral of Eq. (24), i.e.,

$$\frac{1}{2}\left(\frac{d\sigma_\perp}{d\xi}\right)^2 + V(\sigma_\perp) = \mathcal{E} = const\,, \tag{28}$$

where the pseudo-potential $V(\sigma_\perp)$ is

$$V(\sigma_\perp) = \frac{1}{2} K_{NL}\,\sigma_\perp^2 + \eta\,\log\sigma_\perp + \frac{\epsilon^2}{2\,\sigma_\perp^2}\,. \tag{29}$$

and the total conserved energy $\mathcal{E}$ is fixed by the initial conditions for $\sigma_\perp$, i.e., $\sigma_0 = \sigma_\perp(\xi = 0)$ and $\sigma_0' = (d\sigma_\perp/d\xi)_{\xi=0}$.

A. Non-negative K_{NL}

Since η is positive, for $K_{NL} \geq 0$, condition (27) is always fulfilled. Therefore, according to Eq.

(22), the particle beam is localized; in particular, for $n = 0$ (fundamental mode), the transverse density has a pick around $r = 0$. In addition, for $K_{NL} \geq 0$, $V(\sigma_\perp)$ is always confining. Then, any initial condition evolves stably. More specifically, for fixed initial conditions, $\sigma_\perp$ is a periodic function of ξ. This appears as a macroscopic *sausage-like* modulation as a result of the interplay of three concomitant effects: (i) the radial thermal dispersion (spreading), that is associated with the emittance ϵ; (ii) the radial ponderomotive effects of the pre-existing LWF-excited plasma wave, that is associated with the *focusing strength* K_{NL}; (iii) the self-interaction provided by the plasma wake field generated by the beam itself associated with the collective parameter η. We refer this collective response of the beam due to the above complex interaction with the plasma to as *beam self-modulation of the thermal matter wave envelope* (25). Figure 1(a) displays qualitative plots describing the attractive character of $V(\sigma_\perp)$ for sets of parameters in which K_{NL} is non-negative.

B. Negative K_{NL}

If K_{NL} is negative, condition (27) may be no longer fulfilled. Although this condition is satisfied, the analysis of $V(\sigma_\perp)$ shows that there are sets of the parameters corresponding to a locally trapping V as shown qualitatively in the example given in Figure 1(b). In fact, for suitable value of the energy $\mathcal{E}$ (and therefore of the initial conditions), stable oscillations in $\sigma_\perp$ around the (local) minimum are possible (solid plot). Also in this case, they appear as a macroscopic self-modulation of the thermal matter wave envelope driven by the PWF excitation. As $\mathcal{E}$ increases solutions transit from periodic to unbounded ones, which correspond to an unstable evolution. Furthermore, there is also another set of parameters for which $V(\sigma_\perp)$ does not correspond to a trapping potential, even locally (dashed plot). In fact, in this other case, $V(\sigma_\perp)$ does not have local minima to make possible oscillations. Then the thermal matter wave envelope is unstable and evolves in a blowup.

6. Conclusions and remarks

In this paper, we have described the self-modulation of a long, non laminar, relativistic electron (or positron) beam in an unmagnetized plasma. The process is driven by both a 3D, preformed, large amplitude plasma wave, and the self consistent plasma wake field excitation. The analysis has been carried out within the framework of the quantum-like description that is provided by TWM, where the beam as whole can be regarded as a thermal matter wave envelope, whose spatiotemporal evolution is governed by a Zakharov-like pair of equations. In the strongly nonlocal regime, we have analyzed the physical conditions that keep stable the self-modulation of the thermal matter wave envelope, in the form of sausage-like transverse oscillations as well as the instability leading to the beam blowup.

References

[1] Sturrock P 1955 *Static and Dynamic Electron Optics: An Account of Focusing in Lens, Deflector and Accelerator* (Cambridge University Press, London)
[2] Zworykin V, Morton G, Ramberg E, Hillier J and Vance A 1945 *Electron optics and the electron microscope* (John Wiley, New York)
[3] Glaser P 1952 *Grundlagen der Elekronenoptik* (Springer Verlag, Vienna)
[4] Klemperer O and Barnett M 1971 *Electron optics* (Cambridge University Press, London)
[5] Grivet P and Septier A 1972 *Electron optics* (Pergamon, Oxford)
[6] Hawkes P 1972 *Quadrupole optics. Springer tracts in modern physics* (Springer Verlag, Berlin) volume 42
[7] Lawson J D 1988 *The physics of charged-particle beams* (Clarendon, Oxford)
[8] McMorran B, Agrawal A, Anderson I, Herzing A and Lezec, H. 2011 *Science* **331** 192
[9] Jagannathan R, Simon R, Sudarshan E and Mukunda N 1989 *Phys. Lett. A* **134** 457
[10] Jagannathan R 1990 *Phys. Rev. A* **42** 6674
[11] Khan S A and Jagannathan R 1995 *Phys. Rev. E* **51** 2510
[12] Chao A and Tigner M 1998 *Handbook of accelerator physics and engineering* (World Scientific, Singapore)

[13] Fedele R and Man'ko V I 1998 *Phys. Rev. E* **58** 992
[14] Lapostolle P M 1971 *IEEE Trans. Nucl. Sc.* **18** 1101; Sacherer F J 1971 *IEEE Trans. Nucl. Sc.* **18** 1105
[15] Su J J, Katsouleas T, Dawson J M and Fedele R 1990 *Phys. Rev. A* **41** 3321
[16] Fedele R and Wilson E J N 1990 *Nuovo Cim. D* **12** 1497
[17] Fedele R and Jovanović D 2004 *AIP Conf. Proc.* **740** 430-445
[18] Fedele R and Miele G 1991 *Nuovo Cim. D* **13** 1527-1544
[19] Fedele R and Shukla P K 1992 *Phys. Rev. A* **45** 4045-4049
[20] Fedele R and Shukla P K 1992 *A novel approach to the interaction between the plasma wakefield and the driving relativistic electron beam*, Proc. Int. Conf. on Plasma Physics, Innsbruck, Austria, 29 june - 3 July 1992, Vol. II, p.1293-1296, European Physical Society.
[21] De Nicola S, Fedele R, Man'ko V I and Miele G 1995 *Phys. Scr.* **52** 191
[22] Fedele R, Miele G, Palumbo L and Vaccaro V G 1993 *Phys. Lett. A* **179** 407
[23] Fedele R, Galluccio F and Miele G 1994 *Phys. Lett. A* **185** 93
[24] Fedele R, Miele, G and Palumbo L 1994 *Phys. Lett. A* **194** 113
[25] Fedele R, Shukla P K and Vaccaro V G, 1995 *A quantum-like approach to the interaction of relativistic charged particle beams with plasmas*, *J. Physique* **5**, Colloque 6, Suppl. JP II, n.10 p. c6-119 c6-130
[26] Fedele R and Miele G 1992 *Phys. Rev. A* **46** 6634
[27] Fedele R, Galluccio G, Miele G and Man'ko V I 1995 *Phys. Lett. A* **209** 263
[28] Jang J, Cho Y and Kwon H 2007 *Phys. Lett. A* **366** 246
[29] Jang J, Cho Y and Kwon H 2010 *Nucl.Instrum.Meth. A* **624** 578
[30] Fedele R, Tanjia F, De Nicola S, Shukla P K and Jovanović D 2011 *Proc. 38th EPS Conf. on Plasma Physics*, Strasbourg, France 35G, P5.006, ISBN 2-914771-68-1
[31] Tanjia F, De Nicola S, Fedele R, Shukla P K and Jovanović D 2011 *Proc. 38th EPS Conf. on Plasma Physics*, Strasbourg, France 35G, P5.021, ISBN 2-914771-68-1
[32] Anderson D, Fedele R, Vaccaro V, Lisak M, Berntson A and Johansson S 1999 *Phys. Lett. A* **258** 244
[33] Fedele R and Anderson D 2000 *J. Opt. B: Quant. Semicl. Opt.* **2** 207
[34] Johannisson P *et al.* 2004 *Phys. Rev. E* **69** 066501
[35] Chen F F 1984 *Introduction to plasma physics* (Springer, Berlin) second edition
[36] Chen P *et al.* 1985 *Phys. Rev. Lett.* **54** 693-696
[37] Chen P 1987 *Part. Accel.* **20** 171-182
[38] Fedele R, Tanjia F, De Nicola S, Jovanović D and Ronsivalle C 2013 *J. Plasma Phys.* to appear
[39] Jovanović D, Fedele R, Tanjia F, De Nicola S and Belić M 2012 *Eur. Phys. Lett.* **100** 5502-p1 - 5502-p6
[40] Fedele R, Tanjia F, De Nicola S and Jovanović D 2013 *NIMA Proc.* 1st European Advanced Accelerator Concepts Workshop EAAC 2013, Isola d'Elba, 2-7 June 2013, submitted

Perturbative and Exact Results on the Neumann Value for the Nonlinear Schrödinger on the Half-line

A. S. Fokas

Department of Applied Mathematics and Theoretical Physics, University of Cambridge, Cambridge CB3 0WA, UK, and Research Center of Mathematics, Academy of Athens, 11527, Greece.

E-mail: T.Fokas@damtp.cam.ac.uk

J. Lenells

Department of Mathematics and Center for Astrophysics, Space Physics & Engineering Research, Baylor University, One Bear Place #97328, Waco, TX 76798, USA.

E-mail: Jonatan_Lenells@baylor.edu

Abstract. The most challenging problem in the implementation of the so-called *unified transform* to the analysis of the nonlinear Schrödinger equation on the half-line is the characterization of the unknown boundary value in terms of the given initial and boundary conditions. For the so-called *linearizable* boundary conditions this problem can be solved explicitly. Furthermore, for non-linearizable boundary conditions which decay for large t, this problem can be largely bypassed in the sense that the unified transform yields useful asymptotic information for the large t behavior of the solution. However, for the physically important case of periodic boundary conditions it is necessary to characterize the unknown boundary value. Here, we first present a perturbative scheme which can be used to compute explicitly the asymptotic form of the Neumann boundary value in terms of the given τ-periodic Dirichlet datum to any given order in a perturbation expansion. We then discuss briefly an extension of the pioneering results of Boutet de Monvel and co-authors which suggests that if the Dirichlet datum belongs to a large class of particular τ-periodic functions, which includes $\{a\exp(i\omega t)\,|\,a>0,\,\omega\geq a^2\}$, then the large t behavior of the Neumann value is given by a τ-periodic function which can be computed explicitly.

AMS Subject Classification (2000): 37K15, 35Q15.
Keywords: Initial-boundary value problem, long-time asymptotics, nonlinear Schrödinger equation.

1. Introduction

Let $q(x,t)$ satisfy the nonlinear Schrödinger (NLS) on the half-line with a given initial condition,

$$iq_t + q_{xx} - 2\lambda|q|^2q = 0, \qquad 0<x<\infty, \quad t>0, \quad \lambda=\pm 1, \tag{1}$$

$$q(x,0) = q_0(x), \qquad 0<x<\infty, \tag{2}$$

where $q_0(x)$ has sufficient decay as $x\to\infty$.

In the above setting, the so-called unified transform [8] can be used to analyze problems with either a linearizable boundary condition or with a non-linearizable boundary condition which decays for large t.

1.1. Linearizable bounday conditions

Suppose that equations (1) and (2) are supplemented with any of the following boundary conditions:

$$q(0,t)=0, \tag{3a}$$

$$q_x(0,t)=0, \tag{3b}$$

$$q_x(0,t)-\chi q(0,t)=0, \qquad \chi \text{ real constant.} \tag{3c}$$

Then, the unified transform yields a solution with the same level of efficiency as the celebrated inverse scattering transform [8, 9, 11].

The above problems can also be solved by the usual inverse scattering transform after mapping them to certain equivalent problems on the line [1]. However, the unified transform can also be implemented for linearizable boundary value problems for integrable nonlinear PDEs involving a third order spatial derivative—such as the KdV equation—which *cannot* be mapped to appropriate problems on the line.

1.2. Boundary conditions which decay as $t \to \infty$

Suppose that equations (1) and (2) are supplemented with the Dirichlet boundary condition

$$g_0(t)=q(0,t), \quad 0<t<\infty, \tag{4}$$

where $g_0(t)$ vanishes as $t \to \infty$. In this case, the unified transform expresses the solution $q(x,t)$ in terms of the solution of a 2×2 matrix Riemann-Hilbert (RH) problem which is uniquely defined in terms of the *spectral functions* $\{a(k), b(k), A(k), B(k)\}$. The first two of these functions can be obtained in terms of the given initial datum $q_0(x)$, however $\{A(k), B(k)\}$ depend on both $g_0(t)$ and on $g_1(t)=q_x(0,t)$. Thus, in order to compute $q(x,t)$ it is necessary to first characterize the so-called Dirichlet to Neumann map, i.e. to express $g_1(t)$ in terms of $g_0(t)$ and $q_0(x)$. However, since $g_0(t)$ vanishes for large t, using the fact that the above spectral functions are t independent, it is still possible to compute the large t-asymptotics of the above RH problem, and hence the large t-asymptotics of $q(x,t)$ without characterizing fully the spectral functions $A(k)$ and $B(k)$.

1.3. The Dirichlet to Neumann map

Two different approaches for analyzing the Dirichlet to Neumann map were recently presented in [11] and [14]. The formulation in [11] is based on the analysis of the eigenfunctions involved in the definition of $\{A(k), B(k)\}$ (see also [7, 10]), whereas the formulation in [14] is based on an extension of the Gelfand-Levitan-Marchenko approach first introduced in [2].

1.4. τ-periodic boundary condition

Suppose that equations (1) and (2) are supplemented by equation (4), where $g_0(t)$ is a periodic function with period $\tau > 0$.

A perturbative scheme for computing the large t behavior of $g_1(t)$ is introduced in [14]. In particular, it is shown in [14] for the NLS, and in [13] for the mKdV, that if $g_0(t)=\alpha \sin t$, $\alpha \in \mathbb{R}$, then it is possible to obtain the function $g_1(t)$ for the NLS and the functions $\{g_1(t), q_{xx}(0,t)\}$ for the mKdV as a perturbation expansion in the parameter α, up to and including terms of $O(\alpha^3)$. Furthermore, these functions are 2π-periodic, at least up to this order. Unfortunately, the perturbative approach of [14] is quite cumbersome and it is practically impossible to go beyond terms of $O(\alpha^3)$.

In the particular case that

$$g_0(t) = \alpha e^{i\omega t}, \qquad \alpha > 0, \quad \omega \in \mathbb{R}, \tag{5}$$

pioneering results for the focusing NLS have been obtained in a series of papers of Boutet de Monvel and co-authors [3, 4, 5, 6] by employing the unified transform. In particular, it is shown in [5] that for the focusing NLS, $\lambda = -1$, there exists a solution $q(x,t)$, such that

$$g_1(t) \sim c e^{i\omega t}, \qquad t \to \infty, \quad c \in \mathbb{C}, \tag{6}$$

if and only if the triple of constants $\{\alpha, \omega, c\}$ satisfy either

$$\left(c = \pm\alpha\sqrt{\omega - \alpha^2} \text{ and } \omega \geq \alpha^2\right) \quad \text{or} \quad \left(c = i\alpha\sqrt{|\omega| + 2\alpha^2} \text{ and } \omega \leq -6\alpha^2\right). \tag{7}$$

1.5. Outline of the paper

Here we first present a new perturbative scheme for computing the large t asymptotic behavior of $g_1(t)$ in terms of the τ-periodic function $g_0(t)$. As illustrations, we consider the example of the single exponential (6), as well as the example of

$$g_0(t) = \alpha e^{i\omega t} + \beta e^{-i\omega t}, \quad \alpha, \beta \in \mathbb{C}, \quad \omega > 0. \tag{8}$$

The main difference between this new perturbative approach and the one used in [14] is that in the latter approach the analysis was first carried out for all t and then the limit $t \to \infty$ was computed, whereas in the new approach the analysis is carried out directly in the limit $t \to \infty$. This simplifies the relevant algorithm enormously, as illustrated by the fact that for the above two examples we give formulae up to and including terms of order eight.

We also discuss briefly an extension of the formalism introduced in [3, 4, 5, 6], which suggests that using the technique of finite gap integration, it is possible to obtain a large class of τ-periodic functions which have the crucial property that if they are assigned as the Dirichlet data for the NLS on the half line, then the associated function $g_1(t)$ has the property that it asymptotes as $t \to \infty$ to a τ-periodic function and furthermore this function can be computed explicitly.

2. A new perturbative approach

In the context of asymptotically t-periodic data, there are situations where we do not necessarily need to know the Neumann value $g_1(t)$ for all $t > 0$, but it is sufficient to know its asymptotic form $g_1^B(t)$. It is therefore natural to ask the following question: Given the asymptotic form $g_0^B(t)$ of the Dirichlet datum, can we find the asymptotic form $g_1^B(t)$ of the Neumann value? Theorem 2.2 below provides a constructive algorithm for computing the asymptotic form of the Neumann value from the asymptotic form of the Dirichlet data in a perturbative expansion. In other words, it provides an explicit construction of the Dirichlet to Neumann map for asymptotically t-periodic data in the limit of large t and small data.

2.1. Main result

We first define what we mean by a perturbative solution.

Definition 2.1 A *perturbative solution of the NLS equation (1) in the quarter plane* is a sequence of smooth functions $\{q_N(x,t)\}_1^\infty$ defined for $x > 0$ and $t > 0$ with the following properties:

(i) The formal power series

$$q = \epsilon q_1 + \epsilon^2 q_2 + \epsilon^3 q_3 + \cdots \tag{9}$$

satisfies (1) in the quarter plane $\{x > 0, t > 0\}$ to all orders in a perturbative expansion, that is,

$$\begin{aligned}
O(\epsilon): &\quad iq_{1t} + q_{1xx} = 0,\\
O(\epsilon^2): &\quad iq_{2t} + q_{2xx} = 0,\\
O(\epsilon^3): &\quad iq_{3t} + q_{3xx} - 2\lambda|q_1|^2 q_1 = 0,\\
O(\epsilon^4): &\quad iq_{4t} + q_{4xx} - 2\lambda(q_1^2\bar{q}_2 + 2q_1q_2\bar{q}_1) = 0,\\
&\vdots\\
O(\epsilon^N): &\quad iq_{Nt} + q_{Nxx} - 2\lambda\{|q|^2 q\}_N = 0,
\end{aligned}$$

where $\{\cdots\}_N$ denotes the coefficient of ϵ^N of the enclosed expression.

(ii) For each N, $q_N(x,t)$ and all its partial derivatives have continuous extensions to $\{x \geq 0, t \geq 0\}$.

(iii) For each N, $q_N(\cdot, 0)$ belongs to the Schwartz space $\mathcal{S}(\mathbb{R}_+)$.

We can now state our main result.

Theorem 2.2 *Let $\{g_{0N}^B(t), g_{1N}^B(t)\}_1^\infty$ be periodic functions of period $\tau = \frac{2\pi}{\omega} > 0$. Suppose $\{q_N(x,t)\}_1^\infty$ is a perturbative solution of NLS in the quarter plane such that, for each N, the Dirichlet and Neumann boundary values of q_N asymptote towards $g_{0N}^B(t)$ and $g_{1N}^B(t)$ respectively in the sense that*

$$\begin{aligned}
&q_N(0,\cdot) - g_{0N}^B \in \mathcal{S}(\mathbb{R}_+), && N \geq 1,\\
&q_{Nx}(0,t) - g_{1N}^B(t) = O(t^{-3/2}), && t \to \infty, \quad N \geq 1.
\end{aligned} \tag{10}$$

Suppose the set $\{q_N(\cdot,t)|t \geq 0\}$ is bounded in $L^1(\mathbb{R}_+)$ for each $N \geq 1$.

Then the asymptotic Neumann values $\{g_{1N}^B(t)\}_{N=1}^\infty$ can be constructed explicitly from the asymptotic Dirichlet values $\{g_{0N}^B(t)\}_{N=1}^\infty$ as follows. Let $\{a_{N,n}\}_{n=-\infty}^\infty$ denote the Fourier coefficients of g_{0N}^B:

$$g_{0N}^B(t) = \sum_{n=-\infty}^{\infty} a_{N,n} e^{in\omega t}, \qquad N \geq 1. \tag{11}$$

Then the Fourier coefficients $\{c_{N,n}\}_{n=-\infty}^\infty$ of

$$g_{1N}^B(t) = \sum_{n=-\infty}^{\infty} c_{N,n} e^{in\omega t}, \qquad N \geq 1, \tag{12}$$

are given by

$$\begin{aligned}
c_{N,n} = \Bigg\{\Bigg(&- 2i\lambda k \sum_{l,m=-\infty}^{\infty} \bar{a}_l d_m(k) d_{n+l-m}(k) - \lambda \sum_{l,m=-\infty}^{\infty} \bar{c}_l d_m(k) d_{n+l-m}(k)\\
&+ 2\lambda \sum_{l,m=-\infty}^{\infty} \bar{a}_l a_m d_{n+l-m}(k) + 2ika_n\Bigg)\Bigg|_{k=k_1(n)}\Bigg\}_N,
\end{aligned} \tag{13}$$

where

- $a_n = \sum_{N=1}^{\infty} a_{N,n}\epsilon^N$, $c_n = \sum_{N=1}^{\infty} c_{N,n}\epsilon^N$, *and* $d_n(k) = \sum_{N=1}^{\infty} d_{N,n}(k)\epsilon^N$.
- *The coefficients* $d_{N,n}(k)$, $N \geq 1$, $n \in \mathbb{Z}$, *satisfy*

$$d_{N,n}(k) = \frac{1}{4ik^2 + in\omega}\Big\{-2\lambda k \sum_{l,m=-\infty}^{\infty} \bar{a}_l d_m(k) d_{n+l-m}(k)$$
$$+ \lambda i \sum_{l,m=-\infty}^{\infty} \bar{c}_l d_m(k) d_{n+l-m}(k) - 2i\lambda \sum_{l,m=-\infty}^{\infty} \bar{a}_l a_m d_{n+l-m}(k) + 2ka_n + ic_n\Big\}_N,$$
$$n \in \mathbb{Z}, \quad N \geq 1, \quad k \in \partial D_1^0, \tag{14}$$

where $D_1^0 = \{\operatorname{Re} k > 0\} \cap \{\operatorname{Im} k > 0\}$ *denotes the first quadrant of the complex* k*-plane.*
- $k_1(n)$ *denotes the unique root of* $4k^2 + n\omega = 0$ *in* ∂D_1^0, *i.e.*

$$k_1(n) = \begin{cases} \frac{i\sqrt{n\omega}}{2}, & n \geq 0, \\ \frac{\sqrt{-n\omega}}{2}, & n < 0. \end{cases}$$

We refer to [12] for a proof of Theorem 2.2. In the remainder of this section, we explain how the $c_{N,n}$'s are determined from equations (13)-(14) and provide several examples.

2.2. Construction of the $c_{N,n}$*'s*
Equation (13) with $N = 1$ yields

$$c_{1,n} = 2ik_1(n)a_{1,n}, \qquad n \in \mathbb{Z}. \tag{15a}$$

Substituting this into equation (14) with $N = 1$, we find

$$d_{1,n}(k) = \frac{2ka_{1,n} + ic_{1,n}}{4ik^2 + in\omega} = \frac{a_{1,n}}{2i(k + k_1(n))}, \qquad n \in \mathbb{Z}, \quad k \in \partial D_1^0. \tag{15b}$$

Similarly, equations (13) and (14) with $N = 2$ yield

$$c_{2,n} = 2ik_1(n)a_{2,n}, \qquad n \in \mathbb{Z}, \tag{16a}$$
$$d_{2,n}(k) = \frac{a_{2,n}}{2i(k + k_1(n))}, \qquad n \in \mathbb{Z}, \quad k \in \partial D_1^0. \tag{16b}$$

Continuing in this way, equations (13) and (14) with $N = 3$ yield

$$c_{3,n} = \Big(- 2i\lambda k \sum_{l,m=-\infty}^{\infty} \bar{a}_{1,l} d_{1,m}(k) d_{1,n+l-m}(k) - \lambda \sum_{l,m=-\infty}^{\infty} \bar{c}_{1,l} d_{1,m}(k) d_{1,n+l-m}(k)$$
$$+ 2\lambda \sum_{l,m=-\infty}^{\infty} \bar{a}_{1,l} a_{1,m} d_{1,n+l-m}(k) + 2ika_{3,n}\Big)\Big|_{k=k_1(n)}, \qquad n \in \mathbb{Z}, \tag{17a}$$

$$d_{3,n}(k) = \frac{1}{4ik^2 + in\omega}\Big(-2\lambda k \sum_{l,m=-\infty}^{\infty} \bar{a}_{1,l} d_{1,m}(k) d_{1,n+l-m}(k)$$
$$+ \lambda i \sum_{l,m=-\infty}^{\infty} \bar{c}_{1,l} d_{1,m}(k) d_{1,n+l-m}(k) \tag{17b}$$

$$- 2i\lambda \sum_{l,m=-\infty}^{\infty} \bar{a}_{1,l} a_{1,m} d_{1,n+l-m}(k) + 2ka_{3,n} + ic_{3,n} \Big), \qquad n \in \mathbb{Z}, \quad k \in \partial D_1^0.$$

This process can be continued indefinitely. Indeed, suppose we have determined $\{c_{M,n}\}_{n=-\infty}^{\infty}$ and $\{d_{M,n}(k)\}_{n=-\infty}^{\infty}$ for $N \leq M-1$. Then equation (13) with $N = M$ yields

$$c_{M,n} = \Big(F_{Mn}(k) + 2ika_{M,n} \Big) \Big|_{k=k_1(n)}, \qquad n \in \mathbb{Z},$$

where the function $F_{Mn}(k)$ is given in terms of known lower order terms:

$$F_{Mn}(k) = \lambda \sum_{l,m=-\infty}^{\infty} \Big\{ - 2ik\bar{a}_l d_m(k) d_{n+l-m}(k) - \bar{c}_l d_m(k) d_{n+l-m}(k) + 2\bar{a}_l a_m d_{n+l-m}(k) \Big\}_M .$$

We can now use equation (14) with $N = M$ to determine $d_{M,n}(k)$:

$$d_{M,n}(k) = \frac{1}{4ik^2 + in\omega} \big(F_{Mn}(k) + 2ka_{M,n} + ic_{M,n} \big), \qquad n \in \mathbb{Z}.$$

This determines $\{c_{M,n}\}_{n=-\infty}^{\infty}$ and $\{d_{M,n}(k)\}_{n=-\infty}^{\infty}$ for $N = M$ and completes the inductive step.

2.3. Examples

Example 2.3 (Single exponential) Suppose

$$g_0^B(t) = \epsilon e^{i\omega t}, \qquad t \geq 0, \quad \omega > 0.$$

In this case, all coefficients $a_{N,n}$ are zero except for $a_{1,1} = 1$. Equations (15) imply that all the coefficients $c_{1,n}$ and $d_{1,n}(k)$ vanish except for

$$c_{1,1} = -\sqrt{\omega}, \qquad d_{1,1}(k) = -\frac{i}{2k + i\sqrt{\omega}}.$$

Equations (16) yield $c_{2,n} = d_{2,n}(k) = 0$ for all n. Equations (17) imply that all the coefficients $c_{3,n}$ and $d_{3,n}(k)$ vanish except for

$$c_{3,1} = -\frac{\lambda}{2\sqrt{\omega}}, \qquad d_{3,1}(k) = \frac{\lambda}{2(2ik - \sqrt{\omega})^2 \sqrt{\omega}}.$$

Continuing in this way, we find that the nonzero coefficients $c_{N,n}$ with $N \leq 8$ are

$$c_{1,1} = -\sqrt{\omega}, \qquad c_{3,1} = -\frac{\lambda}{2\sqrt{\omega}}, \qquad c_{5,1} = \frac{1}{8\omega^{3/2}}, \qquad c_{7,1} = -\frac{\lambda}{16\omega^{5/2}}.$$

In summary, we have found that

$$g_1^B(t) = -\epsilon \Big(\sqrt{\omega} + \frac{\epsilon^2 \lambda}{2\sqrt{\omega}} - \frac{\epsilon^4}{8\omega^{3/2}} + \frac{\epsilon^6 \lambda}{16\omega^{5/2}} + O(\epsilon^8) \Big) e^{i\omega t}.$$

The summation of this perturbative expansion suggests

$$g_1^B(t) = -\epsilon \sqrt{\omega + \lambda \epsilon^2} e^{i\omega t}, \tag{18}$$

which, upon identifying α and ϵ, is in agreement with (7a) (note that the inequality $\omega \geq \alpha^2$ is automatically satisfied in the perturbative limit $\alpha \to 0$).

Example 2.4 (Single exponential) Suppose

$$g_{01}^B(t) = \epsilon e^{-i\omega t}, \qquad t \geq 0, \quad \omega > 0.$$

In this case, all coefficients $a_{N,n}$ are zero except for $a_{1,-1} = 1$. Proceeding as in the previous example, we find that the nonzero coefficients $c_{N,n}$ with $N \leq 8$ are

$$c_{1,-1} = i\sqrt{\omega}, \qquad c_{3,-1} = -\frac{i\lambda}{\sqrt{\omega}}, \qquad c_{5,-1} = -\frac{i}{2\omega^{3/2}}, \qquad c_{7,-1} = -\frac{i\lambda}{2\omega^{5/2}}.$$

In summary,

$$g_1^B(t) = i\epsilon\Big(\sqrt{\omega} - \frac{\epsilon^2\lambda}{\sqrt{\omega}} - \frac{\epsilon^4}{2\omega^{3/2}} - \frac{\epsilon^6\lambda}{2\omega^{5/2}} + O(\epsilon^8)\Big)e^{-i\omega t}.$$

The summation of this perturbative expansion suggests

$$g_1^B(t) = i\epsilon\sqrt{\omega - 2\lambda\epsilon^2}e^{-i\omega t},$$

which, upon identifying α and ϵ and letting $\omega \to -\omega$, is in agreement with (7b) (note that the inequality $\omega \leq -6\alpha^2$ is automatically satisfied in the perturbative limit $\alpha \to 0$).

Example 2.5 (Sum of exponentials) We consider the case of

$$g_0^B(t) = \epsilon(\alpha e^{i\omega t} + \beta e^{-i\omega t}), \qquad t \geq 0, \tag{19}$$

where $\alpha, \beta \in \mathbb{C}$ and $\omega > 0$ are constants. In this case, all coefficients $a_{N,n}$ vanish except for

$$a_{1,1} = \alpha, \qquad a_{1,-1} = \beta.$$

Equations (15) imply that all the $c_{1,n}$'s and $d_{1,n}(k)$'s vanish except for

$$c_{1,1} = -\alpha\sqrt{\omega}, \quad c_{1,-1} = i\beta\sqrt{\omega}, \quad d_{1,1}(k) = -\frac{i\alpha}{2k + i\sqrt{\omega}}, \quad d_{1,-1}(k) = -\frac{i\beta}{2k + \sqrt{\omega}}.$$

Equations (16) yield $c_{2,n} = d_{2,n}(k) = 0$ for all n. In general, $c_{N,n} = 0$ unless both N and n are odd and $N \geq n$. Continuing in this way, we find that the nonzero coefficients $c_{N,n}$ with $N \leq 8$ are

$$c_{1,1} = -\alpha\sqrt{\omega}, \qquad c_{1,-1} = i\beta\sqrt{\omega}, \tag{20a}$$

$$\begin{aligned} c_{3,3} &= -\frac{i(\sqrt{3} + (-2-i))\alpha^2\bar{\beta}\lambda}{2\sqrt{\omega}}, & c_{3,1} &= -\frac{\alpha\lambda(|\alpha|^2 + 4|\beta|^2)}{2\sqrt{\omega}}, \\ c_{3,-1} &= -\frac{i\beta\lambda(|\beta|^2 + (1-i)|\alpha|^2)}{\sqrt{\omega}}, & c_{3,-3} &= \frac{(\sqrt{3} - 2 - i)\beta^2\bar{\alpha}\lambda}{2\sqrt{\omega}}, \end{aligned} \tag{20b}$$

$$\begin{aligned} c_{5,5} &= \frac{((2+i) - (5+2i)\sqrt{3} - (2+3i)\sqrt{5} + (3+2i)\sqrt{15})\alpha^3\bar{\beta}^2}{16\omega^{3/2}}, \\ c_{5,3} &= -\frac{i\alpha^2\bar{\beta}(((3+6i) + (2+3i)\sqrt{3})|\alpha|^2 + (5+2i)((3+2i) + (2+i)\sqrt{3})|\beta|^2)}{2((12+3i) + (7+2i)\sqrt{3})\omega^{3/2}}, \\ c_{5,1} &= \frac{\alpha(|\alpha|^4 - 4(\sqrt{3} - 6)|\alpha|^2|\beta|^2 + 2(9 - i\sqrt{3})|\beta|^4)}{8\omega^{3/2}}, \end{aligned}$$

$$c_{5,-1} = \frac{\beta((\sqrt{3}+(-2+3i))|\alpha|^4 - 2(\sqrt{3}+(2-5i))|\alpha|^2|\beta|^2 - 2i|\beta|^4)}{4\omega^{3/2}},$$
$$c_{5,-3} = -\frac{\beta^2\bar{\alpha}(((48+9i)+(29+4i)\sqrt{3})|\alpha|^2 + 8((3+5i)+(2+3i)\sqrt{3})|\beta|^2)}{4((21+12i)+(12+7i)\sqrt{3})\omega^{3/2}},$$
$$c_{5,-5} = \frac{((1-2i)-(2-5i)\sqrt{3}-(3-2i)\sqrt{5}+(2-3i)\sqrt{15})\beta^3\bar{\alpha}^2}{16\omega^{3/2}}, \tag{20c}$$

$$\begin{aligned}
c_{7,7} = {}& \frac{\alpha^4\bar{\beta}^3\lambda}{32(56+23\sqrt{7})\omega^{5/2}}\Big\{(-595+147i)+(525-147i)\sqrt{3}+(196+49i)\sqrt{5} \\
&- (223-30i)\sqrt{7}-(189-14i)\sqrt{15}+(192-39i)\sqrt{21}+(67+19i)\sqrt{35} \\
&- (63+i)\sqrt{105}\Big\}, \\
c_{7,5} = {}& \frac{i\alpha^3\bar{\beta}^2\lambda}{8((26+97i)+(15+56i)\sqrt{3})((25+10i)+(11+4i)\sqrt{5})\omega^{5/2}} \\
&\times \Big\{\big((1110+2345i)+(645+1350i)\sqrt{3}+(106+1011i)\sqrt{5} \\
&+ (63+582i)\sqrt{15}\big)|\alpha|^2 + (1+i)\big((4542+903i)+(2623+524i)\sqrt{3} \\
&+ (1612+1359i)\sqrt{5}+(931+786i)\sqrt{15}\big)|\beta|^2\Big\}, \\
c_{7,3} = {}& -\frac{\alpha^2\bar{\beta}\lambda}{96((627+2340i)+(362+1351i)\sqrt{3})\omega^{5/2}}\Big\{\big((42588+1476i) \\
&+ (24588+852i)\sqrt{3}\big)|\alpha|^4 - (6-6i)\big((-25328-35261i) \\
&- (14623+20358i)\sqrt{3}+(3594+1545i)\sqrt{5}+(2075+892i)\sqrt{15}\big)|\alpha|^2|\beta|^2 \\
&+ \big((395885+21770i)+(228564+12569i)\sqrt{3}-(9405+7020i)\sqrt{5} \\
&- (5430+4053i)\sqrt{15}\big)|\beta|^4 - 96i((795+627i)+(459+362i)\sqrt{3})|\alpha|^2|\beta|^2\Big\}, \\
c_{7,1} = {}& -\frac{\alpha\lambda}{48\omega^{5/2}}\Big\{3|\alpha|^6 + 6(4+5\sqrt{3})|\alpha|^4|\beta|^2 + 2\big((162+51i) \\
&- (13+12i)\sqrt{3}\big)|\alpha|^2|\beta|^4 - 6|\alpha|^2|\beta|^2\big((\sqrt{3}-6)|\alpha|^2 \\
&+ (2+2i)(\sqrt{3}+(2+5i))|\beta|^2\big) + 16(15-2i\sqrt{3})|\beta|^6\Big\}, \\
c_{7,-1} = {}& \frac{\beta\lambda}{24\omega^{5/2}}\Big\{3(3-2i\sqrt{3})|\alpha|^6 + \big((39+42i)+(28-45i)\sqrt{3}\big)|\alpha|^4|\beta|^2 \\
&+ ((36-15i)+(23+36i)\sqrt{3})|\alpha|^2|\beta|^4 + 3|\alpha|^2|\beta|^2\big((-2+2i)(\sqrt{3}-6)|\alpha|^2 \\
&- (\sqrt{3}+(2+5i))|\beta|^2\big) - 12i|\beta|^6\Big\}, \\
c_{7,-3} = {}& \frac{(1+i)\beta^2\bar{\alpha}\lambda}{96((2340+8733i)+(1351+5042i)\sqrt{3})\omega^{5/2}}\Big\{3\big((32088+120795i) \\
&+ (18526+69741i)\sqrt{3}+(8106-6393i)\sqrt{5}+(4680-3691i)\sqrt{15}\big)|\alpha|^4 \\
&+ 6i((168629+32048i)+(97358+18503i)\sqrt{3}+(8733+11700i)\sqrt{5} \\
&+ (5042+6755i)\sqrt{15})|\alpha|^2|\beta|^2 - (4-4i)\beta\big(4((265+7718i) \\
&+ (153+4456i)\sqrt{3})|\beta|^2\bar{\beta} - 9((892+3329i)+(515+1922i)\sqrt{3})|\alpha|^2\bar{\beta}\big)\Big\}, \\
c_{7,-5} = {}& -\frac{(1-i)\beta^3\bar{\alpha}^2\lambda}{24((97+26i)+(56+15i)\sqrt{3})((15+40i)+(7+18i)\sqrt{5})\omega^{5/2}}
\end{aligned}$$

$$\times \Big\{3\big((3820+3375i)+(2205+1950i)\sqrt{3}+(1464+1395i)\sqrt{5} + (845+806i)\sqrt{15}\big)|\alpha|^2+(1+i)\big((7755+7962i)+(4466+4602i)\sqrt{3} + (2742+2586i)\sqrt{5}+(1578+1495i)\sqrt{15}\big)|\beta|^2\Big\},$$

$$c_{7,-7} = \frac{\beta^4\bar{\alpha}^3\lambda}{32(56+23\sqrt{7})\omega^{5/2}}\Big\{(-147-595i)+(147+525i)\sqrt{3}-(49-196i)\sqrt{5} - (30+223i)\sqrt{7}-(14+189i)\sqrt{15}+(39+192i)\sqrt{21}-(19-67i)\sqrt{35} + (1-63i)\sqrt{105}\Big\}. \tag{20d}$$

3. Finite-gap solutions and potentially asymptotically admissible pairs

Finally, we propose an approach for generating pairs $\{g_0^B, g_1^B\}$ of periodic functions which can arise as asymptotic Dirichlet and Neumann values of a solution of NLS. We make the following definition.

Definition 3.1 A pair of functions $\{g_0^B(t), g_1^B(t)\}$ is *asymptotically admissible* if there exists a smooth solution $q(x,t)$ of NLS in the quarter plane $\{x>0, t>0\}$, such that

(i) $q(x,t)$ and all its partial derivatives have continuous extensions to $\{x \geq 0, t \geq 0\}$.

(ii) The initial data decay for large x, that is, $q(\cdot, 0) \in \mathcal{S}(\mathbb{R}_+)$.

(iii) As $t \to \infty$, the Dirichlet and Neumann boundary values of q asymptote towards $g_0^B(t)$ and $g_1^B(t)$ respectively, that is,

$$q(0,\cdot) - g_0^B \in \mathcal{S}(\mathbb{R}_+), \qquad q_x(0,t) - g_1^B(t) = O(t^{-3/2}), \qquad t \to \infty. \tag{21}$$

The spectral functions $\{a(k), b(k), A(k), B(k)\}$ associated with the initial and boundary values $\{q_0(x), g_0(t), g_1(t)\}$ of a solution $q(x,t)$ of (1) in the quarter plane are not independent, but satisfy an important relation called the *global relation*. If the Dirichlet and Neumann values satisfy (21) where $\{g_0^B(t), g_1^B(t)\}$ is a pair of smooth periodic functions of period $\tau > 0$, then the global relation requires that the quotient $B(k)/A(k)$ be continuous in a certain region of the complex k-plane. In situations where $B(k)/A(k)$ can be explicitly computed, it is possible to use this continuity condition to impose constraints on $\{g_0^B(t), g_1^B(t)\}$. This approach was employed in [5] to derive the characterization (7) of asymptotically admissible single exponential boundary values for the focusing NLS equation. In [5], the continuity requirement that the global relation imposes on $B(k)/A(k)$ characterizes $\{g_0^B(t), g_1^B(t)\}$ completely.

In general, the nonexplicit nature of the quotient $B(k)/A(k)$ makes it challenging to apply the above approach. However, there exists a large class of pairs $\{g_0^B, g_1^B\}$ for which the quotient $B(k)/A(k)$ is known explicitly. This is the class of finite-gap solutions generated by the Baker-Akhiezer formalism. We propose that potentially asymptotically admissible pairs for the NLS equation can be generated by taking finite-gap solutions for which the data $\{g_0^B, g_1^B\}$ is periodic (not just quasiperiodic) and then enforcing the continuity condition derived from the global relation on $B(k)/A(k)$.

Acknowledgments

The authors acknowledge support from the EPSRC, UK.

References

[1] Biondini G, Fokas A S and Shepelsky D 2013 Comparison of two approaches to the IBVP for the NLS equation on the half-line with Robin boundary conditions, preprint.

[2] Boutet de Monvel A, Fokas A S and Shepelsky D 2003 The analysis of the global relation for the nonlinear Schrödinger equation on the half-line, *Lett. Math. Phys.* **65** 199–212

[3] Boutet de Monvel A, Its A and Kotlyarov V 2007 Long-time asymptotics for the focusing NLS equation with time-periodic boundary condition, *C. R. Math. Acad. Sci. Paris* **345** 615–620.

[4] Boutet de Monvel A, Its A and Kotlyarov V 2009 Long-time asymptotics for the focusing NLS equation with time-periodic boundary condition on the half-line, *Comm. Math. Phys.* **290** 479–522

[5] Boutet de Monvel A , Kotlyarov V and Shepelsky D 2009 Decaying long-time asymptotics for the focusing NLS equation with periodic boundary condition, *Int. Math. Res. Not. IMRN* 547–577

[6] Boutet de Monvel A, Kotlyarov V, Shepelsky D and Zheng C 2010 Initial boundary value problems for integrable systems: towards the long time asymptotics, *Nonlinearity* **23** 2483.

[7] Degasperis A, Manakov S V and Santini P M 2001 On the initial-boundary value problems for soliton equations, *JETP Letters* **74** 481–485.

[8] Fokas A S 1997 A unified transform method for solving linear and certain nonlinear PDEs, *Proc. Roy. Soc. Lond.* A **453** 1411–1443.

[9] Fokas A S 2002 Integrable nonlinear evolution equations on the half-line, *Comm. Math. Phys.* **230** 1–39.

[10] Fokas A S 2005 A generalised Dirichlet to Neumann map for certain nonlinear evolution PDEs, *Comm. Pure Appl. Math.* **LVIII** 639–670.

[11] Fokas A S and Lenells J 2012 The unified method: I Non-linearizable problems on the half-line, *J. Phys. A: Math. Theor.* **45**, 195201.

[12] Fokas A S and Lenells J 2013 The asymptotics of the Neumann value for the nonlinear Schrödinger equation with t-periodic data, preprint.

[13] Hwang G and Fokas A S 2013 The modified Korteweg-de Vries equation on the half-line with a sine-wave as Dirichlet datum, *J. Nonl. Math. Phys.*, to appear.

[14] Lenells J and Fokas, A S 2012 The unified method: II NLS on the half-line with t-periodic boundary conditions, *J. Phys. A: Math. Theor.* **45**, 195202.

Higher order Peregrine breathers, their deformations and multi-rogue waves.

P Gaillard

Université de Bourgogne, 9 Avenue Alain Savary, Campus de Mirande, 21000 Dijon, France

E-mail: pierre.gaillard@u-bourgogne.fr

Abstract. We study the solutions of the one dimensional focusing NLS equation. Here we construct new deformations of the Peregrine breather of order 7 with 12 real parameters. We obtain new families of quasi-rational solutions of the NLS equation. With this method, we construct new patterns of different types of rogue waves. We recover triangular configurations as well as rings isolated. As already seen in the previous studies, one sees appearing for certain values of the parameters, new configurations of concentric rings.

1. Introduction

The nonlinear Schrödinger equation was first solved by Zakharov and Shabat [1] in 1972 by the inverse scattering method. The first expressions of the quasi-rational solutions were given by Peregrine [2] in 1983. From this time, a considerable number of studies were carried out. Eleonski, Akhmediev and Kulagin obtained the first higher order analogue of the Peregrine breather [3] in 1986. Akhmediev et al. [4, 5], constructed other analogues of order 3 and 4, using Darboux transformations.

Rational solutions of the NLS equation have been written in 2010, as a quotient of two Wronskians in [6]. An other representation of the solutions of the NLS equation has been constructed in [7] in 2011, also in terms of a ratio of two Wronskians determinants of order $2N$. In 2012, Guo, Ling and Liu constructed an other representation of the solutions of the focusing NLS equation, as a ratio of two determinants has been given in [8] using generalized Darboux transform.

In the same year, Ohta and Yang [9] have given a new approach where solutions of the focusing NLS equation by means of a determinant representation, obtained from Hirota bilinear method. A the beginning of the year 2012, one obtained a representation in terms of determinants which does not involve limits [10].

The two formulations given in [7, 10] did depend in fact only on two parameters; this remark was first made by V.B. Matveev. Then we found for the order N (for determinants of order 2N), solutions depending on $2N-2$ real parameters.

In this article, we restrict ourself the study to the case of the solutions of NLS of order 7; because of the constraints of the publication, we do not have the space to publish all the deformations. With this new method, we construct news deformations at order 7 with 12 real parameters. The explicit representation in terms of polynomials is found, but is too monstrous to be published. One constructs various drawings to illustrate the evolution of the solutions according to the parameters. One obtains at the same time triangular configurations and ring structures with a maximum of 28 peaks. These deformations are completely new and gives by new patterns a better understanding of the NLS equation.

2. Determinant representation of solutions of NLS equation

We recall the results obtained in [7] and [10]. We consider the focusing NLS equation

$$iv_t + v_{xx} + 2|v|^2 v = 0. \tag{1}$$

In the following, we consider $2N$ parameters λ_ν, $\nu = 1, \dots, 2N$ satisfying the relations

$$0 < \lambda_j < 1, \quad \lambda_{N+j} = -\lambda_j, \quad 1 \leq j \leq N. \tag{2}$$

We define the terms κ_ν, δ_ν, γ_ν by the following equations,

$$\kappa_\nu = 2\sqrt{1-\lambda_\nu^2}, \quad \delta_\nu = \kappa_\nu \lambda_\nu, \quad \gamma_\nu = \sqrt{\frac{1-\lambda_\nu}{1+\lambda_\nu}}, \tag{3}$$

and

$$\kappa_{N+j} = \kappa_j, \quad \delta_{N+j} = -\delta_j, \quad \gamma_{N+j} = 1/\gamma_j, \quad j = 1 \dots N. \tag{4}$$

The terms $x_{r,\nu}$ $(r = 3,\ 1)$ are defined by

$$x_{r,\nu} = (r-1) \ln \frac{\gamma_\nu - i}{\gamma_\nu + i}, \quad 1 \leq j \leq 2N. \tag{5}$$

The parameters e_ν are defined by

$$e_j = ia_j - b_j, \quad e_{N+j} = ia_j + b_j, \quad 1 \leq j \leq N, \tag{6}$$

where a_j and b_j, for $1 \leq j \leq N$ are arbitrary real numbers.
We use the following notations :

$$\begin{aligned} A_\nu &= \kappa_\nu x/2 + i\delta_\nu t - ix_{3,\nu}/2 - ie_\nu/2, \\ B_\nu &= \kappa_\nu x/2 + i\delta_\nu t - ix_{1,\nu}/2 - ie_\nu/2, \end{aligned} \tag{7}$$

for $1 \leq \nu \leq 2N$, with κ_ν, δ_ν, $x_{r,\nu}$ defined in (3), (4) and (5).
The parameters e_ν are defined by (6).
Here, the parameters a_j and b_j, for $1 \leq N$ are chosen in the form

$$a_j = \sum_{k=1}^{N-1} \tilde{a_k} \epsilon^{2k+1} j^{2k+1}, \quad b_j = \sum_{k=1}^{N-1} \tilde{b_k} \epsilon^{2k+1} j^{2k+1}, \quad 1 \leq j \leq N. \tag{8}$$

We consider the following functions :

$$\begin{aligned} f_{4j+1,k} &= \gamma_k^{4j-1} \sin A_k, \quad f_{4j+2,k} = \gamma_k^{4j} \cos A_k, \\ f_{4j+3,k} &= -\gamma_k^{4j+1} \sin A_k, \quad f_{4j+4,k} = -\gamma_k^{4j+2} \cos A_k, \end{aligned} \tag{9}$$

for $1 \leq k \leq N$, and

$$\begin{aligned} f_{4j+1,N+k} &= \gamma_k^{2N-4j-2} \cos A_{N+k}, \quad f_{4j+2,N+k} = -\gamma_k^{2N-4j-3} \sin A_{N+k}, \\ f_{4j+3,N+k} &= -\gamma_k^{2N-4j-4} \cos A_{N+k}, \quad f_{4j+4,N+k} = \gamma_k^{2N-4j-5} \sin A_{N+k}, \end{aligned} \tag{10}$$

for $1 \leq k \leq N$.
We define the functions $g_{j,k}$ for $1 \leq j \leq 2N$, $1 \leq k \leq 2N$ in the same way, we replace only the term A_k by B_k.

$$\begin{array}{ll} g_{4j+1,k} = \gamma_k^{4j-1} \sin B_k, & g_{4j+2,k} = \gamma_k^{4j} \cos B_k, \\ g_{4j+3,k} = -\gamma_k^{4j+1} \sin B_k, & g_{4j+4,k} = -\gamma_k^{4j+2} \cos B_k, \end{array} \tag{11}$$

for $1 \leq k \leq N$, and

$$\begin{array}{ll} g_{4j+1,N+k} = \gamma_k^{2N-4j-2} \cos B_{N+k}, & g_{4j+2,N+k} = -\gamma_k^{2N-4j-3} \sin B_{N+k}, \\ g_{4j+3,N+k} = -\gamma_k^{2N-4j-4} \cos B_{N+k}, & g_{4j+4,N+k} = \gamma_k^{2N-4j-5} \sin B_{N+k}, \end{array} \tag{12}$$

for $1 \leq k \leq N$.
Then we get the following result :

Theorem 2.1 *The function v defined by*

$$v(x,t) = \frac{\det((n_{jk})_{j,k\in[1,2N]})}{\det((d_{jk})_{j,k\in[1,2N]})} e^{2it-i\varphi} \tag{13}$$

is a quasi-rational solution of the NLS equation (1)

$$iv_t + v_{xx} + 2|v|^2 v = 0,$$

depending on $2N-2$ parameters $\tilde{a}_j$, $\tilde{b}_j$, $1 \leq j \leq N-1$, where

$$\begin{array}{l} n_{j1} = f_{j,1}(x,t,0), \quad n_{jk} = \frac{\partial^{2k-2} f_{j,1}}{\partial \epsilon^{2k-2}}(x,t,0), \\ n_{jN+1} = f_{j,N+1}(x,t,0), \quad n_{jN+k} = \frac{\partial^{2k-2} f_{j,N+1}}{\partial \epsilon^{2k-2}}(x,t,0), \\ d_{j1} = g_{j,1}(x,t,0), \quad d_{jk} = \frac{\partial^{2k-2} g_{j,1}}{\partial \epsilon^{2k-2}}(x,t,0), \\ d_{jN+1} = g_{j,N+1}(x,t,0), \quad d_{jN+k} = \frac{\partial^{2k-2} g_{j,N+1}}{\partial \epsilon^{2k-2}}(x,t,0), \\ 2 \leq k \leq N,\ 1 \leq j \leq 2N \end{array} \tag{14}$$

The functions f and g are defined in (9),(10), (11), (12).

We don't have the space to give the proof in this publication. We will give it in an other forthcoming paper.
The solutions of the NLS equation can also be written in the form :

$$v(x,t) = \exp(2it - i\varphi) \times Q(x,t)$$

where $Q(x,t)$ is defined by :

$$Q(x,t) := \frac{\begin{vmatrix} f_{1,1}[0] & \dots & f_{1,1}[N-1] & f_{1,N+1}[0] & \dots & f_{1,N+1}[N-1] \\ f_{2,1}[0] & \dots & f_{2,1}[N-1] & f_{2,N+1}[0] & \dots & f_{2,N+1}[N-1] \\ \vdots & \vdots & \vdots & \vdots & \vdots & \vdots \\ f_{2N,1}[0] & \dots & f_{2N,1}[N-1] & f_{2N,N+1}[0] & \dots & f_{2N,N+1}[N-1] \end{vmatrix}}{\begin{vmatrix} g_{1,1}[0] & \dots & g_{1,1}[N-1] & g_{1,N+1}[0] & \dots & g_{1,N+1}[N-1] \\ g_{2,1}[0] & \dots & g_{2,1}[N-1] & g_{2,N+1}[0] & \dots & g_{2,N+1}[N-1] \\ \vdots & \vdots & \vdots & \vdots & \vdots & \vdots \\ g_{2N,1}[0] & \dots & g_{2N,1}[N-1] & g_{2N,N+1}[0] & \dots & g_{2N,N+1}[N-1] \end{vmatrix}} \tag{15}$$

3. Quasi-rational solutions of order 7 with twelve parameters

Wa have already constructed in [7] solutions for the cases from $N = 1$ until $N = 6$, and in [10] with two parameters.

Because of the length of the expression v of the solution of NLS equation with 12 parameters, we can't give here. We only construct figures to show deformations of the analogue of the seventh Peregrine breather; in the following we will call it for simplicity, the seventh Peregrine breather. Conversely to the study with two parameters given in preceding works [7, 10], we get other type of symmetries in the plots in the (x, t) plane. We give some examples of this fact in the following.

It is important to note the similar role played by the parameters $\tilde{a}_j$ and $\tilde{b}_j$ for a same j; the same configuration of the peaks is obtained. For this reason one will give the figures only for a parameter $\tilde{a}_j$ or $\tilde{b}_j$. On the other hand, to understand the configuration for a value of the parameter, one will give two sights to see the distribution of the peaks.

With different choices of parameters, we obtain all types of configurations : triangles, rings and concentric rings with a maximum of 28 peaks.

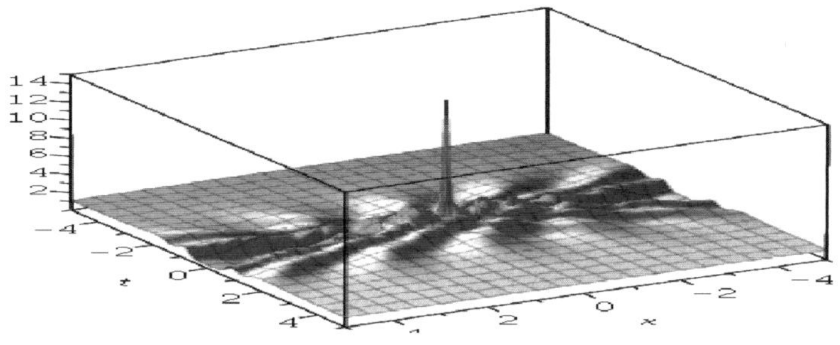

Figure 1. Solution of NLS, N=7; all parameters equal to 0, the Peregrine breather of order 7, P_7.

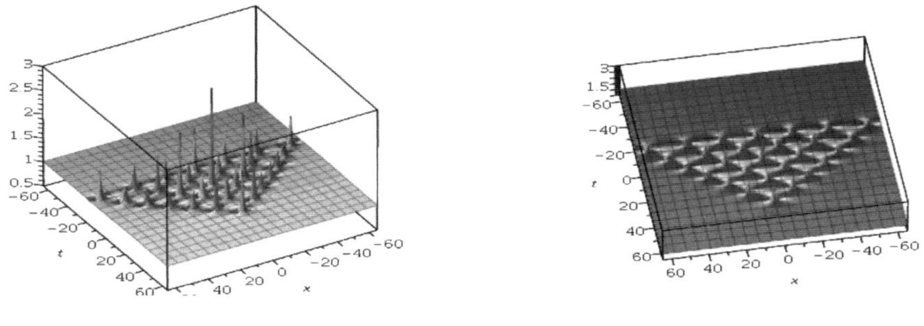

Figure 2. Solution of NLS, N=7; $\tilde{b}_1 = 10^4$; we obtain a regular triangle with 28 peaks; on the right, sight of top.

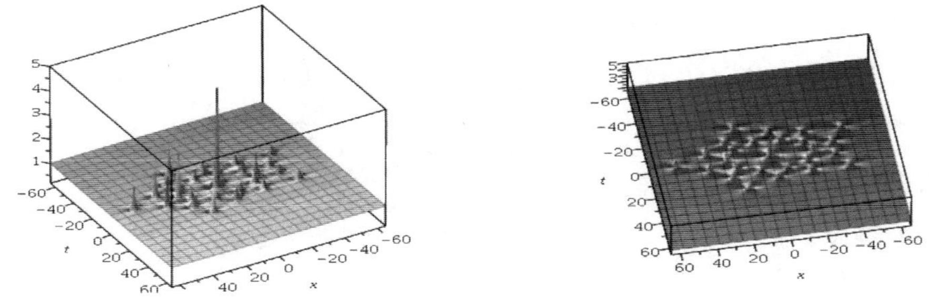

Figure 3. Solution of NLS, N=7; $\tilde{a}_2 = 10^6$, 3 rings with respectively 5, 10, 10 peaks with in the center the Peregrine of order 2, P_2; on the right, sight of top.

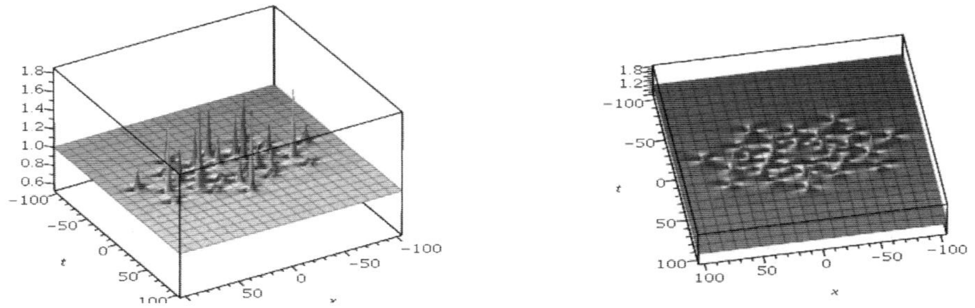

Figure 4. Solution of NLS, N=7; $\tilde{b}_3 = 10^{10}$, 4 rings with 7 peaks on each of them without central peak; on the right, sight of top.

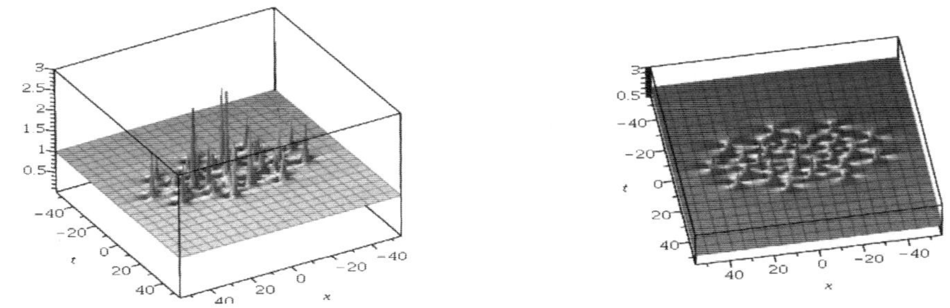

Figure 5. Solution of NLS, N=7; $\tilde{a}_4 = 10^{10}$, 3 rings with 9 peaks on each of them with in the center one peak; on the right, sight of top.

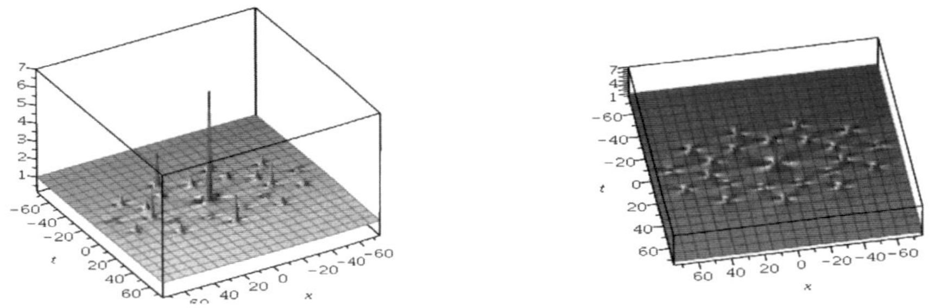

Figure 6. Solution of NLS, N=7; $\tilde{b}_5 = 10^{15}$, 2 rings of 11 peaks with in the center the Peregrine breather of order 3, P_3; on the right, sight of top.

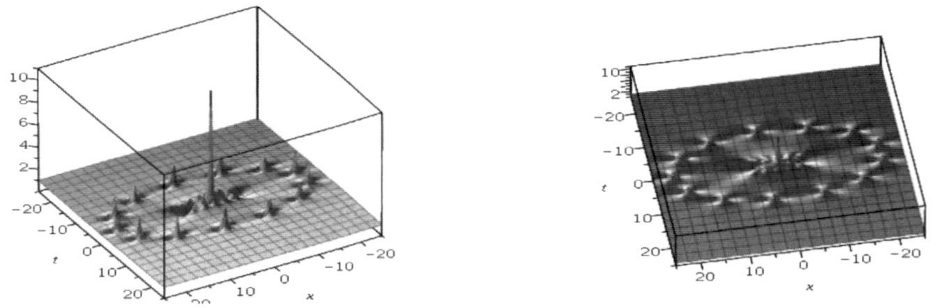

Figure 7. Solution of NLS, N=7; $\tilde{a}_6 = 10^{12}$, $\tilde{b}_6 = -10^8$, a ring with 13 peaks and in the center the Peregrine breather of order 5, P_5; on the right, sight of top.

4. Conclusion

We have constructed explicitly solutions of the NLS equation of order N with $2N - 2$ real parameters. The expressions in terms of polynomials in x and t are too monstrous to be published in this paper.

It is important to note the symmetrical role played by the parameters $\tilde{a}_j$ and $\tilde{b}_j$; the configurations obtained for one of these two parameters $\tilde{a}_j$ or $\tilde{b}_j$ for the index j are the same ones. Thus for each couple $(\tilde{a}_j; \tilde{b}_j)$ we have only built one associated figure for only for one parameter, $\tilde{a}_j \neq 0$, or $\tilde{b}_j \neq 0$.

In the cases $a_1 \neq 0$ or $b_1 \neq 0$ we obtain triangles with a maximum of 28 peaks; for $a_2 \neq 0$ or $b_2 \neq 0$, we have 3 concentric rings with two of them with 10 peaks and an other with 5 peaks with in the center the Peregrine P_2 with 3 peaks. For $a_3 \neq 0$ or $b_3 \neq 0$, we obtain 4 concentric rings without central peak with 7 peaks on each of them. For $a_4 \neq 0$ or $b_4 \neq 0$, we have 3 concentric rings with 9 peaks, with a in the center one peak. For $a_5 \neq 0$ or $b_5 \neq 0$, we obtain 2 concentric rings without central peak with 11 peaks on each of them and the appearance in the center of the Peregrine breather P_3 with 6 peaks. For $a_6 \neq 0$ or $b_6 \neq 0$, we have only one ring with 13 peaks with inside the appearance of the Peregrine breather of order 5 with 15 peaks.

We obtained new patterns in the $(x;t)$ plane, by different choices of these parameters; we recognized rings configurations as already observed in the case of deformations depending on two parameters [7, 10]. We get news triangular shapes and multi-concentric rings.

This study at the order 7 was never still carried out; it is completely new and makes it possible to provide a better understanding of the phenomena of rogue waves.

References

[1] Zakharov V E and Shabat A B 1972 *Sov. Phys. JETP* **34** 62
[2] Peregrine D 1983 *J. Austral. Math. Soc. Ser.* B **25** 16
[3] Akhmediev N, Eleonsky V and Kulagin N 1985 *Sov. Phys. JETP* **62** 894
[4] Akhmediev N, Ankiewicz A and Soto-Crespo J M 2009 *Phys. Rev.* E **80** 026601-1-9
[5] Ankiewicz A, Clarkson P A and Akhmediev N 2010 *J. Phys. A : Math. Theor.* **43** 122002 1
[6] Dubard P, Gaillard P, Klein C and Matveev V B 2010 *Eur. Phys. J. Special Topics* **185** 247
[7] Gaillard P 2011 *J. Phys. A : Meth. Theor.* **44** 1
[8] Guo B, Ling L and Liu Q P 2012 *Phys. Rev.* E **85** 026607-1-9
[9] Ohta Y and Yang J 2012 *Pro. R. Soc.* A **468** 1716
[10] Gaillard P 2013 *J. Math. Phys.* **54** 013504-1-32

Journal of Physics: Conference Series **482** (2014) 012017 doi:10.1088/1742-6596/482/1/012017

Riemann-Hilbert Problems, families of commuting operators and soliton equations

V S Gerdjikov[†] **and A B Yanovski**[‡]

[†]Institute for Nuclear Research and Nuclear Energy
Bulgarian academy of sciences
Tzarigradsko chaussee 72, 1784 Sofia, Bulgaria
[‡] Department of Mathematics and Applied Mathematics
University of Cape Town, 7700 Rondebosch, Cape Town, South Africa

E-mail: gerjikov@inrne.bas.bg, Alexandar.Ianovsky@uct.ac.za

Abstract. We use Riemann-Hilbert Problems with canonical normalization to develop technique for constructing families of commuting operators. As a result we are able to derive new hierarchies of integrable nonlinear evolution equations.

1. Introduction

Let Γ be a contour in the complex λ-plane splitting $\mathbb{C}$ into two parts $\mathbb{C} \equiv \Gamma_+ \cup \Gamma_-$. By multiplicative Riemann-Hilbert problem (RHP)[4, 18] we mean the problem of constructing two functions $\xi^+(\lambda)$ and $\xi^-(\lambda)$ analytic for $\lambda \in \Gamma_+$ and $\lambda \in \Gamma_-$ respectively such that:

$$\xi^+(\lambda) = \xi^-(\lambda)G(\lambda), \qquad \text{for } \lambda \in \Gamma. \tag{1}$$

If the functions $\xi^{\pm}(\lambda)$ are scalar and have no zeroes in their regions of analyticity we can solve the multiplicative RHP by the Plemelj-Sokhotzky formulae [4, 18].

We will consider more general and special RHP for functions $\xi^{\pm}(\vec{x}, t, \lambda)$ taking values in a simple Lie group $\mathcal{G}$ with simple Lie algebra $\mathfrak{g}$.

Let us outline the special properties of our RHP. First, as contour Γ we will choose either the real axis $\mathbb{R}$, or a set of straight lines intersecting at the origin of $\mathbb{C}$. Second, we will assume that the solutions and the sewing function depend on two or more auxiliary variables one of which we will call the time t and the others $\vec{x} = (x_1, \dots, x_p)^T$ will be spatial variables. It is natural that their number p is smaller than the rank of the algebra $r = \operatorname{rank} \mathfrak{g}$. The third special property is that we will specify explicitly the dependence of the sewing function $G(\vec{x}, t, \lambda)$ by:

$$i\frac{\partial G}{\partial t} - \lambda[K, G(\vec{x}, t, \lambda)] = 0, \qquad i\frac{\partial G}{\partial x_s} - \lambda[J_s, G(\vec{x}, t, \lambda)] = 0, \qquad \lambda \in \Gamma, \tag{2}$$

where K and J_s belong to the Cartan subalgebra $\mathfrak{h} \in \mathfrak{g}$.

We shall say that the functions $\xi^{\pm}(\lambda)$ are regular solution of the multiplicative RHP if:

(i) on the contour Γ they satisfy the equation:

$$\xi^+(\vec{x}, t, \lambda) = \xi^-(\vec{x}, t, \lambda)G(\vec{x}, t, \lambda), \qquad \lambda \in \Gamma; \tag{3}$$

(ii) the functions $\xi^{\pm}(\lambda)$ have no singularities or degeneracies for $\lambda \in \mathbb{C}\backslash\Gamma$;

The RHP (3) will have unique solution only after imposing a normalization condition. This in our case is the canonical normalization condition:

$$\lim_{\lambda\to\infty} \xi^{\pm}(\vec{x},t,\lambda) = 1\!\!1. \tag{4}$$

In what follows we will start with an RHP whose sewing functions have special dependence on the auxiliary variables t and x_s and will demonstrate that their solutions are simultaneous fundamental analytic solutions to a set of commuting operators M, L_s. In fact for the simplest nontrivial case when J_s, K are real and $\Gamma \equiv \mathbb{R}$ this has been known for long time [21, 22, 6]. However the RHP was used just for deriving the soliton solutions through the Zakharov-Shabat dressing method [21, 22], see also [14, 10].

The family of operators M, L_s commute provided the first nontrivial coefficient $Q_1(\vec{x},t)$ in the asymptotic expansion of $\xi^{\pm}(\vec{x},t,\lambda)$ (see eq. (6)) satisfies a certain set of nonlinear evolution equations (NLEE). We will demonstrate on several nontrivial examples that this method allows one to derive new types of integrable interactions. Thus we show that the formal approach of Gel'fand and Dickey [3] can be made more precise.

In Section 2 we outline the main idea of constructing the families of commuting operators [7, 8]. The next Section 3 contains several new examples of N-wave type interactions extending the ones found in [12] and their reductions [11, 10, 9]. The last Section contains discussion and conclusions.

2. RHP with canonical normalization

2.1. RHP and Generalized Zakharov-Shabat operators

Consider the simplest nontrivial RHP when $\Gamma \equiv \mathbb{R}$ and the dependence of $G(\vec{x},t,\lambda)$ is provided by eq. (2) with *real* K and $J_s \in \mathfrak{h} \subset \mathfrak{g}$.

Remark 1. *With this choice of K and J_s it is easy to see that*

$$G(\vec{x},t,\lambda) = \exp\left(-iKt - i\sum_{s=1}^{p} x_s J_s\right) G(\vec{0},0,\lambda) \exp\left(iKt + i\sum_{s=1}^{p} x_s J_s\right). \tag{5}$$

We assume that $G(\vec{0},0,\lambda)$ is a smooth bounded function of $\lambda \in \mathbb{R}$ and as a consequence it follows that $G(\vec{x},t,\lambda)$ is a smooth bounded function of $\lambda \in \mathbb{R}$ for all $\vec{x}$ and t.

Remark 2. *The canonical normalization (4) along with the analyticity properties of the solutions ensure that $\xi^{\pm}(\vec{x},\lambda)$ allow asymptotic expansions of the form:*

$$\xi^{\pm}(\vec{x},\lambda) = \exp\left(\mathcal{Q}(\vec{x},t,\lambda)\right), \qquad \mathcal{Q}(\vec{x},t,\lambda) = \sum_{k=1}^{\infty} Q_k(\vec{x},t)\lambda^{-k}. \tag{6}$$

where $Q_k(\vec{x},t)$ are smooth functions of $\vec{x}$ and t vanishing fast enough for $\vec{x}\to\infty$. The rigorous proof of this fact is out of the scope of the present paper.

It is obvious that

$$\mathcal{J}_s(\vec{x},t,\lambda) = \xi^{\pm}(\vec{x},t,\lambda) J_s \hat{\xi}^{\pm}(\vec{x},t,\lambda), \qquad \mathcal{K}(\vec{x},t,\lambda) = \xi^{\pm}(\vec{x},t,\lambda) K \hat{\xi}^{\pm}(\vec{x},t,\lambda), \tag{7}$$

belong to the algebra $\mathfrak{g}$ for any J_s and K from $\mathfrak{g}$ and allow analytic extensions for $\lambda \in \mathbb{C}_{\pm}$; here $\hat{\xi} \equiv \xi^{-1}$. Since K and J_s belong to the Cartan subalgebra $\mathfrak{h}$, then

$$[\mathcal{J}_s(\vec{x},t,\lambda), \mathcal{K}(\vec{x},t,\lambda)] = 0. \tag{8}$$

Theorem 1 (Zakharov-Shabat, 1974). *Let $\xi^{\pm}(x,t,\lambda)$ be solutions to the RHP (3) whose sewing function depends on the auxiliary variables $\vec{x}$ and t as above. Then $\xi^{\pm}(x,t,\lambda)$ are fundamental analytic solutions of the set of differential operators*

$$
\begin{aligned}
L_s\xi^{\pm} \equiv & i\frac{\partial \xi^{\pm}}{\partial x_s} + [J_s, Q_1(\vec{x},t,\lambda)]\xi^{\pm}(\vec{x},t,\lambda) - \lambda[J_s, \xi^{\pm}(\vec{x},t,\lambda)] = 0,\\
M\xi^{\pm} \equiv & i\frac{\partial \xi^{\pm}}{\partial t} + [K, Q_1(\vec{x},t,\lambda)]\xi^{\pm}(\vec{x},t,\lambda) - \lambda[K, \xi^{\pm}(\vec{x},t,\lambda)] = 0.
\end{aligned}
\tag{9}
$$

where $Q_1(\vec{x},\lambda)$ is the first nontrivial coefficient in the asymptotic expansion (6) of $\xi^{\pm}(\vec{x},t,\lambda)$.

Proof. Introduce the functions:

$$
\begin{aligned}
g_s^{\pm}(\vec{x},t,\lambda) &= i\frac{\partial \xi^{\pm}}{\partial x_s}\hat{\xi}^{\pm}(\vec{x},t,\lambda) + \lambda\xi^{\pm}(\vec{x},t,\lambda)J_s\hat{\xi}^{\pm}(\vec{x},t,\lambda),\\
g^{\pm}(\vec{x},t,\lambda) &= i\frac{\partial \xi^{\pm}}{\partial t}\hat{\xi}^{\pm}(\vec{x},t,\lambda) + \lambda\xi^{\pm}(\vec{x},t,\lambda)K\hat{\xi}^{\pm}(\vec{x},t,\lambda),
\end{aligned}
\tag{10}
$$

and using (2) prove that

$$
g_s^{+}(\vec{x},t,\lambda) = g_s^{-}(\vec{x},t,\lambda), \qquad g^{+}(\vec{x},t,\lambda) = g^{-}(\vec{x},t,\lambda), \qquad \lambda \in \Gamma,
\tag{11}
$$

which means that these functions are analytic functions of λ in the whole complex λ-plane. Next we find that for $\lambda \to \infty$:

$$
\lim_{\lambda\to\infty}(g_s^{\pm}(\vec{x},t,\lambda) - \lambda J_s) = -[J_s, Q(\vec{x},t)], \qquad \lim_{\lambda\to\infty} g^{\pm}(\vec{x},t,\lambda) - \lambda K) = -[J_s, Q(\vec{x},t)].
\tag{12}
$$

Then we make use of Liouville theorem to get

$$
\begin{aligned}
g_s^{+}(\vec{x},t,\lambda) &= g_s^{-}(\vec{x},t,\lambda) = \lambda J_s - [J_s, Q_1(\vec{x},t)],\\
g^{+}(\vec{x},t,\lambda) &= g^{-}(\vec{x},t,\lambda) = \lambda K - [K, Q_1(\vec{x},t)].
\end{aligned}
\tag{13}
$$

□

Lemma 1. *The set of operators L_s and M have a common FAS, i.e. they all must commute, that is $Q_1(\vec{x},t)$ satisfies the following NLEE:*

$$
i\left[J_k, \frac{\partial Q_1}{\partial x_s}\right] - i\left[J_s, \frac{\partial Q_1}{\partial x_k}\right] + [[J_s, Q_1(\vec{x},t)], [J_k, Q_1(\vec{x},t)]] = 0,
\tag{14a}
$$

$$
i\left[J_s, \frac{\partial Q_1}{\partial t}\right] - i\left[K, \frac{\partial Q_1}{\partial x_s}\right] + [[J_s, Q_1(\vec{x},t)], [K, Q_1(\vec{x},t)]] = 0,
\tag{14b}
$$

Proof. Follows naturally from the definitions of the operators L_s and M (9).

□

2.2. RHP and Operators of Caudry-Beals-Coifman type

In this Subsection we consider more complicated RHP which is formulated as follows.

- First we introduce the complex valued elements K and J_s of the Cartan subalgebra. The conditions [13]

$$
\operatorname{Im} \lambda\alpha(J) = 0, \qquad \alpha \in \Delta,
\tag{15}
$$

where Δ is the root system of $\mathfrak{g}$, gives a set of M straight lines, or equivalently, a set of $2M$ rays l_ν starting from the origin. We then define the contour as $\Gamma \equiv \bigcup_{\nu=1}^{2M} l_\nu$.

- To each ray l_ν (15) one can associate the subset of roots $\delta_\nu \subset \Delta$ and the corresponding subalgebra $\mathfrak{g}_\nu \subset \mathfrak{g}$. Then the corresponding sewing function takes values in the corresponding subgroup $G_\nu(\vec{x},t,\lambda) \in \mathcal{G}_\nu$ and are bounded functions for $\lambda \in l_\nu$, see remark 1.

The rest of the details are the same as in the previous Subsection. Quite similarly is formulated the generalization of the Zakharov-Shabat theorem. The family of commuting operators formally coincides with the ones in eqs. (9); the difference is that now the Cartan subalgebra elements are complex. As a result however, the operator L in (9) takes the form of a CBC system [2, 1, 13].

2.3. Jets of order k

Another natural generalization consists in formulating the RHP on the complex plane of λ.

$$\xi^+(\vec{x},t,\lambda) = \xi^-(\vec{x},t,\lambda) G(\vec{x},t,\lambda), \qquad \text{for } \lambda^k \in \mathbb{R}, \tag{16}$$

with the canonical normalization (4).

The Zaharov-Shabat method can easily be generalized also for the RHP (16). The result is formulated as

Theorem 2. *Let $\xi^\pm(x,t,\lambda)$ be solutions to the RHP (16) whose sewing function depends on the auxiliary variables $\vec{x}$ and t as follows:*

$$i\frac{\partial G}{\partial t} - \lambda^k [K, G(\vec{x},t,\lambda)] = 0, \qquad i\frac{\partial G}{\partial x_s} - \lambda^k [J_s, G(\vec{x},t,\lambda)] = 0, \qquad \lambda \in \Gamma, \tag{17}$$

where K and J_s belong to the Cartan subalgebra $\mathfrak{h} \in \mathfrak{g}$. Then $\xi^\pm(x,t,\lambda)$ are fundamental analytic solutions of the set of differential operators

$$\begin{aligned} L_s\xi^\pm &\equiv i\frac{\partial \xi^\pm}{\partial x_s} + U_s(\vec{x},t,\lambda)\xi^\pm(\vec{x},t,\lambda) - \lambda^k[J_s, \xi^\pm(\vec{x},t,\lambda)] = 0, \\ M\xi^\pm &\equiv i\frac{\partial \xi^\pm}{\partial t} + V(\vec{x},t,\lambda)\xi^\pm(\vec{x},t,\lambda) - \lambda^k[K, \xi^\pm(\vec{x},t,\lambda)] = 0. \end{aligned} \tag{18}$$

Here $U_s(\vec{x},t,\lambda)$ and $V(\vec{x},t,\lambda)$ are the jets of order k of $\mathcal{J}_s(x,\lambda)$ and $\mathcal{K}(x,\lambda)$, i.e.:

$$\begin{aligned} \lambda^k J_s - U_s(\vec{x},t,\lambda) &= \left(\lambda^k \xi^\pm(\vec{x},t,\lambda) J_l \hat{\xi}^\pm(\vec{x},t,\lambda)\right)_+, \\ \lambda^k K - V(\vec{x},t,\lambda) &= \left(\lambda^k \xi^\pm(\vec{x},t,\lambda) K \hat{\xi}^\pm(\vec{x},t,\lambda)\right)_+. \end{aligned} \tag{19}$$

where the subscript + means that we retain only the nonnegative powers of λ.

Proof. Introduce the functions:

$$\begin{aligned} g_s^\pm(\vec{x},t,\lambda) &= i\frac{\partial \xi^\pm}{\partial x_s}\hat{\xi}^\pm(\vec{x},t,\lambda) + \lambda^k \xi^\pm(\vec{x},t,\lambda) J_s \hat{\xi}^\pm(\vec{x},t,\lambda), \\ g^\pm(\vec{x},t,\lambda) &= i\frac{\partial \xi^\pm}{\partial t}\hat{\xi}^\pm(\vec{x},t,\lambda) + \lambda^k \xi^\pm(\vec{x},t,\lambda) K \hat{\xi}^\pm(\vec{x},t,\lambda), \end{aligned} \tag{20}$$

and using (16) and (17) prove that

$$g_s^+(\vec{x},t,\lambda) = g_s^-(\vec{x},t,\lambda), \qquad g^+(\vec{x},t,\lambda) = g^-(\vec{x},t,\lambda). \tag{21}$$

Thus we find that these functions are analytic functions of λ in the whole complex λ^k-plane. Next we find that:

$$\lim_{\lambda\to\infty} g_s^+(\vec{x},t,\lambda) = \lambda^k J_s, \qquad \lim_{\lambda\to\infty} g^+(\vec{x},t,\lambda) = \lambda^k K. \tag{22}$$

and make use of Liouville theorem to get

$$
\begin{aligned}
g_s^+(\vec{x},t,\lambda) = g_s^-(\vec{x},t,\lambda) &= \lambda^k J_s - \sum_{l=1}^{k} U_{s;l}(\vec{x},t)\lambda^{k-l}, \\
g^+(\vec{x},t,\lambda) = g^-(\vec{x},t,\lambda) &= \lambda^k K - \sum_{l=1}^{k} V_l(\vec{x},t)\lambda^{k-l}.
\end{aligned}
\tag{23}
$$

□

Let us now express $U_s(x) \in \mathfrak{g}$ in terms of the asymptotic coefficients Q_s in eq. (6).

$$
\mathcal{J}_s(\vec{x},t,\lambda) = J_s + \sum_{k=1}^{\infty} \frac{1}{k!}\mathrm{ad}\,_Q^k J_s, \qquad \mathcal{K}(\vec{x},t,\lambda) = K + \sum_{k=1}^{\infty} \frac{1}{k!}\mathrm{ad}\,_Q^k K. \tag{24}
$$

Thus for the first three coefficients of $U_s(\vec{x},t,\lambda)$ we get:

$$
\begin{aligned}
U_{s;1}(\vec{x},t) &= -\mathrm{ad}\,_{Q_1} J_s, \qquad U_{s;2}(\vec{x},t) = -\mathrm{ad}\,_{Q_2} J_s - \frac{1}{2}\mathrm{ad}\,_{Q_1}^2 J_s \\
U_{s;3}(\vec{x},t) &= -\mathrm{ad}\,_{Q_3} J_s - \frac{1}{2}\left(\mathrm{ad}\,_{Q_2}\mathrm{ad}\,_{Q_1} + \mathrm{ad}\,_{Q_1}\mathrm{ad}\,_{Q_2}\right) J_s - \frac{1}{6}\mathrm{ad}\,_{Q_1}^3 J_s,
\end{aligned}
\tag{25}
$$

and similar expressions for $V(\vec{x},t,\lambda)$ with J_s replaced by K.

Lemma 2. *The set of operators L_s and M (18) have a common FAS, i.e. they all must commute, that is, the set of functions $Q_1(\vec{x},t)$, ..., $Q_k(\vec{x},t)$ satisfy the following NLEE:*

$$
\begin{aligned}
i\frac{\partial U_s}{\partial x_j} - i\frac{\partial U_j}{\partial x_s} + [U_s(\vec{x},t,\lambda) - \lambda^k J_s, U_j(\vec{x},t,\lambda) - \lambda^k J_j] &= 0, \\
i\frac{\partial U_s}{\partial t} - i\frac{\partial V}{\partial x_s} + [U_s(\vec{x},t,\lambda) - \lambda^k J_s, V(\vec{x},t,\lambda) - \lambda^k K] &= 0.
\end{aligned}
\tag{26}
$$

Proof. Follows naturally from the definitions of the operators L_s and M (18). □

Remark 3. *Obviously one can use families of operators with different maximal powers of λ. Using them one can derive not only the relevant N-wave type equations, but also the higher order NLEE of the hierarchy.*

Remark 4. *Considering RHP of the form (16) one must take special care of the behavior of the solutions when $\lambda \to 0$. We are not going to discuss here the conditions that are necessary to impose so that the RHP will be properly defined leving it for the future.*

3. Examples of new types of N-wave interactions

The integrability of the well known N-wave equations in two-dimensional space-time was discovered by Zakharov and Manakov [19]. To this end they used the Lax pair

$$
\begin{aligned}
L\chi &\equiv i\frac{\partial \chi}{\partial x} + ([J, Q(x,t)] - \lambda J)\chi(x,t,\lambda) = 0, \\
M\chi &\equiv i\frac{\partial \chi}{\partial t} + ([K, Q(x,t)] - \lambda K)\chi(x,t,\lambda) = -\lambda\chi(x,t,\lambda)K, \\
J &= \mathrm{diag}\,(a_1, a_2, \dots, a_n), \qquad K = \mathrm{diag}\,(b_1, b_2, \dots, b_n),
\end{aligned}
\tag{27}
$$

where the potential $Q(x,t\Delta)$ is a $n \times n$ matrix with $Q_{kk} = 0$. The compatibility of this pair is

$$i\left[J, \frac{\partial Q}{\partial t}\right] - i\left[K, \frac{\partial Q}{\partial x}\right] + [[J, Q(x,t)], [K, Q(x,t)]] = 0. \tag{28}$$

which is a system of $n(n-1)$ equations for the off-diagonal elements of $Q(x,t)$. This system admits the natural reduction $C_0QC_0 = q^\dagger$ where $C_0 = \operatorname{diag}(1, \epsilon_1, \dots \epsilon_{n-1})$ and $\epsilon_k = \pm 1$. After the N-wave system (28) reduces to $n(n-1)/2$ equations for $Q_{km}(x,t)$, $k < m$.

The N-wave are easily generalized to any other simple Lie algebra, see [5]. Indeed, let us consider the simple Lie algebra $\mathfrak{g}$ of rank r with root system Δ and Cartan-Weyl basis H_s, E_α, $\alpha \in \Delta$; assume also that the Cartan generators H_s satisfy $\langle H_s, H_k \rangle = \delta_{sk}$. Then consider the Lax pair

$$\begin{aligned}
&L\chi \equiv i\frac{\partial \chi}{\partial x} + ([J, Q(x,t)] - \lambda J)\chi(x,t,\lambda) = 0, \\
&M\chi \equiv i\frac{\partial \chi}{\partial t} + ([K, Q(x,t)] - \lambda K)\chi(x,t,\lambda) = -\lambda\chi(x,t,\lambda)K, \\
&J = \sum_{s=1}^{r} a_s H_s, \qquad K = \sum_{s=1}^{r} b_s H_s, \qquad Q(x,t) = \sum_{\alpha\in\Delta_+} (q_\alpha(x,t)E_\alpha + p_\alpha(x,t)E_{-\alpha}),
\end{aligned} \tag{29}$$

where $\Delta_+ \subset \Delta$ is the subset of positive roots.

The N-wave equations have as Hamiltonian

$$H_{\rm Nw} = \frac{1}{2i}\int_{x=-\infty}^{\infty} dx\, \left\langle Q, \left[K, \frac{\partial Q}{\partial x}\right]\right\rangle + \frac{1}{3}\int_{x=-\infty}^{\infty} dx\, \langle [J,Q], [Q,[K,Q]]\rangle, \tag{30}$$

where $\langle X, Y\rangle$ is the Killing form between the elements $X, Y \in \mathfrak{g}$. So typically one may say that a given set of NLEE are of N-wave type if: i) they are contain first order derivatives with respect to x and t; ii) the coefficients a_k and b_k are real which physically means that the group velocity of each of these waves is real; and iii) the nonlinearities in the equations are quadratic in the fields q_α.

These types of N-wave equations were known for a long time. A number of their inequivalent $\mathbb{Z}_2$-reductions for the low-rank Lie algebras were described in [11]. Obviously, using the ISM one can show, that in terms of the scattering data of L these equations become *linear* evolution equations. Note that the scattering data of L are determined through the scattering matrix $T(\lambda, t)$ for $\lambda \in \mathbb{R}$ – the continuous spectrum of L, and some additional data characterizing the discrete spectrum of L. The inverse scattering problem for the operator L is best of all reduced to a RHP (3) on the real line.

Our aim in this Section is to demonstrate new examples of qualitatively different N-wave equations which are also integrable.

3.1. CBC systems and 4-wave equations with complex group velocities, $k = 1$.

The 4-wave equations below can be solved by the ISM applied to two operators of CBC type related to the $\mathfrak{g} \simeq so(5)$ algebra. So we consider the Lax pair (29) with

$$J = \operatorname{diag}(a_1, -a_1^*, 0, a_1^*, -a_1), \qquad K = \operatorname{diag}(a_1^*, -a_1, 0, a_1, -a_1^*),$$

$$Q(x,t) = \sum_{\alpha\in\Delta_{1,+}} (q_\alpha E_\alpha + p_\alpha E_{-\alpha}) = \begin{pmatrix} 0 & q_1 & q_2 & q_4 & 0 \\ p_1 & 0 & q_3 & 0 & q_4 \\ p_2 & p_3 & 0 & q_3 & -q_2 \\ p_2 & p_3 & 0 & q_3 & -q_2 \\ p_4 & 0 & p_3 & 0 & q_1 \\ 0 & p_4 & -p_2 & p_1 & 0 \end{pmatrix}. \tag{31}$$

Here a_1 is a complex number that can be assumed to be $a_1 = e^{i\varphi_0}$ and $\Delta_{1,+} \equiv \{e_1 \pm e_2, e_1, e_2\}$ is the set of positive roots of the algebra $so(5)$.

The important difference between (31) and the Lax pair (27) for $\mathfrak{g} \simeq so(5)$ is that now J and K have complex eigenvalues. As a consequence of this the continuous spectrum of L_1 and M_1 fills up the real axis $\mathbb{R}$ and two lines $\mathbb{R}e^{\pm i\varphi_0}$ closing angles $\pm\phi_0$ with $\mathbb{R}$. Again the ISP for the operator L_1 reduces to a RHP, whose solution consists of 6 fundamental analytic solutions $\chi^\nu(x,t,\lambda)$, $\nu = 1,\dots,6$ – one for each of the 6 sectors Ω_ν into which the lines $\mathbb{R}$ and $\mathbb{R}e^{\pm i\varphi_0}$ split the complex λ-plane.

We need in addition a $\mathbb{Z}_2$-reduction which must relate q_j and p_j; several types of such reductions have been described in [11]. However the reduction we will use below

$$C_1 Q(x,t)^\dagger C_1 = Q(x,t), \quad C_1 J^\dagger C_1 = -J, \quad C_1 K^\dagger C_1 = -K, \tag{32}$$

where $C_1 = E_{12} + E_{21} + E_{33} - E_{45} - E_{54}$ corresponds to the Weyl reflection with respect to the root $e_1 - e_2$. This involution leads to a set of algebraic constraints on $q_j(x,t)$ and $p_j(x,t)$. For convenience we introduce

$$\begin{aligned} &p_1(x,t) = w_0(x,t), && q_1(x,t) = w_1(x,t), && q_2(x,t) = w_2(x,t), && q_3(x,t) = w_1(x,t), \\ &q_4(x,t) = w_4(x,t), && p_2(x,t) = w_3^*(x,t), && p_3(x,t) = w_2^*(x,t), && p_4(x,t) = -w_4^*(x,t), \end{aligned} \tag{33}$$

where $w_0(x,t)$ and $w_1(x,t)$ are real-valued functions. Thus instead of the standard 4 complex-valued functions for the 4-wave system (see eqs. (36) and (37) below) we have 2 real and 3 complex-valued functions. The corresponding new 4-wave eqs are:

$$\begin{aligned} &\frac{\partial w_0}{\partial t} - c_0 \frac{\partial w_0}{\partial x} + 2c_0 \sin(\varphi_0)|w_3|^2 = 0, \\ &\frac{\partial w_1}{\partial t} - c_0 \frac{\partial w_1}{\partial x} + 2c_0 \sin(\varphi_0)|w_2|^2 = 0, \\ &\frac{\partial w_2}{\partial t} - c_0 e^{-2i\varphi_0} \frac{\partial w_2}{\partial x} + 2c_0 e^{-2i\varphi_0} \sin(2\varphi_0)(w_2^* w_4 - w_1 w_3) = 0, \\ &\frac{\partial w_3}{\partial t} - c_0 e^{2i\varphi_0} \frac{\partial w_3}{\partial x} - 2c_0 e^{2i\varphi_0} \sin(2\varphi_0)(w_0 w_2 + w_3^* w_4) = 0, \\ &\frac{\partial w_4}{\partial t} + ic_0 \frac{\partial w_4}{\partial x} - 2c_0 \cos(\varphi_0) w_2 w_3 = 0. \end{aligned} \tag{34}$$

The Hamiltonian is given by:

$$\begin{aligned} H_{4\mathrm{w}-1} &= \frac{c_0}{2}\int_{x=-\infty}^{\infty} dx \left\langle Q, \left[K, \frac{\partial Q}{\partial x}\right]\right\rangle - \frac{ic_0}{3}\int_{x=-\infty}^{\infty} dx \ \langle [J,Q],[Q,[K,Q]]\rangle \\ &= c_0 \int_{x=-\infty}^{\infty} dx \ \left(2\cos(\varphi_0)\left(w_1 \frac{\partial w_0}{\partial x} - w_0 \frac{\partial w_1}{\partial x}\right) - i\sin(\varphi_0)\left(w_4^* \frac{\partial w_4}{\partial x} - w_4 \frac{\partial w_4^*}{\partial x}\right)\right. \\ &\left. -e^{-i\varphi_0} w_3^* \frac{\partial w_2}{\partial x} - e^{i\varphi_0} w_3 \frac{\partial w_2^*}{\partial x} - 4\sin(2\varphi_0)\left(w_0|w_2|^2 - w_1|w_3|^2 + w_2 w_3 w_4^* + w_2^* w_3^* w_4\right)\right). \end{aligned} \tag{35}$$

Let us briefly compare the new 4-wave system (34) with the well known 4-wave system (see Chapter 3 of [20]) which are obtained with the generalized Zakharov-Shabat system with real valued $J = a_1 H_1 + a_2 H_2$ and $K = b_1 H_1 + b_2 H_2$ and with the standard reduction $p_k = q_k^*$:

$$\begin{aligned} &i\frac{\partial q_1}{\partial t} - i\frac{b_1 - b_2}{a_1 - a_2}\frac{\partial q_1}{\partial x} - \kappa q_2 q_3^* = 0, && i\frac{\partial q_4}{\partial t} - i\frac{b_1 + b_2}{a_1 + a_2}\frac{\partial q_4}{\partial x} + \kappa q_2 q_3 = 0, \\ &i\frac{\partial q_2}{\partial t} - i\frac{b_1}{a_1}\frac{\partial q_2}{\partial x} + \kappa(q_1 q_3 - q_3^* q_4) = 0, && i\frac{\partial q_1}{\partial t} - i\frac{b_2}{a_2}\frac{\partial q_1}{\partial x} - \kappa(q_2^* q_4 - q_2 q_1^*) = 0, \end{aligned} \tag{36}$$

where $\kappa = a_1b_2 - a_2b_1$. The Hamiltonian is

$$\begin{aligned} H_{4\mathrm{w}-2} = i\int_{x=-\infty}^{\infty} dx\, &\left((b_1-b_2)\left(q_1\frac{\partial q_1^*}{\partial x} - q_1^*\frac{\partial q_1}{\partial x}\right) + (b_1+b_2)\left(q_4^*\frac{\partial q_4}{\partial x} - q_4\frac{\partial q_4^*}{\partial x}\right)\right. \\ &\left. +b_1\left(q_2\frac{\partial q_2^*}{\partial x} - q_2^*\frac{\partial q_2}{\partial x}\right) + b_2\left(q_3\frac{\partial q_3^*}{\partial x} - q_3^*\frac{\partial q_3}{\partial x}\right)\right) \\ &- 2\kappa\int_{x=-\infty}^{\infty} dx\, (q_1q_2^*q_3 - q_1^*q_2q_3^* + q_2q_3q_4^* - q_2^*q_3q_4^*)\,. \end{aligned} \tag{37}$$

The obvious differences are: i) the new 4-wave equation is a system of equations for 3 complex fields w_2, w_3, w_4 and 2 real fields $q_1/a_1, a_1p_1$ (instead of 4 complex fields q_α); ii) the group velocities are now complex (instead of real); iii) the interaction 'strength' between the different waves is different (instead of being equal to κ).

Of course there will be differences between the soliton solutions, but these will be given elsewhere.

3.2. New types of 3-wave interactions, $k=2$.
Let $\mathfrak{g} = sl(3)$ and

$$Q_1(\vec{x},t) = \begin{pmatrix} 0 & u_1 & u_3 \\ -v_1 & 0 & u_2 \\ -v_3 & -v_2 & 0 \end{pmatrix}, \qquad Q_2(\vec{x},t) = \begin{pmatrix} q_{11} & w_1 & w_3 \\ -z_1 & q_{22} & w_2 \\ -z_3 & -z_2 & q_{33} \end{pmatrix}, \tag{38}$$

Fix up $k=2$. Then the Lax pair becomes

$$\begin{aligned} L\xi^\pm &\equiv i\frac{\partial \xi^\pm}{\partial x} + U(x,t,\lambda)\xi^\pm(x,t,\lambda) - \lambda^2[J,\xi^\pm(x,t,\lambda)] = 0, \\ M\xi^\pm &\equiv i\frac{\partial \xi^\pm}{\partial t} + V(x,t,\lambda)\xi^\pm(x,t,\lambda) - \lambda^2[K,\xi^\pm(x,t,\lambda)] = 0, \end{aligned} \tag{39}$$

where

$$\begin{aligned} U \equiv U_2 + \lambda U_1 &= \left([J,Q_2(x)] - \frac{1}{2}[[J,Q_1],Q_1(x)]\right) + \lambda[J,Q_1], \\ V \equiv V_2 + \lambda V_1 &= \left([K,Q_2(x)] - \frac{1}{2}[[K,Q_1],Q_1(x)]\right) + \lambda[K,Q_1]. \end{aligned} \tag{40}$$

Impose a $\mathbb{Z}_2$-reduction of type a) with $A = \operatorname{diag}(1,\epsilon,1)$, $\epsilon^2 = 1$. Thus Q_1 and Q_2 get reduced into:

$$Q_1 = \begin{pmatrix} 0 & u_1 & 0 \\ \epsilon u_1^* & 0 & u_2 \\ 0 & \epsilon u_2^* & 0 \end{pmatrix}, \qquad Q_2 = \begin{pmatrix} 0 & 0 & w_3 \\ 0 & 0 & 0 \\ w_3^* & 0 & 0 \end{pmatrix}, \tag{41}$$

New type of integrable 3-wave equations:

$$\begin{aligned} &i(a_1-a_2)\frac{\partial u_1}{\partial t} - i(b_1-b_2)\frac{\partial u_1}{\partial x} + \epsilon\kappa u_2^*u_3 + \epsilon\frac{\kappa(a_1-a_2)}{(a_1-a_3)}u_1|u_2|^2 = 0, \\ &i(a_2-a_3)\frac{\partial u_2}{\partial t} - i(b_2-b_3)\frac{\partial u_2}{\partial x} + \epsilon\kappa u_1^*u_3 - \epsilon\frac{\kappa(a_2-a_3)}{(a_1-a_3)}|u_1|^2u_2 = 0, \\ &i(a_1-a_3)\frac{\partial u_3}{\partial t} - i(b_1-b_3)\frac{\partial u_3}{\partial x} - \frac{i\kappa}{a_1-a_3}\frac{\partial(u_1u_2)}{\partial x} \\ &+ \epsilon\kappa\left(\frac{a_1-a_2}{a_1-a_3}|u_1|^2 + \frac{a_2-a_3}{a_1-a_3}|u_2|^2\right)u_1u_2 + \epsilon\kappa u_3(|u_1|^2 - |u_2|^2) = 0, \end{aligned} \tag{42}$$

where the interaction constant κ and u_3 are given by:

$$\kappa = \frac{1}{2}\left(a_1 b_3 - a_2 b_3 - a_1 b_2 - b_1 a_3 + b_1 a_2 + b_2 a_3\right), \qquad u_3 = w_3 + \frac{2a_2 - a_1 - a_3}{2(a_1 - a_3)} u_1 u_2. \tag{43}$$

The diagonal terms in the Lax representation are λ-independent. Two of them read:

$$\begin{aligned} &i(a_1 - a_2)\frac{\partial |u_1|^2}{\partial t} - i(b_1 - b_2)\frac{\partial |u_1|^2}{\partial x} - \epsilon\kappa(u_1 u_2 u_3^* - u_1^* u_2^* u_3) = 0, \\ &i(a_2 - a_3)\frac{\partial |u_2|^2}{\partial t} - i(b_2 - b_3)\frac{\partial |u_2|^2}{\partial x} - \epsilon\kappa(u_1 u_2 u_3^* - u_1^* u_2^* u_3) = 0, \end{aligned} \tag{44}$$

These relations are satisfied identically as a consequence of the NLEE.

3.3. New types of 3-wave interactions, $k = 3$.
Our last example of 3-wave interactions is similar to the one, reported in [12] and involves Lax pair, which is cubic in λ:

$$U(x,t,\lambda) = \lambda U_2(x,t) + \lambda^2 U_1(x,t) - \lambda^3 J, \qquad V(x,t,\lambda) = \lambda V_2(x,t) + \lambda^2 V_1(x,t) - \lambda^3 K, \tag{45}$$

where

$$\begin{aligned} &J = \mathrm{diag}\,(a_1, a_2, a_3), \qquad K = \mathrm{diag}\,(b_1, b_2, b_3), \\ &U_1(x,t) = \begin{pmatrix} 0 & 0 & u_{13} \\ u_{21} & 0 & 0 \\ 0 & u_{32} & 0 \end{pmatrix}, \qquad U_2(x,t) = \begin{pmatrix} 0 & u_{12} & 0 \\ 0 & 0 & u_{23} \\ u_{31} & 0 & 0 \end{pmatrix}, \\ &V_1(x,t) = \begin{pmatrix} 0 & 0 & v_{13} \\ v_{21} & 0 & 0 \\ 0 & v_{32} & 0 \end{pmatrix}, \qquad V_1(x,t) = \begin{pmatrix} 0 & v_{12} & 0 \\ 0 & 0 & v_{23} \\ v_{31} & 0 & 0 \end{pmatrix}, \end{aligned} \tag{46}$$

With this choice U and V automatically satisfy the reduction condition [17]

$$C_3 U_k(x,t) C_3^{-1} = \omega^k U_k(x,t), \qquad C_3 V_k(x,t) C_3^{-1} = \omega^k V_k(x,t), \qquad C_3 = \mathrm{diag}\,(1, \omega, \omega^2), \tag{47}$$

with $\omega = \exp(2\pi i/3)$. We can also impose the involution

$$C_2 U_k^\dagger(x,t) C_2^{-1} = \epsilon^{k+1} U_k(x,t), \qquad C_2 V_k^\dagger(x,t) C_2^{-1} = \epsilon^{k+1} V_k(x,t), \qquad C_2 = \begin{pmatrix} 0 & 0 & 1 \\ 0 & \epsilon_1 & 0 \\ 1 & 0 & 0 \end{pmatrix} \tag{48}$$

with $\epsilon = \pm 1, \epsilon_1 = \pm 1$ which gives:

$$\begin{aligned} &u_{12} = \epsilon_1 \epsilon u_{23}^*, \qquad u_{31}^* = \epsilon u_{31}, \qquad u_{13}^* = u_{13}, \qquad u_{21} = \epsilon_1 u_{32}^*, \\ &a_1^* = \epsilon a_3, \qquad a_2 = \epsilon a_2^*, \qquad b_3^* = \epsilon b_1, \qquad b_2^* = \epsilon b_2. \end{aligned} \tag{49}$$

and analogous relations for v_{ij}. The compatibility condition now gives:

Introduce also the coefficients $Q_{1,2}$ which automatically satisfy the above reductions:

$$Q_1(x,t) = \begin{pmatrix} 0 & 0 & R_1 \\ w_1 & 0 & 0 \\ 0 & -\epsilon\epsilon_1 w_1^* & 0 \end{pmatrix}, \qquad Q_2(x,t) = \begin{pmatrix} 0 & w_2 & 0 \\ 0 & 0 & -\epsilon_1 w_2^* \\ R_2 & 0 & 0 \end{pmatrix}, \qquad R_1 = -\epsilon R_1^*, \quad R_2 = -R_2^* \tag{50}$$

Thus we get the following parametrization for U_k and V_k

$$\begin{aligned} U_1(x,t) &= [Q_1(x,t), J], & U_2(x,t) &= [Q_2(x,t), J] + \frac{1}{2}[Q_1(x,t), [Q_1(x,t), J]], \\ V_1(x,t) &= [Q_1(x,t), K], & V_2(x,t) &= [Q_2(x,t), K] + \frac{1}{2}[Q_1(x,t), [Q_1(x,t), K]], \end{aligned} \tag{51}$$

or in components:

$$\begin{aligned} u_{13} &= (a_3 - a_1)R_1, & u_{12} &= (a_2 - a_1)w_2 - \frac{1}{2}\epsilon\epsilon_1(a_1 + a_2 - 2a_3)w_1^* R_1, \\ u_{21} &= (a_1 - a_2)w_1, & u_{23} &= \epsilon_1(a_2 - a_3)w_2^* + \frac{1}{2}(a_2 + a_3 - 2a_1)w_1 R_1, \\ u_{32} &= \epsilon\epsilon_1(a_3 - a_2)w_1^*, & u_{31} &= (a_1 - a_3)R_2 - \frac{1}{2}\epsilon\epsilon_1(a_1 + a_3 - 2a_2)|w_1|^2. \end{aligned} \tag{52}$$

The expressions for v_{jk} differ from (52) only by replacing a_k by b_k.

The corresponding NLEE take the form:

$$\begin{aligned} &\frac{\partial R_1}{\partial t} - \frac{b_1 - b_3}{a_1 - a_3}\frac{\partial R_1}{\partial x} + \frac{\epsilon\epsilon_1}{a_1 - a_3}(u_{12}v_{12}^* - v_{12}u_{12}^*) = 0, & &\frac{\partial u_{12}}{\partial t} - \frac{\partial v_{12}}{\partial x} = 0, \\ &i\frac{\partial w_1}{\partial t} - i\frac{b_1 - b_2}{a_1 - a_2}\frac{\partial w_1}{\partial x} + \frac{\epsilon_1\epsilon}{a_1 - a_2}(u_{12}^* v_{31} - v_{12}^* u_{31}) = 0, & &\frac{\partial u_{31}}{\partial t} - \frac{\partial v_{31}}{\partial x} = 0. \end{aligned} \tag{53}$$

Inserting here the notations from eq. (52) we obtain the four NLEE for the four independent functions w_1, w_2, R_1 and R_2.

4. Conclusions and open questions

We proposed a method for constructing new integrable NLEE based on the use of the RHP with canonical normalization combined with the Mikhailov's reduction group [16, 17]. Obviously we can derive many new classes of such equations making appropriate choices of the: i) the order k of the jets, ii) the simple Lie algebra $\mathfrak{g}$ and iii) the reduction group and its realization as a subgroup of the group of automorphisms of $\mathfrak{g}$.

Since the method is based on the RHP one can apply Zakharov-Shabat dressing method for constructing their explicit (N-soliton) solutions.

The new NLEE are expected to be Hamiltonian. So one must show that the jets $U(\vec{x}, t, \lambda)$ can be viewed as elements of more complicated co-adjoint orbits of the relevant Kac-Moody algebra, generated by the chosen grading of $\mathfrak{f}$. The corresponding Poisson brackets can be derived using the results of Kulish and Reyman [15] and imposing the reduction conditions as constraints.

The last but not least important problem concerns the possible physical applications of these equations.

Acknowledgements

One of us, A. B. Y., is grateful to NRF South Africa for the 2013 incentive research grant.

References

[1] Beals R and Coifman R R 1985 Scattering and inverse scattering for first order systems *Commun. Pure & Appl. Math.* **37**, 39; Beals R and Coifman R R 1984 Inverse scattering and evolution equations *Commun. Pure & Appl. Math.* **38**, 29

[2] Caudrey P J 1980 The inverse problem for the third order equation $u_{xxx} + q(x)u_x + r(x)u = -i\zeta^3 u$ *Phys. Lett. A* **79A** 264 ;
Caudrey P J 1982 The inverse problem for a general $n \times n$ spectral equation. *Physica* **D6**, 56

[3] Dickey L A 1991 *Soliton equations and Hamiltonian systems, Advanced series in Math. Phys* **12** (Singapore, World Scientific)

[4] Gakhov F D 1966 *Boundary value problems.* ed. I. N. Sneddon (Oxford, Pergamon Press)

[5] Gerdjikov V S 1986 Generalised Fourier transforms for the soliton equations. Gauge covariant formulation *Inverse Problems* **2,** n. 1, 51–74

[6] Gerdjikov V S 2002 Algebraic and Analytic Aspects of N-wave Type Equations *Contemporary Mathematics* **301**, 35-68

[7] Gerdjikov V S 2012 On new types of integrable 4-wave interactions. *AIP Conf. Proc.* **1487** 272-279, doi:http://dx.doi.org/10.1063/1.4758968

[8] Gerdjikov V S 2012 Riemann-Hilbert Problems with canonical normalization and families of commuting operators. *Pliska Stud. Math. Bulgar.* **21**, 201–216

[9] Gerdjikov V S and Grahovski G G 2000 Reductions and real forms of hamiltonian systems related to N-wave type equations *Balkan Phys. Lett. BPL* **4**, pp.531-534

[10] Gerdjikov V S, Grahovski G G, Ivanov R I and Kostov N A (2001) N-wave interactions related to simple Lie algebras. $\mathbb{Z}_2$- reductions and soliton solutions *Inverse Problems* **17**, 999-1015

[11] Gerdjikov V S, Grahovski G G ans Kostov N A 2001 Reductions of N-wave interactions related to low–rank simple Lie algebras. I: $\mathbb{Z}_2$- reductions *J. Phys. A: Math & Gen.* **34**, 9425-9461

[12] Gerdjikov V S and Kaup D J 1981 Reductions of 3×3 polynomial bundles and new types of integrable 3–wave interactions in *Nonlinear evolution equations: integrability and spectral methods* ed A P Fordy, A. Degasperis, M. Lakshmanan (Manchester, Manchester University Press) p. 373–380

[13] Gerdjikov V S and Yanovski A B 1994 Completeness of the eigenfunctions for the Caudrey–Beals–Coifman system *J. Math. Phys.* **35**, no. 7, 3687–3725

[14] Ivanov R 2004 On the dressing method for the generalised Zakharov - Shabat system. *Nuclear Physics* **B 694 [PM]**

[15] Kulish P P and Reiman A G 1983 Hamiltonian structure of polynomial bundles *Sci. Notes. LOMI seminars* **123** 67 - 76, Translated in *J. Sov. Math.* **28**, 505-513 (1985).

[16] Mikhailov A 1979 Reduction in the integrable systems. Reduction groups *Lett. JETPh* **32** (1979) 187–192 (In Russian)

[17] Mikhailov A 1981 The reduction problem and inverse scattering method *Physica* **3D** 73–117.

[18] Vekua N P 1967 *Systems of singular integral equations.* Translated from Russian by A. G. Gibs and G. M. Simmons, (Gröningen, P. Noordhoff Ltd., The Netherlands)

[19] Zakharov V E and Manakov S V 1975 The theory of resonant interaction of wave packets in nonlinear media *Sov. Phys. JETP* **69** 1654 (In Russian).

[20] Zakharov V E, Manakov S V and, Novikov S P and Pitaevskii L I 1984 *Theory of solitons: the inverse scattering method.* (New York, Plenum, Consultants Bureau)

[21] Zakharov V E and Shabat A B 1974 A scheme for integrating nonlinear equations of mathematical physics by the method of the inverse scattering transform. I *Funct. Annal. and Appl.* **8**(3) 43–53

[22] Zakharov V E and Shabat A B 1979 A scheme for integrating nonlinear equations of mathematical physics by the method of the inverse scattering transform. II *Funct. Anal. Appl.* **13**(3) 13-23

Euler-Poincaré equations for G-Strands

Darryl D Holm[a], **Rossen I Ivanov**[b]

[a] Department of Mathematics, Imperial College, London SW7 2AZ, UK

[b]Department of Mathematical Sciences, Dublin Institute of Technology, Kevin Street, Dublin 8, Ireland

E-mail: d.holm@imperial.ac.uk, rossen.ivanov@dit.ie

Abstract.
The G-strand equations for a map $\mathbb{R} \times \mathbb{R}$ into a Lie group G are associated to a G-invariant Lagrangian. The Lie group manifold is also the configuration space for the Lagrangian. The G-strand itself is the map $g(t,s) : \mathbb{R} \times \mathbb{R} \to G$, where t and s are the independent variables of the G-strand equations. The Euler-Poincaré reduction of the variational principle leads to a formulation where the dependent variables of the G-strand equations take values in the corresponding Lie algebra $\mathfrak{g}$ and its co-algebra, $\mathfrak{g}^*$ with respect to the pairing provided by the variational derivatives of the Lagrangian.

We review examples of different G-strand constructions, including matrix Lie groups and diffeomorphism group. In some cases the G-strand equations are completely integrable 1+1 Hamiltonian systems that admit soliton solutions.

1. Introduction

We give a brief account of the G-strand construction, which gives rise to equations for a map $\mathbb{R}\times\mathbb{R}$ into a Lie group G associated to a G-invariant Lagrangian. Our presentation reviews our previous works [7, 5, 6, 3, 8] and is aimed to illustrate the G-strand construction with several simple but instructive examples. The following examples are reviewed here:

(i) $SO(3)$-strand equations for the so-called continuous spin chain. The equations reduce to the integrable chiral model in their simplest (bi-invariant) case.

(ii) $SO(3)$ - anisotropic chiral model, which is also integrable,

(iii) Diff($\mathbb{R}$)-strand equations. These equations are in general non-integrable; however they admit solutions in 2 + 1 space-time with singular support (e.g., peakons). Peakon-antipeakon collisions governed by the Diff($\mathbb{R}$)-strand equations can be solved *analytically*, and potentially they can be applied in the theory of image registration.

2. Ingredients of Euler–Poincaré theory for Left G-Invariant Lagrangians

Let G be a Lie group. A map $g(t,s) : \mathbb{R} \times \mathbb{R} \to G$ has two types of tangent vectors, $\dot{g} := g_t \in TG$ and $g' := g_s \in TG$. Assume that the Lagrangian density function $L(g, \dot{g}, g')$ is left G-invariant. The left G–invariance of L permits us to define $l : \mathfrak{g} \times \mathfrak{g} \to \mathbb{R}$ by

$$L(g, \dot{g}, g') = L(g^{-1}g, g^{-1}\dot{g}, g^{-1}g') \equiv l(g^{-1}\dot{g}, g^{-1}g').$$

Conversely, this relation defines for any reduced lagrangian $l = l(\mathsf{u}, \mathsf{v}) : \mathfrak{g} \times \mathfrak{g} \to \mathbb{R}$ a left G-invariant function $L : TG \times TG \to \mathbb{R}$ and a map $g(t,s) : \mathbb{R} \times \mathbb{R} \to G$ such that

$$\mathsf{u}(t,s) := g^{-1}g_t(t,s) = g^{-1}\dot{g}(t,s) \quad \text{and} \quad \mathsf{v}(t,s) := g^{-1}g_s(t,s) = g^{-1}g'(t,s).$$

Lemma 1. *The left-invariant tangent vectors* $\mathsf{u}(t,s)$ *and* $\mathsf{v}(t,s)$ *at the identity of* G *satisfy*

$$\mathsf{v}_t - \mathsf{u}_s = -\,\mathrm{ad}_{\mathsf{u}}\mathsf{v}\,. \tag{1}$$

Proof. The proof is standard and follows from equality of cross derivatives $g_{ts} = g_{st}$.
Equation (1) is usually called a ***zero-curvature relation.*** □

Theorem 2 (Euler-Poincaré theorem for left-invariant Lagrangians).
With the preceding notation, the following two statements are equivalent:

i *Variational principle on* $TG \times TG$ $\delta \int_{t_1}^{t_2} L(g(t,s), \dot g(t,s), g'(t,s))\, ds\, dt = 0$ *holds, for variations* $\delta g(t,s)$ *of* $g(t,s)$ *vanishing at the endpoints in* t *and* s*. The function* $g(t,s)$ *satisfies Euler–Lagrange equations for* L *on* G*, given by*

$$\frac{\partial L}{\partial g} - \frac{\partial}{\partial t}\frac{\partial L}{\partial g_t} - \frac{\partial}{\partial s}\frac{\partial L}{\partial g_s} = 0.$$

ii *The constrained variational principle*[1]

$$\delta \int_{t_1}^{t_2} l(\mathsf{u}(t,s), \mathsf{v}(t,s))\, ds\, dt = 0$$

holds on $\mathfrak{g} \times \mathfrak{g}$*, using variations of* $\mathsf{u} := g^{-1}g_t(t,s)$ *and* $\mathsf{v} := g^{-1}g_s(t,s)$ *of the forms*

$$\delta \mathsf{u} = \dot{\mathsf{w}} + \mathrm{ad}_{\mathsf{u}}\mathsf{w} \quad \textit{and} \quad \delta \mathsf{v} = \mathsf{w}' + \mathrm{ad}_{\mathsf{v}}\mathsf{w}\,,$$

where $\mathsf{w}(t,s) := g^{-1}\delta g \in \mathfrak{g}$ *vanishes at the endpoints. The* ***Euler–Poincaré*** *equations hold on* $\mathfrak{g}^* \times \mathfrak{g}^*$ *(G-**strand equations**)*

$$\frac{d}{dt}\frac{\delta l}{\delta \mathsf{u}} - \mathrm{ad}^*_{\mathsf{u}}\frac{\delta l}{\delta \mathsf{u}} + \frac{d}{ds}\frac{\delta l}{\delta \mathsf{v}} - \mathrm{ad}^*_{\mathsf{v}}\frac{\delta l}{\delta \mathsf{v}} = 0 \quad \& \quad \partial_s \mathsf{u} - \partial_t \mathsf{v} = [\,\mathsf{u},\, \mathsf{v}\,] = \mathrm{ad}_{\mathsf{u}}\mathsf{v}$$

where $(\mathrm{ad}^* : \mathfrak{g} \times \mathfrak{g}^* \to \mathfrak{g}^*)$ *is defined via* $(\mathrm{ad} : \mathfrak{g} \times \mathfrak{g} \to \mathfrak{g})$ *in the dual pairing* $\langle\,\cdot\,,\,\cdot\,\rangle : \mathfrak{g}^* \times \mathfrak{g} \to \mathbb{R}$ *by,*

$$\left\langle \mathrm{ad}^*_{\mathsf{u}}\frac{\delta \ell}{\delta \mathsf{u}}\,,\, \mathsf{v} \right\rangle_{\mathfrak{g}} = \left\langle \frac{\delta \ell}{\delta \mathsf{u}}\,,\, \mathrm{ad}_{\mathsf{u}}\mathsf{v} \right\rangle_{\mathfrak{g}}.$$

In 1901 Poincaré in his famous work proves that, when a Lie algebra acts locally transitively on the configuration space of a Lagrangian mechanical system, the well known Euler-Lagrange equations are equivalent to a new system of differential equations defined on the product of the configuration space with the Lie algebra. These equations are called now in his honor Euler-Poincaré equations. In modern language the contents of the Poincaré's article [12] is presented for example in [4, 11]. English translation of the article [12] can be found as Appendix D in [4].

3. *G*-strand equations on matrix Lie algebras

Denoting $\mathsf{m} := \delta\ell/\delta\mathsf{u}$ and $\mathsf{n} := \delta\ell/\delta\mathsf{v}$ in $\mathfrak{g}^*$, the G-strand equations become

$$\mathsf{m}_t + \mathsf{n}_s - \mathrm{ad}^*_{\mathsf{u}}\mathsf{m} - \mathrm{ad}^*_{\mathsf{v}}\mathsf{n} = 0 \quad \text{and} \quad \partial_t\mathsf{v} - \partial_s\mathsf{u} + \mathrm{ad}_{\mathsf{u}}\mathsf{v} = 0.$$

For G a semisimple *matrix Lie group* and $\mathfrak{g}$ its *matrix Lie algebra* these equations become

$$\begin{aligned} \mathsf{m}_t^T + \mathsf{n}_s^T + \mathrm{ad}_{\mathsf{u}}\mathsf{m}^T + \mathrm{ad}_{\mathsf{v}}\mathsf{n}^T &= 0, \\ \partial_t\mathsf{v} - \partial_s\mathsf{u} + \mathrm{ad}_{\mathsf{u}}\mathsf{v} &= 0 \end{aligned} \tag{2}$$

where the ad-invariant pairing for semisimple matrix Lie algebras is given by

$$\left\langle \mathsf{m}\,,\, \mathsf{n} \right\rangle = \frac{1}{2}\mathrm{tr}(\mathsf{m}^T\mathsf{n}),$$

the transpose gives the map between the algebra and its dual $(\,\cdot\,)^T : \mathfrak{g} \to \mathfrak{g}^*$. For semisimple matrix Lie groups, the adjoint operator is the matrix commutator. Examples are studied in [7, 6, 3].

[1] As with the basic Euler–Poincaré equations, this is not strictly a variational principle in the same sense as the standard Hamilton's principle. It is more like the Lagrange d'Alembert principle, because we impose the stated constraints on the variations allowed.

4. Lie-Poisson Hamiltonian formulation

Legendre transformation of the Lagrangian $\ell(\mathsf{u},\mathsf{v}) : \mathfrak{g}\times\mathfrak{g}\to\mathbb{R}$ yields the Hamiltonian $h(\mathsf{m},\mathsf{v}) : \mathfrak{g}^*\times\mathfrak{g}\to\mathbb{R}$

$$h(\mathsf{m},\mathsf{v}) = \Big\langle \mathsf{m}\,,\,\mathsf{u}\Big\rangle - \ell(\mathsf{u},\mathsf{v})\,. \tag{3}$$

Its partial derivatives imply

$$\frac{\delta l}{\delta \mathsf{u}} = \mathsf{m}\,,\quad \frac{\delta h}{\delta \mathsf{m}} = \mathsf{u} \quad\text{and}\quad \frac{\delta h}{\delta \mathsf{v}} = -\frac{\delta \ell}{\delta \mathsf{v}} = \mathsf{v}.$$

These derivatives allow one to rewrite the Euler-Poincaré equation solely in terms of momentum m as

$$\begin{aligned}\partial_t \mathsf{m} &= \mathrm{ad}^*_{\delta h/\delta \mathsf{m}}\,\mathsf{m} + \partial_s\frac{\delta h}{\delta \mathsf{v}} - \mathrm{ad}^*_{\mathsf{v}}\frac{\delta h}{\delta \mathsf{v}}\,,\\ \partial_t \mathsf{v} &= \partial_s\frac{\delta h}{\delta \mathsf{m}} - \mathrm{ad}_{\delta h/\delta \mathsf{m}}\,\mathsf{v}\,.\end{aligned} \tag{4}$$

Assembling these equations into Lie-Poisson Hamiltonian form gives,

$$\frac{\partial}{\partial t}\begin{bmatrix}\mathsf{m}\\ \mathsf{v}\end{bmatrix} = \begin{bmatrix}\mathrm{ad}^*(\,\cdot\,)\mathsf{m} & \partial_s - \mathrm{ad}^*_{\mathsf{v}}\\ \partial_s + \mathrm{ad}_{\mathsf{v}} & 0\end{bmatrix}\begin{bmatrix}\delta h/\delta \mathsf{m}\\ \delta h/\delta \mathsf{v}\end{bmatrix} \tag{5}$$

The Hamiltonian matrix in equation (5) also appears in the Lie-Poisson brackets for Yang-Mills plasmas, for spin glasses and for perfect complex fluids, such as liquid crystals.

5. Example: The Euler-Poincaré PDEs for the $SO(3)$-strand and the chiral model. The 2-time spatial and body angular velocities on $\mathfrak{so}(3)$

Let us make the following explicit identification:

$$\mathsf{u} = \begin{pmatrix}0 & -u_3 & u_2\\ u_3 & 0 & -u_1\\ -u_2 & u_1 & 0\end{pmatrix} \in \mathfrak{g} \quad\leftrightarrow\quad \mathbf{u} \equiv \begin{pmatrix}u_1\\ u_2\\ u_3\end{pmatrix} \in \mathbb{R}^3 \tag{6}$$

and similarly for $\mathbf{v}$. In terms of the corresponding group element $g(s,t)$, describing rotation, $\mathsf{u}(t,s) = g^{-1}\partial_t g(t,s)$ and $\mathsf{v}(t,s) = g^{-1}\partial_s g(t,s)$ resemble 2 body angular velocities. For $G = SO(3)$ and Lagrangian $\ell(\mathbf{u},\,\mathbf{v}) : \mathbb{R}^3\times\mathbb{R}^3\to\mathbb{R}$, in $1+1$ space-time the Euler-Poincaré equation becomes

$$\frac{\partial}{\partial t}\frac{\delta \ell}{\delta \mathbf{u}} + \mathbf{u}\times\frac{\delta \ell}{\delta \mathbf{u}} = -\left(\frac{\partial}{\partial s}\frac{\delta \ell}{\delta \mathbf{v}} + \mathbf{v}\times\frac{\delta \ell}{\delta \mathbf{v}}\right), \tag{7}$$

and its auxiliary equation becomes

$$\frac{\partial}{\partial t}\mathbf{v} = \frac{\partial}{\partial s}\mathbf{u} + \mathbf{v}\times\mathbf{u}\,. \tag{8}$$

The Hamiltonian form of these equations on $\mathfrak{so}(3)^*$ are obtained from the Legendre transform relations

$$\frac{\delta \ell}{\delta \mathbf{u}} = \mathbf{m}\,,\quad \frac{\delta h}{\delta \mathbf{m}} = \mathbf{u} \quad\text{and}\quad \frac{\delta h}{\delta \mathbf{v}} = -\frac{\delta \ell}{\delta \mathbf{v}}\,.$$

Hence, the Euler-Poincaré equation implies the Lie-Poisson Hamiltonian structure in vector form

$$\partial_t\begin{bmatrix}\mathbf{m}\\ \mathbf{v}\end{bmatrix} = \begin{bmatrix}\mathbf{m}\times & \partial_s + \mathbf{v}\times\\ \partial_s + \mathbf{v}\times & 0\end{bmatrix}\begin{bmatrix}\delta h/\delta \mathbf{m}\\ \delta h/\delta \mathbf{v}\end{bmatrix}.$$

This Poisson structure appears in various other theories, such as complex fluids and filament dynamics. When

$$\ell = \frac{1}{2}\int(\mathbf{u}\cdot A\mathbf{u} + \mathbf{v}\cdot B\mathbf{v})\,ds \tag{9}$$

this is the $SO(3)$ *spin-chain model*, which is in general non-integrable- eq. (7) and (8) give:

$$\frac{\partial}{\partial t}A\mathbf{u} + \mathbf{u}\times A\mathbf{u} + \frac{\partial}{\partial s}B\mathbf{v} + \mathbf{v}\times B\mathbf{v} = 0\,, \tag{10}$$

$$\frac{\partial}{\partial t}\mathbf{v} = \frac{\partial}{\partial s}\mathbf{u} + \mathbf{v}\times\mathbf{u}\,. \tag{11}$$

When $A = -B = 1$, this is the $SO(3)$ *chiral model*, which is an integrable Hamiltonian system.

$$\mathbf{u}_t - \mathbf{v}_s = 0\,, \tag{12}$$

$$\mathbf{v}_t - \mathbf{u}_s + \mathbf{u}\times\mathbf{v} = 0\,. \tag{13}$$

6. Integrability

Some of the G-strands models are well known integrable models. They have a *zero-curvature representation* for two operators L and M of the form

$$L_t - M_s + [L, M] = 0, \tag{14}$$

which is the compatibility condition for a pair of linear equations

$$\psi_s = L\psi, \quad \text{and} \quad \psi_t = M\psi.$$

For the SO(3) chiral model for example these operators are

$$\begin{aligned} L &= \frac{1}{4}\left[(1+\lambda)(\mathsf{u}-\mathsf{v}) - \left(1+\frac{1}{\lambda}\right)(\mathsf{u}+\mathsf{v})\right], \\ M &= -\frac{1}{4}\left[(1+\lambda)(\mathsf{u}-\mathsf{v}) + \left(1+\frac{1}{\lambda}\right)(\mathsf{u}+\mathsf{v})\right]. \end{aligned} \tag{15}$$

Another integrable matrix example: $SO(3)$ anisotropic Chiral model [2]

$$\begin{aligned} &\partial_t\mathbf{v}(t,s) - \partial_s\mathbf{u}(t,s) + \mathbf{u}\times P\mathbf{v} - \mathbf{v}\times P\mathbf{u} = 0\,, \\ &\partial_s\mathbf{v}(t,s) - \partial_t\mathbf{u}(t,s) - \mathbf{v}\times P\mathbf{v} + \mathbf{u}\times P\mathbf{u} = 0\,. \end{aligned} \tag{16}$$

$P = \text{diag}(P_1, P_2, P_3)$ is a constant diagonal matrix. Under the linear change of variables

$$\mathbf{X} = \mathbf{u} - \mathbf{v} \quad \text{and} \quad \mathbf{Y} = -\mathbf{u} - \mathbf{v} \tag{17}$$

equations (16) acquire the form of the following $SO(3)$ anisotropic chiral model,

$$\begin{aligned} &\partial_t\mathbf{X}(t,s) + \partial_s\mathbf{X}(t,s) + \mathbf{X}\times P\mathbf{Y} = 0\,, \\ &\partial_t\mathbf{Y}(t,s) - \partial_s\mathbf{Y}(t,s) + \mathbf{Y}\times P\mathbf{X} = 0\,. \end{aligned} \tag{18}$$

The system (18) represents two *cross-coupled* equations for $\mathbf{X}$ and $\mathbf{Y}$. These equations preserve the magnitudes $|\mathbf{X}|^2$ and $|\mathbf{Y}|^2$, so they allow the further assumption that the vector fields $(\mathbf{X}, \mathbf{Y})$ take values on the product of unit spheres $\mathbb{S}^2\times\mathbb{S}^2 \subset \mathbb{R}^3\times\mathbb{R}^3$. The anisotropic chiral model is an integrable system and its Lax pair in terms of $(\mathbf{u}, \mathbf{v})$ utilizes the following isomorphism between $\mathfrak{so}(3)\oplus\mathfrak{so}(3)$ and $\mathfrak{so}(4)$:

$$A(\mathbf{u},\mathbf{v}) = \begin{pmatrix} 0 & u_3 & -u_2 & v_1 \\ -u_3 & 0 & u_1 & v_2 \\ u_2 & -u_1 & 0 & v_3 \\ -v_1 & -v_2 & -v_3 & 0 \end{pmatrix}. \tag{19}$$

The system (16) can be recovered as a compatibility condition of the operators

$$L = \partial_s - A(\mathbf{v},\mathbf{u})(\lambda\,\mathrm{Id} + J), \quad (20)$$
$$M = \partial_t - A(\mathbf{u},\mathbf{v})(\lambda\,\mathrm{Id} + J), \quad (21)$$

where the diagonal matrix J is defined by

$$J = -\frac{1}{2}\mathrm{diag}(P_1, P_2, P_3, P_1 + P_2 + P_3). \quad (22)$$

This Lax pair is due to Bordag and Yanovski [1]. The $O(3)$ anisotropic chiral model can be derived as an Euler-Poincaré equation from a Lagrangian with quadratic kinetic and potential energy. The details are presented in [7].

Remark 3. *If* $\mathsf{P} = \mathrm{Id}$, *equations (16) recover the* $SO(3)$ *chiral model.*

7. The Diff($\mathbb{R}$)-strand system

The constructions described briefly in the previous sections can be easily generalized in cases where the Lie group is the group of the Diffeomorphisms. Consider Hamiltonian which is a right-invariant bilinear form given by the H^1 Sobolev inner product

$$H(u,v) \equiv \frac{1}{2}\int_{\mathcal{M}} (uv + u_x v_x)dx. \quad (23)$$

The manifold $\mathcal{M}$ is $\mathbb{S}^1$ or in the case when the class of smooth functions vanishing rapidly at $\pm\infty$ is considered, we will allow $\mathcal{M} \equiv \mathbb{R}$.
Let us introduce the notation $u(g(x)) \equiv u \circ g$. If $g(x) \in G$, where $G \equiv \mathrm{Diff}(\mathcal{M})$, then

$$H(u,v) = H(u\circ g, v\circ g)$$

is a right-invariant H^1 metric.
Let us further consider an one-parametric family of diffeomprphisms, $g(x,t)$ by defining the t -evolution as

$$\dot{g} = u(g(x,t),t), \qquad g(x,0) = x, \qquad \text{i.e.} \qquad \dot{g} = u\circ g \in T_gG; \quad (24)$$

$u = \dot{g}\circ g^{-1} \in \mathfrak{g}$, where $\mathfrak{g}$, the corresponding Lie-algebra is the algebra of vector fields, $\mathrm{Vect}(\mathcal{M})$. Now we recall the following result:

Theorem 4. *(A. Kirillov, 1980, [9, 10]) The dual space of* $\mathfrak{g}$ *is a space of distributions but the subspace of local functionals, called the regular dual* $\mathfrak{g}^*$ *is naturally identified with the space of quadratic differentials* $m(x)dx^2$ *on* $\mathcal{M}$. *The pairing is given for any vector field* $u\partial_x \in \mathrm{Vect}(\mathcal{M})$ *by*

$$\langle m dx^2, u\partial_x\rangle = \int_{\mathcal{M}} m(x)u(x)dx$$

The coadjoint action coincides with the action of a diffeomorphism on the quadratic differential:

$$\mathrm{Ad}^*_g: \quad m dx^2 \mapsto m(g)g_x^2 dx^2$$

and

$$\mathrm{ad}^*_u = 2u_x + u\partial_x$$

Indeed, a simple computation shows that

$$\begin{aligned}\langle \mathrm{ad}^*_{u\partial_x} m dx^2, v\partial_x\rangle &= \langle m dx^2, [u\partial_x, v\partial_x]\rangle = \int_{\mathcal{M}} m(u_x v - v_x u)dx = \\ \int_{\mathcal{M}} v(2mu_x + um_x)dx &= \langle (2mu_x + um_x)dx^2, v\partial_x\rangle,\end{aligned}$$

i.e. $\mathrm{ad}^*_u m = 2u_x m + u m_x$.

The Diff($\mathbb{R}$)-strand system arises when we choose $G = \mathrm{Diff}(\mathbb{R})$. For a two-parametric group we have two tangent vectors

$$\partial_t g = u \circ g \quad \text{and} \quad \partial_s g = v \circ g\,,$$

where the symbol $\circ$ denotes composition of functions.
In this right-invariant case, the G-strand PDE system with reduced Lagrangian $\ell(u, v)$ takes the form,

$$\begin{aligned} \frac{\partial}{\partial t}\frac{\delta \ell}{\delta u} + \frac{\partial}{\partial s}\frac{\delta \ell}{\delta v} &= -\,\mathrm{ad}^*_u \frac{\delta \ell}{\delta u} - \mathrm{ad}^*_v \frac{\delta \ell}{\delta v}\,, \\ \frac{\partial v}{\partial t} - \frac{\partial u}{\partial s} &= \mathrm{ad}_u v\,. \end{aligned} \tag{25}$$

Of course, the distinction between the maps $(u, v) : \mathbb{R} \times \mathbb{R} \to \mathfrak{g} \times \mathfrak{g}$ and their pointwise values $(u(t, s), v(t, s)) \in \mathfrak{g} \times \mathfrak{g}$ is clear. Likewise, for the variational derivatives $\delta\ell/\delta u$ and $\delta\ell/\delta v$.

8. The Diff($\mathbb{R}$)-strand Hamiltonian structure

Upon setting $m = \delta\ell/\delta u$ and $n = \delta\ell/\delta v$, the right-invariant Diff($\mathbb{R}$)-strand equations in (25) for maps $\mathbb{R} \times \mathbb{R} \to G = \mathrm{Diff}(\mathbb{R})$ in one spatial dimension may be expressed as a system of two 1+2 PDEs in (t, s, x),

$$\begin{aligned} m_t + n_s &= -\,\mathrm{ad}^*_u m - \mathrm{ad}^*_v n = -(um)_x - mu_x - (vn)_x - nv_x\,, \\ v_t - u_s &= -\,\mathrm{ad}_v u = -uv_x + vu_x\,. \end{aligned} \tag{26}$$

The Hamiltonian structure for these Diff($\mathbb{R}$)-strand equations is obtained by Legendre transforming to

$$h(m, v) = \langle m,\, u\rangle - \ell(u,\, v)\,.$$

One may then write the equations (26) in Lie-Poisson Hamiltonian form as

$$\frac{d}{dt}\begin{bmatrix} m \\ v \end{bmatrix} = \begin{bmatrix} -\,\mathrm{ad}^*(\,\cdot\,)m & \partial_s + \mathrm{ad}^*_v \\ \partial_s - \mathrm{ad}_v & 0 \end{bmatrix} \begin{bmatrix} \delta h/\delta m = u \\ \delta h/\delta v = -\,n \end{bmatrix}. \tag{27}$$

9. Peakon solutions of the Diff($\mathbb{R}$)-strand equations

With the following choice of Lagrangian,

$$\ell(u, v) = \frac{1}{2}\|u\|^2_{H^1} - \frac{1}{2}\|v\|^2_{H^1}\,, \tag{28}$$

the corresponding Hamiltonian is positive-definite and the Diff($\mathbb{R}$)-strand equations (26) admit peakon solutions in *both* momenta

$$m = u - u_{xx} \quad \text{and} \quad n = -(v - v_{xx}),$$

with continuous velocities u and v. This is a two-component generalization of the CH equation.

Theorem 5. *The* Diff($\mathbb{R}$)*-strand equations (26) admit singular solutions expressible as linear superpositions summed over* $a \in \mathbb{Z}$

$$\begin{aligned} m(s,t,x) &= \sum_a M_a(s,t)\delta(x - Q^a(s,t))\,, \\ n(s,t,x) &= \sum_a N_a(s,t)\delta(x - Q^a(s,t))\,, \\ u(s,t,x) &= K * m = \sum_a M_a(s,t)K(x, Q^a)\,, \\ v(s,t,x) &= -K * n = -\sum_a N_a(s,t)K(x, Q^a)\,, \end{aligned} \tag{29}$$

that are peakons *in the case that* $K(x, y) = \frac{1}{2}e^{-|x-y|}$ *is the Green function the inverse Helmholtz operator* $(1 - \partial_x^2)$*:*

$$(1 - \partial_x^2)K(x, 0) = \delta(x)$$

The solution parameters $\{Q^a(s,t), M_a(s,t), N_a(s,t)\}$ with $a \in \mathbb{Z}$ that specify the singular solutions (29) are determined by the following set of evolutionary PDEs in s and t, in which we denote $K^{ab} := K(Q^a, Q^b)$ with integer summation indices $a, b, c, e \in \mathbb{Z}$:

$$\begin{aligned}
\partial_t Q^a(s,t) &= u(Q^a, s, t) = \sum_b M_b(s,t) K^{ab}\,, \\
\partial_s Q^a(s,t) &= v(Q^a, s, t) = -\sum_b N_b(s,t) K^{ab}\,, \\
\partial_t M_a(s,t) &= -\,\partial_s N_a - \sum_c (M_a M_c - N_a N_c) \frac{\partial K^{ac}}{\partial Q^a} \quad \text{(no sum on } a), \\
\partial_t N_a(s,t) &= -\partial_s M_a + \sum_{b,c,e} (N_b M_c - M_b N_c) \frac{\partial K^{ec}}{\partial Q^e} (K^{eb} - K^{cb})(K^{-1})_{ae}\,.
\end{aligned} \tag{30}$$

The last pair of equations in (30) may be solved as a system for the momenta, i.e., Lagrange multipliers (M_a, N_a), then used in the previous pair to update the support set of positions $Q^a(t,s)$.

10. Single-peakon solution of the of the Diff($\mathbb{R}$)-strand system

The single-peakon solution of the Diff($\mathbb{R}$)-strand equations (26) is straightforward to obtain from (30). Combining the equations in (30) for a single peakon shows that $Q^1(s,t)$ satisfies the Laplace equation,

$$(\partial_s^2 - \partial_t^2) Q^1(s,t) = 0\,.$$

Thus, any function $h(s,t)$ that solves the wave equation provides a solution $Q^1 = h(s,t)$. From the first two equations in (30)

$$M_1(s,t) = \frac{1}{K_0} h_t(s,t) \qquad N_1(s,t) = \frac{1}{K_0} h_s(s,t),$$

where $K_0 = K(0,0)$.
The solutions for the single-peakon parameters Q^1, M_1 and N_1 depend only on one function $h(s,t)$, which in turn depends on the (s,t) boundary conditions. The shape of the Green's function comes into the corresponding solutions for the peakon profiles

$$u(s,t,x) = M_1(s,t) K(x, Q^1(s,t))\,, \qquad v(s,t,x) = -N_1(s,t) K(x, Q^1(s,t))\,.$$

11. Peakon-Antipeakon collisions on a Diff($\mathbb{R}$)-strand

Denote the relative spacing $X(s,t) = Q^1 - Q^2$ for the peakons at positions $Q^1(t,s)$ and $Q^2(t,s)$ on the real line and the Green's function $K = K(X)$. Then the first two equations in (30) imply

$$\begin{aligned}
\partial_t X &= (M_1 - M_2)(K_0 - K(X))\,, \\
\partial_s X &= -(N_1 - N_2)(K_0 - K(X))\,,
\end{aligned} \tag{31}$$

where $K_0 = K(0)$.
The second pair of equations in (30) may then be written as

$$\begin{aligned}
\partial_t M_1 &= -\partial_s N_1 - (M_1 M_2 - N_1 N_2) K'(X)\,, \\
\partial_t M_2 &= -\partial_s N_2 + (M_1 M_2 - N_1 N_2) K'(X)\,, \\
\partial_t N_1 &= -\partial_s M_1 + (N_1 M_2 - M_1 N_2) \frac{K_0 - K}{K_0 + K} K'(X)\,, \\
\partial_t N_2 &= -\partial_s M_2 + (N_1 M_2 - M_1 N_2) \frac{K_0 - K}{K_0 + K} K'(X)\,.
\end{aligned} \tag{32}$$

Asymptotically, when the peakons are far apart, the system (32) simplifies, since $\frac{K_0 - K}{K_0 + K} \to 1$ and $K'(X) \to 0$ as $|X| \to \infty$.

The system (32) has two immediate conservation laws obtained from their sums and differences,

$$\begin{aligned} \partial_t(M_1 + M_2) &= -\,\partial_s(N_1 + N_2)\,, \\ \partial_t(N_1 - N_2) &= -\partial_s(M_1 - M_2)\,. \end{aligned} \tag{33}$$

These may be resolved by setting

$$\begin{aligned} M_1 - M_2 &= \frac{\partial_t X}{K_0 - K}\,, \qquad N_1 - N_2 = -\frac{\partial_s X}{K_0 - K}\,, \\ M_1 + M_2 &= \partial_s \phi\,, \qquad N_1 + N_2 = -\,\partial_t \phi\,, \end{aligned} \tag{34}$$

and introducing two potential functions, X and ϕ, for which equality of cross derivatives will now produce the system of equations (31) and (32).

12. A simplification.

A simplification arises if $\phi = 0$, in which case the collision is perfectly antisymmetric, as seen from equation (34). This is the peakon-antipeakon collision, for which the equation for X reduces to

$$(\partial_t^2 - \partial_s^2)X + \frac{K'}{2(K_0 - K)}(X_t^2 - X_s^2) = 0\,. \tag{35}$$

This equation can be easily rearranged to produce a linear equation:

$$(\partial_t^2 - \partial_s^2)F(X) = 0\,, \quad \text{where} \quad F(X) = \int_{X_0}^{X} (K_0 - K(Y))^{-1/2}\, dY\,. \tag{36}$$

When $K(Y) = \frac{1}{2}e^{-|Y|}$, we have

$$F(X) = \sqrt{2}\int_{X_0}^{X} \frac{1}{\sqrt{1 - e^{-|Y|}}}\, dY. \tag{37}$$

We can take for simplicity $X_0 = 0$, this would change $F(X)$ only by a constant. The computation gives

$$F(X) = 2\sqrt{2}\,\mathrm{sign}(X)\cosh^{-1}\left(e^{|X|/2}\right)$$

. Hence the solution $X(t,s)$ can be expressed in terms of any solution $h(t,s)$ of the linear wave equation $(\partial_t^2 - \partial_s^2)h(t,s) = 0$ as

$$X(t,s) = \pm \ln\left(\cosh^2(h(t,s))\right)\,. \tag{38}$$

$h(t,s)$ is any solution of the wave equation.

$$M_1 = -M_2 = \frac{\partial_t X}{2(K_0 - K(X))}\,, \qquad N_1 = -N_2 = -\frac{\partial_s X}{2(K_0 - K(X))}.$$

Complex Diff($\mathbb{R}$)-strand equations

The Diff($\mathbb{R}$)-strands may also be *complexified*. Upon complexifying $(s,t) \in \mathbb{R}^2 \to (z,\bar{z}) \in \mathbb{C}$ where $\bar{z}$ denotes the complex conjugate of z and setting $\partial_z g = u \circ g$ and $\partial_{\bar{z}} g = \bar{u} \circ g$ the Euler-Poincaré G-strand equations in (26) become

$$\begin{aligned} \frac{\partial}{\partial z}\frac{\delta \ell}{\delta u} + \frac{\partial}{\partial \bar{z}}\frac{\delta \ell}{\delta \bar{u}} &= -\,\mathrm{ad}^*_u \frac{\delta \ell}{\delta u} - \mathrm{ad}^*_{\bar{u}}\frac{\delta \ell}{\delta \bar{u}}\,, \\ \frac{\partial \bar{u}}{\partial z} - \frac{\partial u}{\partial \bar{z}} &= \mathrm{ad}_u \bar{u}\,. \end{aligned} \tag{39}$$

Here the Lagrangian ℓ is taken to be real:

$$\ell(u,\bar{u}) = \frac{1}{2}\|\nu\|^2_{H^1} = \frac{1}{2}\int u\,(1 - \partial_x^2)\,\bar{u}\, dx. \tag{40}$$

Upon setting $m = \delta\ell/\delta u$, $\bar{m} = \delta\ell/\delta\bar{u}$, for the real Lagrangian ℓ, equations (39) may be rewritten as

$$
\begin{aligned}
m_z + \bar{m}_{\bar{z}} &= -\,\mathrm{ad}^*_u m - \mathrm{ad}^*_{\bar{u}}\,\bar{m} = -(um)_x - mu_x - (\bar{u}\,\bar{m})_x - \bar{m}\,\bar{u}_x\,, \\
\bar{u}_z - u_{\bar{z}} &= -\,\mathrm{ad}_{\bar{u}}\,u = -u\,\bar{u}_x + \bar{u}\,u_x\,,
\end{aligned}
\tag{41}
$$

where the independent coordinate $x \in \mathbb{R}$ is on the real line, although coordinates $(z, \bar{z}) \in \mathbb{C}$ are complex, as are solutions u, and $m = u - u_{xx}$. This is a possible comlexification of the Camassa-Holm equation. These equations are invariant under two involutions, P and C, where

$$
P : (x, m) \to (-x, -m) \quad \text{and} \quad C : \text{Complex conjugation.}
$$

They admit singular solutions just as before, modulo $\mathbb{R}\times\mathbb{R} \to \mathbb{C}$. For real variables $m = \bar{m}$, $u = \bar{u}$ and real evolution parameter $z = \bar{z} =: t$, they reduce to the CH equation. Their travelling wave solutions and other possible CH complexifications are studied in [5].

Conclusions

The G-strand equations comprise a system of PDEs obtained from the Euler-Poincaré (EP) variational equations for a G-invariant Lagrangian, coupled to an auxiliary *zero-curvature* equation. Once the G-invariant Lagrangian has been specified, the system of G-strand equations in (2) follows automatically in the EP framework. For matrix Lie groups, some of the the G-strand systems are integrable. The single-peakon and the peakon-antipeakon solution of the Diff($\mathbb{R}$)-strand equations (26) depends on a single function of s, t. The *complex* Diff($\mathbb{R}$)-strand equations and their peakon collision solutions have also been solved by elementary means. The stability of the single-peakon solution under perturbations into the full solution space of equations (26) would be an interesting problem for future work.

Acknowledgments

We are grateful for enlightening discussions of this material with F. Gay-Balmaz, T. S. Ratiu and C. Tronci. Work by RII was supported by the Science Foundation Ireland (SFI), under Grant No. 09/RFP/MTH2144. Work by DDH was partially supported by Advanced Grant 267382 FCCA from the European Research Council.

References

[1] Bordag L A and Yanovski A B 1995 *J. Phys. A: Math. Gen.* **28** 4007
[2] Cherednik I 1981 On the integrability of the 2-dimensional asymmetric chiral $O(3)$ field equations and their quantum analogue, *J. Nuc. Phys.* **33** 278 (Russian)
[3] Gay-Balmaz F, Holm D D and Ratiu T S 2013 Integrable G-Strands on semisimple Lie groups *Preprint math-ph 1308.3800*
[4] Holm D D 2011 *Geometric Mechanics II: Rotating, Translating and Rolling* (Singapore: World Scientific - Imperial College Press) 2nd edition
[5] Holm D D and Ivanov R I 2013 *SIGMA* **9** 027 14 pages *(Preprint math.DS 1211.6931)*
[6] Holm D D and Ivanov R I 2013 Matrix G-Strands *Preprint math-ph 1305.4010*
[7] Holm D D, Ivanov R I and Percival J R 2012 *Journal of Nonlinear Science* **22** 517 *(Preprint nlin.CD 1109.4421)*
[8] Holm D D and Lucas A M 2013 Toda lattice G-Strands *Preprint nlin.SI 1306.2984*
[9] Kirillov A 1981 *Funct. Anal. Appl.* **15** 135 (English); Funktsional. Anal. i Prilozhen. **15** 75 (Russian)
[10] Kirillov A 1993 *Contemp. Math.* **145** (Providence, RI: Amer. Math. Soc.) p 33
[11] Marle C-M 2013 *J. Geom. Symm. Phys.* **29** 1
[12] Poincaré H 1901 *C. R. Acad. Sci. Paris* **CXXXII** 369

Journal of Physics: Conference Series **482** (2014) 012019 doi:10.1088/1742-6596/482/1/012019

Stability under persistent perturbation by white noise

L Kalyakin
Institute of mathematics RAS, Chernyshevskii str., 112, Ufa, 450008 Russia
E-mail: klenru@mail.ru

Abstract. Deterministic dynamical system which has an asymptotical stable equilibrium is considered under persistent perturbation by white noise. It is well known that if the perturbation does not vanish in the equilibrium position then there is not Lyapunov's stability. The trajectories of the perturbed system diverge from the equilibrium to arbitrarily large distances with probability 1 in finite time. New concept of stability on a large time interval is discussed. The length of interval agrees the reciprocal quantity of the perturbation parameter. The measure of stability is the expectation of the square distance from the trajectory till the equilibrium position. The method of parabolic equation is applied to both estimate the expectation and prove such stability. The main breakthrough is the barrier function derived for the parabolic equation. The barrier is constructed by using the Lyapunov function of the unperturbed system.

1. Statement of the problem

Let the deterministic system be

$$\frac{d\mathbf{y}}{dT} = \mathbf{a}(\mathbf{y}, T), \qquad \mathbf{y} \in \mathbb{R}^n,\ T > 0. \tag{1}$$

The point $\mathbf{y} = 0$ is equilibrium, which means $\mathbf{a}(0, T) \equiv 0$. Let the perturbed system be the stochastic Ito equation [1, 2, 3]

$$d\mathbf{y} = \mathbf{a}(\mathbf{y}, T)dT + \mu\, B(\mathbf{y}, T)\, d\mathbf{w}(T), \qquad T > 0;\ \ \mathbf{y}|_{T=0} = \mathbf{x};\ \ 0 < \mu^2 \ll 1. \tag{2}$$

Here $\mathbf{w}(T)$ accounts for the standard Brownian motion in $\mathbb{R}^n$ and $B(\mathbf{y}, T)$ is a matrix $n \times n$ with the property $B(0, T) \not\equiv 0$. The solution $\mathbf{y} = \mathbf{y}_\mu(T; \mathbf{x})$ is a stochastic process in $\mathbb{R}^n$ depending on both the initial point $\mathbf{x}$ and a small parameter μ. The solution depends on the matrix B as well.

Let the equilibrium position $\mathbf{y} = 0$ of the system (1) be asymptotically stable in Lyapunov sense. Consider the problem of stability under white noise as follows: *does the trajectory* $\mathbf{y} = \mathbf{y}_\mu(T; \mathbf{x})$ *remain near equilibrium, if the perturbations* $\mu, |\mathbf{x}|$ *are small and the matrix belongs a fixed ball* $||B|| < M$*?*

It is well known, if $B(0, T) \not\equiv 0$, that there is not any Lyapunov type stability with respect to white noise. Almost all trajectories of the perturbed system diverge from the equilibrium to arbitrarily large distances in finite time. The concept of stability must be modified in appropriate way. The similar problem for autonomous systems was considered in [1], see also [4, 5, 6, 7]. We solve the problem for nonautonomous systems by method of parabolic equation. This approach and the obtained results are different from [1].

2. Definition of stability

The measure of stability is the expectation of the square perturbed solution, i.e.

$$\mathbb{E}[(\mathbf{y}_\mu(T;\mathbf{x}))^2] = u_\mu(\mathbf{x},T).$$

It is a function on $\mathbf{x}, T$. In the definition we follow Khasminskii [8], Katc and Krasovskii [9, 10].

Definition 1. *Equilibrium* $\mathbf{y} = 0$ *of the deterministic system* (1) *is weak stable with respect to white noise under the given estimates* $\delta = \delta(\varepsilon)$, $\Delta = \Delta(\varepsilon)$ *on an asymptotically large interval uniformly for* B *from the set* $\mathcal{B}$, *if there exists* $T_0 > 0$, *so that for any* $\varepsilon > 0$ *the expectation along perturbed trajectory is small:* $\mathbb{E}[(\mathbf{y}_\mu(T;\mathbf{x}))^2] < \varepsilon$ *uniformly for all:* $|\mathbf{x}| < \delta(\varepsilon)$, $|\mu| < \Delta(\varepsilon)$, $0 < T < T_0\mu^{-2}$, $B \in \mathcal{B}$.

This definition does not follow the usual mathematical approach [8] by reason of the finite time interval. But it is very useful for physics due to the given estimates and the large time interval under small μ, [9], p.3. We use the expectation instead of probability in the definition. It is not an essential difference due to Chebyshev's inequality [8], p.14, [2], p.65.

3. Reduction to the parabolic equation

Let $f(\mathbf{x})$ be a random function in $\mathbb{R}^n$. The expectation $\mathbb{E}[f(\mathbf{x})]$ is thus a function on $\mathbf{x}$. Consider the parabolic (Kolmogorov) equation

$$\partial_s u + \mathbf{a}(\mathbf{x},s)\partial_{\mathbf{x}} u + \mu^2 \sum_{i,j=1}^{n} \alpha_{i,j}(\mathbf{x},s)\partial_{x_i}\partial_{x_j} u = 0, \quad \mathbf{x} \in \mathbb{R}^n, \quad 0 \le s < T$$

with the specific final condition

$$u|_{s=T} = \mathbb{E}[f(\mathbf{x})].$$

Here the matrix $A = \{\alpha_{i,j}\}$ is determined by $2A = BB^*$. The equations may be reduced to the initial problem by change of time: $s = T - t$. After that the problem takes the form with specific dependence on T, t:

$$L_{T,\mu} u \equiv \partial_t u - \mathbf{a}(\mathbf{x}, T-t)\partial_{\mathbf{x}} u - \mu^2 \sum_{i,j=1}^{n} \alpha_{i,j}(\mathbf{x}, T-t)\partial_{x_i}\partial_{x_j} u = 0, \quad \mathbf{x} \in \mathbb{R}^n, \quad 0 < t \le T; \tag{3}$$

$$u(\mathbf{x}, t; T, \mu)|_{t=0} = \mathbb{E}[f(\mathbf{x})], \quad \mathbf{x} \in \mathbb{R}^n.$$

Connection with the solution $\mathbf{y}_\mu(T;\mathbf{x})$ of the stochastic equation is given by the formula [2, 3]

$$u(\mathbf{x}, t; T, \mu)|_{t=T} = \mathbb{E}[f(\mathbf{y}_\mu(T;\mathbf{x})].$$

In the special case, when $f(\mathbf{y}) = |\mathbf{y}|^2$, the solution of the parabolic problem is taken on the upper boundary: $u(\mathbf{x}, T; T, \mu) = \mathbb{E}[(\mathbf{y}_\mu(T;\mathbf{x}))^2]$ gives the expectation of the quadrate distance from the random trajectory $\mathbf{y} = \mathbf{y}_\mu(T;\mathbf{x})$ till the point $\mathbf{y} = 0$. In such interpretation the function $u(\mathbf{x}, T; T, \mu)$ may be used as a measure of drift of the perturbed trajectory $\mathbf{y}_\mu(T;\mathbf{x})$ from zero. In the case, when the point $\mathbf{y} = 0$ is the equilibrium position of the unperturbed system (1), the magnitude of the function $u(\mathbf{x}, T; T, \mu)$ as $\mathbf{x}, \mu \to 0$ features stability of equilibrium under white noise perturbation.

4. Barrier function

The prove-out of stability is so reduced to estimate the solution $u(\mathbf{x}, t; T, \mu)$ of the parabolic equation near the point $\mathbf{x} = 0$. *Desirable estimate must be uniform with respect to the parameter T on a long interval: $T \in (0, \mathcal{O}(\mu^{-2}))$, $\mu \to 0$.* We construct a barrier function in order to obtain an appropriate estimate.

Definition 2. *Barrier for the equation* (3) *is the function $W(\mathbf{x}, t; T, \mu)$, which possesses continues derivatives involved in differential operator $L_{T,\mu}$ from* (3)*, and the value of the operator is nonnegative $L_{T,\mu} W(\mathbf{x}, t; T, \mu) \geq 0$ in the layer $D^T = \{\mathbf{x} \in \mathbb{R}^n,\ 0 < t \leq T\}$.*

Application of barrier function is based on the maximum principle which says: any solution of the equation (3), which is majorized by barrier at the initial moment $u(\mathbf{x}, t)|_{t=0} \leq W(\mathbf{x}, 0; T, \mu))$, will be majorized at subsequent moments $u(\mathbf{x}, t) \leq W(\mathbf{x}, t; T, \mu))$, $0 < t \leq T$ (see [11], p.14). To construct an appropriate barrier we use a Lyapunov function of the unperturbed system (1).

5. Lyapunov function

Consider the unperturbed system (1). Remind the concept of Lyapunov function [9].

Definition 3. *A function $U(\mathbf{x}, T)$ is called a Lyapunov function for* (1) *at $\mathbf{x} = 0$ if exists a cylinder $D_r = \{(\mathbf{x}, T) :\ |\mathbf{x}| < r,\ T \geq 0\}$ such that 1) $U(\mathbf{x}, T)$ is defined, continuous, and differentiable in D_r; 2) $U(\mathbf{x}) > 0$ if $\mathbf{x} \neq 0$, $U(0) = 0$, and 3) $\big[\partial_T + \mathbf{a}(\mathbf{x}, T)\partial_{\mathbf{x}}\big]U(\mathbf{x}, T) < 0$ in D_r.*

If the system (1) possesses a Lyapunov function of such type, then $x = 0$ is asymptotically stable. Moreover the equilibrium is stable with respect to a persistent smooth deterministic perturbation [9]. A specific of the problem under consideration is that the white noise is not deterministic and the perturbation is not smooth. The perturbed system (2) is not a differential equation in ordinary sense and there is not Lyapunov's stability if $B(0, T) \not\equiv 0$. The trajectories of the stochastic system (2) diverge from the origin to arbitrarily large distances with probability 1 in finite time [1, 2]. But if the diffusion coefficient μ is small then many of trajectories stay near equilibrium for long times. We refer to this property as stability on large interval. The proof of such stability is derived from estimate of the solution of the parabolic equation by means of an appropriate barrier.

Lyapunov function is applied to construct such barrier. The reason is as follows. Let $U(\mathbf{x}, T)$ be a Lyapunov function for (1). Consider the function $V(\mathbf{x}, t; T) = U(\mathbf{x}, T - t)$, $0 \leq t \leq T$. Then for two terms of the parabolic operator the inequality

$$\big[\partial_t - \mathbf{a}(\mathbf{x}, T - t)\partial_{\mathbf{x}}\big]V(\mathbf{x}, t, T) > 0$$

takes place and this suggests to use $V(\mathbf{x}, t; T)$ in the barrier of the equation (3).

There are two obstacle in this way. The first is the diffusion term with a small factor μ^2 in the parabolic operator which gives a negative addend in $L_{T,\mu}V$. The second is the local property of the Lyapunov function while the barrier function must be global. In order to overcome these difficulties we specify the system under consideration.

6. Conditions

Principal restrictions, which are imposed on the differential equations (1), are described by means of conditions on the Lyapunov function $U(\mathbf{x}, T)$. The first part of the conditions has a local nature (in the cylinder D_r) and they are conventional in case of asymptotical stable equilibrium:

$$\partial_t U - \mathbf{a}(\mathbf{x}, T)\partial_{\mathbf{x}} U \geq \gamma U, \tag{4}$$

$$|\mathbf{x}|^2 \leq U(\mathbf{x}, T) \leq M_0 |\mathbf{x}|^2, \quad |\partial_{\mathbf{x}} U|^2 \leq M_1 U, \tag{5}$$

$$\sum_{i,j=1}^{n} a_{i,j}(\mathbf{x},T)\partial_{x_i}\partial_{x_j}U \le M_2; \quad (\mathbf{x},T)\in D_r; \ \gamma, M_k = \text{const} > 0. \tag{6}$$

The local inequality (4) is a fundamental property. In general case there is not any Lyapunov function with global property (4). In what follows we use a function which is given in the whole half-space $D = \{\mathbf{x} \in \mathbb{R}^n, \ T \ge 0\}$ while the condition (4) holds in the subdomain D_r. One can guess that the local Lyapunov function with properties (4)-(6) may be continued in the wide domain D without (4). The second part of the conditions describes the properties of such continuation. They are easily verified if the Lyapunov function is a quadratic form:

$$|\partial_{\mathbf{x}}U|^2 \le M_1 U, \quad L_{T,\mu}U \ge -M_3 U, \quad \mathbf{x},t \in D\setminus D_r, \ M_3 = \text{const} > 0. \tag{7}$$

7. Construction of the barrier function

The barrier function is described by different formulas in different subregions of the layer $D^T = \{\mathbf{x} \in \mathbb{R}^n, \ 0 < t \le T\}$. It is convenient to take the inner subregion delimited by a level surface of the Lyapunov function

$$\mathcal{D}_r^T = \{(\mathbf{x},t) \in D^T : \ U(\mathbf{x}, T-t) < r^2\}, \ r = \text{const} > \text{o}.$$

The outer subregion is the complement $D^T \setminus \mathcal{D}_r^T$. Under condition (5) the boundary $\partial\mathcal{D}_r^T$ is localized in the cylinder layer $\{r/\sqrt{M_0} \le |\mathbf{x}| \le r, \ T > 0\}$.

The first part of the barrier $V_0(\mathbf{x},t;T,\mu)$ is determined by formula:

$$V_0(\mathbf{x},t;T,\mu) = U(\mathbf{x},T-t)\exp(-\alpha_0 t) + \frac{\mu^2 M_2}{\alpha_0}[1-\exp(-\alpha_0 t)], \ \alpha_0 = \text{const} > 0.$$

If we take $\alpha_0 \le \gamma$, then under conditions (4),(5) the barrier property $L_{T,\mu}V_0(\mathbf{x},t;T,\mu) \ge 0$ holds in the subregion $\mathcal{D}_r^T$.

If $r = \infty$ occurs (which means that $U(\mathbf{x},T)$ is a global Lyapunov function), then the function $V_0(\mathbf{x},t;T,\mu)$ is a desirable barrier in D^T. This case corresponds to [8], p.249.

In general case there is not any global Lyapunov function. It means $r < \infty$ and the barrier property for the function $V_0(\mathbf{x},t;T,\mu)$ can violate for large $|\mathbf{x}| > r$. Then we add the exponentially growing function

$$V_2(\mathbf{x},t;T,\mu) = P(U)\exp(\alpha t), \ \ U = U(\mathbf{x},T-t), \ (\mathbf{x},t) \in D^T\setminus\mathcal{D}_\rho^T, \ 0<\rho<r.$$

The factor $P(U)$ is determined by the formulas:

$$P(U) = v + \mu^2\left(m - \frac{\lambda v}{\mu^2+\nu v}\right), \ \ v = [U-\rho^2\chi(U)]\frac{\gamma}{M_1}; \ \ \alpha, m, \lambda, \nu = \text{const} > 0.$$

Here the function $\chi(U) \in C^\infty(\mathbb{R})$ is a smooth steady decreasing patch function with property:

$$\chi(U) \equiv 1, \ U \le \rho^2; \ \chi(U) \equiv 0, \ U \ge r^2.$$

In order to avoid exponentially grows in the neighborhood of the point $\mathbf{x} = 0$, the second part of the barrier is continued in $\mathcal{D}_\rho^T$ by the function

$$V_1(\mathbf{x},t;T,\mu) = m\,\mu^2 \exp(\alpha t + v/\mu^2), \ \ (\mathbf{x},t) \in \overline{\mathcal{D}_\rho^T}.$$

Since the function $v = [U(\mathbf{x},t;T) - \rho^2]\gamma/M_1 < 0$ is negative in $\mathcal{D}_\rho^T$, hence the $V_1(\mathbf{x},t;T,\mu)$ is exponentially small as $\mu \to 0$. In the more narrow domain, where $U(\mathbf{x},t;T) < \rho^2/2$, the estimate $V_1(\mathbf{x},t;T,\mu) = \mathcal{O}(\mu^2)$ is uniform with $(\mathbf{x},t) \in \mathcal{D}_\rho^T$ under $\forall\, T \le \mathcal{O}(\mu^{-2})$. The last property is the main purpose of the construction.

Theorem 1. *Let the Lyapunov function $U(\mathbf{x}, T)$ for the equation* (1) *enjoy the properties* (4)-(7). *Then for all sufficiently small $|\mu| < \mu_0$ there is a barrier function $W(\mathbf{x}, t; T, \mu)$ for the equation* (3)*, which is expressed by the formula*

$$W = V_0(\mathbf{x}, t; T, \mu) + \begin{cases} V_1(\mathbf{x}, t; T, \mu), \ (\mathbf{x}, t) \in \overline{\mathcal{D}^T_\rho}, \\ V_2(\mathbf{x}, t; T, \mu), \ (\mathbf{x}, t) \in D^T \setminus \overline{\mathcal{D}^T_\rho} \end{cases} \tag{8}$$

under appropriate constants $\alpha_0, \alpha, m, \lambda, \nu, \mu_0$ which are not depending on the parameters μ, T.

Proof. The proof of the theorem consists in the fitting of the barrier parameters $\alpha_0, \alpha, m, \lambda, \nu, \mu_0$. There are two requirements which must be satisfied. The first corresponds to smoothness of the function $W(\mathbf{x}, t; T, \mu)$ on the matching surface $\partial\mathcal{D}^T_\rho$ where $v = 0$ and the second is the barrier property $L_{T,\mu}W \geq 0$ in D^T.

The relations $\lambda = 1 - m, \quad \nu = m/2(1-m), \quad 0 < m < 1$ provide a sufficient smooth $W \in C_2(D^T)$. The value m is determined in what follows.

If we take into account the conditions (4),(5), then the estimates

$$L_{T,\mu}V_0 \geq 0, \ L_{T,\mu}V_1 \geq 0$$

in the inner domain $\mathcal{D}^T_\rho$ is derived[1] under appropriate choice of parameters: $\alpha_0 \leq \gamma$, $\alpha \geq \gamma M_2/M_1$. Hence the barrier property is proved in the inner domain $\mathcal{D}^T_\rho$.

In the outer domain $D^T \setminus \mathcal{D}^T_\rho$ the second part of the barrier is taken. It gives

$$L_{T,\mu}V_2 = \Big[\alpha P(U) + \big(\partial_t U - \mathbf{a}\,\partial_{\mathbf{x}} U\big)P'(U) -$$

$$-\mu^2 \sum_{i,j=1}^{n} a_{i,j}(\mathbf{x}, t; T)\big[P'(U)\partial_{x_i}\partial_{x_j}U + P''(U)\partial_{x_i}U\partial_{x_j}U\big]\Big] \exp(\alpha t) + qV_2.$$

In the far subregion $D^T \setminus \overline{\mathcal{D}^T_r}$ the condition (7) is applied. Here $v \equiv U$, and the leading term of the expression $L_{T,\mu}V_2$ is as follows: $\alpha P(U)\exp(\alpha t)$. Since

$$P(U) = U + \mu^2\big(m - \frac{\lambda U}{\mu^2 + \nu U}\big) \geq U + (m - \lambda/\nu) = U + \mu^2(4m - m^2 - 2)/m,$$

then the value m is chosen between root of the quadratic polynomial $4m - m^2 - 2$ under condition $m < 1$. It means $2 - \sqrt{2} < m < 1$. Under this m we have $4m - m^2 - 2 > 0$. Hence the first term in the expression $L_{T,\mu}V_2$ is minorized as follows

$$\alpha P(U)\exp(\alpha t) \geq \alpha U \exp(\alpha t).$$

Due to (7) all remainder terms in the expression

$$L_{T,\mu}W = L_{T,\mu}V_0 + L_{T,\mu}V_2, \ \ (\mathbf{x}, t) \in D^T \setminus \overline{\mathcal{D}^T_r}.$$

are minorized by $-\tilde{M}U\exp(\alpha t)$ with a constant $\tilde{M} > 0$ which is not depending on μ, T. Hence the choice $\alpha \geq \tilde{M}$ provides the inequality $L_{T,\mu}V \geq 0$ in the domain $D^T \setminus \overline{\mathcal{D}^T_r}$.

In the intermediate domain $\mathcal{D}^T_r \setminus \overline{\mathcal{D}^T_\rho}$ the properties $v > 0$ and $\chi'(U) \leq 0$ are used. Due to them the inequalities:

$$P(U) = v - \mu^2\frac{\lambda v}{\mu^2 + \nu v} + m\mu^2 \geq (1-\lambda)v + m\mu^2 = m\,(v + \mu^2) > 0,$$

[1] The inequality $L_{T,\mu}V_0 \geq 0$ takes place in the larger domain $\mathcal{D}^T_r$.

$$P'(U) = \Big[1 - \frac{\mu^4\lambda}{(\mu^2+\nu v)^2}\Big](1-\chi'(U)) \geq 1-\lambda = m$$

are derived. Since in this domain both the inequalities $\rho^2 < U(\mathbf{x},t;T) < r^2$ and the condition (4) take place, hence

$$(\partial_t U - \mathbf{a}\partial_\mathbf{x} U)P'(U) \geq \gamma m \rho^2 = \text{const} > 0, \quad (\mathbf{x},t) \in D_r^T \setminus \overline{\mathcal{D}_\rho^T}.$$

Due to this inequality and conditions (5), (6) the value $L_{T,\mu}V_2 \geq 0$ turns out to be nonnegative for all sufficiently small $|\mu| < \mu_0$. In this step the boundary of the parameter μ_0 is chosen. Remind that $L_{T,\mu}V_0 \geq 0$ in the domain $\mathcal{D}_r^T$. Hence

$$L_{T,\mu}W = L_{T,\mu}V_0 + L_{T,\mu}V_2 \geq 0, \quad (\mathbf{x},t) \in D_r^T \setminus \overline{\mathcal{D}_\rho^T}.$$

So the barrier property is proved in the whole domain D^T.

Corollary 1. *If there exists the solution of the problem* (3) *with initial data* $\mathbb{E}[|\mathbf{x}|^2]$, *then it is majorized by the barrier constructed above:*

$$0 \leq u(\mathbf{x},t;T,\mu) \leq W(\mathbf{x},t;T,\mu), \ 0 < t < T, \ \forall T > 0, \ |\mu| \leq \mu_0.$$

Proof. The property obtained above

$$V_0|_{t=0} = U|_{t=0}, \quad V_1|_{t=0} > 0, \quad V_2|_{t=0} = P(U)|_{t=0} \geq 0$$

provides $W|_{t=0} \geq U|_{t=0} \geq |\mathbf{x}|^2$. Then the desired estimate follows from maximum principle [11].
Note the same barrier is suitable for a set of equations (3) with different matrices $A = \{a_{i,j}\}$, under fixed constants M_2, M_3 in the conditions (6),(7).

8. Stability

Now we came back to the problem of stability with respect to white noise perturbation. In order to discuss stability one has to be sure that the perturbed equation (2) has a global solution i.e. the solution exists on the infinite interval $T \in [0,\infty)$. To ensure this property we have to require additional restrictions on the coefficients. In the most simple approach it is enough to demand a bounded grows of the coefficients with $\mathbf{y}$ at infinity [8, 3]:

$$|\mathbf{a}(\mathbf{y},T)| \leq \text{const}\cdot(1+|\mathbf{y}|), \ \forall \mathbf{y} \in \mathbb{R}^n, \ T > 0, \tag{9}$$

$$||B(\mathbf{y},T)|| =^{def} \max_{i,j}|b_{i,j}(\mathbf{x},t;T)| \leq M\cdot(1+|\mathbf{y}|), \ \forall \mathbf{y} \in \mathbb{R}^n, \ T > 0. \tag{10}$$

Moreover in view of application of the parabolic equation we impose an additional condition: $\mathbf{a}(\mathbf{y},T), B(\mathbf{y},T) \in C_{2,0}(D)$. It ensures smoothness of the expectation $\mathbb{E}[(\mathbf{y}_\mu(T;\mathbf{x})^2]$ with respect to $\mathbf{x}$ on a random trajectory $\mathbf{y}_\mu(T;\mathbf{x})$, [12]. Let the derivatives $\partial_\mathbf{y}B(\mathbf{y},T), \partial^2_\mathbf{y}B(\mathbf{y},T)$ satisfy the conditions (9), (10) as well. To identify a set of matrices of that type we introduce the ball in the linear space $\mathcal{M}^n$ of matrix-fuctions:

$$S_M^{2,0} = \{B(\mathbf{y},T) \in \mathcal{M}^n: \max_{0\leq k\leq 2}\sup_{\mathbf{y},T}||(1+|\mathbf{y}|)^{-1}\partial^k_\mathbf{y}B(\mathbf{y},T)|| \leq M\}. \tag{11}$$

Theorem 2. *Let the deterministic system (1) with the smooth coefficient* $\mathbf{a}(\mathbf{y},T) \in C_{2,0}(D)$ *under condition (9) have the Lyapunov function with the properties (4)-(7). Then for each* $M > 0$ *there exists* $\delta_M, \Delta_M > 0$ *such that the equilibrium* $\mathbf{y} = 0$ *is weak stable with respect to white noise under given estimates* $\delta(\varepsilon) = \delta_M\sqrt{\varepsilon}$, $\Delta(\varepsilon) = \delta_M\sqrt{\varepsilon}$ *on a large interval* $T \in (0, \mathcal{O}(\mu^{-2}))$ *uniformly with matrices* $B(\mathbf{y},T)$ *from the ball* $S^{2,0}_M$.

Proof. The expectation $\mathbb{E}[(\mathbf{y}_\mu(T;\mathbf{x})^2]$ along the random trajectory is a function on $(\mathbf{x},T)$. Due to the condition (11) this function is a solution of the parabolic problem (3) taken on the upper boundary

$$\mathbb{E}[\big(\mathbf{y}_\mu(T;\mathbf{x})\big)^2] = u(\mathbf{x},t;T,\mu)|_{t=T}.$$

Hence it is majorized by barrier constructed above. Due to the properties of the barrier the expectation has the estimate:

$$\mathbb{E}[\big(\mathbf{y}_\mu(T;\mathbf{x})\big)^2] \le M_0|\mathbf{x}|^2 + \mu^2\big[M_3/\alpha_0 + m\big]. \tag{12}$$

From here the property of weak stability follows: $\mathbb{E}[\big(\mathbf{y}_\mu(T;\mathbf{x})\big)^2] < \varepsilon$ under $|\mathbf{x}| < \delta(\varepsilon) = \sqrt{\varepsilon/2M_0}$, $|\mu| < \Delta(\varepsilon) = \sqrt{\varepsilon/2\big[M_3/\alpha_0 + m\big]}$, $0 < T \le R^2/4\alpha\mu^2$.

9. Conclusion

Stability of equilibrium with respect to white noise perturbation on a long time interval was proved in this paper. The length of the interval μ^{-2} corresponds to reciprocal quantity of the noise intensity μ. There is the hypothesis of exponentially large interval $\mathcal{O}\big(\exp(S\mu^{-2})\big)$, $\mu \to 0$, $S = \text{const} > 0$. This conjecture is derived by analogy the formula for the first passage time, which is known in the case of autonomous systems [1]; the nonautonomous systems were not considered. In the approach discussed above the hypothesis is reduced to find an appropriate barrier for the parabolic equation (3).

Acknowledgments

I would like to thank prof. F.S. Nasyrov for helpful comments. I also acknowledge financial support under the RFBR Grant 13-01-00199.

References

[1] Freidlin M I Wentzell A D 1998 *Random Perturbations of Dynamical Systems* 2nd edn. (New York, Heidelberg, Berlin: Springer-Verlag)
[2] Schuss Z 2010 *Theory and applications of stochastic processes* Applied Mathematical Sciences vol 170. (New York, Heidelberg, Berlin: Springer-Verlag)
[3] Oksendal B 1998 *Stochastic Differential Equations. An Introduction with Applications* 5th edn. (New York, Heidelberg, Berlin: Springer-Verlag)
[4] Freidlin M I 1998 *Documenta Mathematica* Extra Volume ICM **III** 223–235
[5] Freidlin M I Weber M 2001 *Stochastic Process. Appl.* **94**(2) 199-239
[6] Freidlin M Weber M 2004 *Probab. Theory Relat. Fields* **128** 441–466.
[7] Cerrai S Freidlin M 2011 *J. Stat. Phys.* **144** 101123
[8] Khasminskii R 2012 *Stochastic Stability of Differential Equations* 2nd edn. (New York, Heidelberg, Berlin: Springer-Verlag)
[9] Krasovskii N N *Stability of Motion* 1963 (Stanford: Stanford University Press)
[10] Kats I Y 1998 *Lyapunov function method in problem of stability and stabilization of the system of random structure* (Ekaterinburg: UrGAPS) (in russian)
[11] Ladyzenskaja O A Solonnikov V A Ural'ceva N N 1968 *Linear and quasi-linear equations of parabolic type* Translations of Mathematical Monographs vol 23 (Providence: American Mathematical Society)
[12] Krylov N V 1980 *Controlled diffusion processes* (New York, Heidelberg, Berlin: Springer-Verlag)

On effective solutions of the nonlinear Schrödinger equation

N Khatiashvili, R Shanidze and D Janjgava

I. Vekua Institute of Applied Mathematics, Iv. Javakhishvili Tbilisi State University, 2, University St. 0186 Tbilisi, Georgia

E-mail: ninakhatia@gmail.com

Abstract. Cubic nonlinear Schrödinger type equation with specific initial-boundary conditions in the infinite domain is considered. The equation is reduced to an equivalent system of partial differential equations and studied in the case of solitary waves. The system is modified by introducing new functions, one of which belongs to the class of functions of negligible fifth order and vanishing at infinity exponentially. For this class of functions the system is reduced to a nonlinear elliptic equation which can be solved analytically, thereby allowing us to present nontrivial approximated solutions of nonlinear Schrödinger equation. These solutions describe a new class of symmetric solitary waves. Graphics of modulus of the corresponding wave function are constructed by using Maple.

1. Introduction

Nonlinear Schrödinger equation describes a lot of physical processes, such as electron plasmatic waves, electromagnetic ion cyclotron waves, waves in cosmic gases, non-linear optics phenomena etc. [1-2, 17-22]. As a result, this equation has been considered by numerous authors and from different points of view (see for example [3-17, 20-22]).

In this paper the nonlinear Schrödinger equation equation is considered in the infinite domain and with specific initial-boundary conditions. We study the case of solitary waves vanishing at infinity and with the modulus of the wave function of the class $0(Re^{-D})$, $D \geq 3$, where D and R are some definite constants. By introducing new variables the equation is modified for this class of waves and non-trivial solutions are derived analytically. Consequently, a new class of localized approximated solutions is obtained.

2. Statement of the problem

In the coordinate system $0xyz$ we consider the following problem.

Problem 1. In the domain $Q_T = R^3 \times \{0 < t < T\}$, find continuous function Ψ satisfying the Schrödinger equation

$$i\frac{\partial \Psi}{\partial t} + \Delta\Psi + \lambda|\Psi|^2\Psi = 0, \tag{1}$$

and the conditions

$$\lim_{|x|+|y|+|z|\to\infty} \Psi = 0, \ |\Psi|_{t=0} = r(x, y, z), \tag{2}$$

where Ψ is a wave function, $\Psi = U + iV$, (U, V are real continuous functions having second order derivatives), λ is some parameter, and $r(x, y, z) \geq 0$ is a function also to be defined.

We will find the solutions of Equation (1) for which $|\Psi|$ is independent of t, vanishing at infinity exponentially, and $|\Psi|^5$ is negligible. This class describes some class of solitary waves. To achieve our goal, we will examine the auxiliary problem first. To this, let us consider the infinite domain

$$G_0 = \{x > 0, y > 0, z > 0\}$$

and the following problem.

Problem 2. In the domain $Q_T^0 = G_0 \times \{0 < t < T\}$, find continuous function Ψ with the modulus of the class $0(Re^{-D}), D \geq 3$, satisfying the equation (1) and the condition (2).

It is obvious that equation (1) is equivalent to the following system of partial differential equations

$$\frac{\partial U}{\partial t} = -\Delta V - \lambda V\,(U^2 + V^2), \qquad \frac{\partial V}{\partial t} = \Delta U + \lambda U\,(U^2 + V^2). \tag{3}$$

Consequently, Problem 2 can be reduced to the following problem.

Problem 3. In the domain $Q_T^0 = G_0 \times \{0 < t < T\}$, find continuous functions U and V of the class $0(Re^{-D}), D \geq 3$, having second order derivatives, satisfying the system (3) and the condition

$$(|U|^2 + |V|^2)_{t=0} = r^2(x, y, z).$$

3. Modification of the Schrodinger equation and its solution

Let us introduce the notation

$$U = r\cos\varphi; \quad V = r\sin\varphi, \tag{4}$$

where $r, r \geq 0$, and φ are unknown functions of the variables x, y, z, t. Taking into account (4), the system (3) becomes [7]

$$\begin{aligned}\frac{\partial r}{\partial t}\cos\varphi - r\sin\varphi\frac{\partial \varphi}{\partial t} &= -\sin\varphi\Delta r - 2\cos\varphi\left(\frac{\partial r}{\partial x}\frac{\partial \varphi}{\partial x} + \frac{\partial r}{\partial y}\frac{\partial \varphi}{\partial y} + \frac{\partial r}{\partial z}\frac{\partial \varphi}{\partial z}\right) + \\ &-r\left\{-\sin\varphi\left(\left(\frac{\partial \varphi}{\partial x}\right)^2 + \left(\frac{\partial \varphi}{\partial y}\right)^2 + \left(\frac{\partial \varphi}{\partial z}\right)^2\right) + \cos\varphi\Delta\varphi\right\} - \lambda r^3\sin\varphi,\end{aligned} \tag{5}$$

$$\begin{aligned}\frac{\partial r}{\partial t}\sin\varphi + r\cos\varphi\frac{\partial \varphi}{\partial t} &= \cos\varphi\Delta r - 2\sin\varphi\left(\frac{\partial r}{\partial x}\frac{\partial \varphi}{\partial x} + \frac{\partial r}{\partial y}\frac{\partial \varphi}{\partial y} + \frac{\partial r}{\partial z}\frac{\partial \varphi}{\partial z}\right) + \\ &-r\left\{\cos\varphi\left(\left(\frac{\partial \varphi}{\partial x}\right)^2 + \left(\frac{\partial \varphi}{\partial y}\right)^2 + \left(\frac{\partial \varphi}{\partial z}\right)^2\right) + \sin\varphi\Delta\varphi\right\} + \lambda r^3\cos\varphi.\end{aligned} \tag{6}$$

After simple transformations, Eqs. (5)-(6) yield [7]

$$\Delta r + \lambda r^3 = r\frac{\partial \varphi}{\partial t} + r\left\{\left(\frac{\partial \varphi}{\partial x}\right)^2 + \left(\frac{\partial \varphi}{\partial y}\right)^2 + \left(\frac{\partial \varphi}{\partial z}\right)^2\right\}, \tag{7}$$

$$\Delta\varphi = -\frac{\partial}{\partial t}\ln r - 2\left\{\frac{\partial}{\partial x}(\ln r)\frac{\partial \varphi}{\partial x} + \frac{\partial}{\partial y}(\ln r)\frac{\partial \varphi}{\partial y} + \frac{\partial}{\partial z}(\ln r)\frac{\partial \varphi}{\partial z}\right\}. \tag{8}$$

Here we consider the case $\varphi = A_0 t + A_1$, where $A_0; A_1; A_0 > 0$; are some constants. From Eqs. (7)-(8) we get

$$\Delta r + \lambda r^3 - A_0 r = 0. \tag{9}$$

We seek the symmetric solutions of (9) in the form

$$r = R \sin \psi, \tag{10}$$

where R is a constant and $\psi(x,y,z)$ is a sufficiently small unknown function such that ψ^5 is negligible (in one dimensional case the exact solution of (9) can be obviously given in terms of elliptic functions [9, 21-22]). By virtue of (10), Equation (9) then turns into a non-linear partial differential equation for $\psi(x,y,z)$, i.e.

$$\cos\psi \Delta\psi - \sin\psi \left\{ \left(\frac{\partial\psi}{\partial x}\right)^2 + \left(\frac{\partial\psi}{\partial y}\right)^2 + \left(\frac{\partial\psi}{\partial z}\right)^2 \right\} + \lambda R^2 \sin^3\psi - A_0 \sin\psi = 0. \tag{11}$$

Taking into the account that ψ^5 is sufficiently small, and putting into (11) the approximated formulas $\sin\psi \approx \psi - \frac{\psi^3}{6}$, and $\cos\psi \approx 1 - \frac{\psi^2}{2} + \frac{\psi^4}{24}$, it then results

$$\left(1 - \frac{\psi^2}{2} + \frac{\psi^4}{24}\right)\Delta\psi - \left(\psi - \frac{\psi^3}{6}\right)\left\{ \left(\frac{\partial\psi}{\partial x}\right)^2 + \left(\frac{\partial\psi}{\partial y}\right)^2 + \left(\frac{\partial\psi}{\partial z}\right)^2 \right\} + \\ + \lambda R^2 \left(\psi - \frac{\psi^3}{6}\right)^3 - A_0\left(\psi - \frac{\psi^3}{6}\right) = 0. \tag{12}$$

For the equation (12), we will solve the following problem.

Problem 3. In the domain G_0 to find symmetric continuous function ψ of the class $0(e^{-D})$, $D \geq 3$, having second order derivatives and satisfying the equation (12).

Let us consider the function

$$\psi = e^{-\alpha x - \beta y - \gamma z - D}, \tag{13}$$

where $D, \alpha, \beta, \gamma > 0$ are some constants satisfying the following conditions

$$\alpha^2 + \beta^2 + \gamma^2 = A_0, \qquad \lambda R^2 = \frac{4}{3} A_0, \tag{14}$$

and D is the constant chosen for the desired accuracy, in such a way that e^{-5D} is negligible (for example for $D = 3$, $e^{-15} \approx 10^{-7}$). It is clear from (13) that ψ belongs to the class $0(e^{-D}), D \geq 3$, and $|\psi| \leq e^{-D}$. Insertion of (13) into the left hand side of (12) yields

$$\left(1 - \frac{\psi^2}{2} + \frac{\psi^4}{24}\right)\Delta\psi - \left(\psi - \frac{\psi^3}{6}\right)\left\{ \left(\frac{\partial\psi}{\partial x}\right)^2 + \left(\frac{\partial\psi}{\partial y}\right)^2 + \left(\frac{\partial\psi}{\partial z}\right)^2 \right\} + \\ + \lambda R^2 \left(\psi - \frac{\psi^3}{6}\right)^3 - A_0\left(\psi - \frac{\psi^3}{6}\right) = A_0\left(-\frac{\psi^5}{72} + \frac{\psi^7}{27} - \frac{\psi^9}{162}\right).$$

Since ψ^5 is negligible we can ignore the rhs term in this equation and the function ψ given by the formula (13) will solve equation (12). Hence we conclude that, if D is chosen accordingly, the function (13) is an effective solution of the equation (12) with the accuracy

$$\left| \left(1 - \frac{\psi^2}{2} + \frac{\psi^4}{24}\right)\Delta\psi - \left(\psi - \frac{\psi^3}{6}\right)\left\{ \left(\frac{\partial\psi}{\partial x}\right)^2 + \left(\frac{\partial\psi}{\partial y}\right)^2 + \left(\frac{\partial\psi}{\partial z}\right)^2 \right\} + \right. \\ \left. + \lambda R^2 \left(\psi - \frac{\psi^3}{6}\right)^3 - A_0\left(\psi - \frac{\psi^3}{6}\right) \right| \leq A_0 \frac{\psi^5}{72}.$$

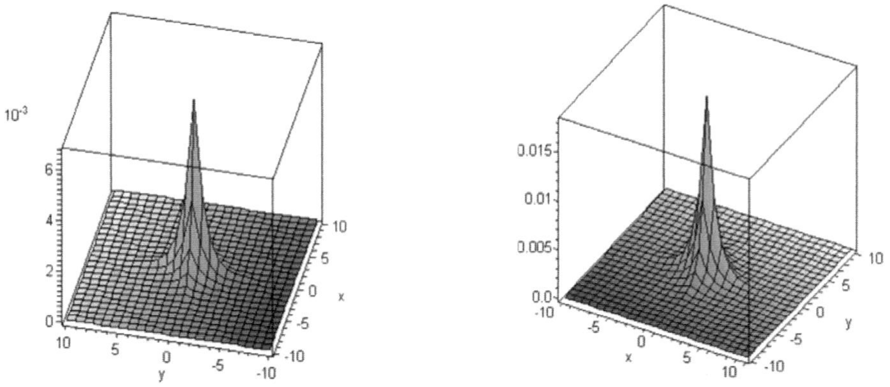

a) b)

Figure 1. Graphics of $|\Psi|$ when $z = 1$, $\alpha = \beta = \gamma = 1, A_0 = 3, \lambda = 4$: a) $D = 4$; b) $D = 3$.

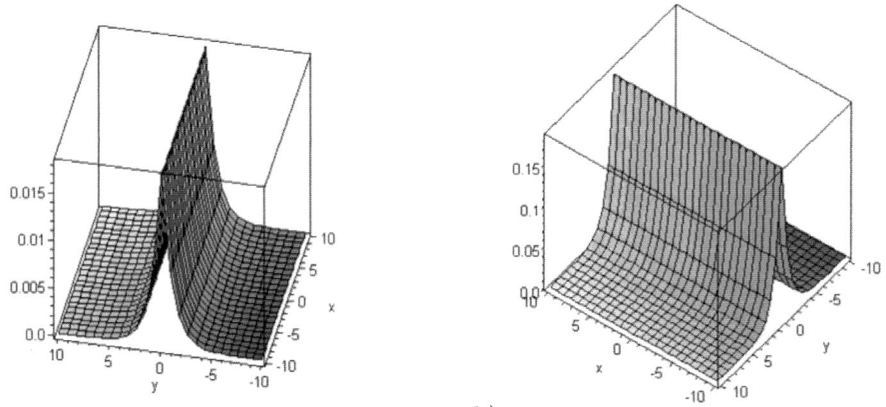

a) b)

Figure 2. Graphics of $|\Psi|$ when $D = 4, \alpha = \gamma = 0, \beta = 1, A_0 = 1$: a) $R = 1, \lambda = 4/3$; b) $R = 10, \lambda = 0.04$.

Consequently, the solution of the Problem 2 will be given by

$$\Psi = Re^{iA_0t+iA_1} \sin e^{-\alpha x-\beta y-\gamma z-D},$$

where R is some constant, e^{-5D} is sufficiently small and the constants $\lambda, \alpha, \beta, \gamma > 0$ satisfy the conditions (14).

Having solved the Problem 2 it is now easy to find the solutions for the Problem 1. By direct verification, we recognize indeed that the symmetric solutions of the Problem 1 will be identified by means of the formula

$$\Psi = Re^{iA_0t+iA_1} \sin e^{-\alpha|x|-\beta|y|-\gamma|z|-D}, \tag{15}$$

where the constants $A_0, \lambda, \alpha, \beta, \gamma > 0$ satisfy the conditions (14). Note that the first order derivatives of the function (15) have discontinuities at the planes $x = 0, y = 0$, and $z = 0$, but their squares and the second order derivatives are continuous at this planes and the equation (12) holds.

4. Conclusions

There exist solitary waves for which the wave function Ψ is of the form (15) and its modulus satisfies the inequality

$$|\Psi| \leq Re^{-D},$$

where R is the amplitude and D is the constant for which e^{-5D} is negligible. For given R and A_0, the constant λ is uniquely defined. The constants R, A_0 and any two of $\alpha, \beta, \gamma > 0$ can be chosen arbitrary. Representive graphics of $r = |\Psi|$ plotted using Maple are given in Figures 1-2.

Acknowledgments The designated project has been fulfilled by financial support of the Georgia Rustaveli Scientific Foundation (Grant #GNSF/ST08/3-395).

References

[1] Alexandrov A, Bogdankevich L and Rukhadze A 1984 *Principles of Plasma Electrodynamics* (Germany, Heidelberg: Springer)
[2] Anderson B J and Hamilton D C 1993 Electromagnetic ion cyclotron waves stimulated by modest magnetospheric compressions *J. Geophys. Res.* **98** 369-382
[3] Bourgain J and Wang W 1998 Construction of blow up solutions for the nonlinear Schrödinger equation with critical nonlinearity *Ann. Scuola Norm. Sup. Pisa CI. Sci.* **25** 197-215
[4] Cazenave T and Wiessler F 1987 *Some Remrks on the Nonlinear Schrödinger Equation in the Critical Case.* In: Nonlinear semigroups, partial differential equations and attractors *Lecture Notes in Math.* **1394** (Berlin, Springer) 18-29
[5] Crosta M, Fratalocchi A and Trillo S 2011 Bistability and instability of dark-antidark solitons in the cubic-quantic nonlinear Schrodinger equation *Phys. Rev.* **84** 763-809
[6] Ginibre J and Velo G 1979 On a class of nonlinear Schrödinger equations. The Cauchy problem, general case *J. Funct. Anal.* **32** 826
[7] Gurevich A V and Shvartsburg A B 1973 *Nonlinear Theory of Radiowave Propagation in the Ionosphere* (Moscow: Science)
[8] Gurevich A V and Krylov A L 1987 Disipationless shock waves in media with positive dispersion *J. Sov. Phys.* **65** 944
[9] Khatiashvili N, Shanidze R and Komurjishvili O 2012 *International J. of Physics and Math. Sciences, CIBTECH* **2** 206-213
[10] Kwong M K 1998 Uniqueness of positive solutions of $\Delta u - u + u^p = 0$ in R^n *Arch. Rati. Mech. Anal.* **105** 243-266
[11] Laptev A, Naboko S and Safronov O 2005 Absolutely continuous spectrum of Schrödinger operators with slowly decaying and oscillating potentials *Commun. Math. Phys.* **253** 611-631
[12] Lions J L 1969 *Qualques methodes de resolution des problemes aux resolution des problems aux limites non lineares* (Paris)
[13] Merle F and Raphael P 2005 Profiles and quantization of the blow up mass for critical Nonlinear Schrödinger equation *Commun. Math. Phys.* **253** 675-704
[14] Perelman G 2001 On the blow up phenomenon for the critical nonlinear Schrödinger equation *Ann. Henri. Poincare* **2** 605-673
[15] Safronov O 2004 On the absolutely continuous spectrum of multi-dimensional Schrödinger operators with slowly decaying potentials *Commun. Math. Phys.* **107**
[16] Simon B 2000 Schrödinger operators in the twenties century *J. Math. Phys.* **41** 3523-35551998
[17] Sulem C and Sulem P 1999 *The nonlinear Schrödinger equation. Self-focusing and wave collapse. Applied mathematical sciences 139* (New York: Springer- Verlag)
[18] Summers D R, Thorne M and Xiao F 1998 Relativistic theory of wave-particle resonant diffusion with application to electron acceleration in the magnetosphere *J.Geophys. Res.* **103** 487-500
[19] Tsintsadze N L, Kaladze T D, Van Dam J, Horton W, Fu X R and Garner T W 2010 Nonlinear dynamics of the electromagnetic ion cyclotron structures in the inner magnetosphere *J. Geophys. Res.* **115** 1-72
[20] Weinstein M 1989 The nonlinear Schrödinger equation singularity formation *Contemp.Math. Providence, RI: Amer. Math. Soc.* **99** 213- 232
[21] Whitham G B 1974 *Linear and Nonlinear Waves* (JOHN Wiley-SONS)
[22] Zakharov V and Shabat A 1972 Exact theory of two-dimensional self-focusing and one-dimensional self-modulation of waves in nonlinear media *Sov. Phys.* **34** 62-69

Gardner's deformations as generators of new integrable systems

Arthemy V Kiselev[1] **and Andrey O Krutov**[2]

[1] Johann Bernoulli Institute for Mathematics and Computer Science, University of Groningen, P.O.Box 407, 9700 AK Groningen, The Netherlands.
[2] Department of Higher Mathematics, Ivanovo State Power University, Rabfakovskaya str. 34, Ivanovo, 153003 Russia.

E-mail: `A.V.Kiselev@rug.nl`, `krutov@math.ispu.ru`

Abstract. We re-address the problem of construction of new infinite-dimensional completely integrable systems on the basis of known ones, and we reveal a working mechanism for such transitions. By splitting the problem's solution in two steps, we explain how the classical technique of Gardner's deformations facilitates – in a regular way – making the first, nontrivial move, in the course of which the drafts of new systems are created (often, of hydrodynamic type). The other step then amounts to higher differential order extensions of symbols in the intermediate hierarchies (e. g., by using the techniques of Dubrovin *et al.* [1, 2] and Ferapontov *et al.* [3, 4]).

The aim of this short note is to further and illustrate a practical concept which was outlined earlier in [5] within this conference series. Namely, we revisit the problem of integrable deformation of a given infinite-dimensional system; the seminal paper was [6]. Much work towards description of the arising moduli spaces has been performed by Dubrovin *et al.* [7], cf. [1, 2]. (It must be recalled that cohomological theories in this context and organization of the moduli spaces are sensitive to the choice of admissible classes of differential functions – e. g., polynomial, rational, or analytic – in which such structures are sought for.) In the world of integrable systems there is a closely related aspect of integrability-preserving transition between (solutions to) systems of PDEs (e. g., via Bäcklund transformations, see [8]; a different approach was developed in [9]). Here one could employ the 'heavy artillery' [8, 10] of jet-bundle techniques for deformation of the Cartan structure elements in coverings over PDEs by using the Frölicher–Nijenhuis bracket, see [11] for illustration.

Having its roots in topological QFT and yet possessing numerous applications elsewhere, a task of extending low-order hierarchies with higher-order symbols remains a topic of particular interest in the field ([7], also [3, 4]). For instance, such is the approach to hydrodynamic-type systems viewed as the weak dispersion limits of larger, initially concealed models. Not limited to the above-mentioned class of evolution equations, this concept suits well for PDE systems of order $\geqslant 2$ whenever those are taken as drafts for the (re)construction of larger models; certain restrictions could be imposed by hand at exactly this moment in order to narrow, e. g., the classes of solutions of the draft systems, cf. [3, 4]. At the same time, there co-exist many schemes for extension of the symbol for a given system (e. g., one follows the perturbative approach of [7] or applies the Lax-pair based techniques from [12]).

From a broader perspective, there arise two natural questions: which systems are proclaimed 'interesting', thus delimiting the sets of start- and endpoints in the proliferation schemes, and where one could take those 'interesting' systems from — or pick the drafts of new interesting PDEs. Leaving now aside the ever-growing supply from Physics or a straightforward idea of ploughing the available lists of already known integrable systems, let us focus on a self-starting, regular algorithm which exploits the classical ideas from geometry of differential equations ([13, 14, 15]).

Specifically, we take the existence of infinitely many integrals of motion as a selection rule for nonlinear evolutionary systems; by default, we shall always assume that the collection of conversed quantities at hand is maximal, that is, it can not be extended within a class of conservation laws with local densities (otherwise, a count of infinities could become risky). As a rule, such systems tend to be bi-Hamiltonian at least in the case when the spatial dimension n is equal to 1, with $x^1 \equiv x$, see [16] and also [17]. Let us note also that a requirement of existence of conserved quantities is, generally speaking, stronger than a 'symmetry integrability' assumption [18]. (However in applications it is often convenient to weaken the former requirement in favour of the latter; we shall profit from the use of both approaches, see Example 1 and Proposition 1 in what follows.)

As soon as we agree to study only those evolutionary systems which admit infinite towers of integrals of motion, it is natural to first ex- and then inspect the existence of a (much better if polynomial) recurrence relation between the integrals' conserved densities [5, 6, 12, 19, 20]. This yields a regular procedure for consecutive calculation of the integrals of motion on the basis of all previously known data by starting from the 'seed' constants. Let us emphasize that such relations between the densities are much more valuable and informative than ordinary recursion operators $R\colon \varphi_i \mapsto \varphi_{i+1}$ for symmetries or say, $R^\dagger\colon \psi_i \mapsto \psi_{i+1}$ for the 'cosymmetries' of evolutionary PDE; in a sense, every algorithm which explicitly produces the densities contains the built-in homotopy formula for reversion of the variational derivative that takes densities to the respective generating functions ψ, cf. [14, § 4.2] and [15].

The classical notion of Gardner's deformation $\mathfrak{m}_\varepsilon\colon \mathcal{E}_\varepsilon \to \mathcal{E}$ for a completely integrable system $\mathcal{E}$ was designed for serving exactly this purpose [6]; in the course of years, it has become the parent structure for a plethora of concepts ranging from the Lax pair to formal τ-function, etc. (for a more general approach to the geometry of Gardner's deformations see [5, 20, 21]). Quite remarkably, this good old construction (see Definition 1 below) also answered the second question which we posed so far: for a system $\mathcal{E}$ which undergoes the deformation, this procedure yields "promising" drafts $\mathcal{E}'$ of "interesting" new systems $\mathcal{E}''$. We sketched this line of reasoning in [5] and we now discuss it in more detail. Our surprising conclusion is that the world of completely integrable systems could be much more 'tense' and regularly organized than it may first seem; for the adjacency relations $\mathcal{E} \to \mathcal{E}''$ spin a web across that set, with topology still to be explored; to the best of our knowledge, a study of the physical sense for a property of two models $\mathcal{E}$ and $\mathcal{E}''$ to be adjacent has not yet begun.

Let us first introduce some notation and conventions. In what follows we consider systems of evolutionary partial differential equations with two independent variables x and t and m-tuples of unknowns $u = {}^t(u^1, \ldots, u^m)$. We denote corresponding dependent variables in extended equation in Gardner's deformation as $\tilde{u}$. We use capital letters for unknowns in extensions of adjoint system.

Definition 1 (Classical Gardner's deformation [5, 6]). Let $\mathcal{E} = \{u_t = f(x, u, u_x, u_{xx}, \ldots, u_k)\}$ be a system of evolution equations (in particular, a completely integrable system). Suppose $\mathcal{E}_\varepsilon = \{\tilde{u}_t = f_\varepsilon(x, \tilde{u}, \tilde{u}_x, \ldots, \tilde{u}_{\tilde{k}}; \varepsilon) \mid f_\varepsilon \in \operatorname{im} \frac{\mathrm{d}}{\mathrm{d}x}\}$ is a deformation of $\mathcal{E}$ such that at each point $\varepsilon \in \mathcal{I}$ of an interval $\mathcal{I} \subseteq \mathbb{R}$ there is the *Miura contraction* $\mathfrak{m}_\varepsilon = \{u = u(\tilde{u}, \tilde{u}_x, \ldots, \varepsilon)\}\colon \mathcal{E}_\varepsilon \to \mathcal{E}$. Then the pair $(\mathcal{E}_\varepsilon, \mathfrak{m}_\varepsilon)$ is the (*classical*) *Gardner deformation* for system $\mathcal{E}$.

Example 1 ([6]). Consider the Koreweg–de Vries equation

$$\mathcal{E}_{\mathrm{KdV}_3} = \{u_{t_3} = -u_{xxx} - 6uu_x\}. \tag{1}$$

Its deformation found by Gardner is as follows: the extended equation

$$\mathcal{E}_{\mathrm{KdV}_3;\varepsilon} = \left\{\tilde{u}_{t_3} = -(\tilde{u}_{xx} + 3\tilde{u}^2 - 2\varepsilon^2\tilde{u}^3)_x\right\}, \tag{2a}$$

is mapped to the initial KdV at $\varepsilon = 0$ by the Miura contraction

$$\mathfrak{m}_\varepsilon = \left\{u = \tilde{u} - \varepsilon\tilde{u}_x - \varepsilon^2\tilde{u}^2\right\} : \mathcal{E}_{\mathrm{KdV}_3;\varepsilon} \to \mathcal{E}_{\mathrm{KdV}}. \tag{2b}$$

By expanding the left- and right-hand sides of (2b) in ε, one obtains a well-known recurrence relation for densities ω_k of the integrals of motion $\int_{-\infty}^{+\infty} \omega_k \,\mathrm{d}x$ for (1),

$$\omega_0 = u, \qquad \omega_1 = u_x, \qquad \omega_k = \frac{\mathrm{d}}{\mathrm{d}x}\omega_{k-1} + \sum_{i+j=k-2} \omega_i\omega_j, \qquad k \geqslant 2.$$

We say that the coefficient $\varphi(\tilde{u}, \tilde{u}_x, \ldots)$ of the highest power of ε in the right-hand side of a *polynomial* (in ε) Gardner's extension $\mathcal{E}_\varepsilon$ determines the *adjoint* system $\mathcal{E}' = \{\tilde{u}_{t_k} = \varphi\}$.

Example 2. The evolution equation

$$\tilde{u}_{t_3} = -6\tilde{u}^2\tilde{u}_x \tag{3}$$

is adjoint to the Korteweg–de Vries equation (1).

We notice that the adjoint systems are often dispersionless, although this is not always the case (cf. [5] and [19]). Let us now address the natural problem of extension of the new, adjoint equation – and its hierarchy which appears by construction of Gardner's deformation – by adding terms with higher-order derivatives (in particular, by switching on the dispersion in $\mathcal{E}'$). There are many techniques for solving this problem: a straightforward computational algorithm, which does not require that the adjoint system be hydrodynamic type, is illustrated in what follows by using the 'symmetry integrability' approach (3) and software [23]; we then report on the second iteration of such proliferation scheme and discuss the third and other steps to follow.

Example 3. It is readily seen that equation (3) is the second element in the infinite hierarchy of adjoint systems (corresponding to Gardner's extensions of higher KdV flows, with the Miura contraction $\mathfrak{m}_\varepsilon$ common for all of them), which is

$$\begin{array}{ll}
\vdots & \vdots \\
\mathcal{E}'_{\mathrm{KdV}_2} = \{\tilde{u}_{t_5} = \tilde{u}^4\tilde{u}_x\}, & \varphi_5 = \tilde{u}^4\tilde{u}_x, \\
\mathcal{E}'_{\mathrm{KdV}_1} = \{\tilde{u}_{t_3} = -6\tilde{u}^2\tilde{u}_x\}, & \varphi_3 = -6\tilde{u}^2\tilde{u}_x, \\
\mathcal{E}'_{\mathrm{KdV}_0} = \{\tilde{u}_{t_1} = \tilde{u}_x\}, & \varphi_1 = \tilde{u}_x.
\end{array}$$

Clearly, the scaling weights are not uniquely determined for the dependent variables in the adjoint hierarchy; for definition, set $[U] = 1$, $[\mathrm{d}/\mathrm{d}x] = 1$, and $[\mathrm{d}/\mathrm{d}t_k] = (k-1)[U] + [\mathrm{d}/\mathrm{d}x] = k$, which is consistent with the dynamics. For all $k \in \mathbb{N}$, let us now list all scaling-homogeneous differential polynomials f_k of weights $[f_k] = [U] + [\mathrm{d}/\mathrm{d}t_k] = k+1$ with undetermined coefficients, excluding at once those terms which are already contained in the respective right-hand sides of the adjoint hierarchy $\mathcal{E}'_{\text{KdV}_k}$. For instance, we let

$$\begin{aligned}
&\vdots\\
f_5 &= q_5U^6 + q_6U^3U_{xx} + q_7U^2U_{xxx} + q_8U^2(U_x)^2 + q_9UU_{4x} + q_{10}UU_{xx}U_x + q_{11}U_{5x} + q_{12}U_{xxx}U_x\\
&\quad + q_{13}(U_{xx})^2 + q_{14}(U_x)^3,\\
f_3 &= q_1U^4 + q_2UU_{xx} + q_3U_{xxx} + q_4(U_x)^2,\\
f_1 &= 0.
\end{aligned}$$

The ansatz for the full hierarchy is thus

$$\begin{aligned}
U_{t_k} &= \varphi_k + f_k,\\
&\dots\\
U_{t_5} &= \varphi_5 + f_5 = U^4U_x + q_5U^6 + q_6U^3U_{xx} + q_7U^2U_{xxx} + q_8U^2(U_x)^2 + q_9UU_{4x} + q_{10}UU_{xx}U_x\\
&\qquad + q_{11}U_{5x} + q_{12}U_{xxx}U_x + q_{13}(U_{xx})^2 + q_{14}(U_x)^3,\\
U_{t_3} &= \varphi_3 + f_3 = -6U^2U_x + q_1U^4 + q_2UU_{xx} + q_3U_{xxx} + q_4(U_x)^2,\\
U_{t_1} &= \varphi_1 + f_1 = U_x.
\end{aligned}$$

By solving the determining system of algebraic equations $(U_{t_i})_{t_j} = (U_{t_j})_{t_i}$ upon the undetermined coefficients q_α and then taking its nontrivial solution (if any), we obtain a new, dispersionful hierarchy (which is symmetry integrable by construction). Specifically, for (1) and its adjoint (3) the solution is

$$\begin{aligned}
&\vdots\\
U_{t_5} &= \tfrac{1}{30}\left(U_{4x} + 6U^5 + 10U^2U_{xx} + 10U(U_x)^2\right)_x\\
U_{t_3} &= -U_{xxx} - 6U^2U_x,\\
U_{t_1} &= U_x,
\end{aligned}$$

which is none other than the hierarchy of modified Korteweg–de Vries equation.

We now approach the main result of this report. Consider the second term in one of the two towers of the Kaup–Boussinesq hierarchy,

$$u_{t_2} = -v_{xxx} + 4(uv)_x, \tag{4a}$$
$$v_{t_2} = -u_x + 4vv_x. \tag{4b}$$

Its Gardner deformation is known from [20]: let us recall that the Miura contraction is

$$u = \tilde{u} + \varepsilon\big(\tilde{u}_x - 2\tilde{v}\tilde{v}_x\big) + \varepsilon^2\big(4\tilde{u}\tilde{v}^2 - \tilde{u}^2 - \tilde{v}_x^2\big) + 4\varepsilon^3\tilde{u}\tilde{v}\tilde{v}_x - 4\varepsilon^4\tilde{u}^2\tilde{v}^2, \tag{5a}$$
$$v = \tilde{v} + \varepsilon\tilde{v}_x - 2\varepsilon^2\tilde{u}\tilde{v}, \tag{5b}$$

and the parametric-extended equations are as follows,

$$\tilde{u}_{t_2} = \tilde{v}_{xxx} + 4\big(\tilde{v}\tilde{u}\big)_x - 2\varepsilon\big(\tilde{v}\tilde{u}_x\big)_x - 4\varepsilon^2\big(\tilde{v}\tilde{u}^2\big)_x, \tag{6a}$$
$$\tilde{v}_{t_2} = -\tilde{u}_x + 4\tilde{v}\tilde{v}_x + 2\varepsilon\big(\tilde{v}\tilde{v}_x\big)_x - 4\varepsilon^2\big(\tilde{v}^2\tilde{u}\big)_x. \tag{6b}$$

By definition, the adjoint system is

$$\tilde{u}_{t_k} = (\tilde{u}^k\tilde{v}^{k-1})_x, \qquad \tilde{v}_{t_k} = (\tilde{u}^{k-1}\tilde{v}^k)_x. \tag{7}$$

Proposition 1. *Let us require that the dispersionful extension of (7) itself is an infinite-dimensional integrable system and that it is scaling-invariant with respect to the weights* $[U] = [V] = \frac{1}{2}$ *and* $[\mathrm{d}/\mathrm{d}t_k] = k$ *for* $k \in \mathbb{N}$. *Then there is a unique solution to the extension problem:*

$$\begin{aligned} &\vdots && \vdots \\ U_{t_3} &= (U_{xx} + 6UU_xV + 6U^3V^2)_x, & V_{t_3} &= (V_{xx} - 6UVV_x + 6U^2U^3)_x, \\ U_{t_2} &= -U_{xx} + 2(U^2V)_x, & V_{t_2} &= -V_{xx} + 2(UV^2)_x, \\ U_{t_1} &= U_x, & V_{t_1} &= V_x. \end{aligned}$$

This is the Kaup–Newell hierarchy [22].

It would be quite logical to iterate the reasoning by first constructing a Gardner's deformation – or several such deformations – for the Kaup–Newell system, and then by extending the available adjoint system(s). However, this algorithmically simple problem appears unexpectedly complex as far as computations are concerned. Specifically, by using the analytic software [23] we obtain a 'no-go' result: there is no Gardner's deformation for the Kaup–Newell equation under the following set of assumption:

- we supposed that the deformation $(\mathcal{E}_\varepsilon, \mathfrak{m}_\varepsilon)$ is polynomial in ε and differential polynomial in $\tilde{U}$ and $\tilde{V}$;
- we let such deformations be scaling homogeneous with respect to the weights $[\tilde{V}] = [\tilde{U}] = \frac{1}{2}$ and $[\varepsilon] = -1\frac{1}{2}$;
- the polynomial Ansätze for Gardner's deformations were bounded by using $\deg_\varepsilon \mathfrak{m}_\varepsilon \leqslant 5$ and $\deg_\varepsilon \mathcal{E}_\varepsilon \leqslant 10$ (here we note that $\max(\deg_\varepsilon \mathfrak{m}_\varepsilon) = 2 \times \max(\deg_\varepsilon \mathcal{E}_\varepsilon)$ for the Kaup–Newell system).

Let as also note that the extended equation $\mathcal{E}_\varepsilon$ for the Kaup–Newell system can depend on derivative of U and U with respect to x of orders up to but not exceeding *two*.[1] We expect that the Kaup–Newell system can be Gardner deformed strictly outside the class of differential polynomials (but can not be deformed within such class of functions).

It therefore remains an open problem to find Gardner's deformation(s) for the Kaup–Newell system and extend the arising adjoint equations(s) so that new, higher-order completely integrable hierarchies are attained.

We conclude that Gardner's deformations of infinite-dimensional completely integrable systems can be effectively used not only through their '$\mathfrak{m}_\varepsilon$-parts,' which encode the recurrence relations between conserved densities, but – viewed via their '$\mathcal{E}_\varepsilon$-parts' – as a source of new completely integrable systems, or draft approximations to larger systems for which the integrability is retained.

The reproduction process is self-starting. Moreover, whenever there is a Gardner deformation for the new hierarchy, one could attempt another iteration. This scheme yields the oriented graph whose vertices are integrable systems and whose edges associate new such systems to the ones at their starting points. We emphasize that the degree of a vertex can be greater than two, meaning that a given system admits several deformations (cf. [5, 19, 20]), and that, in principle, multiple edges may occur. A study of topology of such graph and its correlation with the structure of moduli spaces for higher perturbations of low-order models is a challenging open problem.

[1] The proof is as follows: Consider the determining equation $(\mathfrak{m}_\varepsilon)_t = \mathcal{E}(\mathfrak{m}_\varepsilon)$ for Gardner's deformation and calculate the differential orders of both sides; by the chain rule, this yields that $\operatorname{ord}_x(\mathfrak{m}_\varepsilon) + \operatorname{ord}_x(\mathcal{E}_\varepsilon) = \operatorname{ord}_x(\mathcal{E}) + \operatorname{ord}_x(\mathfrak{m}_\varepsilon)$, which implies a rough estimate $\operatorname{ord}_x(\mathcal{E}_\varepsilon) = \operatorname{ord}_x(\mathcal{E})$.

Acknowledgments

The first author thanks the Organizing committee of international workshop 'Physics and Mathematics of nonlinear phenomena' (June 22 – 29, 2013; Gallipoli, Italy) for stimulating discussions and partial financial support. The research of A.V.K. was supported in part by JBI RUG project 103511 (Groningen); A.O.K. was supported by ISPU scholarship for young scientists. A part of this research was done while the first author was visiting at the IHÉS (Bures-sur-Yvette); the financial support and hospitality of this institution are gratefully acknowledged.

References

[1] Dubrovin B, Liu Si-Qi and Zhang Y 2006 *Commun. Pure Appl. Math.* **59** 559–615 (*Preprint* math.DG/0410027)
[2] Dubrovin B 2006 *Commun. Math. Phys.* **267** 117–39 (*Preprint* math-ph/0510032)
[3] Ferapontov E V and Moro A 2009 *J. Phys.* A: *Math. Theor.* **42** 035211 (*Preprint* nlin.SI/0510032)
[4] Ferapontov E V, Moro A and Novikov V S 2009 *J. Phys.* A: *Math. Theor.* **42** 345205 (*Preprint* nlin.SI/0903.3586)
[5] Kiselev A V 2007 *Theor. Math. Phys.* **152** 963–76 (*Preprint* nlin.SI/0610072)
[6] Miura R M, Gardner C S and Kruskal M D 1968 *J. Math. Phys.* **9** 1204–9
[7] Dubrovin B A 1996 *Integrable systems and quantum groups* (*Lect. Notes in Math.* vol 1620) (Berlin: Springer) pp 120–348
[8] Igonin S and Krasil'shchik J S 2002 *Adv. Stud. Pure Math.* **37** 99–114 (*Preprint* nlin.SI/0010040)
[9] Borisov A B and Zykov S A 1998 *Theor. Math. Phys.* **115** 530–41
[10] Igonin S, Kersten P H M and Krasil'shchik I S 2003 *Differential Geom. Appl.* **19** 319–42 (*Preprint* math.DG/0301344)
[11] Kiselev A V and Krutov A O 2013 On the (non)removability of spectral parameters in $\mathbb{Z}_2$-graded zero-curvature representations and its applications *Preprint* math.DG/1301.7143
[12] Kupershmidt B A 1983 *Proc. Roy. Irish Acad.* **A83** 45–74
[13] Bocharov A V, Verbovetsky A V, Vinogradov A M *et al.* 1999 *Symmetries and conservation laws for differential equations of mathematical physics* ed Krasil'shchik I S and Vinogradov A M (Providence, RI: AMS) p 333
[14] Kiselev A V 2012 The twelve lectures in the (non)commutative geometry of differential equations, *Preprint* IHÉS/M/12/10 p 140
[15] Olver P J 1993 *Applications of Lie groups to differential equations* Grad. Texts in Math. vol 107 (2nd ed) (NY: Springer–Verlag)
[16] Magri F 1978 *J. Math. Phys.* **19** 1156–62
[17] Sergyeyev A 2004 *Acta Appl. Math.* **83** 183–97 (*Preprint* nlin.SI/0310012)
[18] Mikhailov A V, Shabat A B and Sokolov V V 1991 *The symmetry approach to classification of integrable equations* (*What is integrability?*) ed Zakharov V E (Berlin: Springer-Verlag) pp 115–84
[19] Karasu A and Kiselev A V 2006 *J. Phys.* A: *Math. Gen.* **39**:11453–60 (*Preprint* nlin.SI/0603029)
[20] Hussin V, Kiselev A V, Krutov A O and Wolf T 2010 *J. Math. Phys.* **51** 083507 (*Preprint* nlin.SI/0911.2681)
[21] Kiselev A V and Krutov A O 2012 *J. Math. Phys.* **53** 103511 (*Preprint* nlin.SI/1108.2211)
[22] Kaup D J and Newell A C 1978 *J. Math. Phys.* **19** 798–801
[23] Kiselev A V, Wolf T 2007 *Comput. Phys. Commun.* **177** 315–28 (*Preprint* nlin.SI/0609065)

Novel Hierarchies & Hidden Dimensions in Integrable Field Models: Theory & Application

Anjan Kundu

Theory Division, Saha Institute of Nuclear Physics
Calcutta, INDIA

E-mail: anjan.kundusaha.ac.in

Abstract. Focusing on the problem of finding the right number of infinite series of conserved charges in integrable models with infinite degrees of freedom, we explore the well known nonlinear Scrödinger (NLS) equation discovering novel sets of charges with hidden integrable hierarchies. This prompts us to construct a new integrable NLS equation in $2+1$-dimensions. Few important applications of our result including a 2D analytic model for the ocean rogue wave are reported.

1. Introduction

Theory and application of nonlinear integrable systems has a long and rich history starting from the KdV equation and its application to shallow water waves to the NLS, sine-Gordon, derivative NLS, Toda chain etc. stretching from the discrete to the continuum models, with corresponding applications to various fields. At the same time the classical theory was developed into the quantum domain with the discovery of new tools like Yang-Baxter equation Bethe ansatz and construction of quantum integrable models like spin chains, t-J model, Hubbard model etc. [1, 2].

The notion of integrability is intimately related to the symmetries of the system, which following the Nöther's theorem are linked to the conserved quantities. To be precise, the system to be integrable the total number of the associated conserved charges C_n must coincide with the degrees of its freedom.

For mechanical systems with finite N degrees of freedom the definition of integrability becomes straightforward, since such a system becomes integrable, when the total number of the associated conserved quantities: $C_n, \ n = 1, 2, .., M, \ M$ matches with N. If $M < N$, the system might be partially integrable, while for $M > N$ it becomes a superintegrable system. Examples of such integrable systems with finite degrees of freedom may be given by the following classic cases.

I. *Kepler Problem*
Degrees of freedom of this 3D problem is $N = 3$, while the total number of conserved quantities is $M = 4$, which makes this well known problem a superintegrable system.

II. *Euler, Lagrange, Kovalevskaya Top* : These classic top problems posses $N = 4$ degrees of freedom, while as Kovalevskaya has shown, the total number of conserved charges of the system is also $M = 4$. This confirms the integrability of the system, which has been tested by Kovalevskaya through the Painlevé type integrability criteria.

III. *Toda chain* This nonlinear N coupled oscillator type problem on a chain has N degrees of freedom with $M = N$ number of conserved charges, making the system completely integrable.

However the problem may arise when the above concept of integrability is applied to the field theoretic models having infinite degrees of freedom with $N \to \infty$. Extending the above criterion of integrability to the systems with infinite degrees of freedom, we naturally demand the existence of infinite number of independent conserved charges C_n, $n = 1, 2, \ldots$, with $M \to \infty$. This very criterion of integrability is applied to all known integrable field models in $1 + 1$-dimensions like KdV, NLS, derivative NLS, sine-Gordon models etc. However a big, not well answered question for the field theoretic models is that, is there any guarantee that these two infinities: $N \to \infty$ and $M \to \infty$ would coincide, that is the infinite degrees of freedom for such systems would match the number of associated conserved quantities. Nevertheless, in $(1 + 1)$-dimensional integrable field models like in the NLS equation

$$iq_t - (q_{xx} + 2|q|^2 q) = 0, \tag{1}$$

all conserved charges: C_n, $n = 1, 2, \ldots$ associated with the model is believed to be well known, with known integrable hierarchies, derived from these charges. On the other hand, since an infinity added to another infinity would make again an infinity, there might exist more than one sets of charges, yet undiscovered, for the true integrability of a field model. Though this might be an universal problem concerning the integrability of all models with infinite degrees of freedom, here we will focus only on the NLS field model (1) .

Our aim therefore is to focus on this less discussed but extremely crucial conceptual problem regarding the integrable NLS model, asking some important questions like, can there be new sets of charges yet undiscovered for the well known NLS model, can there be hidden space-dimensions underlying the standard $1 + 1$-dimensional NLS equation.

2. Generation of charges

Since the conserved charges are intimately linked to the integrability of a system, we present, for going deep into such properties, first a simple but systematic method for generating the full set of finite number of charges for the NLS equation, based on the Lax operator method, which is also an important signature of an integrable system. The space-Lax equation related to an nonlinear integrable equation can be given by an auxiliary linear system

$$\Phi_x = U_1(\lambda)\Phi, \tag{2}$$

where the Lax operator U_1 for the NLS model can be expressed in the AKNS form

$$U_1 = i(\lambda\sigma^3 + U^{(0)}), \; U^{(0)} = \begin{pmatrix} 0 & q \\ q^* & 0 \end{pmatrix}, \tag{3}$$

where q, q^* are the basic NLS fields, λ is the spectral parameter.

Scale dimension: Before proceeding further we would focus on an important concept of scale dimension (or inverse length dimension), which is usually not much emphasised in the literature. Note that all terms in any object should be of the same scale dimension with the dimension of ∂_x being L^{-1}, which we define as scale dimension 1. Therefore looking at the Lax equation (2) we may conclude that the scale dimension of U_1 should be 1, with that of q, q^* as well as of λ must also be 1. We shall actively use this concept below to identify the higher scale dimensions of different objects of our interest.

Journal of Physics: Conference Series **482** (2014) 012022 doi:10.1088/1742-6596/482/1/012022

By taking the Jost matrix function as $\Phi(\lambda, x) = (\phi, \tilde{\phi})$, the infinite set of conserved charges: $C_n, \ n = 1, 2, \ldots$ expressed as $C_n = \int dx \rho_n, \ n = 1, 2, \ldots$ can be linked to

$$\phi(x, \lambda) = e^{\int^x \rho(\lambda, x')dx'}, \ \rho(\lambda, x) = \sum_{n=1}^{\infty} \rho_n(x)\lambda^{-n} \tag{4}$$

resulting to the generation of charges through

$$\ln\phi(x = \infty, \lambda) = \sum_{n=1}^{\infty} C_n \lambda^{-n}. \tag{5}$$

Therefore constructing Riccati type equation from the Lax equation (2) one can systematically generate the well known conserved quantities for the NLS model as $C_n = \int dx \rho_n, \ n = 1, 2, \ldots$ where

$$\rho_1 = |q|^2, \quad \rho_2 = i(q_x^* q - q^* q_x), \quad \rho_3 = q_x^* q_x + |q|^4, \tag{6}$$

and so on. Taking for example $H = C_3$, one can derive the well known NLS equation (1) from $iq_t = \{q, H\}$, using the PB relation

$$\{q(x), q^*(x')\} = i\delta(x - x'), \tag{7}$$

where time $t = t_2$ is of scale dimension 2 (one can easily check this using the concept of scale dimension, defined above.)

For $H = C_4$ with the same PB we get the higher NLS equation

$$iq_t - (q_{xxx} + 6|q|^2 q_x) = 0, \ c.c. = 0,$$

with time $t = t_3$ having scale dimension 3. Thus one derives the integrable hierarchy of the NLS equation taking the conserved charges $C_n, n = 1, 2, \ldots$ as the respective Hamiltonian of the system: $H = C_n, \ n = 1, 2, \ldots$ with the corresponding time : $t \equiv t_n$.

We have noticed above an interesting fact, which has not been focused much, is that the NLS field $q(x, t_1, t_2, t_3, \cdots)$ involved in the integrable hierarchy contains multi-dimensional time having different scale dimensions (see similar observation for the Toda chain in Suris [3]). We intend to exploit this known though less discussed fact to the deepest extend, asking the intriguing question whether we can revert the concept of multi-dimensional time to construct an integrable NLS equation with multi-dimensional space $x_m, \ m = 1, 2, \ldots$.

Another nontrivial possibility which will be our primary concern here also seems to have been ignored in the existing literature. To understand this point, let us focus first on the fact that, the well known conserved charges of the NLS model, as we have shown above, can be derived from the Lax operator U_1 in the AKNS form (3) having the scale dimension 1. It is worth recalling that not only the NLS, but also all other well known integrable systems, e.g., KdV, mKdv, sine-Gordon, Toda chain etc. have their conserved charges and the integrable hierarchies constructed from the same Lax operator U_1 in the AKNS form of scale dimension 1 (with different conjugate field r).

Therefore, we raise a nontrivial question asking whether one can use a different Lax operator with higher scale dimension for generating new series of conserved quantities and hence new integrable hierarchy for the same model, e.g. NLS model. For concreteness we use a Lax operator with scale dimension 2 for the NLS equation, choosing it as $U_2 \sim V_{NLS}$ with Lax equation

$$\begin{aligned} \Phi_y &= U_2(\lambda)\Phi, \ y \equiv x_2, \\ U_2 &= i\left(2\lambda^2\sigma^3 + \lambda U^{(0)} - (U^{(0)})^2 - i\sigma^3 U_x^{(0)}\right) \end{aligned} \tag{8}$$

where $U^{(0)}$ is the same scale dimension 1 object as defined in (3). It is not difficult to see that U_2 containing λ^2 etc. is of scale dimension 2.

3. New charges, hidden hierarchies and novel 2D NLS model

Repeating now the same procedure for finding the well known infinite set of charges from the Lax equation (2) with Lax operator U_1 of scale dimension 1, as shown above, for the Lax equation (8) associated with U_2 of scale dimension 2, we can derive a new set of infinite number of conserved charges. In explicit form they can be given as

$$\begin{aligned} C_n^{(2)} &= \int dy \rho_n^{(2)}, \text{where} \\ \rho_1^{(2)} &= i(q_x^* q - q^* q_x), & (9) \\ \rho_2^{(2)} &= i q_y^* q + q_x^* q_x + |q|^4 + c.c., \quad \rho_3^{(2)} = q_y^* q_x, & (10) \\ \rho_4^{(2)} &= i q_{xy}^* q_x + q_y^* q_y - i|q|^2(q^* q_y - q_y^* q) - 2|q|^2 \, q_x^* q_x \\ &+ (q^{*2} q_x^2 + q_x^{*2} q^2) + c.c., & (11) \end{aligned}$$

etc. It is important to note that, all the above densities of conserved charges $\rho_n^{(2)}(x, y)$ are dependent on both the space dimensions x, y as a consequence of the fact that the basic fields q, q^* of this NLS system are now dependent also on the 2-dimensional space coordinates. These hidden space dimensions are manifestation of a dual picture of the multi-dimensional time in the NLS field discussed above.

Taking these new conserved quantities as Hamiltonians one can find therefore a new form of integrable hierarchy:

$$i q_{t_n} = \{q, H\}_{(2)}, H = C_n^{(2)}, \quad n = 1, 2, \dots \qquad (12)$$

hidden in the NLS system .

3.1. Novel PB structure

The above hierarchy (12) could be derived when we use a new kind of PB structure

$$\{q(y), q_x^*(y')\}_{(2)} = i\delta(y - y'). \qquad (13)$$

For comparison recall the standard bosonic type of PB: $\{q(x), q^*(x')\} = i\delta(x - x')$ valid for the well known NLS model. It is remarkable that this completely new type of PB (13) can be derived from the Yang-Baxter equation (YBE)

$$\{U_2, \otimes \tilde{U}_2\}_{(2)} = [r, U_2 \otimes I + I \otimes \tilde{U}_2] \qquad (14)$$

related to the Lax operator U_2, but associated again with the same rational r as in case of 1D NLS model. Note that the validity of the YBE confirms also the complete integrability of the 2D NLS system. We wish to elaborate on this point including its quantum generalisation elsewhere [4]

3.2. Novel 2D integrable NLS equation

If we choose as Hamiltonian $H = C_4^{(2)}$, in the above 2D NLS hierarchy (12) $i q_{t_4} = \{q, H\}_{(2)}$, using the PB structure (13) we can arrive at a novel $(2 + 1)$-dimensional integrable NLS equation given as

$$i q_t + q_{xy} + 2iq(q q_x^* - q^* q_x) = 0 \qquad (15)$$

Note that while the well known 2D extension of the NLS equation is a nonintegrable system with unstable soliton solution, our 2D NLS model (15) is a completely integrable system exhibiting exact soliton solutions. Some specific properties and details of this novel model can be found in our recent reports [5, 6, 7].

4. Hierarchy of hierarchies in the NLS system

Now the question remains to be addressed whether this novel integrable hierarchy of a 2D integrable NLS equation is really independent from the well known NLS hierarchy, to which we will return below.

Recall that the compatibility of the linear system (2) with the Lax pair (U_1, V_2) leading to the flatness condition

$$U_{1t} - V_{2x} + [U_1, V_2] = 0$$

gives the 1D NLS equation (1). The NLS Hierarchy similarly is generated from the Lax pair $U_1, V_n)$, $n = 1, 2, 3, \ldots$, which can be presented in the (x, t) plain with a vertical line with a single jump x_1 along the x-direction while a jump of $t_1, t_2, \ldots, t_n$ in the t-direction making the vertical line $(1, n)$, $n = 1, 2, \ldots$ (See first column in Fig. 1).

Now we ask some important questions, which seem to have been never put forward, as: *Is this hierarchy complete? Can there be other hidden hierarchies?*

To address these questions let us take first the case of the integrable hierarchy of the 2D-NLS equatio we have already discussed above having the Lax equations $\Phi_y = U_2\Phi$, $\Phi_{t_n} = V_n\Phi$, with the Lax pairs $(U_2, V_n), n = 1, 2, \ldots$, yielding the hierarchy from the flatness condition

$$U_{2t} - V_{nx} + [U_2, V_n] = 0.$$

In similarity with the well known NLS hierarchy representing the first column in the x, t plane, the present case would naturally represent the second column with a double jump x_2 along the x-direction, while jumps of $t_1, t_2, \ldots, t_n$ along the t-direction making the next vertical line $(2, n)$, $n = 1, 2, \ldots$ (See second vertical column in Fig. 1)

interchanging $U_m \leftrightarrow V_n$ in the above scheme one can derive similarly dual hierarchies along the horizontal lines in the x, t plane with the generation of infinite sets of dual charges $J_m^{(n)}$.

Therefore, exploring these hidden hierarchies we can conclude finally that, different Lax pairs (U_m, V_n), $n = 1, 2, \ldots;\ m = 1, 2, \ldots$
through their flatness condition

$$U_{nt_m} - V_{mx_n} + [U_n, V_m] = 0$$

would generate complete hierarchies covering all vertical and horizontal lines at all 2D lattice points (n, m) in the x, t plane (see Fig. 1).

As shown in the figure:

1. $(m, n) = 1, 2\ldots$ with $m < n$ describes the *vertical lines* with novel Integrable hierarchies with conserved charges $C_n^{(m)}$.
2. The diagonal line $n = m$ corresponds to *linear equations* (since nonlinear part of these equations $[U_n, V_n] = 0$).
3. Horizontal lines with $m > n$ correspond to the *dual hierarchies* with dual charges (or currents) $J_m^{(n)} = \int dt_n j_m^{(n)}$.

Now we address to the independence of all these hidden integrable hierarchies with likely conclusions that:

Conserved charges $C_n^{(m)}$ (from U_m) commute within each hierarchy & hence they are independent (with fixed m).

Dual charges $J_m^{(n)}$ (generated by V_n) commute within each hierarchy & hence they are independent (with fixed n).

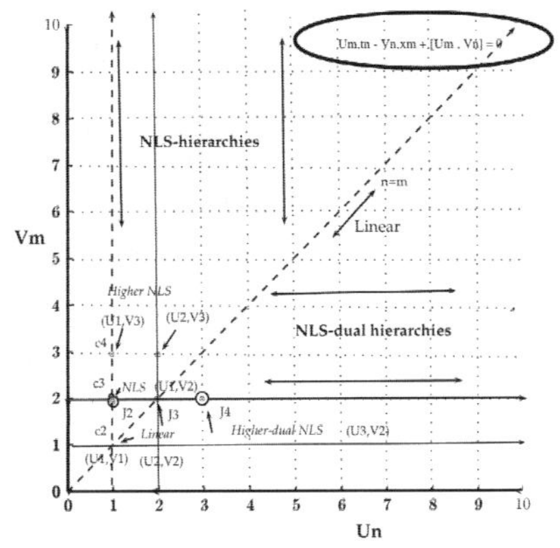

Figure 1. Complete integrable hierarchies of the NLS system

However, these charges (dual charges) within a hierarchy are likely to be related to the charges belonging to other hierarchies with different m (different n for dual charges), due to ther underlying continuity relations

$$\partial_{t_n} \rho_n^{(m)} = \partial_{x_m} j_m^{(n)}$$

Nevertheless the hidden hierarchies we found here yield novel equations in higher dimension with applicable interest.

5. Applications

We briefly present now various possibilities of application of our result, more elaborate account of which can be found in our recent work [6, 8].

5.1. Non-periodic models with nontrivial boundaries

The usual integrable models with conserved charges considered above are valid only for periodic or trivial boundary conditions. However if we consider a more general integrable model in the interval $[x_0, , x_1]$ with nontrivial boundary conditions, then interestingly the standard conserved charges are no longer conserved, while the actual conserved charges $\tilde{C}_k$ take the form

$$\tilde{C}_k = C_k + I_k^{BC},$$

where the boundary contribution is expressed through the dual charges $J_k =\equiv J_k^{(1)}$ that we have generated above through the time-Lax operator V_2 related to the dual hierarchy of the NLS equation in the form

$$I_k^{BC} = J_k(x_0, t) - J_k(x_1, t), \ k = 1, 2, \ldots$$

5.2. Integrable defect contribution through dual charge

Models with defect point at $x = x_d$, preserving their integrability become popular in recent years. To find the infinite set of conserved quantities of the whole system including the defect point the defect contribution can be given as

$$\tilde{C}_k = C_k^- + C_k^+ + I_k^{def}.$$

The defect contribution, as we show, can again be expressed through the dual charges J_k at the defect point as

$$I_k^{def} = J_k(x_d^+) - J_k(x_d^-), \; x_d^\pm = \; (x_d \pm \epsilon)_{|\epsilon \to 0}, \; k = 1, 2, \ldots$$

with $\pm$ denoting the left/right limits from the defect point. The details of this method with application to the Toda chain with an integrable defect site can be found in our recent work [8].

5.3. Ocean rough wave model through integrable 2D NLS

Oceanic rogue wave, one of the mysteries of nature has not been understood and modelled yet satisfactorily. The rogue waves are 2D surface waves, which can appear suddenly on a relatively calm ocean, could attain enormously large amplitude (17-30 meters) and disappear fast without a trace. Such monster wave usually comes as a single event and preceded by strange hole waves. Nonlinearity, modulational instability and ocean current are supposed to be the crucial factors in the formation of ocean rogue waves.

However strangely, the analytic rogue wave mode s available so far, are only in one space dimension and known as the Peregrine breather [9] and its higher order extensions, while the ocean rogue waves being surface waves are truly two-dimensional objects. Our aim therefore is to propose an analytic 2D surface rogue wave model based on a solution of our 2D integrable NLS equation (15) , controlled by the ocean current represented by a term $U_c(x,t)q_x$ (see for details our report [6]). Our proposed rogue wave model is given by a 2D dynamical lump solution

$$P(2d)(x,y,t) = e^{4iy}(-1 + \frac{1 - i4y}{\alpha x^2 + \mu t^2 + 4y^2 + c}) \qquad (16)$$

with tunable amplitude & steepness (due to the existence of free parameters c, α, which are absent in the available rogue wave model: Peregrine breather.)

5.4. Real time behaviour of our 2D rogue wave model

The lump solution (16) constructed by us shows close similarity with the dynamical behaviour of the observed ocean rogue wave as evident from its real time dynamics as described below.

1) In distant past ($t \to -\infty$) the solution turns into a plane wave e^{4ix}.

2) At $t = 0$ the solution becomes a high and steep 2D lump peaked at $x = y = 0$, accompanied by hole states.

3) It disappears again into the plane wave e^{4ix} in future ($t \to +\infty$)

More vivid description of the dynamics of the model is given graphically Fig. 2a-c as snap-shots of the 2D wave at different moments of time.

6. Concluding remarks

Focusing on the problem of ambiguity in determining the right number of conserved charges in integrable field models with infinite degrees of freedom, we have found novel infinite sets of charges and dual charges together with the hidden hierarchies in the integrable NLS model. We have also discovered hidden possibilities of constructing higher dimensional integrable NLS model within this framework reverting the concept of multi-dimensional time. However these equations, as in other $2 + 1$ dimensional known integrable equations are accompanied by constraint equations. As a concrete example we have constructed a novel $2 + 1$ dimensional integrable NLS equation with direct application to model ocean surface rogue wave. Other applications of our result include an easy and explicit way of constructing defect contribution

Journal of Physics: Conference Series **482** (2014) 012022 doi:10.1088/1742-6596/482/1/012022

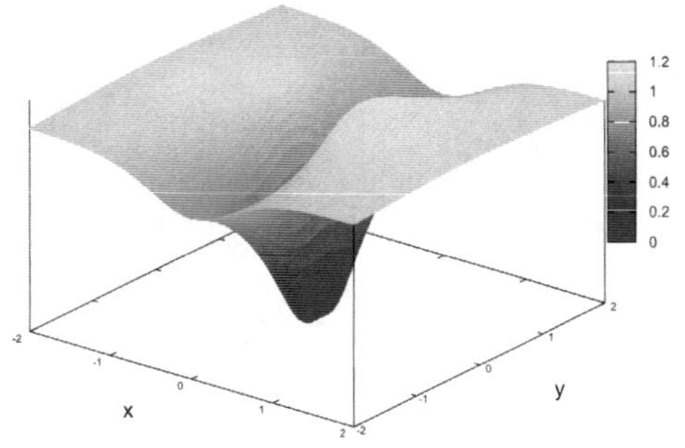

Figure 2. Creation of 2D hole at time $t_h = -0.83$, as told in marine-lores.

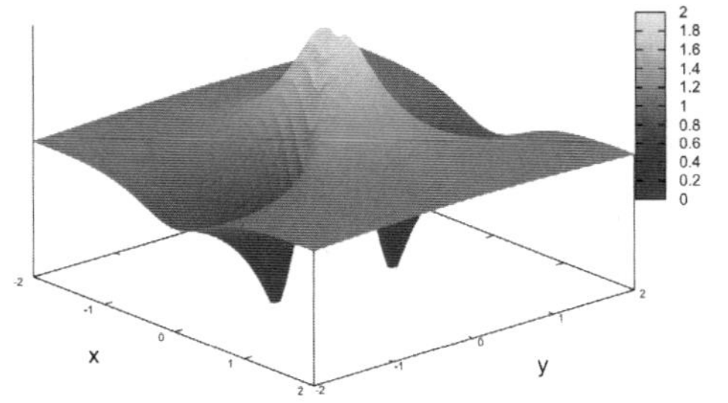

Figure 3. The hole splits into two and drifted away from the centre (at $t = -0.40$).

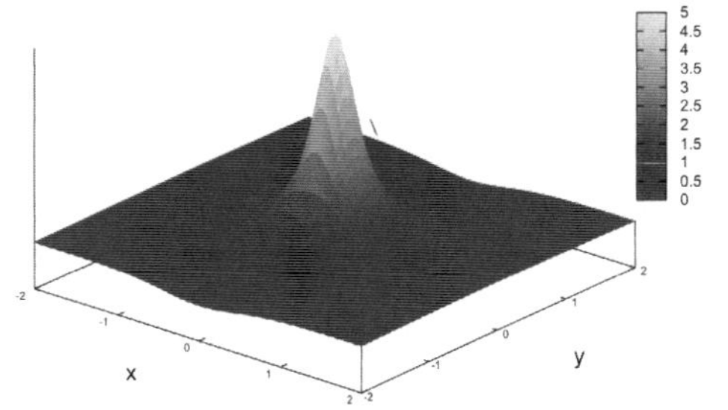

Figure 4. The full grown surface rogue wave at $t = 0.0$.

in integrable defect models and boundary contributions in integrable models with nontrivial boundary conditions, both expressed through dual charges.

Since our result is loaded with new ideas, we hope to develop them for new applications as well as extension to cover quantum integrable models.

Acknowledgement The author acknowleges with thanks the support of ICTP, Trieste as a Senior Associate of the Centre, the help of Dr. Tapan Naskar with graph and latex and expresses his sincere gratitude to Prof Zakharov for stimulatig discussions.

Reference

[1] Ablowitz M, Segur H 1981 *Solon's and the Inverse Scattering Transform*, (SIAM, Philadelphia)
[2] P. Kulish and E. K. Sklyanin, Lect. Notes. Phys. (ed. J. Hietarinta et al, Springer, Berlin, 1982), **151**, 61
[3] Suris Ye B 2013 Variational Formulation of Commuting Hamiltonian flows: Multi-time Lagrangian 1-forms (*Preprint* math-ph/1212.3314)
[4] Kundu A 2014 *Under preperation*
[5] Kundu A 2012 Unraveling hidden hierarchies and dual structures in an integrable field model (*Preprint* nonlin.SI/1201.0627)
[6] Kundu A, Mukherjee A and Naskar T 2013 Modelling rogue waves through exact dynamical lump soliton controlled ocean currents (*Preprint* nonlin.SI/1204.0916)
[7] Kundu A and Mukherjee A 2013 Novel integrable higher-dimensional nonlinear Schroedinger equation: properties, solutions, applications (*Preprint* nonlin.SI/1305.4023)
[8] Caudrelier V and Kundu A 2013 Defect Toda chain (to be submitted for publication)
[9] Peregrine D H 1083 *Austral. Math. Soc.* B **25** 16

Quantum corrections to quasi-periodic solution of Sine-Gordon model and periodic solution of ϕ^4 model

G Kwiatkowski[1] and S Leble[2]

Gdansk University of Technology, ul. Narutowicza 11/12, 80-952, Gdansk, Poland

E-mail: gkwiatkowski@mifgate.pg.gda.pl[1], leble@mifgate.pg.gda.pl[2]

Abstract. Analytical form of quantum corrections to quasi-periodic solution of Sine-Gordon model and periodic solution of ϕ^4 model is obtained through zeta function regularisation with account of all rest variables of a d-dimensional theory. Qualitative dependence of quantum corrections on parameters of the classical systems is also evaluated for a much broader class of potentials $u(x) = b^2 f(bx) + C$ with b and C as arbitrary real constants.

1. Introduction

Since the seminal work of Dashen et al. [1], where the determinant approximation of path integrals proposed by Maslov [2] was first used for quantisation of a ϕ^4 kink, there is a constant interest in semiclassical quantisation of nonlinear field theories. Over the years many different methods of obtaining quantum corrections to energy basing either on the Feynman form of propagation operator (see [3]), or alternative form of path integral formulation of quantum mechanics proposed by Garrod [4] were developed. A year after publications of Dashen et al. Korepin and Faddeev calculated energy corrections for the static Sine-Gordon soliton [5] in $1 + 1$ dimensions. In the late seventies generalised zeta function was introduced as a powerful tool for regularisation procedure. This development allowed Konoplich to quantise the Sine-Gordon soliton in a general d-dimensional space [6] as well as account for non-zero temperature [7]. At the same time Hawking used generalized zeta function to define field quantisation on curved manifolds [8]. Many other methods and variations of existing ones were developed in subsequent years including direct mode summation [9] and use of contour integrals for zeta function construction [10, 11], which allowed Pawellek to quantise periodic solutions of Sine-Gordon and ϕ^4 models [12, 13] without inclusion of any rest variables. It is of note, that the final integrals obtained by Pawellek have a very similar form to those derived by Bordag in [9].

Our aim is to quantise the quasi-periodic solution of Sine-Gordon system as well as the periodic solution of ϕ^4 model using zeta function regularisation in form used by Konoplich [6] in order to examine the influence of rest variables on the energy corrections in those cases, since results for the kink solutions of respective models show a very strong dependence on the overall number of dimensions [6, 14, 15]. Additionally we provide general qualitative analysis of semiclassical quantisation for a large class of potentials, which provides an interesting new insight into effects of scaling of the classical system onto its quantum counterpart as well as the meaning of the mass scale and its dependence on physical parameters.

2. Semiclassical corrections to energy

For the purpose of this publication we use quantisation procedure presented by Maslov in [2] with regularisation procedure proposed by Konoplich in [6]. The scheme is based on the expansion of the action (with unitless variables $x_n = \frac{x'_n}{a}$ and $t = \frac{t'}{T}$)

$$S(\psi, T) = Ta^d \int_0^1 \int_{\mathbb{R}\times[0,l]^{d-1}} \left(\frac{M}{2T^2} \left(\frac{\partial \psi}{\partial t} \right)^2 - \frac{G}{2a^2} \sum_{n=1}^{d} \left(\frac{\partial \psi}{\partial x_n} \right)^2 - V(\psi) \right) \prod_{n=1}^{d} dx_n dt \tag{1}$$

in path integral formulation of the propagator

$$\langle \phi | e^{-\frac{i}{\hbar}TH} | \phi \rangle = \int_{C^{0,T}_{\phi,\phi}} D\psi e^{\frac{i}{\hbar}S(\psi,T)} \tag{2}$$

in a Taylor series around the classical solution φ and cutting it at the first non-trivial term with T as an arbitrary time period. The series is cut at the first nontrivial element (second derivative) and energy corrections can be formally represented as

$$\Delta E = -\frac{\hbar}{iT} \ln\left(\det\left[L\right]\right), \tag{3}$$

with L as the second functional derivative of the Lagrangian taken at the classical solution

$$L = -\frac{iTa^d}{2\pi\hbar r^2} \left(-\frac{M}{T^2}\frac{\partial^2}{\partial t^2} + \frac{G}{a^2} \sum_{n=1}^{d} \frac{\partial^2}{\partial x_n^2} - V''(\varphi) \right) \tag{4}$$

Necessary regularisation consists of two steps: subtraction of analogous expression for classical vacuum solution

$$\Delta E = -\frac{\hbar}{iT} \ln\left(\frac{\det\left[L\right]}{\det\left[L_0\right]} \right), \tag{5}$$

and a choice of a multiplicative constant r^2 in L and L_0 operators (the so called mass scale) connected to the norm of base functions used in (2), which cuts logarithmic divergences. On purely mathematical level r^2 can be viewed as a free parameter of the theory (which is a reason for keeping it unspecified in works of Konoplich [6]). Yet, to obtain physically relevant results one has to find a way of fitting its value. This problem will be further discussed in section 3. Similar issues might arise with the L_0 for fields spanning over a finite domain, where any choice of the constant potential leads to finite results. However, known results for fields spanning over infinite domains suggest, that the lowest eigenvalue of L_0 should coincide with start of the unbound states band. The above regularisation is realised by means of generalised zeta function

$$\Delta E = -\frac{\hbar}{iT} \lim_{s\to 0_+} \frac{\partial}{\partial s} \frac{1}{\Gamma(s)} \int_0^\infty \tau^{s-1} \int_{[0,1]\times\mathbb{R}\times[0,l]^{d-1}} (g_L(\tau, \vec{x}, \vec{x}) - g_{L_0}(\tau, \vec{x}, \vec{x})) d\vec{x} d\tau, \tag{6}$$

with $\vec{x}$ covering all variables of the classical system including time and g_L as a Green function of the following equation

$$\left(\frac{\partial}{\partial \tau} + L \right) g_L(\tau, \vec{x}, \vec{x}_0) = \delta(\tau)\delta(\vec{x} - \vec{x}_0) \tag{7}$$

For convenience we will also define

$$\gamma_L(\tau) = \int_{[0,1]\times\mathbb{R}\times[0,l]^{d-1}} g_L(\tau, \vec{x}, \vec{x}) d\vec{x}, \tag{8}$$

$$\gamma(\tau) = \int_{[0,1]\times\mathbb{R}\times[0,l]^{d-1}} (g_L(\tau, \vec{x}, \vec{x}) - g_{L_0}(\tau, \vec{x}, \vec{x}))d\vec{x} \tag{9}$$

and

$$\zeta(s) = \frac{1}{\Gamma(s)} \int_0^\infty \tau^{s-1}\gamma(\tau)d\tau. \tag{10}$$

Since the classical fields considered in this publication depend on a single spatial variable, we will use the fact, that for an operator L expressable as a sum of operators working on independent variables ($L = \sum_i L_n$), heat equation Green function can be constructed as a product of Green functions for L_n (see [16]). For this purpose we will define

$$L_1 = A\left(\frac{\partial^2}{\partial x_1^2} - \frac{a^2 V''(\varphi(x_1))}{G}\right) \tag{11}$$

$$L_2 = -\frac{A}{c^2}\frac{\partial^2}{\partial t^2} \tag{12}$$

$$L_3 = A\sum_{n=2}^{d} \frac{\partial^2}{\partial x_n^2} \tag{13}$$

with

$$A = -\frac{iTGa^{d-2}}{2\pi\hbar r^2} \tag{14}$$

$$c^2 = \frac{GT^2}{Ma^2} \tag{15}$$

and can readily write

$$\gamma_{L_2} = \sqrt{\frac{c^2}{4\pi A\tau}} \tag{16}$$

$$\gamma_{L_3} = \left(-\frac{l}{4\pi A\tau}\right)^{\frac{d-1}{2}} \tag{17}$$

3. General results

All potentials considered in following sections have many similarities, so it is useful to study general properites of a family of potentials of form

$$u(x) = b^2 f(bx) + C \tag{18}$$

where b and C are some real constants and f is an arbitrary integrable function. Our goal will be to extract as much information from (6) as possible without explicitly solving the Green function problem. This will be most useful for quasi-periodic solution of Sine-Gordon and periodic solution of ϕ^4 model, where exact analytic solutions are difficult to obtain. Let us begin with the Green function equation for L_1 with (18) as the potential

$$\left(\frac{\partial}{\partial\tau} + A\left(\frac{\partial^2}{\partial x^2} - b^2 f(bx) - C\right)\right) g_{L_1}(\tau, x, x_0) = \delta(\tau)\delta(x - x_0) \tag{19}$$

We will now try to remove b from the equation by a series of substitutions

$$x_b = bx \tag{20}$$

will give us

$$\left(\frac{\partial}{\partial \tau} + Ab^2\left(\frac{\partial^2}{\partial x_b^2} - f(x_b) - \frac{C}{b^2}\right)\right) g_{L_1}(\tau, x_b, x_{b,0}) = b\delta(\tau)\delta(x_b - x_{b,0}) \tag{21}$$

Next we rescale τ

$$\tau = \frac{\tau_b}{|A|b^2} \tag{22}$$

obtaining

$$\left(\frac{\partial}{\partial \tau_b} - i\left(\frac{\partial^2}{\partial x_b^2} - f(x_b) - \frac{C}{b^2}\right)\right) g_{L_1}(\tau_b, x_b, x_{b,0}) = b\delta(\tau_b)\delta(x_b - x_{b,0}) \tag{23}$$

From here we can extract both b and C by rescaling the Green function

$$g_{L_1}(\tau_b, x_b, x_{b,0}) = be^{\frac{C}{b^2}\tau_b} g_{L_1,b}(\tau_b, x_b, x_{b,0}) \tag{24}$$

after inserting that form into (23), we find that $g_{L_1,b}(\tau_b, x_b, x_{b,0})$ solves the equation

$$\left(\frac{\partial}{\partial \tau_b} - i\left(\frac{\partial^2}{\partial x_b^2} - f(x_b)\right)\right) g_{L_1,b}(\tau_b, x_b, x_{b,0}) = \delta(\tau_b)\delta(x_b - x_{b,0}) \tag{25}$$

The same procedure has to be done for the Green function of operator L_0 - both rescaling of the x and τ variables as well as potential shift by $\frac{C}{b^2}$ (if done after rescaling of τ) even if the constant potential in L_0 has a different value than $\frac{C}{b^2}$. As for the Green functions (or γ functions (8)) of L_2 and L_3 operators, it is most convenient to express them through τ_b as well

$$\gamma_{L_2}(\tau_b) = \sqrt{\frac{ic^2b^2}{4\pi\tau_b}} \tag{26}$$

$$\gamma_{L_3}(\tau_b) = \left(\frac{l^2b^2}{i4\pi\tau_b}\right)^{\frac{d-1}{2}} \tag{27}$$

With this information we can reproduce the formula for energy corrections remembering to change the integration over x to integration over x_b.

$$\Delta E = \Re\left(\lim_{s\to 0}\frac{\partial}{\partial s}\frac{\hbar c b^d l^{d-1}}{2^{d+1}\pi^{\frac{d}{2}}T(|A|b^2)^{-s}\Gamma(s)}\int_0^\infty i^{1-\frac{d}{2}}\tau_b^{s-\frac{d+2}{2}}e^{\frac{C}{b^2}\tau_b}\int_{\mathbb{R}}\left(g_{L_1,b}(\tau_b, x_b, x_b) - g_{L_0,b}(\tau_b, x_b, x_b)\right)dx_b d\tau_b\right) \tag{28}$$

Interestingly, in all cases considered in this work C is proportional to b^2, so we can rightfully substitute

$$C = C_b b^2 \tag{29}$$

and obtain

$$\Delta E = \Re\left(\lim_{s\to 0}\frac{\partial}{\partial s}\frac{\hbar c b^d l^{d-1}}{2^{d+1}\pi^{\frac{d}{2}}T(|A|b^2)^{-s}\Gamma(s)}\int_0^\infty i^{1-\frac{d}{2}}\tau_b^{s-\frac{d+2}{2}}e^{C_b\tau_b}\int_{\mathbb{R}}\left(g_{L_1,b}(\tau_b, x_b, x_b) - g_{L_0,b}(\tau_b, x_b, x_b)\right)dx_b d\tau_b\right) \tag{30}$$

Unfortunately, extracting C_b out of the Mellin transform is not a trivial task. Nevertheless at this point we have most of physical parameters of the classical system extracted out of the Green function. Apart from qualitative estimations of quantum corrections, it helps us in choosing the renormalisation parameter r^2, since the logarithmic divergences will arise from differentiation of exponential components of the zeta function. This does not concern divergence in T parameter only - it is also reasonable to assume, that if the classical solution vanishes, corrections should vanish as well. This means, that r^2 should contain b^2 The same argument can be used to cut divergences in all physically relevant parameters, thus it seems valid to propose for the renormalisation factor to cancel the whole $(|A|b^2)^{-s}$ term. Yet, there is no strict way of choosing the value of r^2, so in practice we will fit its value to recover well known results in the cases obtainable by different methods.

At this point it is also worth noting, that in our chosen dimensionless variables c is linearly dependent on T, so it cancels out the T component in the denominator. This would not be the case, if we didn't include the kinetic energy component in the action integral and later on in calculation of energy corrections. Combined with the strong dependence of the result on the number of spatial dimensions included it indicates that one should never omit any components of the action integral even if it would be valid for the classical solution.

One of the most important findings going beyond the chosen class of potentials, if we consider a $\tau_A = |A|\tau$ scaling, is that quantum corrections to energy don't depend on the scale of the classical system. If we were to amplify the action integral by a constant factor, it would have no impact on the quantum corrections.

4. Cnoidal waves

We will consider periodic solutions of two well known models (Sine-Gordon and ϕ^4). Considering the findings of the previous section, we will present them in a simplified form:

$$S(\psi, T) = \int_0^1 \int_{\mathbb{R}\times[0,l]^{d-1}} \left(\frac{1}{2c^2}\left(\frac{\partial\psi}{\partial t}\right)^2 - \sum_{n=1}^{d}\left(\frac{\partial\psi}{\partial x_n}\right)^2 - V(\psi)\right)\prod_{n=1}^{d} dx_n dt \tag{31}$$

with

$$V_{SG}(\psi) = m^2(1-\cos(\psi)) \tag{32}$$

for Sine-Gordon and

$$V_{\phi^4} = \frac{m^2}{2V^2}\psi^4 - \frac{m^2}{2}\psi^2 + \frac{m^2V^2}{8} \tag{33}$$

for ϕ^4 model. Static, periodic solution of respective systems:

$$\psi_{SG}(x) = 2\arcsin(k\,\mathrm{sn}(mx;k)) + \pi \tag{34}$$

$$\psi_{\phi^4}(x) = \sqrt{\frac{k^2}{1+k^2}}V\,\mathrm{sn}\left(\frac{m}{\sqrt{1+k^2}}x;k\right) \tag{35}$$

They will give following potential for the L_1 operator:

$$V''_{SG}(\psi_{SG}(x)) = u_{SG}(x) = m^2(2k^2 - 1 - 2k^2\mathrm{cn}^2(mx;k)) \tag{36}$$

$$u_{\phi^4}(x) = m^2\left(5k^2 - 1 - 6k^2\mathrm{cn}^2\left(\frac{m}{\sqrt{1+k^2}}x;k\right)\right) \tag{37}$$

After the scaling procedure from previous section (with $b = m$ for Sine-Gordon and $b = \frac{m}{\sqrt{1+k^2}}$ for ϕ^4 model) we calculated respective Laplace transforms of Green function diagonals using the method described in [17]:

$$G_{SG}(p,z) = i\frac{ip - k^2 z}{2\sqrt{ip(ip-(k^2-1))(k^2-ip)}} \tag{38}$$

$$G_{\phi^4}(p,z) = i\frac{-p^2+9k^2(1+k^2(-1+z))z-3(ip+k^2pz)}{2\sqrt{(3-ip)ip(ip-3k^2)((ip-1-k^2)^2-4(1-k^2+k^4))}} \tag{39}$$

with $z = \mathrm{cn}^2(x_b;k)$. At this point we perform the vacuum cut-off with

$$G_0(p) = \frac{i}{2\sqrt{C-ip}} \tag{40}$$

with the constant C chosen separately for both systems to coincide with the only root of the polynomial in the denominator, which does not become degenerate in the $k \to 1$ limit, which corresponds to the boundary of the unbound state band of our potential. Next we integrate the function over half-period of the classical solution ($2\mathcal{K}(k)$) in both cases with $\mathcal{K}$ as complete elliptic integral of the first kind to obtain Laplace transform of the γ function

$$\hat{\gamma}(p) = \int_{-\mathcal{K}(k)}^{\mathcal{K}(k)} (iG(ip,x_b) - iG_0(ip))\,dx_b \tag{41}$$

$$\hat{\gamma}_{SG}(p) = i\frac{-\mathcal{E}(k)+\left(-1+k^2+ip-ip\sqrt{\frac{1-k^2}{ip}+1}\right)\mathcal{K}(k)}{\sqrt{(ip+1-k^2)(k^2-ip)ip}} \tag{42}$$

$$\begin{aligned}\hat{\gamma}_{\phi^4}(p) = i&\frac{3(1+k^2-ip)\mathcal{E}(k)+(-3+k^2(3-3ip)-p^2)\mathcal{K}(k)}{\sqrt{(3k^2-ip)p(3i+p)((ip-1-k^2)^2-4(1-k^2+k^4))}}+\\ i&\frac{\mathcal{K}(k)}{\sqrt{(-1-k^2-2\sqrt{1-k^2+k^4}+ip)}}\end{aligned} \tag{43}$$

where $\mathcal{E}$ is the complete elliptic integral of the second kind. We now have to perform an inverse Laplace transform of obtained functions, which unfortunately proves to be beyond our reach at the moment. However, we managed to obtain approximate results by expanding problematic parts of $\hat{\gamma}$ functions in a power series around $k = 1$ treating elliptic integrals as independent parameters. We will showcase the procedure on the Sine-Gordon case.

$$\begin{aligned}\hat{\gamma}_{SG}(p) \approx -i\frac{\mathcal{E}(k)}{ip\sqrt{1-ip}} + i(k-1)\frac{\mathcal{E}(k)(1-2ip)+3\mathcal{K}(k)(p^2+ip)}{(-1+ip)p^2\sqrt{1-ip}}\\ -i(k-1)^2\frac{\mathcal{E}(k)(-3+7ip+5p^2+2ip^3)+\mathcal{K}(k)(7ip+17p^2-7ip^3+3p^4)}{(-1+ip)^2p^3\sqrt{ip-1}}+\ldots\end{aligned} \tag{44}$$

Inverse Laplace transform of this function is easily obtained, since any function of form $\frac{\prod_{j=1}^{N}(p-a_j)}{\prod_{j=1}^{N}(p-b_j)\sqrt{b_0-p}}$ can be rewritten as a sum of $\frac{1}{(p-b_j)^n\sqrt{b_0-p}}$ for which inverse Laplace transforms are well known. We will obtain

$$\begin{aligned}\gamma_{SG}(\tau_b) = -\mathcal{E}(k)\mathrm{Erf}(\sqrt{i\tau_b}) + (k-1)\left(\tfrac{1}{\sqrt{\pi}}e^{-i\tau_b}\mathcal{E}(k)\sqrt{i\tau_b}+\tfrac{1}{2}(6\mathcal{K}(k)-\mathcal{E}(k)(1+2i\tau_b))\mathrm{Erf}(\sqrt{i\tau_b})\right)\\ -\tfrac{1}{8}(k-1)^2\left(\tfrac{2}{\sqrt{\pi}}e^{-i\tau_b}\sqrt{i\tau_b}(8\mathcal{K}(k)(1+i\tau_b)+5\mathcal{E}(k)(1+2i\tau_b))+\right.\\ \left.\mathrm{Erf}(\sqrt{i\tau_b})(4\mathcal{K}(k)(1-14i\tau_b)+\mathcal{E}(k)(-5+4i\tau_b+12\tau_b^2))\right)+\ldots\end{aligned} \tag{45}$$

With this result we can readily use formula (30) to obtain energy corrections assuming value of r^2 as noted in the previous section

$$\Delta E_1(k) = \frac{\hbar cm}{4T\pi}\left(-4\mathcal{E}(k)+12(k-1)\mathcal{K}(k)-\frac{(k-1)^2}{8}(80\mathcal{K}(k)-10\mathcal{E}(k))+\ldots\right) \tag{46}$$

$$\begin{aligned}\Delta E_2(k) = \tfrac{\hbar cm^2 l}{8T\pi}\left(2\mathcal{E}(k)+(k-1)(-6\mathcal{K}(k)+\mathcal{E}(k)(-\ln(4)))-\right.\\ \left.\tfrac{(k-1)^2}{8}(\mathcal{K}(k)(-20+28(-\ln(4)))+\mathcal{E}(k)(8-2(-\ln(4))))+\ldots\right)\end{aligned} \tag{47}$$

$$\Delta E_3(k) = \frac{\hbar c m^3 l^2}{16T\pi^2}\left(\tfrac{10}{9}\mathcal{E}(k) + (k-1)\left(-\tfrac{15}{3}\mathcal{K}(k) + \mathcal{E}(k)\left(-5+\tfrac{5}{9}\right)\right) - \frac{(k-1)^2}{8}\left(\mathcal{K}(k)\left(-\tfrac{20}{9} - 128\right) + \mathcal{E}(k)\left(10 + \tfrac{25}{9}\right)\right) + \ldots\right) \tag{48}$$

The $d = 1$ case in the $k \to 1$ limit coincides with other quantisation methods [5, 12]. Using the same method we also obtained energy corrections for the ϕ^4 cnoidal wave with r^2 chosen in a way, that ensures $4|A|b^2 = 1$ in order to reproduce the results of Daschen et al. [1] for the $k \to 1$ limit in the $d = 1$ case.

$$\Delta E_1(k) = -\frac{\hbar m c}{4T\pi\sqrt{1+k^2}}\Big(2\mathcal{E}(k)(6 - \tfrac{\pi}{\sqrt{3}}) + (k-1)\Big(-\tfrac{2\pi}{\sqrt{3}}\mathcal{K}(k) - 6(\mathcal{E}(k) - 2\mathcal{K}(k))\Big) + (k-1)^2\Big(-15\mathcal{K}(k) + \tfrac{3}{2}\mathcal{E}(k) + \tfrac{\pi}{\sqrt{3}}(\mathcal{K}(k) - 2\mathcal{E}(k))\Big) + \ldots\Big) \tag{49}$$

$$\Delta E_2(k) = -\frac{\hbar l c m^2}{8T\pi(1+k^2)}\left(-\mathcal{E}(k)(12 + 3\ln(3)) - (k-1)\tfrac{3}{2}(\mathcal{E}(k) + 2\mathcal{K}(k))(4 + \ln(3)) + (k-1)^2\tfrac{3}{8}(\mathcal{E}(k)(6 - 33\mathrm{arccoth}(2)) + \mathcal{K}(k)(-40 + 6\ln(3))) + \ldots\right) \tag{50}$$

$$\Delta E_3(k) = -\frac{\hbar l^2 m^3 c}{48\pi^2 T(1+k^2)^{\frac{3}{2}}}\left(4\mathcal{E}(k)(\sqrt{3}\pi - 18) + (k-1)4(\sqrt{3}(\mathcal{K}(k) + \mathcal{E}(k))\pi - 9(2\mathcal{K}(k) + \mathcal{E}(k))) - (k-1)^2(2\mathcal{K}(k)(7 + 2\sqrt{3}\pi) - \mathcal{E}(k)(14\sqrt{3}\pi - 27)) + \ldots\right) \tag{51}$$

It is important to note, that in the $k \to 1$ limit the results converge to those obtained by Konoplich in [6]. As yet we were unable to obtain convergence radius for the power series approximation and in this sense the results are incomplete, we are however able to calculate as many terms in this series as needed. It is evident, that the dependence of the results on the elliptic parameter (which is connected to the period of classical solutions) is heavily influenced by the number of spatial dimensions included in the quantisation procedure.

5. Conclusions

If we carry all physical constants through the calculations, it becomes evident, that quantum corrections to energy are independent of the energy scale of the classical system. If we were to multiply the classical action integral by any constant, it would not affect the corrections due to the way parameter A is cut by regularisation. Even if our choice of the regularisation coefficient r^2 was incorrect, scaling of the classical Hamiltonian would at most result in a logarithmic change in quantum corrections. In a way it coincides with the intuition, that quantum effects should only be noticeable in small scale systems.

The zeta-function regularisation scheme is incomplete in the sense, that it doesn't give clear method of choosing the regularisation coefficient r^2. The choice of vacuum cutoff in the case of fields over a finite interval is not necessarily straightforward as well. We can solve the problem by comparing the energy corrections for cases solved by other methods and extending the results to those otherwise unattainable as we did for the ϕ^4 and Sine-Gordon models.

Energy corrections show strong dependence on the overall number of dimensions of the classical system. It would be of interest to calculate energy corrections without the continuum approximation to research the system's geometry effect on energy in semiclassical regime.

References

[1] R F Dashen, B Hasslacher and A Neveu, Phys. Rev. D 10, p. 4130 (1974).
[2] V P Maslov, Zhurnal Vychislitelnoj Matematiki i Matematicheskoj Fiziki 1, p. 638 (1961).
V P Maslov, Teor. i Mat. Fiz. 2, p. 30 (1970).
[3] R P Feynman, Reviews of Modern Physics 20, 2, p. 367 (1948).
[4] C Garrod, Reviews of Modern Physics 38, 3, p. 483 (1966).
[5] W E Korepin and L D Faddeev, Teor. i Mat. Fiz. 25, 2, p. 147 (1975)
[6] R V Konoplich, Teoreticheskaya i Matematicheskaya Fizika 73, 3, p. 379 (1987).
[7] R V Konoplich, Teor. Mat. Fiz. 78, 3, p. 444 (1989).

[8] S W Hawking, Commun. Math. Phys. 55, 2, p. 133 (1977).
[9] M Bordag, J. Phys. A Math. Gen. 28, 3, p. 755 (1995).
[10] K Kirsten and A J McKane, Annals. Phys. 308, 2, p. 502 (2003).
[11] K Kirsten and P Loya, Am. J. Phys. 76, 1 p. 60 (2008).
[12] M Pawellek, Nuclear Physics B 810, 3, p. 527 (2009).
[13] M Pawellek, arXiv:0802.0710v1 [hep-th] (6 Feb 2008).
[14] G Kwiatkowski and S Leble, Physics Letters A 376, 1-2, p. 991 (2012).
[15] G Kwiatkowski and S Leble, Central European Journal of Physics (Jul 2013).
[16] S Leble, Theoretical and Mathematical Physics 160, 1, p. 976 (2009).
[17] G Kwiatkowski and S Leble, Applied Mathematics and Computation 219, 11, p. 6084 (2013).

Journal of Physics: Conference Series **482** (2014) 012024 doi:10.1088/1742-6596/482/1/012024

Nonlinear nuclear equation of state and thermodynamical instabilities in warm and dense nuclear matter

A Lavagno[1,3], G Gervino[2,3], D Pigato[1,3]

[1]Department of Applied Science and Technology, Politecnico di Torino, I-10129 Torino, Italy
[2]Dipartimento di Fisica, Università di Torino, I-10126 Torino, Italy
[3]INFN, Sezione di Torino, I-10125 Torino, Italy

Abstract.
We study a nonlinear nuclear equation of state in the framework of a relativistic mean field theory. We investigate the possible thermodynamic instability in a warm and dense asymmetric nuclear medium where a phase transition from nucleonic matter to resonance dominated Δ matter can take place. Such a phase transition is characterized by both mechanical instability (fluctuations on the baryon density) and by chemical-diffusive instability (fluctuations on the isospin concentration) in asymmetric nuclear matter. Similarly to the liquid-gas phase transition, the nucleonic and the Δ-matter phase have a different isospin density in the mixed phase. In the liquid-gas phase transition, the process of producing a larger neutron excess in the gas phase is referred to as isospin fractionation. A similar effects can occur in the nucleon-Δ matter phase transition due essentially to a negative Δ-particles excess in asymmetric nuclear matter. In this context, we investigate also the effects of power law effects, due to the possible presence of nonextensive statistical mechanics effects.

1. Introduction

Because nuclei are made of neutrons and protons, the nuclear liquid-gas phase transition is in a binary system where one has to deal with two independent proton and neutron chemical potentials for baryon number and electric charge conservation. Taking into account of this important property, a very detailed study of Müller and Serot [1] focused on the main thermodynamic properties of asymmetric nuclear matter in the framework of a relativistic mean field model.

A relevant aspect of a system with two conserved charges (baryon and isospin numbers) is that the phase transition is of second order from the viewpoint of Ehrenfests definition. At variance with the so-called Maxwell construction for one conserved charge, the pressure is not constant in the mixed phase and therefore the incompressibility does not vanish [1, 2]. Such feature plays a crucial role in the structure and in the possible hadron-quark phase transition in compact star objects [3]. Moreover, for a binary system with two phases, the binodal coexistence surface is two dimensional and the instabilities in the mixed liquid-gas phase arise from fluctuations in the proton concentration (chemical instability) and in the baryon density (mechanical instability).

In this article, we study a nonlinear hadronic EOS at finite temperature and density by means of a relativistic mean-field model with the inclusion Δ-isobars and by requiring the

Gibbs conditions on the global conservation of baryon number and net electric charge. In this context, let us observe that, for the range of temperatures and baryon densities considered in this investigation ($T \leq 50$ and $\rho_B \leq 3\rho_0$), the contribution of strange hadron particles can be neglected in a good approximation due to their very low concentration.

The main goal of this paper is twofold. First, we are going to show that, for an asymmetric warm and dense nuclear medium, the possible Δ-matter phase transition is characterized by mechanical and chemical-diffusive instabilities. Similarly to the liquid-gas phase transition, chemical instabilities play a crucial role in the characterization of the phase transition and can imply a very different electric charge fraction Z/A in the coexisting phases during the phase transition. Second, we study the influence of nonextensive statistical effects on the thermodynamical instabilities in warm and asymmetric nuclear matter and we investigate how the phase diagram of the nuclear liquid-gas phase transition can be modified in the framework of nonextensive statistical mechanics [4].

2. Nonlinear hadronic equation of state

The relativistic mean-field model (RMF) is widely successful used for describing the properties of finite nuclei as well as hot and dense nuclear matter [3, 5, 6].

In the RMF model the Lagrangian density for nucleons can be written as

$$\begin{aligned}\mathcal{L}_{\rm N} &= \overline{\psi}_N\,[i\,\gamma_\mu\,\partial^\mu-(M_N-g_{\sigma N}\,\sigma)-g_{\omega N}\,\gamma_\mu\,\omega^\mu-g_{\rho N}\,\gamma_\mu\,\vec{t}\cdot\vec{\rho}^{\,\mu}]\,\psi_N+\frac{1}{2}(\partial_\mu\sigma\partial^\mu\sigma-m_\sigma^2\sigma^2)\\ &\quad -U(\sigma)+\frac{1}{2}\,m_\omega^2\,\omega_\mu\omega^\mu+\frac{1}{4}\,c\,(g_{\omega N}^2\,\omega_\mu\omega^\mu)^2+\frac{1}{2}\,m_\rho^2\,\vec{\rho}_\mu\cdot\vec{\rho}^{\,\mu}-\frac{1}{4}F_{\mu\nu}F^{\mu\nu}-\frac{1}{4}\vec{G}_{\mu\nu}\vec{G}^{\mu\nu}\,,\end{aligned} \quad (1)$$

where $M_N = 939$ MeV is the nucleon vacuum mass and $\vec{t}$ is the isospin operator which acts on the nucleon. The field strength tensors for the vector mesons are given by the usual expressions $F_{\mu\nu} \equiv \partial_\mu\omega_\nu - \partial_\nu\omega_\mu$, $\vec{G}_{\mu\nu} \equiv \partial_\mu\vec{\rho}_\nu - \partial_\nu\vec{\rho}_\mu$, and $U(\sigma)$ is the nonlinear potential of σ meson

$$U(\sigma)=\frac{1}{3}a(g_{\sigma N}\sigma)^3+\frac{1}{4}b(g_{\sigma N}\sigma)^4\,, \quad (2)$$

usually introduced to achieve a reasonable compression modulus for equilibrium normal nuclear matter. In the following, the meson-nucleon coupling constants and the other parameters (a, b, c) of the EOS will be fixed to the parameters set marked as TM1 of Ref. [5].

In a regime of finite values of temperature and density, a state of high-density resonance matter may be formed and the $\Delta(1232)$-isobar degrees of freedom are expected to play a central role. In particular, the formation of resonances matter contributes essentially to baryon stopping, hadronic flow effects and enhanced strangeness [7].

The Lagrangian density concerning the Δ-isobars can be then expressed as [8]

$$\mathcal{L}_\Delta=\overline{\psi}_{\Delta\,\nu}\,[i\gamma_\mu\partial^\mu-(M_\Delta-g_{\sigma\Delta}\sigma)-g_{\omega\Delta}\gamma_\mu\omega^\mu]\psi_\Delta^\nu\,, \quad (3)$$

where ψ_Δ^ν is the Rarita-Schwinger spinor for the Δ-isobars (Δ^{++}, Δ^+, Δ^0, Δ^-). Due to the uncertainty on the meson-Δ coupling constants, we limit ourselves to consider only the coupling with the σ and ω meson fields, more of which are explored in the literature.

Because we are going to describe a finite temperature and density asymmetric nuclear matter, we have to require the conservation of two "charges": baryon number (B) and electric charge (C) (as already remarked, we neglect the contribution of strange hadrons, because a tiny amount of strangeness can be produced in the range of temperature and density explored in this study). As a consequence, the system is described by two independent chemical potentials: μ_B and

μ_C, the baryon and the electric charge chemical potential, respectively. Therefore, the chemical potential of particle of index i can be written as

$$\mu_i = b_i\, \mu_B + c_i\, \mu_C\,, \tag{4}$$

where b_i and c_i are, respectively, the baryon and the electric charge quantum numbers of the ith hadron.

The thermodynamical quantities can be obtained from the baryon grand potential Ω_B in the standard way. More explicitly, the baryon pressure $P_B = -\Omega_B/V$ and the energy density can be written in the following nonlinear equations

$$\begin{aligned} P_B &= \frac{1}{3}\sum_i \gamma_i \int \frac{\mathrm{d}^3k}{(2\pi)^3}\, \frac{k^2}{E_i^*(k)}\, [n_i(k)+\overline{n}_i(k)] - \frac{1}{2}\, m_\sigma^2\, \sigma^2 - U(\sigma) && (5)\\ &+ \frac{1}{2}\, m_\omega^2\, \omega^2 + \frac{1}{4}\, c\,(g_{\omega N}\, \omega)^4 + \frac{1}{2}\, m_\rho^2\, \rho^2, \\ \epsilon_B &= \sum_i \gamma_i \int \frac{\mathrm{d}^3k}{(2\pi)^3}\, E_i^*(k)\, [n_i(k)+\overline{n}_i(k)] + \frac{1}{2}\, m_\sigma^2\, \sigma^2 + U(\sigma) && (6)\\ &+ \frac{1}{2}\, m_\omega^2\, \omega^2 + \frac{3}{4}\, c\,(g_{\omega N}\, \omega)^4 + \frac{1}{2}\, m_\rho^2\, \rho^2\,. \end{aligned}$$

Let us observe that in the presence of nonlinear statistical mechanics effects, the nuclear equation of state will be modified as discussed in Ref. [9].

3. Phase transition and stability conditions

We are dealing with the study of a multi-component system at finite temperature and density with two conserved charges: baryon number and electric charge. For such a system, the Helmholtz free energy density F can be written as

$$F(T,\rho_B,\rho_C) = -P(T,\mu_B,\mu_C) + \mu_B\rho_B + \mu_C\rho_C\,, \tag{7}$$

with

$$\mu_B = \left(\frac{\partial F}{\partial \rho_B}\right)_{T,\rho_C}, \qquad \mu_C = \left(\frac{\partial F}{\partial \rho_C}\right)_{T,\rho_B}. \tag{8}$$

In a system with N different particles, the particle chemical potentials are expressed as the linear combination of the two independent chemical potentials μ_B and μ_C and, as a consequence, $\sum_{i=1}^N \mu_i\rho_i = \mu_B\rho_B + \mu_C\rho_C$.

Assuming the presence of two phases (denoted as I and II, respectively), the system is stable against the separation in two phases if the free energy of a single phase is lower than the free energy in all two phases configuration. The phase coexistence is given by the Gibbs conditions

$$\mu_B^I = \mu_B^{II}\,, \qquad \mu_C^I = \mu_C^{II}\,, \tag{9}$$
$$P^I(T,\mu_B,\mu_C) = P^{II}(T,\mu_B,\mu_C)\,. \tag{10}$$

Therefore, at a given baryon density ρ_B and at a given net electric charge density $\rho_C = y\,\rho_B$ (with $y = Z/A$), the chemical potentials μ_B are μ_C are univocally determined. An important feature of this conditions is that, unlike the case of a single conserved charge, the pressure in the mixed phase is not constant and, although the total ρ_B and ρ_C are fixed, baryon and charge densities can be different in the two phases. For such a system in thermal equilibrium, the possible phase transition can be characterized by mechanical (fluctuations in the baryon density) and chemical instabilities (fluctuations in the electric charge density). As usual the condition of the mechanical stability implies

$$\rho_B \left(\frac{\partial P}{\partial \rho_B}\right)_{T,\rho_C} > 0\,. \tag{11}$$

By introducing the notation $\mu_{i,j} = (\partial\mu_i/\partial\rho_j)_{T,P}$ (with $i,j = B,C$), the chemical stability for a process at constant P and T can be expressed with the following conditions [10]

$$\rho_B\,\mu_{B,B} + \rho_C\,\mu_{C,B} = 0\,, \tag{12}$$

$$\rho_B\,\mu_{B,C} + \rho_C\,\mu_{C,C} = 0\,. \tag{13}$$

Whenever the above stability conditions are not respected, the system becomes unstable and the phase transition take place. The coexistence line of a system with one conserved charge becomes in this case a two dimensional surface in (T,P,y) space, enclosing the region where mechanical and diffusive instabilities occur.

By increasing the temperature and the baryon density during the high energy heavy ion collisions ($T \approx 50$ MeV and $\rho_B \geq \rho_0$), a multi-particle system with Δ-isobar and pion degrees of freedom may take place.

In analogy with the liquid-gas case, we are going to investigate the existence of a possible phase transition in the nuclear medium by studying the presence of instabilities (mechanical and/or chemical) in the system. The chemical stability condition is satisfied if [10]

$$\left(\frac{\partial\mu_C}{\partial y}\right)_{T,P} > 0 \;\;\text{or}\;\; \begin{cases} \left(\dfrac{\partial\mu_B}{\partial y}\right)_{T,P} < 0\,, & \text{if } y > 0\,, \\[2ex] \left(\dfrac{\partial\mu_B}{\partial y}\right)_{T,P} > 0\,, & \text{if } y < 0\,. \end{cases} \tag{14}$$

4. Results and discussion

4.1. Δ-matter phase transition

In Fig. 1, we report the pressure as a function of the baryon density at $T = 30$ MeV (left panel) and $T = 50$ MeV (right panel), for different values of $y = Z/A$ and $x_{\sigma\Delta} = 1.3$. In this context it is interesting to observe that at $T = 50$ MeV and below $y = 0.3$, the system becomes mechanically stable, but, in a similar manner to the liquid-gas case, is chemically unstable.

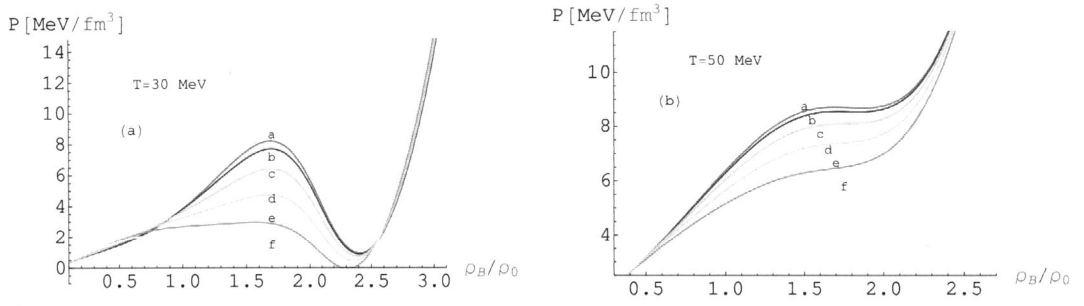

Figure 1. Pressure as a function of baryon density for $T = 30$ MeV (left panel) and $T = 50$ MeV (right panel) with $x_{\sigma\Delta} = 1.3$. Labels a to f correspond to $y = 0.5, 0.4, 0.3, 0.2, 0.1, 0$, respectively.

From the analysis of the chemical potential isobars, we are able to construct the binodal surface relative to the nucleon-Δ matter phase transition.

At lower temperatures, the mixed-phase region becomes more relevant at higher values of y. This feature can be seen in Fig. 2 where the binodal section (left panel) and the isothermal pressure as a function of the baryon density (right panel) is reported at $T = 40$ MeV and $x_{\sigma\Delta} = 1.3$. The Gibbs construction corresponds to the curve from A to C; the isothermal curves in B and D (with $y_B \neq y_D$) are also reported. In this case, we assume that the system is initially prepared in the low-density (nucleonic) phase with $y = 0.3$, corresponding to point A. During the compression each phase evolves following the corresponding curve up to points C and D, where the system leaves the instability region in the Δ-matter phase.

The right branch (at lower density) corresponds to the initial phase (I), where the dominant component of the system is given by nucleons. The left branch (II) is related to the final phase at higher densities, where the system is composed primarily by Δ-isobar degrees of freedom (Δ-dominant phase). In presence of Δ-isobars the phase coexistence region results very different from what obtained in the liquid-gas case, in particular it extends up to regions of negative electric charge fraction.

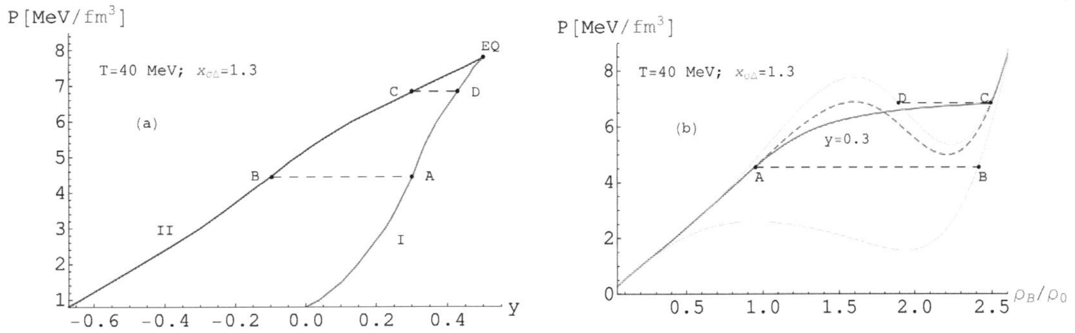

Figure 2. Left panel: binodal section at T= 40 MeV and $x_{\sigma\Delta} = 1.3$, with the point of equal equilibrium in evidence. Right panel: the corresponding isothermal curves, with the Gibbs construction (curve from the point A to C) at $y = 0.3$ and the isotherms of points B and D shown.

Therefore, similarly to the liquid-gas phase transition, the nucleonic and the Δ-matter phase have a different electric charge fraction in the mixed phase. The electric charge fraction in the nucleonic phase reflects a system with higher values of y than the Δ-matter phase. In the liquid-gas phase transition, the process of producing a larger neutron excess in the gas phase is referred to as isospin fractionation. A similar effects can occur in the nucleon-Δ matter phase transition due essentially to a Δ^- excess in the Δ-matter phase with lower values of y.

Many effects discussed in this paper are more evident at low values of y, obtainable, in principle, with radioactive ion beam facilities. On the other hand, it is rather unlikely, at least in the near future, that neutron rich nuclei can be accelerated to energies larger than a few GeV per nucleon. However, some precursor signals of the considered instabilities could be observed even in collisions of stable nuclei at intermediate energies. For example, in Ref. [11], the simulation of the reaction ^{238}U+^{238}U (average $y = 0.39$), at 1 A GeV and semicentral impact parameter $b = 7$ fm, shows that rather exotic nuclear matter can be formed in a transient time of the order of 10 fm/c, with a baryon density up to $3\,\rho_0$, $T \leq 50 \div 60$ MeV and $y \approx 0.35 \div 0.40$. Such conditions would agree fully with the results for the nucleon-Δ mixed phase region.

4.2. Nonextensive nuclear liquid-gas phase transition

Recently, there has been increasing evidence that the nonlinear and nonextensive statistical mechanics, proposed by Tsallis, can be considered as an appropriate basis to deal with physical phenomena where strong dynamical correlations, long-range interactions and microscopic memory effects take place [4, 12]. A considerable variety of physical applications involve a quantitative agreement between experimental data and theoretical models based on Tsallis thermostatistics. In particular, in the last years there has been a growing interest in high energy physics applications of nonextensive statistics and several authors have outlined the possibility that experimental observations in relativistic heavy ion collisions can reflect nonextensive statistical behaviors [13, 14, 15, 16, 17].

Nonextensive statistical mechanics introduced by Tsallis consists of a generalization of the common Boltzmann-Gibbs statistical mechanics and it is based upon the introduction of entropy [4]

$$S_q[f] = \frac{1}{q-1}\left(1 - \int [f(\mathbf{x})]^q \, d\Omega\right) , \quad \left(\int f(\mathbf{x}) \, d\Omega = 1\right) , \tag{15}$$

where $f(\mathbf{x})$ stands for a normalized probability distribution, $\mathbf{x}$ and $d\Omega$ denoting, respectively, a generic point and the volume element in the corresponding phase space (here and in the following we set the Boltzmann and the Planck constant equal to unity). The real parameter q determines the degree of non-additivity exhibited by the entropy form (15) which reduces to the standard Boltzmann-Gibbs entropy in the limit $q \to 1$. By means of maximizing the entropy S_q, under appropriate constraints, it is possible to obtain a probability distribution (or particle distribution) which generalized, in the classical limit, the Maxwell-Boltzmann distribution. The nonextensive classical distribution can be seen as a superposition of the Boltzmann one with different temperatures which has a mean value corresponding to the temperature appearing in the Tsallis distribution [13].

We are going to study the effects of nonextensive statistical effects on the presence of nuclear instabilities at low temperature and subnuclear densities, in the framework of the previous introduced nuclear EOS. In the following we will focus our investigation for small deviations from the standard Boltzmann-Gibbs statistical mechanics and for values $q > 1$, because these values were obtained in several phenomenological studies in high energy heavy ion collisions (see, for example, Ref.s [18, 19, 20, 21, 22]).

In Fig. 3, we report the pressure as a function of baryon density ρ_B (in units of the nuclear saturation density ρ_0) for various values of the electric charge fraction y at fixed temperature $T = 10$ MeV. In the left panel, a comparison of the equation of state in the presence of ($q = 1.05$, dashed lines) and in the absence ($q = 1$, continuous lines) of the nonextensive statistical effects is shown. In the right panel, for $q = 1.05$ only, the continuous lines correspond to the solution obtained with the Gibbs construction in the mixed phase, whereas the dashed lines are without correction.

Let us observe that in presence of small deviation from the standard Boltzmann-Gibbs statistics, the nuclear EOS appears stiffer, with higher values of pressure at fixed baryon density. As we will see, this implies a greater value of the first transition density and a reduction of second transition density in the mixed phase with respect to the standard case ($q = 1$). This feature results in significant changes in the nuclear incompressibility and may be particularly important in identifying the presence of nonextensive effects in heavy-ion collision experiments. Moreover, it appears evident that for a proton fraction $y > 0.2$ a mechanical instability is present, whereas for $y < 0.2$ the system becomes unstable only under chemical-diffusive instability.

Finally, in Fig. 4, we report the phase diagram with evidence of the coexistence regions of the liquid-gas phase transition for $y = 0.3$ and 0.5, in the presence ($q = 1.05$) and in the absence ($q = 1$) of nonextensive statistical effects. As expected, the relevance of nonextensive

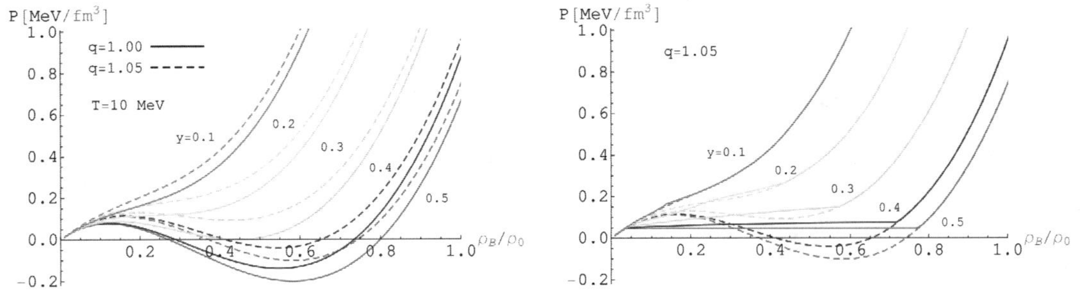

Figure 3. Pressure as a function of baryon density (in units of the nuclear saturation density ρ_0) for various values of the proton fraction y and at $T = 10$ MeV. In the left panel, the dashed lines are related to the entropic value $q = 1.05$ in comparison to the standard case ($q = 1$), corresponding to the continuous line. In the right panel, the continuous (dashed) lines correspond to the solution obtained with (without) the Gibbs construction for $q = 1.05$.

statistical mechanics increases by increasing the temperature and implies a reduction of the critical temperature T_c and a reduction of the baryon density range involved in the mixed phase, with an enhancement of the first critical density and a reduction of the second critical density, with respect to the standard (extensive) case.

Therefore, several features of the nuclear liquid-gas phase transition turn out to be sensibly modified also for small deviations from the standard Boltzmann-Gibbs statistics. In asymmetric nuclear matter, instabilities that produce a liquid-gas phase separation arise from fluctuations in the proton fraction (chemical instability), rather than from fluctuations in the baryon density (mechanical instability). In the presence of nonlinear and nonextensive statistical effects, the relevance of the chemical instability becomes more pronounced; it appears an effective reduction of the dynamical instability region and the so-called isospin fractionation effect turns out to be reduced, with fewer protons in the liquid phase and more proton in the gas phase, with respect to the standard case.

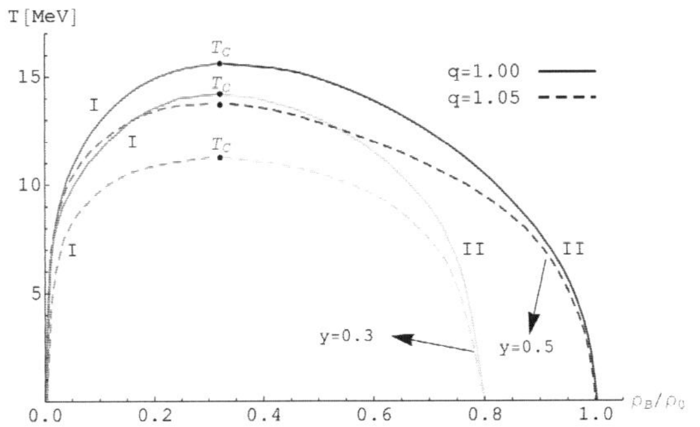

Figure 4. Phase diagram of the liquid-gas phase transition for asymmetric ($y = 0.3$) and symmetric ($y = 0.5$) nuclear matter and for different values of q. The lines labeled with I and II, delimitate the first and second critical densities of the coexistence regions, respectively.

References

[1] Müller H and Serot B D 1995 *Phys. Rev.* C **52** 2072
[2] Bonanno L, Drago A and Lavagno A 2007 *Phys. Rev. Lett.* **99** 242301
[3] Glendenning N K 1992 *Phys. Rev.* D **46** 1274
[4] Tsallis C 2009 *Introduction to Nonextensive Statistical Mechanics* (Springer-Verlag)
[5] Lavagno A 2010 *Phys. Rev.* C **81** 044909

[6] Lavagno A 2013 *Eur. Phys. J.* A **49** 102
[7] Hofmann M, Mattiello R, Sorge H, Stocker H and Greiner W 1995 *Phys. Rev.* C **51** 2095
[8] Li Z, Mao G, Zhuo Y and Greiner W 1997 *Phys. Rev.* C **56** 1570
[9] Lavagno A and Pigato D 2013 *Physica* A **392** 5164
[10] Lavagno A and Pigato D 2012 *Phys. Rev.* C **86** 024917
[11] Di Toro M *et al.* 2006 *Nucl. Phys.* A **775** 102
[12] Lavagno A, Scarfone A M and Swamy P Narayana 2006 *Eur. Phys. J.* C **47** 253
[13] Wilk G and Wlodarczyk Z 2009 Eur. Phys. J. A 40 299
[14] Biró T S, Purcsel G and Ürmössy K 2009 *Eur. Phys. J.* A **40** 325
[15] Alberico W M and Lavagno A 2009 *Eur. Phys. J.* A **40** 313
[16] Lavagno A and Pigato D 2012 *J. Phys. G: Nucl. Part. Phys.* **39** 125106
[17] Quarati P *et al.* 1997 *Nucl. Phys.* A **621** 345c
[18] Cleymans J and Worku D 2012 *J. Phys. G: Nucl. Part. Phys.* **39** 025006
[19] Chinellato D D, Takahashi T and Bediaga I 2010 *J. Phys. G: Nucl. Part. Phys.* **37** 094042
[20] Alberico W M et al. 2008 *Physica* A **387** 467
[21] Lavagno A, Pigato D and Quarati P 2010 *J. Phys. G: Nucl. Part. Phys.* **37** 115102
[22] Lavagno A and Pigato D 2011 *Eur. Phys. J* A **47** 52

Integrable zero-range potentials in a plane.

S Leble
Gdansk University of Technology, ul. Narutowicza 11/12, 80-952, Gdansk, Poland
E-mail: leble@mifgate.pg.gda.pl

Abstract. We examine general statements in the Wronskian representation of Darboux transformations for plane zero-range potentials. Such expressions naturally contain scattering problem solution. We also apply Abel theorem to Wronskians for differential equations and link it to chain equations for Darboux transforms to fix conditions for further development of the underlying distribution concept. Moutard transformations give a convenient insight into the problem that allows one to formulate general assertions and give complete description of a point potential in a limit of long waves.

1. Introduction

Solvable models play a basic role as in classical/quantum mechanics [1] as in electrodynamics [2, 3] and statistical physics [4]. In this rather poor set of models, a significant place is occupied by the ones based on Zero-Range Potentials (ZRP), whose history started from the seminal work of E. Fermi [5]. Their mathematical significance in the context of extension theory was established in [6], see the review [7] and consideration in arbitrary dimensions [3]. The book [8] covers as multicenter applications of the theory of ZRPs as its multichannels generalizations. Very recent paper [9] studies two integrable models: it consider ZRP as perturbations of harmonic oscillator. Interesting applications in solid state physics can be found in [10].

Integrability concept is directly related to Darboux transformations (DT) theory [11, 12]. The zero range potentials were first considered by means of this theory in [13]. The approach introduces so-called generalized ZRPs (compared to spherically symmetrical - partial s-wave ZRP). Its distributional nature was examined and extended to two-dimensional space in [14]. An important development of the theory from a point of view of DT or dressing of a many-center potential and applications to electron-molecular scattering problems can be found in [15]. Dressing of ZRP also gives useful finite range potentials that could be widely applied (see e.g. [16]). To explain the idea, let us address the one-dimensional spectral problem at $x \in (-\infty, \infty)$

$$-\psi_{xx} + u\psi = \lambda\psi, \tag{1}$$

which is covariant with respect to DT

$$\psi_1 = (\partial - \sigma)\psi \tag{2}$$

if σ is a solution of the Riccati equation

$$\sigma' + \sigma^2 - \kappa^2 = u. \tag{3}$$

Substitution $\sigma = (\ln\phi)' = \phi'/\phi$ links it to the solution of the spectral problem (1) with $\lambda = -\kappa^2$. It defines the dressed potential

$$u_1 = u - 2\sigma'. \quad (4)$$

Consider a dressing of the ZRP represented by the distribution $-\alpha\delta(x)$, whose presence is equivalent to boundary condition

$$\lim_{\epsilon\to 0}[\psi(\epsilon) - \psi(-\epsilon)] = -\alpha\psi(0). \quad (5)$$

We should rely upon linear independent generalized solutions of the problem (1) with the ZRP:

$$\psi_1 = \sin(kxsgn(x) + \eta), \quad \psi_2 = \sin(kx), \qquad \lambda = k^2,\ k\cot\eta = -\alpha, \quad (6)$$

whose derivatives are also distributions. For $\alpha > 0$, the chosen potential implies one bound state with eigenfunction

$$\phi = \exp[-\alpha xsgnx].$$

A dressing with such prop function deletes the bound state and leads to the ZRP

$$u_1 = \alpha\delta(x).$$

The boundary condition (5) also changes. This simple example explains how the algorithm enlarges the space of integrable potentials. Other combinations of eigenfunctions generate ZRP with different properties.

Separation of variables on three-dimensional Schrödinger equation with spherical symmetry gives a radial differential equation for partial waves $\psi_l(r) = \frac{1}{r}\phi_l(r)$. Precisely, the function $\phi_l(r)$ satisfies the equation

$$\left[-\frac{1}{2}\frac{d^2}{dr^2} + \frac{l(l+1)}{2r^2} + u_l - E\right]\phi_l(r) = 0. \quad (7)$$

The covariance of this equation with respect to DT

$$\psi_{l1} = \frac{1}{\sqrt{k^2 + \kappa_l^2}}\left(\frac{d}{dr} - \sigma_l\right)\phi_l(r) \quad (8)$$

combines two solutions of the spectral problem (7) with $E = k^2$ and $E_l = -\kappa_l^2$, so that

$$\sigma_l = \ln'\phi_l(r, i\kappa_l). \quad (9)$$

The transformation was considered in [13, 15] for the quantum problem of scattering on integrable potential, by applying N-th order DT and taking spherical Hankel functions with specific parameters κ_m as prop functions. The potentials are expressed in terms of Wronskians, which go to Vandermond determinants at infinity. Zero logarithmic derivative of the Wronskians corresponds to generalized ZRP's.

The construction is similar to the one given in this article (see Section 2), where the Wronskians are computed and the solution of the scattering problem is given in explicit form. For example, in the case of l=1, the direct application of Rayleigh's formulas for Bessel functions yields

$$\phi_{1,i} = \sqrt{\frac{\pi}{2\kappa_i r}}H_{\frac{3}{2}}(\kappa_i r) = \frac{i}{\kappa_i^2 r^2}e^{i\kappa_i r}(i\kappa_i r - 1). \quad (10)$$

The Wronskian determinant in this case, by direct evaluation, is equal to:

$$W = e^{ir(\kappa_3+\kappa_2+\kappa_1)}(\kappa_2-\kappa_3)(\kappa_1-\kappa_3)(\kappa_1-\kappa_2)\frac{\kappa_1\kappa_2+\kappa_1\kappa_3+\kappa_2\kappa_3 - ir\kappa_1\kappa_2\kappa_3}{r\kappa_1^2\kappa_2^2\kappa_3^2}. \quad (11)$$

Its derivative $ln_{rr}W$ is zero if $\kappa_1\kappa_2 + \kappa_1\kappa_3 + \kappa_2\kappa_3 = 0$, and ZRP hence follows.

Section 2 is devoted to novel results: explicit form of the boundary conditions in 2d, analog of the three-dimensional generalized ZRP, is derived by the iterated DT. One can compare resulting formulas from [3] with those derived here.

Section 3 contains general formalism, based on separated variables, Abel theorem for Wronskians and DT chain equations that allow to avoid direct differentiations in some physically significant cases, including novel ones.

In the last Section, a Moutard transformation (MT) is applied to new alternative construction of ZRP for the wave, Helmholtz and Schrödinger equations in two dimensions. Iterated MT may produce locally dense set of such ZRP [14].

2. Two-dimensional ZRP for cylindrical symmetry case

To develop the method described in the introduction and show the dressing origin and generalizations of the results for abundant two-dimensional problems of, e.g., [17], let us consider the Schrödinger equation in atomic units for xy plane

$$-\frac{1}{2}(\frac{\partial^2\psi}{\partial^2 x} + \frac{\partial^2\psi}{\partial^2 y}) + U(\vec{r})\psi = E\psi. \tag{12}$$

As in the three-dimensional case [13], by reducing to a one-dimensional problem via separation of variables, ZRP may be introduced whose explicit form depends on the problem symmetry. Equation (12) for $U = 0$ is equivalent to the Helmholtz equation $(-\triangle + k^2)\psi(\vec{\rho}) = 0$, $E = 2k^2$, which, like its inhomogeneous counterpart $(-\triangle + u(\rho) + k^2)\psi(\vec{\rho}) = 0$ where $u = 2U$, finds application in electrodynamics. For a cylindric symmetry, going to polar coordinates $x = \rho\cos\phi, y = \rho\sin\phi$, we have

$$\Delta = \frac{d^2}{d\rho^2} + \frac{1}{\rho}\frac{d}{d\rho} + \frac{1}{\rho^2}\frac{d^2}{d\phi^2}. \tag{13}$$

The separation of variables $\exp[i\nu\phi]R$ yields R either as the solution $R(\rho) = Y_v(k\rho), ...$ of the Bessel equation

$$[\frac{d^2}{d(k\rho)^2} + \frac{1}{k\rho}\frac{d}{d\,(k\rho)} + 1 - \frac{\nu^2}{(k\rho)^2}]R = 0, \tag{14}$$

if $E = k^2 > 0$, or, when $E = \kappa^2 < 0$, as the solution of the modified Bessel equation

$$-[\frac{d^2}{d\rho^2} + \frac{1}{\rho}\frac{d}{d\rho} - \frac{\nu^2}{\rho^2}]R = -\kappa^2 R, \tag{15}$$

with the feature of the singular behaviour of $R = K_v(\kappa z)$ at $\rho = 0$. The special case of $E = 0$ and analogue separation of variables $\exp[i\nu\phi]R$ yields

$$[\frac{d^2}{d\rho^2} + \frac{1}{\rho}\frac{d}{d\rho} - \frac{\nu^2}{\rho^2}]R(\rho) = 0. \tag{16}$$

It is the Euler equation with the basic solutions $R = x^{\pm\nu}$.

Let us apply the DT technique to these equations following the scheme of [13, 15]. For the sake of convenience in the application of Crum formulas (see again e.g. [13]), let us transform the equations via $R = \rho^{-1/2}a\,(\rho)$. This gives

$$-\frac{\partial^2 a\,(\rho)}{\partial(k\rho)^2} + \frac{-1+\nu^2}{4(k\rho)^2}a\,(\rho) - a\,(\rho) = 0. \tag{17}$$

Hence, for a seed solution $a_\nu(\rho) = \rho^{1/2}R(k\rho)$ and a prop functions $c_m(\rho) = \rho^{1/2}R(\kappa_m\rho)$, the N-fold transform of $a = \rho^{1/2}R$ is identified by means of the dressed potential

$$u^{[N]}(\rho) = u(\rho) - 2\frac{d^2}{d\rho^2}\ln W(\rho^{1/2}R_1(\rho), ..., \rho^{1/2}R_N(\rho)), \quad (18)$$

and the dressed solution

$$S_\nu(\rho)^{[N]} = \frac{W(\rho^{1/2}R_1(\rho), ..., \rho^{1/2}R_N(\rho), \rho^{1/2}S_\nu(\rho))}{\rho^{1/2}W(\rho^{1/2}R_1(\rho), ..., \rho^{1/2}R_N(\rho))}. \quad (19)$$

Having in mind the scattering problem application as it is explained in the Introduction (see also [13]), it is convenient to use Hankel functions $H_\nu^{(1,2)}(\kappa_i\rho) = J_v(z) \pm iY_\nu(z)$, and express the general solution of the radial equation as

$$\psi_\nu = C(H_\nu^{(1)}(k\rho)s_\nu + H_\nu^{(2)}(k\rho)), \quad (20)$$

s_ν being a scattering matrix. The convenient choice for the dressing scheme is

$$S_\nu = \sqrt{\frac{\pi}{2k}}J_v(k\rho), \qquad R_m = \sqrt{\frac{\pi}{2\kappa_m}}H_\nu^{(1)}(\kappa_m\rho). \quad (21)$$

Substitution of asymptotes at infinity (for integer ν),

$$J_v(z) \sim \left[\sqrt{\frac{2}{\pi z}}\cos\left(z - \frac{\nu\pi}{2} - \frac{\pi}{4}\right)\right]_{z=k\rho} = \frac{\sqrt{2}}{\sqrt{\pi k\rho}}\left[\cos\left(k\rho - \frac{1}{2}\pi\nu - \frac{1}{4}\pi\right)\right], \quad (22)$$

$$Y_\nu(z) = \frac{J_v(z)\cos(\nu\pi) - J_{-v}(z)}{\sin(\nu\pi)} \sim \sqrt{\frac{2}{\pi k\rho}}\sin\left(k\rho - \frac{\nu\pi}{2} - \frac{\pi}{4}\right),$$

into (19) yields the solution

$$S^{[N]}(\rho) \sim \frac{C}{\sqrt{\rho}}\left(\frac{V(\kappa_1, ..., \kappa_N, ik)}{V(\kappa_1, ..., \kappa_N)}\exp[ik\rho] - \frac{V(\kappa_1, ..., \kappa_N, -ik)}{V(\kappa_1, ..., \kappa_N)}\exp[-ik\rho]\right), \quad (23)$$

which solves the scattering problem on the potentials of the form (18). The dressed zero potential (18) goes to

$$u^{[N]}(\rho) = -2\frac{d^2}{d\rho^2}\ln(V(\kappa_1, ..., \kappa_N)\exp[i\sum\kappa_m\rho]), \quad (24)$$

which has also zero value, hence defining ZRP.

Let us consider asymptotes at zero, having in mind $Y_0(z) \sim \frac{2}{\pi}\ln z$, and restrict ourselves to the case $\nu = n > 0$; then $Y_n(z) \sim \frac{2^n(n-1)!}{\pi}z^{-n}$ and $J_n(z) \sim \frac{1}{2^n n!}z^n$.

The boundary condition which defines ZRP is extracted from the asymptotic solutions monomials behaviour

$$\partial^{2n}(z^n\psi_n)|_{z=0} = \frac{1}{2^n n!} = (z^n\psi_n)|_{z=0} = \alpha_n\frac{2^n(n-1)!}{\pi}, \quad (25)$$

where α_n is the partial scattering length and κ_m are roots of the algebraic equation $k^N = i\alpha_n$, that guarantee the relation $s_n = \prod_{m=1}^{m=N}\frac{ik-\kappa_m}{ik+\kappa_m}$.

3. General remarks: one-dimensional case.

For a problem with symmetry, after separation of variables one is led to a one-dimensional description. The integrable potentials are effectively generated in terms of the Wronskian representation naturally appearing within the DT theory [11, 20]. To evaluate the Wronskians it is convenient to use Abel theorem [18]:

Theorem 1 *Let* $\psi_i(x) \in C^n, \partial = \partial/\partial x$. *If*

$$(p_0\partial^N + p_1\partial^{N-1} + ...)\psi_i(x) = 0,$$

the Wronskian derivative is given by

$$W'(\psi_1, ..., \psi_N) = \frac{p_1}{p_0} W. \tag{26}$$

Consider now the second order Schrödinger operator $L = -\partial^2 + u = (-\partial + \sigma)(\partial + \sigma)$ for a one-dimensional problem based on Equation (1), no matter what interval is of interest. The iterated Darboux transform (in a slightly different notation compared to Introduction)

$$\psi_{i+1} = (\partial - \sigma_i)\psi_i = \left(\partial - \frac{\varphi_i'}{\varphi_i}\right)\psi = \left(\varphi_i \partial \varphi_i^{-1}\right)\psi_i, \tag{27}$$

$\psi_i \in R_\lambda, \varphi_i \in R_{\mu_i}$, combines eigenspaces of L. For potentials one has

$$u_{i+1} = u_i - 2\sigma_i', \tag{28}$$

if the so-called Miura constraint

$$\sigma_i' + {\sigma_i}^2 + \mu_i = u_i \tag{29}$$

holds. The combining of DT (28) and Miura link (29) yields the chain equation (e.g. in [20])

$$(\sigma_i + \sigma_{i+1})' = \sigma_i^2 - \sigma_{i+1}^2 + \mu_i - \mu_{i+1}. \tag{30}$$

N-fold DT [11, 20] have the property:

$$(\partial - \sigma_N) \cdot ... \cdot (\partial - \sigma_1)\phi_i = 0,$$

where $\phi_i \in R_{\mu_i}$. Expanding the l.h.s. yields

$$\left(\partial^N - \sum_{i=1}^{i=N} \sigma_i \partial^{N-1} + ...\right)\phi_i = 0,$$

The application of the Crum result and the Abel theorem, Eq. (26), gives

$$u[N] = u - 2\ln_{xx} W[N] = u - 2\left(\frac{W'[N]}{W[N]}\right)' = u - 2\,(\frac{p_1}{p_0})' = u + 2\sum_{i=1}^{i=N} \sigma_i', \tag{31}$$

because $p_0 = 1,\ p_1 = -\sum_{i=1}^{i=N} \sigma_i$. This gives the necessary condition of the extra ZRP

$$\sum_{i=1}^{i=N} \sigma_i = const, \tag{32}$$

and an expression for the Wronskian

$$W = \exp[-\int \sum_{i=1}^{i=N} \sigma_i dx]. \tag{33}$$

Further, by insertion of $\sigma_i = \phi_i'/\phi_i = (\ln \phi_i)'$ into (33), one obtains

$$W = \exp[-\int \sum_{i=1}^{i=N} \ln_x \phi_i dx] = \exp[-\ln \prod_{i=1}^{i=N} \phi_i] = \prod_{i=1}^{i=N} \phi_i^{-1}. \tag{34}$$

N-fold application of (28) thus leads to

$$\sum_{i=1}^{i=N} \sigma_i = \frac{1}{2} \int (u_{N+1} - u_1) dx + C. \tag{35}$$

This formula results in a constant Wronskian via (33) and may be applied in scattering or eigenvalue problems if $u_i = u_0$ or $u_{N+1} - u_1 = 0$. For the cases studied in the previous section, $u_0 = \frac{-1+\nu^2}{4(k\rho)^2}, \rho \in [0, \infty)$ and $u_0 = \frac{l(l+1)}{2r^2}$, $r \in [0, \infty)$ [13].

There are many interesting applications for an operator

$$u_0 = -\frac{d}{dx} x \frac{d}{dx} + (\frac{x}{4} + \frac{m^2}{x}), \tag{36}$$

arising after separation of variables in the three-dimensional Shrödinger operator [19]. We can supply it with extra zero-range potential. The variable x here is directly linked with parabolic coordinates, and solutions describe Stark effect and particles with spin via Pauli equation. The ZRP construction is based on the scheme presented here and solutions of the correspondent equation with u_0, given in the Fock textbook [19].

In the alternative language of "superpotentials" σ_i, we base ourselves on Eq. (30). Developing it in a sequence of equalities

$$(\sigma_{i+1} + \sigma_{i+2})' = \sigma_{i+1}^2 - \sigma_{i+2}^2 + \mu_{i+1} - \mu_{i+2}, \\ \dots\dots \tag{37}$$

a relation convenient for evaluation of Wronskian via (33) can be derived:

$$(\sigma_1 + 2\sigma_2 + ...2\sigma_N + \sigma_{N+1})' = \sigma_1^2 - \sigma_{N+1}^2 + \mu_1 - \mu_{N+1}. \tag{38}$$

As a test, this reproduces (35) if the link (29) is taken into account. Notice that the relevant case of periodic closure (details in e.g. [20]) $\sigma_{N+1} = \sigma_1$, i.e. $(\ln \phi_{N+1})' = (\ln \phi_1)'$ and hence $\mu_1 - \mu_{N+1} = 0$, is concerned with the constraint

$$\sum_{i=1}^{i=N} \sigma_i' = 0, \tag{39}$$

that again gives ZRP as extra potential. The solutions of the closed chain (30) on $x \in (-\infty, \infty)$ provide finite-gap potentials and the corresponding eigenfunctions of continuous spectrum [21]. A combination of finite-gap potential with ZRP on the axis $x \in (-\infty, \infty)$, implemented directly by (5), leads to eigen functions similar to (6). It is obtained by direct application of N-fold DT in Wronskian form similar to (19). The results generalize [9] for the case of delta-perturbation of periodic finite-gap potential. The result may also be obtained as a limit with respect to soliton dressing on finite-gap background [21]. Another option, defined through the relations $\sigma_{N+1} = \sigma_1 + C$ and $\sigma_1^2 - \sigma_{N+1}^2 = -2\sigma_1 C - C^2$, furnishes

$$\sum_{i=1}^{i=N} \sigma_i' = \frac{\mu_1 - \mu_N}{2} - \sigma_1 C - \frac{C^2}{2}. \tag{40}$$

Plugging this into (33), one has

$$W = \phi_1^C \exp[-(\frac{\mu_1 - \mu_N}{2} - \frac{C^2}{2})x], \tag{41}$$

i.e

$$-2\ln_{xx} W = \mu_N - \mu_1 + 2C\sigma_1 + C^2 \tag{42}$$

that gives simple expression of either iterated potential or Wronskians.

Consider one more example of a chain condition, useful in the context of ZRP theory. Let us put $u_{i+1} = a_i u_i$, then

$$\begin{aligned} \sigma_i' = \tfrac{1}{2}(u_i - u_{i+1}) &= \tfrac{1}{2}u_i(1 - a_i), \\ u_1 &= a_0 u_0, \\ u_2 = a_1 u_1 &= a_1 a_0 u_0, \\ &\dots \\ u_{N+1} = a_N u_N &= \prod_{i=0}^{N} a_i u_0. \end{aligned} \tag{43}$$

Consider an example with $u_i = b_i/x^2$, which means

$$b_i = \prod_{j=0}^{i} a_j. \tag{44}$$

On the other hand, $\sigma_i' = \frac{b_i}{2x^2}(1-a_i)$, so that $\sigma_i = \int \frac{b_i}{2x^2}(1-a_i)dx = \frac{(a_i-1)b_i}{2x} + c_i$. Miura constraint (29) implies

$$\frac{b_i}{2x^2}(1 - a_i) + [\frac{(a_i - 1)b_i}{2x} + c_i]^2 + \mu_i = b_i/x^2. \tag{45}$$

Euler equation case (16) (zero energy in quantum mechanics, long wave limit in wave theory) is recovered for $\mu_i = c_i = 0$, or

$$\frac{b_i}{2}(1 - a_i) + [\frac{(a_i - 1)^2 b_i^2}{2^2}] - b_i = 0, (1 - a_i) + [\frac{(a_i^2 - 2a_i + 1)b_i}{2}] - 2 = 0. \tag{46}$$

By virtue of (44), then

$$(a_i^2 - 2a_i + 1)\prod_{j=0}^{i} a_j = 2(1 + a_i). \tag{47}$$

The condition is a recurrent algebraic equation for a_i; for instance, if $i = 1$ we have a cubic equation for a_1

$$(a_1^2 - 2a_1 + 1)a_1 a_0 = 2(1 + a_1).$$

In general Bessel function case ($Z_n(z), z = k\rho$ in 2d), a similar transformation changes the parameters l (3d) or ν (2d case). It is known that the case is concerned with recurrent formulas

$$Z_{n+1} = \left(-\frac{d}{dz} + \frac{n}{z}\right) Z_n, \qquad Z_{n-1} = \left(\frac{d}{dz} + \frac{n}{z}\right) Z_n, \tag{48}$$

leading to a set of iterated DTs

$$Z_0 = \prod_{i=1}^{i=n} \left(\frac{d}{dz} + \frac{i}{z}\right) Z_n \tag{49}$$

that link different potentials (u_0 in our notation).

4. General algorithm in 2d. Moutard transformation.

The Moutard equation

$$\psi_{\sigma\tau} + u(\sigma,\tau)\psi = 0 \tag{50}$$

(which has the obvious connection with the Schrödinger, wave and Helmholtz equations for complex σ, τ) is covariant with respect to the Moutard transformation. The transformed coefficient (potential in mathematical physics) is given by

$$u[1] = u - 2(\log\varphi)_{\sigma\tau} = -u + \varphi_\sigma\varphi_\tau/\varphi^2 \ . \tag{51}$$

ZRP chain is generated under condition

$$(\log\varphi)_{\sigma\tau} = 0, \tag{52}$$

or, equivalently, when $\log\varphi = \Phi(\sigma) + \Psi(\tau)$ with arbitrary Φ and Ψ. So, if φ and ψ are different solutions of (50), the solution $\psi[1]$ of the twin equation with $(\psi, u) \to (\psi[1], u[1])$ is

$$\psi[1] = \psi - \varphi\Omega(\varphi,\psi)/\Omega(\varphi,\varphi) \tag{53}$$

where Ω is the integral of the exact differential form

$$d\Omega = \varphi\psi_\sigma d\sigma + \psi\varphi_\tau d\tau. \tag{54}$$

We specify now results from [20] in the following convenient form:

Theorem 2 *The N-times iterated potential built on the zero background is*

$$u = 6(\ln\Delta)_{\sigma\tau} \ , \tag{55}$$

where $\Delta = \det[\Delta_{ik}]$ and,

$$\Delta_{ik} = \int d\Omega(\phi_k,\phi_i) + C_{ik} \ , \quad C_{ik} + C_{ki} = \phi_k(0)\phi_i(0) \ ,$$

This choice fix the constants of integration C_{ik}. The integrand is determined by

$$\Omega(\phi_k,\phi_i) = -2\int[\delta_1 d\sigma + \delta_2 d\tau] \ ,$$

with $\delta_1 = \phi_k\phi_{i\sigma}$ and $\delta_2 = \phi_{k\tau}\phi_i$.

Journal of Physics: Conference Series **482** (2014) 012025 doi:10.1088/1742-6596/482/1/012025

Proof directly follows from results published in [20]. Investigation of asymptotes proves multikink structure of $(\ln\Delta)_x$ for the solutions

$$\phi_k = A_k \exp(a_k\sigma + c_k) + B_k \exp(b_k\tau), \tag{56}$$

which means a multisoliton content of u.

If we perform integration from $-\infty$, to σ, τ, for kinks with the choice of constants $a_i > 0$, $b_i > 0, c_k > 0$, we obtain

$$\Delta_{ik} = [\alpha_{ik} p_i p_k \exp(\chi_i + \chi_k) + p_i \exp(\chi_i) + \beta_{ik}] \exp[(b_i + b_k)\tau] + C_{ik}, \tag{57}$$

where $\chi_i = a_i\sigma - b_i\tau + c_i$, $\alpha_{ik} = \frac{a_i}{a_i+a_k}$,$\beta_{ik} = \frac{b_i}{b_i+b_k}$, $\xi_k = a_k\sigma + \phi_k$,$A_i/B_i = p_i$. Due to the impossibility to expand Δ_{ik} as a superposition of exponents with the opposite powers like for the multisoliton determinant representation for the KP equation, a special asymptotic calculation technique was elaborated in [20]. A zero limit of the parameters that corresponds to the width of solitons results in delta - functions that form a network in $\xi\eta$ plane.

There is a way to define ZRP based on direct map of the equation (1) (equivalent to (50) for complex variables), taking into account a symmetry and class of eventual singularities [14]. For example, integrating (1) over a disk S inside a circumference L of small radius ϵ and using Gauss theorem, one has an identity that may be considered as a definition of $\delta_2(\rho, \phi)$, if the rest terms exist in some class of ψ.

An important paper of E. Ganzha, on local completeness of iterated Moutard transformations [22], allows to built coordinate systems by means of the link with Ribokur transformations and solutions of the corresponding Lame equations. The generalization is straightforward if the coordinate curve is closed [14]. In a similar way, the consideration in a vicinity of arbitrary coordinate line may lead to a ZRP in the spirit of the theorem in this paper expressed by means of Equation (55).

5. Conclusion

Integrability means also that the ZRPs may be considered in the context of evolution by Lax pair, therefore serving as a basis for point particles motion description. The technical difference is in the accounting of the σ_t term in Miura link. Note, that Goursat transformation and its generalizations open a way to the definition of ZRP for matrix spectral problems, arisen, e.g., in Pauli equation [14]. Application of the theory conventionally makes use of the Green function of a correspondent problem [2], and in the 2d case this needs special consideration [17]. The theory of ZRP perturbation of periodic finite-gap potential may be useful in models of solid state physics that would account point defects of a crystal.

References

[1] Albeverio S *et al*, 1988. Solvable Models in Quantum Mechanics (Springer-Verlag, New York)
[2] Szmytkowski R, and Szmytkowski Cz 1999 Scalar wave diffraction from zero-range scatterers, *Journal of Mathematical Chemistry* **26** 243
[3] Krajewska K, Kaminski J and Wódkiewicz 2010 Zero-range interaction in arbitrary dimensions and in the presence of external forces *Optics Communications* **283** 843
[4] Craig T 2013 Integrable Models in Statistical Physics and Their Universality, *Bernoulli Lecture*
[5] Fermi E 1934 Sopra lo spostamento per pressione delle righe elevante delle serie spettrali. *Nuovo Cimento* **11** 157
[6] Berezin F A and Faddeev L D 1961 Note on the Schrödinger equation with singular regularly potential *Dokl.AN SSSR*, **137** 1011
[7] Pavlov B S 1987 Extension theory and explicitly solvable models. *Usp. Mat. Nauk* **42** 99
[8] Demkov Yu and Ostrovskij V 1988, Zero-range potentials and their applications in Atomic Physics (Plenum Press, NY-London)

[9] Albeverio S, Fassari S and Rinaldi F 2013 A remarkable spectral feature of the Schrödinger Hamiltonian of the harmonic oscillator perturbed by an attractive delta-interaction centered at the origin: double degeneracy and level crossing. *J. Phys. A: Math. Theor.* **46** 385305

[10] Farrell A and van Zyl. B 2009 Universality of the energy spectrum for two interacting monically trapped ultra-cold atoms in one and two dimensions (*Preprint* arXiv:0911.3121v1 [cond-mat.quant-gas]

[11] Matveev V B 2000 Darboux transformations, covariance theorems and integrable systems, *L. D. Faddeev's Seminar on Mathematical Physics, Amer. Math. Soc. Transl.*, RI, Series 2, **201**, 179-209.

[12] Leble S 2005 On necessary covariance conditions of one-field Lax pair, *Theor. Math. Phys.* **144**, 985

[13] Leble S and Yalunin S 2002, A dressing of zero-range potentials and electron-molecule scattering problem at low energies, arXiv:quant-ph/0210133v1; Zero-range potentials in multi-channel diatomic molecule scattering *Phys. Lett. A* **306**, 35.

[14] Leble S 2012 Pseudopotentials via Moutard transformations and differential geometry. *Geometric Methods in Physics: XXX Workshop, (Bialowieza 2011) Trends in Mathematics*, ed P. Kielanowski S.T. Ali, A. Odzijewicz, M. Schlichenmaier, Th. Voronov-Eds.,(Springer-Verlag)

[15] Leble S and Yalunin S 2005 A dressing of zero-range potentials and electron–molecule scattering problem at low energies, *Phys. Lett. A* **339**, 83.

[16] Ponomarev D and Leble S 2011 Molecular zero-range potential method and its application to cyclic structure, (*Preprint* arXiv:1101.0439v1). Leble S and Ponomarev D 2011 Dressing of zero-range potentials into realistic molecular potentials of finite range *TASK Quaterly* **14** 1001

[17] 2005 de Llano M, Salazar A, and Sol M, Two-dimensional delta potential wells and condensed-matter physics, *Revista Mexicana de Fizica* **51** (6) 626

[18] Ince E 1958 *Ordinary differential equations*, (Dover)

[19] Fock V A 1978 *Fundamentals of quantum mechanics* (Mir publishing)

[20] Doktorov E and Leble S 2006 *Dressing method in mathematical physics* (Springer)

[21] Belokolos E, Bobenko A, Enolsky V, Its A and Matveev V B 1991 *Algebraic-Geometrical Approach to Nonlinear Evolution Equations* (Springer)

[22] Ganzha E 2000 On the approximation of solutions of some 2+1-dimensional integrable systems by Bäcklund transformations *Sibirsk. Mat. Zh.* **41:3** 541

Resonant dispersive Benney and Broer-Kaup systems in 2+1 dimensions

Jyh-Hao Lee[1] and Oktay K Pashaev[2]

[1] Institute of Mathematics, Academia Sinica, Taipei, 10617 Taiwan

[2] Department of Mathematics, Izmir Institute of Technology, Urla-Izmir, 35430 Turkey

E-mail: [1] `leejh@math.sinica.edu.tw`

E-mail: [2] `oktaypashaev@iyte.edu.tr`

Abstract. We represent the Benney system of dispersionless hydrodynamic equations as NLS type infinite system of equations with quantum potential. We show that negative dispersive deformation of this system is an integrable system including vector generalization of Resonant NLS and 2+1 dimensional nonlocal Resonant NLS. We obtain bilinear form and soliton solutions in these systems and find the resonant character of soliton interaction. Equivalent vector Broer-Kaup system and non-local 2+1 dimensional nonlocal Broer-Kaup equation are constructed.

1. Introduction

Resonant interaction of solitons in KP-II system has attracted much attention recently. This type of solitons can be constructed from the second and the third flow of the $SL(2, R)$ AKNS hierarchy [1]. The second flow represented by the Reaction-Diffusion (RD) equation has envelope soliton form as the Resonant NLS (RNLS) equation [2]. The resonant interaction of envelope solitons by creation and annihilation processes has been then studied in several nonlinear Schrödinger type models of RNLS type [3]. As was shown in [4], any envelope soliton equation admits the resonant counterpart. This process of recovering dispersion of integrable models from the dispersionless equations always has the resonant option. In the present paper we study dispersionless hydrodynamic system of Benney. Positive dispersive version of this system was considered by Zakharov [5], and soliton solutions was found in [6]. Here we are going to study the negative dispersion equation and resonant phenomena for it.

2. Benney system in the Schrödinger representation

The system of Benney equations

$$\frac{\partial u}{\partial t} + u\frac{\partial u}{\partial x} + \frac{\partial h}{\partial x} = 0, \tag{1}$$

$$\frac{\partial \eta}{\partial t} + \frac{\partial}{\partial x}(\eta u) = 0, \tag{2}$$

$$h = \int_0^1 \eta(x, \xi, t)\, d\xi, \tag{3}$$

as a descriptive of long shallow water waves has been reduced to an infinite system of two-dimensional hydrodynamic equations [5]. Here the horizontal velocity $u(x, \xi, t)$ is a function of layers in the z-direction, enumerated by parameter ξ $(0 < \xi < 1)$ and the fluid surface shape is $h(x,t)$. The Benney system (1), (2), (3) can be rewritten as nonlinear Schrodinger type wave equation by the Madelung transform to complex wave function

$$\psi(x,\xi,t) = \sqrt{\eta} e^{-i\int_{-\infty}^{x} u dx}. \tag{4}$$

Then we get the NLS type infinite system of equations with the quantum potential(QPNLS)

$$i\psi_t = \frac{1}{2}\psi_{xx} - h\psi - \frac{1}{2}\frac{|\psi|_{xx}}{|\psi|}\psi, \tag{5}$$

where

$$h = \int_0^1 |\psi(x,\xi,t)|^2 d\xi. \tag{6}$$

This Madelung transform is some kind of complex Cole-Hopf transformation

$$i\frac{\psi_x}{\psi} = u + iv, \tag{7}$$

where the imaginary part of complex velocity, $v = (\ln\sqrt{\eta})_x$, sometimes is called the quantum velocity. The Benney system rewritten in this wave form is the Hamiltonian system

$$\frac{\partial}{\partial t}\psi(x,\xi,t) = i\frac{\delta H}{\delta\bar{\psi}(x,\xi,t)}, \quad -\frac{\partial}{\partial t}\bar{\psi}(x,\xi,t) = i\frac{\delta H}{\delta\psi(x,\xi,t)}, \tag{8}$$

with

$$H = \frac{1}{2}\int_{-\infty}^{\infty} dx [\int_a^b d\xi\, (\bar{\psi}_x\psi_x - |\psi|_x|\psi|_x) + (\int_a^b d\xi\, |\psi|^2)^2]. \tag{9}$$

2.0.1. Discrete layers In the special case, when the flow is divided into N layers,

$$\eta(x,\xi,t) = \sum_{k=1}^{N}\eta_k(x,t)\delta(\xi - \xi_k) \Rightarrow h(x,t) = \sum_{k=1}^{N}\eta_k(x,t) \tag{10}$$

the Benney system reduces to a multicomponent (vector) hydrodynamic system. This system via Madelung transform $\psi_k = \sqrt{\eta_k} e^{-i\int_{-\infty}^{x} u_k dx}$ to N-complex wave functions $\psi(x,\xi_k,t) \equiv \psi_k(x,t)$, $k = 1,2,...,N$, gives the vector NLS equation with quantum potential

$$i\psi_t^a = \frac{1}{2}\psi_{xx}^a - \left(\sum_{b=1}^{n}|\psi^b|^2\right)\psi^a - \frac{1}{2}\frac{|\psi^a|_{xx}}{|\psi^a|}\psi^a. \tag{11}$$

2.0.2. Dispersive deformation and RNLS It is noted that in (5) the dispersion term and the quantum potential term have equal but opposite sign strength. If this condition is broken then we have more general system

$$i\psi_t = \frac{1}{2}\psi_{xx} - h\psi - \frac{1}{2}(1-\hbar^2)\frac{|\psi|_{xx}}{|\psi|}\psi, \tag{12}$$

$$h = \int_a^b |\psi(x,\xi,t)|^2 d\xi, \tag{13}$$

where $\hbar$ is deformation parameter. This system is a nonlocal 2+1-dimensional generalization of the Resonant NLS equation [2]. For given ψ in (4) we can relate new wave function

$$\chi(x,\xi,t) = \sqrt{\eta}e^{-\frac{i}{\hbar}\int_{-\infty}^{x} udx}, \tag{14}$$

so that the system (12),(13) becomes the infinite system of NLS equations in 2+1 dimensions with non-local self-interaction considered in [5]

$$i\hbar\chi_t = -\frac{\hbar^2}{2}\chi_{xx} - h\chi, \qquad h = \int_a^b |\chi(x,\xi,t)|^2 d\xi. \tag{15}$$

For discrete set of parameter $\xi_1, \xi_2, ..., \xi_N$ the wave functions $\chi^k(x,t) = \chi(x,\xi_k,t)$ satisfy the vector $U(N)$ NLS equation, and for $N \to \infty$, the infinite component vector $U(\infty)$ NLS. Recently in [6] this equation has been considered as nonlocal NLS in 2+1 dimensions with a domain of integration as a whole real line, $-\infty < \xi < \infty$, and the bilinear method of Hirota was applied to solve it for N - soliton solutions localized in plane.

3. Negative dispersion and RNLS

Dispersive deformation described by system (12), (13) is not unique. Indeed the semiclassical expansion of a solution is going in according to parameter $\hbar^2$. Analyticity of this expansion implies that semiclassical series should be analytic in full disk in complex plane with radius $\hbar$. This poses intriguing question of behavior of this expansion inside of the disk. By replacing $\hbar \to i\hbar$ then we get

$$i\psi_t = \frac{1}{2}\psi_{xx} - h\psi - \frac{1}{2}(1+\hbar^2)\frac{|\psi|_{xx}}{|\psi|}\psi, \tag{16}$$

where

$$h = \int_a^b |\psi(x,\xi,t)|^2 d\xi. \tag{17}$$

Both systems (12), (13) and (16), (17) correspond to dispersive Benney system with positive and negative dispersions correspondingly:

$$\frac{\partial u}{\partial t} + u\frac{\partial u}{\partial x} + \frac{\partial h}{\partial x} = \pm\frac{\hbar^2}{2}\left(\frac{(\sqrt{\eta})_{xx}}{\sqrt{\eta}}\right)_x, \tag{18}$$

$$\frac{\partial \eta}{\partial t} + \frac{\partial}{\partial x}(\eta u) = 0, \tag{19}$$

$$h = \int_a^b \eta(x,\xi,t)\, d\xi. \tag{20}$$

Positive dispersion case was treated in details in [5]. Here we are going to study the negative dispersion case. We have several motivations for this.

The negative dispersion hydrodynamic system (18), (19), (20) for particular case $\eta(x,\xi,t) = \eta(x,t)\delta(\xi-\xi_0)$ of one layer, and with $h = \eta$, describes propagation of long magneto-acoustic waves in cold plasma, moving across the magnetic field, in shallow water approximation [9], [10]. Parameter $\hbar = 2\beta$ then is rescaling parameter for this approximation. Reaction-diffusion form of this system has been derived from the Jackiw-Teitelboim model of low dimensional gravity [11], [12], [2]. It has subsequently been derived in plasma physics and an auto-Bäcklund transformation was constructed in that context in [8]. The resonant NLS equation as a theoretical capillarity model was discussed in [7]. In addition, it turns out that a wide class of NLS equations with underlying Hamiltonian structure may be reduced to consideration of the resonant NLS equation [3].

4. Nonlocal Reaction-Diffusion System

The negative dispersion system (16), (17), can not be reduced to NLS equation and solvability of the model is not evident. However, if instead of Madelung transform (4), we introduce the couple of real functions,

$$e^+(x,\xi,t) = \sqrt{\eta}\, e^{-\int^x u dx'}, \quad e^-(x,\xi,t) = \sqrt{\eta}\, e^{+\int^x u dx'} \tag{21}$$

then the system (18), (19), (20) for negative dispersion case can be rewritten as the couple of nonlocal reaction-diffusion equations

$$e_t^+ = \frac{1}{2} e_{xx}^+ + h e^+, \tag{22}$$

$$-e_t^- = \frac{1}{2} e_{xx}^- + h e^-, \tag{23}$$

$$h = \int_a^b e^+ e^- \, d\xi. \tag{24}$$

When integration domain is given by the set of discrete points, for $e^{\pm}(x, \xi_a, t) \equiv e^{\pm(a)}(x,t)$ we are getting the vector reaction-diffusion system

$$e_t^{+(a)} = \frac{1}{2} e_{xx}^{+(a)} + \left(\sum_{b=1}^{n} e^{+(b)} e^{-(b)} \right) e^{+(a)}, \tag{25}$$

$$-e_t^{-(a)} = \frac{1}{2} e_{xx}^{-(a)} + \left(\sum_{b=1}^{n} e^{+(b)} e^{-(b)} \right) e^{-(a)}. \tag{26}$$

4.1. Linear Problem

The linear problem in Dirac's notations is

$$\Phi_x = U\Phi, \quad \Phi_t = V\Phi, \tag{27}$$

where

$$\Phi = \begin{pmatrix} \varphi_0 \\ |\varphi> \end{pmatrix}, \tag{28}$$

$$U = \begin{pmatrix} -i\lambda & <e^-| \\ |e^+> & i\lambda I \end{pmatrix}, \tag{29}$$

$$V = \begin{pmatrix} -i\lambda^2 - \frac{i}{2} <e^-|e^+> & \lambda <e^-| + \frac{i}{2} <e_x^-| \\ \lambda|e^+> - \frac{i}{2}|e_x^+> & i\lambda^2 I + \frac{i}{2}|e^+><e^-| \end{pmatrix}. \tag{30}$$

Compatibility condition for this linear system is equivalent to general vector form of the Reaction-Diffusion (RD) system of equations

$$|e_t^+> = \frac{1}{2}|e_{xx}^+> + <e^-|e^+> |e^+>, \tag{31}$$

$$-<e_t^-| = \frac{1}{2} <e_{xx}^-| + <e^-|e^+><e^-|. \tag{32}$$

Here most general real symmetric bilinear form is mixture of discrete and continuous spectrum of ξ:

$$< e^-|e^+ >= \sum_n e_n^+(x,t)e_n^-(x,t) + \int_a^b e^+(x,\xi,t)e^-(x,\xi,t)\, d\xi,$$

corresponding to the unit identity

$$\sum_n |\xi_n >< \xi_n| + \int_a^b |\xi > \, d\xi \, < \xi| = I,$$

and giving the vector discrete-continuous system of RD equations.

4.2. Conserved quantities

The mass distribution density along ξ direction, $\mu(\xi) = dM/d\xi$, so that $dM = \mu(\xi)d\xi$, is conserved quantity for arbitrary ξ:

$$\mu(\xi) = \int_{-\infty}^{\infty} dx\, e^+e^-. \tag{33}$$

Hence the total mass

$$M = \int_a^b d\xi\, \mu(\xi) = \int_a^b d\xi \int_{-\infty}^{\infty} dx\, e^+e^- \tag{34}$$

is conserved as well. Conserved quantities are also the total momentum and the Hamiltonian function respectively

$$P = \int_a^b d\xi \int_{-\infty}^{\infty} dx\, (e^+e_x^- - e_x^+e^-), \tag{35}$$

$$H = \frac{1}{2}\int_{-\infty}^{\infty} dx \left[\int_a^b d\xi e_x^+ e_x^- - \left(\int_a^b d\xi e^+ e^-\right)^2\right]. \tag{36}$$

4.3. Hamiltonian structure

$$e_t^+ = -\frac{\delta H}{\delta e^-}, \quad e_t^- = +\frac{\delta H}{\delta e^+}, \tag{37}$$

or by Poisson bracket

$$\{A, B\} = \int_{-\infty}^{\infty}\int_a^b \left(\frac{\delta A}{\delta e^+}\frac{\delta B}{\delta e^-} - \frac{\delta A}{\delta e^-}\frac{\delta B}{\delta e^+}\right), \tag{38}$$

$$e_t^+ = \{H, e^+\}, \quad e_t^- = \{H, e^-\}. \tag{39}$$

5. Bilinear Form

For discrete spectrum of ξ we have the vector RD system

$$e_{n_t}^+ = e_{n\,xx}^+ + \left(\sum_{m=1}^{\infty} e_m^+ e_m^-\right) e_n^+, \tag{40}$$

$$-e^-{}_{nt} = e_{n\,xx}^- + \left(\sum_{m=1}^{\infty} e_m^+ e_m^-\right) e_n^-, \tag{41}$$

where $< \xi_n|e^+ >= e_n^+(x,t)$, $< e^-|\xi_n >= e_n^-(x,t)$. For continuous case we have the nonlocal reaction-diffusion system

$$e_t^+ = e_{xx}^+ + he^+, \tag{42}$$
$$-e_t^- = e_{xx}^- + he^-, \tag{43}$$
$$h = \int_a^b e^+e^- d\xi. \tag{44}$$

were for convenience to compare with RNLS we drop coefficient $1/2$ in front of dispersion. Here $< \xi|e^+ >= e^+(x,\xi,t)$, $< e^-|\xi >= e^-(x,\xi,t)$. In addition, to treat most general situation we consider integration domain between points a and b, so that in particular case $a = 0$, $b = 1$ we recover the original structure of Benney system. Moreover, for improper integral $a \to -\infty$, $b \to \infty$, we have 2+1 dimensional nonlocal problem similar to the one considered in [6].

We represent two real functions $e^\pm(x,\xi,t)$ in terms of three real functions

$$e^+(x,\xi,t) = \frac{g^+(x,\xi,t)}{f(x,t)}, \quad e^-(x,\xi,t) = \frac{g^-(x,\xi,t)}{f(x,t)}. \tag{45}$$

Then substituting it to the above system (42)-(44) gives bilinear equations

$$(D_t - D_x^2)(g^+ \cdot f) = 0, \tag{46}$$
$$(D_t + D_x^2)(g^- \cdot f) = 0, \tag{47}$$
$$D_x^2(f \cdot f) = \int_a^b g^+g^- \, d\xi. \tag{48}$$

We have following particular cases of this bilinear form. When g^+ and g^- are independent of ξ the system reduces to the one considered in [2]. If integration region contains only the discrete set of points we have bilinear form for the vector reaction-diffusion system (25), (26)

$$(D_t - D_x^2)(g^{+(a)} \cdot f) = 0, \tag{49}$$
$$(D_t + D_x^2)(g^{-(a)} \cdot f) = 0, \tag{50}$$
$$D_x^2(f \cdot f) = \sum_{a=1}^n g^{+(a)}g^{-(a)}. \tag{51}$$

where $a = 1, ..., n$ and

$$e^{+(a)}(x,t) = \frac{g^{+(a)}(x,t)}{f(x,t)}, \quad e^{-(a)}(x,t) = \frac{g^{-(a)}(x,t)}{f(x,t)}. \tag{52}$$

From (48) by dividing on f^2 we have identity

$$\int_a^b e^+(x,\xi,t)\, e^-(x,\xi,t)\, d\xi = 2(\ln f)_{xx}. \tag{53}$$

For the scalar case it gives

$$e^+(x,t)\, e^-(x,t) = 2(\ln f)_{xx}, \tag{54}$$

and for the vector case

$$\sum_{a=1}^n e^{+(a)}(x,t)\, e^{-(a)}(x,t) = 2(\ln f)_{xx}. \tag{55}$$

5.1. One Soliton Solution

By Hirota perturbation $g^{\pm} = \epsilon g_1^{\pm} + \epsilon^3 g_3^{\pm} + ...$, $f = 1 + \epsilon^2 f_2 + \epsilon^4 f_4 + ...$ we get one dissipaton solution

$$g_1^{\pm} = a_1^{\pm}(\xi)\, e^{\eta_1^{\pm}}, \quad f = 1 + \frac{e^{\eta_1^+ + \eta_1^-}}{(k_1^+ + k_1^-)^2} \int_a^b a_1^+ a_1^- \, d\xi, \tag{56}$$

where $\eta_1^{\pm} = k_1^{\pm} x \pm (k_1^{\pm})^2 t + \eta_1^{\pm(0)}$. We notice that constants $\eta_1^{\pm(0)}$ can be absorbed to functions $a_1^{\pm}(\xi)$, but for comparison our results with RNLS case it would be convenient to use it in the above form. Then particular choice of $a_1^{\pm}(\xi) = 1$ gives reduction to the RNLS case. For one dissipaton solution we have

$$e^+ = \frac{a_1^+(\xi)\, e^{\eta_1^+}}{1 + \frac{e^{\eta_1^+ + \eta_1^-}}{(k_1^+ + k_1^-)^2} \int_a^b a_1^+ a_1^- \, d\xi}, \quad e^- = \frac{a_1^-(\xi)\, e^{\eta_1^-}}{1 + \frac{e^{\eta_1^+ + \eta_1^-}}{(k_1^+ + k_1^-)^2} \int_a^b a_1^+ a_1^- \, d\xi}, \tag{57}$$

with

$$e^+ e^- = \frac{a_1^+(\xi) a_1^-(\xi)}{4\gamma_{11}} \frac{(k_1^+ + k_1^-)^2}{\cosh^2 \frac{\eta_1^+ + \eta_1^- + \phi_{11} + \ln \gamma_{11}}{2}}, \tag{58}$$

where $\gamma_{11} = \int_a^b a_1^+(\xi) a_1^-(\xi) d\xi$ and $e^{\phi_{11}} = (k_1^+ + k_1^-)^{-2}$. In terms of amplitude and velocity variables $k = \frac{k_1^+ + k_1^-}{2}$, $v = k_1^- - k_1^+$, we have soliton

$$e^+ e^- = \frac{a_1^+(\xi) a_1^-(\xi)}{\gamma_{11}} \frac{k^2}{\cosh^2 k(x - vt - x_0)}, \tag{59}$$

where $x_0 = -(\phi_{11} + \ln \gamma_{11})/2k$.

5.1.1. Scalar Dissipaton

In particular case when $a_1^{\pm}$ are independent of ξ, we can choose $a_1^{\pm} = 1 \to \gamma_{11} = 1$, and we get one soliton solution of RNLS [2].

5.1.2. Vector Dissipaton

For N discrete values of parameter ξ: $\xi_1, ..., \xi_N$ we have solution

$$g_1^{\pm(a)}(x,t) = a_1^{\pm(a)} e^{\eta_1^{\pm}}, \quad (a = 1, ..., N), \tag{60}$$

$$f = 1 + \frac{e^{\eta_1^+ + \eta_1^-}}{(k_1^+ + k_1^-)^2} \sum_{a=1}^{N} a_1^{+(a)} a_1^{-(a)}, \tag{61}$$

where $a_1^{\pm(a)} \equiv a_1^{\pm}(\xi_a)$, $(a = 1, ..., n)$. It gives the vector dissipaton

$$e^{\pm(a)} = \frac{a_1^{\pm(a)}\, e^{\eta_1^{\pm}}}{1 + \frac{e^{\eta_1^+ + \eta_1^-}}{(k_1^+ + k_1^-)^2} \sum_{b=1}^{N} a_1^{+(b)} a_1^{-(b)}}, \tag{62}$$

with soliton shape of the partial mass density

$$e^{+(a)} e^{-(a)} = \frac{a_1^{+(a)} a_1^{-(a)}}{\gamma_{11}} \frac{k^2}{\cosh^2 k(x - vt - x_0)}, \tag{63}$$

where $x_0 = -(\phi_{11} + \ln \sum_{a=1}^{n} a_1^{+(a)} a_1^{-(a)})/2k$, $\gamma_{11} \equiv \sum_{a=1}^{n} a_1^{+(a)} a_1^{-(a)}$ and the soliton shape of the total mass density

$$\sum_{a=1}^{n} e^{+(a)} e^{-(a)} = \frac{k^2}{\cosh^2 k(x - vt - x_0)}. \tag{64}$$

5.2. Two Soliton Solution

Continuing Hirota's perturbation we have two dissipaton solution

$$
\begin{aligned}
g^+ &= a_1^+(\xi)e^{\eta_1^+} + a_2^+(\xi)e^{\eta_2^+} \\
&+(k_1^+ - k_2^+)\left[\frac{a_1^+(\xi)\gamma_{21}}{k_{11}^{+-}(k_{21}^{+-})^2} - \frac{a_2^+(\xi)\gamma_{11}}{k_{21}^{+-}(k_{11}^{+-})^2}\right]e^{\eta_1^+ + \eta_1^- + \eta_2^+} \\
&+(k_1^+ - k_2^+)\left[\frac{a_1^+(\xi)\gamma_{22}}{k_{12}^{+-}(k_{22}^{+-})^2} - \frac{a_2^+(\xi)\gamma_{12}}{k_{22}^{+-}(k_{12}^{+-})^2}\right]e^{\eta_2^+ + \eta_2^- + \eta_1^+},
\end{aligned}
\tag{65}
$$

$$
\begin{aligned}
g^- &= a_1^-(\xi)e^{\eta_1^-} + a_2^-(\xi)e^{\eta_2^-} \\
&+(k_1^- - k_2^-)\left[\frac{a_1^-(\xi)\gamma_{12}}{k_{11}^{+-}(k_{12}^{+-})^2} - \frac{a_2^-(\xi)\gamma_{11}}{k_{12}^{+-}(k_{11}^{+-})^2}\right]e^{\eta_1^+ + \eta_1^- + \eta_2^-} \\
&+(k_1^- - k_2^-)\left[\frac{a_1^-(\xi)\gamma_{22}}{k_{21}^{+-}(k_{22}^{+-})^2} - \frac{a_2^-(\xi)\gamma_{21}}{k_{22}^{+-}(k_{21}^{+-})^2}\right]e^{\eta_2^+ + \eta_2^- + \eta_1^-},
\end{aligned}
\tag{66}
$$

and

$$
\begin{aligned}
f &= 1 + \gamma_{11}\frac{e^{\eta_1^+ + \eta_1^-}}{(k_{11}^{+-})^2} + \gamma_{12}\frac{e^{\eta_1^+ + \eta_2^-}}{(k_{12}^{+-})^2} + \gamma_{21}\frac{e^{\eta_2^+ + \eta_1^-}}{(k_{21}^{+-})^2} + \gamma_{22}\frac{e^{\eta_2^+ + \eta_2^-}}{(k_{22}^{+-})^2} \\
&+ \frac{(k_{11}^{+-}k_{22}^{+-} - k_{12}^{+-}k_{21}^{+-})(\gamma_{12}\gamma_{21}k_{11}^{+-}k_{22}^{+-} - \gamma_{11}\gamma_{22}k_{12}^{+-}k_{21}^{+-})}{(k_{11}^{+-}k_{12}^{+-}k_{21}^{+-}k_{22}^{+-})^2}e^{\eta_1^+ + \eta_1^- + \eta_2^+ + \eta_2^-},
\end{aligned}
\tag{67}
$$

where $\eta_i^{\pm} = k_i^{\pm}x \pm (k_i^{\pm})^2 t + \eta_i^{\pm(0)}$, and constants $k_{ij}^{+-} \equiv k_i^+ + k_j^-$, $i,j = 1,2$ and $\gamma_{ij} \equiv \int_a^b a_i^+(\xi)a_j^-(\xi)\,d\xi$.

If $\frac{\partial}{\partial\xi}e^{\pm} = 0$, we have reduction to the reaction-diffusion system and the above two-dissipaton solution reduces to the one obtained in [2]. But as was shown in that paper, interaction of two dissipatons has resonant character. This is why now we can describe the resonant soliton dynamics for the 2+1 dimensional model in the continuous case, and for vector RNLS in the discrete case. For vector reduction in above formulas we substitute $a_i^{\pm}(\xi_a) = a_i^{\pm(a)}$, $(i = 1,2)$, $(a = 1,...,n)$, and $\gamma_{ij} = \sum_{a=1}^{n} a_i^{+(a)}a_j^{-(a)}$.

6. Soliton as non-relativistic particle

For one soliton solution (57) we have the mass density

$$
\mu(\xi) = \frac{a_1^+(\xi)a_1^-(\xi)}{\gamma_{11}}2\,|k| \tag{68}
$$

and the momentum density

$$
\pi(\xi) = \mu(\xi)\,v, \tag{69}
$$

where $\pi(\xi) = dP/d\xi$ and $dP = \pi(\xi)d\xi$. The total mass and total momentum of the soliton are

$$
M = 2\,|k|, \quad P = M\,v. \tag{70}
$$

Then for energy $E = -4H$ of one soliton we have

$$
E = \frac{Mv^2}{2} + \frac{1}{6}M^3. \tag{71}
$$

6.0.1. Resonance conditions The positive rest energy of soliton $E_0 = \frac{1}{6}M^3$, is the reason why the resonant soliton behavior takes place for this system. Decay of a soliton at rest on the pair of solitons with positive energies is allowed only if the rest energy is positive. Due to inequality $\frac{1}{6}M^3 = \frac{1}{6}(M_1+M_2)^3 > \frac{1}{6}M_1^3 + \frac{1}{6}M_2^3$ it allows creation of two solitons. The first three conservation laws for the decaying process are

$$M = M_1 + M_2, \quad P = P_1 + P_2, \quad E = E_1 + E_2. \tag{72}$$

By substituting asymptotic form of created solitons, we find that the defect of mass vanishes $\Delta M = M - (M_1 + M_2) = 0$ and velocity of the decaying soliton is just the center-of-mass velocity $v = (M_1 v_1 + M_2 v_2)/(M_1 + M_2)$. From energy conservation law we find resonance constraint on velocities as

$$|v_1 - v_2| = M_1 + M_2. \tag{73}$$

As an example, let us illustrate decay of soliton in a rest to two solitons. Then $v_2 = -v_1 M_1/M_2$ and $v_1^2 = M_2^2$, $v_2^2 = M_1^2$. In particular case of equal solitons $|v_1| = |v_2|$, $M_1 = M_2$ it gives $k_1^+ = k_2^+$ and $k_1^- = k_2^-$. It is easy to see that under this constraint the last term in (67) vanishes and two soliton solution reduces to the one soliton.

7. Distribution in ξ direction

Functions $a_{1,2}^{\pm}(\xi)$ are arbitrary integrable functions determining the mass distribution density $\mu(\xi)$ along the ξ direction according to (33). By choosing these functions we have different situations. If for one soliton solution (68) we consider the discrete set of mass distribution

$$\mu(\xi) = \sum_{k=1}^{N} \mu_k \delta(\xi - \xi_k), \tag{74}$$

then one soliton solution becomes vector soliton. We can also describe continuous and localized distribution of soliton mass by considering delta sequence $\delta_n(\xi)$, $\lim_{n\to\infty} \delta_n(\xi) = \delta(\xi)$:

$$\int_a^b \delta_n(\xi)\, d\xi = 1. \tag{75}$$

Below we list some examples (with $\alpha = const.$, $(a, b) = (-\infty, \infty)$):

$$a^{\pm}(\xi) = \sqrt{\gamma_{11}\frac{n}{2}}\frac{e^{\pm\alpha}}{\cosh n\xi} \to \delta_n(\xi) = \frac{n}{2\cosh^2 n\xi}; \tag{76}$$

$$a_1^{\pm} = \sqrt{\gamma_{11}\frac{n}{\pi}}\frac{1}{\sqrt{1+n^2\xi^2}} \to \delta_n(\xi) = \frac{n}{\pi}\frac{1}{1+n^2\xi^2}; \tag{77}$$

$$a_1^{\pm} = \sqrt{\gamma_{11}\frac{n}{\sqrt{\pi}}}\, e^{-n^2\xi^2/2} \to \delta_n(\xi) = \frac{n}{\sqrt{\pi}} e^{-n^2\xi^2}. \tag{78}$$

For the set of solutions localized in ξ direction we can take also superposition of these distributions. Since the mass density $\mu(\xi)$ is conserved quantity, the mass distribution will not change with time. However, as we have seen above, only total momentum and energy are conserved. So it opens possibility to have momentum and energy exchange between localized distributions in ξ direction. For discrete distribution it would leads to the exchange between vector soliton components. Combination of this phenomena with resonant character of soliton interaction in our model promises quite interesting structure and is under investigation.

8. Non-Madelung Fluid Form

In addition to the Madelung fluid representation (18), (19), (20) the system (42)-(44) can be rewritten in terms of another hydrodynamical variables. The density of fluid is the same

$$\rho(x,\xi,t) = e^+e^-, \tag{79}$$

but the velocity field called the drift velocity is derived by the Cole-Hopf type formula

$$v(x,\xi,t) = \frac{e^+_x}{e^+}. \tag{80}$$

In terms of these variables we have nonlocal 2+1 dimensional version of the Broer-Kaup system

$$v_t = (v_x + v^2 + h)_x, \tag{81}$$

$$\rho_t + \rho_{xx} = (\rho v)_x, \tag{82}$$

$$h = \int_a^b \rho\, d\xi. \tag{83}$$

Dispersionless limit of this hydrodynamic system is the Benney system (1), (2), (3). Moreover, in particular case when ρ is independent of ξ , so that $h(x,t) = \rho(x,t)$, the system reduces to the Broer-Kaup system

$$v_t = (v_x + v^2 + \rho)_x, \tag{84}$$

$$\rho_t + \rho_{xx} = (\rho v)_x. \tag{85}$$

For one soliton solution, velocity

$$v = (\ln e^+)_x = \frac{k_1^+ - k_1^- e^{\eta_1^+ + \eta_1^- + \phi_{11} + \ln\gamma_{11}}}{1 + e^{\eta_1^+ + \eta_1^- + \phi_{11} + \ln\gamma_{11}}} \tag{86}$$

is shock-soliton

$$v = -\frac{v}{2} - k \tanh k(x - vt - x_0), \tag{87}$$

with soliton profile for the density

$$\rho(x,\xi,t) = \frac{a_1^+(\xi)a_1^-(\xi)}{\gamma_{11}} \frac{k^2}{\cosh^2 k(x - vt - x_0)}. \tag{88}$$

By choosing proper distribution in ξ direction we can localize this soliton in (x,ξ) plane.

8.1. Acknowledgments

This work was supported by Institute of Mathematics, Academia Sinica and Izmir Institute of Technology. O.K. Pashaev was supported also by TUBITAK Grant (Project TBAG No. 110T679).

References

[1] Pashaev O K and Francisco M L Y 2005 *Theor Math Phys* **144** 1022-29
[2] Pashaev O K and Lee J H 2002 *Mod. Phys.Lett.* A **17** 1601-19
[3] Pashaev O K, Lee J H and Rogers C, 2008 *J. Phys. A. Math Theor* **41** 452001-9
[4] Pashaev O K 2012 *Theor Math Phys* **172** 1147-59
[5] Zakharov V E 1981 *Physica* D **3** 193-202
[6] Maruno K I and Y. Ohta Y 2008 *Phys. Lett.* A **372** 4446
[7] Rogers C and Schief W K 1999 *Nuovo Cimento* **114** 1409-12
[8] Lee J H, Pashaev O K, Rogers C and Schief W K, 2007 *J. Plasma Physics* **73** 257-72
[9] Gurevich A V and Meshcherkin A P 1984 *Soviet Phys. JETP* **60** 73240
[10] El G A, Khodorovskii V V and Tyurina A V 2004 *Phys. Lett.* A **333** 33440
[11] Pashaev O K 1997 *Nucl. Phys. B Proc. Suppl.* **57** 33841
[12] Martina L, Pashaev O K and Soliani G 1997 *Class. Quantum Grav.* **14** 317986

A macroscopic system with undamped periodic compressional oscillations

F Leyvraz[1,2] and F Calogero[3,4]

[1] Instituto de Ciencias Físicas, UNAM, Cuernavaca, Mexico
[2]Centro Internacional de Ciencias, Cuernavaca, Mexico
[3] Dipartimento di Fisica, Università di Roma "La Sapienza", Italy
[4] Istituto Nazionale di Fisica Nucleare, Sezione di Roma, Italy

E-mail: leyvraz@fis.unam.mx,francesco.calogero@roma1.infn.it, francesco.calogero@uniroma1.it

Abstract. A system consisting of an arbitrary number of particles of equal masses interacting via an arbitrary potential of homogeneity degree -2 and confined by an isotropic harmonic potential has the property of sustaining undamped isochronous compressional oscillations, as has been shown earlier. In this paper, we review and generalize this finding, and also the concept of thermodynamic equilibrium for such systems. It turns out that these compressional oscillations are *adiabatic*, and that correspondingly, the temperature varies when the size of the system does (in the specific case stated above, this dependence is one of inverse proportionality). It is also shown that some of these results extend to the quantal case.

1. Introduction

Oscillations arise in a large variety of physical systems. When the systems contain few particles or are in some other way highly idealized, it may happen that the oscillation does not decay. However, whenever the system contains a macroscopic number of particles, and is capable of displaying irreversible behaviour, one expects that any macroscopic oscillation arising in such a system will eventually be damped. In the following, we show analytically how systems of an arbitrary (finite) number of particles obeying classical Hamiltonian dynamics may display an *undamped* finite amplitude compressional oscillation. By this we mean that the oscillation amplitude never decays, even though its amplitude is arbitrary. Thus the size of such a system varies with time. As we shall show in greater detail, so do also various thermodynamic variables characterizing it, such as temperature and pressure.

This leads us to the most peculiar feature of these systems: they never reach thermal equilibrium. They are thus a counterexample to the generally accepted principle that all (sufficiently chaotic) systems must somehow relax to equilibrium, see for example [1]. Note that these systems are *not* integrable. Rather, their dynamics separate in a chaotic part, describing the motion of all degrees of freedom except two. These two are collective in nature and satisfy an equation of motion which does not couple to the other degrees of freedom. Approach to equilibrium has been extensively studied in recent years, see for example [2, 3]. Presumably, our systems satisfy the "relaxation theorem" proved there, if one previously factors out the oscillatory motion, as discussed in Section 3.

Let us tersely describe in informal terms the class of systems we consider. A quite particular case is described as follows: an arbitrary number of equal-mass particles moving according to classical (i.e., nonrelativistic) dynamics, are confined by a common, external isotropic harmonic potential centered at the origin (representing a container with "soft" walls). These particles further interact among themselves via conservative repulsive forces scaling as the inverse cube of distance (but which are otherwise *arbitrary*: such as, for instance, pairwise forces with *arbitrary* coupling constants). The main peculiarity of this system, first pointed out in [4] (for similar results obtained by a related approach, see [5]) is that the dynamical and thermodynamical behaviors of this system display the peculiar features mentioned above: in particular, for generic initial data the radius of gyration of the system oscillates isochronously, that is, with a fixed frequency independent of the initial data. As we shall see, this oscillation is *adiabatic*, and implies a corresponding variation of all other thermodynamic variables. Perhaps the most remarkable aspect of this finding is its validity for systems living in spaces of *arbitrary* dimension d (including, of course, $d = 3$): indeed, for $d \geq 2$ the many-body system under consideration is *not* integrable, as is readily seen by numerical simulations, which show clearly a non-vanishing maximal Lyapunov exponent. This system is also generally *not* integrable in the $d = 1$ case, also in the case with two-body forces unless these are restricted to feature the same coupling constant for every particle pair [6, 7]. One therefore expects that such general systems should evolve towards thermodynamic equilibrium. We shall see that there is, in fact, a sense in which this presumably happens, and numerics presented in [8, 9] certainly seems to confirm this.

It should be pointed out that the results stated above were published by us in [10] without realizing that these same results had been shown a decade earlier in [4, 8] by D. Lynden-Bell and R. M. Lynden-Bell, as was kindly pointed out to us by M. Kiessling. We have since submitted another paper recognizing this fact [11]. Indeed, the underlying property of systems having homogeneous interactions of degree -2 relies on a certain kind of dilation symmetry presumably first pointed out in [5].

In the following Section 2 the many-body problem treated in this paper is described in the context of classical (Hamiltonian and Newtonian) dynamics and the notation used throughout is introduced. In Section 2.2 we describe the main results of this paper: we obtain a new derivation of the results obtained in [4, 8] which lead to a rather extensive generalization of their claims to a broader class of Hamiltonians than those described in the preceding paragraph. In particular, an extension to time-dependent systems is discussed, which permits to analyse methods for acting externally on such systems. In Section 3 the thermodynamics of these systems is described and several of its peculiar features are discussed. In Section 4 we discuss the systems' behaviour with respect to the second law of thermodynamics, making use of the time-dependent Hamiltonians discussed earlier. In Section 5 we shortly discuss the extension of our results to quantum mechanics.

2. The many-body problem: notation and main results

2.1. Preliminaries and notation

The fundamental building blocks of the Hamiltonians we shall be considering are the following

$$Q \equiv \sum_{n=1}^{N} r_n^2 = \sum_{n=1}^{N}\sum_{j=1}^{d} x_{nj}^2 \,, \tag{1a}$$

$$D \equiv D\left(\underline{\vec{r}}, \underline{\vec{p}}\right) = \sum_{n=1}^{N} \vec{r}_n \cdot \vec{p}_n \,, \tag{1b}$$

$$H_0(\underline{\vec{r}}; \underline{\vec{p}}) = \frac{1}{2}\sum_{n=1}^{N} p_n^2 + V^{(-2)}\left(\underline{\vec{r}}\right) \,. \tag{1c}$$

Notation 2.1. Here and hereafter the (arbitrary) positive integer N denotes the number of (point) particles; the index n (and likewise m, see below) ranges generally from 1 to N; the "coordinates" d-vector $\vec{r}_n \equiv (x_{n1}, ..., x_{nd})$ identifies the position in d-dimensional space (with d an arbitrary positive integer) of the n-th particle and depends of course on the "time" variable t, $\vec{r}_n \equiv \vec{r}_n(t)$ (but often the time dependence of the various quantities is not explicitly displayed); the notation $\underline{\vec{r}}$ denotes the $N \times d$-matrix with elements x_{nj} where $n = 1, ..., N$ and $j = 1, ..., d$; likewise for the "momentum" (or "velocity") N-vector $\vec{p}_n \equiv (p_{n1}, ..., p_{nd}) \equiv \vec{p}_n(t)$ and the $N \times d$-matrix $\underline{\vec{p}} \equiv \underline{\vec{p}}(t)$; of course $p_n^2 = \vec{p}_n \cdot \vec{p}_n \equiv \sum_{j=1}^{d} p_{nj}^2$ and likewise $r_n^2 = \vec{r}_n \cdot \vec{r}_n \equiv \sum_{j=1}^{d} x_{nj}^2$; superimposed dots indicate time-differentiations; and we trust the rest of the notation to be self-evident (note that for simplicity we set all particle masses to *unity*). Throughout this paper we work of course with real functions of real variables.

We always assume the potential $V^{(-2)}(\underline{\vec{r}})$ to be *nonnegative* for all values of its Nd arguments,

$$V^{(-2)}(\underline{\vec{r}}) \geq 0 \; , \tag{2}$$

and to scale as the *inverse-square* of its arguments,

$$V^{(-2)}(\alpha \underline{\vec{r}}) = \alpha^{-2} V^{(-2)}(\underline{\vec{r}}) \; . \tag{3}$$

This scaling entails that the forces and the potential energy diverge when the particle positions all tend to zero. Moreover, whenever the potential energy has a singularity, it diverges to *positive* infinity; hence, for any initial conditions, the moving particles are kept away by a finite amount from any singularity. Hence, as long as attention is restricted to the real domain (corresponding to the physical situation), *all* solutions $\underline{\vec{r}}(t)$ and $\underline{\vec{p}}(t)$ of the Hamiltonian and Newtonian equations written above are uniquely defined and finite for *all* time.

A subcase of the class of potentials we just described is given as follows

$$V^{(-2)}(\underline{\vec{r}}) = \frac{1}{4} \sum_{n,m=1; m \neq n}^{N} \frac{g_{nm}^2}{r_{nm}^2} \; , \tag{4a}$$

entailing

$$\vec{F}_n^{(-3)}(\underline{\vec{r}}) = \sum_{m=1; \, m \neq n}^{N} \frac{g_{nm}^2 \vec{r}_{nm}}{r_{nm}^4} \; . \tag{4b}$$

Of course in these formulas

$$\vec{r}_{nm} \equiv \vec{r}_n - \vec{r}_m \; , \qquad r_{nm}^2 \equiv \vec{r}_{nm} \cdot \vec{r}_{nm} = \sum_{j=1}^{d} (x_{nj} - x_{mj})^2 \; . \tag{4c}$$

This case is mentioned here because of its physically more interesting character. Note that the "coupling constants" g_{nm}^2 may be different (for different pairs of interacting particles; of course $g_{nm}^2 \equiv g_{mn}^2$, see (4a)), but they should all be nonnegative, $g_{nm}^2 \geq 0$ (as suggested by their notation), to avoid that the corresponding (singular) interaction be attractive. But the restriction (4) to pair interactions is not required for the validity of our main results.

2.2. Derivation of the main results

The crucial remark is now the following: the various quantities H_0, Q and D close as follows under Poisson brackets:

$$\{D, H_0\} = -2H_0 \; , \tag{5a}$$

$$\{D, Q\} = 2Q \; , \tag{5b}$$

$$\{H_0, Q\} = 2D \; . \tag{5c}$$

The Casimir operator Ξ corresponding to the Lie algebra defined by (5), is given by

$$\Xi = QH_0 - \frac{D^2}{2} , \tag{6}$$

which clearly commutes with Q, D and H_0.

From this follows that any Hamiltonian of the form

$$H(\underline{\vec{r}}; \underline{\vec{p}}; t) = \Phi\left(H_0, Q, D; t\right) \tag{7}$$

has a dynamics for the collective variables H_0, Q and D given by

$$\dot{Q} = \{H, Q\} = 2Q\frac{\partial\Phi}{\partial D} + 2D\frac{\partial\Phi}{\partial H_0} , \tag{8a}$$

$$\dot{D} = \{H, D\} = -2Q\frac{\partial\Phi}{\partial Q} + 2H_0\frac{\partial\Phi}{\partial H_0} , \tag{8b}$$

$$\dot{H}_0 = \{H, H_0\} = -2D\frac{\partial\Phi}{\partial Q} - 2H_0\frac{\partial\Phi}{\partial D} . \tag{8c}$$

These equations always have Ξ as an integral of motion, whether the system be time-dependent or not. In the time-independent case, these equations are additionally an *integrable* Hamiltonian system, since they possess the energy as an additional integral of motion apart from Ξ. In the following, we shall limit ourselves to the time-independent case, returning to the general case in Section 4, where we shall discuss the possibility of using such time-dependent variations of the Hamiltonians to describe compression and expansion of the system, whether adiabatic or sudden.

Whenever the orbit is bound, we therefore have systems for which the variable Q, which describes the size of the system, varies periodically. The simplest case is the following:

$$H(\underline{\vec{r}}; \underline{\vec{p}}) = H_0 + \frac{\Omega^2 Q}{2} = \frac{1}{2}\sum_{n=1}^{N}\left(p_n^2 + \Omega^2 r_n^2\right) + V^{(-2)}\left(\underline{\vec{r}}\right) , \tag{9a}$$

$$\Omega \equiv \Omega\left(N\right) = \omega N^{-1/d} , \tag{9b}$$

which was discovered in a slightly different manner in [4] and further discussed in [8, 9]. A somewhat more general case but still yielding "reasonable" equations of motion is

$$H(\underline{\vec{r}}; \underline{\vec{p}}) = H_0 + \Psi\left(Q, D\right) , \tag{10}$$

for which the equations of motion can be interpreted as the ones corresponding to H_0 modified by a harmonic potential and a dilational dynamics (determined by D) the parameters of which are determined by the collective variables Q and D. The equations of motion for the collective variables are

$$\dot{Q} = 2D + 2Q\frac{\partial\Psi}{\partial D} , \tag{11a}$$

$$\dot{D} = 2H_0 + 2Q\frac{\partial\Psi}{\partial Q} , \tag{11b}$$

$$\dot{H}_0 = -2D\frac{\partial\Psi}{\partial Q} - 2H_0\frac{\partial\Psi}{\partial D} \tag{11c}$$

and Newton's equations read

$$\ddot{\vec{r}}_i = -\frac{\partial V^{(-2)}(\underline{\vec{r}})}{\partial\vec{r}_i} + 2\vec{r}_i\left[2\left(\frac{\partial\Psi}{\partial D}\right)^2 - \frac{\partial\Psi}{\partial Q} + \frac{\partial^2\Psi}{\partial D^2}\dot{D} + \frac{\partial^2\Psi}{\partial D\partial Q}\dot{Q}\right] . \tag{12}$$

These equations of motion are readily interpreted as describing particles interacting both via a general interaction potential of homogeneity degree -2 and oscillating in a self-consistent harmonic potential, the strength of which is determined by the values of the collective coordinates H_0, D and Q, which themselves evolve according to (11).

If the interaction potential $V^{(-2)}(\underline{\vec{r}})$ is additionally assumed to be both rotationally and translationally invariant, as most physically relevant interactions should be, then the above argument can be considerably extended. Indeed, then the components of d-dimensional angular momentum, the position and momentum of the center of mass, D, Q and H_0 all yield a closed set of commutation relations among themselves. It can then easily be checked that the most general Hamiltonian in this enlarged set of variables needs no more be integrable even in the time-independent case, but bounded motions of the various quantities can still be generated. We may thus have a non-integrable dynamics for the collective variables D and Q, these being then coupled to angular motion as well as center-of-mass motion. While these possibilities are no doubt of interest, we leave their exploration to future work and limit ourselves in the following to the case in which the position and momentum of the center of mass as well as the total angular momentum are all zero.

3. Thermodynamics

In the following, we aim at making thermodynamic statements concerning systems of the type we have just been describing. Normally, thermodynamics only describes states of equilibrium, that is, states such that the values of macroscopic observables are constant up to fluctuations of relative order $1/\sqrt{N}$.

For a system which is strictly speaking out of equilibrium, devising an equivalent of entropy is quite a challenging problem. However, we shall not need to discuss such issues or to develop techniques for systems out of equilibrium. What happens in the systems described above is that there is an additional conserved quantity Ξ in addition to the total energy of the system given by the Hamiltonian (7), or more specifically (10), as was shown in section 2, see (6). To take into account the existence of Ξ, it is standard to introduce a *restricted ensemble*, that is, to fix the value of Ξ. There are, of course, different, but essentially equivalent, possibilities, such as using a generalized canonical ensemble. In the following we shall only consider an ensemble of the microcanonical type, both with respect to the energy and to Ξ. The reason for this is that, for the systems we consider, the standard theorems concerning equivalence of ensembles derived, say, in [12] do not apply in dimensions above two, as interaction potentials satisfying (3) are then long-range. The microcanonical ensemble, however, is the only one appropriate to describe a perfectly isolated system such as the one we have been describing.

But if we consider such a restricted ensemble, two related problems arise: since the region of phase space in which E and Ξ take prescribed values is the part of phase space corresponding to an oscillation of a given amplitude, the system does not explore this region uniformly. Rather, it runs through various parts of this region, corresponding to different values of Q and D as the oscillation proceeds. This has the additional consequence that macroscopic variables such as D and Q generally vary over a macroscopic range of values in an ensemble characterized by fixing only the value of E and Ξ, contrary to the basic property of thermodynamic equilibrium.

Since, however, the oscillation is always *slow* in the thermodynamic limit, that is, the period of the oscillation is always much larger than the collision time, we may ask whether it is not possible to describe what happens in terms of ensembles where the values of Q and D are fixed separately. We shall see that this is indeed possible.

Let us therefore consider such an ensemble, restricted in the values of both Q and D. The partition function is then given by

$$W_r(E,Q,D;N) = \int d\underline{\vec{r}}\, d\underline{\vec{p}}\, \delta\big(E - \Phi\left[H_0\left(\vec{p}_n,\vec{r}_n\right),Q,D\right]\big) \times \delta\left(Q-\sum_{n=1}^{N} r_n^2\right)\delta\left(D-\sum_{n=1}^{N} r_n p_n\right) . \tag{13}$$

Note that the parameter ω, see (9b), is not mentioned explicitly: this corresponds to the fact that it corresponds physically to a pressure, and we are fixing the volume via the value of Q.

The following remark is elementary: out of the two equations

$$E = \Phi\left(H_0,D,Q\right) , \tag{14a}$$

$$\Xi = QH_0 - \frac{D^2}{2} , \tag{14b}$$

we may in general eliminate H_0 to obtain a relationship between E and Ξ at given values of Q and D. Out of this relationship, we may finally express D as a function of Ξ, E and Q, say as

$$D = \mathcal{D}(E,\Xi,Q). \tag{15}$$

We can thus express the restricted partition function in terms of E, Ξ and Q: we define

$$\overline{W}_r(E,\Xi,Q;N) = W\left[E,Q,\mathcal{D}(E,\Xi,Q);N\right] . \tag{16}$$

Let us now assume that we have started an ensemble of restricted systems all at the same values $Q(0)$ and $D(0)$ of Q and D and distributed according to the corresponding generalized microcanonical ensemble having the partition function $\overline{W}_r(E,\Xi(0),Q(0),N)$. We now ask what happens when the restriction is simultaneously lifted for all systems of the ensemble. They then start to oscillate and all the systems belonging to the ensemble have at time t the same values of $Q(t)$ and $\Xi(t) = \Xi(0)$ defined by (8). Since the systems belonging to the ensemble all obey a dynamics defined by a Hamiltonian, Liouville's theorem holds. From this follows that, since the systems were originally uniformly distributed with respect to the Liouville measure restricted to the values $Q(0)$ and $D(0)$ of Q and D respectively, they will remain thus uniformly distributed for all times with respect to the Liouville measure similarly restricted to the values $Q(t)$ and $D(t)$.

Hence, defining restricted thermodynamic equilibrium as the one determined by the restricted generalized microcanonical ensemble for fixed values of Q and D as described above, it follows from the considerations just made that, if the system starts out in a restricted equilibrium, it stays in such an equilibrium forever.

Since it follows from (14) that the restricted partition function $\overline{W}_r$ depends on E, Ξ, Q and N, and since, of those four variables, only Q varies in the course of an oscillation, it follows that $\overline{W}_r$ can in fact only depend on Ξ and N:

$$W_r(E,Q,D;N) = \overline{W}_r(E,\Xi,Q;N) = \Psi_r(E,\Xi,N), \tag{17}$$

where $\Psi_r(K,N)$ is defined by the last equality.

It thus follows that the oscillation can be described as an *adiabatic* oscillation, that is, it has constant entropy. Here we only need a quite mild generalization of the concept of entropy, and by no means require to consider true non-equilibrium properties. We are dealing, in fact, not with a true non-equilibrium system, but with a system in ordinary thermodynamic equilibrium, coupled with a hydrodynamic oscillation. The nature of the coupling is, however, peculiar: no energy exchange takes place between the oscillation and the microscopic motion. Indeed, no irreversible processes that arise in the microscopic motion can damp the collective oscillation. That such processes may well occur is clear enough: if initially, for example, a temperature or pressure gradient were present, the fact that the restricted system tends to equilibrium tells us that these gradients will dissipate. Such relaxation processes will presumably satisfy the relaxation theorem described in [2, 3] and have in fact been observed in numerical work, see [8]. However, as we have seen, the dynamics of the macroscopic oscillation is unaffected by anything that happens at the microscopic level, so the dissipation of initial temperature gradients will take place without causing any damping of the macroscopic oscillation.

It thus follows that the entropy of the system only depends on Ξ:

$$s(E,Q,D) = \frac{1}{N}\ln \overline{W}_r(E,\Xi,Q;N) = \frac{1}{N}\ln \Psi_r(E,\Xi;N) = \sigma(E,\Xi) \tag{18}$$

where the final equality *defines* the function $\sigma(E,\Xi)$. The fact that the N dependence vanishes in the thermodynamic limit is standard, see for example [12]. The temperature is thus given by

$$\frac{1}{T} = N\left(\frac{\partial s(E,Q,D)}{\partial E}\right)_{Q,D} = N\left[\left(\frac{\partial \sigma(E,\Xi)}{\partial E}\right)_{\Xi} + \left(\frac{\partial \sigma(E,\Xi)}{\partial \Xi}\right)_{E}\left(\frac{\partial \Xi}{\partial E}\right)_{Q,D}\right] \tag{19}$$

which means that the temperature will generally depend on Q and D, that is, it will vary during the oscillation, leading to the main conclusion of this paper, that such systems do not approach a physical state of equilibrium.

This leads to the remarkable fact that the systems described here do not tend to thermodynamic equilibrium, since the temperature oscillates indefinitely. Ordinarily, relaxation to equilibrium is shown to follow from appropriate ergodicity and mixing properties [1], which can then presumably be assumed in the case of sufficiently complex systems. What we see here is an instance of a system that behaves in a somewhat peculiar and unexpected manner. It has indeed a conservation law distinct from energy, which contradicts the traditional statement of ergodicity. However, in such cases, it generally suffices to limit consideration to a surface in which the additional conservation law has a fixed value to recover the usual ergodic results and relaxation to equilibrium. This is, for example, what happens if angular momentum is conserved. It is also a useful approach, say, when adiabatically conserved quantities exist [13]. Here, however, the situation is different, because two macroscopic variables, namely Q and D, follow a separate, non-trivial, conservative dynamics uninfluenced by the other degrees of freedom; on the other hand, these two variables do affect the microscopic motion, at least to the extent of affecting the system temperature.

Finally, let us note that, since no damping of the radial oscillation described above takes place, it follows that compression and dilation of the system cannot be accompanied by any dissipative processes. Specifically, this implies that the bulk viscosity of the class of systems we consider must vanish. While this is an immediate consequence of our work, this does not seem to have been noticed before. It is, of course, a straightforward consequence of the Green–Kubo formula for the bulk viscosity [14] in the case of systems such as the ones described by (9).

4. Generalization to a time-dependent Hamiltonian

In this section we investigate whether a system which fails to approach thermodynamic equilibrium in the manner described above might conceivably provide a counterexample to the second law of thermodynamics. In the sense that this law states the impossibility of perpetual motion machines of the second kind, we found no way to violate the second law using the above class of systems; nor do we believe this to be possible. What we show below is that the simplest attempts to obtain energy out of this class of systems remain consistent with the second law.

The class of systems described here presumably provides an interesting test of any proofs of the second law, such as the one provided in [15].

The Hamiltonian (7) in its most general form provides various possibilities of acting on the system in a mechanical manner by an external agency. Indeed, allowing the external harmonic potential in (9) to depend on time, one can compress or expand the system, either suddenly, adiabatically or following any other protocol. Similarly, one can use such a potential to extract work out of an existing oscillation, while decreasing its amplitude. The Hamiltonian (7) in its most general form presumably allows many possibilities of acting on the system in a mechanical manner. Of course, one cannot describe in this way processes involving heat flow, such as isothermal expansion or compression, or putting the system in contact with a heat bath at a given temperature.

We now describe a mechanical cyclic process as follows: we start from a system without an oscillation and at thermodynamic equilibrium in the usual sense. The system is subjected to a time evolution $H(t)$, where the Hamiltonian is only time-dependent for $0 \leq t \leq T$, such that $H(0) = H(T)$ and $H(t)$ is at all times of the form (7). Such a cyclic process can, as is readily seen, describe fairly arbitrary mechanical operations on the system, but no operations involving heat transfer. Let us further assume that the system returns to its initial state, that is, the initial values of Q and D are recovered at the end of the cycle. Since Ξ is conserved also for the case of a time-dependent Hamiltonian, the fact that Q and D return to their original values means that H_0 does so as well. Since the final Hamiltonian is the same as the initial one, the total energy has remained constant. Since by definition no transfer of heat took place, the total work done on the system in such a cyclic evolution vanishes. This means, of course, that no violations of the second law can be obtained in this manner, consistently with the result proved in [15].

Thus, for example, if starting from an initial condition in which the system is in equilibrium with a frequency ω_1 one suddenly decreases the frequency to $\omega_2 < \omega_1$, an oscillation will be set in motion. It is then possible to get work by damping the oscillation, and finally one may get back to the initial configuration by an adiabatic compression. From the above result we see that the total work performed on the system is zero. This means that, in such a free expansion, contrary to what is well-known in the case of the ideal gas, the entropy does not grow irreversibly.

5. The quantum case

In this section we show that our main finding remains true in the *quantal* context; i. e., also in that context the time evolution of the effective volume occupied by the N-body system whose dynamics is determined by the Hamiltonian (9) oscillates isochronously without any damping.

A straightforward way to show this is by working in the Heisenberg representation. The Heisenberg equations of motion for the observables corresponding to the classical quantities Q and D then reduce to (8), which must, however, be interpreted as operator equations. In the specific case where the Hamiltonian is of the form (9), these equations reduce to

$$\dot{Q} = 2D \,, \tag{20a}$$

$$\dot{D} = -2\Omega^2 Q + 2H \,. \tag{20b}$$

Since H is a constant, the equations for D and $Q - H/\Omega^2$ are *linear*, so that their classical solution can be immediately reinterpreted as an operator solution in quantum mechanics. A more detailed proof in the Schrödinger representation can be found in [10].

The generalization to Hamiltonians of more general form, such as, for example, (7) or (10) is not so straightforward, since the Heisenberg equations of motion are not linear any more, and therefore the classical solutions for the system of ODE's have no relation to the corresponding solution of the operator equations.

6. Conclusions

To summarize, we have studied a rather general class of systems including those in which particles of equal masses interact with each other according to a potential of degree of homogeneity -2 and these particles are additionally confined by a harmonic external potential: these have the property of showing *undamped* compressional oscillations whenever the initial conditions deviate from thermodynamic equilibrium. These oscillations are adiabatic, and lead therefore to variations in temperature and pressure. If the system starts out in (restricted) thermodynamic equilibrium at a given stage of the oscillation, it remains in equilibrium, meaning that no pressure or temperature gradients can be generated. Further, the oscillation remains uncoupled from the microscopic motion, in the sense that dissipative processes arising in the system cannot affect the amplitude of the oscillation. In other words, energy cannot pass from the microscopic degrees of freedom to the collective variables characterizing the oscillation or viceversa. On the other hand, as we have seen, the oscillation modifies the temperature of the system, so that the system really does not approach thermal equilibrium.

The reason for the phenomenon of undamped oscillations is that a set of three collective coordinates, of which the radius of gyration of the system is one, form a closed algebra under Poisson brackets, and that the full Hamiltonian is a function of these.

Finally, it was shown that the main results remain valid in the quantal context. The extension of the thermodynamic considerations to the quantum case has not been done. We do not expect that it should present any significant difficulties.

We also generalized the approach to include a fairly general time-dependent Hamiltonian, so that we could model mechanical manipulations on the system. This allowed to show, for example, that it is possible to reverse a free expansion made by suddenly decreasing the strength of the confining harmonic potential. As expected, it was not possible to display any counterexamples to the second law of thermodynamics.

Acknowledgements

We wish to acknowledge with thanks the hospitality, extended many times, to one of us (FC) by the Centro Internacional de Ciencias (CIC) in Cuernavaca, and to the other one of us (FL) by the Physics Department of the University of Rome "La Sapienza". This paper was completed during the Scientific Gathering on "Integrable systems—continuous and discrete—and the transition to chaos" hosted by CIC in November-December 2012. FL also wishes to acknowledge the financial support of the following projects: Projects CONACyT 154586 and UNAM DGAPA PAPIIT RR113311.

7. References

[1] Dorfman J R 1999 *An Introduction to Chaos in Nonequilibrium Statistical Mechanics*, (Cambridge Lecture Notes in Physics)

[2] Evans D J, Williams S R and Searles D J 2011 On the entropy of relaxing deterministic systems *J. Chem. Phys.* **135** 194107 (6 pages)

[3] Reid J C, Evans D J and Searles D J 2012 Communication: Beyond Boltzmann's H-theorem: demonstration of the relaxation theorem for a non-monotonic approach to equilibrium *J. Chem. Phys.* **136** 021101 (4 pages)

[4] Lynden-Bell D and Lynden-Bell R M 1999 Exact general solutions to extraordinary N-body problems *Proc. Roy. Soc. Lond.* A **455** 475–489

[5] Bobylev A V, Dukes P, Illner R and Victory H D 1997 On Vlasov–Manev equations. I: Foundations, properties, and nonglobal existence *J. Stat. Phys.* **88** 885–911

[6] Calogero F 1971 Solution of the one-dimensional N-body problems with quadratic and/or inversely quadratic pair potentials *J. Math. Phys.* **12** 419–436

[7] Calogero F 2001 *Classical many-body problems amenable to exact treatments* (Lecture Notes in Physics Monograph **m 66**, Springer)

[8] Lynden-Bell D and Lynden-Bell R M 2004 Relaxation to a perpetually pulsating equiibrium *J. Stat. Phys.* **117** 199–209

[9] Pichon C, Lynden-Bell D, Pichon J, Lynden-Bell R M 2007 Lattice melting and rotation in perpetually pulsating equilibria *Physical Review E* **75** 011125 (2007), 11 pages

[10] Calogero F and Leyvraz F 2013 A Macroscopic System with Undamped Periodic Compressional Oscillations *J. Stat. Phys.* **151** 922–937

[11] Calogero F and Leyvraz F Many-body problem with quadratic and/or inversely-quadratic potentials in one- and more-dimensional spaces: some retrospective remarks (submitted to *J. Stat. Phys.*)

[12] Ruelle D 1999 *Statistical Mechanics: Rigorous Results* (World Scientific, Singapore)

[13] Leyvraz F, Firpo M-C and Ruffo S 2002 Inhomogeneous quasi-stationary states in the antiferromagnetic mean-field XY model J. Phys. A Math. Gen. **35** 4413–4427

[14] Gubbins K E 1972 *Statistical Mechanics*, vol. 1, Singer ed., pp.194–253, The Chemical Society, Burlington House

[15] Allahverdyan A E and Nieuwenheuizen Th M 2002 A mathematical theorem as the basis for the second law: Thomson's formulation applied to equilibrium Physica **A305** 542–552

Haantjes Manifolds

F Magri

Dipartimento di Matematica ed Applicazioni, Università di Milano Bicocca, 20125 Milano, Italy

E-mail: franco.magri@unimib.it

Abstract. The aim of this paper is to introduce a new category of manifolds, called Haantjes manifolds, and to show their interest by a few selected examples.

1. Introduction

Many years ago, at the first conference in this series held in Lecce in 1979, I had the occasion to present a simple model of integrable systems, henceforth called bihamiltonian systems. They were defined on manifolds called bihamiltonian manifolds. In this conference I wish to present an evolution of the concept of bihamiltonian manifold called Haantjes manifold. I believe that the study of the new class of manifolds may enlight a range of topics even outsides the domain of the theory of integrable systems. Among them I list:

- Topological Field Theories
- Frobenius manifolds
- Singularity Theory
- Quantum cohomology
- Coisotropic deformations of associative algebras
- Quasi-Hamiltonian systems of hydrodynamic type
- Whitham equations
- Orthogonal coordinates in $\mathbf{R^n}$.

In this paper the focus is on the relation between Haantjes manifolds and the theory of singularities.

To introduce the concept of Haantjes manifold, I need to recall that a bihamiltonian system is a vector field X on a bihamiltonian manifold M which admits two compatible Hamiltonian decompositions

$$X = Pdh = Qdk.$$

If one of the Poisson bivectors P and Q is invertible, say Q, this decomposition endows the manifold M with three distinguished geometrical objects: the vector field X, the 1-form $\theta = dh$, and the recursion operator $K = PQ^{-1}$. These geometrical objects satisfy five basic conditions

- $Torsion(K) = 0$
- $d\theta = 0$

- $dK\theta = 0$
- $Lie_X(K) = 0$
- $Lie_X(\theta) = 0$.

Acting on the vector field X and the 1-form θ, the tensor field K generate two infinite chains of vector fields and 1-forms respectively :

$$\theta_{k+1} = K\theta_k, \qquad \theta_0 = \theta,$$
$$X_{k+1} = KX_k, \qquad X_0 = X.$$

They are called the *Lenard chains* associated with the given bihamiltonian system. Owing to the five conditions listed above, these chains enjoy the remarkable property that the 1-forms θ_k are closed, and that the vector fields X_k commute. Locally the 1-forms θ_k are the differentials of certain functions I_k. These functions turn out to be in involution with respect to both the Poisson brackets associated with the Poisson bivectors P and Q. Consequently the vector fields X_k and the functions I_k are symmetries and integrals of motion of X. They are the additional informations which permit to integrate the given bihamiltonian vector field X .

The concept of Haantjes manifold arises when one accepts to slighty modify the previous scheme, allowing the recursion operator to *possess torsion* but in a controlled manner. This idea has been presented two years ago at the conference on *Geometrical Methods in Mathematical Physics* organized by Boris Dubrovin in Moscow. In that occasion I suggested to replace the conditions of the classical theory of bihamiltonian manifolds by the new conditions

- $Haantjes(K) = 0$
- $d\theta = 0$
- $dK\theta = 0$
- $\theta(Torsion(K)) = 0$
- $Lie_X(K) = 0$
- $Lie_X(\theta) = 0$,

on the basis of their significance for the theory of Frobenius manifolds and $WDVV$ equations [1]. In this conference I wish to continue the exploration of the new geometrical setting. In particular, I am interested to describe the changes which are undergone by the theory of Lenard chains .

First I collect the definitions which are relevant to the present discussion.

Definition 1. (**Haantjes manifold**) *A manifold M is a Haantjes manifold if it is endowed with a distinguished 1-form θ and with a distinguished vector-valued 1-form K which satisfy the four conditions*

- $Haantjes(K) = 0$
- $d\theta = 0$
- $dK\theta = 0$
- $\theta(Torsion(K)) = 0$.

If moreover there exists a vector field X obeying the conditions

- $Lie_X(K) = 0$
- $Lie_X(\theta) = 0$.

the manifold is said to be a Haantjes manifold with symmetry.

On a Haantjes manifold there are two distinguished tensor fields of type $(1,1)$: the identity maps Id and K. They form the core of a Lenard chain. Any family of tensor fields which prolong this elementary core and enjoy the same properties of Id and K is a Lenard chain of the Haantjes manifold.

Definition 2. *(Lenard chain) A Lenard chain of lenght p on a Haantjes manifold of dimension n is a family of p vector-valued 1-forms K_j which contains $K_0 = Id$ and $K_1 = K$, and satisfies the conditions*

- $Haantjes(K_j) = 0$
- $[K_j, K_l] = 0$
- $dK_j\theta = 0$
- $\theta(Torsion(K_j)) = 0.$

This extension of the concept of Lenard chain, from the case of bihamiltonian manifolds to the case of Haantjes manifolds, is of great use, since it enlarges significantly the field of application of these chains, but it has a cost. In the case $Torsion(K) = 0$ there was a simple and general rule to construct the family of tensor fields K_j: to take the powers of K. This rule is lost on Haantjes manifolds. In the new setting, the family of tensor fields K_j must be constructed in each example according to a rule which accounts for the specific form of the torsion of K. Despite this difficulty, Lenard chains on Haantjes manifolds are still quite worth of interest. This is due to a subtle property enjoyed by them, which I will describe next. I need two preliminary remarks.

Lemma 1. *Consider two tensor fields K_1 and K_2 on a manifold M (not necessarily a Haantjes manifold), and assume that they commute, that their Haantjes tensors vanish, and that their eigenvalues α_j and β_j are real and distinct. Then any 1-form θ which annihilates the torsion of K_1 and K_2 separately, annihilates the torsion of $K_1 + K_2$ as well (or of any linear combination $\lambda_1 K_1 + \lambda_2 K_2$ with real coefficients).*

Lemma 2. *On a manifold M (not necessarily a Haantjes manifold) consider a closed 1-form θ and a tensor field K, possibly with non vanishing Haantjes tensor. Assume that θ annihilates the torsion of K, and that K maps θ into a closed 1-form. Then, the 1-form $K^2\theta$ is closed as well.*

Let make a comment on the last Lemma. By assumption the 1-forms θ and $K\theta$ are closed , as always in the theory of Lenard chains. The condition $\theta(Torsion(K)) = 0$ then entails that the 1-form $K^2\theta$ is closed as well. But if one try to go a step further, one fails. The next 1-form $K^3\theta$ is, usually, not longer closed. So the Lenard chain breaks at the third iteration , and cannot be further prolonged. This is the difference between the case of bihamiltonian manifolds with respect to the case of Haantjes manifolds. The triple of 1-forms $(\theta, K\theta, K^2\theta)$ will be referred to as the *short Lenard chain* generated by θ and K.

Proof. I will omit the proof of the second Lemma, which is a rephrasing of the concept of Nijenhuis torsion. To prove the first Lemma, I make use of the concept of canonical coordinates. It is well-known that the vanishing of the Haantjes torsion of K_1, combined with the assumption that its eigenvalues α_j are real and distinct , entails the existence of a special system of coordinates $(u_1, u_2, \ldots, u_n)$ which diagonalize K_1 [2]. In these coordinates the torsion of K_1 has the following simple form

$$T_{K_1}\left(\frac{\partial}{\partial u_j}, \frac{\partial}{\partial u_k}\right) = (\alpha_j - \alpha_k)\left(\frac{\partial \alpha_k}{\partial u_j}\frac{\partial}{\partial u_k} - \frac{\partial \alpha_j}{\partial u_k}\frac{\partial}{\partial u_j}\right).$$

Therefore, the 2-form $\theta(Torsion(K_1))$ has components

$$\theta(Torsion(K_1))_{j,k} = (\alpha_j - \alpha_k)\left(\frac{\partial \alpha_k}{\partial u_j}\theta_k - \frac{\partial \alpha_j}{\partial u_k}\theta_j\right),$$

where θ_k are the components of the 1-form θ. Since the eigenvalues of K_1 are distinct, the 1-form θ verifies the condition $\theta(Torsion(K_1)) = 0$ iff

$$\frac{\partial \alpha_k}{\partial u_j}\theta_k - \frac{\partial \alpha_j}{\partial u_k}\theta_j = 0.$$

The same condition holds for K_2:

$$\frac{\partial \beta_k}{\partial u_j}\theta_k - \frac{\partial \beta_j}{\partial u_k}\theta_j = 0.$$

Let us now compute the 2-form $\theta(Torsion(K_1 + K_2))$. It vanishes iff

$$(\beta_j - \beta_k)\left(\frac{\partial \alpha_k}{\partial u_j}\theta_k - \frac{\partial \alpha_j}{\partial u_k}\theta_j\right) + (\alpha_j - \alpha_k)\left(\frac{\partial \beta_k}{\partial u_j}\theta_k - \frac{\partial \beta_j}{\partial u_k}\theta_j\right) = 0.$$

This condition is manifestly a linear combination of the conditions verified separately by K_1 and K_2. Therefore, if θ annihilates the torsion of K_1 and K_2, and if K_1 and K_2 have distinct eigenvalues and satisfy the Haantjes condition, θ annihilates also the torsion of $K_1 + K_2$. □

The important consequence of this result comes to light when, on a Haantjes manifold endowed with a Lenard chain K_j of lenght p, one considers the second generation of iterared 1-forms

$$\theta_{jl} = K_j K_l \theta.$$

This family is referred to as the *square* associated to the Lenard chain of 1-forms.

Proposition 1. *All the forms of the square are closed . So, any Lenard chain of lenght p on a Haantjes manifold of dimension n generates a square of $1/2p(p+1)$ locally exact 1-forms.*

Proof. Consider the linear pencil $K(\lambda) = \lambda_1 K_1 + \cdots + \lambda_p K_p$ of all the tensor fields of the Lenard chain. By the first Lemma its torsion is annihilated by the 1-form θ. By the second Lemma the 1-form $K(\lambda)^2\theta$ is closed for any possible choice of the parameters of the pencil. This form depends quadratically on the parameters. Hence the coefficient of the product of λ_j and λ_k must vanish. This coefficient is the exterior differential of the form θ_{jl} of the square. Therefore, under the assumptions listed above, all the 1-forms of the square are closed. □

The case of major interest is, of course, the case when p equals the dimension of the manifold M. One can show that this case is related to $WDVV$ equations [1] and to the concept of (dispersionless) Hirota *tau*-function. The study of this case is, however, outside the scopes of this note. In this paper I will limit myself to show, in a concrete example, that Lenard chains are related to the theory of singularities of type E_6.

2. Two examples of Haantjes manifolds

In 1951 Albert Nijenhuis [3] noticed that the eigenvalues α_j of a tensor field K with vanishing torsion

$$Torsion(K) = 0$$

may be used as canonical coordinates for K in the domain where they are real, distinct, and functionally independent. Thus

$$K\frac{\partial}{\partial \alpha_j} = \alpha_j \frac{\partial}{\partial \alpha_j}$$

in this domain. If instead of the eigenvalues α_j one uses the coefficients of the characteristic polynomial

$$Q(\alpha) = \alpha^n + a_n\alpha^{n-1} + \cdots + a_1$$

as coordinates, one finds for K the cyclic representation

$$\begin{aligned}
K\frac{\partial}{\partial a_1} &= \frac{\partial}{\partial a_2}, \\
K\frac{\partial}{\partial a_2} &= \frac{\partial}{\partial a_3}, \\
&\vdots \\
K\frac{\partial}{\partial a_{n-1}} &= \frac{\partial}{\partial a_n}, \\
K\frac{\partial}{\partial a_n} &= -\left(a_1\frac{\partial}{\partial a_1} + a_2\frac{\partial}{\partial a_2} + \cdots + a_n\frac{\partial}{\partial a_n}\right).
\end{aligned}$$

The idea pursued in this section is to look for tensor fields K verifying the weaker condition

$$Haantjes(K) = 0$$

inside the class of tensor fields defined by the equations

$$\begin{aligned}
K\frac{\partial}{\partial a_1} &= \frac{\partial}{\partial a_2}, \\
K\frac{\partial}{\partial a_2} &= \frac{\partial}{\partial a_3}, \\
&\vdots \\
K\frac{\partial}{\partial a_{n-1}} &= \frac{\partial}{\partial a_n}, \\
K\frac{\partial}{\partial a_n} &= -\left(A_1\frac{\partial}{\partial a_1} + A_2\frac{\partial}{\partial a_2} + \cdots + A_n\frac{\partial}{\partial a_n}\right),
\end{aligned}$$

inspired by the example studied by Nijenhuis. This class depends on the choice of n arbitrary functions $(A_1, A_2, \cdots, A_n)$ of the cyclic coordinates $(a_1, a_2, \cdots, a_n)$. Since these functions are the coefficients of the characteristic polynomial of K

$$Q(\alpha) = \alpha^n + A_n\alpha^{n-1} + \cdots + A_1,$$

the new class may be tought of as a gentle deformation of the tensor field of Nijenhuis. The deformation process changes the characteristic polynomial of K, but keeps unchanged the cyclic coordinates. The vector field

$$A = A_1\frac{\partial}{\partial a_1} + \cdots + A_n\frac{\partial}{\partial a_n}$$

is accordingly referred to as the *characteristic vector field* of K. In the example of Nijenhuis the characteristic vector field is the Euler vector field

$$A = -\left(a_1\frac{\partial}{\partial a_1} + a_2\frac{\partial}{\partial a_2} + \cdots + a_n\frac{\partial}{\partial a_n}\right).$$

The study of the Haantjes torsion of K shows that there are other two elementary solutions of the Haantjes condition besides that found by Nijenhuis. They are selected by demanding that the components of the characteristic vector field A depend *linearly* on the cyclic coordinates. These new solutions are :

$$A = -\frac{1}{n+1}\left[a_2\frac{\partial}{\partial a_1} + 2a_3\frac{\partial}{\partial a_2} + \cdots + (n-1)a_n\frac{\partial}{\partial a_{n-1}}\right],$$
$$B = -\frac{1}{n-1}\left[-a_1\frac{\partial}{\partial a_1} + a_3\frac{\partial}{\partial a_3} + \cdots + (n-3)a_{n-1}\frac{\partial}{\partial a_{n-1}} + (n-2)a_n\frac{\partial}{\partial a_n}\right].$$

The corresponding tensor fields will be denoted by L and M respectively. They inherit several interesting properties from the tensor field of Nijenhuis. A few of these properties will be presented below.

Eigencovectors. I notice first that the 1-form

$$\eta = \lambda(da_1 + \alpha da_2 + \alpha^2 da_3 + \cdots + \alpha^{n-1}da_n)$$

is an eigencovector of K, corresponding to the eigenvalue α, for any choice of the arbitrary function λ . This property follows from the cyclic form of K, and holds for any choice of the arbitrary functions A_j.The function λ serves to normalize the eigencovector . When the Haantjes torsion of K vanishes, and when the eigenvalues α_j are real and distinct, the function λ may be chosen so that the 1-form η is exact :

$$\eta = du.$$

The function u is the canonical coordinate associated with the eigenvalue α. The construction of the canonical coordinates is usually a difficult problem. It becomes simpler when the tensor field K has symmetries. The following result is stated without proof.

Lemma 3. *Let X be a symmetry of K, and demand that*

$$\eta(X) = 1.$$

Then the 1-form η is exact.

This result is used to select the normalization factor λ.

Canonical coordinates. To find the symmetries of L and M, I look first at the symmetries of the corresponding characteristic vector fields. Since the coordinates a_1 and a_2 are cyclic for A and B respectively, the conclusion is that the vector fields

$$X = \frac{\partial}{\partial a_1}, \qquad Y = \frac{\partial}{\partial a_2}$$

are symmetries of L and M respectively. Let us use these symmetries to normalize the eigenvector forms. The normalization rule leads to

$$\lambda_L = 1 \qquad \text{and} \qquad \lambda_M = \alpha^{-1},$$

Journal of Physics: Conference Series **482** (2014) 012028 doi:10.1088/1742-6596/482/1/012028

so that the normalized eigencovectors are

$$\eta_L = da_1 + \alpha da_2 + \alpha^2 da_3 + \cdots + \alpha^{n-1} da_n,$$
$$\eta_M = \alpha^{-1} da_1 + da_2 + \alpha^1 da_3 + \cdots + \alpha^{n-2} da_n.$$

The canonical coordinates follow then in two steps. A simple integration by parts allows to write immediately η as a total differential up to a multiple of the differential $d\alpha$ of the eigenvalue α. The extra term is subsequently eliminated by means of the characteristic polynomial of K. One arrives in this way to the following expressions

$$u_L = a_1 + a_2\alpha + \cdots + a_n\alpha^{n-1} + \alpha^{n+1},$$
$$u_M = a_1\alpha^{-1} + a_2 + a_3\alpha + \cdots + a_n\alpha^{n-2} + \alpha^{n-1},$$

for the canonical coordinates of L and M respectively. These coordinates are special deformations of the canonical coordinates

$$u = \alpha^n$$

of the tensor field studied by Nijenhuis. They are called miniversal deformations of type A_n and D_n respectively.

Torsion and annihilating 1-form . The tensors fields L and M have torsion, but their torsion is very mild since they are gentle deformations of the torsionless tensor field considered by Nijenhuis. For L one finds that all the components of the torsion vanish except

$$T_L\left(\frac{\partial}{\partial a_j}, \frac{\partial}{\partial a_n}\right) = \frac{\partial}{\partial a_j},$$

for $j = 1, 2, \cdots, n-1$. Similarly , for M one finds that all the components of the torsion vanish except

$$T_M\left(\frac{\partial}{\partial a_j}, \frac{\partial}{\partial a_n}\right) = \frac{\partial}{\partial a_{j+1}}$$

for $j = 1, 2, \cdots, n-1$. These outcomes make it simple to construct the 1-forms annihilating the torsion of L and M respectively. They are:

$$\theta_L = da_n, \qquad \theta_M = da_1.$$

These forms are exact and invariant along the the symmetry vector fields X and Y introduced above.

Haantjes manifolds with symmetry. To check the last condition, that is that the 1-forms $L\theta_L$ and $M\theta_M$ are exact, it is expedient to pass to represent the cyclic tensor fields K on the dual basis da_j. The equations

$$\begin{aligned} K da_n &= da_{n-1} + A_n da_n, \\ K da_{n-1} &= da_{n-2} + A_{n-1} da_n, \\ &\vdots \\ K da_2 &= da_1 + A_2 da_n, \\ K da_1 &= A_1 da_n \end{aligned}$$

hold for any tensor field of the class considered at the beginning of this section. By inserting in these equations the explicit form of the characteristic vector fields of L and M respectively, one may check that the required property holds in both cases. The conclusion is that among the deformations of the bihamiltonian manifold studied by Nijenhuis in 1951 there are two Haantjes manifolds with symmetry. The canonical coordinates of these manifolds are the miniversal deformations of singularities of type A_n and D_n respectively [4].

This is a first instance of the connection between Haantjes manifolds and the Singularity Theory. The outcome of this example may be deepened by pursuing the study of Lenard chains. The second example dealt with in the next section goes one step in this direction. It is meant to suggest that the geometrical framework of Haantjes manifolds is sufficiently wide to encompass the Frobenius structure of the orbit spaces of Coxeter groups [5].

3. An example of Lenard chain

The class of tensor fields studied in the previous section depended on a single characteristic vector field. The obvious extension is to consider now a class of tensor fields depending on two characteristic vector fields. This class is defined by the equations

$$\begin{aligned}
L\frac{\partial}{\partial a_1} &= \frac{\partial}{\partial a_2},\\
L\frac{\partial}{\partial a_2} &= \frac{\partial}{\partial a_3},\\
&\vdots\\
L\frac{\partial}{\partial a_{n-1}} &= \frac{\partial}{\partial a_n},\\
L\frac{\partial}{\partial a_n} &= A,\\
L\frac{\partial}{\partial a_{n+1}} &= \frac{\partial}{\partial a_{n+2}},\\
L\frac{\partial}{\partial a_{n+2}} &= \frac{\partial}{\partial a_{n+3}},\\
&\vdots\\
L\frac{\partial}{\partial a_{n+m-1}} &= \frac{\partial}{\partial a_{n+m}},\\
L\frac{\partial}{\partial a_{n+m}} &= B,
\end{aligned}$$

where A and B are unspecified vector fields to be properly determined afterward. The new class will be referred to as the *gluing* of two Nijenhuis operators (for a different but related notion of gluing see [6]).

In this section I sketch the study of this type of tensor fields on a manifold of low dimension. Specifically, I consider the class of tensor fields defined by the equations

$$L\frac{\partial}{\partial a_1} = \frac{\partial}{\partial a_2}, \qquad L\frac{\partial}{\partial a_2} = \frac{\partial}{\partial a_3}, \qquad L\frac{\partial}{\partial a_3} = A,$$
$$L\frac{\partial}{\partial a_4} = \frac{\partial}{\partial a_5}, \qquad L\frac{\partial}{\partial a_5} = \frac{\partial}{\partial a_6}, \qquad L\frac{\partial}{\partial a_6} = B,$$

on a manifold of dimension six. The aim is to construct a concrete example of Lenard chain. The process is rather articulate, and I split it in seven steps for convenience.

The intertwining tensor M. In view of the study of the spectral properties of the tensor field L it is quite useful to introduce a second tensor field M, called the intertwining tensor field associated with L. It is defined as the unique tensor field which commute with L, and that intertwines the generator $\frac{\partial}{\partial a_1}$ of the first cyclic chain of L to the generator $\frac{\partial}{\partial a_4}$ of the second cyclic chain. On the basis associated with the cyclic coordinates a_j, this tensor field admits the following representation:

$$M\frac{\partial}{\partial a_1} = \frac{\partial}{\partial a_4}, \qquad M\frac{\partial}{\partial a_2} = \frac{\partial}{\partial a_5}, \qquad M\frac{\partial}{\partial a_3} = \frac{\partial}{\partial a_6},$$
$$M\frac{\partial}{\partial a_4} = C, \qquad M\frac{\partial}{\partial a_5} = D, \qquad M\frac{\partial}{\partial a_6} = E.$$

The three vector fields (C, D, E), referred to as the characteristic vector fields of M, are well determined rational functions of the characteristic vector fields (A, B) of L. I omit to give their explicit representation here, since later on we shall have the occasion to characterize them in a more effective way.

Eigencovectors of L. Since the characteristic vector fields (A, B) determine completely the tensor fields L and M, they fix their eigenvalues α_j and β_j as well. The study of the eigenvalue problems

$$L\eta = \alpha\eta, \qquad M\eta = \beta\eta,$$

leads to two important conclusions. The first is that the eigenvalues of L and M are related to their characteristic vector fields through the equations

$$\begin{aligned}
(A_1 + \alpha A_2 + \alpha^2 A_3) + \beta(A_4 + \alpha A_5 + \alpha^2 A_6) &= \alpha^3, \\
(B_1 + \alpha B_2 + \alpha^2 B_3) + \beta(B_4 + \alpha B_5 + \alpha^2 B_6) &= \alpha^3\beta, \\
(C_1 + \alpha C_2 + \alpha^2 C_3) + \beta(C_4 + \alpha C_5 + \alpha^2 C_6) &= \beta^2, \\
(D_1 + \alpha D_2 + \alpha^2 D_3) + \beta(D_4 + \alpha D_5 + \alpha^2 D_6) &= \alpha\beta^2, \\
(E_1 + \alpha E_2 + \alpha^2 E_3) + \beta(E_4 + \alpha E_5 + \alpha^2 E_6) &= \alpha^2\beta^2.
\end{aligned}$$

The second is that the eigencovector η related to the generic pair of eigenvalues α and β is given by

$$\eta = \lambda(da_1 + \alpha da_2 + \alpha^2 da_3 + \beta da_4 + \alpha\beta da_5 + \alpha^2\beta da_6).$$

It is defined, as usual, up to the arbitrary scaling factor λ. Having in mind to study the Haantjes condition for the tensor fields L and M, it is worth to notice that the correspondance between characteristic vector fields and eigenvalues is bijective , at least generically in an open dense domain. This property allows to introduce a second representation of L and M in terms of their eigenvalues, conceived as arbitrarily functions of the cyclic coordinates.

Haantjes conditions on the eigenvalues. To impose the Haantjes conditions upon the tensor fields L and M amounts to choose the normalizing factor λ and the eigenvalues α and β in such a way to make the 1-form η exact. This problem is easily solved by an integration by parts. Choose $\lambda = 1$ and write η in the form

$$\eta = d(a_1 + \alpha a_2 + \alpha^2 a_3 + \beta a_4 + \alpha\beta a_5 + \alpha^2\beta a_6) - (a_2 + 2\alpha a_3 + \beta a_5 + 2\alpha\beta a_6)d\alpha - (a_4 + \alpha a_5 + \alpha^2 a_6)d\beta.$$

To make this form exact it is sufficient to impose the constraints

$$\begin{aligned}
-(a_2 + 2\alpha a_3 + \beta a_5 + 2\alpha\beta a_6) &= \alpha^3, \\
-(a_4 + \alpha a_5 + \alpha^2 a_6) &= \beta^2.
\end{aligned}$$

on the eigenvalues of L and M. This is the form of the Haantjes conditions on the eigenvalues.

Haantjes conditions on the characteristic vector fields. The final form of the Haantjes conditions is on the characteristic vector fields , seen as functions of the cyclic coordinate instead than as functions of the eigenvalues. This final form is attained by a process of elimination. One has to eliminate the eigenvalues α and β among the five equations which give the characteristic vector fields as functions of the eigenvalues , and the two equations which give the eigenvalues as functions of the cyclic coordinates. It is convenient to split the process of elimination in two parts. First I give the explicit representation of the characteristic vector fields A and C, which depend quite simply from the cyclic coordinates. Their form is:

- *Field* A :

$$A_1 = -a_2, \qquad A_2 = -2a_3, \qquad A_3 = 0,$$
$$A_4 = -a_5, \qquad A_5 = -2a_6, \qquad A_6 = 0;$$

- *Field* C :

$$C_1 = -a_2, \qquad C_2 = -2a_3, \qquad C_3 = 0,$$
$$C_4 = 0, \qquad C_5 = 0, \qquad C_6 = 0.$$

Then I give the expression of the components of the other characteristic vector fields as functions of the components of A and C. The expression are a little more cumbersome:

- *Field* B :

$$B_1 = A_4C_1 + A_5(0 + A_1C_3), \qquad B_2 = A_4C_2 + A_5(C_1 + A_2C_3),$$
$$B_3 = A_4C_3 + A_5(C_2 + A_3C_3), \qquad B_4 = A_1 + A_4A_5C_3,$$
$$B_5 = A_2 + A_5A_5C_3, \qquad B_6 = A_3;$$

- *Field* D :

$$D_1 = 0 + A_1C_3, \qquad D_2 = C_1 + A_2C_3,$$
$$D_3 = C_2 + A_3C_3, \qquad D_4 = A_4C_3,$$
$$D_5 = A_5C_3, \qquad D_6 = A_6C_3;$$

- *Field* E :

$$E_1 = A_1C_2 + C_3(0 + A_1A_3), \qquad E_2 = A_2C_2 + C_3(A_1 + A_2A_3),$$
$$E_3 = A_3C_2 + C_3(A_2 + A_3A_3) + C_1, \qquad E_4 = A_4C_2 + C_3(A_4 + 0),$$
$$E_5 = A_5C_2 + C_3(A_4 + A_5), \qquad E_6 = A_6C_2 + C_3(A_5 + 0).$$

Canonical coordinates. At this point we dispose of all the informations concerning the tensor fields L and M. This allows to compute, for instance, their canonical coordinates. From the final form of the eigencovector η one may see that the canonical coordinate associated with the pair of eigenvalues α and β of L and M is

$$u = a_1 + \alpha a_2 + \alpha^2 a_3 + \beta a_4 + \alpha\beta a_5 + \alpha^2\beta a_6 + 1/4\alpha^4 + 1/3\beta^3.$$

This formula shows that the tensor field L is related to the theory of singularities of type E_6 [4].

The Haantjes manifold. To complete the construction of the Haantjes manifold it remains to find a 1-form θ which is exact and annihilates the torsion of L, and such that the iterated 1-form $L\theta$ be furthermore a second exact 1-form. A look at the action of L on the dual basis da_j provides quickly this form. Indeed the 1-form

$$\theta = da_6$$

verifies all the required conditions. So the pair (L, θ) defines a Haantjes manifold of dimension six.

The Lenard chain. Let us set, as usual, $K_0 = Id$ and $K_1 = L$. To construct a Lenard chain one has to find, inside the ring of tensor fields commuting with L and M, other four tensor fields (K_2, K_3, K_4, K_5) satisfying the same conditions verified by K_0 and K_1. One looks for polynomials in L and M, and one selects these polynomials by imposing the conditions $dK\theta = 0$ and $dK^2\theta = 0$ characteristic of the theory of short Lenard chains. I omit the details of the computation, giving only the final result. The Haantjes manifold defined by the pair (L, θ) is endowed with a Lenard chain formed by the tensor fields:

$$\begin{aligned}
K_0 &= Id, \\
K_1 &= L, \\
K_2 &= M - 1/24a_6^2 Id, \\
K_3 &= L^2 + 1/4a_6 M + (1/4a_3 + 1/144a_6^3) Id, \\
K_4 &= LM + 1/12a_6^2 L - 1/12a_5a_6 Id, \\
K_5 &= L^2M - 1/8a_6^2L^2 + (1/4a_3 + 1/288a_6^3)M - 1/2a_5a_6L + \\
&\quad + (-1/12a_4a_6 + 1/96a_3a_6^2 - 1/24a_5^2 + 1/3456a_6^5)\, Id.
\end{aligned}$$

It is a special Lenard chain, since it possesses an additional symmetry. Indeed the tensor fields K_j and the 1-form θ are invariant along the vector field $\frac{\partial}{\partial a_1}$ which served as starting point of the present construction. It is therefore a Lenard chain with symmetry. This property has far reaching consequences, but their study go outside the limits of the present paper.

Acknowledgments

First, I would like to thank the organizers for the opportunity to present these results at the conference on Physics and Mathematics of Nonlinear Phenomena 2013. Secondly, I owe special thanks to Boris Konopeltchenko: without his strong impulse and friendly support , this research would have never been carried on. It is a part of a longstanding and unfinished collaboration with him. Thirdly, I want to thank Boris Dubrovin, since his work on Frobenius manifolds has been a great source of inspiration for this research. Finally thanks are also due to Federico Rossi, a former PhD student in Mathematics at the University of Milano Bicocca, for the help in the computation of the Lenard chains of E_6, E_7, and E_8.

References

[1] Magri F 2012 Recursion Operators and Frobenius Manifolds *SIGMA* **8** 076
[2] Haantjes J 1955 On X_m-forming sets of eigenvectors *Indag. Math.* **17** 158
[3] Nijenhuis A 1951 X_{n-1} Forming sets of eigenvectors *Indag. Math.* **13** 200
[4] Hertling C 2002 *Frobenius Manifolds and moduli spaces for singularities* Cambridge Tracts in Mathematics **151** (Cambridge: Cambridge University Press)
[5] Dubrovin B 1998 Differential Geometry of the Space of Orbits of Coxeter Groups *Surveys in Differential Geometry* **4** 181
[6] Konopelchenko B and Magri F 2007 Coisotropic deformations of associative algebras and dispersionless integrable hierarchies *Comm. Math. Phys.* **274** n.3 627

Integrable dispersionless PDEs arising as commutation condition of pairs of vector fields

S. V. Manakov[1] **and P. M. Santini**[2]

[1] Landau Institute for Theoretical Physics, Chernogolovka, Russia.
[2] Department of Physics, University of Roma "La Sapienza" and INFN, Sezione di Roma, Roma, Italy

E-mail: [2] paolo.santini@roma1.infn.it

Abstract.
In this paper we review some results about the theory of integrable dispersionless PDEs arising as commutation condition of pairs of one-parameter families of vector fields, developed by the authors during the last years. We review, in particular, the basic formal aspects of a novel Inverse Spectral Transform including, as inverse problem, a nonlinear Riemann - Hilbert (NRH) problem, allowing one i) to solve the Cauchy problem for the target PDE; ii) to construct classes of RH spectral data for which the NRH problem is exactly solvable, corresponding to distinguished examples of exact implicit solutions of the target PDE; iii) to construct the longtime behavior of the solutions of such PDE; iv) to establish in a simple way if a localized initial datum breaks at finite time and, if so, to study analytically how the multidimensional wave breaks. We also comment on the existence of recursion operators and Backlünd - Darboux transformations for integrable dispersionless PDEs.

1. Introduction

Waves propagating in weakly nonlinear and dispersive media are well described by integrable soliton equations, like the Korteweg - de Vries [1], the Nonlinear Scrhödinger [2] equations and their integrable $(2+1)$ dimensional generalizations, the Kadomtsev - Petviashvili [3] and Davey - Stewartson [4] equations respectively. The Inverse Spectral Transform (IST), introduced by Gardner, Green, Kruskal and Miura [5], is the spectral method allowing one to solve the Cauchy problem for such PDEs, predicting that a localized disturbance evolves into a number of soliton pulses + radiation, and solitons arise as an exact balance between nonlinearity and dispersion [6]-[9]. It is known that, apart from exceptional cases, soliton PDEs do not generalize naturally to more than $(2+1)$ dimensions; therefore, in the context of soliton equations, integrability is a property of low dimensional PDEs. There is another important class of integrable PDEs, the so-called dispersionless PDEs (dPDEs), or PDEs of hydrodynamic type, arising in various problems of Mathematical Physics and intensively studied in the recent literature (see, f.i., [10] - [34]). Since integrable dPDEs arise from the condition of commutation $[\hat{L}_1, \hat{L}_2] = 0$ of pairs of one-parameter families of vector fields, implying the existence of common zero energy eigenfunctions:

$$[\hat{L}_1, \hat{L}_2] = 0 \quad \Rightarrow \quad \hat{L}_j \psi = 0, \quad j = 1, 2, \tag{1}$$

they can be in an arbitrary number of dimensions [10]. Due to the lack of dispersion, these multidimensional PDEs may or may not exhibit a gradient catastrophe at finite time, and their integrability gives a unique chance to study analytically such a mechanism. Also with this motivation, a novel IST for vector fields, significantly different from that of soliton PDEs [6, 7, 9], has been recently constructed [35, 36, 37] i) to solve the Cauchy problem for dPDEs, ii) obtain the longtime behavior of solutions, iii) costruct distinguished classes of exact implicit solutions, iv) establish if, due to the lack of dispersion, the nonlinearity of the PDE is "strong enough" to cause the gradient catastrophe of localized multidimensional disturbances, and v) study analytically the breaking mechanism [35]-[43].

This paper is the written rendition of the talk presented by one of the authors (PMS) at the PMNP 2013 Conference, in a special session dedicated to the memory of Sergey V. Manakov, and it contains some of the results obtained by the authors in the period 2005-2011 (in §2 and §3), before the premature death of Sergey, with some additional formulas and considerations (in §4) not written before. This paper is dedicated to the memory of Sergey, a great scientist and a loyal friend.

2. Two basic examples

Here we consider, as illustrative examples, two integrable dPDEs; the first is associated with a Lax pair of vector fields not containing the partial derivative with respect to the spectral parameter λ, the second containing it.

1. The N-vector PDE in $(N+4)$ dimensions [36]:

$$\vec{u}_{t_1 z_2} - \vec{u}_{t_2 z_1} + (\vec{u}_{z_1} \cdot \nabla_{\vec{x}})\, \vec{u}_{z_2} - (\vec{u}_{z_2} \cdot \nabla_{\vec{x}})\, \vec{u}_{z_1} = \vec{0}, \tag{2}$$

equivalent to the commutativity condition $[\hat{L}_1, \hat{L}_2] = \hat{0}$ of the $(N+1)$ dimensional vector fields

$$\hat{L}_i = \partial_{t_i} + \lambda \partial_{z_i} + \vec{u}_{z_i} \cdot \nabla_{\vec{x}}, \qquad i = 1, 2, \tag{3}$$

where $\vec{u}(t_1, t_2, z_1, z_2, \vec{x}) \in \mathbb{R}^N$, $\vec{x} = (x_1, \dots, x_N) \in \mathbb{R}^N$ and $\nabla_{\vec{x}} = (\partial_{x_1}, .., \partial_{x_N})$, together with its deepest Hamiltonian reduction, the scalar PDE in $(2M+4)$ dimensions (for N even and $M = N/2$) and its Hamiltonian Lax pair [44]:

$$\begin{aligned} &\theta_{t_2 z_1} - \theta_{t_1 z_2} + \{\theta_{z_1}, \theta_{z_2}\}_{\vec{x}} = c(t_1, t_2, z_1, z_2), \\ &\hat{L}_j = \partial_{t_j} + \lambda \partial_{z_j} + \{\theta_{z_j}, \cdot\}, \quad j = 1, 2, \\ &\{f, g\}_{\vec{x}} \equiv \sum_{k=1}^{M} (f_{x_k} g_{x_{M+k}} - f_{x_{M+k}} g_{x_k}), \end{aligned} \tag{4}$$

where c is an arbitrary function of its arguments, reducing, for $N = 2$, to the well-known first and second heavenly equations of Plebanski [45]

$$\begin{aligned} &\{\theta_{z_1}, \theta_{z_2}\}_{x_1, x_2} = c(z_1, z_2), \quad \text{first heavenly equation} \\ &\theta_{t_2 x_1} - \theta_{t_1 x_2} + \{\theta_{x_1}, \theta_{x_2}\}_{x_1, x_2} = c(t_1, t_2), \quad \text{second heavenly equation,} \end{aligned} \tag{5}$$

obtained assuming respectively that θ does not depend on $t_j, j = 1, 2$ and that $z_j = x_j, j = 1, 2$.

2. The following system of two nonlinear PDEs in $2+1$ dimensions and its Lax pair of two - dimensional vector fields [37]:

$$\begin{aligned} &u_{xt} + u_{yy} + (u u_x)_x + v_x u_{xy} - v_y u_{xx} = 0, \\ &v_{xt} + v_{yy} + u v_{xx} + v_x v_{xy} - v_y v_{xx} = 0, \\ &\hat{L}_1 \equiv \partial_y + (\lambda + v_x)\partial_x - u_x \partial_\lambda, \\ &\hat{L}_2 \equiv \partial_t + (\lambda^2 + \lambda v_x + u - v_y)\partial_x + (-\lambda u_x + u_y)\partial_\lambda, \end{aligned} \tag{6}$$

describing a fairly general Einstein - Weyl metric [46]. This system reduces, for $v = 0$, to the celebrated dispersionless Kadomtsev - Petviashvili (dKP) equation and its Lax pair of Hamiltonian two-dimensional vector fields:

$$\begin{array}{l} u_{xt} + u_{yy} + (uu_x)_x = 0, \quad u = u(x,y,t) \in \mathbb{R}, \quad x,y,t \in \mathbb{R}, \\ \hat{L}_1 \equiv \partial_y + \lambda\partial_x - u_x\partial_\lambda = \partial_y + \lambda\partial_x + \{\frac{\lambda^2}{2} + u, \cdot\}_{\lambda,x}, \\ \hat{L}_2 \equiv \partial_t + (\lambda^2 + u)\partial_x + (-\lambda u_x + u_y)\partial_\lambda = \partial_t + \{\frac{\lambda^3}{3} + \lambda u - \partial_x^{-1}u_y, \cdot\}_{\lambda,x}, \\ \{f,g\}_{\lambda,x} \equiv f_\lambda g_t - f_t g_\lambda, \end{array} \tag{7}$$

describing the evolution of weakly nonlinear, nearly one-dimensional waves in Nature, in the absence of dispersion and dissipation [47],[48],[43], and, for $u = 0$, to the Pavlov equation [28] and its Lax pair [29] of non Hamiltonian one-dimensional vector fields:

$$\begin{array}{l} v_{xt} + v_{yy} + v_x v_{xy} - v_y v_{xx} = 0, \quad v = v(x,y,t) \in \mathbb{R}, \quad x,y,t \in \mathbb{R}, \\ \hat{L}_1 \equiv \partial_y + (\lambda + v_x)\partial_x, \\ \hat{L}_2 \equiv \partial_t + (\lambda^2 + \lambda v_x - v_y)\partial_x. \end{array} \tag{8}$$

3. The IST for vector fields

Since the Lax pair of integrable dPDEs is made of n-dimensional vector fields, Hamiltonian in some cases, the zero - energy eigenfunctions satisfy the following basic properties.
1) *The space of eigenfunctions is a ring.* If $f_1, f_2, \dots$ are solutions of the Lax pair equations (1), then an arbitrary differentiable function $F(f_1,, f_2, \dots)$ of them is also a solution of (1). A basis of this ring consists of n independent eigenfunctions.
2) *In the Hamiltonian reduction, the space of eigenfunctions is also a Lie algebra, whose Lie bracket is the natural Poisson bracket.* If $f_1, f_2, \dots$ are solutions of the Lax pair (1), then the Poisson bracket of any two of them $\{f_i, f_j\}$ is also a solution of (1).
For the sake of concreteness, we concentrate on the IST for the system (6) [35, 36, 37, 38]. Assuming that $\vec{u} = (u,v)^T \to \vec{0}$ as $x^2 + y^2 \to \infty$, we have that $\vec{\xi} = (\xi, \lambda)^T$, $\xi \equiv x - \lambda y$ is the basic eigenfunction of $\hat{L}_1$ at $x^2 + y^2 \to \infty$. We also assume that $\vec{u} = (u,v)^T \in \mathbb{R}^2$; it follows that, for $\lambda \in \mathbb{R}$, the vector fields are real.

The Direct Problem A basic role in the IST for real vector fields is played by a suitable basis of real (Jost) eigenfunctions $\vec{\phi}_\pm(x,y,\lambda)$, defined for $\lambda \in \mathbb{R}$, the solutions of equations $\hat{L}_1\vec{\phi}_\pm = \vec{0}$ satisfying the boundary conditions

$$\vec{\phi}_\pm(x,y,\lambda) \to \vec{\xi}, \quad y \to \pm\infty, \quad \lambda \in \mathbb{R}, \tag{9}$$

intimately related to the system of real ODEs

$$\frac{dx}{dy} = \lambda + v_x(x,y), \quad \frac{d\lambda}{dy} = -u_x(x,y) \tag{10}$$

defining the characteristics of $\hat{L}_1$. If the potentials (u,v) are sufficiently regular, the solution $(x(y), \lambda(y))$ of the ODE (10) exists unique globally in the (time) variable y, with the following free particle asymptotic behavior

$$x(y) \to \lambda_\pm y + x_\pm, \quad \lambda(y) \to \lambda_\pm, \quad y \to \pm\infty, \tag{11}$$

reducing to the asymptotics

$$x(y) \to \lambda y + x_\pm, \quad \lambda(y) = \lambda = \text{constant}, \quad y \to \pm\infty, \tag{12}$$

in the Pavlov reduction $u = 0$. Once the asymptotics $\lambda_{\pm}, x_{\pm}$ are constructed in terms of the initial data $x_0 = x(y_0), \lambda_0 = \lambda(y_0)$ of the ODE: $\lambda_{\pm}(x_0, y_0, \lambda_0), x_{\pm}(x_0, y_0, \lambda_0)$, the real eigenfunctions $\vec{\phi}_{\pm}$, that are particular constants of motion of the ODE, are given by

$$\vec{\phi}_{\pm}(x_0, y_0, \lambda_0) = (x_{\pm}(x_0, y_0, \lambda_0), \lambda_{\pm}(x_0, y_0, \lambda_0)). \tag{13}$$

Another important ingredient of the formalism is given by the complex eigenfunction $\vec{\psi}$, defined by the asymptotics

$$\vec{\psi}(y, \vec{x}, \lambda) \sim \vec{\xi}, \quad x^2 + y^2 \to \infty, \quad \lambda \notin \mathbb{R}, \tag{14}$$

analytic for $\lambda \notin \mathbb{R}$, having continuous boundary values $\vec{\psi}^{\pm}(x, y, \lambda)$, $\lambda \in \mathbb{R}$ from above and below the real λ axis, with the following asymptotics for large complex λ:

$$\begin{array}{l} \vec{\psi}^{\pm}(x, y, \lambda) = \vec{\xi} + \frac{1}{\lambda}\vec{U}(x, y) + \vec{O}\left(\frac{1}{\lambda^2}\right), \quad |\lambda| >> 1, \\ \vec{U}(x, y) \equiv \begin{pmatrix} -yu(x, y) - v(x, y) \\ u(x, y) \end{pmatrix}. \end{array} \tag{15}$$

The analyticity properties of $\vec{\psi}^{\pm}$ and of their $y \to \pm\infty$ limits follow from those of the analytic Green's functions $G^{\pm}$ of the undressed operator $\partial_y + \lambda\partial_x$

$$G^{\pm}(x, y, \lambda) = \pm \left(2\pi i(x - (\lambda \pm i\epsilon)y)\right)^{-1}, \tag{16}$$

exhibiting the following asymptotics for $y \to \pm\infty$:

$$\begin{array}{l} G^{\pm}(x - x', y - y', \lambda) \to \pm \left(2\pi i(\xi - \xi' \mp i\epsilon)\right)^{-1}, \quad y \to +\infty, \\ G^{\pm}(x - x', y - y', \lambda) \to \pm \left(2\pi i(\xi - \xi' \pm i\epsilon)\right)^{-1}, \quad y \to -\infty, \end{array} \tag{17}$$

and entailing that: the $y = +\infty$ asymptotics of $\vec{\psi}^{+}$ and $\vec{\psi}^{-}$ are analytic respectively in the lower and upper halves of the complex plane ξ, while the $y = -\infty$ asymptotics of $\vec{\psi}^{+}$ and $\vec{\psi}^{-}$ are analytic respectively in the upper and lower halves of the complex plane ξ. Similar features have been observed first in [49]. In addition $\vec{\psi}^{\pm} - \vec{\xi} = \vec{O}(\xi^{-1})$ as $|\xi| \gg 1$.

Scattering and spectral data. The $y = +\infty$ limit of $\vec{\phi}_{-}$ defines the natural (y - time) *scattering vector* $\vec{\sigma}$ for $\hat{L}_1$:

$$\lim_{y \to +\infty} \vec{\phi}_{-}(x, y, \lambda) \equiv \vec{\mathcal{S}}(\vec{\xi}) = \vec{\xi} + \vec{\sigma}(\vec{\xi}). \tag{18}$$

Since the space of eigenfunctions is a ring, the eigenfunctions $\vec{\psi}^{\pm}$ for $\lambda \in \mathbb{R}$ can be expressed in terms of the real eigenfunctions $\vec{\phi}_{\pm}$, and this expression defines the *spectral data* $\vec{\chi}^{\pm}_{\beta}(\xi, \lambda)$:

$$\begin{array}{l} \vec{\psi}^{\pm}(x, y, \lambda) = \vec{\mathcal{K}}^{\pm}_{-}(\vec{\phi}_{-}(x, y, \lambda)) = \vec{\mathcal{K}}^{\mp}_{+}(\vec{\phi}_{+}(x, y, \lambda)), \quad \lambda \in \mathbb{R}, \\ \vec{\mathcal{K}}^{\pm}_{\beta}(\vec{\xi}) \equiv \vec{\xi} + \vec{\chi}^{\pm}_{\beta}(\vec{\xi}), \quad \vec{\xi} = (\xi, \lambda), \end{array} \tag{19}$$

where $\vec{\chi}^{+}_{\beta}(\vec{\xi})$ and $\vec{\chi}^{-}_{\beta}(\vec{\xi})$ are analytic wrt the first argument ξ respectively in the upper and lower halves of the complex ξ - plane, as a consequence of equations (17) and of the above analyticity properties of $\vec{\psi}^{\pm}$ as $y \to \pm\infty$. We remark tha equations (19) for $\vec{\psi}^{-}$ can be omitted taking account of the symmetry properties coming from the reality of the potentials:

$$(u, v) \in \mathbb{R}^2 \;\Rightarrow\; \vec{\phi}_{\pm} \in \mathbb{R}^2, \; \vec{\psi}^{-} = \overline{\vec{\psi}^{+}}, \;\; \vec{\sigma} \in \mathbb{R}^2, \; \vec{\chi}^{-}_{\alpha} = \overline{\vec{\chi}^{+}_{\alpha}}, \;\; \lambda \in \mathbb{R}. \tag{20}$$

Evaluating the second of equations (19) for $\vec{\psi}^+$ at $y = +\infty$, one obtains the following linear Riemann - Hilbert (RH) problem with a shift:

$$\begin{array}{l} \vec{\sigma}(\xi,\lambda) + \vec{\chi}_-^{+}(\vec{\xi} + \vec{\sigma}(\xi,\lambda)) - \vec{\chi}_+^{-}(\xi,\lambda) = \vec{0}, \\ |\vec{\chi}_\beta^{\pm}(\xi,\lambda)| = O(\xi^{-1}), \quad \xi \sim \infty \end{array} \tag{21}$$

equivalent to a linear Fredholm integral equation [50], allowing one to uniquely construct the spectral data $\vec{\chi}_-^{+}$ and $\vec{\chi}_+^{-}$ from the scattering data $\vec{\sigma}$, under the hypothesis that the mapping $\xi \to \xi + \sigma_1(\xi,\lambda)$ be invertible.
Soon after the introduction of the RH problem (21) in [35], an alternative construction was proposed, based on the linear integral equations [36, 37]

$$\begin{array}{l} \tilde{\vec{\chi}}_-^{+}(\vec{\omega}) + \theta(\omega_1)\left(\tilde{\vec{\sigma}}(\vec{\omega}) + \int_{\mathbb{R}^2} \tilde{\vec{\chi}}_-^{+}(\vec{\eta})Q(\vec{\eta},\vec{\omega})d\vec{\eta}\right) = \vec{0}, \\ \tilde{\vec{\chi}}_-^{-}(\vec{\omega}) + \theta(-\omega_1)\left(\tilde{\vec{\sigma}}(\vec{\omega}) + \int_{\mathbb{R}^2} \tilde{\vec{\chi}}_-^{-}(\vec{\eta})Q(\vec{\eta},\vec{\omega})d\vec{\eta}\right) = \vec{0}, \end{array} \tag{22}$$

involving the Fourier transforms $\tilde{\vec{\sigma}}$ and $\tilde{\vec{\chi}}_\beta^{\pm}$ of $\vec{\sigma}$ and $\vec{\chi}_\beta^{\pm}$:

$$\tilde{\vec{\sigma}}(\vec{\omega}) = \int_{\mathbb{R}^2} \vec{\sigma}(\vec{\xi})e^{-i\vec{\omega}\cdot\vec{\xi}}d\vec{\xi}, \quad \tilde{\vec{\chi}}_\beta^{\pm}(\vec{\omega}) = \int_{\mathbb{R}^2} \vec{\chi}_\beta^{\pm}(\vec{\xi})e^{-i\vec{\omega}\cdot\vec{\xi}}d\vec{\xi} \tag{23}$$

and the scalar kernel:

$$Q(\vec{\eta},\vec{\omega}) = \frac{1}{(2\pi)^2}\int_{\mathbb{R}^2} e^{i(\vec{\eta}-\vec{\omega})\cdot\vec{\xi}}[e^{i\vec{\eta}\cdot\vec{\sigma}(\vec{\xi})} - 1]d\vec{\xi}. \tag{24}$$

To construct, say, (22) for $\tilde{\vec{\chi}}_-^{+}$, one applies the integral operator $\int_{\mathbb{R}} d\xi e^{-i\vec{\omega}\cdot\vec{\xi}}\cdot$ for $\omega_1 > 0$ to equation (21), using the above analyticity properties and the Fourier representations of $\vec{\chi}^{\pm}$ and $\vec{\sigma}$.

We remark that, in the Pavlov reduction $u = 0$, $\phi_{2\pm} = \psi_2^{\pm} = \lambda$, implying that $\sigma_2 = \chi_2^{\pm} = 0$, while in the Hamiltonian dKP reduction $v = 0$ [37]:

$$\{\phi_{\pm 1}, \phi_{\pm 2}\}_{x,\lambda} = \{\psi_1^{\pm}, \psi_2^{\pm}\}_{x,\lambda} = 1, \tag{25}$$

and, consequently:

$$\{\mathcal{S}_1, \mathcal{S}_2\}_{\vec{\xi}} = \{\mathcal{K}_1^{\pm}, \mathcal{K}_2^{\pm}\}_{\vec{\xi}} = 1. \tag{26}$$

Summarizing, from the initial data $\vec{u}(x,y,0) = (u(x,y,0), v(x,y,0))$ one constructs the real eigenfunctions $\vec{\phi}_-$ and then the scattering data $\vec{\sigma}(\xi,\lambda)$ through the solution of the ODE system (10). From the scattering data one constructs the spectral data $\vec{\chi}^{\pm}$ through the solution of the RH problem with shift (21). The main difficulties to make the above formalism rigorous are associated with the proof of the existence of the analytic eigenfunctions, and of their limits on the real λ axis from above and below. While such a proof exists [51] for the Pavlov reduction $u = 0$, for the dKP reduction the existence of the analytic eigenfunctions is proven, at the moment, only sufficiently far from the real λ axis [52]. We remark that, while the construction of the spectral data from the scattering data through the RH problem with shift (21) does not present difficulties, the one that makes use of the integral equations (22) may not be easy, due to the behavior of its kernel, as pointed out in [51].

Two inverse problems The first inversion (the reconstruction of $\vec{\phi}_-$ from the spectral data $\vec{\chi}_-^{+}$) is provided by the nonlinear integral equation

$$\vec{\phi}_-(x,y,\lambda) + H_\lambda \vec{\chi}_{-I}^{+}(\vec{\phi}_-(x,y,\lambda) + \vec{\chi}_{-R}^{+}(\vec{\phi}_-(x,y,\lambda)) = \vec{\xi}, \tag{27}$$

where $\vec{\chi}^{+}_{-R}$ and $\vec{\chi}^{+}_{-I}$ are the real and imaginary parts of $\vec{\chi}^{+}_{-}$, and H_λ is the Hilbert transform operator wrt λ

$$H_\lambda f(\lambda) = \frac{1}{\pi} PV \int_{-\infty}^{\infty} \frac{f(\lambda')}{\lambda - \lambda'} d\lambda'. \tag{28}$$

Equation (27) expresses the fact that the RHS of (19) for $\vec{\psi}^{+}$ is the boundary value of a function analytic in the upper half λ plane.

Once $\vec{\phi}_{-}$ is reconstructed from $\vec{\chi}^{+}_{-}$ solving the nonlinear integral equation (27), equations (19) give $\vec{\psi}^{\pm}$, and (u,v) is finally reconstructed from

$$\begin{aligned} u(x,y) &= \lim_{\lambda\to\infty} \left(\lambda(\psi_2^-(x,y,\lambda) - \lambda\right), \\ v(x,y) &= -yu - \lim_{\lambda\to\infty} \left(\lambda(\psi_1^-(x,y,\lambda) - (x - \lambda y)\right), \end{aligned} \tag{29}$$

consequence of (15). This inversion procedure was first introduced in [53].

A second inverse problem can be obtained eliminating the real eigenfunctions from the first of equations (19) for $\vec{\psi}^{\pm}$, obtaing a 2 vector nonlinear RH (NRH) problem on the real line:

$$\begin{aligned} &\psi_1^+(\lambda) = \mathcal{R}_1\left(\psi_1^-(\lambda), \psi_2^-(\lambda)\right), \quad \lambda \in \mathbb{R}, \\ &\psi_2^+(\lambda) = \mathcal{R}_2\left(\psi_1^-(\lambda), \psi_2^-(\lambda)\right), \\ &\psi_1^+(\lambda) = -y\lambda + x + O(\lambda^{-1}), \quad \psi_2^+(\lambda) = \lambda + O(\lambda^{-1}), \quad \lambda \sim \infty. \end{aligned} \tag{30}$$

or, in vector form:

$$\vec{\psi}^{+}(\lambda) = \vec{\mathcal{R}}\left(\vec{\psi}^{-}(\lambda)\right), \quad \lambda \in \mathbb{R}, \tag{31}$$

for the RH data $\vec{\mathcal{R}}$, constructed, via algebraic manipulation, from the spectral data. Once the analytic eigenfunctions are reconstructed through the solution of the NRH problem (30), the solution of the nonlinear PDE (6) is obtained from (29).

We remark that, in the two basic reductions, the RH data are constrained as follows:

$$\begin{aligned} &\mathcal{R}_2(\zeta_1,\zeta_2) = \zeta_2, \quad \text{Pavlov reduction}, \\ &\{\mathcal{R}_1,\mathcal{R}_2\}_{\zeta_1,\zeta_2} = 1, \quad \text{dKP reduction}. \end{aligned} \tag{32}$$

Evolution of the spectral data. The evolution of the scattering, spectral, and RH data is described by the following simple formula [37, 38]:

$$\Sigma_1(\xi,\lambda,t) = \Sigma_1(\xi - \lambda^2 t, \lambda, 0) \tag{33}$$

for the Pavlov equation, and

$$\begin{aligned} \Sigma_1(\xi,\lambda,t) &= t\left(\Sigma_2(\xi - \lambda^2 t,\lambda,0)\right)^2 + \Sigma_1(\xi - \lambda^2 t,\lambda,0), \\ \Sigma_2(\xi,\lambda,t) &= \Sigma_2(\xi - \lambda^2 t,\lambda,0) \end{aligned} \tag{34}$$

for the dKP equation.

We remark that, from the eigenfunctions $\vec{\phi}_{\pm}, \vec{\psi}^{\pm}$ of $\hat{L}_1$, one can constructs the common eigenfunctions $\vec{\Phi}_{\pm}, \vec{\Psi}^{\pm}$ of $\hat{L}_1$ and $\hat{L}_2$ through the formulae

$$\begin{aligned} &\Phi_{\pm 1} = \phi_{\pm 1} - t\left(\phi_{\pm 2}\right)^2, \quad \Phi_{\pm 2} = \phi_{\pm 2}, \\ &\Psi_1^{\pm} = \psi_1^{\pm} - t\left(\psi_2^{\pm}\right)^2, \quad \Psi_2^{\pm} = \psi_2^{\pm}, \end{aligned} \tag{35}$$

and the inverse (dressing) problem for the common eigenfunctions reads as follows.

Nonlinear Riemann - Hilbert dressing [37, 38, 39]. Let $\vec{\Psi}^{\pm}(\lambda)$ be the solutions of the following 2 vector NRH problem on the line

$$\vec{\Psi}^{+}(\lambda) = \vec{\mathcal{R}}\left(\vec{\Psi}^{-}(\lambda)\right), \quad \lambda \in \mathbb{R}, \tag{36}$$

with the normalization

$$\vec{\Psi}^{\pm}(\lambda) = \begin{pmatrix} -t\lambda^2 - y\lambda + x - 2ut \\ \lambda \end{pmatrix} + \vec{O}(\lambda^{-1}), \quad \lambda \sim \infty, \tag{37}$$

for the RH data $\vec{\mathcal{R}}(\vec{\zeta}) = (\mathcal{R}_1(\vec{\zeta}), \mathcal{R}_2(\vec{\zeta}))$, $\vec{\zeta} \in \mathbb{C}^2$. Then $\vec{\Psi}^{\pm}(\lambda)$ are eigenfunctions of $\hat{L}_j$ $j = 1, 2$: $\hat{L}_j \vec{\Psi}^{\pm} = \vec{0}$, $j = 1, 2$, and (u, v), constructed through the analogue of formulae (29):

$$\begin{aligned} u(x, y) &= \lim_{\lambda \to \infty} \left(\lambda \left[\Psi_2^-(x, y, \lambda) - \lambda\right]\right), \\ v(x, y) &= -yu - \lim_{\lambda \to \infty} \left(\lambda \left[\Psi_1^- + t(\Psi_2^{\pm})^2 - (x - \lambda y)\right]\right), \end{aligned} \tag{38}$$

are solutions of the nonlinear system (6).

If the spectral data $\vec{\mathcal{R}}(\vec{\zeta})$ satisfy the reality constraint

$$\vec{\mathcal{R}}(\overline{\vec{\mathcal{R}}(\bar{\vec{\zeta}})}) = \vec{\zeta}, \quad \forall \vec{\zeta} \in \mathbb{C}^2, \tag{39}$$

then the solutions u, v are real: $u, v \in \mathbb{R}$.
If the transformation $\vec{\zeta} \to \vec{\mathcal{R}}(\vec{\zeta})$ is simplectic:

$$\{\mathcal{R}_1, \mathcal{R}_2\}_{\zeta_1, \zeta_2} = 1, \tag{40}$$

then $v = 0$, $\vec{\Psi}^{\pm}(\lambda)$ are common eigenfunctions of the Hamiltonian vector fields (7), and u satisfies the dKP equation (7).
If $\mathcal{R}_2(\zeta_1, \zeta_2) = \zeta_2$, then $\Psi_2^{\pm}(\lambda) = \lambda$, $u = 0$ and v satisfies the Pavlov equation (8).

From the integral equations

$$\begin{aligned} \Psi_1^-(\lambda) &= -\lambda^2 t - \lambda y + x - 2ut + \frac{1}{2\pi i}\int_{\mathbb{R}} \frac{d\lambda'}{\lambda' - (\lambda - i0)} R_1\left(\Psi_1^-(\lambda'), \Psi_2^-(\lambda')\right), \\ \Psi_2^-(\lambda) &= \lambda + \frac{1}{2\pi i}\int_{\mathbb{R}} \frac{d\lambda'}{\lambda' - (\lambda - i0)} R_2\left(\Psi_1^-(\lambda'), \Psi_2^-(\lambda')\right), \\ \mathcal{R}_j(\zeta_1, \zeta_2) &\equiv \zeta_j + R_j(\zeta_1, \zeta_2), \quad j = 1, 2 \end{aligned} \tag{41}$$

characterizing the solutions of the RH problem (36), and from the definition (38), one obtains the following spectral characterization of the solution u:

$$u = F\left(x - 2ut, y, t\right) \in \mathbb{R}, \tag{42}$$

where the spectral function F, defined by

$$F\left(\xi, y, t\right) = -\int_{\mathbb{R}} \frac{d\lambda}{2\pi i} R_2\left(\Psi_1^-(\lambda; \xi, y, t), \Psi_2^-(\lambda; \xi, y, t)\right), \tag{43}$$

is connected to the initial data via the direct problem [37].

We remark that, while the IST of most of soliton PDEs provides a spectral representation of the solution involving, as parameters, the space-time coordinates, the inverse problem of system (6) provides a spectral representation (42), (43) of the solution involving, as parameter, also the solution u itself, in the combination $(x - 2ut)$. This is the spectral mechanism for the breaking of a generic localized initial condition at finite time t. This spectral mechanism is absent for the Pavlov equation (8) ($u = 0$), having the same linear part as dKP, but different nonlinear part.

The NRH dressing provides a very convenient tool i) to study the longtime behavior of the solutions of an integrable dispersionless PDE [39, 40, 41]; ii) to easily detect if a localized initial disturbance evolving according to such a PDE goes through a gradient catastrophe at finite time (f.i., there is no gradient catastrophe for the second heavenly equation (5) [36, 40], while there is a gradient catastrophe [41] for the dispersionless 2D Toda (or Boyer - Finley) equation $u_{xx} + u_{yy} = (exp(u))_{tt}$ [54, 55]); iii) to investigate analytically the wave breaking mechanism of such multidimensional waves (see Figure 1) [39, 41]; iv) to construct classes of RH data giving rise to exactly solvable NRH problems, and to distinguished exact implicit solutions of the dispersionless PDEs through an algorithmic approach [42, 56, 39, 40, 41]; v) to detect integrable differential reductions of the associated hierarchy of PDEs [57, 58], like the Dunajski interpolating equation $v_{xt}+v_{yy}+cv_xv_{xx}+v_xv_{xy}-v_yv_{xx} = 0$ [46], an integrable PDE interpolating between the dKP and the Pavlov equations, corresponding to the reduction $u = cv_x$ of system (6).

We also remark that the mechanism responsible for this feature, in our example, is that the vector field $\hat{L}_2$ is quadratic in λ and, at the same time, it contains the partial differential operator ∂_λ. Due to that, the unknown field u is present in the normalization of the eigenfunctions of the vector field. It is easy to see that these properties are shared by the whole dKP hierarchy, associated with time operators involving higher powers of λ.

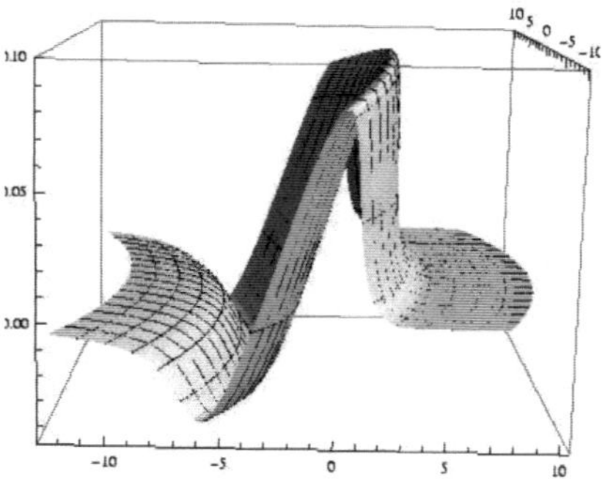

Figure 1. A small localized profile evolves according to the dKP equation into a parabolic wave front and breaks in a point of it. A detail of the parabolic wave front generated by a Gaussian initial profile around the breaking point at the breaking time, obtained from the analytical formula describing the wave breaking mechanism in the longtime regime [39, 43].

4. Recursion operators and hierarchies of dispersionless PDEs

Soliton PDEs like the KdV and NLS equations possess infinitely many symmetries (and constants of motions), organized in integrable hierarchies of PDEs. These commuting flows are generated by Nijenhuis (or hereditary) operators, and possess a multi-Hamiltonian structure [59, 60]. This pictures generalizes to integrable soliton PDEs in $(2 + 1)$ dimensions [61]. Soliton equations possess also discrete symmetries: the Backlünd - Darboux transformations, allowing one to construct solutions from solutions through a recursive procedure [7, 8]. Do we have a similar picture for multidimensional dispersionless PDEs?

The picture is the exactly same for example (2). We first observe that the vector equation (2) and its Lax operators (3) are nothing but the self-dual Yang - Mills (SDYM) equation [9] and its integrability structure

$$\begin{aligned} &[L_1, L_2] = 0 \quad \Rightarrow \quad U_{t_1z_2} - U_{t_2z_1} + [U_{z_1}, U_{z_2}] = 0, \\ &L_i \equiv \partial_{t_i} + \lambda\partial_{z_i} + U_{z_i}, \quad i = 1, 2, \end{aligned} \tag{44}$$

corresponding to the Lie algebra of N dimensional and λ independent vector fields:

$$U \to \vec{u} \cdot \nabla_{\vec{x}}, \quad \vec{u} = \vec{u}(\vec{x}, t_1, t_2, z_1, z_2). \tag{45}$$

Since the SDYM equation plays a role of master equation in the theory of integrable soliton PDEs [9], equation (2) should play an analogous role in the theory of integrable multidimensional dPDEs arising as commutativity condition of pairs of one parameter families of vector fields not containing derivatives with respect to the parameter λ [44].

It is known that the recursion operator [62]:

$$R = \Theta_2 \left(\Theta_1\right)^{-1}, \quad (\text{or} \;\; R' = \left(\Theta_1\right)^{-1} \Theta_2), \tag{46}$$

defined as factorization of the bi-Hamiltonian structure

$$\begin{array}{l} \Theta_1 = \partial_{z_1}, \quad \Theta_2 = \partial_{t_1} + ad(U_{z_1}), \\ ad(g) f \equiv [g, f], \end{array} \tag{47}$$

generates the hierarchy of symmetries of the SDYM equation. Since this operator contains the ad operator, it is perfectly well defined also in the vector field case (45) (the commutator of two vector fields is a vector field), giving rise to the following recursion operator and bi-Hamiltonian structures of the PDE (2):

$$\begin{array}{l} \check{R}' = \left(\check{\Theta}_1\right)^{-1} \check{\Theta}_2, \\ \check{\Theta}_1 = \partial_{z_1}, \quad \check{\Theta}_2 \vec{f} = \vec{f}_{t_1} + (\vec{u}_{z_1} \cdot \nabla_{\vec{x}}) \vec{f} - \left(\vec{f} \cdot \nabla_{\vec{x}}\right) \vec{u}_{z_1} \end{array} \tag{48}$$

Also in this vector field setting, $\check{\Theta}_1$, $\check{\Theta}_2$ is a compatible pair of Hamiltonian operators and $\check{R}$ is a Nijenhuis or hereditary operator, generating commuting symmetries of the PDE (2). The recursion operator $\check{R}'$ in (48) was first derived in [63] through different considerations (see also [44] for a third derivation of (48) and of the hierarchy of integrable flows associated with (2)).

If we try, instead, to construct the Backlünd transformations of (2) starting from the recursion operator $\tilde{R}$ of the Backlünd transformations of the SDYM equation [62]:

$$\begin{array}{l} \tilde{R} = \tilde{\Theta}_2 \left(\Theta_1\right)^{-1}, \\ \Theta_1 = \partial_{z_1}, \quad \tilde{\Theta}_2 = \partial_{t_1} + \tilde{ad}(U_{z_1}) \end{array} \tag{49}$$

depending on the generalized commutator

$$\tilde{ad}(U_{z_1}) f \equiv \tilde{U}_{z_1} f - f U_{z_1} \tag{50}$$

(where U and $\tilde{U}$ are two solutions of SDYM), one fails, since this generalized commutator of vector fields is not a vector field. For the same reason, one cannot use the construction of the Darboux transformations of the SDYM equation, defined by the generalized commutator equation $\tilde{L}_1 \mathcal{D} = \mathcal{D} L_1$, where L_1 is defined in (44), $\tilde{L}_1 = \partial_{t_1} + \lambda \partial_{z_1} + \tilde{U}_{z_1}$, and $\mathcal{D}$ is the unknown Darboux operator.

These elementary considerations give a simple explanation of the difficulty in constructing Darboux - Backlünd transformations of integrable dispersionless PDEs.

If the Lax pair is made of vector fields containing the operator ∂_λ, like in the examples (6), (7), no results are known so far concerning the recursion operator generating the hierarchy of commuting flows. However, elegant alternative characterizations of such hierarchies in the case of Hamiltonian vector fields can be found, f.i., in [14]-[20], and, in the general case, f.i., in [57, 58].

References

[1] D. J. Korteweg and G. de Vries 1895 *Philosophical Magazine* **39** (240) 422.
[2] V. E. Zakharov and A. B. Shabat 1973 Sov Phys *JETP* **37** 823.
[3] B. B. Kadomtsev and V. I. Petviashvili 1970 *Sov. Phys. Dokl.* **15** 539.
[4] A. Davey and K. Stewartson 1974 *Proc. R. Soc. A* **338** 101110.
[5] C. S. Gardner, J. M. Greene, M. D. Kruskal and R. M. Miura 1967 *Phys. Rev. Lett.* **19** 1095.
[6] V. E. Zakharov, S. V. Manakov, S. P. Novikov and L. P. Pitaevsky *Theory of solitons* Plenum Press, New York, 1984.
[7] M. J. Ablowitz and H. Segur *Solitons and the Inverse Scattering Transform* SIAM 1981.
[8] F. Calogero and A. Degasperis *Spectral Transform and Solitons* North-Holland Publishing Company, 1982.
[9] M. J. Ablowitz and P. A. Clarkson *Solitons, nonlinear evolution equations and Inverse Scattering*, London Math. Society Lecture Note 1991.
[10] V. E. Zakharov and A. B. Shabat 1979 *Functional Anal. Appl.* **13**, 166.
[11] V. E. Zakharov 1982 *Integrable systems in multidimensional spaces* in Lecture Notes in Physics Springer-Verlag Berlin **153** 190.
[12] Y. Kodama and J. Gibbons 1990 *Integrability of the dispersionless KP hierarchy* Proc. 4th Workshop on Nonlinear and Turbulent Processes in Physics, World Scientific, Singapore.
[13] I. M. Krichever 1988 *Functional Analysis and Its Applications* **22** 200.
[14] I. M. Krichever 1992 *Comm. Math. Phys.* **143**(2) 415.
[15] K. Takasaki 1991 *Area preserving diffeomorphisms and nonlinear integrable systems* in "Turku 1991, Proceedings, Topological and geometrical methods in field theory" 383.
[16] K. Takasaki and T. Takebe 1991 *Lett. Math. Phys.* **23** 205.
[17] K. Takasaki K and T. Takebe 1992 *Int. J. Mod. Phys. A Suppl.* **1B** 889.
[18] V. E. Zakharov 1994 *Dispersionless limit of integrable systems in 2+1 dimensions Singular Limits of Dispersive Waves*, edited by N. M. Ercolani et al., Plenum Press, New York.
[19] I. M. Krichever 1994 *Comm. Pure Appl. Math.* **47**, 437.
[20] K. Takasaki and T. Takebe 1995 *Rev. Math. Phys.* **7**, 743.
[21] M. Dunajski and L. J. Mason 2000 *Comm. Math. Phys.* **213**, 641.
[22] M. Dunajski and L. J. Mason 2003 *J. Math. Phys.* **44**, 3430.
[23] M. Dunajski, L. J. Mason and P. Tod 2001 *J. Geom. Phys.* **37** 63.
[24] M. Dunajski and K. P. Tod 2002 Einstein-Weyl spaces and dispersionless Kadomtsev-Petviashvili equation from Painlevé I and II *Preprint arXiv:nlin.SI/0204043.*
[25] B. Konopelchenko, L. Martinez Alonso and O. Ragnisco 2001 *J.Phys. A: Math. Gen.* **34** 10209.
[26] L. Martinez Alonso and A. B. Shabat 2003 *J. Nonlinear Math. Phys.* 10(2) 229.
[27] F. Guil, M. Manas and L. Martinez Alonso 2003 *J. Phys. A: Math. Gen.* **36** 6457.
[28] M. V. Pavlov 2003 *J. Math. Phys.* **44** 4134.
[29] M. Dunajski 2004 *J. Geom. Phys.* **51** 126.
[30] E. V. Ferapontov and K. R. Khusnutdinova 2004 *Comm. Math. Phys.* **248** 187.
[31] F. Neyzi, Y. Nutku and M. B. Sheftel 2005 *J. Phys. A: Math. Gen.* **38** 8473.
[32] L. V. Bogdanov and B. G. Konopelchenko 2005 *Phys. Lett. A* **345** 137.
[33] B. Konopelchenko and F. Magri 2007 *Comm. Math. Phys.* **274**, 627.
[34] B. Konopelchenko and F. Magri 2007 *Theoretical and Mathematical Physics* **151** (3) 803.
[35] S. V. Manakov and P. M. Santini 2005 Inverse Scattering Problem for Vector Fields and the Heavenly Equation *Preprint arXiv:nlin/0512043.*
[36] S. V. Manakov and P. M. Santini 2006 *Physics Letters A* **359** 613.
[37] S. V. Manakov and P. M. Santini 2006 *JETP Letters* **83**, No 10, 462.
[38] S. V. Manakov and P. M. Santini 2007 *Theor. Math. Phys.* **152**(1) 1004.
[39] S. V. Manakov and P. M. Santini 2008 *J. Phys. A: Math. Theor.* **41** 055204 (23pp).
[40] S. V. Manakov and P. M. Santini 2009 *J. Phys. A: Math. Theor.* **42** 404013 (16pp).
[41] S. V. Manakov and P. M. Santini 2009 *J. Phys. A: Math. Theor.* **42** 095203 (16pp).
[42] S.V. Manakov and P.M. Santini 2011 *J. Phys. A: Math. Theor.* **44** 345203.
[43] S. V. Manakov and P. M. Santini 2011 *J. Phys. A* **44** 405203.
[44] P. M. Santini 2013 On the integrable vector nonlinear PDE $\vec{u}_{t_1 z_2} - \vec{u}_{t_2 z_1} + (\vec{u}_{z_1} \cdot \nabla_{\vec{x}})\, \vec{u}_{z_2} - (\vec{u}_{z_2} \cdot \nabla_{\vec{x}})\, \vec{u}_{z_1} = \vec{0}$ in arbitrary dimensions: recursion and bi-Hamiltonian operators, and solvable nonlinear Riemann problems *Preprint to be submitted.*
[45] J. F. Plebanski 1975 *J. Math. Phys.* **16** 2395.
[46] M. Dunajski 2008 Interpolating dispersionless integrable systems *Preprint arXiv:0804.1234.*
[47] R. Timman 1962 Unsteady motion in transonic flow *Symposium Transsonicum* Aachen 1962. Ed. K. Oswatitsch, Springer 394.

[48] E. A. Zobolotskaya and R. V. Kokhlov 1969 *Sov. Phys. Acoust.* **15** 1 35.
[49] S. V. Manakov and V. E. Zakharov 1981 *Letters in Mathematical Physics* **5** 247.
[50] F. D. Gakhov 1990 *Boundary value problems*, Dover, New York.
[51] P. G. Grinevich, P. M. Santini and D. Wu The Cauchy problem for the Pavlov equation 2013 *Preprint, to be submitted.*
[52] P. G. Grinevich and P. M. Santini 2013 *J. Differential Equations* 2013 **255** 1469.
[53] S. V. Manakov 1981 *Physica D* **3** 420.
[54] J. D. Finley and J. F. Plebanski 1979 *J. Math. Phys.* **20** 1938.
[55] C. Boyer and J. D. Finley 1982 *J. Math. Phys.* 23 1126.
[56] L. Bogdanov, V. Dryuma and S. V. Manakov 2007 *J. Phys. A: Math. Theor.* **40** 14383.
[57] L. V. Bogdanov 2010 *J. Phys. A: Math. Theor.* **43** 115206.
[58] L. V. Bogdanov 2011 *Theor. Math. Phys.* **167** (3) 705.
[59] F. Magri 1978 *J. Math. Phys.* **19** 1156-1162.
[60] B. Fuchssteiner and A. S. Fokas 1981 *Physica D* **4** 47.
[61] P. M. Santini and P. M. Fokas 1988 *Comm. Math. Phys.* **115** 375.
[62] M. Bruschi, D. Levi and O. Ragnisco 1982 *Lett. Nuovo Cimento* **33** 263.
[63] M. Marvan and A. Sergyeyev 2012 *Inverse Problems* **28** 025011 (12pp).

Ring localized structures in nonlinear shallow water wave dynamics

A. Mannan[1,6], **R. Fedele**[2,6], **M. Onorato**[3], **S. De Nicola**[4,6] **and D. Jovanović**[5,6]

[1] Dipartimento di Matematica e Fisica, Seconda Università degli Studi di Napoli, Sede di Caserta, Caserta, Italy
[2] Dipartimento di Fisica, Università di Napoli "Federico II", Napoli, Italy
[3] Dipartimento di Fisica, Università di Torino, Torino, Italy
[4] CNR-SPIN Napoli, Napoli, Italy
[5] Institute of Physics, University of Belgrade, Belgrade, Serbia
[6] INFN Sezione di Napoli, Napoli, Italy

E-mail: abdul.mannan@na.infn.it

Abstract. The nonlinear dynamics of the concentric shallow water waves is described by means of the *cylindrical* Korteweg-de Vries equation, often referred to as the *concentric* Korteweg-de Vries equation (cKdVE). By using the mapping that transforms a cKdVE into the standard one – hereafter also referred to as the *planar* Korteweg-de Vries equation (KdVE) – the spatiotemporal evolution of a cylindrical surface water wave, corresponding to a *tilted* cylindrical bright soliton, is described. The usual representation of a tilted soliton is 'non-physical'; here the cylindrical coordinate and the retarded time play the role of time-like and space-like variables, respectively. However, we show that, when we express such analytical solution of the cKdVE in the appropriate representation in terms of the two horizontal space coordinates, say X and Y, and the 'true' time, say T, this non-physical character disappears. The analysis is then carried out numerically to consider the surface water wave evolution corresponding to initially localized structures with standard boundary conditions, such as bright soliton, Gaussian and Lorentzian profiles. A comparison among those profiles is finally presented.

1. Introduction

Large amplitude wave propagation is encountered in almost all physical areas, ranging from nano- to astro-physical scales [1, 2, 3, 4, 5]. Remarkably, in such diverse areas, most of the nonlinear physical processes involved in the nonlinear wave propagation are modeled by similar nonlinear partial differential equations. The physics of extreme events, or simply the *extreme physics*, is a very recently-developed interdisciplinary research area, based on a very wide platform of phenomena and methodologies that are imported from the diverse disciplines. For instance, ultra-strong wave intensities can be artificially produced by means of very advanced techniques, such as the use of ultra-short (duration in femto-seconds and, in the near future, atto-seconds and ultra strong ($10^{20} - 10^{22}$ W/cm^2 intensities) laser pulses [6, 7] in plasmas to achieve the particle acceleration to very high energies (plasma-based accelerator schemes) [6, 7, 8, 9] or the controlled thermonuclear inertial fusion [10]. Ultra-strong wave intensities can be also naturally generated, such as in a number of astrophysical conditions [11, 12] or in ocean wave physics [13]. The latter is essentially the physics of the surface gravity waves, ranging from shallow to deep water regimes, and it is probably the oldest discipline deeply related to the extreme physics [14], such as the one involving rogue wave [15] and tsunami generation [16]. However, water wave physics, that includes also the study of artificially-produced nonlinear wave phenomena, is the subject for which several mathematical techniques to analyze and solving nonlinear partial differential equations have been developed for the first time and, later on, transferred to the other disciplines.

In this paper, we describe the dynamics of concentric large amplitude surface gravity waves in shallow water regime, in the horizontal space coordinates. This is done both analytically and numerically. In particular, the analytical description is done by using the mapping that relates the analytical solutions of the concentric (or cylindrical) Korteweg-de Vries equation (cKdVE) to the ones of the standard (or planar) Korteweg-de Vries equation (KdVE). Then, we analyze numerically the spatiotemporal evolution of cylindrical localized surface gravity waves with different initial profiles (f. i., bright soliton, Gaussian and Lorentzian). In the next section, we briefly review the fluid model typically adopted to obtain both the families of the KdVE and cKdVE describing the nonlinear water waves in the shallow water regime. In section 3, we briefly present the transformation that allows to go from cKdVE to KdVE and vice versa. Then the analytical nonlinear shallow water solutions of cKdVE with *tilted* boundary conditions are found: they are the image of the localized shallow water solutions of the planar KdVE by means of the above mapping. In section 4, we find concentric numerical solutions of the cKdVE, corresponding to bright, Gaussian and Lorentzian initial profiles. All the solutions are first represented in the retarded time and the radial space coordinates (ζ and s, respectively). Once these variables are expressed in the horizontal ones, X and Y, all the above solutions assume the form of ring localized structures (vortex state) characterized by a cylindrical propagation which is associated with a decreasing of the amplitude as $1/s$. Finally, conclusions and remarks are given in section 5.

2. Model equations

We assume the model of water corresponding to an incompressible, inviscid fluid of constant density ρ with the acceleration of gravity g and a zero surface tension. We also assume that the water moves over an impermeable bed with constant depth and with a constant pressure (atmospheric pressure) at the free surface. To write the governing equations of our system, we refer for procedure and details to the previous literature [f. i., see [17, 18, 19], and references therein]. These consist in the Euler's equations of motion (separated into vertical and perpendicular components with respect to the direction of the gravity acceleration vector $\mathbf{g}$) and mass conservation equation. These can be cast as follows

$$\frac{\partial \mathbf{v}_\perp}{\partial t} + \epsilon(\mathbf{V}\cdot\nabla)\mathbf{v}_\perp = -\nabla_\perp p, \tag{1}$$

$$\delta^2\left[\frac{\partial w}{\partial t} + \epsilon(\mathbf{V}\cdot\nabla)w\right] = -\frac{\partial p}{\partial z}, \tag{2}$$

$$\nabla\cdot\mathbf{V} = 0, \tag{3}$$

where: z is the vertical component of the position vector normalized to the constant unperturbed depth h_0; $\nabla = \nabla_\perp + \hat{z}\partial/\partial z$ is the gradient operator, $\nabla_\perp$ being its perpendicular component; p denotes the dimensionless pressure which measures the deviation from the hydrostatic pressure distribution; $\mathbf{V} = (\mathbf{v}_\perp, w)$ is the fluid velocity, $\mathbf{v}_\perp$ and w being the perpendicular component normalized by $\sqrt{gh_0}$ and the vertical component normalized by $h_0\sqrt{gh_0}/\lambda$, respectively, with λ wavelength of the wave; t is the time variable normalized by $\lambda/\sqrt{gh_0}$. Note that, by denoting by $\mathbf{x}_\perp$ as the perpendicular component of the position vector normalized by λ, $z = h(\mathbf{x}_\perp, t)/h_0$ represents the surface normalized by h_0, h being the actual depth. Equations (1)-(3) have been written by introducing some parameters that fix the scales of the problem. First of all, we have defined the parameter $\delta = h_0/\lambda$ which estimates the shallowness of the water and introduced an arbitrarily large perturbation in height, namely $h(\mathbf{x}_\perp, t) = h_0 + \Delta h(\mathbf{x}_\perp, t)$. Then, denoting by a the maximum value of Δh, we have defined the dimensionless quantity $\eta(\mathbf{x}_\perp, t) = \Delta h(\mathbf{x}_\perp, t)/a$, which represents the profile of the perturbation, usually referred to as the *elevation* of the surface wave. Therefore, the expression for h can be easily rewritten, i.e., $h(\mathbf{x}_\perp, t) = h_0 + a\eta(\mathbf{x}_\perp, t)$ and another fundamental parameter has been introduced, i.e., $\epsilon = a/h_0$, that gives the measure of the nonlinearity (i.e. estimates how large is the amplitude).

Note that, if $\epsilon = 0$ no wave is present. Consequently, as $\epsilon \to 0$, the disturbances associated with the passage of the wave vanish. However, when ϵ increases the amplitude of the perturbations becomes larger, so that $\epsilon \ll 1$ describes the linear regime, whilst the values closer to 1 describe the nonlinear regime. It turns out, that, in order to be consistent with the choice of the ϵ value, it is required that we scale our physical quantity with respect to ϵ as follows: $(\mathbf{v}_\perp, w, p) \to \epsilon(\mathbf{v}_\perp, w, p)$. Equations (1)-(3) have to be implemented by the following boundary conditions with

$$p = \eta \quad and \quad w = \frac{\partial \eta}{\partial t} + \epsilon(\mathbf{v}_\perp \cdot \nabla_\perp)\eta \quad on \quad z = 1 + \epsilon\eta \tag{4}$$

and

$$w = 0 \quad on \quad z = 0. \tag{5}$$

Boundary condition (4) corresponds to impose that the pressure on the wave surface be the atmospheric pressure and the vertical velocity component is given by the total derivative of the elevation, whilst the boundary condition (5) fixes to 0 the vertical velocity component at the bottom.

2.1. Korteweg-de Vries equation

To derive the KdVE, we assume that a fluid disturbance is propagating in a narrow channel with $\mathbf{x}_\perp = (x, y)$ and $\mathbf{v}_\perp = (u_x, 0)$ (absence of $y-$component of the transverse velocity field $\mathbf{v}_\perp$), and that all quantities depends only on x, z and t. Moreover, the effects of surface tension are assumed to be negligible. We further introduce the new variables $\mathbf{X}_\perp$, T and W, by the following scaling

$$\mathbf{x}_\perp = \frac{\delta}{\sqrt{\epsilon}}\mathbf{X}_\perp \equiv \frac{\delta}{\sqrt{\epsilon}}(X, Y), \quad t = \frac{\delta}{\sqrt{\epsilon}}T, \quad w = \frac{\sqrt{\epsilon}}{\delta}W\,. \tag{6}$$

By substituting this scaling in eqs. (1)-(5), we obtain the governing system of equations which is similar to the former one, but where δ^2 is replaced by ϵ. For disturbances that are propagating in the positive X-direction, we introduce the following stretched coordinates:

$$\xi = X - \tau, \qquad \tau = \epsilon T\,. \tag{7}$$

Note that τ describes the long time scale. We restrict our study to a weakly nonlinear evolution equation and perform a Taylor expansion in the small parameter ϵ of η, u_x, W, and p. After some tedious calculations (for details, see Ref. [17]), we finally obtain an equation for η_0 (the surface elevation at the leading order):

$$\frac{\partial \eta_0}{\partial \tau} + \frac{3}{2}\eta_0\frac{\partial \eta_0}{\partial \xi} + \frac{1}{6}\frac{\partial^3 \eta_0}{\partial \xi^3} = 0. \tag{8}$$

Equation (8), usually referred to as the Korteweg-de Vries equation, governs the spatio-temporal evolution of surface waves in shallow water.

2.2. Cylindrical Korteweg-de Vries equation

In cylindrical coordinates (r, θ, z), with the z-axis pointing upwards in the vertical direction, the flow region is bounded below by solid bottom at $z = 0$, and above by a free surface, say $z = \eta(r, t)$, which measures the displacement from the undisturbed free surface at $z = 0$. The fluid is supposed to be unbounded in the radial direction $(0 < r < \infty)$. Here, the cylindrical coordinates are dimensionless as in the previous case. The perpendicular components $\mathbf{x}_\perp$ of the position vector $\mathbf{x}$ is here represented by $\mathbf{x}_\perp = (r, \theta)$. For simplicity, we consider long waves having a vertical axis of symmetry at all values of the time t (cylindrical symmetry). So that, the governing equations describe surface waves with a circular wave front. Therefore, all the quantities are independent of θ and the perpendicular component of the velocity field is only radial, say u_r. To determine the cylindrical KdVE, we follow a procedure similar to the one used for the KdVE case. Thus, we introduce the stretched coordinates, ζ (retarded time) and s, and new variables H, P, U, W, respectively, i.e.

$$\zeta = \frac{\epsilon^2}{\delta^2}(r - t), \quad s = \frac{\epsilon^6}{\delta^4} r, \quad (\eta, p, u_r) = \frac{\epsilon^3}{\delta^2}(H, P, U), \quad w = \frac{\epsilon^5}{\delta^4} W. \tag{9}$$

in the governing eqs. (1)-(5). Following the weakly nonlinear analysis performed in the one dimensional case (for details see Ref. [17]), we get for the leading order term, say H_0, the following equation:

$$\frac{\partial H_0}{\partial s} + \frac{3}{2} H_0 \frac{\partial H_0}{\partial \zeta} + \frac{1}{6}\frac{\partial^3 H_0}{\partial \zeta^3} + \frac{H_0}{2s} = 0. \tag{10}$$

Equation (10) is usually referred to as the concentric Korteweg-de Vries equation.

3. Analytical soliton solutions for cylindrical surface gravity waves

In order to describe analytically the spatiotemporal evolution of the surface gravity waves in the form of solitons, we use the connection between eqs. (8) and (10). Let us take their more general forms, where the coefficients 3/2 and 1/6 are replaced by arbitrary coefficients A and B, respectively, that are assumed to be real numbers. For simplicity of the notation, we introduce also the following replacements: $\eta_0 \to u$ and $H_0 \to v$. Then, we consider the following planar and cylindrical Korteweg-de Vries equations, respectively:

$$\frac{\partial u}{\partial \tau} + B\, u \frac{\partial u}{\partial \xi} + A \frac{\partial^3 u}{\partial \xi^3} = 0 \tag{11}$$

and

$$\frac{\partial v}{\partial s} + B\, v \frac{\partial v}{\partial \zeta} + A \frac{\partial^3 v}{\partial \zeta^3} + \frac{v}{2s} = 0\,. \tag{12}$$

According to the definitions given in the previous sections, all the quantities involved in these equations are dimensionless. It is easy to see that eqs. (11) and (12) are related by the following transformations [20, 21]:

$$\mathcal{T} : \begin{cases} \xi = s^{-1/2}\zeta, \quad \tau = -2s^{-1/2} \\ v(\zeta, s) = s^{-1}\left[u\left(\xi = s^{-1/2}\zeta, \tau = -2s^{-1/2}\right) + \zeta/2B\right] \end{cases} \tag{13}$$

and

$$\mathcal{T}' : \begin{cases} \zeta = -2\xi/\tau, \quad s = 4/\tau^2 \\ u(\xi, \tau) = \tau^{-1}\left[4\tau^{-1} v\left(\zeta = -2\xi/\tau,\; s = 4/\tau^2\right) + \xi/B\right] \end{cases} . \tag{14}$$

Note that $\mathcal{T}$ transforms (11) into (12) whilst $\mathcal{T}'$ transforms (12) into (11). A transformation similar to $\mathcal{T}'$ have been first introduced in Ref. [22].

Remarkably, similar transformations have been found to transform the planar to the cylindrical Kadomtsev-Petviashvili equations [23].
To find an analytical solutions of the cKdVE, we consider here the classical planar bright soliton:

$$u(\xi,\tau) = u_m \operatorname{sech}^2\left[\sqrt{\frac{Bu_m}{12A}}\,(\xi - V_0\tau)\right], \tag{15}$$

where $B\,u_m/A > 0$, and $V_0 = B\,u_m/3$.
By using the transformation (13), we find the solutions of the concentric equation (12) corresponding to the bright soliton of the planar KdVE. By definition, we may refer these concentric solutions to as the *cylindrical (or concentric) bright soliton*, with tilted boundary conditions. Therefore, we have:

$$v(\zeta,s) = \frac{1}{B\,s}\left\{\frac{1}{2}\zeta + B\,u_m \operatorname{sech}^2\left[\sqrt{\frac{B\,u_m}{12A\,s}}\,(\zeta + 2V_0)\right]\right\}, \tag{16}$$

where, according to (13) and (14), the standard boundary conditions of solitons (15) are transformed into *tilted* boundary conditions by means of transformation $\mathcal{T}$. The analytical

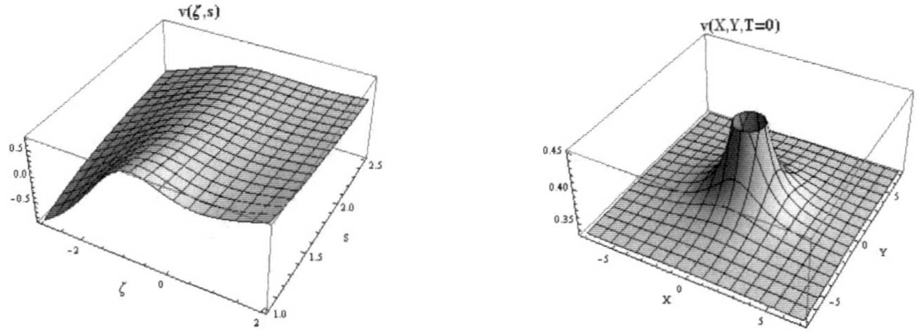

Figure 1. 3D spatial profile of the height v of the analytical initial bright soliton in true space (ζ,s) and X-Y plane for $u_m = 1.0$, $A = 1/6$, and $B = 3/2$.

concentric single-soliton solutions are displayed in figures 1 and 2. The plot on the left side of figure 1 represents tilted solitons as a function of ζ and s. The 3D plot on the right side of figure 1 shows the corresponding solutions in the $X - Y$ space at $T = 0$ (note that the cylindrical coordinate $s \propto \sqrt{X^2 + Y^2}$ corresponds to the time-like variable for cKdVE). Figure 2 shows the temporal evolution (in $T \propto t$) of the 3D plots in the $X - Y$ plane for a tilted-bright soliton. We observe that the initial profiles ($T = 0$) *radiate* circularly. This is due to the non-stationary features of the tilted-solitons with respect to the artificial time s.

4. Numerical solutions of the cylindrical case

We consider three different initial conditions: the bright soliton case given by eq. (17), the Gaussian and the Lorentzian profiles given by eqs. (18) and (19), respectively. The parameter s_i in each case governs the width of the three distributions at the initial time.

$$v(\zeta) = \frac{u_m}{s_i}\operatorname{sech}^2\left[\zeta\sqrt{\frac{Bu_m}{12As_i}}\right], \tag{17}$$

$$v(\zeta) = \frac{u_m}{s_i}\exp\left[-\left(\frac{\zeta}{s_i}\right)^2\right], \tag{18}$$

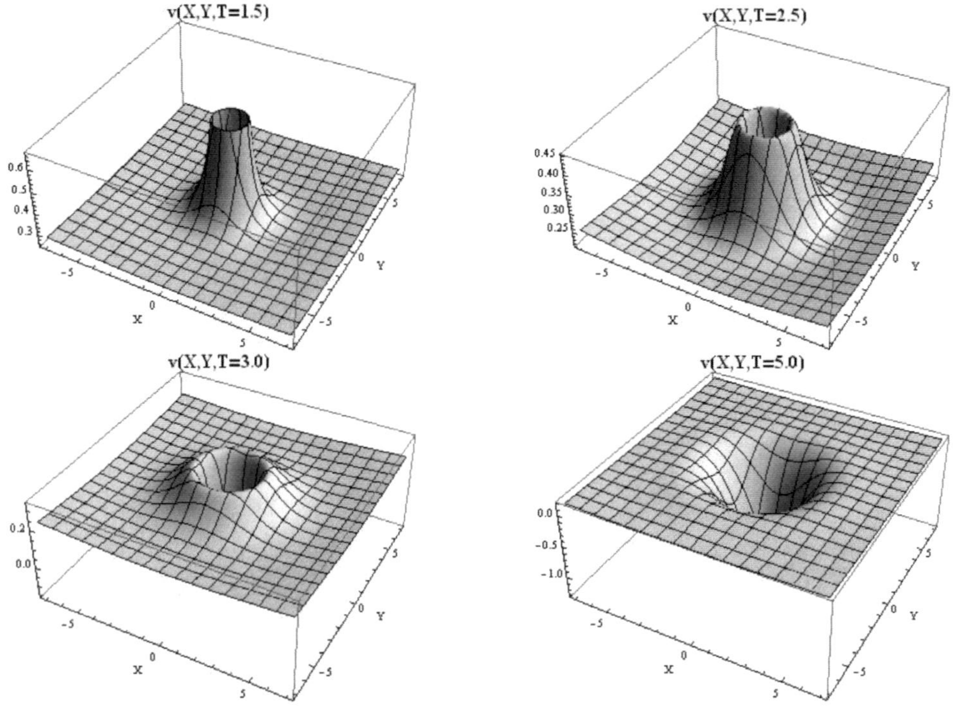

Figure 2. 3D spatial profile of the height v of the analytical bright soliton in X-Y plane for $u_m = 1.0$, $A = 1/6$, and $B = 3/2$.

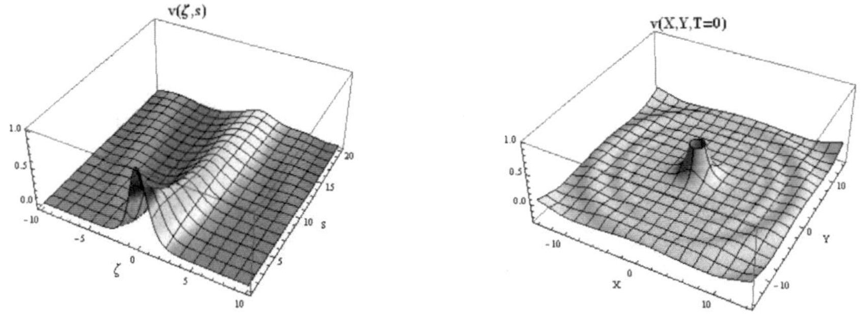

Figure 3. 3D spatial profile of the height v of the initial bright soliton in true space (ζ,s) and X-Y space for $u_m = 1.0$, $A = 1/6$, and $B = 3/2$.

$$v(\zeta) \;=\; \frac{u_m/s_i}{1+(\zeta/s_i)^2}. \tag{19}$$

The plot on the left side of figure 3 displays the numerical calculation of the 3D spatial profile of the height of the cylindrical bright soliton in true space. The numerical results have been obtained by integrating the eq. (12) with standard boundary conditions. The corresponding representation of the 3D profile of the height is represented in the real $X - Y$ plane at the true time $T = 0$ and it is shown on the right side of figure 3 for the same values of the parameters $u_m = 1.0$, $A = 1/6$, and $B = 3/2$. Figure 4 shows the time evolution of the 3D spatial profile of the height of the cylindrical bright soliton in $X - Y$ plane for increasing the values of the true time T. The amplitude of the localized initial profile decreases with increasing the time

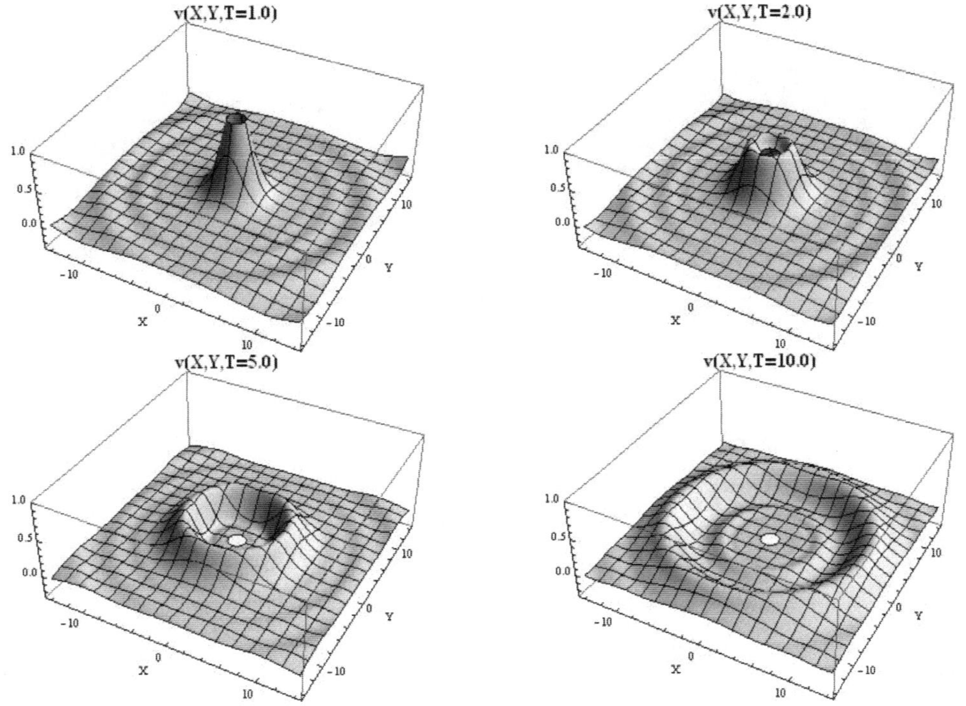

Figure 4. 3D spatial profile of the height v of the bright soliton in X-Y space for $u_m = 1.0$, $A = 1/6$, and $B = 3/2$.

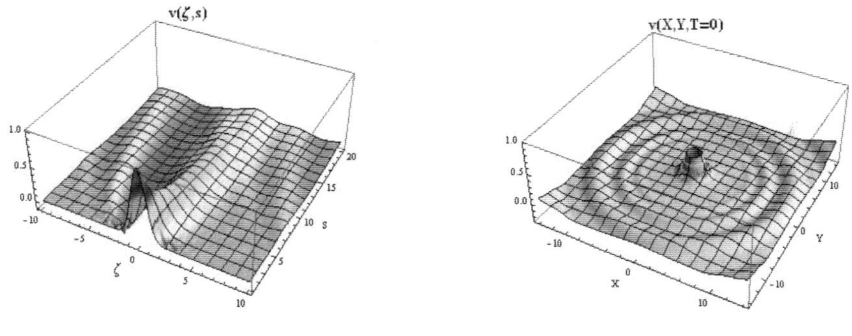

Figure 5. 3D spatial profile of the height v of the initial Gaussian case in true space (ζ,s) and X-Y space for $u_m = 1.0$, $A = 1/6$, and $B = 3/2$.

T while the waves are radiated in a form of circular front in the $X - Y$ plane. Figures 5, 6, 7, and 8 show the results of the numerical integration of eq. (12) for Gaussian and Lorentzian initial conditions given by the eqs. (18) and (19), respectively. It can be clearly seen that the initial peaked amplitude decreases rapidly with increasing time T. Both the initial profiles tend to radiate in circular front with increasing radii. However the radiation emitted in the case of the initial Lorentzian profile appeared to be less pronounced compared to the Gaussian case and resulting in the formation of circular front comparable to the bright soliton case.

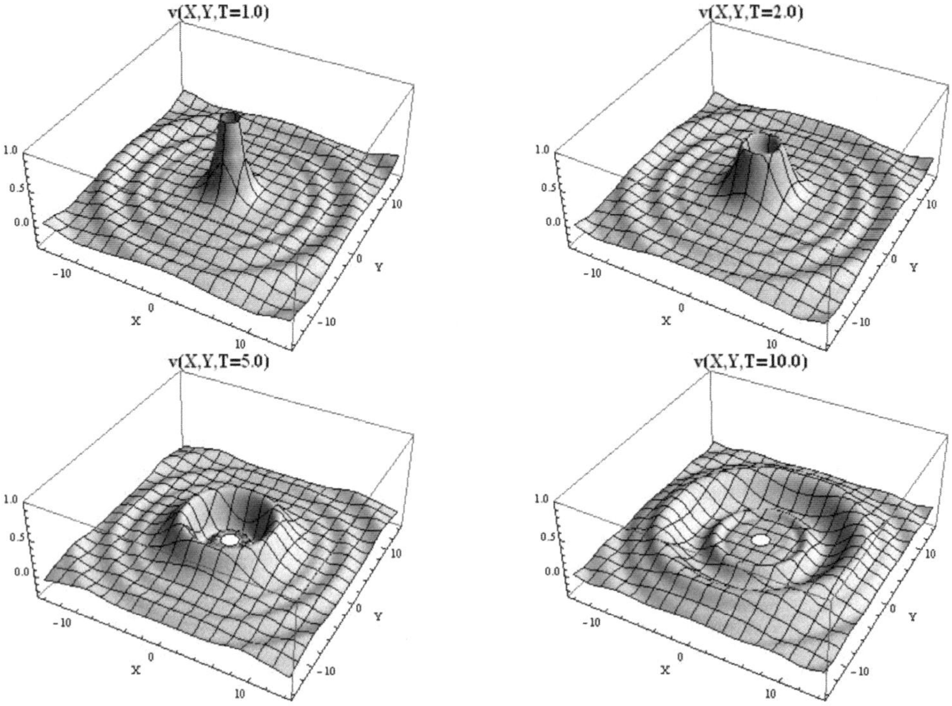

Figure 6. 3D spatial profile of the height v of the Gaussian profile in X-Y space for $u_m = 1.0$, $A = 1/6$, and $B = 3/2$.

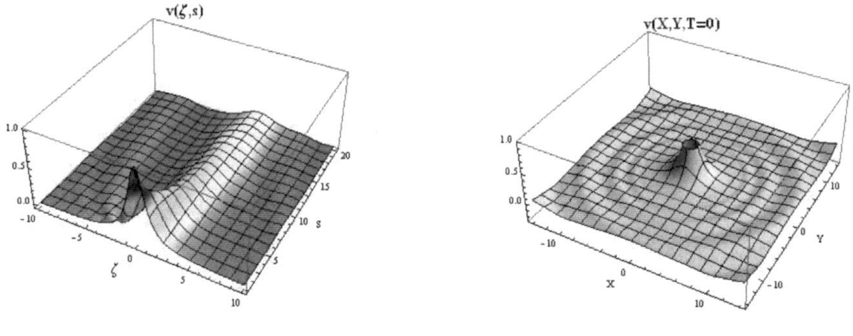

Figure 7. 3D spatial profile of the height v of the initial Lorentzian case in true space (ζ,s) and X-Y plane for $u_m = 1.0$, $A = 1/6$, and $B = 3/2$.

5. Conclusions and remarks

In this paper, we have studied both analytically and numerically, the nonlinear dynamics of the concentric surface gravity waves in the form of localized solutions of the cylindrical KdVE. We have also reviewed the previous approaches, by using a transformation that relates the solutions of the planar (or standard) KdVE to those of the cKdVE. The analytical localized solution (in the $\zeta - s$ domain) in the form of bright tilted solitons has been found. These solutions have also been represented in the real $X - Y$ space and in the true time T, where the spatio-temporal evolution of the tilted solitons ($\zeta - s$ representation) is transformed to X-Y space with an initially given ($T = 0$) soliton profile. The "temporal" behavior with respect to the time-like variable s shows that the localized shape of the analytical solutions reduces as s increases, vanishing

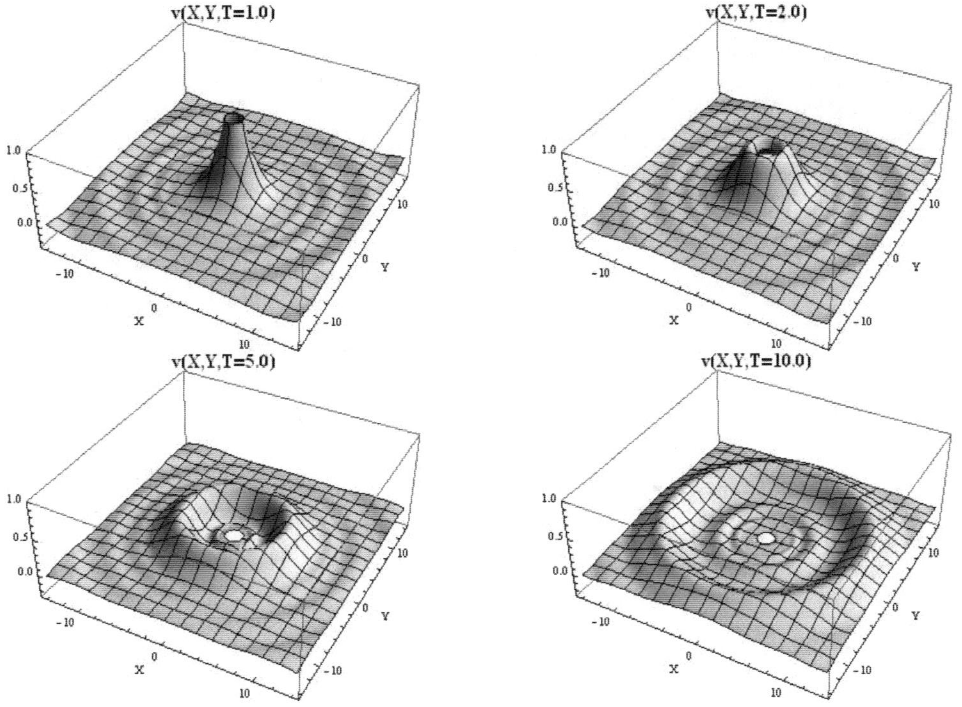

Figure 8. 3D spatial profile of the height v of the Lorentzian profile in X-Y plane for $u_m = 1.0$, $A = 1/6$, and $B = 3/2$.

asymptotically; correspondingly, the tilted character of these solutions reduces as s increases, as well, for asymptotic values of s. Numerical localized solutions with standard boundary conditions corresponding to single-pulse initial profiles, has been found as function of ζ and s. They show a temporal behavior (with respect to s) of the local amplitude similar to the one shown by the analytical localized solutions. Also here this behavior is transformed to $X - Y$ space with an initially given localized profile ($T = 0$). Furthermore, we observe that the analytical solutions of the cKdVE (with *tilted* character) and the numerical ones (with *standard* character of the boundary conditions), both given in $\zeta - s$ space, when transformed to the $(X - Y)$ space, show very similar behavior during the propagation. In particular they exhibit similar features of ring localized structures (in particular, ring bright solitons) that are typically associated with vortex states. Remarkably, this physical circumstance was observed in an irrotational medium.

Finally, we would like to stress that, in this preliminary investigation, we have analysed only the behaviour of localized structures that assume the form of a single pulse in the $\zeta - s$ domain. However, a more detailed analysis, that takes into account both analytical and numerical localized structures, in the form of multi-pulses (in particular, multi-solitons), in $\zeta - s$ domain, is under way. In this space, the simple typical feature exhibited by the *planar* multi-pulse structures, for which the higher pulses have the higher speed, becomes more complex or not easily recognizable for concentric waves, and in the tilted analytical solutions is *apparently* lost. Furthermore, once such structures are transformed to the $X - Y$ space, this nonlinear feature seems to be hidden by the ring character of the waves in this representation. In a forthcoming paper, we analyse in detail all these aspects.

References

[1] Witham G B 1974 *Linear and nonlinear waves* (John Wiley, New York)

[2] Calogero F and Degasperis A 1982 *Spectral transform and solitons* (North-Holland, Amsterdam) vol. I
[3] Dodd R K, Eilbeck J C, Gibbon J D and Morris H C 1982 *Soliton and nonlinear wave equations* (Academic Press, London)
[4] Karpman V I 1975 *Nonlinear Waves in Dispersive Media* (Pergamon Press, Oxford)
[5] Sulem C and Sulem P L 1999 *The Nonlinear Schrödinger Equation: Self-focusing and Wave Collapse* (Springer, New York)
[6] Koke S *et al.* 2010 *Nature Photonics* **4** 462
[7] Jovanović D, Fedele R, Tanjia F, De Nicola S and Gizzi L A 2012 *Eur. Phys. J. D* **66** 328
[8] Esarey E, Sprangle P, Krall J and Ting A 1996 *IEEE Trans. Plasma Sci.* **24** 253, and references therein
[9] Bingham R, Mendosça J T and Shukla P K 2004 *Plasma Phys. Control. Fusion* **46** R1, and references therein
[10] Atzeni S and Meyer-ter-Vehn J 2004 *The Physics of Inertial Fusion* (Oxford University Press, New York)
[11] http://www.astro.cornell.edu/extreme-physics-and-astrophysics-of-compact-objects.html
[12] http://www.eli-beams.eu/eli-events/relativistic-laboratory-astrophysics-with-extreme-power-lasers/
[13] Kharif C, Pelinovsky E and Slunyaev A 2009 *Rogue Waves in the Ocean* (Springer, New York)
[14] Nott J 2006 *Extreme Events: A Physical Reconstruction and Risk Assessment* (Cambridge University Press, Cambridge)
[15] Onorato M *et al.* 2013 *Physics Reports* **528** 47-89
[16] Synolakis C E and Bernard E N 2006 *Phil. Trans. R. Soc. A* **364** 2231-2265
[17] Johnson R S 1997 *A Modern Introduction to the Mathematical Theory of Water Waves* (Cambridge University Press, Cambridge)
[18] Johnson R S 2003 *J. Nonlinear Math. Phys.* **10** 72
[19] Johnson R S 2012 *J. Nonlinear Math. Phys.* **19** 1240012
[20] Fedele R *et al.* 2008 *Frontier in Modern Plasma Physics. In: AIP Conf. Proc.*, **1061** (eds. P. K. Shukla, B. Eliasson, and L. Stenflo), p. 273
[21] Fedele R *et al.* 2009 *New Developments in Nonlinear Plasma Physics. In: AIP Conf. Proc.*, **1188** (eds. P. K. Shukla, B. Eliasson, and L. Stenflo), p. 365
[22] Hirota R 1979 *Phys. Lett. A* **71** 393
[23] Khusnutdinova K R, Klein C, Matveev V B and Smirnov A O 2013 *Chaos* **23** 013126, and references therein

Magnetic domains and waves in the Skyrme - Faddeev model

L Martina[1,2], **M.V. Pavlov**[3]

[1] Dipartimento di Matematica e Fisica "Ennio De Giorgi", Universitådel Salento Italy
[2] INFN Sezione di Lecce, Italy
[3] Lebedev Physical Institute of Russian Academy of Sciences, Moscow, Leninskij Prospeckt, 53

E-mail: martina@le.infn.it

Abstract. The Skyrme-Faddeev model admits exact analytical non localized solutions, which describe magnetic domain wall solutions when multivalued singularities appear or, differently, always regular periodic nonlinear waves, which may degenerate into linear spin waves or solitonic structures. Here both classes of solutions are derived and discussed and a general discusssion about the existence of integrable subsectors of the model is addressed.

1. Introduction

The present contribution is mainly based on the developments of the results reported in the recent paper [1], where several reductions of the Skyrme-Faddeev model [2] were solved implicitly or in terms of special functions. In particular we consider the 4-dimensional relativistic ($g_{\mu\nu} = \mathrm{diag}\,(+,-,-,-)$) Skyrme–Faddeev model [6] for the unimodular three vector $\phi \in \mathcal{S}^2$ given by the Lagrangian density

$$\mathcal{L} = \frac{1}{32\pi^2}\left(\partial_\mu \phi \cdot \partial^\mu \phi - \frac{\lambda}{4}\left(\partial_\mu \phi \times \partial_\nu \phi\right) \cdot \left(\partial^\mu \phi \times \partial^\nu \phi\right)\right) - \kappa\left(1 - \phi \cdot \phi\right), \tag{1.1}$$

where $\lambda > 0$ is a scaling parameter, determining the breaking of the conformal symmetry, and κ is a Lagrangian multiplier, implementing the constraint on ϕ. The considered reductions lead to extended infinite energy solutions in the 3-dimensional space, which are rarely considered in the current literature on the subject, that being focused on the hopfion solutions [2] - [13], mainly by numerical integration of the equations of motion. Thus, there are at least three different motivations to the present approach: i) find analytical solutions to the system by exploiting its hidden symmetries, ii) give a suitable mathematical description of the of the extended phases observed in experimentsin ferromagnets and multiferroincs [14, 15, 16, 17], iii) find a method in order to extract as as we can from the general model special subsectors completely integrables. Moreover, numerical investigations concerning the anelastic scattering of hopfions show the excitation of wave-like modes disperding the energy, which are far to be studied. From our view point, a strong connection with the d'Alembert-Eikonal system has been obtained. This observation leads to find large classes of solutions of the Cauchy problem. Here special classes are described and interpreted as magnetic domains for the considered model. In Sec. 3 we will consider certain exact analytic periodic spin waves solutions for the Skyrme-Faddeev model, given in terms of elliptic integrals of third kind, specifying their behavior in some particular limits. In general the obtained solutions are quasi-periodic, but periodic dispersive wave-train solutions, one can show that the Lagrangian can be averaged by the Whitham method. This provides a Lagrangian for a set of parameters, describing the modulation of periodic waves in terms of a quasilinear system of partial derivatives of the first order. Finally, we remark that the imposed constraints is similar to the restrictions imposed by [18] in order to select an integrable sector of the general non integrable system.

2. Domain-Walls

First, let os observe that the geometric constraint $\phi \cdot \phi = 1$ can be realized in several way, for instance by resorting to the stereographic projection of the sphere on the complex plane, but loosing some evident symmetries here it seems useful to introduce the polar representation

$$\phi = (\sin w \cos u, \sin w \sin u, \cos w)\,, \tag{2.1}$$

where w and u are suitable functions on the variables $(x^0, \ldots, x^3)$ to be determined. The Lagrangian (1.1) becomes

$$\mathcal{L}_p = \frac{1}{32\pi^2}\left\{w_\mu w^\mu + \sin^2 w\left[u_\mu u^\mu - \frac{\lambda}{2}\left(w_\mu w^\mu u_\nu u^\nu - w_\mu w_\nu u^\mu u^\nu\right)\right]\right\}, \tag{2.2}$$

and the Euler–Lagrange equations read

$$\partial_\mu w^\mu = \tfrac{1}{2}\sin(2w)u_\nu u^\nu + \tfrac{\lambda}{2}\sin w\; u_\nu\, \partial_\mu[\sin w(w^\mu u^\nu - w^\nu u^\mu)], \tag{2.3}$$

$$w_\mu u^\mu \sin(2w) + \sin^2 w[\partial_\mu u^\mu + \tfrac{\lambda}{2} w_\nu \partial_\mu(u^\mu w^\nu - u^\nu w^\mu)] = 0. \tag{2.4}$$

The first equation can be interpreted as a quadratic constraint among first and second derivatives of the function u, while the second equation is a linear combination of first and second derivatives of the function u. In both cases the coefficients contain trigonometric functions and derivatives of w.

Such an asymmetry suggest to deal in different way u and w For instance, A first observation comes from the assumption $w = const$, which drastically reduces a d'Alembert equation constrained by the homogeneous Eikonal equations, i.e.

$$\partial_\mu u^\mu = 0, \qquad u_\nu u^\nu = 0. \tag{2.5}$$

This overdetermined system was investigated in many papers [19, 20] and for which the general solution was given in the implicit form

$$G\left(u, A_\mu\left(u\right)x^\mu, B_\mu\left(u\right)x^\mu\right) = 0, \qquad A_\mu A^\mu = B_\mu B^\mu = A_\mu B^\mu = 0, \tag{2.6}$$

with G, A_μ and B_μ arbitrary real regular functions. The process to provide explicit form for u may suggest that multi-valued functions appear. In fact, for sake of simplicity, let us impose $B_\mu \equiv 0$ and restrict to two spacial dimensions coordinated by (x^1, x^2). Then, under the suitable assumptions of regularity of the function G, one can write the solution in the form $u = F\left(x^0 - n_i\left(u\right)_i x^i\right)$, where F and n_i are differentiable functions, such that $\sum_{i=1}^{2} n_i^2 = 1$. Thus, setting $n_1\left(u\right) = \cos\left(f\left(u\right)\right)$, for a suitable function f, one can introduce the auxiliary variable $y = \cos\left(f\left(u\right) - z\right)$, being $z = \arccos\left(\frac{x^1}{r}\right)$ and $r = \sqrt{(x^1)^2 + (x^2)^2}$. Thus the equation defining u is reduced to

$$\arccos y = f\left(F\left(x^0 - ry\right)\right) - z. \tag{2.7}$$

For suitable bounded continuous on $\mathbf{R}$ functions f and F, the above equation admits continuous solutions in (x^0, x^1, x^2) variables, possibly multiple coincident or not. If F goes to finite value at infinity, asymptotic values, may depending on the direction, are taken by u. This value is preserved in time evolution, while other localized structures, caustics or singularities, move at constant velocity along the characteristics. Such a behavior suggests the existence of solutions domain wall in the Skyrme-Faddeev model. However, such a reduction has few of interest, for two reasons: i) asymptotics in time lead to the same boundary value and the solution becomes uniform, ii) the energy density is vanishing everywhere. So one has to look to less drastic reduction, for example imposing that

$$w_\mu u^\mu = 0, \qquad u_\nu u^\nu = \alpha \quad . \tag{2.8}$$

Then, the system (2.4) reduces to the separable set of equations

$$\partial_\mu u^\mu = 0, \qquad u_\nu u^\nu = \alpha, \tag{2.9}$$

$$w_\mu u^\mu = 0, \qquad \partial_\mu w^\mu = \frac{\alpha}{2}\frac{\sin(2w)}{1-\frac{\lambda\alpha}{2}\sin^2 w}(1+\frac{\lambda}{2}w^\mu w_\mu), \tag{2.10}$$

being the last one highly nonlinear for the w field, but depending on the general solution of the d'Alembert-Eikonal system (2.9). For that one the general solution is given in the implicit form by [20]

$$u = A_\mu(\tau)x^\mu + R_1(\tau), \tag{2.11}$$

$$B_\mu(\tau)x^\mu + R_2(\tau) = 0, \tag{2.12}$$

$$A_\mu A^\mu = \alpha, \quad A_\mu B^\mu = A'_\mu B^\mu = B_\mu B^\mu = 0, \tag{2.13}$$

where the function τ is implicitly defined by the (2.12), the real differentiable functions A_μ, B_μ satisfy the constraints (2.13) and, finally, the function R_i are differentiable up to second order at least.

If in (2.9) it results that $\alpha < 0$, setting $\alpha = -\eta^2$ the solution can be put in the form

$$u = x_k A_k(\tau) + A_0(\tau), \quad t = x_k B_k(\tau) + B_0(\tau), \tag{2.14}$$

$$(A_i) = \eta\hat{\mathbf{A}}\left(f(\tau), g(\tau)\right)\right) \left(\hat{\mathbf{A}}^2 = 1\right), \quad (B_i) = \hat{\mathbf{B}} = \pm\frac{\hat{\mathbf{A}}\times\hat{\mathbf{A}}'}{|\hat{\mathbf{A}}\times\hat{\mathbf{A}}'|}\left[f(\tau), g(\tau)\right)] \tag{2.15}$$

where $f(\tau)$, $g(\tau)$ (the $[\ldots]$ in (2.15) indicate the dependency also on their first derivatives), $A_0(\tau)$ and $B_0(\tau)$ are arbitrary differentiable functions. Thus, choosing them large classes of solutions can be found for the d'Alembert-Eikonal system (2.9). To give some insight on it, one may consider the simplest situation in which $\hat{\mathbf{A}}$ rotate uniformly, tracing a maximal circle on the unit sphere embedded into an auxiliary 3-space. The circle axis is given by the constant $\hat{\mathbf{B}}$. Then, accordingly with the properties of invertibility of $B_0(\tau)$, at least locally τ is a function of the linear combination $t - \hat{\mathbf{B}}\cdot\mathbf{x}$. Neglecting for sake of clarity the function A_0 in (2.14), the function u oscillates, increasing its amplitude like r. But the oscillations are strongly dependent on the direction and, in particular, they move in the space according to the time dependency of τ seen above. Moreover, similar static solutions exist. If $B_0(\tau)$ is not an injective continuous function, caustics and singularities of the wave front may appear. We know that in general the singularities of the wave fronts are classified by the Coxeter groups [21].

Similar considerations concern a different reduction, simply obtained from (2.4) by setting to zero the coefficients of all functions of w. After some work one prove that the set of the independent constraints reduces to the overdetermined quasilinear system of the first order in (u^μ, w^μ) fields (we will name it the reduced Skyrme–Faddeev system)

$$\partial_\mu w^\mu = 0, \qquad w_\mu w^\mu = -\epsilon^2, \tag{2.16}$$

$$u_\mu w^\mu = 0, \qquad u_\nu\partial_\mu(w^\mu u^\nu - w^\nu u^\mu) = 0, \tag{2.17}$$

where $\epsilon^2 = \frac{2}{\lambda}$ for notational clarity. Now the d'Alembert-Eikonal system (2.16) has the general solution (2.11)-(2.13) after the change $u \to w$ and $\alpha \to -\epsilon^2$. The first equation in (2.17) is the orthogonality condition among the gradients of the two fields, and the second one is a quadratic differential constraint among the derivatives of the function u. Actually it can be rewritten as two divergence terms (namely $\partial_\mu (u_\nu u^\nu w^\mu)$ and $\partial_\mu (u_\nu w^\nu u^\mu)$) balanced by a contribution of the form $(w^\mu u^\nu - w^\nu u^\mu) \partial_\mu u_\nu$. However, the second divergence is vanishing, because of the orthogonality condition, while the balancing contribution is the trace of the product of an antisymmetric tensor with a symmetric one, then it vanishes. Finally, recalling that w satisfies the d'Alembert equation, the equation we considered can be concisely written as

$$a_\mu w^\mu = 0 \quad \text{with} \quad a = u^\nu u_\nu, \tag{2.18}$$

in complete analogy with the first equation in (2.17).

Cross differentiation of those equations and systematic substitution of the x^0-derivatives

$$w_0 = \sqrt{w_m^2 - \epsilon^2},\ u_0 = \frac{u_k w_k}{\sqrt{w_m^2 - \epsilon^2}}, \tag{2.19}$$

lead to a set of compatibility conditions. For the d'Alembert-Eikonal equations it is well known [20] that such a set is finite and any further compatibility condition is satisfied only by the Monge-Ampére equation Det $[w_{ij}] = 0$,. On the other hand, the compatibility for the remaining equations leads to the two quadratic constraints

$$(w_s^2 - \epsilon^2) u_m u_k w_{km} + (u_k w_k)^2 w_{mm} = 2 u_s w_s u_m w_k w_{km}, \tag{2.20}$$

$$4 u_k w_k u_s w_{sp} (w_m w_{pm} - w_p w_{mm}) + 2 (u_s w_m w_{sm})^2 + \tag{2.21}$$
$$(u_s w_s)^2 (w_{mm} w_{pp} - w_{pm}^2) = 2 (w_p^2 - \epsilon^2)(u_s w_{sm})^2. \tag{2.22}$$

Such constraints can be simplified and possibly solved, if one can find a first order linear system of the form

$$u_0 = A u_1, \quad u_2 = B u_1, \quad u_3 = C u_1, \tag{2.23}$$

where the functions A, B, C depend on first and second order derivative of field variable w only. This formulation is similar to look at solutions of the Skyrme-Faddeev system (2.4) by the method of the hydrodynamic reductions [22, 23], involving a finite number of Riemann invariants. This can be certainty done in two space dimensions, where the equations of the form (2.23)

$$u_0 = \frac{w_1 w_{11} + w_2 w_{12}}{w_0 w_{11}} u_1, \quad u_2 = \frac{w_{12}}{w_{11}} u_1, \tag{2.24}$$

allow to identically satisfy the constraints (2.20) - (2.22). Moreover, it is easy to prove also that a solution of the system (2.24) is given by

$$u = F[w_1, w_2] \tag{2.25}$$

where F is an arbitrary real differentiable function of its arguments, depending on the initial data. Finally, by direct computation the equation (2.18) is identically satisfied, because of the vanishing of the quantity a. Thus, we have the general solution for the reduced Skyrme-Faddeev system in 2 space dimensions in terms of the four arbitrary functions involved in the solution of the d'Alembert-Eikonal system.

In three space dimensions, a similar analysis is much more complicated. In fact, substitution of (2.23) into (2.20) and (2.22) one is lead to an algebraic system of 4th degree. Thus, in principle the functions A, B, C in (2.23) can be explicitly determined, but their long expressions are intractable. On the other hand, from the orthogonality condition and (2.19), it is easy to prove that a class of solutions for u is given by $u = F[w_1, w_2, w_3]$, being F an arbitrary differentiable real function, constrained by the second equation in (2.17), which is not longer an identity as in 2-dimensions.

A direct substitution of a general solution for w, in the form (2.14)-(2.15), into the orthogonality condition (2.17-i) leads to the linear PDE

$$[X_m(B'_m(\tau)A_p(\tau) - A'_m(\tau)B_p(\tau)) + B'_0(\tau)A_p(\tau) - A'_0(\tau)B_p(\tau)]u_{X_p} = 0, \tag{2.26}$$

having used the coordinate transformation

$$(x_m B'_m(\tau) + B'_0(\tau))d\tau = dt - B_k(\tau)dx_k, \; X_k = x_k. \tag{2.27}$$

By the methods of characteristics it is equivalent to a 3 ODE system, for the unknown $X_k(s)$, being s an auxiliary variable. Depending the coefficient only on τ, one has a constant coefficient linear system nilpotent of order 2, because of the peculiar symmetry of the constraints (2.14)-(2.15). Thus, the solution is linear in s, from which one can extract two integrals of motion, in terms of which express the solution of (2.26). Then a general solution for the complete system of equations is determined by a sole function of a single variable only and the formula $u = F[w_1, w_2, w_3]$, is certainly redundant.

Finally, Let us notice that the eikonal - like condition was used in [18] to reduce the Skyrme-Faddeev model.

3. Waves

Now, we look at the invariant solutions of any 2-dimensional sub-algebra of the translational symmetries of the Skyrme-Faddeev model. Actually, by using the adjoint action of the space-time rotational subgroup, only 3 parameters will be left arbitrary. The invariant reduction is of the form

$$w = \Theta[\theta], \; u = \Phi[\theta] + \tilde{\theta}, \quad \text{where } \theta = \alpha_\mu x^\mu, \tilde{\theta} = \beta_\mu x^\mu \tag{3.1}$$

in which one distinguishes θ as the *phase* from the *pseudo-phase* $\tilde{\theta}$. Thus the equations of motion reduce to the announced 3-parametric family

$$\left[2B_3 - \frac{\lambda}{4}\mathcal{B}\sin^2\Theta\right]\Theta_{\theta\theta} = \sin 2\Theta\left(\frac{\lambda}{8}\mathcal{B}\,\Theta_\theta^2 + B_3\Phi_\theta^2 + B_2\Phi_\theta + B_1\right) \tag{3.2}$$

$$2B_3\sin^2\Theta\,\Phi_{\theta\theta} + \Theta_\theta\sin 2\Theta\,(2B_3\Phi_\theta + B_2) = 0, \tag{3.3}$$

where $B_1 = -\beta_\mu\beta^\mu$, $B_2 = -2\alpha_\mu\beta^\mu$ $B_3 = -\alpha_\mu\alpha^\mu$ and $\mathcal{B} = B_2^2 - 4B_1B_3$. These equations provide trigonometric plane waves, for instance by setting $B_3 = 0$ and $B_2 \neq 0$.

To obtain the general situation, one uses the expression of the energy-stress tensor $T^{\mu\nu} = (w^\nu\partial_{w_\mu} + u^\nu\partial_{u_\mu})\mathcal{L}_p - g^{\mu\nu}\mathcal{L}_p$ and build the conserved quantities $\mathcal{E}^\mu = T^{\mu\nu}\alpha_\nu$, given by

$$\mathcal{E}^0 = \frac{-1}{32\pi^2}\left\{B_3\alpha_0\Theta_\theta^2 + \sin^2\Theta\left[2\vec{\alpha}\cdot\vec{\beta}\,\beta_0 + \left(B_1 - 2\vec{\beta}^2\right)\alpha_0\right.\right. \tag{3.4}$$

$$\left.\left.+B_3\left(2\beta_0 + \alpha_0\Phi_\theta\right)\Phi_\theta - \frac{\lambda\mathcal{B}}{8}\alpha_0\Theta_\theta^2\right]\right\}, \tag{3.5}$$

$$\mathcal{E}^i = \frac{-1}{32\pi^2}\left\{B_3\alpha_i\Theta_\theta^2 + \sin^2\Theta\left[B_2\beta_i - B_1\alpha_i + B_3\left(2\beta_i + \alpha_i\Phi_\theta\right)\Phi_\theta - \frac{\lambda\mathcal{B}}{8}\alpha_i\Theta_\theta^2\right]\right\}. \tag{3.6}$$

These equations can be used to find an expression of Θ_θ^2 and Φ_θ. Precisely, assuming $B_3 \neq 0$ one finds

$$\Theta_\theta^2 = \frac{8B_3\left(B_1 \sin^2\Theta + U_3\right) - 2B_2^2\left(\sin^2\Theta + U_2^2 \csc^2\Theta\right)}{B_3\left(8B_3 - \lambda\mathcal{B}\sin^2\Theta\right)} \tag{3.7}$$

$$\Phi_\theta = -\frac{B_2\left(U_2 \csc^2(\Theta) + 1\right)}{2B_3}, \tag{3.8}$$

where the U_i's are two constants completely defining the quantities in (3.5) by the expressions $\mathcal{E}^\mu = U_3\alpha_\mu + \frac{B_2U_2}{2}\left(\frac{B_2}{B_3}\alpha_\mu - 2\beta_\mu\right)$. The equation (3.7) can be set in algebraic form by the transformation

$$\Theta = \arcsin\sqrt{\psi}, \tag{3.9}$$

yielding the equation

$$\psi_\theta^2 = \frac{64(\psi-1)\left(\psi - A_1\right)\left(\psi - A_2\right)}{\lambda^2\mathcal{B}\psi_1\left(\psi_1 - \psi\right)}, \quad (0 \leq \psi \leq 1) \tag{3.10}$$

where one has defined the constants $A_{1,2} = \frac{2B_3U_3 \pm \sqrt{4B_3^2U_3^2 - \mathcal{B}U_2^2}}{\mathcal{B}}$ and $\psi_1 = \frac{8B_3}{\lambda\mathcal{B}}$, related 1-1 to the values of the integrals of motion. Assuming for instance $0 < A_1 < A_2 < 1$, by a continuous variation of ψ_1 one obtains that only one oscillating solution exists for ψ, bounded between two of the three zeros of the numerator in (3.10), even if real unbounded solution may appear or complex ones. Analytically, equation (3.10) can be integrated in terms of incomplete elliptic integrals of the third kind. Precisely, by introducing a parametric variable Z one obtains the parametric form

$$\begin{aligned}
\theta(\psi) &= \theta_0 + \frac{1}{4}\sqrt{\frac{\mathcal{B}\lambda^2\psi_1\left(\psi_1 - A_1\right)^2}{\left(A_1 - 1\right)\left(A_2 - \psi_1\right)}}\Pi\left[\frac{A_1 - A_2}{\psi_1 - A_2}; Z \middle| \frac{\left(\psi_1 - 1\right)\left(A_1 - A_2\right)}{\left(A_1 - 1\right)\left(\psi_1 - A_2\right)}\right], \\
\psi &= -\frac{A_2\psi_1 \sin^2 Z + A_1\left(\psi_1 \cos^2 Z - A_2\right)}{A_1 \sin^2 Z + A_2 \cos^2 Z + \psi_1}
\end{aligned} \tag{3.11}$$

which can be expressed in terms of Weierstrass $\mathcal{P}$ function. Furthermore, from (3.8) the function Φ can be expressed again in terms of incomplete elliptic integrals, namely

$$\begin{aligned}
\Phi &= -\frac{B_2U_2}{2B_3}\left[\int \frac{d\theta}{\psi(\theta)} + \theta\right] + \Phi_0 = \\
&- \frac{s_1}{2\psi_1}\left[\sqrt{\frac{2\psi_1\left(A_1 - \psi_1\right)^2\left(B_1\lambda\psi_1 + 2\right)}{\left(A_1 - 1\right)\left(A_2 - \psi_1\right)}}\Pi\left(\frac{A_2 - A_1}{A_2 - \psi_1}; Z \middle| \frac{\left(A_1 - A_2\right)\left(\psi_1 - 1\right)}{\left(A_1 - 1\right)\left(\psi_1 - A_2\right)}\right)\right. \\
&\left. + 2s_2\sqrt{\frac{A_2\psi_1\left(A_1 - \psi_1\right)^2}{A_1\left(A_1 - 1\right)\left(A_2 - \psi_1\right)}}\Pi\left(\frac{\left(A_1 - A_2\right)\psi_1}{A_1\left(\psi_1 - A_2\right)}; Z \middle| \frac{\left(A_1 - A_2\right)\left(\psi_1 - 1\right)}{\left(A_1 - 1\right)\left(\psi_1 - A_2\right)}\right)\right],
\end{aligned} \tag{3.12}$$

where $s_1 = \text{sign } B_2$, $s_2 = \text{sign } U_2$.

This general solution leads to three different linear harmonic branches in the limit of ψ_1 is approaching A_1, A_2 and 1 respectively. For instance, one has

$$\psi_1 \to 1: \quad \psi \to \frac{1}{2}\left(\left(A_1 - A_2\right)\cos\left(\frac{8}{\sqrt{\mathcal{B}\lambda}}\theta\right) + A_1 + A_2\right), \tag{3.13}$$

from which the dispersion relation for such waves can be explicitly derived. The wavelength is affected by the presence of the pseudo-phase by the factor in front of θ. At the opposite, one can notice from the general solution above that the length-wave can be made very large when $A_2 \to 1$ and $\psi_1 \to \infty$. In this limit the solution can be expressed in terms of elementary hyperbolic functions. A similar situation occurs when $A_1 = A_2$, then the the elliptic module is 0 and the solutions are given in terms of trigonometric functions. Thus one can conjecture that varying in a suitable way all the parameters, a set of periodic solutions can be found. This suggests to consider slowly deformations of them. The lowest order of modulation approximation is found by substituting the multi-parametric family of periodic solutions (3.10) (see also (3.7) and (3.8)) into the Skyrme-Faddeev Lagrangian (see (1.1) and (2.2)), introducing an averaged Lagrangian, say $L(\gamma, \omega, \vec{\beta}, \vec{k})$, depending on the slowly changing parameters as new dynamic variables $\gamma, \omega, \vec{\beta}, \vec{k}$, which correspond to the derivatives with respect to space-time variables of phase θ and pseudo-phase $\tilde{\theta}$, now not necessarily linear as in (3.1). It means that $\omega = -\theta_{X^0}, k_i = \theta_{X^i}$ and $\gamma = -\tilde{\theta}_{X^0}, \beta_i = \tilde{\theta}_{X^i}$, where X^0, X^1, X^2, X^3 are the so called "slow" variables in comparison with "fast" variables x^0, x^1, x^2, x^3 (see detail in [24]).

Thus, one immediately derives the four-dimensional quasilinear system (below we use $\partial_k \equiv \partial_{X^k}$)

$$\partial_0 L_\omega = \partial_i L_{k^i}, \quad \partial_0 L_\gamma = \partial_i L_{\beta^i}, \tag{3.14}$$

with the compatibility conditions

$$\begin{array}{rclcl} \partial_0 k^1 + \partial_i \omega &=& 0, &,& \partial_j k^i = \partial_i k^j \quad i \neq j, \\ \partial_0 \beta^i + \partial_i \gamma &=& 0, &,& \partial_j \beta^i = \partial_i \beta^j \quad i \neq j. \end{array} \tag{3.15}$$

After the substitution of the family of periodic solutions (3.7) and (3.8) into the Lagrangian density (2.2), one obtains

$$\hat{\mathcal{L}}_p = \sin^2(\Theta)\left(-\frac{1}{2}\lambda\left(\frac{B_2^2}{4} - B_1B_3\right)\Theta_\theta^2 + B_3\Phi_\theta^2 + B_2\Phi_\theta + B_1\right) + B_3\Theta_\theta^2, \tag{3.16}$$

which is a function only on θ. Thus, performing an integration over a finite space-like region, contributions from θ-independent coordinates are just time-independent finite multiplicative factors. Then, on a period of the wave, one leads to the Lagrangian

$$L \equiv \frac{1}{2\pi}\oint \hat{\mathcal{L}}_p \, d\theta, \tag{3.17}$$

which, generalizing the standard Whitham approach, is supplemented by the introduction of two natural normalizations (or constraints)

$$\oint d\theta = 2\pi, \qquad <\Phi_\theta> = \oint \Phi d\theta = 2\pi m, \tag{3.18}$$

where the integer "m" is the number of rotations of the vector $\vec{\phi}$ around a value determined by a given pseudo-phase $\tilde{\theta}$. This situation is very similar to to the spin wave configurations called cyclon and extra-cyclon in multiferroic materials ([17]). Then the corresponding averaged Lagrangian is

$$L = \left(B_1 - \frac{B_2^2}{4B_3}\right)\left(A_1 + A_2 + W\sqrt{\frac{\lambda}{2}B_3}\right) + \frac{B_2 + 2mB_3}{2B_3}\sqrt{A_1A_2(B_2^2 - 4B_1B_3)}, \tag{3.19}$$

where we introduced the function

$$W = \frac{1}{2\pi}\oint \sqrt{\frac{(\psi - A_1)(\psi - A_2)(\psi - \psi_1)}{1 - \psi}}\frac{d\psi}{\psi}. \tag{3.20}$$

One can immediately check that two equations $L_{A_1} = 0$ and $L_{A_2} = 0$ (see [24]) coincide with normalizations (3.18), while the Euler–Lagrange equations lead to four dimensional quasilinear system of the first order (3.15).

4. Conclusions

We have found several reductions of the Skyrme-Faddeev model, which can be solved or implicitly or via special functions. We have shown the strong connection with the d'Alembert-Eikonal system, which allows to find large classes of solutions. Even in the more complicated system (2.16)-(2.18), we found large classes of solution in terms of two arbitrary functions. Finally, we have shown that the overdetermined system for the field u can be formulated as a first order system of linear pde's, which can be solved by the "method of hydrodynamic reductions ". However, the analysis of the special constraint 2.18) deserves several technical complexities and for this only particular solutions are given. Only, we remark that the constraint is a further restriction on the subsector of the solution space described in [18]. In Sec. 3 we have found exact analytic periodic spin waves for the Skyrme-Faddeev model, determined in terms of elliptic integrals of third kind. Assuming that there exist such periodic dispersive wave-train solutions, we have shown that the Lagrangian can be averaged by the Whitham method. This provides a Lagrangian for a set of parameters, describing the modulation of periodic waves in terms of a quasilinear system of partial derivatives of the first order

Acknowledgments

The work was partially supported by the PRIN "Geometric Methods in the Theory of Nonlinear Waves and their Applications" of the Italian MIUR and by the INFN - Sezione of Lecce under the project LE41. MVP's work was partially supported by the RF Government grant #2010-220-01-077, ag. #11.G34.31.0005, by the grant of Presidium of RAS "Fundamental Problems of Nonlinear Dynamics" and by the RFBR grant 11-01-00197. MVP acknowledges the Department of Matematica e Fisica "E. De Giorgi" for the warm hospitality.

References

[1] L. Martina, M.V. Pavlov and S. Zykov: Waves in the Skyrme–Faddeev model and Integrable reductions, J. Phys. A: Math. Theor. 46 275201 (2013)

[2] L. D. Faddeev: "Quantization of solitons", Princeton preprint IAS-75-QS70 (1975) ; L. D. Faddeev: "Some comments on the many-dimensional solitons" , Lett. Math. Phys. **1** n. 4, (1976) 289-293

[3] Faddeev L D and Niemi A J 2007 Spin-Charge Separation, Conformal Covariance and the SU(2) Yang–Mills Theory *Nucl. Phys.* **B 776** 38

[4] De Vega H J 1978 Closed Vortices and the Hopf index in classical field theory *Phys. Rev.* **D 18** 2945

[5] Vakulenko A F and Kapitanski L V 1979 Stability of solitons in S2 in the nonlinear σ-model *Dokl. Akad. Nauk USSR* **246** 840

[6] Kundu A and Rybakov Y P 1982 Closed-vortex-type solitons with Hopf index *J. Phys. A: Math. Gen.* **15** 269

[7] Gladikowski J and Hellmund M 1997 Static solitons with nonzero Hopf number *Phys. Rev.* **D 56** 5194

[8] Faddeev L D and Niemi A J 1997 Stable knot-like structures in classical field theory *Nature* **387** 58

[9] Battye R A and Sutcliffe P. M. 1999 Solitons, Links and Knots", *Proc. R. Soc. Lond.* **A 455** 4305

[10] Ward R S 1999 Hopf solitons on S^3 and R^3 *Nonlinearity* **12** 241

[11] Hietarinta J and Salo P 1999 Faddeev–Hopf knots: dynamics of linked un-knots", *Phys. Lett.* **B 451** 60;Hietarinta J and Salo P 2000 Ground state in the Faddeev–Skyrme model *Phys. Rev.* **D 62**, 081701

[12] Sutcliffe P 2007 Knots in the Skyrme–Faddeev mode *Proc. R. Soc.* **A 463** 3001

[13] Martina L, Martone G I and Zykov S 2012 Symmetry reductions of the Skyrme – Faddeev model *Acta Math. Appl.* **122** 323

[14] Yu X Z *et al.* 2010 Real-space observation of a two-dimensional skyrmion crystal *Nature* 465, 901-904.

[15] Romming N *et al.* 2013 Writing and Deleting Single Magnetic Skyrmions *Science* **341** 636-639

[16] Okamura Y *et al.* 2013 Microwave magnetoelectric effect via skyrmion resonance modes in a helimagnetic multiferroic 2013 *Nature Comm.*

[17] Rovillain P *et al.* 2010 Electric-field control of spin waves at room temperature in multiferroic $BiFeO_3$ *Nature materials* **9** 975

[18] Alvarez O, Ferreira L A and Sanchez-Guillen J 2009 Integrable theories and loop spaces: fundamentals, applications and new developments *Int.J.Mod.Phys.* **A24** 1825-1888.

[19] Cieciura G and Grundland A M 1984 A certain class of solutions of the nonlinear wave equation *J. Math. Phys.* **2** 3460-3469

[20] Fushchich V I, Zhdanov R Z and Revenko I V 1991 General solutions of the nonlinear wave equation and of the eikonal equation *Ukr. Mat. Z.* **43** 1471-1487; Zhdanov R Z, Revenko I V and Fushchich V I 1995 On the general solution of the d'Alembert equation with a nonlinear eikonal constraint and its applications *J. Math. Phys.* **36** 7109

[21] Arnold V I 1990 *Singularities of caustics and wave fronts* (Dordrecht, Kluwer Acad. Pub.)

[22] Ferapontov E V and Pavlov M V 2003 Hydrodynamic reductions of the heavenly equation *Class. Quantum Grav.* **20** 2429-2441.

[23] Ferapontov E V, Khusnutdinova K R and Tsarev S P 2006 On a class of three-dimensional integrable Lagrangians *Comm. Math. Phys.* **261** no. 1, 225–243

[24] Whitham G B 1974 *Linear and Nonlinear Waves* (Wiley, New York)

Spectral realization of the Riemann zeros by quantizing $H = w(x)(p + \ell_p^2/p)$: the Lie-Noether symmetry approach

M C Nucci

Dipartimento di Matematica e Informatica, Università degli Studi di Perugia & INFN Sezione Perugia, 06123 Perugia, Italy

E-mail: `mariaclara.nucci@unipg.it`

Abstract. If t_n are the heights of the Riemann zeros $1/2 + it_n$, an old idea, attributed to Hilbert and Polya [6], stated that the Riemann hypothesis would be proved if the t_n could be shown to be eigenvalues of a self-adjoint operator. In 1986 Berry [1] conjectured that t_n could instead be the eigenvalues of a deterministic quantum system with a chaotic classical counterpart and in 1999 Berry and Keating [3] proposed the Hamiltonian $H = xp$, with x and p the position and momentum of a one-dimensional particle, respectively. This was proven not to be the correct Hamiltonian since it yields a continuum spectrum [23] and therefore a more general Hamiltonian $H = w(x)(p + \ell_p^2/p)$ was proposed [25], [4], [24] and different expressions of the function $w(x)$ were considered [25], [24], [16] although none of them yielding exactly t_n. We show that the quantization by means of Lie and Noether symmetries [18], [19], [20], [7] of the Lagrangian equation corresponding to the Hamiltonian H yields straightforwardly the Schrödinger equation and clearly explains why either the continuum or the discrete spectrum is obtained. Therefore we infer that suitable Lie and Noether symmetries of the classical Lagrangian corresponding to H should be searched in order to alleviate one of Berry's quantum obsessions [2].

Keywords: Lagrangian; Jacobi last multiplier; Lie symmetry; Noether symmetry; Classical quantization.

PACS numbers: 02.30.Xx, 02.20.Sv, 45.20.Jj, 03.65.-w

1. Introduction

In [2] Michael Berry wrote:

Why should a physicist be concerned with the zeros of the Riemann zeta function $\zeta(s)$, and in particular the Riemann hypothesis, according to which all complex zeros of $\zeta(s)$ have $Res = \frac{1}{2}$?

Not for the reasons that motivate mathematicians, e.g. fluctuations in the distribution of primes. Rather, my interest grew from the study of quantum systems whose classical counterparts possess chaotic trajectories.

One hundred years ago Polya and Hilbert suggested that in order to prove the Riemann hypothesis one has to find a self-adjoint operator whose spectrum contains the imaginary part of the nontrivial Riemann zeros [6]. Michael Berry suggested the existence of a classical Hamiltonian whose quantum version would realize the Polya-Hilbert conjecture. This conjectured Hamiltonian must satisfy the following conditions [25]:

(i) be chaotic, with isolated periodic orbits related to the prime numbers,

(ii) break time reversal symmetry, to agree with the Gaussian unitary ensemble statistics,

(iii) be quasi-one dimensional.

The first classical Hamiltonian proposed by Berry, i.e.

$$H_{cl} = xp$$

was then shown by Berry and Keating [3] not to fulfill the requirement (i).

Sierra and collaborators [25], [24] and also Berry and Keating [4] have proposed different modifications of the xp Hamiltonian in order to have bounded classical trajectories and a discrete quantum spectrum. The first such Hamiltonian was [25]

$$H_S = x\left(p + \frac{\ell_p^2}{p}\right). \tag{1}$$

Others followed, in particular

$$H_B = \cosh\left(\frac{x}{R}\right)\left(p + \frac{\ell_p^2}{p}\right). \tag{2}$$

In the present paper we study the Lagrangian equations that derive from those Hamiltonians. In particular, we find their Lie point symmetries, their Jacobi last multipliers, various Lagrangians and the Noether symmetries that they admit. Finally we use those Noether symmetries to straightforwardly construct the Schrödinger equations [18], [19], [20], [7].

The paper is organized in the following way. In the next section we recall the properties of the Jacobi last multiplier, its connection with Lagrangians of second-order equations [12], [27], and the link with Lie symmetries [14], [15], [5]. In section 3 we study the Hamiltonian (1) and in section 4 the Hamiltonian (48). The last section contains some final remarks.

2. Jacobi last multiplier, Lie symmetries, Lagrangians

The method of the Jacobi Last Multiplier [9]-[12] provides a means to determine all the solutions of the partial differential equation

$$\mathcal{A}f = \sum_{i=1}^{n} a_i(x_1,\dots,x_n)\frac{\partial f}{\partial x_i} = 0 \tag{3}$$

or its equivalent associated Lagrange's system

$$\frac{\mathrm{d}x_1}{a_1} = \frac{\mathrm{d}x_2}{a_2} = \dots = \frac{\mathrm{d}x_n}{a_n}. \tag{4}$$

In fact, if one knows the JLM and all but one of the solutions, then the last solution can be obtained by a quadrature. The JLM M is given by

$$\frac{\partial(f,\omega_1,\omega_2,\dots,\omega_{n-1})}{\partial(x_1,x_2,\dots,x_n)} = M\mathcal{A}f, \tag{5}$$

where

$$\frac{\partial(f,\omega_1,\omega_2,\ldots,\omega_{n-1})}{\partial(x_1,x_2,\ldots,x_n)} = \det\begin{bmatrix} \frac{\partial f}{\partial x_1} & \cdots & \frac{\partial f}{\partial x_n} \\ \frac{\partial \omega_1}{\partial x_1} & & \frac{\partial \omega_1}{\partial x_n} \\ \vdots & & \vdots \\ \frac{\partial \omega_{n-1}}{\partial x_1} & \cdots & \frac{\partial \omega_{n-1}}{\partial x_n} \end{bmatrix} = 0 \tag{6}$$

and $\omega_1,\ldots,\omega_{n-1}$ are $n-1$ solutions of (3) or, equivalently, first integrals of (4) independent of each other. This means that M is a function of the variables $(x_1,\ldots,x_n)$ and depends on the chosen $n-1$ solutions, in the sense that it varies as they vary. The essential properties of the JLM are:

(a) If one selects a different set of $n-1$ independent solutions $\eta_1,\ldots,\eta_{n-1}$ of equation (3), then the corresponding JLM N is linked to M by the relationship:

$$N = M\frac{\partial(\eta_1,\ldots,\eta_{n-1})}{\partial(\omega_1,\ldots,\omega_{n-1})}.$$

(b) Given a non-singular transformation of variables

$$\tau: \quad (x_1,x_2,\ldots,x_n) \longrightarrow (x'_1,x'_2,\ldots,x'_n),$$

then the JLM M' of $\mathcal{A}'F=0$ is given by:

$$M' = M\frac{\partial(x_1,x_2,\ldots,x_n)}{\partial(x'_1,x'_2,\ldots,x'_n)},$$

where M obviously comes from the $n-1$ solutions of $\mathcal{A}F=0$ which correspond to those chosen for $\mathcal{A}'F=0$ through the inverse transformation τ^{-1}.

(c) One can prove that each JLM M is a solution of the following linear partial differential equation:

$$\sum_{i=1}^{n}\frac{\partial(Ma_i)}{\partial x_i} = 0, \tag{7}$$

or equivalently:

$$\frac{\mathrm{d}}{\mathrm{d}t}(\log M) + \sum_{i=1}^{n}\frac{\partial a_i}{\partial x_i} = 0; \tag{8}$$

viceversa every solution M of this equation is a JLM.

(d) If one knows two JLMs M_1 and M_2 of equation (3), then their ratio is a solution ω of (3), or, equivalently, a first integral of (4). Naturally the ratio may be quite trivial, namely a constant. Viceversa the product of a multiplier M_1 times any solution ω yields another JLM $M_2 = M_1\omega$.

Since the existence of a solution/first integral is consequent upon the existence of symmetry, an alternate formulation in terms of symmetries was provided by Lie [14], [15]. A clear treatment of the formulation in terms of solutions/first integrals and symmetries is given by Bianchi [5]. If we know $n-1$ symmetries of (3)/(4), say

$$\Gamma_i = \sum_{j=1}^{n}\xi_{ij}(x_1,\ldots,x_n)\partial_{x_j}, \quad i=1,n-1, \tag{9}$$

then a JLM is given by $M = \Delta^{-1}$, provided that $\Delta \neq 0$, where

$$\Delta = \det \begin{bmatrix} a_1 & \cdots & a_n \\ \xi_{1,1} & & \xi_{1,n} \\ \vdots & & \vdots \\ \xi_{n-1,1} & \cdots & \xi_{n-1,n} \end{bmatrix}. \tag{10}$$

Another property of the JLM is its (almost forgotten) relationship with the Lagrangian, $L = L(t, q, \dot{q})$, for any second-order equation

$$\ddot{q} = F(t, q, \dot{q}) \tag{11}$$

i.e. [12] (Lecture 10)[1], [27]

$$M = \frac{\partial^2 L}{\partial \dot{q}^2} \tag{12}$$

where $M = M(t, q, \dot{q})$ satisfies the following equation

$$\frac{\mathrm{d}}{\mathrm{d}t}(\log M) + \frac{\partial F}{\partial \dot{q}} = 0. \tag{13}$$

Then equation (11) becomes the Euler-Lagrange equation:

$$-\frac{\mathrm{d}}{\mathrm{d}t}\left(\frac{\partial L}{\partial \dot{q}}\right) + \frac{\partial L}{\partial q} = 0. \tag{14}$$

The proof is based on taking the derivative of (14) with respect to $\dot{q}$ and showing that this yields (13). If one knows a JLM, then L can be easily obtained by a double integration, i.e.:

$$L = \int \left(\int M \, \mathrm{d}\dot{q} \right) \mathrm{d}\dot{q} + f_1(t, q)\dot{q} + f_2(t, q), \tag{15}$$

where f_1 and f_2 are functions of t and q which have to satisfy a single partial differential equation related to (11) [21]. As it was shown in [22], f_1, f_2 are related to the gauge function $g = g(t, q)$. In fact, we may assume

$$\begin{aligned} f_1 &= \frac{\partial g}{\partial q} \\ f_2 &= \frac{\partial g}{\partial t} + f_3(t, q) \end{aligned} \tag{16}$$

where f_3 has to satisfy the mentioned partial differential equation and g is obviously arbitrary.

In [21] it was shown that if one knows several (at least two) Lie symmetries of the second-order differential equation (11), i.e.

$$\Gamma_j = V_j(t, q)\partial_t + G_j(t, q)\partial_q, \quad j = 1, r, \tag{17}$$

then many Jacobi Last Multipliers could be derived by means of (10), i.e.

$$\frac{1}{M_{nm}} = \Delta_{nm} = \det \begin{bmatrix} 1 & \dot{q} & F(t, q, \dot{q}) \\ V_n & G_n & \dfrac{\mathrm{d}G_n}{\mathrm{d}t} - \dot{q}\dfrac{\mathrm{d}V_n}{\mathrm{d}t} \\ V_m & G_m & \dfrac{\mathrm{d}G_m}{\mathrm{d}t} - \dot{q}\dfrac{\mathrm{d}V_m}{\mathrm{d}t} \end{bmatrix}, \tag{18}$$

with $(n, m = 1, r)$, and therefore many Lagrangians can be obtained by means of (15).

[1] *Jacobi's Lectures on Dynamics* are finally available in English [13].

Journal of Physics: Conference Series **482** (2014) 012032 doi:10.1088/1742-6596/482/1/012032

3. The case $H = x(p + \ell_p^2/p)$

Sierra has proposed the following Hamiltonian [25], [24]:

$$H_S = x\left(p + \frac{\ell_p^2}{p}\right) \tag{19}$$

This Hamiltonian yields the Lagrangian equation:

$$\ddot{x} = -4x + 6\dot{x} - \frac{\dot{x}^2}{x}. \tag{20}$$

This equations admits an eight-dimensional Lie point symmetry algebra $sl(3, \mathbb{R})$ generated by

$$\begin{gathered}\Gamma_1 = \exp(-2t)x^2(\partial_t + 2x\partial_x), \quad \Gamma_2 = \exp(-4t)x^2(\partial_t + x\partial_x)\\ \Gamma_3 = \exp(2t)(\partial_t + 2x\partial_x), \quad \Gamma_4 = \partial_t, \quad \Gamma_5 = \exp(-2t)(\partial_t + x\partial_x),\\ \Gamma_6 = x\partial_x, \quad \Gamma_7 = \frac{\exp(4t)}{x}\partial_x, \quad \Gamma_8 = \frac{\exp(2t)}{x}\partial_x,\end{gathered} \tag{21}$$

which implies that equation (20) is linearizable by means of a point transformation[2] [15]. In order to find the linearizing transformation we have to look for an abelian intransitive subalgebra of $sl(3, \mathbb{R})$ and, following Lie's classification of two-dimensional algebras in the real plane [15], we have to transform it into the canonical form

$$\partial_{\tilde{x}}, \quad \tilde{t}\partial_{\tilde{x}} \tag{22}$$

with $\tilde{x}$ and $\tilde{t}$ the new dependent and independent variables, respectively. We have found that such subalgebra[3] is that generated by Γ_7 and Γ_8. Then it is easy to derive that

$$\tilde{x} = \tfrac{1}{2}\exp(-2t)x^2, \quad \tilde{t} = \exp(2t) \tag{23}$$

and equation (20) is transformed into the free-particle equation, i.e.:

$$\frac{\mathrm{d}^2\tilde{x}}{\mathrm{d}\tilde{t}^2} = 0. \tag{24}$$

Several JLM can be obtained by means (18) and two of the eight symmetries (21). Consequently several Lagrangians can be derived from (15). In particular from Γ_7 and Γ_8 comes the Lagrangian

$$L_{78} = -\exp(-6t)\left(\frac{1}{4}\dot{x}^2x^2 - \tfrac{1}{2}x^4\right) \tag{25}$$

that admits the following five Noether symmetries [17]

$$\Gamma_3, \quad \Gamma_4 + \frac{3}{2}\Gamma_6 = \partial_t + \frac{3}{2}x\partial_x, \quad \Gamma_5, \quad \Gamma_7, \quad \Gamma_8. \tag{26}$$

If we quantize by preserving those Noether symmetries then the following Schrödinger equation is obtained:

$$4\mathrm{i}\exp(-2t)x^4\Phi_t + 4\exp(4t)x^2\Phi_{xx} - \left(2\exp(-8t)x^8 + 4\mathrm{i}\exp(-2t)x^4 + 3\exp(4t)\right)\Phi = 0 \tag{27}$$

[2] We prefer to cite the original work by Sophus Lie, although many textbooks/papers on Lie symmetries report his findings.

[3] We may find other such subalgebras but they are obviously related by a point transformation since there exist only one abelian intransitive subalgebra of $sl(3, \mathbb{R})$ [26]. For example, two other abelian intransitive subalgebras are those generated by $< \Gamma_1, \Gamma_3 >$, and $< \Gamma_2, \Gamma_5 >$, respectively.

and its spectrum is obviously a continuum [24]. Indeed equation (27) admits the following five Lie symmetries

$$\begin{aligned}
X_1 &= \Gamma_3 + \Phi\left(-2\exp(2t) + \frac{\mathrm{i}}{2}x^4\exp(-4t)\right)\partial_\Phi, \\
X_2 &= \Gamma_4 + \frac{3}{2}\Gamma_6, \\
X_3 &= \Gamma_5 - \Phi\left(\frac{1}{2}\exp(-2t) + \frac{\mathrm{i}}{4}x^4\exp(-8t)\right)\partial_\Phi, \\
X_4 &= \Gamma_7 + \Phi\left(-\frac{1}{2x^2}\exp(4t) + \mathrm{i}x^2\exp(-2t)\right)\partial_\Phi, \\
X_5 &= \Gamma_8 + \Phi\left(-\frac{1}{2x^2}\exp(2t) + \frac{\mathrm{i}}{2}x^2\exp(-4t)\right)\partial_\Phi,
\end{aligned} \tag{28}$$

and also $\Phi\partial_\Phi$, and $\phi(t,x)\partial_\Phi$ with ϕ any solution of (27).

From Γ_1 and Γ_3 comes another Lagrangian

$$L_{13} = \frac{1}{4x(2x-\dot{x})} \tag{29}$$

that admits the following five Noether symmetries [17]

$$\Gamma_1, \quad \Gamma_2, \quad \Gamma_3, \quad \Gamma_4, \quad \Gamma_7. \tag{30}$$

If we quantizing by requiring the preservation of the five Noether symmetries (30) then the following parabolic equation is obtained

$$\Omega_{tt} + 4x\Omega_{xt} + 4x^2\Omega_{xx} + 4x\Omega_x - \Omega = 0 \tag{31}$$

Its Lie symmetries are generated by[4]

$$\hat{X}_1 = \Gamma_1 + x^2\Omega\exp(-2t)\partial_\Omega, \quad \hat{X}_2 = \Gamma_2, \quad \hat{X}_3 = \Gamma_3 + \Omega\exp(2t)\partial_\Omega, \quad \hat{X}_4 = \Gamma_4, \quad \hat{X}_5 = \Gamma_7, \tag{32}$$

and can be reduced to its canonical form by the change of independent variable $x = \zeta\exp(2t)$, i.e.

$$\Omega_{tt} - \Omega = 0 \tag{33}$$

that can be solved to give

$$\Omega(t,\zeta) = f_1(\zeta)\exp(t) + f_2(\zeta)\exp(-t), \tag{34}$$

with f_1, f_2 arbitrary functions of ζ. Equation (33) is not surprising. Indeed equation (20) can be transformed into

$$\ddot{z} = 2(b+3)\dot{z} - (b+2)(b+4)z \tag{35}$$

by means of the transformation

$$z = x^2\exp(bt) \tag{36}$$

where b is an arbitrary constant. Substituting $b = -3$ into (35) yields

$$\ddot{z} = z, \tag{37}$$

[4] Also $\Omega\partial_\Omega$, $\omega(t,x)\partial_\Omega$ with ω any solution of (31).

which corresponds exactly to (33).

The Lagrangian associated with the Hamiltonian (1) is given by[5] [24]:

$$L_S = -2\ell_p\sqrt{x(x-\dot{x})} \tag{38}$$

This Lagrangian admits three Noether symmetries only, i.e.:

$$\Gamma_4, \quad \Gamma_5, \quad \Gamma_8. \tag{39}$$

These symmetries generate the following family of equations:

$$\ddot{x} = -4x + 6\dot{x} - \frac{\dot{x}^2}{x} + A(x-\dot{x})\sqrt{x(x-\dot{x})} \tag{40}$$

where A is an arbitrary constant. The three symmetries (39) yield three JLM and consequently three Lagrangians. In particular Γ_5 and Γ_8 yields the JLM

$$M_{58} = \frac{\sqrt{x}}{A(x-\dot{x})^{3/2}} \tag{41}$$

by means of (18), and the Lagrangian

$$L_{58} = -2\ell_p\sqrt{x(x-\dot{x})} + \frac{A}{4}\ell_p x^2, \tag{42}$$

by means of (15). Lagrangian L_{58} coincides with L_S if $A = 0$, but in this case M_{58} does not exist since in (18) Δ_58 is equal to zero. This means that the JLM that yields the Lagrangian L_S does not come from the two symmetries Γ_5 and Γ_8. If $A \neq 0$ then L_{58} admits only two Noether symmetries, namely Γ_4 and Γ_5.

If we quantizing by requiring the preservation of the three Noether symmetries (39) then the following parabolic equation is obtained

$$4\Psi_{tt} + 8x\Psi_{xt} + 4x^2\Psi_{xx} - 4\Psi_t - 3\Psi = 0 \tag{43}$$

Its Lie symmetries are generated by[6]

$$Y_1 = \Gamma_4, \quad Y_2 = \Gamma_5 - \frac{1}{2}\exp(2t)\Psi\partial_\Psi, \quad Y_3 = \Gamma_8 \tag{44}$$

and can be reduced to its canonical form by the change of independent variable $x = \xi\exp(t)$, i.e.

$$4\Psi_{tt} - 4\Psi_t - 3\Psi = 0 \tag{45}$$

that can be solved to give

$$\Psi(t,\xi) = F_1(\xi)\exp(3t/2) + F_2(\xi)\exp(-t/2), \tag{46}$$

with F_1, F_2 arbitrary functions of ξ. Equation (45) is also not surprising. Substituting $b = -5/2$ into (35) yields

$$\ddot{z} = \dot{z} + \frac{3}{4}z, \tag{47}$$

which corresponds exactly to (45).

[5] The reality of L_S implies that $\dot{x} < x$.

[6] Also $\Psi\partial_\Psi$, $\psi(t,x)\partial_\Psi$ with ψ any solution of (43).

4. The case $H = \cosh\left(\frac{x}{R}\right)\left(p + \frac{\ell_p^2}{p}\right)$

Another classical Hamiltonian proposed by Sierra in [24] is

$$H_B = \cosh\left(\frac{x}{R}\right)\left(p + \frac{\ell_p^2}{p}\right) \tag{48}$$

This Hamiltonian yields the Lagrangian equation:

$$\ddot{x} = \frac{1}{R}\tanh\left(\frac{x}{R}\right)\left(-4\cosh^2\left(\frac{x}{R}\right) + 6\dot{x}\cosh\left(\frac{x}{R}\right) - \dot{x}^2\right) \tag{49}$$

Since this equation admits an eight-dimensional Lie symmetry algebra then it is linearizable by means of an abelian intransitive subalgebra and we found that the linearizing transformation is

$$\begin{aligned} \tilde{t} &= -\tan\left(\frac{2}{R}\left(-t + R\arctan\left(e^{x/R}\right)\right)\right), \\ \tilde{x} &= \frac{1 - e^{2x/R}}{4\cos(2t/R)(1 - e^{2x/R}) + 8\sin(2t/R)e^{x/R}} \end{aligned} \tag{50}$$

that yields the free-particle equation (24). The Lagrangian that derives from the JLM obtained by means of the abelian intransitive subalgebra is

$$L = \frac{1}{2R\cosh\left(\frac{x}{R}\right)\left(2\cosh\left(\frac{x}{R}\right) - \dot{x}\right)} \tag{51}$$

and admits five Noether symmetries, while the Lagrangian L_B corresponding to the Hamiltonian H_B (48), i.e.

$$L_B = -2\ell_p\sqrt{\cosh\left(\frac{x}{R}\right)^2 - \dot{x}\cosh\left(\frac{x}{R}\right)} \tag{52}$$

admits just three Noether symmetries that actually generate the following family of equations of second order

$$\begin{aligned} \ddot{x} &= A\sqrt{\cosh\left(\frac{x}{R}\right)\left(\dot{x} - \cosh\left(\frac{x}{R}\right)\right)}\left(\dot{x} - \cosh\left(\frac{x}{R}\right)\right) \\ &\quad + \frac{1}{R}\tanh\left(\frac{x}{R}\right)\left(-4\cosh^2\left(\frac{x}{R}\right) + 6\dot{x}\cosh\left(\frac{x}{R}\right) - \dot{x}^2\right) \end{aligned} \tag{53}$$

with A an arbitary constant. Here we do not pursue this case any further since the discussion is analogous to that of the previous section but the formula are quite lenghtier. We observe that the linearizing transformation (50) is singular and resembles somehow the transformation between the free-particle and the linear harmonic oscillator: a similar instance recurs also in the case of a Liénard-type nonlinear oscillator [7]. We will address this issue in future work.

5. Final Remarks

We have shown that Lie symmetries may explain what goes wrong/right if one takes a classical (even non-physical) problem into the realm of quantum mechanics.
In fact, the application of Lie symmetries yield that:

- the Lagrangian equation (20) admits an eight dimensional Lie symmetry algebra, therefore is linearizable and can be transformed to either a free-particle by a t-dependent transformation (23) or a variety of linear equations (35);

- there exist many Lagrangians for equation (20) but only one, up to a representation of the two-dimensional abelian intransitive (Type II) subalgebra, admits the maximum number (five) of Noether symmetries;
- those five symmetries lead to the derivation of a time-dependent Schrödinger equation (27) with a continuum spectrum;
- the Hamiltonian (1) introduced by Sierra corresponds to a t-independent Lagrangian (38) which admits three Noether symmetries $\Gamma_4, \Gamma_5, \Gamma_8$: those three symmetries generate the complete symmetry group of the family of equations (40);
- those three symmetries are not admitted by the Lagrangian (42) of equation (40) unless $A = 0$: the Lagrangian (42) can be obtained through the JLM coming from the two symmetries Γ_5 and Γ_8 but not if $A = 0$
- those three symmetries lead to the Schrödinger equation (43) that admits a continuum spectrum.

Indeed we may have transformed one of Berry's quantum obsessions [2] into a Lie symmetry obsession.

Acknowledgments

MCN acknowledges the support of the Italian Ministry of University and Scientific Research through PRIN 2010-2011, Prot. 2010JJ4KPA_004.

References

[1] Berry MV (1986) Riemann's zeta function: a model for quantum chaos? in *Quantum Chaos and Statistical Nuclear Physics* ed T H Seligman and H Nishioka vol 263, pp 1–17
[2] Berry MV (2008) Three quantum obsessions *Nonlinearity* **21** T19–T26
[3] Berry MV and Keating JP (1999) $H = xp$ and the Riemann zeros, in *Supersymmetry and Trace Formulae: Chaos and Disorder* ed J P Keating and I V Lerner, Plenum, New York, pp 355–367
[4] Berry MV and Keating JP (2011) A compact hamiltonian with the same asymptotic mean spectral density as the Riemann zeros *J. Phys. A: Math. Theor.* **44** 285203
[5] Bianchi L (1918) *Lezioni sulla teoria dei gruppi continui finiti di trasformazioni* (Pisa, Enrico Spoerri)
[6] Edwards HM (1974) *Riemann's Zeta Function* (New York, Academic Press)
[7] Gubbiotti G and Nucci MC (2013) Noether symmetries and the quantization of a Lienard-type nonlinear oscillator *arXiv:1307.3803*
[8] Jacobi CGJ (1842) Sur un noveau principe de la mécanique analytique *Comptes Rendus du Académie des Sciences de Paris* **15**, 202–205
[9] Jacobi CGJ (1844) Sul principio dell'ultimo moltiplicatore, e suo uso come nuovo principio generale di meccanica *Giornale Arcadico di Scienze, Lettere ed Arti* **99** 129–146
[10] Jacobi CGJ (1844) Theoria novi multiplicatoris systemati æquationum differentialum vulgarium applicandi: Pars I *J. Reine Angew. Math.* **27** 199–268
[11] Jacobi CGJ (1845) Theoria novi multiplicatoris systemati æquationum differentialum vulgarium applicandi: Pars II *J. Reine Angew. Math.* **29** 213–279 and 333–376
[12] Jacobi CGJ (1886) *Vorlesungen über Dynamik. Nebst fünf hinterlassenen Abhandlungen desselben herausgegeben von A Clebsch* (Berinl, Druck und Verlag von Georg Reimer)
[13] Jacobi CGJ (2009) *Jacobi's Lectures on Dynamics*, tr. by K. Balagangadharan, ed. by Biswarup Banerjee, Hindustan Book Agency, available through the American Mathematical Society
[14] Lie S (1874) Veralgemeinerung und neue Verwerthung der Jacobischen Multiplicator-Theorie *Forhandlinger I Videnskabs-Selskabet I Christiania* 255–274
[15] Lie S (1912) *Vorlesungen über Differentialgleichungen mit bekannten infinitesimalen Transformationen* (Leipzig: Teubner)
[16] Molina-Vilaplana J and Sierra G (2012) An xp model on AdS2 spacetime *arXiv:1212.2436v2*
[17] Noether E (1918) Invariante Variationsprobleme *Nachr. d. König Gesellsch. d. Wiss. zu Göttingen Math-phys. Klasse* 235–257
[18] Nucci MC (2011) Quantization of classical mechanics: shall we Lie? *Theor.Math. Phys.* **168** 997–1004

[19] Nucci MC (2012) From Lagrangian to Quantum Mechanics with Symmetries *J. Phys.: Conf. Ser.* **380** 012008
[20] Nucci MC (2013) Quantizing preserving Noether symmetries *J. Nonlinear Math. Phys.* **20** 451–463
[21] Nucci MC and Leach PGL (2007) Lagrangians galore *J. Math. Phys.* **48** 123510
[22] Nucci MC and Leach PGL (2008) Gauge variant symmetries for the Schrödinger equation *Il Nuovo Cimento B* **123** 93–101
[23] Sierra G (2007) $H = xp$ with interaction and the Riemann zeros *Nucl. Phys. B* **776** 327–364
[24] Sierra G (2012) General covariant xp models and the Riemann zeros *J. Phys. A:Math. Theor.* **45** 055209
[25] Sierra G and Rodriguez-Laguna J (2011) The H = xp model revisited and the Riemann zeros *Phys. Rev.Lett.* **106** 200201
[26] Winternitz P (2004) Subalgebras of Lie Algebras. Example of $sl(3,\mathbb{R})$ *Centre de Recherches Mathématiques CRM Proceedings and Lecture Notes* **34** 215–227
[27] Whittaker ET (1999) *A Treatise on the Analytical Dynamics of Particles and Rigid Bodies* (Cambridge, Cambridge University Press)

Two-circles theorem, q-periodic functions and entangled qubit states

Oktay K. Pashaev

Department of Mathematics, Izmir Institute of Technology, Gulbahce Campus, Izmir, Urla 35430, Turkey

E-mail: oktaypashaev@iyte.edu.tr

Abstract. For arbitrary hydrodynamic flow in circular annulus we introduce the two circle theorem, allowing us to construct the flow from a given one in infinite plane. Our construction is based on q-periodic analytic functions for complex potential, leading to fixed scale-invariant complex velocity, where q is determined by geometry of the region. Self-similar fractal structure of the flow with q-periodic modulation as solution of q-difference equation is studied. For one point vortex problem in circular annulus by fixing singular points we find solution in terms of q-elementary functions. Considering image points in complex plane as a phase space for qubit coherent states we construct Fibonacci and Lucas type entangled N-qubit states. Complex Fibonacci curve related to this construction shows reach set of geometric patterns.

1. Introduction

1.1. Method of Images

The problem of vortex images in circular annulus was formulated by Poincare [1] and solved by q-elementary functions in [2]. It was shown that all images are ordered in geometric progression and are determined by singularities of q-logarithmmic and q-exponential functions. By this approach we solved one and two vortex problems in circular annulus [3] and found regular polygon type N vortex configurations [2]. In the present paper we are generalizing these results by studying arbitrary hydrodynamic flow in annular domain bounded by two concentric circles. We formulate general two circle theorem and find self-similar structure of the flow. Then we describe the method of images implications for quantum qubit states.

1.2. Two dimensional stationary flow

Incompressibility condition $div\,\vec{u} = 0$ gives $u_1 = \varphi_x$, $u_2 = \varphi_y$, where $\varphi(x, y)$ is the velocity potential. From irrotational condition $rot\,\vec{u} = 0$ we have $u_1 = \psi_y$, $u_2 = -\psi_x$, where $\psi(x, y)$ is the stream function. They satisfy the Cauchy-Riemann equations $\varphi_x = \psi_y$, $\varphi_y = -\psi_x$. Then the complex potential $f(z) = \varphi + i\psi$ is analytic function $\frac{\partial}{\partial \bar{z}} f(z) = 0$, $z = x + iy$, while the complex velocity $\bar{V}(z) = \frac{\partial}{\partial z} f(z)$ is anti-analytic.

1.2.1. Point vortex The point vortex corresponds to complex potential with logarithmic singularity $f(z) = \frac{\Gamma}{2\pi i} Log(z - z_0)$ and complex velocity with simple pole singularity $\bar{V} = \frac{\Gamma}{2\pi i} \frac{1}{z - z_0}$. Residue at this singularity determines the vortex strength $\Gamma = \oint \vec{u} d\vec{s}$.

1.3. Hydrodynamic flow in bounded domain
For a domain bounded by curve C the problem is to find analytic function $F(z)$ with boundary condition

$$\Im F|_C = \psi|_C = 0,$$

where ψ is the stream function. Then the normal velocity vanishes at boundary $v_n|_C = 0$. For one circle C: $z\bar{z} = r_1^2$ the last condition gives $(\bar{V}(z)z + V(\bar{z})\bar{z})|_C = 0$.

1.4. Milne-Thomson's one circle theorem
For a given flow in plane with complex potential $f(z)$, introduction of boundary circle at the origin C: $|z| = r$ produces the flow with complex potential [4]

$$F(z) = f(z) + \bar{f}\left(\frac{r^2}{z}\right)$$

and complex velocity

$$\bar{V}(z) = u_1 - iu_2 = \bar{v}(z) - \frac{r^2}{z^2}v\left(\frac{r^2}{z}\right).$$

1.4.1. Circle theorem for point vortex For the vortex at z_0 with strength $(\Gamma = -2\pi\kappa)$ it gives

$$\bar{V}(z) = \frac{i\kappa}{z - z_0} - \frac{i\kappa}{z - \frac{r_1^2}{\bar{z}_0}} + \frac{i\kappa}{z},$$

where the second term describing vortex $-\kappa$ at the inverse point $r_1^2/\bar{z}_0$ to the circle, is the vortex image. The last term as vortex $+\kappa$ at the origin cannot be fixed by the boundary.

1.4.2. Symmetry The above theorem admits the exterior - interior symmetry. If we denote $g(z) \equiv \bar{f}\left(\frac{r^2}{z}\right)$ - potential of interior flow, then $\bar{g}\left(\frac{r^2}{z}\right) = f(z)$ becomes potential of exterior flow so that

$$F(z) = f(z) + \bar{f}\left(\frac{r^2}{z}\right) = g(z) + \bar{g}\left(\frac{r^2}{z}\right)$$

This symmetry allows one to extend results to the problem inside of circular domain.

1.5. Double connected domain
Due to the Riemann mapping theorem, the unit disk can be mapped conformally onto any simply connected region in the plane. This is why the Milne-Thomson one circle theorem allows one to solve the problem for wide class of contours in simply connected domain.

However extension to multiply connected domains is not straightforward. A doubly connected domain can be conformally mapped to annular region $r_1 < |z| < r_2$ uniquely up to linear map,

$$r_2/r_1 = r_2'/r_1' \equiv \sqrt{q}$$

2. Two circles theorem
For annular domain, $r_1 < |z| < r_2$, between two concentric circles $C_1 : |z| = r_1$, $C_2 : |z| = r_2$ the complex potential is

$$F_q(z) = f_q(z) + \bar{f}_q\left(\frac{r^2}{z}\right),$$

Journal of Physics: Conference Series **482** (2014) 012033 doi:10.1088/1742-6596/482/1/012033

where $q = \frac{r_2^2}{r_1^2}$, $f_q(z) = \sum_{n=-\infty}^{\infty} f(q^n z)$ is flow in even annular domains, $\bar{f}_q\left(\frac{r^2}{z}\right) = \sum_{n=-\infty}^{\infty} \bar{f}\left(q^n \frac{r^2}{z}\right)$ is the flow in odd annular domains (see Figure 1).

Proof: By shifting summation index it is easy to show that [7] $Im\, F(z)|_{C_1} = 0$ and $Im\, F(z)|_{C_2} = 0$.

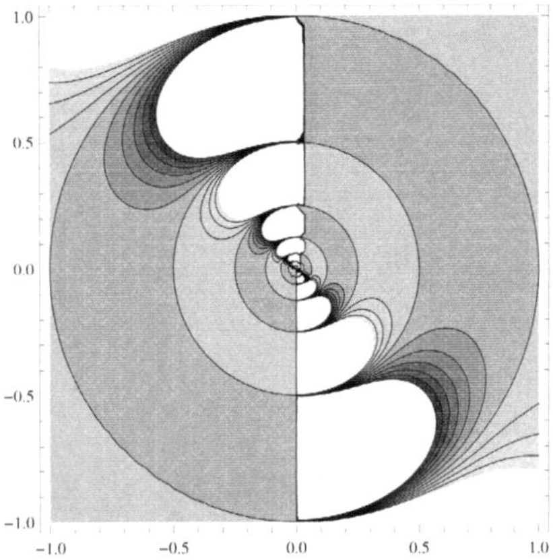

Figure 1. q-periodic flow.

2.1. *q-periodicity*

From above definition follows that $f_q(qz) = f_q(z)$ which implies that complex potential is q-periodic function $F_q(qz) = F_q(z)$ and

$$D_z f_q(z) \equiv \frac{f_q(qz) - f_q(z)}{(q-1)z} = 0. \qquad (1)$$

Corresponding complex velocity

$$\bar{V}(z) = \sum_{n=-\infty}^{\infty} q^n \bar{v}(q^n z) - \frac{r_1^2}{z^2} \sum_{n=-\infty}^{\infty} q^n v(q^n r_1^2 z),$$

then is the q-scale invariant function

$$\bar{V}(qz) = q^{-1}\bar{V}(z).$$

2.2. *Jackson complex q-Integral Representation*

By definition

$$\int f(z) d_q z \equiv (1-q) z \sum_{n=0}^{\infty} q^n f(q^n z)$$

we get the next representation for our two circle potential

$$F(z) = \frac{1}{1-q} \int \frac{f(z) + \bar{f}(r_1^2/z)}{z} d_q z + \frac{q}{q-1} \int \frac{f(z) + \bar{f}(r_1^2/z)}{z} d_{1/q} z - f(z) - \bar{f}\left(\frac{r_1^2}{z}\right).$$

3. q scale-invariant analytic fractal

For any complex valued function of two variables $f(x,y)$ we define q-differential $d_q f(x,y) = f(qx,qy) - f(x,y)$, expressible by complex q-derivatives (see definition in (1)) as

$$d_q f(x,y) = d_q z\, D_z f(x,y) + d_q \bar{z}\, q^{z\frac{d}{dz}}\, D_{\bar{z}} f(x,y).$$

If function $f(x,y)$ is complex analytic function $\partial_{\bar{z}} f = 0$ then $D_{\bar{z}} f = 0$ and $d_q f(x,y) = d_q z\, D_z f(x,y)$. If in addition the analytic function is scale invariant $f(qz) = q^d f(z)$ then it satisfies the q-difference equation

$$D_z f(z) = \frac{f(qz) - f(z)}{(q-1)z} = \frac{(q^d - 1)}{(q-1)z} f(z) \tag{2}$$

or

$$z D_z f(z) = [d]_q f(z), \tag{3}$$

where $[n]_q = \frac{q^d - 1}{q-1}$ is q-number. General solution of this equation has the following form

$$f(z) = z^d A_q(z), \tag{4}$$

where $A_q(qz) = A_q(z)$ is q-periodic function playing the role of q-periodic modulation.

3.1. Weierstrass-Mandelbrot Function

Well known classical example of continuous but anywhere non-differentiable function

$$W(t) = \sum_{n=-\infty}^{\infty} q^{-nd}(1 - \cos q^n t)$$

$0 < d < 1$, $q > 1$ is a fractal with dimension $2-d$. It is self-similar $W(qt) = q^d W(t)$ and satisfies q-difference equation $t D_t W(t) = [d]_q W(t)$ similar to considered above [5]. Decomposition $W(t) = t^d A_q(t)$, then gives q-periodic part $A_q(qt) = A_q(t)$ as

$$A_q(t) = \sum_{n=-\infty}^{\infty} q^{-nd} t^{-d} (1 - e^{iq^n t}).$$

Analytic extension of this result to complex domain as in previous section gives analytic fractal in plane.

3.2. Elliptic functions

Let $F(qz) = F(z)$ is analytic q-periodic function. We take arbitrary positive number ω and define pure imaginary number ω' [9] so that $q = e^{2\pi i \frac{\omega'}{\omega}}$ and new function

$$F(z) \equiv \phi\left(\frac{\omega}{\pi i} \ln z\right).$$

Then due to q-periodicity this function is periodic $\phi(u + 2\omega') = \phi(u)$ where $u \equiv \frac{\omega}{\pi i} \ln z$. Another period is related with rotation on 2π around origin $\phi(u + 2\omega) = \phi(u)$. Thus $\phi(u)$ is double-periodic function with periods 2ω, $2\omega'$. To construct this function we need to fix its singular points.

3.3. q-periodic flow

Simplest q-periodic function is

$$f(z) = z^{\frac{2\pi i}{\ln q}} = e^{\frac{2\pi i}{\ln q}\ln z},$$

the principal branch of which is analytic in complex plain with a branch cut along the negative real axis, and $f(qz) = f(z)$. In polar coordinates $z = re^{i\varphi}$ it gives the flow

$$f(z) = e^{-\frac{2\pi}{\ln q}\varphi}(\cos\frac{2\pi}{\ln q}\ln r + i\sin\frac{2\pi}{\ln q}\ln r)$$

with the stream function

$$\psi(x,y) = e^{-\frac{2\pi}{\ln q}\varphi}\sin\left(\frac{2\pi}{\ln q}\ln r\right). \tag{5}$$

The flow has self-similar velocity

$$\bar{V}(z) = \frac{df}{dz} = \frac{2\pi i}{\ln q}\frac{1}{z}e^{\frac{2\pi i}{\ln q}\ln z}$$

and can be interpreted as q-periodically modulated point vortex at origin with strength $\Gamma = -4\pi^2/\ln q$. It is plotted in Figure 1.

The stream function (5) vanishes $\psi|_C = 0$ when $\sin\frac{2\pi}{\ln q}\ln r = 0$, and $\frac{2\pi}{\ln q}\ln r = \pi n$, $n = 0, \pm1, \pm2...$, which determines the set of concentric circles with radiuses ordered in geometric series $r_n = q^{\frac{n}{2}}$. The flow is circular along these circles (see Figure 1). For $n = 0$ we have unit circle at the origin. Then every pair of circles $r_{+k} = q^{k/2}$ and $r_{-k} = 1/q^{k/2}$ is symmetric with respect to the unit circle $r_{-k}\, r_{+k} = 1$. Moreover, for any chosen circle with radius $r_s = q^{s/2}$, exists infinite set of symmetric circles with radiuses $R_k = r_s\, q^{k/2}$ and $r_k = r_s/q^{k/2}$ so that $R_k\, r_k = r_s$. In addition every circle R has the symmetric one r with respect to r_s. Indeed, let $R = q^{n/2}$. Then exist integer numbers $k = n - s$ and $m = s - k$ such that $R = q^{\frac{s+k}{2}}$ and $r = q^{\frac{s-k}{2}}$ satisfy equation $R\,r = r_s$.

From above consideration follows that $f(z) = z^{\frac{2\pi i}{\ln q}}$ describes the hydrodynamic flow in circular annulus with radiuses $r = 1$ and $R = \sqrt{q}$, so that $q = R^2/r^2$. In fact it describes also the flow in any annulus with $r = q^{n/2}$ and $R = q^{m/2}$, but simplest one is when $n = 0$ and $m = 1$. Taking power of this function we get more general set of functions $f^{(N)}(z) = z^{\frac{2\pi i}{\ln q}N}$, where $N \in Z$. This function is q-periodic and in addition $q^{1/N}$-periodic. The stream function vanishes on circles C: $|z|^2 = q^{n/N}$, $n = 0, \pm1, \pm2, ...$, which are symmetric with respect to unit circle. For the set of numbers n multiple N:$n = kN$, where $k = 1, 2...$, circles $r_{+k} = q^{k/2N}$ and $r_{-N} = q^{-k/2N}$ are symmetric and coincides with $N = 1$ case. With growing N we have more and more dense set of geometric circles, splitting the plane on circular annuluses. Combining superposition of these functions we have more general q- periodic function

$$F(z) = \sum_{N=-\infty}^{\infty} a_N\, z^{\frac{2\pi i}{\ln q}N} \tag{6}$$

describing flow in circular annulus $r = 1$, $R = \sqrt{q}$.

3.4. Application to vortex in annular domain

By our Two Circle Theorem for one vortex in annular domain we find

$$F(z) = \frac{\Gamma}{2\pi i}\sum_{n=-\infty}^{\infty}\ln\frac{z - z_0 q^n}{z - \frac{r_1^2}{\bar{z}_0}q^n}.$$

This function is q-periodic $F(qz) = F(z)$ and can be represented in terms of double periodic functions $F(z) \equiv \varphi(\ln(z))$. Explicit expression in terms of elliptic functions was given in [2].

The corresponding velocity potential

$$\bar{V}(z) = \frac{\Gamma}{2\pi i}\sum_{n=-\infty}^{\infty}\left[\frac{1}{z - z_0 q^n} - \frac{1}{z - \frac{r_1^2}{\bar{z}_0}q^n}\right] \tag{7}$$

is scale-invariant analytic function $\bar{V}(qz) = q^{-1}\bar{V}(z)$ and can be represented as $\bar{V}(z) = \frac{\Gamma}{2\pi i\, z}A_q(z)$, where $A(qx, qy) = A(x, y)$ is q-periodic function of two arguments. The last representation shows that the problem looks like q-periodic modulation of one vortex at the origin. Then from our solution we have q-periodic function as

$$A(z) = \sum_{n=-\infty}^{\infty}\left[\frac{z}{z - z_0 q^n} - \frac{z}{z - \frac{r_1^2}{\bar{z}_0}q^n}\right].$$

3.4.1. Pole singularities and vortex images This expression has pole singularities at $z_0\, q^n$, and at symmetric points $\frac{r_1^2}{\bar{z}_0}\, q^n$, where $n = 0, \pm 1, \pm 2, \ldots$. For $q > 1$, $0 < |z| < q$ by q-logarithm function [2]

$$\ln_q(1+z) = \sum_{n=1}^{\infty}\frac{(-1)^{n-1}z^n}{[n]} = (q-1)\sum_{n=1}^{\infty}\frac{z}{q^n + z},$$

with infinite set of pole singularities at $z = -q^n$, n = 1,2,... then we get expression

$$A(z) = \frac{z}{z - z_0} + \frac{1}{(q-1)}\left[\ln_q\left(1 - \frac{z}{z_0}\right) - \ln_q\left(1 - \frac{z\bar{z}_0}{r_1^2}\right) + \ln_q\left(1 - \frac{r_2^2}{z\bar{z}_0}\right) - \ln_q\left(1 - \frac{z_0}{z}\right)\right].$$

4. q-scale invariant quantum fractal

As known in Fock-Bargmann representation any vector of quantum state $|\psi>$ is associated in one-to-one correspondence with entire analytic function $\psi(z)$. This function is a wave function

$$< z|\psi >= e^{-|z|^2/2}\psi(\bar{z})$$

in the Glauber coherent state basis $|z>$. The dilatation operator q^N acts on this function as

$$q^{z\frac{d}{dz}}\psi(z) = \psi(qz).$$

It was shown in [10], [11] that this operator produces the q-deformation process associated with fractal generation process and is equivalent to the squeezing transformation. The q-number operator $[N]_q = \frac{q^N - 1}{q-1}$ in this representation becomes $[z\frac{d}{dz}]_q = \frac{q^{z\frac{d}{dz}} - 1}{q-1} = zD_z$ so that q-difference equation (3) for scale invariant wave function $\psi(z)$, represents the eigenvalue equation for scale invariant quantum state $|\psi>$,

$$[N]_q|\psi >= [n]_q|\psi > .$$

The general solution of this equation (4) can be considered as descriptive of quantum q-scale invariant fractal. This implies that with hydrodynamic problems discussed in the first part of this paper we can associate quantum states in the Fock-Bargmann representation. Then to the flow in bounded domain determined by the set of images, we introduce the "image" quantum states. These states for one and two circles are given by pairs of symmetric quantum states. One set of images is given by q-deformation of state $|\psi>$: $|q\psi>$, $|q^2\psi>$,... and corresponds to even number of inversions in the circles. Another set of images is related with odd number of inversions: $|\frac{1}{\bar{\psi}}>$, $|q\frac{1}{\bar{\psi}}>$,... To understand structure of elementary inversion state, in next section we study quantum "image" states for $SU(2)$ coherent states, in stereographic projections of qubits. The full construction of q-scale-invariant quantum states as in two circle theorem is under investigation.

5. Mobius transformation and qubit

In this part we apply method of images to construct a set of multiple qubit coherent states. Qubit as a unit of quantum information, can be ascribed to arbitrary point in extended complex plane. Linear transformation

$$\begin{pmatrix} w_1 \\ w_2 \end{pmatrix} = \begin{pmatrix} a & b \\ c & d \end{pmatrix} \begin{pmatrix} \psi_1 \\ \psi_2 \end{pmatrix}$$

acts on homogeneous coordinates $\psi = \psi_1/\psi_2$, $w = w_1/w_2$ by Möbius Transformation

$$w = \frac{a\psi + b}{c\psi + d}, \quad ad - bc \neq 0.$$

5.1. *Qubit and coherent state*

In terms of this coordinates for quantum qubit state we have

$$|\psi> = \begin{pmatrix} \psi_1 \\ \psi_2 \end{pmatrix} \Rightarrow \frac{1}{\sqrt{1+|\psi|^2}} \begin{pmatrix} 1 \\ \psi \end{pmatrix}.$$

This dermines spin $\frac{1}{2}$ generalized coherent state in terms of stereographic projection of the Bloch sphere to complex plane $\psi \in C$, $\psi = \tan\frac{\theta}{2}e^{i\varphi}$,

$$|\theta, \varphi> = \cos\frac{\theta}{2}|0> + \sin\frac{\theta}{2}e^{i\varphi}|1>.$$

5.2. *Symmetric points and symmetric qubits*

5.2.1. Symmetric inverse points Points in complex plane ψ and $1/\bar{\psi}$ are symmetrical points with respect to unit circle C around the origin $|\psi| = 1$. As we have seen in the first part, they correspond to positions of hydrodynamic vortex and its image in unit circle bounded domain. Under Möbius transformation generalized circles and corresponding symmetrical points transform to generalized circles and related symmetric points. These symmetric points as projections of $M(x, y, z)$ and $M^*(x, y, -z)$ inverse points on Bloch sphere determine the pair of inverse symmetric qubits $|\theta, \varphi>$ and $|\pi - \theta, \varphi>$ and hence the couple of coherent states

$$|\psi> = \frac{|0> + \psi|1>}{\sqrt{1+|\psi|^2}}, \quad |\frac{1}{\bar{\psi}}> = \frac{\bar{\psi}|0> + |1>}{\sqrt{1+|\psi|^2}}.$$

The unitary Möbius transformation acting on these states produces another couple of symmetric coherent states.

5.2.2. Negative symmetric inverse points Symmetric coherent states are not orthogonal. However if we consider the pair of negative symmetric points ψ and $-\frac{1}{\bar{\psi}}$ being projections of $M(x, y, z)$ and $M^*(-x, -y, -z)$ antipodal points on Bloch sphere, then we have the pair of qubits $|\theta, \varphi>$ and $|\pi - \theta, \varphi>$ and coherent states

$$|\psi> = \frac{|0> + \psi|1>}{\sqrt{1+|\psi|^2}}, \quad |-\frac{1}{\bar{\psi}}> = \frac{-\bar{\psi}|0> + |1>}{\sqrt{1+|\psi|^2}}.$$

These coherent states are orthogonal $< -\frac{1}{\bar{\psi}}|\psi >= 0$. Since Möbius transformation transforms also the pair of negative inverse points (antipodal points) to couple of negative inverse points, the corresponding antipodal quantum states transform in a similar way.

6. Two qubit coherent states

For $\psi \to \infty$ our antipodal points give 0 and ∞ as (negative inverse)symmetric points and corresponding quantum states become orthogonal computational basis states $|1>$ and $|0>$ correspondingly. Then motivated by Bell's states, we can construct the set of orthonormal two qubit coherent states [7], [8]

$$|P_\pm >= \frac{1}{\sqrt{2}}\left(|\psi > |\psi > \pm| - \frac{1}{\bar{\psi}} > | - \frac{1}{\bar{\psi}} >\right),$$

$$|G_\pm >= \frac{1}{\sqrt{2}}\left(|\psi > | - \frac{1}{\bar{\psi}} > \pm| - \frac{1}{\bar{\psi}} > |\psi >\right).$$

These states are maximally entangled states and the average energy distributions for spin model in these states are surfaces with characteristic extreme points [8].

7. N qubit coherent states

Tensor product of N qubit coherent states can be expanded to computational basis

$$|\psi >^N \equiv \prod_{n=1}^{N} \otimes(|0> + \psi|1>)^n = |00...0> + \psi(|10...0> + ...|00...1>) \quad (8)$$

$$+\psi^2(|11...0> + ...|00...11>) + ...\psi^N|11...1>. \quad (9)$$

It has form of N-degree polynomial with vector coefficients.

7.1. Fibonacci type states

Combination of N qubit states in the form similar to Binet formula gives entangled state

$$|\mathcal{F}_N(\psi) >= \frac{|\psi >^N - | - \psi^* >^N}{\psi - (-\psi^*)} = F_1(\alpha, \beta)(|10...0> + ... + |00...1>) \quad (10)$$

$$+F_2(\alpha, \beta)(|11...0> + ... + |0...11>) + ... + F_N(\alpha, \beta)|11...1>, \quad (11)$$

where coefficients $F_n(\alpha, \beta)$ are complex Fibonacci polynomials.

7.2. Lucas type states

Similar way we get another N qubit entangled state

$$|\mathcal{L}_N(\psi) >= |\psi >^N + | - \psi^* >^N = |00...0> + L_1(\alpha, \beta)(|10...0> + ... + |00...1>) \quad (12)$$

$$+L_2(\alpha, \beta)(|11...0> + ... + |0...11>) + ... + L_N(\alpha, \beta)|11...1>, \quad (13)$$

where coefficients $L_n(\alpha, \beta)$ are complex Lucas polynomials.

8. Complex Fibonacci polynomials

Negative symmetric points with respect to unit circle, ψ and $-\frac{1}{\bar{\psi}}$, are roots of equation

$$\xi^2 = (\psi - \frac{1}{\bar{\psi}})\xi + \frac{\psi}{\bar{\psi}}.$$

Substituting $\psi = |\psi|e^{i\phi}$ and $\xi = \eta e^{i\phi}$ we get $\eta^2 = a\eta + 1$, $a = |\psi| - |\psi|^{-1}$, $\eta^n = \eta F_n(a) + F_{n-1}(a)$ where Fibonacci polynomials are $F_1(a) = 1$, $F_2(a) = a$,

$$F_{n+1}(a) = aF_n(a) + F_{n-1}(a).$$

8.1. Binet representation and q-calculus

Fibonacci polynomials can be interpreted as q-numbers with Binet formula

$$F_n(\eta) = \frac{\eta^n - (-\eta)^{-n}}{\eta - (-\eta)^{-1}} = [n]_{\eta,-\eta^{-1}}$$

for negative-symmetrical points as bases [6]. In particular case when $a = 1$ and $|\psi| = \varphi = \frac{1+\sqrt{5}}{2}$ is the Golden Ratio, we have Fibonacci numbers as q-numbers

$$F_n = \frac{\varphi^n - (-\varphi)^{-n}}{\varphi - (-\varphi)^{-1}} = [n]_F.$$

8.1.1. Complex Fibonacci and Lucas polynomials Recursion formula for complex Fibonacci polynomials is

$$F_{n+1}(\alpha, \beta) = \alpha F_n(\alpha, \beta) + \beta F_{n-1}(\alpha, \beta),$$

where $\alpha = \psi - \bar{\psi}^{-1}$, $\beta = \psi/\bar{\psi}$. They are related to real Fibonacci polynomials by

$$F_n(\alpha, \beta) = F_n(a)e^{i\phi(n-1)},$$

where $\phi = arg\,\psi$. In a similar way for complex Lucas polynomials we have

$$L_n(\alpha, \beta) = \psi^n + (-\bar{\psi})^{-n}$$

and

$$L_n(\alpha, \beta) = L_n(a)e^{i\phi n}.$$

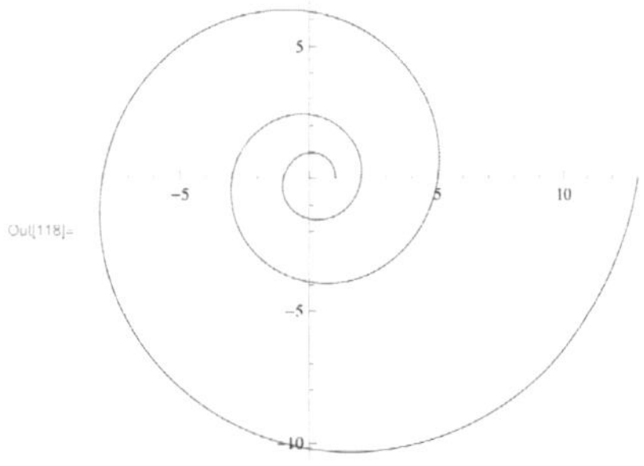

Figure 2. Complex Fibonacci spiral: $\phi = \pi$.

8.2. Complex Fibonacci curve

We can draw the set of these polynomials as points in complex plane. For $a = 1 \rightarrow \varphi = \frac{1+\sqrt{5}}{2}$ we introduce the Fibonacci spiral curve

$$\mathcal{F}_(t) = F(t)e^{i\phi(t-1)} = \frac{\varphi^t - (-\varphi)^{-t}}{\sqrt{5}}e^{i\phi(t-1)}.$$

It is parametric curve with

$$x(t) = \frac{1}{\sqrt{5}}[e^{t\ln\varphi}\cos(\phi(t-1)) - e^{-t\ln\varphi}\cos(\pi t + \phi(t-1))],$$

$$y(t) = \frac{1}{\sqrt{5}}[e^{t\ln\varphi}\sin(\phi(t-1)) - e^{-t\ln\varphi}\sin(\pi t + \phi(t-1))].$$

We draw these spiral curve on Figure 2 for $\phi = \pi$. It is rotating anticlockwise in such a way that intersect x axis at points distant from origin as Fibonacci sequence. To have more fast rotation around the curve In Figures 3-8 we choose $\phi = 100\pi$. Then to produce Fibonacci sequence after every 50 rotation, the curve shows reach set of patterns with interesting shapes.

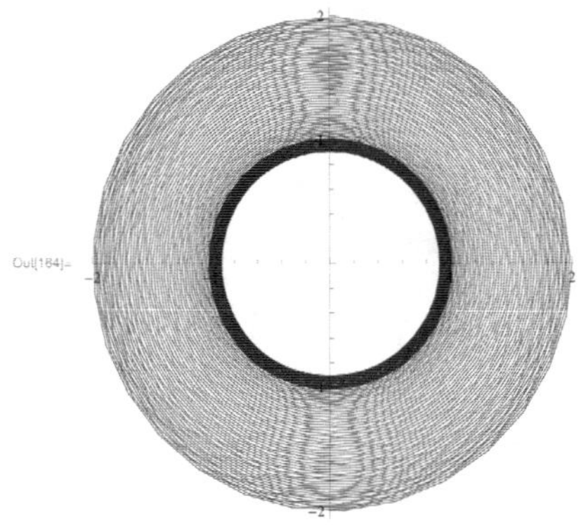

Figure 3. Complex Fibonacci spiral $\phi = 100\pi$: t=3.

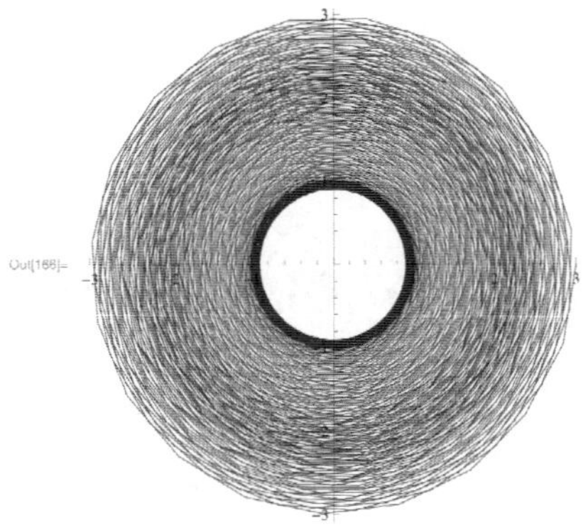

Figure 4. Complex Fibonacci spiral $\phi = 100\pi$: t=4.

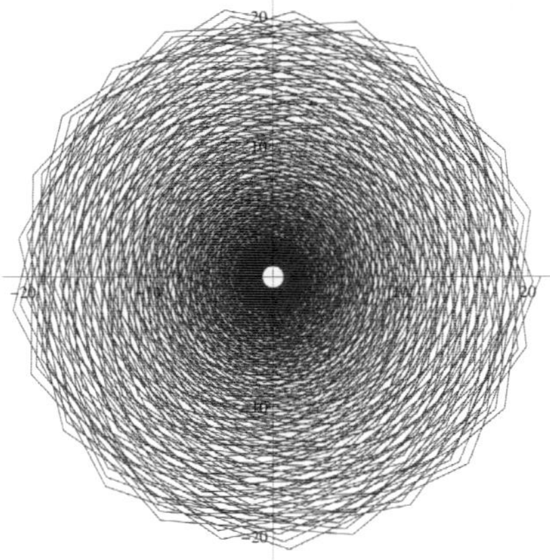

Figure 5. Complex Fibonacci spiral $\phi = 100\pi$: t=8.

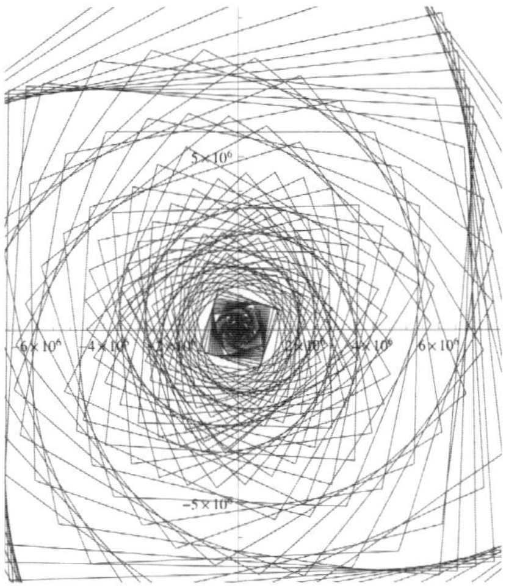

Figure 6. Complex Fibonacci spiral $\phi = 100\pi$: t=50.

8.3. Acknowledgments
This work was supported by Izmir Institute of Technology and by TUBITAK Grant (Project TBAG No. 110T679).

References

[1] Poincare H 1893 *Theorie des Tourbillions* (Paris: Georges Carre)
[2] Pashaev O K and Yilmaz O 2008 *J. Phys. A. Math Theor* **41** 135207
[3] Pashaev O K and Yilmaz O 2011 *J. Phys. A. Math Theor* **44** 185501
[4] Milne-Thomson L M 1968 *Theoretical Hydrodynamics* (London: Macmillan)
[5] Erzan A and Eckmann J P O 1997 *Phys. Rev. Lett.* **78** 3245-8
[6] Pashaev O K and Nalci S 2012 *J. Phys. A. Math Theor* **45** 015303
[7] Pashaev O K 2012 *J. Phys.: Conf. Series* **343** 012093
[8] Pashaev O K and Gurkan N 2012 *New Journal of Physics* **14** 063007
[9] Akhiezer N I 1990 *Elements of the Theory of Elliptic Functions* (Moscow: AMS)
[10] Vitiello G 2008 Topological defects, fractals and the structure of quantum field theory *Preprint* hep-th/0807.2164v1; 2012 math-ph/1206.1854v1

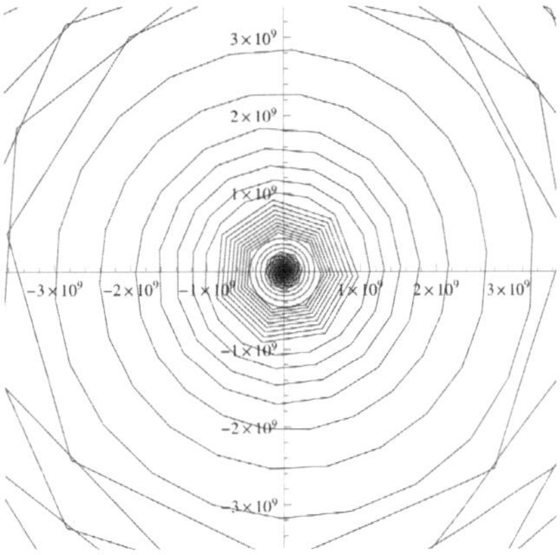

Figure 7. Complex Fibonacci spiral $\phi = 100\pi$: t=70.

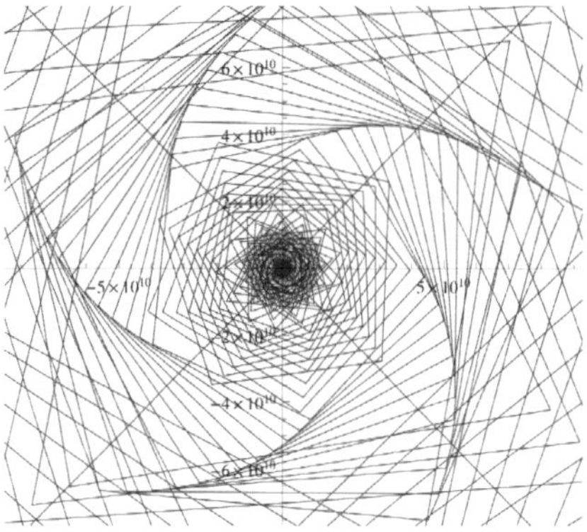

Figure 8. Complex Fibonacci spiral $\phi = 100\pi$: t=80.

[11] Vitiello G 2012 Fractals, coherent states and self-similarity induced noncommutative geometry *Preprint* math-ph/1206.1854v1

Complete bifurcation analysis of DC-DC converters under current mode control

D Pikulin
Leading researcher, Riga Technical University, Riga, Latvia

E-mail: dmitrijs.pikulins@rtu.lv

Abstract. The purpose of this research is to investigate to what extend application of novel method of complete bifurcation groups to the analysis of global dynamics of piecewise-smooth hybrid systems enables one to highlight new nonlinear effects before periodic and chaotic regimes. Results include the construction of complete one and two-parameter bifurcation diagrams, detection of various types of bifurcation groups and investigation of their interactions, localization of rare attractors, and the investigation of different principles of birth of chaotic attractors. Effectiveness of the approach is illustrated in respect to one of the most widely used switching systems–boost converter under current mode control operating in continuous current mode.

1. Introduction

Bifurcations of piecewise - smooth hybrid dynamical systems have received much attention in recent years due to some unexpected, and even unique, phenomena observed in this kind of systems [1-6, 9]. As a matter of fact, the numerical and analytical tools that can be used to study these bifurcations are still very incomplete [1, 2, 9] and a need for an efficient methods and software tools that can handle a large variety of nonlinear problems associated with piecewise-smooth systems therefore remains. The brute-force approach is a general method used for calculating bifurcation diagrams of stable steady-state solutions of arbitrary systems. In contrast, this paper presents the improvement of innovative approach, called Method of Complete Bifurcation Groups (MCBG) [8, 10], allowing the complete bifurcation analysis of highly nonlinear systems, considering stable and unstable periodic orbits. The purpose of this research is to study the possibilities of application of MCBG to the analysis of dynamics of piecewise-smooth hybrid systems, constructing the complete one-parametric bifurcation diagrams and two-parametric diagram (bifurcation map) for simple switching converter – boost one.

2. The discrete model of the current mode controlled boost converter with compensating ramp

The schematic diagram of the conventional current controlled boost converter operating in continuous conductions mode (CCM – when the inductor current never falls to zero) with a control circuitry including a comparator and a RS flip-flop is shown in the figure 1.

The comparator likens the inductor current to a reference current I_{ref} in order to control the state of switching element *S*. Thus the circuit could be described by two different structures, depending on whether the switch is open (*OFF*) or closed (*ON*). When the switch *S* is in the *ON* position, the diode *D* is reverse biased and non-conducting. Assuming that the resistance of inductor *L* is negligible, the current i_L rises almost linearly (see figure 1(b)) and the energy is accumulated in the magnetic field of the inductor. The switch *S* is opened when the inductor current reaches the reference value I_{ref}, then the voltage Ldi_L/dt is induced in the inductor in order to maintain the current flow. This forward biases the

Journal of Physics: Conference Series **482** (2014) 012034 doi:10.1088/1742-6596/482/1/012034

diode and the current decays almost linearly, transferring the energy from the inductor to the capacitor *C*. The switch is closed each time the clock pulse with period *T* arrives [1].

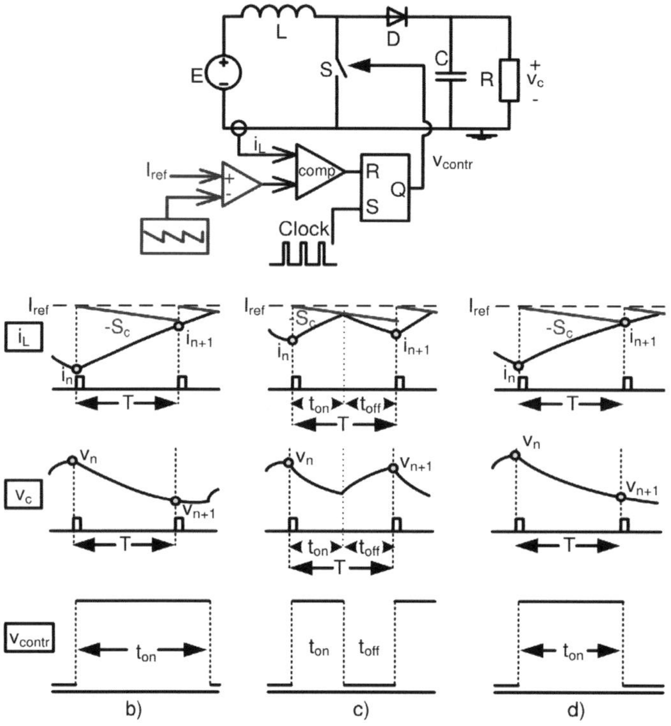

Figure 1. (a) The current controlled boost converter; **(b)-(d)** typical waveforms of inductor current and the control signal for the switch.

It is well known and documented fact that almost all current mode controlled switching converters are capable of exhibiting subharmonic modes of operation [1]. Thus, in order to avoid the occurrence of undesirable operating regimes, special compensating ramp is added to the reference current. This compensating signal is widely integrated in the DC-DC converter control chips and is taken into account while defining the model of the boost converter under study.

The operation of boost converter could be described by two systems of differential equations, accordingly to the two possible switching structures. However, it is now a common practice to use the discrete-time models, which provide the simple tool for the direct construction of bifurcation diagrams as well as analytical investigation of various types of nonlinear phenomena. The discrete-time model of the boost converter under study is the function that relates the voltage and the current vector at one instant $(i_{(n+1)}, v_{(n+1)})$ to the vector at previous instant $(i_{(n)}, v_{(n)})$; the sequence of time instants here is clearly defined by the arrival of clock pulses. The boost converter is unusual in that the discrete-time model can be derived in closed form without approximations [1, 6]. It could be deduced that there are two ways in which the state can move from one clock instant to the next. A clock pulse may arrive before the inductor current reaches the I_{ref} (figure 1(b)). In that case, the obtained discrete-time model is the following:

$$v_{n+1}=v_n e^{-T/RC} \tag{1}$$

$$i_{n+1}=i_n+ET/L \tag{2}$$

Journal of Physics: Conference Series **482** (2014) 012034 doi:10.1088/1742-6596/482/1/012034

On the other hand, if the inductor current reaches I_{ref} before the arrival of the next clock pulse (figure 1(c)) the map would include the *ON* and *OFF* intervals. In such case, the discrete time mapping is as follows:

$$v_{n+1}=e^{-mt_2}[K_1\cos(\mu t_2)+K_2\sin(\mu t_2)]+E \qquad (3)$$

$$i_{n+1}=e^{-mt_2}[C[-m(K_1\cos(\mu t_2)+K_2\sin(\mu t_2))+\mu(-K_1\sin(\mu t_2)+K_2\cos(\mu t_2))]+ \\ +(K_1\cos(\mu t_2)+K_2\sin(\mu t_2))/R]+E/R, \qquad (4)$$

where $K_1=v_n e^{-2mt_1}-E$, $K_2=[I_{ref}/C-m(v_n e^{-2mt_1}+E)]/\mu$, $t_1=(I_{ref}-i_n)/(E/L+S_c)$, $t_2=T-t_1$, $S_c=V_{p-p}/T$, $\mu=\sqrt{(p^2-m^2)}$, $m=1/(2RC)$ and $p=1/\sqrt{(LC)}$.

The borderline between the two regimes described by maps (1)-(2) and (3)-(4) is given by the value I_{border} for which the inductor current reaches the reference value exactly at the arrival of the next clock pulse (figure 1(d)):

$$I_{border}=I_{ref}-T(E/L+S_c) \qquad (5)$$

The model presented in this section is used to obtain the complete bifurcation diagrams and provide all necessary investigation of phenomena observed in the current controlled DC-DC boost converter.

3. The concepts of the Method of Complete Bifurcation Groups (MCBG)

The Method of Complete Bifurcation Groups, developed in Riga Technical University in a group under guidance of Professor M.V. Zakrzhevsky [10], includes the complex of approaches used for the analysis of highly nonlinear dynamical systems. Precisely, it implies the following procedures:

- for fixed parameters of the system: detection of all stable and unstable periodic regimes and bifurcation subgroups with so called unstable periodic infinitiums (UPI – the area in which only unstable periodic orbits of definite bifurcation group exist [8,10]) on plane of states, construction of regimes' basins of attraction;
- for varied system parameters: construction of complete one parameter bifurcation diagrams and bifurcation maps.

The continuation along a solution branch of selected regime (not parameter) is of special importance in the MCBG. This approach allows finding new unknown before stable regimes in widely used dynamical models of strongly nonlinear oscillatory systems.

One-parameter bifurcation analysis within the MCBG includes following procedures:

- the construction of periodic skeleton: all stable and unstable periodic solutions at fixed value of bifurcation parameter are found, the number and types of bifurcation groups is identified;
- on the basis of obtained periodic skeleton, using the numerical continuation, the branches of bifurcation diagram, corresponding to each found group are constructed. In many cases, on the branches of diagram corresponding to unstable regimes, the greater or smaller stable periodic solutions are found (rare attractors – RA);
- the complete bifurcation diagram is composed, including stability estimation procedure, defining (analytically of numerically) appropriate monodromy matrixes for all found regimes and determining their multipliers;
- the revision of the diagram allows detection of different types of RA and UPI.

In the forthcoming Sections, the described procedure will be applied to the model (1)-(4) and appropriate complete bifurcation diagrams for boost DC-DC converter under study will be constructed.

4. The complete bifurcation analysis of the boost converter

At the beginning of the investigation, the two-parameter bifurcation diagram (map) is constructed, dividing the two parameter space to different regions of periodic and chaotic behavior. The reference current and the output capacitance are selected accordingly as primary and secondary bifurcation parameters. The constructed bifurcation map was obtained for the boost converters with parameters defined in Table 1. The map shows that for small values of output capacitance, the boost converter exhibits period doubling (PD) bifurcation cascade, chaotic mode of operation and a great variety of subharmonic windows. However, the increment of the secondary bifurcation parameter leads to the contraction of area of the first period doubling cascade with the subsequent transition to robust chaotic operation (without periodic windows and coexisting attractors).

Table 1. Parameters of the boost converter.

Parameter	Value
Switching period, T	100 μs
Inductance, L	1.5 mH
Load resistance, R	40 Ω
Output capacitance, C	2-20 μF
Input voltage, E	5 V
Reference current, I_{ref}	0.2…1.5 A
Compensating ramp, Sc	0

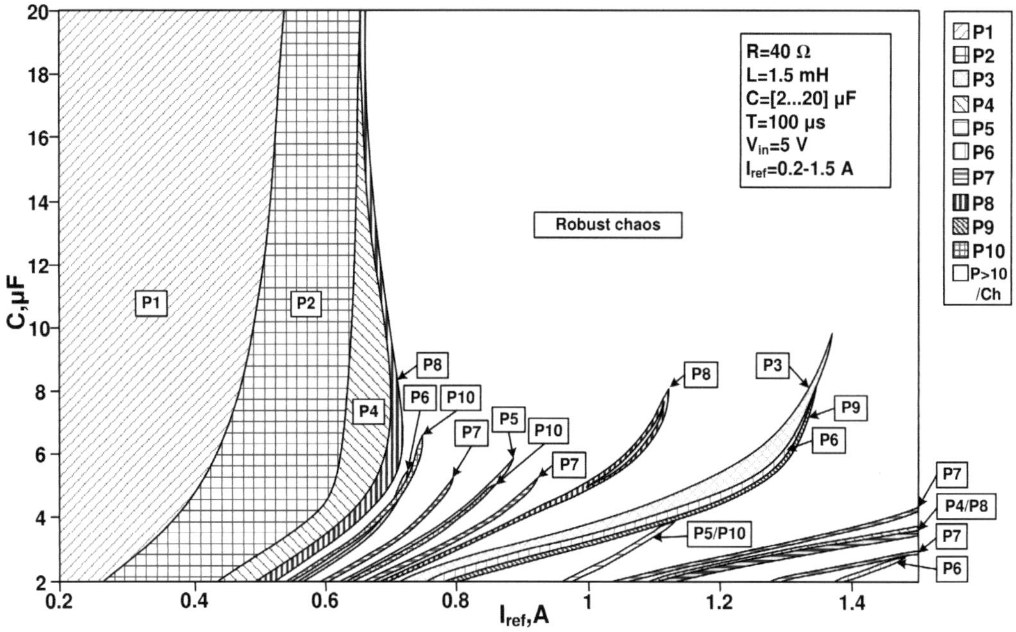

Figure 2. Two parameter bifurcation diagram of the boost converter.

To provide the detailed analysis of the dynamics of boost converter, the complete bifurcation diagrams are constructed as the cross-sections of the bifurcation map. The first complete bifurcation diagram (for $C=5\ \mu F$) depicted in Figure 3 (where dark lines represent stable and light lines represent unstable periodic regimes), shows that for the small values of reference current the system operates in the stable period-1 (P1) regime (within the 1T bifurcation group), that loses its stability through smooth period doubling bifurcation at $I_{ref}\approx 0.42\ A$. Increasing the primary bifurcation parameter (I_{ref}), the classical period doubling bifurcation cascade and the subsequent chaotization is observed (which is connected to the formation of the UPI-1). Further increment of I_{ref} leads to the occurrence of the 6T bifurcation group that also includes the smooth period doubling cascade ending with the chaotic mode of operation. Figure 4 shows the amplified fragment of the complete bifurcation diagram, exhibiting the sudden disappearance and the subsequent appearance of the chaotic attractor defined by 6T bifurcation group. The described phenomenon could not be interpreted in terms of the loss of local stability of definite periodic regime as it is connected to the appearance of different types of global crises. The type of the crisis could be simply detected, studying the unstable branches of bifurcation diagram, constructed within the MCBG.

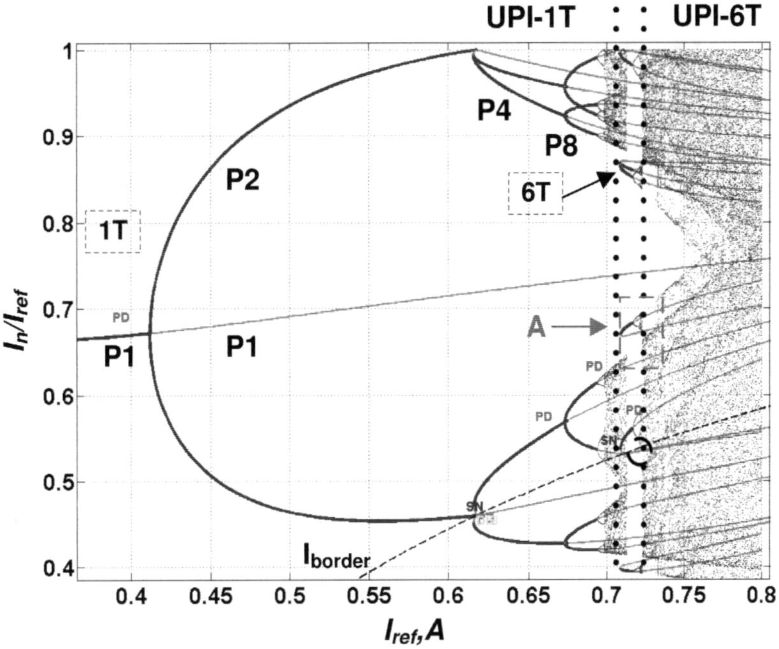

Figure 3. The complete bifurcation diagram for the boost converter $C= 5\ \mu F$.

Figure 4 shows that, at the point of "disappearance" of the chaotic attractor, the unstable branch of P6 regime touches the attractor that becomes unstable through the boundary crisis. It is important to note, that the unstable chaotic attractor of 6T bifurcation group still exists, and becomes stable through the internal crisis when the unstable branch of P6 touches the 1T chaotic attractor (at $I_{ref}\approx 0.73\ A$), forming the global chaotic dynamics of the system.

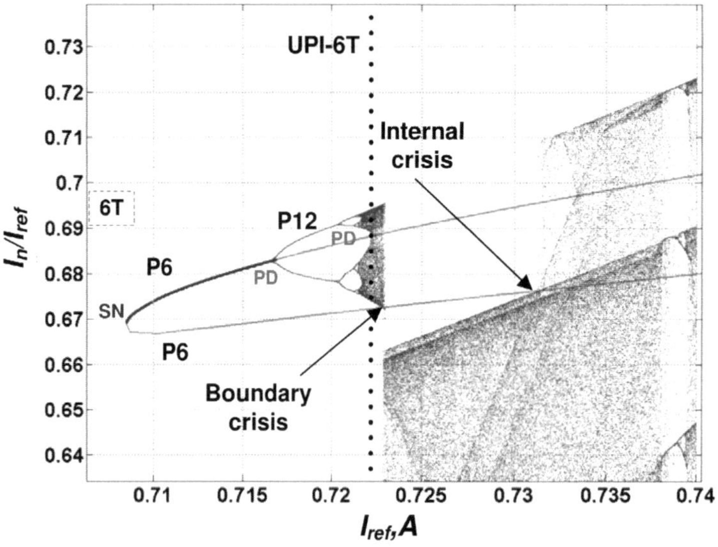

Figure 4. The magnified fragment A of the complete bifurcation diagram in the figure 3.

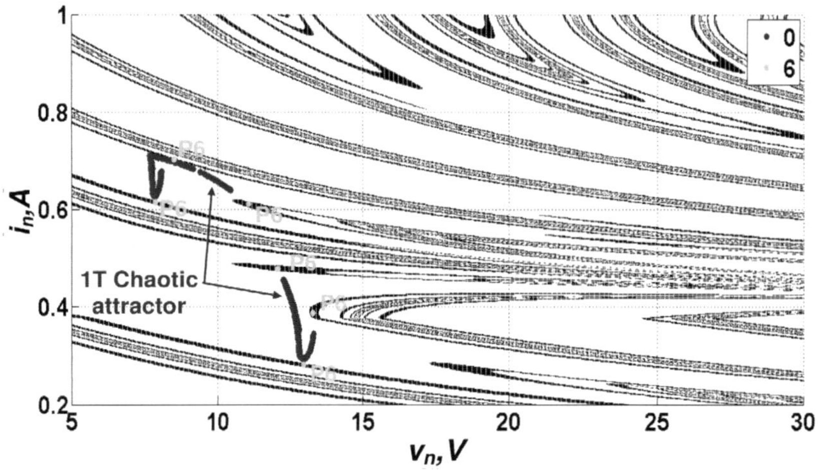

Figure 5. The basins of attraction of P6 and chaotic regimes for $I_{ref}=0.71\ A;\ C=5\ \mu F$.

The dynamics of each bifurcation group could be studied separately; however, the global dynamics of the boost converter is defined by the interaction of various bifurcation groups. The bifurcation diagram, depicted in the figure 3, shows that for certain parameter values the 1T chaotic attractor coexists with the P6 regime of 6T bifurcation group, defining the region of multistability, in which each regime has its own basin of attraction. The mentioned attractors along with the appropriate basins of attraction are shown in the figure 5 (black area represents the basin of attraction of P6 regime, and the white area – that of 1T chaotic regime). It could be observed that the border of the basin of attraction is rather complex and the occurrence of external noise -or even small deviation in system parameters- could lead to skipping between chaotic and subharmonic modes of operation.

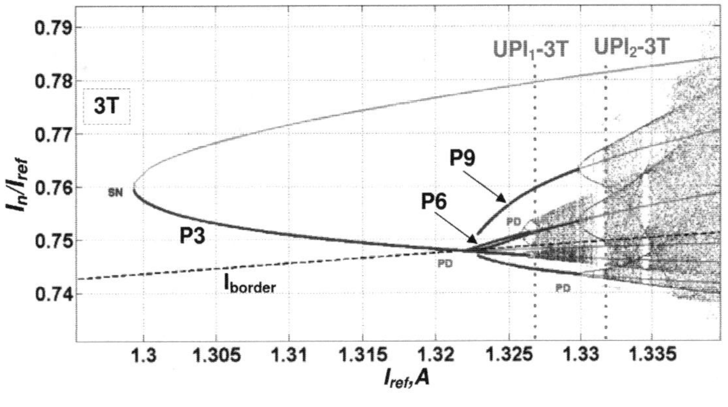

Figure 6. Coexistence of P6 and P9 regimes, both born from single P3 regime for *C= 7 μF.*

The complete bifurcation diagrams, obtained for the greater values of the output capacitance in general have structure similar to one shown in the figure 3, so for the sake of simplicity only the most atypical bifurcation groups will be shown.

The increment of the output capacitance leads to the growth of the influence of the borderline, defined in (5), on the global dynamics of the converter. For example figure 6 shows the 3T bifurcation group within which the first non-smooth period doubling bifurcation caused by the collision with I_{border} leads to the appearance of two stable coexisting P6 and P9 regimes with their own chaotic attractors (UPI$_1$-3T and UPI$_2$-3T). In fact the observed bifurcation is simultaneous period doubling and period tripling, leading to very complicated dynamics of the boost converter.

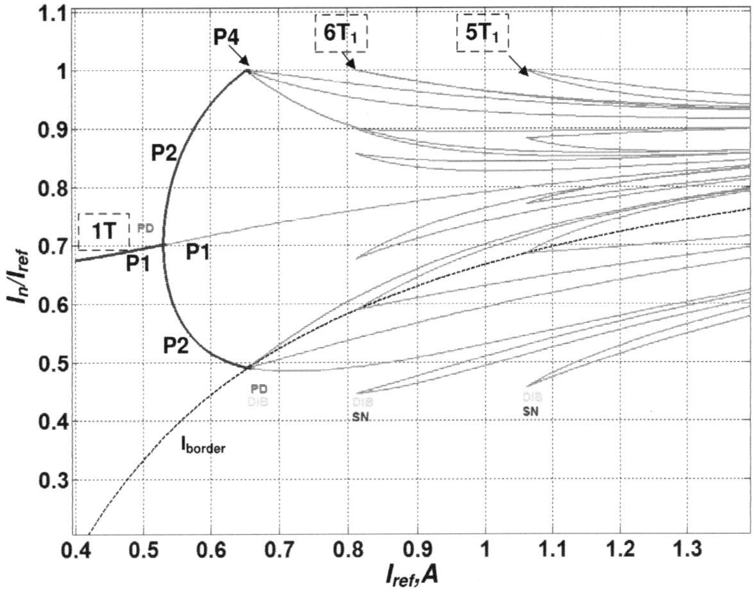

Figure 7. The complete bifurcation diagram for the boost converter for *C= 17 μF.*

Another example of the effect of the pronounced borderline is shown in the figure 7, where after the non-smooth period doubling bifurcation no stable orbits occur (see the region of robust chaos in figure 2). All bifurcation groups (see for example $6T_1$ and $5T_1$) are formed by uncommon border collision, which leads to the jump of the characteristic multiplier across the whole unity circle, causing the simultaneous occurrence of period doubling and saddle-node (SN) bifurcation (see figure 8). The only analogous of this "strange bifurcation" up to authors knowledge is described in [7], within the study of complex behavior of mechanical systems.

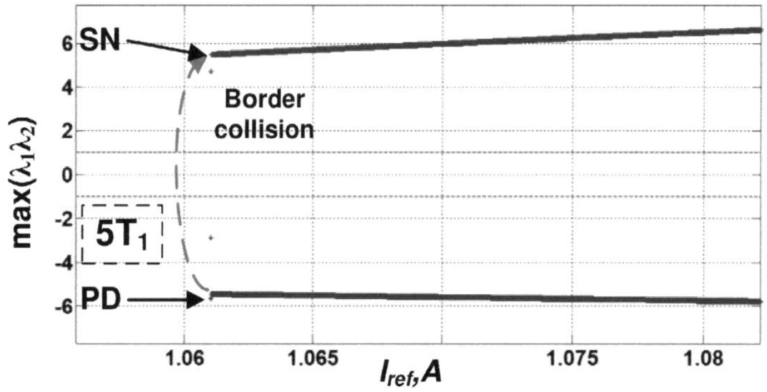

Figure 8. The dependence of the maximal multiplier of P5 regime within $5T_1$ bifurcation group on parameter for $C= 17\ \mu F$.

Thus it has been shown, that for the small values of output capacitance the smooth bifurcation dominate in the boost converter. However, for greater values of C the collision with the borderline (5) may lead to the manifestation of sudden "strange" bifurcations and the appearance of rather uncommon types of coexisting bifurcation groups.

5. Conclusions

Nowadays the typical design objective in majority of engineering systems stable operation with wide and predictable stability margins. Switching mode DC–DC converters often do not fit cleanly into such a framework. The use of switching elements makes this kind of systems highly nonlinear and the prediction of their instabilities is a challenging task.

It has been demonstrated in this communication that the MCBG could be efficiently applied to investigation of the dynamics of switching DC–DC converters in order to provide the researcher and designer with necessary information about the domains of existence and stability of period-1 and subharmonic operation, as well as the appearance of smooth and various types of unordinary non-smooth bifurcations in this kind of circuits.

The application of MCBG to the analysis of complex behavior of switching converters demonstrates potential for general applicability of pronounced technique to exploration of dynamics of switching (hybrid) systems described by set of difference or differential equations. The development in this direction may bring us to a more detailed understanding of the complex dynamics of even more complicated systems with several control loops and sophisticated switching borders, using the concepts of complete bifurcation groups and rare attractors.

Acknowledgements

Travel costs and participation fee for this conference are financially supported by ERDF project „ The development of international cooperation, projects and capacities in science and technology at Riga Technical University" Nr.2DP/2.1.1.2.0/10/APIA/VIAA/003.

References

[1] Banerjee S, Verghese G C 2001 *Nonlinear Phenomena in Power Electronics . Attractors, Bifurcations, Chaos, and Nonlinear Control* (IEEE Press)

[2] di Bernardo M, Budd C J, Champneys A R and Kowalczyk P 2008 *Piecewise-Smooth Dynamical Systems: Theory and Application* (Berlin: Springer)

[3] di.Bernardo M, Fosas E, Olivar G, Vasca F 1997 Secondary bifurcations and high-periodic orbits in voltage controlled buck converter *International Journal of Bifurcation and Chaos* **12** pp 2755-2771

[4] Chan W C Y, Tse C K 1996 A universal bifurcation path conjectured from current-programmed switching converters *Proceedings of International Symposium on Nonlinear Theory and Its Applications* (Japan) pp 121-124

[5] Fosas E and Olivar G 1996 Study of chaos in buck converter *IEEE Transactions on Circuits and Systems Part I* **43** pp 13-25

[6] Hammil D C, Deane J H B and Jefferies D J 1992 Modeling of chaotic dc/dc converters by iterative nonlinear mappings *IEEE Transactions on Circuits and Systems Part I* **35** pp 25-36.

[7] Leine R I 2000 *Bifurcations in Discontinuous Mechanical Systems of Filippov-Type* Ph.D. thesis (Eindhoven University of Technology)

[8] Schukin I, Zakrzhevsky M 2008 Application of Software SPRING and Method of Complete Bifurcation Groups for the Bifurcation Analysis of Nonlinear Dynamical *System Journal of Vibroengineering* **10** pp. 510-518

[9] Tse C K 2003 *Complex Behavior of Switching Power Converters (*Boca Raton: CRC Press)

[10] Zakrzhevsky M 2008 New Concepts of Nonlinear Dynamics: Complete Bifurcation Groups, Protuberances, Unstable Periodic Infinitiums and Rare Attractors *Journal of Vibroengineering* **10** pp 421-441

Topological phase states of the $SU(3)$ QCD

Alexander P Protogenov[1,2], **Evgueni V Chulkov**[2,3,4], **Jeffrey C Y Teo**[5]

[1]Institute of Applied Physics, Nizhny Novgorod 603950, Russia
[2] Donostia International Physics Center (DIPC), 20018 San Sebastián/Donostia, Spain
[3] Departamento de Física de Materiales UPV/EHU; Centro de Física de Materiales CFM - MPC and Centro Mixto CSIC-UPV/EHU, Facultad de Ciencias Químicas, Universidad del Pais Vasco/Euskal Herriko Unibertsitatea, 20080 San Sebastián/Donostia, Spain
[4] Tomsk State University, Tomsk 634050, Russia
[5]Department of Physics, University of Illinois at Urbana-Champaign, Urbana IL 61801, USA

E-mail: alprot@appl.sci-nnov.ru

Abstract. We consider the topologically nontrivial phase states and the corresponding topological defects in the $SU(3)$ d-dimensional quantum chromodynamics (QCD). The homotopy groups for topological classes of such defects are calculated explicitly. We have shown that the three nontrivial groups are $\pi_3 SU(3) = \mathbb{Z}$, $\pi_5 SU(3) = \mathbb{Z}$, and $\pi_6 SU(3) = \mathbb{Z}_6$ if $3 \leq d \leq 6$. The latter result means that we are dealing exactly with six topologically different phase states. The topological invariants for d=3,5,6 are described in detail.

Introduction. - Topological invariants of field configurations are the fundamental objects in the quantum field theory and condensed matter physics, which classify topological defects and possible phase states [1, 2]. The well-known examples of topological field distributions are vortices, hedgehogs, and instantons. They are a direct consequence of the nontrivial homotopy groups $\pi_n \mathbb{S}^n = \mathbb{Z}$ for the spheres $\mathbb{S}^n$ with $n = 1, 2$, $\pi_3 SU(2) = \mathbb{Z}$, for the spatial dimensions $d = 1, 2, 3$, respectively. Here, $\mathbb{Z}$ is a group of integers. Recent progress in the theory of topologically ordered phase states [3, 4] is associated with the classification of the systems, in which the D-dimensional surface $\mathbb{S}^D$ surrounds a defect in d-dimensional topological insulators or superconductors [5, 6, 7, 8] and $D \neq d$. In this case, the first nontrivial example $\pi_3 \mathbb{S}^2 = \mathbb{Z}$ is the well-known Hopf mapping of the three-dimensional sphere $\mathbb{S}^3$ into the two-dimensional one $\mathbb{S}^2$. The corresponding topological invariant Q is called the helicity in magnetohydrodynamics or the Abelian Chern-Simons action in the $(2 + 1)$-dimensional topological field theory. The integer Q means the knotting degree of the field distributions and determines, in particular, the lower bound [9, 10] of the energy of the two-component Ginzburg-Landau model expressed [11] in the form of the Skyrme-Faddeev-Niemi model [12, 13]. In this $O(3)$ σ-model, the $U(1)$ two-form $dA = \mathbf{n}[\mathrm{d}\mathbf{n} \wedge \mathrm{d}\mathbf{n}]$ is parametrized the Hopf invariant $Q = \frac{1}{16\pi^2} \int_{\mathbb{S}^3} A \wedge dA \in \mathbb{Z}$ by the unit 3d vector $\mathbf{n}$ which maps the base space $\mathbb{S}^3$ into $\mathbb{S}^2$. The target sphere $\mathbb{S}^2$ of the map is topologically equivalent to the coset $SU(2)/U(1) \cong \mathbb{S}^2$. The $\mathbf{n}$-field is also a relevant on-shell variable [15, 14, 16, 18] in the infrared limit of the $SU(2)$ QCD.

In this paper, we use the $SU(3)$ group instead of the $SU(2)$ one. The change in the value of the rank is due to several reasons. Primarily, to the three colors of the QCD. From the point of view of the knot theory, this choice is also due to an attempt to extend the low-dimensional topology of the standard knot theory to higher dimensions of the $SU(3)$ QCD target space

d	0	1	2	3	4	5	6	7	8	9	10
$\pi_d F_2$	0	0	$\mathbb{Z}\times\mathbb{Z}$	$\mathbb{Z}$	0	$\mathbb{Z}$	$\mathbb{Z}_6$	0	$\mathbb{Z}_{12}$	$\mathbb{Z}_3$	$\mathbb{Z}_{30}$

Table 1. A list of the homotopy groups $\pi_d F_2$ for dimensions $d \leq 10$.

d	0	1	2	3	4	5	6	7
$\pi_d CP^2$	0	0	$\mathbb{Z}$	0	0	$\mathbb{Z}$	$\mathbb{Z}_2$	$\mathbb{Z}_2$

Table 2. A list of the homotopy groups $\pi_d CP^2$ for dimensions $d \leq 7$.

[17]. One approach to this problem is to use many-valued functionals [20] in accordance with the conjecture given in Ref. [21]. Another elegant method is based on the results obtained in Ref. [22]. However, it is more expedient to describe the target spaces of the $SU(3)$ QCD as a generalization of the $SU(2)$ target sphere $\mathbb{S}^2$.

Flag space. - To generalize the $SU(2)$ group target space to the $SU(3)$ one, we consider the coset $SU(3)/(U(1) \times U(1)) = F_2$ that is now the flag space F_2 [23, 24]. The remained freedom of the maps is the dimension of the base space having a nontrivial homotopy group. For simplicity, we restrict our consideration to the spheres $\mathbb{S}^n$ as the base spaces. Therefore, we focus on such n of the homotopy groups $\pi_n F_2$, which yields nontrivial results. For comparison, in addition to the maximal torus $U(1)^2$ of $SU(3)$ that results in the general orbit F_2 [25], we calculate the homotopy groups of the degenerate orbit CP^2, which are equivalent to the coset space $SU(3)/U(2) = SU(3)/\left(SU(2) \times U(1)\right) = CP^2$.

It should be noted that we are restricted to the framework of the homotopy group approach. Therefore, we would like to determine the constraints on the type of the possible topological phase states and topological defects only. We will describe the geometry of the flag space F_2 and topological features in the last two sections. The results of our calculations are presented in two tables.

It is seen in Table I that the nontrivial homotopy groups for $d \leq 6$ are $\pi_2 F_2$, $\pi_3 F_2$, $\pi_5 F_2$, and $\pi_6 F_2$.

(i) It is known [24] that nontriviality of $\pi_2 F_2 = \mathbb{Z} \times \mathbb{Z}$ accounts for the presence of two different monopoles in the theory (cf. $\pi_2 CP^2 = \mathbb{Z}$ in Table II which means that we deal with a monopole of one type). The second homotopy group is nontrivial due to the fact that the simply connected flag space F_2 is a compact symplectic manifold.

(ii) The integers in the RHS of $\pi_3 F_2 = \mathbb{Z}$ have the meaning of $SU(3)$ instanton topological charges because $\pi_3 F_2 = \pi_3 SU(3) = \pi_3 SU(2)$.

(iii) The integers in the RHS of $\pi_5 F_2 = \mathbb{Z}$ describe some textures and the corresponding phases. The nature of these textures is difficult to guess now.

(iv) The most interesting answer $\pi_6 F_2 = \mathbb{Z}_6$ means that there are only six phase states with the labels $\{0, 1, \ldots 5\}$. They are usually ordered as three quark doublets. We can topologically distinct the quark states because of the gauge invariant coupling of the fermions to the gauge potential. This takes place on the scales where we can consider the six-dimensional base space as the sphere $\mathbb{S}^6$. Note that some additional parameters of the $(3+1)$d gauge theory can add dimensions in order to have finally 6 dimensions of the base space [31]. We encounter these phenomena in some topologically ordered phases of condensed matter [4]. In our case, the best natural choice for the interpretation of the base space $\mathbb{S}^6$ corresponds to the standard six-dimensional space-momentum phase space. We are free also to interpret the six-dimensional compact space $\mathbb{S}^6$ as a complement to the $(3+1)$-dimensional space-time, but the previous suggestion is much better.

Gauge fields on the flag space. - Let us describe the flag space F_2 in detail to explain in particular at the end, why we addressed to homotopy theory approach. It is a compact Kähler manifold which is a homogeneous nonsymmetric space of dimension $\dim F_2 = 6$. Since the flag manifold F_2 is the Kähler one, it possesses the complex local coordinates w_α, $\alpha = 1, 2, 3$, the Hermitian Riemanian metric, $ds^2 = g_{\alpha\bar{\beta}} dw^\alpha d\bar{w}^\beta$, and the closed two-form (field strength) $\Omega_K = i g_{\alpha\bar{\beta}} dw^\alpha \wedge d\bar{w}^\beta$, i.e., $d\Omega_K = 0$. Here, $d = \partial + \bar{\partial} = dw_\alpha \frac{\partial}{\partial w_\alpha} + d\bar{w}_\beta \frac{\partial}{\partial \bar{w}_\beta}$ denotes the exterior derivative, while the operators ∂ and $\bar{\partial}$ are called the Dolbeault operators.

According to the Poincaré lemma, any closed form Ω_K is *locally* exact, i.e., $\Omega_K = d\omega$, where ω is the gauge potential. The condition $d\Omega_K = 0$ is equivalent to $g_{\alpha\bar{\beta}} = \frac{\partial}{\partial w^\alpha} \frac{\partial}{\partial \bar{w}^\beta} K$, where $K = K(w, \bar{w})$ is the Kähler potential:

$$K(w, \bar{w}) = \ln[(\Delta_1(w, \bar{w}))^m (\Delta_2(w, \bar{w}))^n] , \tag{1}$$

$$\Delta_1(w, \bar{w}) = 1 + |w_1|^2 + |w_2|^2 , \tag{2}$$

$$\Delta_2(w, \bar{w}) = 1 + |w_3|^2 + |w_2 - w_1 w_3|^2 . \tag{3}$$

By means of three complex variables w_α, the flag space F_2 is realized as a set of triangular matrices of the form

$$\begin{pmatrix} 1 & w_1 & w_2 \\ 0 & 1 & w_3 \\ 0 & 0 & 1 \end{pmatrix}^t \in F_2 = SU(3)/U(1)^2 . \tag{4}$$

The Kähler one-form and the two-form are $\omega = \frac{i}{2}(\partial - \bar{\partial})K, \Omega_K = i\partial\bar{\partial}K$. The explicit forms of the gauge potential ω and the field strength Ω_K are given by

$$\begin{aligned} \omega &= im \frac{w_1 d\bar{w}_1 + w_2 d\bar{w}_2}{\Delta_1(w, \bar{w})} + in \frac{w_3 d\bar{w}_3}{\Delta_2(w, \bar{w})} \\ &+ in \frac{w_2 - w_1 w_3)(d\bar{w}_2 - \bar{w}_1 d\bar{w}_3 - \bar{w}_3 d\bar{w}_1}{\Delta_2(w, \bar{w})} , \end{aligned} \tag{5}$$

$$\begin{aligned} \Omega_K &= d\omega = im(\Delta_1)^{-2}[(1 + |w_1|^2) dw_2 \wedge d\bar{w}_2 \\ &- \bar{w}_2 w_1 dw_2 \wedge d\bar{w}_1 \\ &- w_2 \bar{w}_1 dw_1 \wedge d\bar{w}_2 + (1 + |w_2|^2) dw_1 \wedge d\bar{w}_1] \\ &+ in(\Delta_2)^{-2}[\Delta_1 dw_3 \wedge d\bar{w}_3 \\ &- (w_1 + \bar{w}_3 w_2) dw_3 \wedge (d\bar{w}_2 - \bar{w}_3 d\bar{w}_1) \\ &- (\bar{w}_1 + w_3 \bar{w}_2)(dw_2 - w_3 dw_1) \wedge d\bar{w}_3 \\ &+ (1 + |w_3|^2)(dw_2 - w_3 dw_1) \\ &(d\bar{w}_2 - \bar{w}_3 d\bar{w}_1)] . \end{aligned} \tag{6}$$

Calculation of the Poincaré polynomial $P_{F_2}(t) = \sum_{i=0}^{6} b_i t^i$ of F_2 (see [26, 27]) with the Betti numbers b_i yields $P_{F_2}(t) = 1 + 2t^2 + 2t^4 + t^6$, i.e., $b_0 = b_6 = 1, b_2 = b_4 = 2$. We see that the cohomology class is not zero because all even Betti numbers are nonzero.

The Kähler potential for CP^2 is given by

$$K(w, \bar{w}) = \ln[(\Delta_1)^m], \tag{7}$$

which is obtained as a special case of F_2 by specifying the coordinate $w_3 = 0$ and the parameter $n = 0$ in Eq. (1). Hence, we have

$$\omega = im\frac{w_1 d\bar{w}_1 + w_2 d\bar{w}_2}{\Delta_1(w,\bar{w})} \tag{8}$$

up to the total derivative and

$$\begin{aligned} \Omega_K &= d\omega = im(\Delta_1)^{-2}[(1+|w_1|^2)dw_2 \wedge d\bar{w}_2 \\ &-\bar{w}_2 w_1 dw_2 \wedge d\bar{w}_1 \\ &-w_2 \bar{w}_1 dw_1 \wedge d\bar{w}_2 \\ &+(1+|w_2|^2)dw_1 \wedge d\bar{w}_1]. \end{aligned} \tag{9}$$

This should be compared to the case $F_1 = CP^1$, where

$$e^{ws_+} = \begin{pmatrix} 1 & w \\ 0 & 1 \end{pmatrix} \in F_1 = CP^1 = SU(2)/U(1) \cong \mathbb{S}^2 \tag{10}$$

and $s_+ = \begin{pmatrix} 0 & 1 \\ 0 & 0 \end{pmatrix}$. Here, the complex variable w is the CP^1 variable written as $w = e^{i\phi}\cot\frac{\theta}{2}$ in terms of the polar coordinates of the unit vector $\mathbf{n}$ in $\mathbb{S}^2$. The results for $SU(2)$ are well-known: $K(w,\bar{w}) = m\ln[(1+|w|^2)], \omega = im\frac{wd\bar{w}}{1+|w|^2}$, and $\Omega_K = ig_{w\bar{w}}dw \wedge d\bar{w} = im\frac{dw\wedge d\bar{w}}{(1+|w|^2)^2}$.

It is seen from these equations that the degenerate orbit CP^2 is the four-dimensional feature inside the six-dimensional flag space F_2. Therefore, two-form (6) is closed, but not *globally* exact. One can say that Ω_K is an element of the second cohomology group of F_2. This means that we cannot define the gauge connection ω everywhere in F_2, because the one-form ω is not well defined on the manifold. This is the reason why it is difficult to directly determine the Hopf-like topological invariants in the F_2 case.

Topological invariants . - Let us proceed with the analysis of topological invariants by calculating the homotopy groups of $SU(3)$. The flag space $F_2 = SU(3)/U(1)^2$ is the base space of the $U(1)^2$-fiber bundle $SU(3) \to F_2$. We have the following exact sequences:

$$\begin{aligned} &0 \to \pi_d\left(U(1)^2\right) \to \pi_d SU(3) \overset{\cong}{\to} \pi_d F_2 \\ &\to \pi_d\left(U(1)^2\right) \to 0, \qquad \text{for } d \geq 3, \end{aligned} \tag{11}$$

$$\begin{aligned} &0 \to \pi_2 SU(3) \to \pi_2 F_2 \to \pi_1\left(U(1)^2\right) \\ &\to \pi_1 SU(3) \to \pi_1 F_2 \to 0, \end{aligned} \tag{12}$$

where $\pi_1\left(U(1)^2\right) = \mathbb{Z}\times\mathbb{Z}$ and $\pi_1 SU(3) = \pi_2 SU(3) = 0$. Thus, we have

$$\begin{cases} \pi_0 F_2 = \pi_1 F_2 = 0, \\ \pi_2 F_2 = \mathbb{Z}\times\mathbb{Z}, \\ \pi_d F_2 = \pi_d SU(3), \qquad \text{for } d \geq 3. \end{cases} \tag{13}$$

We summarized the results in Table I. It presents the nontrivial homotopy groups of F_2, which are in accord with previous studies [28] (see also Ref. [29]). For completeness and comparison, we also have shown a list of the homotopy groups $\pi_d CP^2$ for $d \leq 7$ in Table II.

Table I is proved by the results of Ref. [30] and references therein. In particular, Ref. [30] presents two theorems that account for the 5th and 6th homotopy groups of SU(3). *Theorem 1*: $\pi_{2n-1}U(N) = \mathbb{Z}$ for $N \geq n$ and *Theorem 2*: $\pi_{2n}U(n) = \mathbb{Z}_{n!}$ for $n \geq 2$. We will study now the 3rd, 5th, and 6th homotopy groups in more detail.

1. *The* 3rd *homotopy group of SU(3).* The exact sequence of the fibration $SU(3) \to SU(3)/SU(2) \cong \mathbb{S}^5$ is

$$\pi_{d+1}\mathbb{S}^5 \to \pi_d SU(2) \xrightarrow{i_*} \pi_d SU(3) \to \pi_d \mathbb{S}^5\,. \tag{14}$$

Let $d = 3$ and, since $\pi_4\mathbb{S}^5 = \pi_3\mathbb{S}^5 = 0$, let the inclusion i: $SU(2) \hookrightarrow SU(3)$ induce an isomorphism i_*: $\pi_3 SU(2) \xrightarrow{\cong} \pi_3 SU(3)$. A generator for $\pi_3 SU(2)$ is given by

$$g_2(\mathbf{r}) = r^0\mathbf{1} + ir^j\sigma_j, \tag{15}$$

$$\text{for } \mathbf{r} = (r^0, r^1, r^2, r^3) \text{ and } \quad |\mathbf{r}| = 1\,, \tag{16}$$

where $\mathbf{1}$ is the identity matrix and σ_j are the Pauli matrices. Thus, the generator for $\pi_3 SU(3)$ is

$$g_3(\mathbf{r}) = \begin{pmatrix} 1 & 0 & 0 \\ 0 & r^0 + ir^3 & ir^1 + r^2 \\ 0 & ir^1 - r^2 & r^0 - ir^3 \end{pmatrix}. \tag{17}$$

Given any continuous function g: $\mathbb{S}^3 \to SU(3)$, the topological invariant, i.e., the winding degree $[g] \in \pi_3 SU(3) = \mathbb{Z}$, is determined by the integral formula

$$[g] = \frac{1}{24\pi^2}\int_{\mathbb{S}^3} \mathrm{Tr}\left[(gdg^\dagger)^3\right] =$$

$$\frac{1}{24\pi^2}\int_{\mathbb{S}^3} d^3x\varepsilon_{\mu\nu\lambda}\,\mathrm{Tr}(g^\dagger\partial_\nu g g^\dagger\partial_\mu g g^\dagger\partial_\lambda g)\,. \tag{18}$$

2. *The* 5th *homotopy group of SU(3).* Using exact sequence (14), we have

$$\pi_5 SU(2) \to \pi_5 SU(3) \to \pi_5\mathbb{S}^5$$

$$\to \pi_4 SU(2) \to \pi_4 SU(3)\,. \tag{19}$$

It is known that $\pi_5 SU(2) = \pi_5\mathbb{S}^3 = \mathbb{Z}_2$ and $\pi_5 SU(3) = \mathbb{Z}$ (from theorem 1), and thus the first arrow must be the zero map ($\mathbb{Z}_2 \xrightarrow{0} \mathbb{Z}$). We also know that $\pi_4 SU(3) = 0$, as the homotopy group $\pi_4 SU(N) = 0$ stabilizes after $N \geq 3$. Finally, we need $\pi_4 SU(2) = \pi_4\mathbb{S}^3 = \mathbb{Z}_2$. We now have

$$0 \to \pi_5 SU(3) \xrightarrow{\times 2} \pi_5\mathbb{S}^5 \to \mathbb{Z}_2 \to 0\,. \tag{20}$$

This means that given g: $\pi_5\mathbb{S}^5 \to SU(3)$,

$$g(\mathbf{r}) = \begin{pmatrix} | & | & | \\ \mathbf{u}_1(\mathbf{r}) & \mathbf{u}_2(\mathbf{r}) & \mathbf{u}_3(\mathbf{r}) \\ | & | & | \end{pmatrix}, \quad \text{for } \mathbf{r} \in \mathbb{S}^5, \tag{21}$$

the vector $\mathbf{u}_1$: $\mathbb{S}^5 \to \mathbb{S}^5$ has an even winding degree, namely, the winding degree $[\mathbf{u}_1] = 2 \times$ winding degree$[g]$.

The exact sequence for the fibration $SU(N+1) \to SU(N+1)/SU(N) = \mathbb{S}^{2N+1}$ shows that $\pi_5 SU(N) = \mathbb{Z}$ stabilizes after $N \geq 3$. Thus, the winding degree of g can be also deduced by the usual formula

$$[g] = \frac{1}{480\pi^3 i} \int\limits_{\mathbb{S}^5} \mathrm{Tr}\left[(gdg^\dagger)^5\right] . \tag{22}$$

A particular generator of $\pi_5 SU(3)$ can be found in Ref. [30].

3. *The* 6th *homotopy group of SU(3).* Exact sequence (14) yields

$$\pi_7 \mathbb{S}^5 \to \pi_6 SU(2) \to \pi_6 SU(3)$$

$$\xrightarrow{0} \pi_6 \mathbb{S}^5 \xrightarrow{\cong} \pi_5 SU(2) , \tag{23}$$

where $\pi_7 \mathbb{S}^5 = \mathbb{Z}_2$ and $\pi_6 SU(2) = \mathbb{Z}_{12}$. It turns out that $\pi_6 SU(3) = \mathbb{Z}_{3!} = \mathbb{Z}_6$. A generator for $\pi_6 SU(3)$ can be found in [30] in page 6.

Conclusion. - In conclusion, we focus on the nontrivial homotopy groups for $d \leq 6$ $\pi_2 F_2$, $\pi_3 F_2$, $\pi_5 F_2$, and $\pi_6 F_2$ considered so far for the spheres $\mathbb{S}^2$ as the base space. The generalization $\mathbb{S}^2 \to T^n$, where T^n is the n-dimensional torus, is an interesting and more complicated extension even in the $SU(2)$ case. The result of calculations [32] of the mapping class groups in the last case with $T^3 = S^1 \times S^1 \times S^1$ leads to the linear superposition of the topological invariants beginning from the first Chern class to the Hopf invariant (see also [12]). Similar behavior also takes place in the case of $T^3 = \mathbb{S}^2 \times \mathbb{S}^1$. The classification problems of the mappings $T^n \to F_2$ are still totally open.

Up to now, we did not pay any attention to the relation between the existence of strong interaction in the system and homotopy group results. It is well known in the condensed matter community that nontrivial answers $\mathbb{Z}_2$ or $\mathbb{Z}$ for the topological invariant of non-interacting systems change drastically in particular to $\mathbb{Z}_8$ in the case of the interacting system [33]. Considering from this point of view the result $\pi_6(F_2) = \mathbb{Z}_6$, one can say that we deal here with the significant interaction as it takes place in our QCD system.

Thorough understanding of the role of the flag space F_2 in the $SU(3)$ gauge theory is related to the search for an analog of the Hopf number, i.e., the linking number of pullbacks on a space $\mathbb{M}$ of two arbitrary "points" on a target space $\mathbb{N}$ of the map $\mathbb{M} \to \mathbb{N}$. Such an analog can have the form of pre-images of the target points in the codimension two [22]. This could take place if $\mathbb{M} = \mathbb{S}^3$ and $\mathbb{N}$ is the $2d$ complement of CP^2 with respect to the whole space F_2. This is an open question, which is difficult to answer without knowing the details of the map. We leave the problem of describing the details of this map for future work.

Acknowledgments

We are grateful to V.I. Arnold, J.E. Avron, J. Bernatska, L.D. Faddeev, P.G. Grinevich, C.L. Kane, Y. Hatsugai, A.W.W. Ludwig, I.A. Taimanov, G. E. Volovik for fruitful discussions. The coauthors (A.P., J.T.) especially thank the organizers of the 28th Jerusalem Winter School in Theoretical Physics A. Stern and S.-C. Zhang for hospitality at the Israel Institute for Advanced Studies where a part of the work was done. This work was supported in part by the RFBR Grant No. 13-02-12110.

[1] Mermin N D 1979 *Rev. Mod. Phys.* **51** 591
[2] Volovik G E 2003 *The Universe in a Helium Droplet* (Clarendon: Oxford)
[3] Teo J.C.Y. and Kane C L 2010 *Phys. Rev. Lett.* **104** 046401

[4] Teo J C Y and Kane C L 2010 *Phys. Rev.* B **82** 115120
[5] Hasan M Z and Kane C L 2010 *Rev. Mod. Phys.* **82** 3045
[6] Qi X-L and Zhang S-C 2011 *Rev. Mod. Phys.* **83** 1057
[7] Schnyder A P, Ryu S, Furusaki A and Ludwig A W W 2008 *Phys. Rev.* B **78** 195125; *AIP Conf. Proc.* 2009 **1134** 10; *New J. Phys.* 2010 **12** 065010
[8] Kitaev A. 2009 *AIP Conf. Proc.* **1134** 22 (2009)
[9] Vakulenko A F and Kapitanskii L V 1979 *Sov. Phys. Dokl.* **24** 433
[10] Protogenov A P and Verbus V A 2002 *JETP Lett.* **76** 53
[11] Babaev E, Faddeev L. D. and Niemi A J 2002 *Phys. Rev.* B **65** 100512
[12] Protogenov A P 2006 *Physics-Uspekhi* **49** 667
[13] Jäykka J, Hietarinta J and Salo P 2008 *Phys. Rev.* B **77** 094509
[14] Cho Y M 1980 *Phys. Rev.* D **21** 1080; 1981 *Phys. Rev. Lett.* **46** 302
[15] Duan Y S and Ge M L 1979 *Sinica Sci.* **11** 1072
[16] Faddeev L D and Niemi A J 1999 *Phys. Rev. Lett.* **82** 1624
[17] The problem in this form has been formulated by L. Faddeev in Ref. [18] (see also Ref. [19]).
[18] Faddeev L 2001 *Philos. Trans. R. Soc. London* A **359** 1399
[19] Bolokhov T A and Faddeev L D 2004 *Theor. and Math. Phys.* **139** 679
[20] Novikov S P 1984 *Uspekhi Mat. Nauk* **39** 97
[21] V.I. Arnold has suggested in 2004 to one (AP) of us to use the Novikov's analytic generalized Hopf invariant [20] for description of the knots in the $SU(3)$ case.
[22] A. Ranicki A 1998 *A High-dimensional Knot Theory. Algebraic Surgery in Codimension 2* (New York: Springer)
[23] Picken R F 1990 *J. Math. Phys.* **31** 616
[24] Kondo K-I and Taira Y 2000 *Prog. Theor. Phys.* **104** 1189
[25] Bernatska J and P. Holod P 2008 Geometry and topology of coadjoint orbits of semisimple Lie groups *Preprint math.RT/0801.2913*
[26] Borel A 1955 *Bull. Am. Math. Soc.* **67** 397
[27] Boya L J, Perelomov A M and M. Santander M 2001 *J. Math. Phys.* **42** 5130
[28] Mimura M and Toda H 1963 *J. Math. Kyoto Univ.* **3** 217
[29] Deligne P, Griffiths Ph, Morgan J and Sullivan D, 1975 *Invent. Math.* **29** 245
[30] Puettmann T and Rigas R 2003 *Comment. Math. Helv.* **78**, 648
[31] Zubkov M A and Volovik G E 2012 *Nucl. Phys.* B **860** 295
[32] Pontrjagin L S 1941 *Mat. Sbornik (Recueil Mathematique N. S.)* **9** 331
[33] Fidkowski L and Kitaev A The effects of interactions on the topological classification of free fermion systems *Preprint cond-mat/0904.2197*

Journal of Physics: Conference Series **482** (2014) 012036 doi:10.1088/1742-6596/482/1/012036

New exact solutions for a coupled KdV–like model

Marianna Ruggieri[1] **and Maria Paola Speciale**[2]
[1]Faculty of Engineering and Architecture, Kore University of Enna, Italy
[2]Department of Mathematics and Computer Science, University of Messina, Italy
E-mail: marianna.ruggieri@unikore.it
E-mail: mpspeciale@unime.it

Abstract. An extension of a recent method is applied in order to construct new explicit exact solutions for a system of coupled KdV-like equations.

1. Introduction

Recently, see [1]-[3], we have studied the system of Korteweg–de Vries (KdV) equations

$$\begin{aligned} &u_t + u_{xxx} + 2uu_x + 2e_1 vv_x + e_2(u_x v + uv_x) + e_3 v_{xxx} = 0, \\ &c_1 v_t + v_{xxx} + 2vv_x + c_2 v_x + c_3[e_1(u_x v + uv_x) + 2e_2 uu_x + e_3 u_{xxx}] = 0, \end{aligned} \tag{1}$$

coupled through both dispersive and nonlinear terms.

The system was originally developed by Gear and Grimshaw [4] to model the strong interaction of two-dimensional, long, internal gravity waves propagating on neighboring pycnoclines in a stratified fluid.

In (1), $u(t, x)$ and $v(t, x)$ are the dependent variables and represent the desplacement from the horizontal of the fluid interfaces; subscripts denote partial derivatives with respect to the indicated independent variables, whereas e_i $(i = 1, 2, 3)$ and c_j $(j = 1, 2, 3)$ are constant parameters. It should be noted that c_1 is positive and finite (regardless of its magnitude).

In real physics, the linear dispersion relations are generally different for the two layers of the fluids, especially for different fluids; therefore, we are interested to find solutions for different choices of the parameters in order to capture the different linear dispersion relations of the fluids.

Some mathematical questions as existence and uniqueness of the solution for the model equations (1) have been studied in [5], where it is shown that the Cauchy problem is globally well-posed in suitably strong function spaces. Moreover, Bona and Saut have also stated that the system (1) is susceptible of experiencing the dispersive blow-up phenomenon. In [6], a study of the integrability of models for coupled KdV systems with different linear dispersion relations has been performed. A study by means of quasi self-adjointness can be found in [2], while, following the guidelines of [7], an approximate symmetry reduction of the model (1) is performed in [8].

In this paper, we devise an extension of an algebraic method to uniformly construct a series of explicit exact solutions for the model.

2. Methodology

The soliton theory have been studied extensively and a lot of interesting results have been found for many coupled systems of partial differential equations. It is known that systems of coupled KdV-like equations admit soliton solutions - see for example [9] -[12] and references therein for a review - moreover, in [13] the author discusses the problem how the coupled two-component KdV systems are connected with the generalized two component Harry Dym systems, while in [14] periodic wave solutions to a coupled KdV equations with variable coefficients are obtained. The literature is also very rich in interesting results obtained on generalized systems of coupled KdV-like equations by Fordy and his co-workers [15],[16]. In [17], an extended algebraic method is used for constructing exact traveling wave solutions for some coupled nonlinear evolution equations.

These facts are confirmed by the symmetry classification performed in [1], by which it is clear that starting from the Principal Lie Algebra of (1) spanned by the operators

$$X_1 = \frac{\partial}{\partial t}, \quad X_2 = \frac{\partial}{\partial x}, \tag{2}$$

the model admits traveling wave solutions

$$u(x,t) = U(z), \qquad v(x,t) = V(z), \qquad z = x - ct, \qquad (c \text{ constant}),$$

where the functions U and V satisfy the following system of ODEs:

$$U''' + e_3 V''' + U'(e_2 V + 2U + c) + V'(2 e_1 V + e_2 U) = 0 \tag{3}$$

$$c_3 e_3 U''' + V''' + c_3(2e_2 U' + e_1 V')U + V'(c_2 + c\, c_1) + (c_3 e_1 U' + 2V')V = 0, \tag{4}$$

where the prime $'$ denotes the derivative with respect to z.

Due to the strong nonlinearity of the system of generalized KdV equations (1), in order to characterize a complete hierarchy of traveling wave solutions admitted by the model at hand for some values of the parameters e_i and c_i $(i = 1, 2, 3)$, we use in [3] an algebraic method proposed recently [18].

According to the procedure outlined in [18], we introduce a function $\phi = \phi(z)$ which is a solution of the first-order ODE

$$\phi' = \epsilon \sqrt{\sum_j^r \lambda_j \phi^j} \tag{5}$$

where $\epsilon = \pm 1$. Then the derivatives with respect to z become the derivatives with respect to the variable ϕ as follows

$$\frac{d}{dz} \to \epsilon \sqrt{\sum_j^r \lambda_j \phi^j}\, \frac{d}{d\phi}, \quad \frac{d^2}{dz^2} \to \epsilon^2 \frac{1}{2}\left[\sum_j^r j\, \lambda_j \phi^{j-1} \frac{d}{d\phi} + \sum_j^r \lambda_j \phi^j \frac{d^2}{d\phi^2}\right], \ldots.$$

and write the solution of (4),

$$u(x,t) = U(z) = \sum_i^{n_1} \alpha_i \phi^i(z), \quad v(x,t) = V(z) = \sum_i^{n_2} \beta_i \phi^i(z), \tag{6}$$

as a sum of powers of $\phi(z)$, α_i and β_i being arbitrary constants. Balancing the highest derivative terms with the nonlinear terms we give a relation between positive integers n_1, n_2 and r: $n_2 = n_1 = n$ and $r = n + 2$.

Substituting the expansion of $U(z)$ and $V(z)$ into (4), and setting the coefficients of all powers ϕ^j and $\phi^j\sqrt{\sum_j^r \lambda_j\phi^j}$ to zero, we get a system of algebraic equations, from which the constants λ_j, α_i and β_i ($j = 0 \ldots n+2$ and $i = 0 \ldots n$) can be found explicitly.

In the following, we ask if the model (1) admits not only the classical traveling wave solutions but also a general solution $\phi(z)$ of (5), which depends of a variable z of the form

$$z = x - f(t), \tag{7}$$

where $f(t)$ is a suitable function. Of course, when $f(t)$ is a linear function of t, we recover the results of [1, 3] in which the complete group classification of the model and some exact solutions in closed form have been found. In fact, in the case of PDEs, Lie symmetries may lead to the determination of invariant solutions of a variety of models and to transformations into new equivalent and more convenient forms (some examples in different fields can be found in [19]-[24]).

Now we ask if, generalizing the procedure introduced by Fan, we can obtain a more general class of solutions admitted by the model which includes the soliton solution as a particular case. We suppose that the solution of (1) is of the form

$$u(x,t) = U(z) + g_1(t), \quad v(x,t) = V(z) + g_2(t), \tag{8}$$

where $z = x - f(t)$. Substituting (8) into the system (1), in order to obtain a system of ODEs, we have that

- $g_1(t)$ and $g_2(t)$ must be linear function of t linked by constraints

$$c_1(2g_1(t) + e_2g_2(t)) - (c_3e_1g_1(t) + 2g_2(t)) = 0, \tag{9}$$

$$e_2g_1(t) + 2e_1g_2(t) = 0, \quad 2c_3(e_2g_1(t) + e_1g_2(t)) = 0, \tag{10}$$

  satisfied for suitable subclasses of the model ($c_3 = 0$ or $e_1 = 0$ or $e_2 = 0$, or $e_1 = e_2 = 0$).
- $f(t)$ quadratic and linked to $g_1(t)$, $g_2(t)$ by the relation

$$c_1f'(t) = c_3e_1g_1(t) + 2g_2(t). \tag{11}$$

When we introduce the function $\phi(z)$, with $z = x - f(t)$, as solution of (5), we assume

$$U(z) = \sum_i^{n_1} \alpha_i\phi^i(z) + f_1(z), \quad V(z) = \sum_i^{n_2} \beta_i\phi^i(z) + f_2(z) \tag{12}$$

$f_1(z)$ and $f_2(z)$ being functions of their argument.

Setting the coefficients the different powers $\phi^j\sqrt{\sum_j^r \lambda_j\phi^j}$ to zero we get a set of constraints for $f_1(z)$ and $f_2(z)$ that must be compatible with the algebraic system obtained by setting the coefficients of the powers ϕ^j to zero:

$$(2\alpha_2 + e_2\beta_2)f_1(z) + (e_2\alpha_2 + 2e_1\beta_2)f_2(z) = A, \tag{13}$$

$$(2\alpha_1 + e_2\beta_1)f_1(z) + (e_2\alpha_1 + 2e_1\beta_1)f_2(z) = B, \tag{14}$$

$$c_3(2e_2\alpha_2 + e_1\beta_2)f_1(z) + (c_3e_1\alpha_2 + 2\beta_2)f_2(z) = C, \tag{15}$$

$$c_3(2e_2\alpha_1 + e_1\beta_1)f_1(z) + (c_3e_1\alpha_1 + 2\beta_1)f_2(z) = D, \tag{16}$$

$$(2\alpha_0 + e_2\beta_0)f_1'(z) + (e_2\alpha_0 + 2e_1\beta_0)f_2'(z) \tag{17}$$

$$+[(f_1(z)^2 + e_2f_1(z)f_2(z) + e_1f_2(z)^2)]' + k_1 + f_1'''(z) + e_3f_2'''(z) = 0, \tag{18}$$

$$c_3(2e_2\alpha_0 + e_1\beta_0)f_1'(z) + (c_3e_1\alpha_0 + 2\beta_0 + c_2)f_2'(z) \tag{19}$$

$$+[c_3(e_2f_1(z)^2 + e_1f_2(z)f_1(z)) + f_2(z)^2]' + c_1k_3 + c_3e_3f_1'''(z) + f_2'''(z) = 0 \tag{20}$$

where $g_1(t) = k_1 t + k_2$ and $g_2(t) = k_3 t + k_4$, and A, B, C and D are constants. Through the substitution of (8) into (1) and balancing the highest derivative terms with the nonlinear terms in (4) we yield the following set of algebraic equations

$$\alpha_2^2 + (e_2\alpha_2 + e_1\beta_2)\beta_2 + 6(\alpha_2 + e_3\beta_2)\lambda_4 = 0, \tag{21}$$
$$(2\alpha_1 + e_2\beta_1)\alpha_2 + (e_2\alpha_1 + 2e_1\beta_1)\beta_2 + 5(\alpha_2 + e_3\beta_2)\lambda_3 + 2(\alpha_1 + e_3\beta_1)\lambda_4 = 0, \tag{22}$$
$$2((2\alpha_0 + e_2\beta_0)\alpha_2 + (e_2\alpha_0 + 2e_1\beta_0)\beta_2) + (2\alpha_1^2 + e_2\alpha_1\beta_1 + e_1\beta_1^2) + 2A \tag{23}$$
$$+8(\alpha_2 + e_3\beta_2)\lambda_2 + 3(\alpha_1 + e_3\beta_1)\lambda_3 = 0, \tag{24}$$
$$(2\alpha_0 + e_2\beta_0)\alpha_1 + (e_2\alpha_0 + 2e_1\beta_0)\beta_1 + 3(\alpha_2 + e_3\beta_2)\lambda_1 + (\alpha_1 + e_3\beta_1)\lambda_2 + B = 0, \tag{25}$$
$$\beta_2^2 + c_3(e_2\alpha_2 + e_1\beta_2)\alpha_2 + 6(c_3 e_3\alpha_2 + \beta_2)\lambda_4 = 0, \tag{26}$$
$$c_3(2e_2\alpha_1 + e_1\beta_1)\alpha_2 + (c_3 e_1\alpha_1 + 2\beta_1)\beta_2 + 5(\beta_2 + c_3 e_3\alpha_2)\lambda_3 + 2(\beta_1 + c_3 e_3\alpha_1)\lambda_4 = 0, \tag{27}$$
$$c_3(2e_2\alpha_0 + e_1\beta_0)\alpha_2 + (c_3 e_1\alpha_0 + 2\beta_0 + c_2)\beta_2 + (c_3 e_1\alpha_1\beta_1 + 2\beta_1^2 + c_3 e_2\alpha_1^2) + 2c_3 C \tag{28}$$
$$+4(\beta_2 + c_3 e_3\alpha_2)\lambda_2 + 3(\beta_1 + c_3 e_3\alpha_1)\lambda_3 = 0, \tag{29}$$
$$(2\beta_0 + c_3 e_1\alpha_0 + c_2)\beta_1 + c_3(2e_2\alpha_0 + e_1\beta_0)\alpha_1 + 3(\beta_2 + c_2 e_3\alpha_2)\lambda_1 \tag{30}$$
$$+(\beta_1 + c_3 e_3\alpha_1)\lambda_2 + D = 0. \tag{31}$$

The only possible choices which can be compatible with the method proposed by Fan are $e_1 = 0$ or $e_2 = 0$; the cases $c_3 = 0$ or the particular subcase $e_1 = e_2 = 0$ do not allow for having a compatible generalized solution.

If $e_2 = 0$, $e_1 = 2\frac{c_1}{c_3}$, $g_1(t) = k_1 t + k_2$ and $g_2(t) = 0$ and $f_1(z) = \delta f_2(z)$ with $\delta \neq 0, 1$ (for $\delta = 0$ or $\delta = 1$ an incompatibility with algebraic relations since f_1 and $f_2(z)$ turn out to be constant), the algebraic system provides the explicit solution for $f_2(z)$:

$$f_2(z) = \frac{(K_1 - K_2)\cosh(2z\sqrt{\lambda_2}) + (K_1 + K_2)\sinh(2z\sqrt{\lambda_2})}{2\sqrt{\lambda_2}} + k_1\frac{z}{4\delta\lambda_2(1-\delta)} + K_3, \tag{32}$$

K_1,K_2, K_3 being constants with $\lambda_2 \neq 0$; moreover, for $\lambda_3 = \lambda_4 = 0$, the following expressions of parameters arise:

$$\alpha_2 = 2\delta^2\beta_2, \quad \alpha_1 = 2\delta^2\beta_1, \quad \alpha_0 = \beta_0\delta + 2\lambda_2(\delta - 1), \quad \beta_1 = \tfrac{\lambda_1}{\lambda_2}\beta_2, \tag{33}$$
$$c_3 = \tfrac{1}{\delta^3}, \quad c_1 = -\tfrac{1}{2\delta}, \quad e_3 = -\delta^2, \quad c_2 = 2\lambda_2\tfrac{\delta-1}{\delta}. \tag{34}$$

Now, the solution of (5) reads as:

$$\phi(z) = \frac{-2\lambda_1 + (1 + \lambda_1^2 - 4\lambda_0\lambda_2)\cosh(z\sqrt{\lambda_2}) \mp (1 - \lambda_1^2 + 4\lambda_0\lambda_2)\sinh(z\sqrt{\lambda_2})}{4\lambda_2}. \tag{35}$$

This class of solutions admitted by the model includes a series of "generalized traveling wave solutions" such as: soliton, rational, triangular generalized solutions etc.. We use the term "generalized" because the independent variable has the form $z = x - k_2 t - 2\,k_1 t^2$.

In particular, if we assume $\lambda_2 < 0$ (this is the case when $\lambda_2 = -\frac{1}{2}$, $\lambda_1 = -\frac{1}{4}$, $\lambda_0 = 0$), we obtain:

$$\phi(z) = -\frac{1}{2}\sin^2\left(\frac{z}{2\sqrt{2}}\right). \tag{36}$$

Therefore, we are able to write:

$$U(z) = k_1\tfrac{z}{2(\delta-1)} + \beta_2\delta\tfrac{\cos(\sqrt{2}z)-1}{32} + \delta\tfrac{K_1}{\sqrt{2}}\sin(\sqrt{2}z) + \delta\beta_0 - \delta + 1, \tag{37}$$

$$V(z) = k_1\tfrac{z}{2\delta(\delta-1)} + \beta_2\tfrac{\cos(\sqrt{2}z)-1}{32} + \tfrac{K_1}{\sqrt{2}}\sin(\sqrt{2}z) + \beta_0, \tag{38}$$

Journal of Physics: Conference Series **482** (2014) 012036 doi:10.1088/1742-6596/482/1/012036

for $K_2 = K_1$ and $K_3 = 0$.

The choices $e_1 = 0$, $e_2 = 2$ and $g_2(t) = k_3 t + k_4$ with $c_1 f_2(z) = \delta f_1(z)$, lead us to a family of solutions of the model (1) whose profiles are specular to the previous ones.

For suiable values of β_0, β_2, δ and k_1 we obtain the classical traveling wave solution ($k_1 = 0$) and the "generalized traveling waves" that are shown in the Figures 1 and 2.

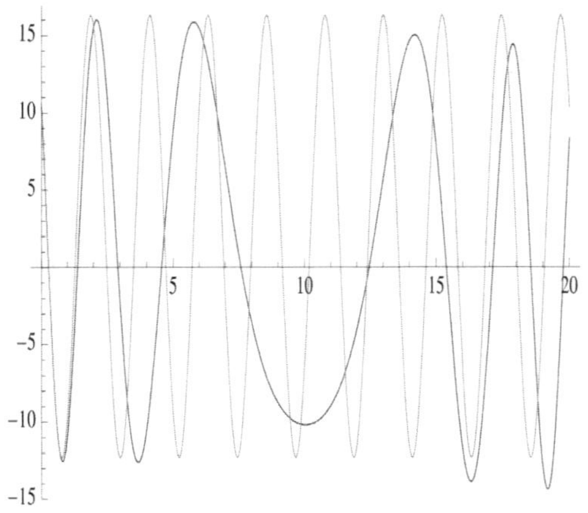

Figure 1. 2D view of the solution (38) for fixed value of the independent variable with $u(x,t)$ versus $t \in [0, 20]$; when $\beta_0 = 0$, $\beta_2 = 32$, $k_2 = 1/5$, $k_1 = 1/20$, $K_1 = 10$, $\delta = 2$. We have in orange the classical solitary wave and in blue the "generalized traveling wave".

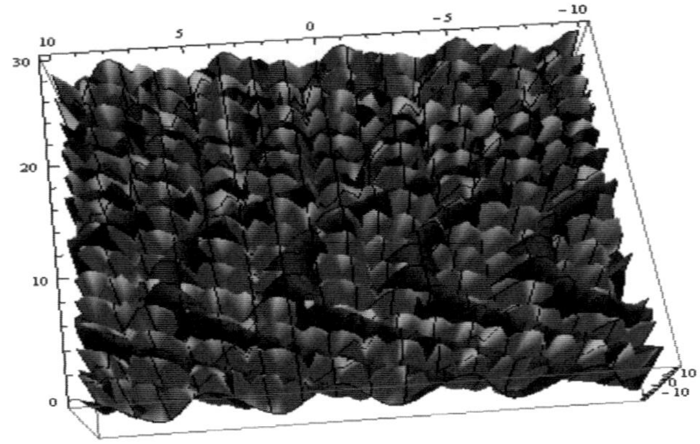

Figure 2. 3D view of the solution (38) when $x \in [-10, 10]$ and $t \in [0, 30]$; with $\beta_0 = 0$, $\beta_2 = 32$, $k_2 = 1/5$, $k_1 = 1/20$, $K_1 = 10$, $\delta = 2$. In the picture we have in orange the classical solitary wave and in blue the "generalized traveling wave".

[1] Ruggieri M. and Speciale M. P., 2013, *Similarity Reduction and Closed Form Solutions for a Model Derived from Two Layer Fluids*, in press on Advances in Difference Equations, Springer.

[2] Ruggieri M. and Speciale M. P., 2013 *Quasi Self-adjoint Coupled KdV-like Equations*, In Numerical Analysis and Applied Mathematics, AIP Conference Proceedings, **1558**, doi: 10.1063/1.4825730, 1220-1223.

[3] Ruggieri M. and Speciale M. P., 2013, *On a hierarchy of traveling wave solutions in a shallow stratified fluid*, In Numerical Analysis and Applied Mathematics, AIP Conference Proceedings **1558**, doi: 10.1063/1.4825873, 1793-1796.

[4] Gear J. A. and Grimshaw R., 1984, *Weak and strong interactions between internal solitary waves*, Stud. Appl. Math (70), **3**, 235–258.

[5] Bona J. L. *et al.*, 1992, *A Model System for Strong Interaction Between Internal Solitary Waves*, Commun. Math. Phys., **143**, 287–313.

[6] Bin T. *et al.*, 2006, *A New Coupled KdV Equation: Painlevé Test*, Commun. Theor. Phys., **45**, 965–968.

[7] Ruggieri M. and Valenti A., 2013, *Approximate symmetries in nonlinear viscoelastic media,* Boundary Value Problems, **143**, doi: 10.1186/1687-2770-2013-143.

[8] Ruggieri M. and Speciale M. P., 2013, *Approximate Analysis of a generalized KdV-like Equations*, to appear.

[9] Hu H.C. and Lou S.Y., 2004, Exact solutions of Bogoyavlenskii coupled KdV equations, Commun. Theor. Phys. 42, 485–487.

[10] Fan E. and Hon Y., 2003, A series of traveling wave solutions for two variant Boussinesq equations in shallow water waves, Chaos Solitons Fractals 15, 559–566.

[11] Hirota R. and Satsuma J., 1981, Soliton solutions of a coupled Kortewegde Vries equation, Phys. Lett. A 85, 407–408.

[12] Hirota R. and Satsuma J., 1982, A coupled KdV equation is one case of the four-reduction of the KP hierarchy, J. Phys. Soc. Japan 51, 3390–3397.

[13] Popowicz Z. 2012, *Two-Componet Coupled KdV Equations and its Connection with the Generalized Harry Dym Equations*, arXiv:1210.5822v1.

[14] Zhou Y. *et al.*, 2003, *Periodic wave solutions to a coupled KdV equations with variable coefficients*, Physics Letters A, **308**, 3136.

[15] Dodd R. and Fordy A., 1982, *On the integrability of a system of coupled KdV equations*, Physics Letters A, **89**, 168-170.

[16] Antonowicz M. and Fordy A. P., 1987, *Coupled KdV equations with multi-Hamiltonian structures*, Physica D: Nonlinear Phenomena **28**, 3, 345-357.

[17] Seadawy A.R. and Rashidy El, 2013, Traveling wave solutions for some coupled nonlinear evolution, Mathematical and Computer Modelling equations, 57, 1371-1379.

[18] Fan, E.G., 2003, *An algebraic method for finding a series of exact solutions to integrable and nonintegrable nonlinear equations*, Journal of Physics A, **36**, 7009–7026.

[19] Ruggieri M. and Valenti A., 2011, *Exact solutions for a nonlinear model of dissipative media*, Journal of Mathematical Physics, **52**, 043520.

[20] Ruggieri M. and Valenti A., 2009, *Symmetries and reduction techniques for dissipative models*, Journal of Mathematical Physics, **50**, 063506 - 063506-9.

[21] M. Ruggieri, 2012, *Kink solutions for a class of generalized dissipative equations*, Abstract and Applied Analysis, **2012**, 237135-7.

[22] Oliveri F. and Speciale M. P., 2012, *Equivalence Transformations of Quasilinear First Order Systems and Reduction to Autonomous and Homogeneous Form*, Acta Applicandae Mathematicae, **122**, 447–460.

[23] Margheriti L. and Speciale M. P., 2011,*Unsteady solutions of Euler equations generated by steady solutions*, Acta applicandae mathematicae, **113**, 3, 289-303.

[24] Oliveri F. and Speciale M. P., 2013,*Reduction of balance laws to conservation laws by means of equivalence transformations*, Journal of Mathematical Physics, **54**, 041506.

Conservation laws for a model derived from two layer-fluids

Marianna Ruggieri[1] **and Maria Paola Speciale**[2]
[1]Faculty of Engineering and Architecture, Kore University of Enna, Italy
[2]Department of Mathematics and Computer Science, University of Messina, Italy
E-mail: marianna.ruggieri@unikore.it
E-mail: mpspeciale@unime.it

Abstract. In this paper, using a recent approach for finding conservation laws, based on Lie symmetries, we establish the conservation laws for a model admitting quasi self-adjoint equations.

1. Introduction

Conservation laws are very important in the construction of mathematical models described by partial differential equations. Recently, N.H. Ibragimov [1] developed a new method which generalizes Noether's theorem. In particular, it allows one to associate conservation laws with symmetries of partial differential equations not having classical Lagrangian. In the present paper, we apply this method to a class of generalized system of KdV–like equations

$$\begin{aligned} &u_t + u_{xxx} + 2uu_x + 2e_1vv_x + e_2(u_xv + uv_x) + e_3v_{xxx} = 0, \\ &c_1v_t + v_{xxx} + 2vv_x + c_2v_x + c_3[e_1(u_xv + uv_x) + 2e_2uu_x + e_3u_{xxx}] = 0, \end{aligned} \tag{1}$$

coupled through both dispersive and nonlinear terms, which was derived by Gear and Grimshaw [2]; $u(t,x)$ and $v(t,x)$ are the dependent variables and represent the displacement from the horizontal of the fluid interfaces, subscripts denote partial derivatives with respect to the indicated independent variables; moreover, e_i $(i = 1,2,3)$ and c_j $(j = 1,2,3)$ are arbitrary constants. The complete group classification of (1) and some exact solutions of physical relevance in closed form have been derived in [3, 4], and a complete hierarchy of traveling wave solutions admitted by the model at hand have been characterized. An approximate study of the model (1), following the guidelines of [5], can be found in [6].

By using a general theory introduced by Ibragimov in [1, 7, 8], we present the general concept of nonlinear self-adjointness in order to construct all conservation laws of the system (1), using the admitted Lie point symmetries. So, the construction of conservation laws demonstrates a practical significance of the nonlinear self-adjointness.

The plan of this paper is the following. In Section 2, we report the complete point symmetry classification of (1) obtained in [3]; in Section 3, we introduce the quasi self-adjoint forms associated to (1), and, finally, in Section 4, by applying the general theorem proved in [1, 7, 8], we determine all conservation laws associated with symmetries for all nonlinearly self-adjoint differential subsystems of (1) determined in Section 3.

2. Lie Symmetries

Essentially, all the fundamental equations describing physical phenomena are nonlinear; such nonlinearity is often responsible for the mathematical difficulties in finding exact solutions. Lie symmetries, in general, may lead to the determination of invariant solutions of a variety of initial and boundary value problems in many disciplines of the applied sciences (some recent examples can be found in [9]–[14]).

In [3], we have proved that system (1) has a two-dimensional Algebra of point symmetries spanned by

$$X_1 = \frac{\partial}{\partial t}, \quad X_2 = \frac{\partial}{\partial x}; \tag{2}$$

this algebra extends to a three–dimensional or to a four–dimensional one according to special choices of the involved parameters.

- **Case 1**
  If one of the parameters c_3, e_1, e_2 is zero we have

$$X_3 = 3k\,t\frac{\partial}{\partial t} + \left[kx - 2c_2(e_2^2 - 4e_1)\,t\right]\frac{\partial}{\partial x} - 2\,(k\,u - 2e_1c_2)\frac{\partial}{\partial u} - 2\,(k\,v + e_2c_2)\frac{\partial}{\partial v}, \tag{3}$$

with the position $k = 2(e_2 - c_3e_1^2) - c_1(e_2^2 - 4e_1)$. A special case arises when both k and c_2 are equal to zero:

$$\acute{X}_3 = 3\,t\frac{\partial}{\partial t} + x\frac{\partial}{\partial x} - 2\,u\frac{\partial}{\partial u} - 2\,v\frac{\partial}{\partial v}, \qquad \acute{X}_4 = -(e_2^2 - 4e_1)\,t\frac{\partial}{\partial x} + 2e_1\frac{\partial}{\partial u} - e_2\frac{\partial}{\partial v}. \tag{4}$$

- **Case 2**
  When $c_1 = 1$ and $e_3 = 0$, we get $c_3 = \frac{e_2^2 - e_2 - 2 - 4e_1}{e_1(2e_1 + e_2)}$ $(e_2 \neq -2e_1, e_1 \neq 0)$, along with the following constraint on the parameters

$$e_2^3 - (2 + e_1)e_2^2 + 4e_1e_2 + (4e_1 + 1)e_1 = 0, \tag{5}$$

we have the operator

$$\begin{aligned}\tilde{X}_3 = {} & 3e_2\,t\frac{\partial}{\partial t} + \left(\frac{2\,k_0\,c_2 t}{\Delta} + e_2 x\right)\frac{\partial}{\partial x} \\ & +2\left[\frac{c_2e_1k_1}{\Delta} + \frac{3\,e_1e_2k_2}{\Delta}v - e_2\left(\frac{3\,e_2k_3}{\Delta} + 1\right)u\right]\frac{\partial}{\partial u} \\ & +2e_2\left[\frac{c_2k_4}{\Delta} + \frac{3k_5}{\Delta}u + \left(\frac{3e_1k_3}{\Delta} - 1\right)v\right]\frac{\partial}{\partial v},\end{aligned} \tag{6}$$

with Δ and $k_i = k_i(e_1, e_2)$ $(i = 1, \dots, 5)$ given by

$$\begin{aligned} k_0 &= (2e_1 + e_2)(24e_1^2 - 18e_1e_2^2 + 10e_1e_2 + 8e_1 + 3e_2^4 - 4e_2^3 - 5e_2^2), \\ k_1 &= (2e_1 + e_2)(12e_1 - 3e_2^2 + 5e_2 + 4), \\ k_2 &= (2e_1 + e_2)(e_2^2 - e_2 - 4e_1 - 2), \\ k_3 &= (1 - e_2)(e_2^2 - e_2 - 4e_1 - 2), \\ k_4 &= (2e_1 + e_2)(3e_2^2 - 4e_2 - 12e_1 - 5), \\ k_5 &= (e_2^2 - 2e_2 - 4e_1 - 1)(e_2^2 - e_2 - 4e_1 - 2), \\ \Delta &= 3(e_1 - e_2)(e_2 - 1)(e_2^2 - e_2 - 4e_1 - 2), \end{aligned}$$

where $\Delta \neq 0$; on the contrary, if $\Delta = 0$, we fall in other cases of the classification (if $e_1 = e_2$ we fall in the next case 4, if $e_2 = 1$ or $e_2^2 - e_2 - 4e_1 - 2 = 0$ we recover the next case 3).

- **Case 3** A singularity arises when $e_2 = -2e_1$, that leads us to consider the special values $e_2 = 1$, $e_1 = -\frac{1}{2}$, $c_3 = 2(5 \pm 3\sqrt{3})$; then, the related operator reads

$$\begin{aligned}\bar{X}_3 &= 3\bar{k}\,t\frac{\partial}{\partial t} + \left[\bar{k}x - 24c_2(5c_3+4)\,t\right]\frac{\partial}{\partial x}\\ &-2\left(2c_2(8+7c_3) + (\bar{k} - 6c_3(c_3-4))u - 18c_3v\right)\frac{\partial}{\partial u}\\ &+2\left(-16c_2(1+2c_3) + 12(c_3+2)c_3u - (\bar{k}\,v - 3c_3(c_3-4))v\right)\frac{\partial}{\partial v},\end{aligned} \tag{7}$$

where $\bar{k} = -7c_3^2 - 4c_3 - 16$.

- **Case 4**
Finally, another singularity arises if $e_1 = 0$ and $e_2 = 0$, whereupon we have

$$\hat{X}_3 = 3t\frac{\partial}{\partial t} + x\frac{\partial}{\partial x} - 2u\frac{\partial}{\partial u} - (c_2 + 2v)\frac{\partial}{\partial v}, \qquad \hat{X}_4 = 2t\frac{\partial}{\partial x} + \frac{\partial}{\partial u} + c_1\frac{\partial}{\partial v}. \tag{8}$$

3. Adjoint system

The general concept of nonlinear self-adjointness [7]–[8] was introduced to construct conservation laws associated with Lie symmetries of differential equations.
Following the guidelines of [7]-[8], we introduced the *formal Lagrangian* for the system (1)

$$\mathcal{L} = \phi_1 F_1 + \phi_2 F_2, \tag{9}$$

where F_1 and F_2 are

$$F_1 \equiv u_t + u_{xxx} + 2uu_x + 2e_1vv_x + e_2(uv)_x + e_3v_{xxx}, \tag{10}$$
$$F_2 \equiv c_1v_t + v_{xxx} + 2vv_x + c_2v_x + c_3[e_1(uv)_x + 2e_2uu_x + e_3u_{xxx}], \tag{11}$$

and ϕ_i $(i = 1, 2)$ are two new dependent variables. In agreement with [1, 8] the *adjoint system* for (1) is defined as

$$F_1^\star = \frac{\delta\mathcal{L}}{\delta u} = 0, \quad F_2^\star = \frac{\delta\mathcal{L}}{\delta v} = 0 \tag{12}$$

with $\frac{\delta}{\delta u^i} = \frac{\delta}{\delta u^i} - D_{x_j}\frac{\partial}{\partial u^i_{x_j}} + D_{x_j}D_{x_k}\frac{\partial}{\partial u^i_{x_jx_k}}\ldots$ $(i, j, k = 1, 2)$, where $u^1 = u$, $u^2 = v$, $x_1 = t$, $x_2 = x$. We recall that the system of PDEs (1) is said to be *quasi self-adjoint* if

$$F_1^\star|_{\phi_1=\psi_1(u,v),\phi_2=\psi_2(u,v)} = F_1\lambda_{11} + F_2\lambda_{12}, \quad F_2^\star|_{\phi_1=\psi_1(u,v),\phi_2=\psi_2(u,v)} = F_1\lambda_{21} + F_2\lambda_{22}, \tag{13}$$

with regular undetermined coefficients λ_{ij} $(i, j = 1, 2)$.

Using the differential consequences of (13), we obtain the following system

$$\lambda_{11} + \psi_u^1 = \lambda_{22} + \psi_v^2 = \lambda_{12} + \frac{\psi_v^1}{c_1} = \lambda_{21} + \psi_u^2 c_1 = 0, \tag{14}$$
$$c_3e_3\left(-\frac{\psi_v^1}{c_1} + \psi_u^2\right) = \frac{(c_1-1)}{c_1}\psi_v^1 + e_3(c_3\psi_v^2 - \psi_u^1) = 0, \tag{15}$$
$$\psi_{vv}^1 + c_3e_3\psi_{vv}^2 = \psi_{uv}^1 + c_3e_3\psi_{uv}^2 = \psi_{uu}^1 + c_3e_3\psi_{uu}^2 = 0, \tag{16}$$
$$\psi_{uuv}^1 + c_3e_3\psi_{uuv}^2 = \psi_{uvv}^1 + c_3e_3\psi_{uvv}^2 = \psi_{uuu}^1 + c_3e_3\psi_{uuu}^2 = \psi_{vvv}^1 + c_3e_3\psi_{vvv}^2 = 0, \tag{17}$$
$$c_3e_3\psi_v^2 - e_3\psi_u^1 + (c_1-1)\psi_u^2 = e_3(\psi_v^1 - c_1\psi_u^2) = 0, \tag{18}$$
$$e_3\psi_{vv}^1 + \psi_{vv}^2 = e_3\psi_{uu}^1 + \psi_{uu}^2 = e_3\psi_{uv}^1 + \psi_{uv}^2 = 0, \tag{19}$$

$$
\begin{aligned}
&e_3\psi^1_{uvv}+\psi^2_{uvv}=0,\quad e_3\psi^1_{uuv}+\psi^2_{uuv}=0,\quad e_3\psi^1_{uuu}+\psi^2_{uuu}=0,\quad e_3\psi^1_{vvv}+\psi^2_{vvv}=0, &(20)\\
&(c_2-2c_1u+c_3e_1u+2v-c_1e_2v)\psi^1_v-c_1c_3(2e_2u+e_1v)\psi^2_v+c_1(e_2u+2e_1v)\psi^1_u=0, &(21)\\
&c_3(2e_2u+e_1v)(\psi^1_v-c_1\psi^2_u)=0, &(22)\\
&(c_2-2c_1u+c_3e_1u+2v-c_1e_2v)\psi^2_u-c_3(2e_2u+e_1v)\psi^2_v+(e_2u+2e_1v)\psi^1_u=0, &(23)\\
&(e_2u+2e_1v)(\psi^1_v-c_1\psi^2_u)=0. &(24)
\end{aligned}
$$

From these relations we get

$$\lambda_{11}=-\psi^1_u\quad \lambda_{22}=-\psi^2_v,\quad \lambda_{12}=-\frac{\psi^1_v}{c_1},\quad \lambda_{21}=-\psi^2_u c_1, \tag{25}$$

and, by integration (16) and (17),

$$\psi^1=-c_3e_3\psi^2+k_1v+k_2u+k_3, \tag{26}$$

with k_1,k_2 and k_3 arbitrary constants, and ψ^2 subjected to the following constraints

$$
\begin{aligned}
&c_3e_3(c_3e_3\psi^2_v+c_1\psi^2_u-k_1)=0,\quad (c_1-1)k_1-c_1e_3k_2+c_3e_3(\psi^2_v+c_1e_3\psi^2_u)=0,\\
&(c_3e_3^2-1)\psi^2_{uvv}=0,\quad (c_3e_3^2-1)\psi^2_{uuv}=0, (c_3e_3^2-1)\psi^2_{uuu}=0,\\
&(1-c_3e_3^2)\psi^2_{vvv},\quad (c_3e_3^2-1)\psi^2_{vv}=0,\quad (c_3e_3^2-1)\psi^2_{uv}=0,\quad (c_3e_3^2-1)\psi^2_{uu}=0,\\
&e_3k_2-c_3e_3\psi^2_v-(c_1+c_3e_3^2-1)\psi^2_u=0,\quad e_3(k_1-c_3e_3\psi^2_v-c_1\psi^2_u)=0\\
&c_3(2e_2u+e_1v)(k_1-c_3e_3\psi^2_v-c_1\psi^2_u)=0\\
&c_1(e_2u+2e_1v)(k_2-c_3e_3\psi^2_u)-c_3(e_3(c_2-(2c_1-c_3e_1)u+(2-c_1e_2)v+c_1(2e_2u+e_1v))\psi^2_v\\
&\quad-((2c_1-c_3e_1)u-(2-c_1e_2)v-c_2)k_1=0\\
&(e_2u+2e_1v)(c_3e_3\psi^2_v+c_1\psi^2_u-k_1)=0\\
&(e_2u+2e_1v)(k_2-c_3e_3\psi^2_u)-c_3(2e_2u+e_1v)\psi^2_v+((2c_1-c_3e_1)u-(2-c_1e_2)v-c_2)\psi^2_u=0.
\end{aligned}
\tag{27}
$$

Then, we can state the following result.
Theorem: The system (1) is quasi self-adjoint if and only if the constant parameters c_i, e_j, $(i,j=1,2,3)$ satisfy the conditions (27).

Note that, for clarity of further developments, we have review some constants which appear in [15] according to the symmetry classification of (1) reported in the previous section.

In order to obtain solutions to the previous conditions, we must distinguish different classes of subsystems.

Subclass 1. If $c_3e_3^2-1\neq 0$ we get $\psi^2=h_1u+h_2v+h_3$ with h_1, h_2, h_3 constants. When one of the constants e_1, e_2 and c_3 is vanishing we get

$$k_1=c_1h_1+c_3e_3h_2,\quad h_1=e_3\frac{c_3h_2-k_2}{1-c_1-c_3e_3^2};$$

in particular, we have:

- If $e_1=0$ we get

$$k_2=\frac{c_3(-2c_1e_3+e_2(-2+2c_1+c_3e_3^2))h_2}{(-1+c_1)e_2-2c_1e_3}$$

 associated to one of the following constraints on parameters:
 - $e_3=0$, whereupon it is $\psi^1=k_3+2c_3h_2u$, $\psi^2=h_3+h_2v$;
 - $h_2=0$, whereupon $\psi^1=k_3-e_3c_3h_3$, $\psi^2=h_3$;

– $c_2 = 0$ and $e_2 = \frac{2}{c_1}$, whereupon

$$\psi^1 = k_3 - e_3 c_3 h_3 + c_3 h_2 \frac{(4(c_1-1) - 2c_1^2 e_3)u + 2e_3 v}{2(c_1-1) - 2c_1^2 e_3},$$
$$\psi^2 = h_3 + \frac{2h_2 c_3 e_3}{2(c_1-1) - 2c_1^2 e_3} u + h_2 v.$$

In this case a singulariy arises when $e_3 = 0$ and $c_1 = 1$, so that we obtain

$$\psi^1 = k_3 + k_2 u + (k_2 - 2c_3 h_2)v, \quad \psi^2 = h_3 + (k_2 - 2c_3 h_2)u + h_2 v. \tag{28}$$

- If $e_2 = 0$ we get

$$k_2 = -c_3 h_2 \frac{(1-c_1)e_1 - e_3(2 - c_3 e_1 e_3)}{2((c_1-1)e_1 + e_3)}$$

along with one of the following constraints on parameters:
  – $e_3 = 0$, whereupon we get $\psi^1 = k_3 + \frac{c_3 h_2}{2} u$, $\psi^2 = h_3 + h_2 v$;
  – $h_2 = 0$ and $\psi^1 = k_3 - e_3 c_3 h_3$, $\psi^2 = h_3$;
  – $c_2 = 0$ and $e_1 = \frac{2c_1}{c_3}$, whereupon

$$\psi^1 = k_3 - e_3 c_3 h_3 + c_3 h_2 \frac{(c_3 e_3 - c_1)u + c1^2(u - e_3 v)}{2(c_1-1)c_1 + c_3 e_3},$$
$$\psi^2 = h_3 - \frac{c_3 e_3 c_1 h_2}{2(c_1-1)c_1 + c_3 e_3} u + h_2 v.$$

In this case a singulariy arises if $e_3 = 0$ and $c_1 = 1$, so that we recover

$$\psi^1 = k_3 + (2\frac{k_2}{c_3} - h_2)u + k_2 v, \quad \psi^2 = h_3 + h_2 u + (2\frac{k_2}{c_3} - h_2)v. \tag{29}$$

- If $c_3 = 0$ we have:
  – if $h_1 = 0$ then $\psi^1 = k_3$, $\psi^2 = h_3 + h_2 v$;
  – if $c_2 = 0$ and $e_1 = \frac{e_2(c_1 e_2 - 2)}{4c_1}$, $e_3 = \frac{(c_1-1)e_2}{2c_1}$, $k_2 = 2\frac{k_1}{e_2}$ it is

$$\psi^1 = k_3 + k_1\left(\frac{2}{e_2}u + v\right), \quad \psi^2 = h_3 + \frac{k_1}{c_1}u + h_2 v. \tag{30}$$

A special subclass arises when $e_1 = e_2 = 0$, whence we get:

- if $e_3 = 0$ then $\psi^1 = k_3 + k_2 u$, $\psi^2 = h_3 + h_2 v$;
- if $k_2 = c_3 h_2$ then $\psi^1 = k_3 - c_3 e_3 h_3 + c_3 h_2 u$, $\psi^2 = h_3 + h_2 v$.

Subclass 2. If $c_3 = \frac{1}{e_3^2}$ $(e_3 \neq 0)$ we get $k_2 = \frac{k_1}{e_3}$ and

$$\psi^2 = e_3 k_1 v + F(\omega),$$

$F(\omega)$ being an arbitrary function of $\omega = (u - e_3 c_1 v)$.

- If $c_2 = 0$ and $c_1 = \frac{e_1^2 - 3e_1 e_2 e_3 + 2e_2 e_3^2}{e_3(2e_1 e_3 - 3e_1 e_2 + e_2^2 e_3)}$ we get an explicit form for the arbitary function $F(\omega)$ that leads us to write:

$$\psi^1 = k_3 + \frac{k_1}{e_3}u - k_1 \frac{(c_1 e_2 e_3 - e_1)(u - c_1 e_3 v)}{e_3 \Delta_1} - \frac{h_3}{e_3}, \tag{31}$$
$$\psi^2 = e_3 k_1 v + k_1 \frac{(c_1 e_2 e_3 - e_1)(u - c_1 e_3 v)}{\Delta_1} + h_3, \tag{32}$$

with $\Delta_1 = 2[(c_1-1)e_1 + e_3 - c_1 e_3(e_2 - c_1 e_2 + c_1 e_3)]$ and h_3 arbitrary constant.

- If $\Delta_1 = 0$ we get $e_1 = e_3 \frac{e_3 c_1^2 - 1}{2(c_1-1)}$, $e_2 = \frac{e_3 c_1^2 - 1}{2(c_1-1)c_1}$, and we obtain:

$$\psi^1 = k_3 + \frac{k_1}{e_3} u - e_2 k_1 \frac{((3c_1 - 1)e_2 - 2c_1 e_3)(u - c_1 e_3 v) - c_2 e_3 g(u,v)}{e_3((3c_1-1)e_2 - 2c_1 e_3)^2} - \frac{h_3}{e_3}, \quad (33)$$

$$\psi^2 = e_3 k_1 v + e_2 k_1 \frac{((3c_1 - 1)e_2 - 2c_1 e_3)(u - c_1 e_3 v) - c_2 e_3 g(u,v)}{((3c_1-1)e_2 - 2c_1 e_3)^2} + h_3 \quad (34)$$

with $g(u,v) = \ln[(e_2 - 3c_1 e_2 + 2c_1 e_3)(u - c_1 e_3 v) - c_2 e_3]$ and h_3 arbitrary constant.

- If $e_1 = e_2 = 0$ we get $F(\omega) = h_3$, h_3 arbirtary constant, and

$$\psi^1 = k_3 - \frac{h_3}{c_3} + \frac{k_1}{e_3} u, \quad \psi^2 = e_3 k_1 v + h_3. \quad (35)$$

Remark: The results on quasi self-adjointness are in agreement with the complete symmetry classification, according to which for all subclasses of the system (1) that admit a quasi self-adjoint form, we find subalgebras of three and four dimensional algebras.

4. Conservation laws

In order to determine some conservation laws of system (1), according to [7], once we know the explicit expression of the lagrangian form $\mathcal{L}$ in terms of ψ^1, ψ^2 and λ_{ij} reported in the previous section, we must consider the associated Lie algebra $\mathcal{L}$ of system (1), that, in general, is the set of vector fields of the form

$$X = \xi^1 \frac{\partial}{\partial t} + \xi^2 \frac{\partial}{\partial x} + \eta^1 \frac{\partial}{\partial u} + \eta^2 \frac{\partial}{\partial v} \quad (36)$$

leading to a conservation law for (1) constructed by the following formula:

$$C^i = \xi^i \mathcal{L} + W^\alpha \left[\frac{\partial \mathcal{L}}{\partial u_i^\alpha} - D_j \left(\frac{\partial \mathcal{L}}{\partial u_{ij}^\alpha} \right) + D_j D_k \left(\frac{\partial \mathcal{L}}{\partial u_{ijk}^\alpha} \right) - \dots \right]$$
$$+ D_j(W^\alpha) \left[\frac{\partial \mathcal{L}}{\partial u_{ij}^\alpha} - D_k \left(\frac{\partial \mathcal{L}}{\partial u_{ijk}^\alpha} \right) + \dots \right] + D_j D_k(W^\alpha) \left[\frac{\partial \mathcal{L}}{\partial u_{ijk}^\alpha} - \dots \right], \quad (37)$$

where $i, \alpha = 1, 2$; in particular,

$$W^1 = \eta^1 - \xi^1 u_t - \xi^2 u_x, \quad (38)$$
$$W^2 = \eta^2 - \xi^1 v_t - \xi^2 v_x \quad (39)$$

and $\mathcal{L}$ is the *formal Lagrangian* defined in (9), and η^1, η^2, ξ^1 and ξ^2 have an explicit form according to one of previous Cases 1, 2, 3, 4. (*i.e.*, according to the operators X_3, $\acute{X}_3$,$\acute{X}_4$, ...).

Then, let us construct all the conserved vectors associated with all the Lie point symmetries of system (1) using the algorithm given in (37).

In order to find new conservation laws we must merge the results of adjoint analysis in group classification, and we obtain the following simplified form of C^1 and C^2,

$$C^1 = \xi^1 \mathcal{L} + \psi^1 W^1 + c_1 \psi^2 W^2 =$$
$$= \psi^1 \eta^1 + c_1 \psi^2 \eta^2 + \xi^1 (\psi^1 P_x^1 + \psi^2 P_x^2) - \xi^2 (\psi^1 u_x + c_1 \psi^2 v_x) \quad (40)$$
$$C^2 = \xi^2 \mathcal{L} + (\psi^1 (2u + e_2 v) + c_3 \psi^2 (2e_2 u + e_1 v) + G_{xx}^1) W^1$$
$$(\psi^1 (e_2 u + 2e_1 v) + \psi^2 (c_2 + c_3 e_1 u + 2v) + G_{xx}^2) W^2$$
$$- G_x^1 W_x^1 - G_x^2 W_x^2 + G^1 W_{xx}^1 + G^2 W_{xx}^2, \quad (41)$$

where

$$P^1 = u_{xx} + u^2 + e_1 v^2 + e_2 uv + e_3 v_{xx}, \quad P^2 = v_{xx} + v^2 + c_2 v + c_3(e_1 uv + e_2 u^2 + e_3 u_{xx}), \quad (42)$$

along with the positions

$$G^1 = \psi^1 + c_3 e_3 \psi^2, \quad G^2 = \psi^2 + e_3 \psi^1. \quad (43)$$

Independently of the subclass and its associated operator, we are able to write C^1 and C^2 in the following forms

$$C^1 = D_x(\frac{1}{2}\xi^1\theta^1 - \xi^2\theta^2) + \gamma^1, \qquad C^2 = -D_t(\frac{1}{2}\xi^1\theta^1 - \xi^2\theta^2) + \gamma^2, \quad (44)$$

being γ^1 and γ^2 suitable functions of $u, v, u_t, v_t, u_x, v_x, u_{xx}, v_{xx}$ whose form will be specialized according to the particular values of the parameters. Obviously, the first terms of fluxes (44) disappear and the only contributions are given by γ^1 and γ^2 where

$$\theta^1 = (\psi^1 + (k_3 - c_3 e_3 h_3))u - c_1(\psi^2 + h_3)v + uv(k_1 - c_1 h_1 - c_3 e_3 h_2), \quad \theta^2 = \psi^1 P^1 + \psi^2 P^2.$$

4.1. Conservation Laws of Subclass 1

Some cases of subclass 1 have operators X_3 whose generators are

$$\xi^1 = 3kt, \quad \xi^2 = 2c_2(4e_1 - e_2^2)t + kx, \quad \eta^1 = 2(c_2 e_1 - ku), \quad \eta^2 = -(c_2 e_2 + kv),$$

where $k = 2(e_2 - c_3 e_1^2) - c_1(e_2^2 - 4e_1)$.

- If $e_1 = 0$:
  - when $e_3 = 0$ with $\psi^1 = k_3 + 2c_3 h_2 u$, $\psi^2 = h_3 + h_2 v$ we obtain

$$\begin{aligned}\gamma^1 &= -k(k_3 + 3c_3 h_2 u)u - c_1(2c_2 e_2 h_2 + kh_3)v - \frac{3}{2}c_1 k h_2 v^2 + 2c_1 c_2 e_2 h_3,\\ \gamma^2 &= -k(k_3 + 6c_3 h_2 u)P^1 - (2c_2 e_2 h_2 + kh_3)P^2 - 3kh_2 v P^2 - 2c_2^2 e_2 h_3 +\\ &\quad kh_2(c_3(2u^3 + 3e_2 u^2 v + 3u_x^2) + v^3 + \frac{3}{2}(c_2 v^2 + v_x^2)),\end{aligned}$$

    and $\mathcal{L} = -k(k_3 + 6c_3 h_2 u)F^1 - (2c_2 e_2 h_2 + k(h_3 + 3h_2 v))F^2$;
  - when $h_2 = 0$, ψ^1 and ψ^2 are constants and as a consequence we get only the trivial form.
  - In the case $c_2 = 0$, it is $e_2 = \frac{2}{c_1}$ with

$$\begin{aligned}\psi^1 &= k_3 - e_3 c_3 h_3 + c_3 h_2 \frac{(4(c_1 - 1) - 2c_1^2 e_3)u + 2e_3 v}{2(c_1 - 1) - 2c_1^2 e_3},\\ \psi^2 &= h_3 + \frac{2h_2 c_3 e_3}{2(c_1 - 1) - 2c_1^2 e_3}u + h_2 v.\end{aligned}$$

    Since $k = c_2 = 0$, we have to consider, according to the Lie classification admitted by the system (see [3]), the operators

$$\acute{X}_3 = 3t\frac{\partial}{\partial t} + x\frac{\partial}{\partial x} - 2u\frac{\partial}{\partial u} - 2v\frac{\partial}{\partial v} \quad \acute{X}_4 = -(e_2^2 - 4e_1)t\frac{\partial}{\partial x} + 2e_1\frac{\partial}{\partial u} - e_2\frac{\partial}{\partial v}; \quad (45)$$

now, by taking a linear combination $a_1\acute{X}_3 + a_2\acute{X}_4$ (but it is sufficient to consider the operator X_3 with $k = a_1$ and $2c_2 = a_2$), we get

$$\gamma^1 = (a_1(c_3e_3h_3 - k_3) - 2a_2h_1 - 3a_1c_1h_1v - \frac{3}{2}a_1(k_2 - c_3e_3h_1)u)u - (2a_2h_2 + a_1c_1(h_3 + \frac{3}{2}h_2v))v,$$

$$\gamma^2 = (a_1(c_3e_3h_3 - k_3) - 2a_2h_1 - 3a_1c_1h_1v - 3a_1(k_2 - c_3e_3h_1)u)P^1 + (2\frac{a_2}{c_1}h_2 + a_1(h_3 + 3h_2v))P^2 + a_1h_1(2\frac{c_1-1}{e_3} - (c_1^2 - e_2c_3))u^3 + \frac{3}{2}a_1((c_1-1)\frac{h_1}{e_3} + c_3h_2)u^2v + 3a_1h_1v^2u + h_2v^3 + a_1(3(c_3e_3h_2 + c_1h_1)u_xv_x + \frac{3}{2}(k_2u_x^2 + (h_2 + c_1h_1e_3)v_x^2)),$$

and $\mathcal{L} = (a_1(c_3e_3h_3 - k_3) - 2a_2h_1 - 3a_1c_1h_1v) - 3a_1(k_2 - c_3e_3h_1)u)F_1 - (2\frac{a_2}{c_1}h_2 + a_1(h_3 + 3h_2v) - 3a_1h_1u)F_2$.

- If $e_2 = 0$:
  - when $e_3 = 0$ with $\psi^1 = k_3 + \frac{c_3h_2}{2}u$, $\psi^2 = h_3 + h_2v$ we get

$$\gamma^1 = (2c_2c_3e_1h_2 - k(k_3 + \frac{3}{4}c_3h_2u))u - c_1k(h_3 + \frac{3}{2}h_2v) + 4c_2e_1k_3,$$

$$\gamma^2 = (2c_2c_3e_1h_2 - k(k_3 + \frac{3}{2}c_3h_2u))P^1 - k(h_3 + 3h_2v)P^2 + \frac{k}{2}h_2(c_3(u^3 + 3e_1uv^2 + \frac{3}{2}u_x^2) + 2v^3 + 3c_2v^2 + 3v_x^2),$$

and $\mathcal{L} = (2c_2c_3e_1h_2 - k(k_3 + \frac{3}{2}c_3h_2u))F^1 - k(h_3 + 3h_2v)F^2$;
  - when $h_2 = 0$, ψ^1 and ψ^2 are constants and we get only the trivial form.
  - In the case $c_2 = 0$, $e_1 = \frac{2c_1}{c_3}$ with

$$\psi^1 = k_3 - e_3c_3h_3 + c_3h_2\frac{(c_3e_3 - c_1)u + c1^2(u - e_3v)}{2(c_1-1)c_1 + c_3e_3},$$

$$\psi^2 = h_3 - \frac{c_3e_3c_1h_2}{2(c_1-1)c_1 + c_3e_3}u + h_2v.$$

we get $k = c_2 = 0$ and, as before, we again consider the previous generators by taking $k = a_1$ and $2c_2 = a_2$ as coefficients of the linear combination $a_1\acute{X}_3 + a_2\acute{X}_4$ so that

$$\gamma^1 = (2a_2e_1(c_3e_3h_1 - k_2) + a_1(k_3 - c_3e_3h_3) - 3a_1(\frac{1}{2}(k_2 - c_3e_3h_1)u + c_1h_1v)u + c_1(4a_2e_1h_1 - a_1h_3 - \frac{3}{2}a_1h_2v)v + 2a_2e_1(e_3c_3h_3 - k_3),$$

$$\gamma^2 = (2a_2e_1(e_3c_3h_1 - k_2) + a_1(k_3 - c_3e_3h_3) - 3a_1((k_2 - c_3e_3h_1)u + c_1h_1v)P^1 + (4a_2e_1h_1 - a_1h_3 - 3a_1h_2v)P^2 - a_1(h_1(\frac{1-c_1}{e_3} - \frac{c_3}{c_1})u^3 + (h_2 + 2c_1^2\frac{h_1}{c_3})v^3) - \frac{3}{2}a_1e_2(h_1(2\frac{c_1-1}{e_3} + (c_1^2 + \frac{c_3}{c_1})) + c_3h_2)u^2v - 3a_1h_1(1 + 2\frac{c_1(c_1-1)}{c_3e_3})v^2u + a_1(3k_1u_xv_x + \frac{3}{2}(h_2 + c_1\frac{h_1}{e_3}v_x^2 + k_2u_x^2)),$$

$\mathcal{L} = (2a_2e_1(k_2 - c_3e_3h_1) - a_1(k_3 - c_3e_3h_3) - 3a_1(c_1h_1v + (k_2 - c_3e_3h_1)u)F_1 + (4a_2e_1h_1 - a_1h_3 - 3a_1h_2v - 3a_1h_1u))F_2$.

- If $c_3 = 0$:
  - when $h_1 = 0$, it is $\psi^1 = k_3$, $\psi^2 = h_3 + h_2 v$ and we have

$$\gamma^1 = -kk_3 u - c_1(2c_2e_2h_2 + k(h_3 + \frac{3}{2}h_2 v))v - 2c_2(c_1e_2h_3 - 2e_1k_3),$$
$$\gamma^2 = -kk_3P^1 - (2c_2e_2h_2 + k(h_3 + 3h_2v))P^2 + kh_2(\frac{3}{2}(v^2c_2 + v_x^2) + v^3),$$

  and $\mathcal{L} = -kk_3F^1 - (2c_2h_2 + k(h_3 + 3h_2v))F^2$;
  - when $c_2 = 0$ and $e_1 = \frac{e_2(c_1e_2-2)}{4c_1}$, $e_3 = \frac{(c_1-1)e_2}{2c_1}$, $k_2 = 2\frac{k_1}{e_2}$ with

$$\psi^1 = k_3 + k_1\left(\frac{2}{e_2}u + v\right), \quad \psi^2 = h_3 + \frac{k_1}{c_1}u + h_2 v. \tag{46}$$

  it results $k = c_2 = 0$, and, by considering the same linear combination of operators as above, we get

$$\gamma^1 = -\frac{1}{e_2}(a_2(e_2^2 - 4e_1)k_1 + a_1e_2k_3 + 3a_1k_1u)u + a_2(c_1e_2h_3 - 2e_1k_3) -$$
$$((a_2c_1e_2h_2 + a_1c_1h_3 - 2a_2e_1k_1 + 3a_1k_1u) + \frac{3}{2}a_1c_1e_2h_2v)v,$$
$$\gamma^2 = -\frac{1}{e_2}(a_2(e_2^2 - 4e_1)k_1 + a_1e_2k_3 + 6a_1k_1u)P^1 -$$
$$((a_2e_2h_2 + a_1h_3 - 2a_2\frac{e_1}{c_1}k_1 + 3\frac{a_1}{c_1}k_1u) + 3a_1h_2v)P^2 +$$
$$\frac{1}{e_2}(a_1(2k_1u^3 + 3e_2k_1u^2v + \frac{3}{2}e_2^2k_1uv^2 + e_2(h_2 + e_1k_1)v^3) +$$
$$\frac{a_1}{e_2}(3k_1u_x(u_x + e_2v_x) + \frac{3}{2}e_2(h_2 + e_3k_1)v_x^2),$$

  $\mathcal{L} = -\frac{1}{e_2}((e_2^2 - 4e_1)a_2k_1 + e_2a_1k_3 + 6a_1k_1u - 3e_2a_1k_1v)F^1 + (a_2e_2h_2 + a_1h_3 - 2a_2\frac{e_1}{c_1}k_1 + 3\frac{a_1}{c_1}k_1u + 3a_1h_2v)F^2$.
- If $e_1 = e_2 = 0$, the operators associated to this subclass are $\hat{X}_3$ and $\hat{X}_4$ [3]

$$\hat{X}_3 = 3t\frac{\partial}{\partial t} + x\frac{\partial}{\partial x} - 2u\frac{\partial}{\partial u} - (c_2 + 2v)\frac{\partial}{\partial v}, \quad \hat{X}_4 = 2t\frac{\partial}{\partial x} + \frac{\partial}{\partial u} + c_1\frac{\partial}{\partial v}. \tag{47}$$

  and, by considering a linear combination of the operators, the generators read:

$$\xi_1 = 3a_1t, \quad \xi_2 = a_1x + 2a_2t, \quad \eta_1 = -2a_1u + a_2, \quad \eta_2 = -a_1(c_2 + 2v) + c_1a_2;$$

  - when $e_3 = 0$, $\psi^1 = k_3 + k_2u$, $\psi^2 = h_3 + h_2v$ so that we get

$$\gamma^1 = (a_2k_2 - a_1k_3 - \frac{3}{2}a_1k_2u)u + c_1((a_1c_2 - a_2h_1)h_2 - a_1h_3 + \frac{3}{2}a_1h_2v)v +$$
$$c_1(a_1c_2 - a_2c_1)h_3 - a_2k_3,$$
$$\gamma^2 = (a_2k_2 - a_1k_3 - 3a_1k_2u)P^1 + ((a_2c_1 - a_1c_2)h_2 - a_1h_3 - 3a_1h_2v)P^2 +$$
$$a_1(k_2(u^3 + u_x^2) + h_2(v^3 + \frac{3}{2}(c_2h_2v^2 + h_2v_x^2)) - c_2(a_1c_2 - a_2c_1)h_3,$$

  and $\mathcal{L} = (a_2k_2 - a_1k_3 - 3a_1k_2u)F_1 - ((a_2c_1 - a_1c_2)h_2 - a_1h_3 + 3a_1h_2v)F_2$;

– when $k_2 = c_3h_2$, $k_1 = c_3e_3h_2$ and $\psi^1 = k_3 - c_3e_3h_3 + c_3h_2u$, $\psi^2 = h_3 + h_2v$ we obtain

$$\begin{aligned}
\gamma^1 &= (c_3(a_2h_2 + a_1e_3h_3) - a_1k_3 - \frac{3}{2}a_1c_3h_2u)u - \\
&\quad c_1(a_1c_2h_2 - a_1(c_2h_2 + h_3) - \frac{3}{2}h_2v)v - a_1c_1c_2h_3 + a_2(c_1^2h_3 - c_3e_3h_3 + k_3), \\
\gamma^2 &= (c_3(a_2h_2 + a_1e_3h_3) - a_1k_3 - 3a_1c_3h_2u)P^1 + \\
&\quad (a_2c_1h_2 - a_1c_2h_2 + a_1h_3 - 3a_1h_2v)P^2 + \\
&\quad a_1h_2(c_3(u^3 + \frac{3}{2}u_x^2 + 3e_3u_xv_x) + v^3 + \frac{3}{2}(c_2v^2 + v_x^2)) + c_2(a_1c_2 - a_2c_1)h_3,
\end{aligned}$$

$$\mathcal{L} = (c_3(a_2h_2 + a_1e_3h_3) - a_1k_3 - 3a_1c_3h_2u)F^1 + (a_2c_1h_2 - a_1(c_2h_2 + h_3) - 3a_1h_2v)F^2.$$

4.2. Subclass 2 associated with its symmetries

For this subclass, characterized by the relation $c_3 = \frac{1}{e_3^2}$, there is no need of distinguishing particular values of parameters, and only the principal Lie algebra, generated by the operators X_1 and X_2 given in (2) is admitted by the system. In this case, the contributions γ^1 and γ^2 (44) are vanishing, and, as a consequence, both C^1 and C^2 vanish too.

In the particular case of this subclass with $e_1 = e_2 = 0$, the operators admitted are $\hat{X}_3$ and $\hat{X}_4$, and we recover a subcase of subclass 1.

5. Conclusions

Using the new outstanding concepts of quasi self-adjoint equations and the general theorem on conservation laws, developed recently [8], nonlinear self-adjointness and conservation laws for the system (1) have been discussed. We underline that the property of a differential equation to be quasi self-adjoint is important for constructing conservation laws associated with symmetries of the differential equation.

[1] Ibragimov N.H., 2007, Quasi-self-adjoint differential equations, Archives of ALGA, Vol. 4, 55–60.
[2] Gear J. A. and Grimshaw R., 1984, *Weak and strong interactions between internal solitary waves*, Stud. Appl. Math (70), **3**, 235–258.
[3] Ruggieri M. and Speciale M. P., 2013, *Similarity Reduction and Closed Form Solutions for a Model Derived from Two Layer Fluids*, Advances in Difference Equations, Springer.
[4] Ruggieri M. and Speciale M. P., 2013, *On a hierarchy of traveling wave solutions in a shallow stratified fluid*, In AIP Conference Proceedings **1558**, doi: 10.1063/1.4825873, 1793-1796.
[5] Ruggieri M. and Valenti A., 2013, Approximate symmetries in nonlinear viscoelastic media, Boundary Value Problems, doi: 10.1186/1687-2770-2013-143.
[6] Ruggieri M. and Speciale M. P., 2013, Approximate Analysis of KdV-like Equations, to appear.
[7] Ibragimov N. H., 2007, A new conservation theorem, J. Math. Anal. Appl. 333, 311–328.
[8] Ibragimov N.H., 2006, Integrating factors, adjoint equations and Lagrangians, J. Math. Anal. Appl. 318, 742–757.
[9] Oliveri F. and Speciale M. P., 2013, *Reduction of balance laws to conservation laws by means of equivalence transformations*, Journal of Mathematical Physics, **54**, 041506.
[10] Oliveri F. and Speciale M. P., 2012, *Equivalence Transformations of Quasilinear First Order Systems and Reduction to Autonomous and Homogeneous Form*, Acta Applicandae Mathematicae, **122**, 447 – 460.
[11] Ruggieri M. and Valenti A., 2011, *Exact solutions for a nonlinear model of dissipative media*, Journal of Mathematical Physics, **52**, 043520.
[12] Ruggieri M. and Valenti A., 2009, *Symmetries and reduction techniques for dissipative models* Journal of Mathematical Physics, **50**, 063506 - 063506-9.
[13] Margheriti L. and Speciale M. P., 2011,*Unsteady solutions of Euler equations generated by steady solutions*, Acta Applicandae Mathematicae, **113**, 3, 289-303.
[14] Ruggieri M., 2012, Kink solutions for class of generalized dissipative equations, *Abstract and Applied Analysis*, Article ID 237135, 7 pages, doi:10.1155/2012/237135.
[15] Ruggieri M. and Speciale M. P., 2013 *Quasi Self-adjoint Coupled KdV-like Equations*, In Numerical Analysis and Applied Mathematics, AIP Conference Proceedings, **1558**, doi: 10.1063/1.4825730, 1220-1223.

Building Generalized Lax Integrable KdV and MKdV Equations with Spatiotemporally Varying Coefficients

M Russo and S R Choudhury

Department of Mathematics, University of Central Florida, Orlando, FL 32816-1364 USA

E-mail: choudhur@cs.ucf.edu

Abstract. We present a technique based on extended Lax Pairs to derive variable-coefficient generalizations of various Lax-integrable NLPDE hierarchies. As illustrative examples, we consider generalized KdV equations, and three variants of generalized MKdV equations. It is demonstrated that the technique yields Lax- or S-integrable NLPDEs with both time- AND space-dependent coefficients which are thus more general than almost all cases considered earlier via other methods such as the Painlevé Test, Bell Polynomials, and various similarity methods. Some solutions are also presented for the generalized KdV equation derived here by the use of the Painlevé singular manifold method. Current and future work is centered on generalizing other integrable hierarchies of NLPDEs similarly, and deriving various integrability properties such as solutions, Backlund Transformations, and hierarchies of conservation laws for these new integrable systems with variable coefficients.

Key Words: Generalizing Lax or S-integrable equations, spatially and temporally-dependent coefficients, solutions.

1. Introduction

Variable Coefficient Korteweg de Vries (vcKdV) and Modified Korteweg de Vries (MKdV)equations have a long history dating from their derivation in various applications[1]-[10].

Almost all studies, including those which derived exact solutions by a variety of techniques, as well as those which considered integrable sub-cases and various integrability properties by methods such as Painlevé analysis, Hirota's method, and Bell Polynomials treat vcKdV equations with coefficients which are functions of the time only. For instance, for generalized variable coefficient NLS (vcNLS) equations, a particular coefficient is usually taken to be a function of x[11], as has also been sometimes done for vcMKdV equations[12]. The papers [13]-[14] are somewhat of an exception in that they treat vcNLS equations having coefficients with general x and t dependences. Variational principles, solutions, and other integrability properties have also been considered for some of the above variable coefficient NLPDEs in cases with time-dependent coefficients.

In applications, the coefficients of vcKdV equations may include spatial dependence, in addition to the temporal variations that have been extensively considered using a variety of techniques. Both for this reason, as well as for their mathematical interest, extending integrable hierarchies of nonlinear PDEs (NLPDEs) to include *both* spatial and temporal dependence of the coefficients is worthwhile.

Given the above, we consider a direct method for deriving the integrability conditions of both a general form of variable-coefficient MKdV (vcMKdV) equation, as well as a general, variable-coefficient KdV (vcKdV) equation. In both cases,the coefficients are allowed to vary in space AND time.

The method used is based on directly establishing Lax integrability (or S-integrability to use the technical term) as detailed in the following section. As such, it is rather general, although subject to the ensuing equations being solvable. We should stress that the computer algebra involved is quite challenging, and an order of magnitude beyond that encountered for integrable, constant coefficient NLPDEs.

However, there is an additional important proviso or qualifier that needs to be stressed. In order that Lax integrability ensues from the compatibility of the Lax Pair, i.e., the compatibility condition(s) for the Lax Pair contain the NLPDE under question, a form for each entry in the Lax Pair must be guessed *a priori*. Since there is no strict algorithm for doing this, one must rely on something like inspired guesswork. The inspiration comes from basing the Lax Pair on the known one for the existing constant coefficient version of the NLPDE under investigation. However, non-trivial guesswork must now go into adding additional terms and coefficients in each entry of this Lax Pair, and also generalizing some of the constants (related to the spectral parameter of the constant coefficient NLPDE) to functions.

We shall explain the above paragraph, which may seem rather dense at the first reading, in more detail in the context of the actual examples in the following two sections.

The outline of this paper is as follows. In Section 2, we briefly review the Lax Pair method and its modifications for variable-coefficient NLPDEs, and then apply it to three classes of generalized vcMKdV equation. Section 3 considers an analogous treatment of a generalized vcKdV equation. Some solutions of the generalized vcKdV equations of Section 3 are then derived in Section 4 via the use of truncated Painlevé expansions. Section 5 briefly considers the results and conclusions of the paper, and discusses the natural next steps or directions for further work.

2. Extended Lax Pair method and application to three generalized vcMKdV equations

In the Lax pair method [15] - [16] for solving and determining the integrability conditions for nonlinear partial differential equations (NLPDEs) a pair of $n \times n$ matrices, **U** and **V** needs to be derived or constructed. The key component of this construction is that the integrable nonlinear PDE under consideration must be contained in, or result from, the compatibility of the following two linear Lax equations (the Lax Pair)

$$\Phi_x = U\Phi \tag{1}$$

$$\Phi_t = V\Phi \tag{2}$$

where Φ is an eigenfunction, and **U** and **V** are the time-evolution and spatial-evolution (eigenvalue problem) matrices. From the cross-derivative condition (i.e. $\Phi_{xt} = \Phi_{tx}$) we get

$$U_t - V_x + [U, V] = \dot{0} \tag{3}$$

known as the zero-curvature condition where $\dot{0}$ is contingent on $v(x, t)$ being a solution to the nonlinear PDE. A Darboux transformation can then be applied to the linear system to obtain solutions from known solutions and other integrability properties of the integrable NLPDE.

We first consider the following three variants of generalized variable-coefficient MKdV (vcMKdV) equations:

$$v_t + a_1 v_{xxx} + a_2 v^2 v_x = 0 \tag{4}$$
$$v_t + b_1 v_{xxxxx} + b_2 v^2 v_{xxx} + b_3 v v_x v_{xx} + b_4 v_x^3 + b_5 v^4 v_x = 0 \tag{5}$$
$$\begin{aligned} v_t + c_1 v_{xxxxxxx} + c_2 v^2 v_{xxxxx} + c_4 v v_{xx} v_{xxx} + c_5 v_x^2 v_{xxx} + \\ + c_6 v_x v_{xx}^2 + c_7 v^4 v_{xxx} + c_8 v^3 v_x v_{xx} + c_9 v^2 v_x^3 + c_{10} v^6 v_x = 0. \end{aligned} \tag{6}$$

These equations, which we shall always call the physical (or field) NLPDEs to distinguish them from the many other NLPDEs we encounter, will be Lax-integrable or S-integrable if we can find a Lax pair whose compatibility condition (3) contains them.

One expands the Lax pair **U** and **V** in powers of v and its derivatives with unknown functions as coefficients. This results in a VERY LARGE system of coupled NLPDEs for the variable coefficient functions. Upon solving these (and a solution is not guaranteed, and may prove to be impossible to obtain in general for some physical NLPDEs), we simultaneously obtain the Lax pair and integrability conditions on the a_i, b_i, and c_i for which (4) - (6) are Lax-integrable.

The results are given in the following three subsections. The details of derivations are omitted for the sake of readability.

2.1. *Conditions on the a_i*

$$\begin{aligned} 6a_1 a_{2x}^3 - 6a_1 a_2 a_{2x} a_{2xx} + a_1 a_2^2 a_{2xxx} - \frac{K_t}{K} a_2^3 + a_2^2 a_{2t} - a_2^3 a_{1xxx} + \\ v + 3a_{1xx} a_2^2 a_{2x} - 6a_{1x} a_2 a_{2x}^2 + 3a_{1x} a_2^2 a_{2xx} = 0 \end{aligned} \tag{7}$$

where $K(t)$ is an arbitrary function of t.

2.2. *Conditions on the b_i*

$$b_3 = 2(b_2 + b_4), \qquad b_5 = H_1(t)(2b_2 - b_4). \tag{8}$$

where $H_1(t)$ is an arbitrary function of t and b_2 or b_4 are considered psuedo-arbitrary. We also do not display one more condition which we omit because it is very long. Now, once b_2 or b_4 is given all other coefficients can be found. The latter equation may look a bit daunting but is quite easily managed with the aid of a CAS (in this case MAPLE) once b_2 and b_4 are given. One example is presented here:

2.3. *Conditions on the c_i*

$$c_4 = -10c_2 + 5c_3 - 4c_5 + 2c_6 \tag{9}$$
$$c_8 = \frac{2}{3} c_9 + 4c_7 \tag{10}$$
$$c_{10x} c_6 - c_6 c_{10x} - 2c_5 c_{10x} + 2c_{5x} c_{10} + c_{10x} c_3 - c_{3x} c_{10} = 0 \tag{11}$$
$$-12c_{10x} c_7 + 12c_{7x} c_{10} + c_{10x} c_9 - c_{9x} c_{10} = 0 \tag{12}$$
$$5Gc_{10}^3 - 2Hc_{10x} c_{7x} c_{10} + Hc_{10}^2 c_{7xx} + 2Hc_{10x}^2 c_7 - Hc_7 c_{10} c_{10xx} = 0 \tag{13}$$
$$\begin{aligned} 14c_3 c_{10x}^2 - 7c_3 c_{10} c_{10xx} - 60c_2 c_{10x}^2 + 30c_2 c_{10} c_{10xx} + 60c_{2x} c_{10} c_{10x} - 30c_{2xx} c_{10}^2 + \\ +4c_{5x} c_{10} c_{10x} - 14c_{3x} c_{10} c_{10x} + 7c_{3xx} c_{10}^2 - 4c_5 c_{10x}^2 + 2c_5 c_{10} c_{10xx} - 2c_{5xx} c_{10}^2 = 0 \end{aligned} \tag{14}$$

$$\begin{aligned} &18c_3 c_{10x}^2 - 18c_3 c_{10} c_{10x} c_{10xx} + 3a_3 c_{10}^3 c_{10xxx} - 18c_{3x} c_{10} c_{10x}^2 + 9c_{3x} c_{10}^2 c_{10xx} - 60c_2 c_{10x}^3 \\ &+60c_2 c_{10} c_{10x} c_{10xx} - 10c_2 c_{10}^2 c_{10xxx} + 60c_{2x} c_{10} c_{10x}^2 - 30c_{2x} c_{10}^2 c_{10xx} - 30c_{2xx} c_{10}^2 c_{10x} \\ &+10c_{2xxx} c_{10}^3 + 6c_{5x} c_{10} c_{10x}^2 - 3c_{5x} c_{10}^2 c_{10xx} - 3c_{5xx} c_{10}^2 c_{10x} + 9c_{3xx} c_{10}^2 c_{10x} - 3c_{3xxx} c_{10}^3 \\ &-6c_5 c_{10x}^3 + c_5 c_{10} c_{10x} c_{10xx} - c_5 c_{10}^2 c_{10xxx} + c_{5xxx} c_{10}^3 = 0 \end{aligned} \tag{15}$$

$$
\begin{aligned}
&c_{1xxxxxxx}c_{10}^{7} + 5040c_{1x}c_{10}c_{10x}^{6} - 7c_{1xxxxxx}c_{10}^{6}c_{10x} - c_{1}c_{10}^{6}c_{10xxxxxxx} - 630c_{1x}c_{10}^{4}c_{10xx}^{3} \\
&+140c_{1x}c_{10}^{5}c_{10xxx}^{2} - 7c_{1x}c_{10}^{6}c_{10xxxxxx} - 2520c_{1xx}c_{10}^{2}c_{10x}^{5} - 21c_{1xx}c_{10}^{6}c_{10xxxxx} \\
&+840c_{1xxx}c_{10}^{3}c_{10x}^{4} + 210c_{1xxx}c_{10}^{5}c_{10xx}^{2} - 35c_{1xxx}c_{10}^{6}c_{10xxxx} - 210c_{1xxxx}c_{10}^{4}c_{10x}^{3} \\
&-35c_{1xxxx}c_{10}^{6}c_{10xxx} + 42c_{1xxxxx}c_{10}^{5}c_{10x}^{2} - 21c_{1xxxxx}c_{10}^{6}c_{10xx} + 5040c_{1}c_{10}^{3}c_{10x}^{2}c_{10xx}c_{10xxx} \\
&-630c_{1}c_{10}^{4}c_{10x}c_{10xx}c_{10xxxx} - 2520c_{1x}c_{10}^{4}c_{10x}c_{10xx}c_{10xxx} + \frac{H_t}{H}c_{10}^{7} - c_{10}^{6}c_{10t} - 5040c_{1}c_{10x}^{7} \\
&+15210c_{1}c_{10}c_{10x}^{5}c_{10xx} - 4200c_{1}c_{10}^{2}c_{10x}^{4}c_{10xxx} + 840c_{1}c_{10}^{3}c_{10x}^{3}c_{10xxxx} - 630c_{1}c_{10}^{4}c_{10xx}^{2}c_{10xxx} \\
&-126c_{1}c_{10}^{4}c_{10x}^{2}c_{10xxxxx} + 70c_{1}c_{10}^{5}c_{10xxx}c_{10xxxx} + 42c_{1}c_{10}^{5}c_{10xx}c_{10xxxxx} \\
&+14c_{1}c_{10}^{5}c_{10x}c_{10xxxxxx} - 12600c_{1}c_{10}^{2}c_{10x}^{3}c_{10xx}^{2} + 2520c_{1}c_{10}^{3}c_{10x}c_{10xx}^{3} - 420c_{1}c_{10}^{4}c_{10x}c_{10xxx}^{2} \\
&-12600c_{1x}c_{10}^{2}c_{10x}^{4}c_{10xx} + 7560c_{1x}c_{10}^{3}c_{10x}^{2}c_{10xx}^{2} + 3360c_{1x}c_{10}^{3}c_{10x}^{3}c_{10xxx} \\
&-630c_{1x}c_{10}^{4}c_{10x}^{2}c_{10xxxx} + 210c_{1x}c_{10}^{5}c_{10xx}c_{10xxxx} + 84c_{1x}c_{10}^{5}c_{10x}c_{10xxxxx} \\
&+5040c_{1xx}c_{10}^{3}c_{10x}^{3}c_{10xx} - 1890c_{1xx}c_{10}^{4}c_{10x}c_{10xx}^{2} - 1260c_{1xx}c_{10}^{4}c_{10x}^{2}c_{10xxx} \\
&+420c_{1xx}c_{10}^{5}c_{10xx}c_{10xxx} + 210c_{1xx}c_{10}^{5}c_{10x}c_{10xxxx} - 1260c_{1xxx}c_{10}^{4}c_{10x}^{2}c_{10xx} \\
&+280c_{1xxx}c_{10}^{5}c_{10x}c_{10xxx} + 210c_{1xxxx}c_{10}^{5}c_{10x}c_{10xx} = 0
\end{aligned} \tag{16}
$$

$$
\begin{aligned}
&-240c_{3}c_{10x}^{5} + 720c_{2}c_{10x}^{5} + 120c_{5}c_{10x}^{5} + 40c_{3xxx}c_{10}^{3}c_{10x}^{2} - 20c_{3xxx}c_{10}^{4}c_{10xx} + 6c_{2}c_{10}^{4}c_{10xxxxx} \\
&-180c_{2x}c_{10}^{3}c_{10xx}^{2} + 30c_{2x}c_{10}^{4}c_{10xxxx} + 360c_{2xx}c_{10}^{2}c_{10x}^{3} + 60c_{2xx}c_{10}^{4}c_{10xxx} - 120c_{2xxx}c_{10}^{3}c_{10x}^{2} \\
&+60c_{2xxx}c_{10}^{4}c_{10xx} - 30c_{5x}c_{10}^{3}c_{10xx}^{2} + 5c_{5x}c_{10}^{4}c_{10xxxx} + 60c_{5xx}c_{10}^{2}c_{10x}^{3} + 10c_{5xx}c_{10}^{4}c_{10xxx} \\
&-20c_{5xxx}c_{10}^{3}c_{10x}^{2} + 10c_{5xxx}c_{10}^{4}c_{10xx} + c_{5}c_{10}^{4}c_{10xxxxx} - 120c_{3xx}c_{10}^{2}c_{10x}^{3} - 20c_{3xx}c_{10}^{4}c_{10xxx} \\
&-2c_{3}c_{10}^{4}c_{10xxxxx} + 60c_{3x}c_{10}^{3}c_{10xx}^{2} - 10c_{3x}c_{10}^{4}c_{10xxxx} + 30c_{2xxxx}c_{10}^{4}c_{10x} + 5c_{5xxxx}c_{10}^{4}c_{10x} \\
&-10c_{3xxxx}c_{10}^{4}c_{10x} - 6c_{2xxxxx}c_{10}^{5} + 2c_{3xxxxx}c_{10}^{5} - c_{5xxxxx}c_{10}^{5} - 720c_{2x}c_{10}c_{10x}^{4} - 120c_{5x}c_{10}c_{10x}^{4} \\
&+240c_{3x}c_{10}c_{10x}^{4} + 480c_{3}c_{10}c_{10x}^{3}c_{10xx} - 180c_{3}c_{10}^{2}c_{10x}c_{10xx}^{2} - 120c_{3}c_{10}^{2}c_{10x}^{2}c_{10xxx} \\
&+40c_{3}c_{10}^{3}c_{10xx}c_{10xxx} + 20c_{3}c_{10}^{3}c_{10x}c_{10xxxx} - 360c_{3x}c_{10}^{2}c_{10x}^{2}c_{10xx} + 80c_{3x}c_{10}^{3}c_{10x}c_{10xxx} \\
&+120c_{3xx}c_{10}^{3}c_{10x}c_{10xx} - 1440c_{2}c_{10}c_{10x}^{3}c_{10xx} + 540c_{2}c_{10}^{2}c_{10x}c_{10xx}^{2} + 360c_{2}c_{10}^{2}c_{10x}^{2}c_{10xxx} \\
&-120c_{2}c_{10}^{3}c_{10xx}c_{10xxx} - 60c_{2}c_{10}^{3}c_{10x}c_{10xxxx} + 1080c_{2x}c_{10}^{2}c_{10x}^{2}c_{10xx} - 240c_{2x}c_{10}^{3}c_{10x}c_{10xxx} \\
&-360c_{2xx}c_{10}^{3}c_{10x}c_{10xx} + 180c_{5x}c_{10}^{2}c_{10x}^{2}c_{10xx} - 40c_{5x}c_{10}^{3}c_{10x}c_{10xxx} - 60c_{5xx}c_{10}^{3}c_{10x}c_{10xx} \\
&-240c_{5}c_{10}c_{10x}^{3}c_{10xx} + 90c_{5}c_{10}^{2}c_{10x}c_{10xx}^{2} + 60c_{5}c_{10}^{2}c_{10x}^{2}c_{10xxx} \\
&-20c_{5}c_{10}^{3}c_{10xx}c_{10xxx} - 10c_{5}c_{10}^{3}c_{10x}c_{10xxxx} = 0
\end{aligned} \tag{17}
$$

where $G(t)$ and $H(t)$ are arbitrary functions of t.

3. Generalized vcKdV Equations

Here, we will applying the technique of the last section in exactly the same fashion to generalized vcKdV equations, but will omit the details for the sake of brevity. Please note that *the coefficients* a_i, *as well as the quantities labeled* c_i *in this section are totally distinct or different from those given the same symbols in the previous section. All equations in this section are thus to be read independently of those in the previous one.*

Consider the generalized vcKdV equation in the form

$$
u_t + a_1uu_{xxx} + a_2u_xu_{xx} + a_3u^2u_x + a_4uu_x + a_5u_{xxx} + a_6u_{xxxxx} + a_7u + a_8u_x = 0 \tag{18}
$$

As before, we consider the generalized variable-coefficient KDV equation to be integrable if we can find a Lax pair which satisfies (3). In the method given in [14] one expands the Lax pair $\mathbf{U}$ and $\mathbf{V}$ in powers of u its derivatives with unknown function coefficients and require (3) to be equivalent to the nonlinear system. This results in a system of coupled PDEs for the unknown coefficients for which upon solving we simultaneously obtain the Lax pair and integrability conditions on the a_i. The results, for which the details are similar to the previous section and so are omitted here, are as follows

$$a_{1-4} = \frac{H_{1-4}}{p}, \qquad a_7 = \frac{p_t + (pa_5)_{xxx} + (pa_6)_{xxxxx} + (pa_8)_x}{p} \tag{19}$$

where $H(t)$ is an arbitrary function of t, c_1, c_2 are arbitrary constants and a_1, a_5, a_6, a_8 and $p(\neq 0)$ are taken to be arbitrary functions of x and t. This form helps to give integrability conditions in specific cases. For example an integrable variable-coefficient KDV equation would require that $H_{1-3}(t) = a_{6-8}(x,t) = 0$. For $a_7(x,t) = 0$ we would need to further require that (through a little algebraic manipulation) the following be satisfied

$$\left(\frac{H(t)}{a_1}\right)_t + H(t)\left(\frac{a_5}{a_1}\right)_{xxx} = 0 \tag{20}$$

Note that with the choices $p(x,t) = e^{\int m(t)dt}$, $H_1(t) = H_2(t) = \frac{1}{2\mu_2}H_4(t) = a(t)e^{\int m(t)dt}$, $H_3(t) = \mu_1 e^{2\int m(t)dt}$, $a_5(x,t) = 5\mu_2 a_6(x,t) = \frac{\mu_2 a(t)}{\mu_1 e^{\int m(t)dt}}$ and $a_8(x,t) = n(t)$, we see that $a_7(x,t) = m(t)$ and thus we get back exactly the integrability conditions found in earlier papers on generalized vcKdV systems.

4. Painleve Analysis Method

Given a nonlinear partial differential equation in $(n+1)$-dimensions, without specifying initial or boundary conditions, we may find a solution about a movable singular manifold $\phi - \phi_0 = 0$ as an infinite Laurent series given by

$$u(x_1, \ldots, x_n, t) = \phi^{-\alpha} \sum_{m=0}^{\infty} u_m \phi^m. \tag{21}$$

Note that when $m \in (\mathbb{Q} - \mathbb{Z})$ (21) is more commonly known as a Puiseux series. One can avoid dealing with Puiseux series if proper substitutions are made, as we will see a little later on. Plugging this infinite series into the NLPDE yields a recurrence relation for the u_m's. As with most series-type solution methods for NLPDEs we will seek a solution to our NLPDE as (21) truncated at the constant term. Plugging this truncated series into our original NLPDE and collecting terms in decreasing order of ϕ will give us a set of determining equations for our unknown coefficients $u_0, \ldots, u_\alpha$ known as the Painleve-Backlund equations. We now define new functions

$$C_0(x_0, \ldots, x_n, t) = \frac{\phi_t}{\phi_{x_0}} \tag{22}$$

$$C_1(x_0, \ldots, x_n, t) = \frac{\phi_{x_1}}{\phi_{x_0}} \tag{23}$$

$$\vdots \tag{24}$$

$$C_n(x_0, \ldots, x_n, t) = \frac{\phi_{x_n}}{\phi_{x_0}} \tag{25}$$

$$V(x_0, \ldots, x_n, t) = \frac{\phi_{x_0 x_0}}{\phi_{x_0}} \tag{26}$$

which will allow us to eliminate all derivatives of ϕ other than ϕ_{x_0}. For simplicity it is common to allow $C_i(x_0, \ldots, x_n, t)$ and $V(x_0, \ldots, x_n, t)$ to be constants, thereby reducing a system of PDEs (more than likely nonlinear) in $\{C_i(x,t), V(x,t)\}$ to an algebraic system in $\{C_i, V\}$ for $(i = 0, \ldots, n)$.

4.1. Exactly Solvable Examples

Consider the following example

$$\begin{aligned}&u_t + \frac{H_1}{F(\frac{x}{C}+t)}uu_{xxx} + \left(\frac{c_1H_3 - 2H_1 + 1}{F(\frac{x}{C}+t)}\right)u_xu_{xx} + \frac{H_3}{F(\frac{x}{C}+t)}u^2u_x + \frac{H_4(t)}{F(\frac{x}{C}+t)}uu_x \\ &+ \left(\frac{c_1(c_1H_3(t) + 2H_4 - 2H_1)}{8F(\frac{x}{C}+t)}\right)u_{xxx} + \left(\frac{c_1(c_1H_3 - 2)}{40F(\frac{x}{C}+t)}\right)u_{xxxxx} \\ &+ \left(\frac{(5c_1(H_1 - H_4) - c_1 - 3c_1^2H_3 - 20F(\frac{x}{C}+t)C}{20F(\frac{x}{C}+t)}\right)u_x = 0 \end{aligned} \tag{27}$$

Note that in this example we have $a_7 = 0$. The leading order analysis yields $\alpha = 2$. Therefore we seek a solution of the form

$$u(x,t) = \frac{u_0}{\phi(x,t)^2} + \frac{u_1}{\phi(x,t)} + u_2(x,t) \tag{28}$$

This forms an auto-Backlund transformation. For simplicity we will allow our initial solution $u_2(x,t)$ to be 0. Plugging this into our pde yields the following determining equations for $\phi(x,t), u_0(x,t), u_1(x,t), V$ and C_1 :

$$O(\phi^{-7}) \; : \; 2u_0\phi_x[(9c_1^2H_3 + 18c_1)\phi_x^4 + (6H_3c_1 + 6)u_0\phi_x^2 + H_3u_0^2] = 0 \tag{29}$$

$$\begin{aligned} O(\phi^{-6}) \; : \; & -10H_1u_0u_1\phi_x^3 - 5H_3u_0^2u_1\phi_x + 60c_1u_0\phi_x^3\phi_{xx} + 15c_1^2H_3u_{0x}\phi_x^4 - 3c_1^2H_3u_1\phi_x^5 \\ & +10H_1u_0^2\phi_x\phi_{xx} - 10H_1u_0u_{0x}\phi_x^2 + 30c_1^2H_3u_0\phi_x^3\phi_{xx} + 14c_1H_3u_0u_{0x}\phi_x^2 \\ & +4c_1H_3u_0^2\phi_x\phi_{xx} - 10c_1H_3u_0u_1\phi_x^3 + 30c_1u_{0x}\phi_x^4 - 6c_1u_1\phi_x^4 + 14u_0u_{0x}\phi_x^2 \\ & +4u_0^2\phi_x\phi_{xx} - 10u_0u_1\phi_x^3 + H_3u_0^2u_{0x} = 0 \end{aligned} \tag{30}$$

$$\begin{aligned} O(\phi^{-5}) \; : \; & H_3^2u_0^2u_{1x} - 6H_4u_0\phi_x^3 - 6u_0\phi_x^3(c_1H_3 - 2H_1 + 1) + 3u_0\phi_x^3 \\ & +6H_1u_0\phi_x^3(c_1H_3 - 2H_1 + 1) - 4c_1H_3^2u_0u_1\phi_x\phi_{xx} - 16H_1H_3u_0u_1\phi_x\phi_{xx} \\ & +9c_1^2H_3^2u_0\phi_x\phi_{xx}^2 + 4H_3u_{0x}^2\phi_x + 2H_3u_1^2\phi_x^3 - 6H_1H_3u_{0x}u_1\phi_x^2 \\ & +14H_1H_3u_0u_{1x}\phi_x^2 - 10c_1H_3^2u_0u_{1x}\phi_x^2 - 6c_1H_3^2u_{0x}u_1\phi_x^2 + 36c_1H_3u_{0x}\phi_x^2\phi_{xx} \\ & +12c_1H_3u_0\phi_x^2\phi_{xxx} + 2H_1H_3u_0u_{0x}\phi_{xx} - 4H_3u_0u_1\phi_x\phi_{xx} - 6c_1^2H_3^2u_1\phi_x^3\phi_{xx} \\ & +6c_1^2H_3^2u_0\phi_x^2\phi_{xxx} + 18c_1^2H_3^2u_{0x}\phi_x^2\phi_{xx} + 2c_1H_3^2u_0u_{0x}\phi_{xx} + 2H_1H_3u_0u_{0xx}\phi_x \\ & -12c_1H_3u_1\phi_x^3\phi_{xx} + 2c_1H_3^2u_0u_{0xx}\phi_x + 18c_1H_3u_0\phi_x\phi_{xx}^2 + 2H_1H_3u_1^2\phi_x^3 \\ & +4H_3^2u_0u_1^2\phi_x + 2H_3H_4u_0^2\phi_x - 6c_1H_3u_{1x}\phi_x^4 - 3c_1^2H_3^2u_{1x}\phi_x^4 - 8H_1H_3u_{0x}^2\phi_x \\ & -10H_3u_0u_{1x}\phi_x^2 + 2c_1H_3^2u_1^2\phi_x^3 - 6H_3u_{0x}u_1\phi_x^2 + 2H_3u_{0xx}\phi_x + 4c_1H_3^2u_{0x}^2\phi_x \\ & +6H_4u_0\phi_x^3(c_1H_3 - 2H_1 + 1) + 2H_1H_3u_0^2\phi_{xxx} + 12c_1H_3u_{0xx}\phi_x^3 \\ & +6c_1^2H_3^2u_{0xx}\phi_x^3 - 2H_3^2u_0u_{0x}u_1 + 12H_1H_4u_0\phi_x^3 + 2H_3u_0u_{0x}\phi_{xx} \\ & +3u_0\phi_x^3(c_1H_3 - 2H_1 + 1)^2 - 6H_1u_0\phi_x^3 = 0 \end{aligned} \tag{31}$$

Upon solving the $O(\phi^{-7})$ and $O(\phi^{-6})$ equations for u_0 and u_1 respectively we find that

$$u_0(x,t) = -3c_1\phi_x^2, \qquad u_1(x,t) = 3c_1\phi_{xx} \tag{32}$$

which lends itself nicely to a representation of the solution as $u(x,t) = 3c_1\log[\phi(x,t)]_{xx}$. Further with the choice $V = 1$ the choices for coefficients the remaining orders of ϕ are identically satisfied. Now solving the system for ϕ given in the previous section we find that $\phi(x,t) = c_2 + c_3e^{x+Ct}$. Therefore we have the solution

$$u(x,t) = \frac{3c_1c_2c_3e^{x+Ct}}{(c_2 + c_3e^{x+Ct})^2} \tag{33}$$

which for the selection $c_2 = c_3$ reduces to the solution $u(x,t) = \frac{3c_1}{4}\text{sech}^2\left(\frac{x}{2} + \frac{Ct}{2}\right)$

The next example is similar to the first however in this case we don't have $a_7 = 0$ and we will not force the u_2 term to be the trivial solution. We thus consider the following example

$$\begin{aligned} u_t &+ \frac{10H_1\xi(t)}{F(\int\eta(t)dt + x)}uu_{xxx} + \frac{2(3+2H_1)\xi(t)}{F(\int\eta(t)dt + x)}u_x u_{xx} + \\ &+ \frac{6H_1 - 1}{F(\int\eta(t)dt + x)}u^2 u_x + \frac{10H_4\xi(t)}{F(\int\eta(t)dt + x)}uu_x + \frac{10H_5\xi(t)}{F(\int\eta(t)dt + x)}u_{xxx} + \\ &+ \frac{4(3H_1+2)\xi(t)^2}{5F(\int\eta(t)dt + x)}u_{xxxxx} - \left(\frac{1}{\xi(t)}\right)' F\left(\int\eta(t)dt + x\right)u + \\ &+ \left(H_6(t) + \frac{\xi(t)^2(c_2^4(8H_1 - 3) - 2500c_1(H_4 + 30c_1H_1 - 5c_1))}{5F(\int\eta(t)dt + x)}\right)u_x = 0 \end{aligned} \tag{34}$$

where $\xi(t) = H_5/(10c_1H_1 - 10c_1 + H_4)$ and $H_1(t), H_4(t), H_5(t)$ and $\eta(t)$ are arbitrary functions of t and c_1, c_2 are arbitrary constants. As with our last example the leading order analysis yields $\alpha = 2$. Unlike our last example we will not force the u_2 term to be 0 initially. The first orders of ϕ which determine the u_i are as follows :

$$O(\phi^{-7}): -576(3H_1+2)\xi(t)^2u_0\phi_x^5 - 2(6H_1-1)u_0^3\phi_x - 24H_1u_0^2\phi_x^3 - 24(3+2H_1)\xi(t)u_0^2\phi_x^3 = 0 \tag{35}$$

$$\begin{aligned} O(\phi^{-6}): \quad & 1440H_1H_5^2\phi_x^4u_{0x} - 288H_1H_5^2\phi_x^5u_1 + 600c_1^2H_1^3u_0^2u_{0x} - 1300c_1^2H_1^2u_0^2u_{0x} \\ & +800c_1^2H_1u_0^2u_{0x} + 6H_1H_4^2u_0^2u_{0x} + 500c_1^2\phi_xu_0^2u_1 + 20c_1H_4u_0^2u_{0x} + 5H_4^2\phi_xu_0^2u_1 \\ & -840c_1H_5\phi_x^2u_0u_{0x} + 84H_4H_5\phi_x^2u_0u_{0x} + 600c_1H_5\phi_x^3u_0u_1 - 60H_4H_5\phi_x^3u_0u_1 \\ & -3000c_1^2H_1^3\phi_xu_0^2u_1 + 6500c_1^2H_1^2\phi_xu_0^2u_1 + 120c_1H_1^2H_4u_0^2u_{0x} - 4000c_1^2H_1\phi_xu_0^2u_1 \\ & -140c_1H_1H_4u_0^2u_{0x} - 30H_1H_4^2\phi_xu_0^2u_1 - 100c_1H_4\phi_xu_0^2u_1 - 192H_5^2\phi_x^5u_1 \\ & -100c_1^2u_0^2u_{0x} - H_4^2u_0^2u_{0x} + 960H_5^2\phi_x^4u_{0x} + 1920H_5^2\phi_x^3\phi_{xx}u_0 \\ & +2880H_1H_5^2\phi_x^3\phi_{xx}u_0 - 240c_1H_5\phi_x\phi_{xx}u_0^2 + 24H_4H_5\phi_x\phi_{xx}u_0^2 \\ & +1960c_1H_1^2H_5\phi_x\phi_{xx}u_0^2 - 1720c_1H_1H_5\phi_x\phi_{xx}u_0^2 + 196H_1H_4H_5\phi_x\phi_{xx}u_0^2 \\ & +2360c_1H_1^2H_5\phi_x^2u_0u_{0x} - 1520c_1H_1H_5\phi_x^2u_0u_{0x} + 236H_1H_4H_5\phi_x^2u_0u_{0x} \\ & -3400c_1H_1^2H_5\phi_x^3u_0u_1 + 2800c_1H_1H_5\phi_x^3u_0u_1 - 340H_1H_4H_5\phi_x^3u_0u_1 \\ & -600c_1H_1^2H_4\phi_xu_0^2u_1 + 700c_1H_1H_4\phi_xu_0^2u_1 = 0 \end{aligned} \tag{36}$$

and another equation at $O(\phi^{-5})$ which we omit for reasons of length.

Upon solving the $O(\phi^{-7}), O(\phi^{-6})$ and $O(\phi^{-5})$ equations for u_0, u_1 and u_2 respectively we find that

$$u_0(x,t) = -\frac{12H_5\phi_x^2}{10c_1H_1 - 10c_1 + H_4} = -12\xi(t)\phi_x^2 \tag{37}$$

$$u_1(x,t) = \frac{12H_5\phi_{xx}}{10c_1H_1 - 10c_1 + H_4} = 12\xi(t)\phi_{xx} \tag{38}$$

$$u_2(x,t) = -\frac{(4\phi_x\phi_{xxx} - 50c_1\phi_x^2 - 3\phi_{xx}^2)H_5}{(10c_1H_1 - 10c_1 + H_4)\phi_x^2} = \frac{(4\phi_x\phi_{xxx} - 50c_1\phi_x^2 - 3\phi_{xx}^2)\xi(t)}{\phi_x^2} \tag{39}$$

which similarly lends itself nicely to a representation of the solution as

$$u(x,t) = 12\xi(t)[\log(\phi(x,t))]_{xx} + u_2(x,t)$$

Further, if we let $C(x,t) = B(t)$ and once again $V(x,t) = 1$ the choices for coefficients reduce the remaining orders of ϕ to an identically satisfied system. Solving the determining equations for $\phi(x,t)$ we have that $\phi(x,t) = c_2 + c_3 e^{\int VB(t)dt+Vx}$. Therefore we have the solution

$$u(x,t) = -\frac{\xi(t)\left(c_2^2(V^2-50c_1) - 10c_2c_3(V^2+10c_1)e^{\int VB(t)dt+Vx} + c_3^2(V^2-50c_1)e^{2\int VB(t)dt+2Vx}\right)}{\left(c_2 + c_3 e^{\int VB(t)dt+Vx}\right)^2} \tag{40}$$

5. Conclusions and Future Work

We have used a direct method to obtain very significantly extended Lax- or S-integrable families of generalized KdV and MKdV equations with coefficients which may in general vary in both space and time. Some solutions for the generalized inhomogeneous KdV equations have also been presented here.

Future work will address the derivation of additional solutions by various methods, as well as detailed investigations of other integrability properties of these novel integrable inhomogeneous NLPDEs such as Backlund Transformations and conservation laws.

References

[1] Grimshaw R and Pudjaprasetya S 2004 *Stud. Appl. Math.* **112** 271
[2] Ei G, Grimshaw R and Kamchatov A 2007 *J. Fluid Mech.* **585** 213
[3] Das Sharma G and Sarma M 2000 *Phys. Plasmas* **7** 3964
[4] Fan E G 2002 *Phys. Lett* A **294** 26
[5] Grimshaw R 1979 *Proc. Roy. Soc.* A **368** 359
[6] Joshi N 1987 *Phys. Lett.* A **125** 456
[7] Zhang Y C 2006 *J. Phys.* A **39**14353
[8] Fan E G 2011 *Phys. Lett.* A **375** 493
[9] Hu X and Chen Y 2012 *J. Nonlin. Math. Phys.* **19** 1250002
[10] Tian S F and Zhang H Q 2013 *Stud. Appl. Math.*, doi 10.1111/sapm.12026, 13 pages
[11] Suslov S K (*Preprint* arXiv: 1012.3661v3 [math-ph])
[12] 2011 Sun Z Y, Gao Y T, Liu Y and Yu X 2011 *Phys. Rev* E **84** 026606
[13] He J and Li Y 2010 *Stud. Appl. Math.*, **126** 1
[14] Al Khawaja U 2010 A comparative analysis of Painlevé, Lax Pair, and Similarity Transformation methods in obtaining the integrability conditions of nonlinear Schrodinger equations *J. Math. Phys.* **51** 053506
[15] Ablowitz M J and Segur H 1981 *Solitons and the IST* (Philadelphia: SIAM)
[16] Drazin P G and Johnson R S 1989 *Solitons: an introduction* (Cambridge: Cambridge Univ. Press.)

Asymptotic behavior of a generalized Burgers equation solutions on a finite interval

Alexey Samokhin

Dept. of Math., Moscow State Technical University of Civil Aviation, 20 Kronshtadtsky blvd., Moscow, 125493, Russia

E-mail: samohinalexey@gmail.com

Abstract. The article is concerned with the study of asymptotic behavior of solutions of the Burgers equation and its generalizations with initial value — boundary problem on a finite interval, with constant boundary conditions. Since these equations take a dissipation into account, it is naturally to presuppose that any initial profile will evolve to an invariant time-independent solution with the same boundary values. Yet the answer happens to be slightly more complex. There are three possibilities: the initial profile may regularly decay to an invariant solution; or a Heaviside-type gap develops through a dispersive shock and multi-oscillations; or, exotically, an asymptotic limit is a 'frozen multi-oscillation' piecewise-differentiable solution, composed of different smooth invariant solutions.

1. Introduction

The Burgers equation

$$u_t(x,t) = \varepsilon^2 u_{xx}(x,t) - u(x,t)u_x(x,t). \tag{1}$$

is related to the viscous medium whose oscillations it describes. This viscosity dampens oscillations except for stationary solutions which are invariant for some subalgebra of the full symmetry algebra of the equation. While studying the equation on the whole line only bounded solutions are usually taken into account since only they have a physical meaning. It is not the case for a finite interval as an unbounded solution may still remain bounded within an interval. Thus we obtain a wider choices of invariant solutions and asymptotics and, consequently, some new effects. The generalized Burgers equation here is of the form

$$u_t(x,t) = \varepsilon^2 u_{xx}(x,t) - \alpha u^n(x,t)u_x(x,t). \tag{2}$$

We consider initial value - boundary problem (IVBP) for the Burgers equation on a finite interval:

$$u(x,0) = f(x), \quad u(a,t) = l(t), \quad u(b,t) = r(t), \quad x \in [a,b]. \tag{3}$$

The case of constant boundary conditions $u(a,t) = A, \quad u(b,t) = B$ and the related asymptotics are of a special interest here.

Some of our results are similar to those of Dubrovin *et al* [1, 2, 3] dealing with a formation of dispersive shocks in a class of Hamiltonian dispersive regularizations of the quasi-linear transport equation. For the Burgers equation the shocks resulting in breaks (and preceded by a multi-oscillation) do develops for some IVBPs; some other IVBPs lead to a monotonic convergence to an invariant solutions. One new possibility for the asymptotic profile is a class of frozen multi-oscillating solutions.

This paper is a continuation of [4, 5]. Numeric results are obtained via the Maple *PDETools* package.

2. Solutions and symmetries

The Burgers (1) stationary solutions are:

$$\begin{aligned} u(x,t) &= c, &(4)\\ u(x,t) &= -\varepsilon^2 k\tanh(kx+c), &(5)\\ u(x,t) &= -\varepsilon^2 k\coth(kx+c), &(6)\\ u(x,t) &= \varepsilon k\tan(\frac{kx+c}{\varepsilon}), &(7)\\ u(x,t) &= \frac{k\varepsilon^2}{kx+c}. &(8)\end{aligned}$$

For the generalization (2) stationary solutions are given by

$$\begin{aligned} x &= C_1+\varepsilon^2(n+1)\int\frac{dy}{C_2+\alpha y^{n+1}},\\ y &= C.\end{aligned}$$

These solutions are invariant under the t-translation symmetry.

Consider a simple IVBP for (1) of the form (3)

$$u(x,0)=f(x),\quad u(0,t)=A,\quad u(1,t)=B,\quad A,B\in\mathbb{R}. \tag{9}$$

Taking the dissipation into the account it is naturally to presuppose that at $t\to\infty$ we get $u(x,t)\to y_{AB}(x))$ where $y_{AB}(x))$ is a unique stationary solution corresponding to the ordinary differential problem $y''-2yy'=0,\quad y(0)=A,\quad y(1)=B$.

Such solutions do exist and the first conjecture was that this limit does not depend on the initial profile $f(x)$.

Note that only bounded solutions (they are, incidentally, non-decreasing) are of interest if (1) is considered on the whole line $x\in\mathbb{R}$. But on $x\in[a,b]$ anyone of the above list suits, providing the singularity is not on the interval. The same is true for a more general (2).

3. Stability of invariant solutions

A solution of the (generalized) Burgers equation

$$u_t=u_{xx}-\alpha u^n u_x \tag{10}$$

with zero boundary conditions

$$u(t,a)=u(t,b)=0,\quad u(0,x)|_{[a,b]}=f(x) \tag{11}$$

monotonically tends to zero as $t\to\infty$ in $\mathcal{L}^2$ norm since

$$\frac{\partial}{\partial t}\int_a^b u^2dx=\int_a^b 2uu_tdx=2\int_a^b u(u_{xx}-\alpha u^n u_x)dx \;= \tag{12}$$

$$2\int_a^b udu_x+\left.\frac{-\alpha}{n+1}u^{n+1}\right|_a^b=2uu_x|_a^b-2\int_a^b u_x^2dx=-2\int_a^b u_x^2dx\;\leqslant\;0$$

The greater u_x the faster the convergence.

When the boundary conditions are non-zero but constant

$$u(0,x)|_{[a,b]} = f(x) \quad u(t,a) = f(a) = A, \quad u(t,b) = f(b) = B, \tag{13}$$

one may expect the solution to converge to the respective stationary invariant solution, ie, to $\mu(x)$,

$$\mu_{xx} - \alpha\mu^n\mu_x = 0, \quad \mu(a) = A, \quad \mu(b) = B \tag{14}$$

Such a solution exists and is of one of the above listed forms depending on the combination of A and B.

The answer to this hypothesis is complex.

Let us see how evolves the difference between u and the solution of (14). Denote $\nu(t,x) = u(t,x) - \mu(x)$, ie, $u(t,x) = \nu(t,x) + \mu(x)$. Substituting the latter into (10) we get

$$\begin{aligned} u_t = (\nu(t,x)+\mu(x))_t = \nu(t,x)_t = u_{xx} - \alpha u^n u_x \quad = \\ (\nu(t,x)+\mu(x))_{xx} - \alpha(\nu(t,x)+\mu(x))^n(\nu(t,x)+\mu(x))_x. \end{aligned} \tag{15}$$

In the case $n = 1$ and $\alpha = 2$ it equals $\nu_{xx} - 2\nu\nu_x + [\mu_{xx} - 2\mu\mu_x] - 2\{\nu_x\mu + \nu\mu_x\}$ The expression in square brackets equals zero. So

$$\nu_t = \nu_{xx} - 2\nu\nu_x - 2(\nu\mu)_x. \tag{16}$$

Boundary conditions for ν are zero by definition.

We evaluate the rate of ν by analogy with (12):

$$\begin{aligned} \langle\nu_t\rangle|_{\mathcal{L}^2} = \frac{\partial}{\partial t}\int_a^b \nu^2 dx = \int_a^b 2\nu\nu_t dx = 2\int_a^b \nu(\nu_{xx} - 2\nu\nu_x - 2(\nu\mu)_x)dx \quad = \\ 2\int_a^b \nu d\nu_x - \frac{4}{3}\nu^3\Big|_a^b - 4\int_a^b \nu d(\nu\mu) = 2\nu\nu_x|_a^b - 2\int_a^b \nu_x^2 dx - 4\nu(\nu\mu)|_a^b + 4\int_a^b \mu\nu\nu_x dx \quad = \\ -2\int_a^b \nu_x^2 dx + 2\int_a^b \mu d\nu^2 = -2\int_a^b \nu_x^2 dx + 2\int_a^b \mu_x\nu^2 dx - 2\,\nu^2\mu\Big|_a^b = -2\int_a^b \left(\nu_x^2 + \mu_x\nu^2\right)dx. \end{aligned} \tag{17}$$

Thus the monotony of $\mathcal{L}^2$-convergence is not automatically guaranteed; but it surely takes place, for instance in the case $\mu_x \geqslant 0$ (the case of the increasing initial profile, which agrees with the characteristics method). In the case $n > 1$ the corresponding conditions are somewhat less transparent; for instance when $n = 2$

$$\frac{\partial}{\partial t}\int_a^b \nu^2 dx = -2\int_a^b \left(\nu_x^2 - \mu_x\nu^2(\mu_x - \nu_x)\right)dx.$$

It follows that $\mu_x(\mu_x - \nu_x) > 0$ guarantees decay: if such conditions are satisfied the deviation ν decays to zero. When the inequality $\langle\nu_t\rangle|_{\mathcal{L}^2} \geqslant 0$ fails (eg, for decreasing initial profile) the difference ν doesn't necessarily tend to zero. Usually the evolution ends in catastrophe or decay, but sometimes and somehow it stabilizes half-way.

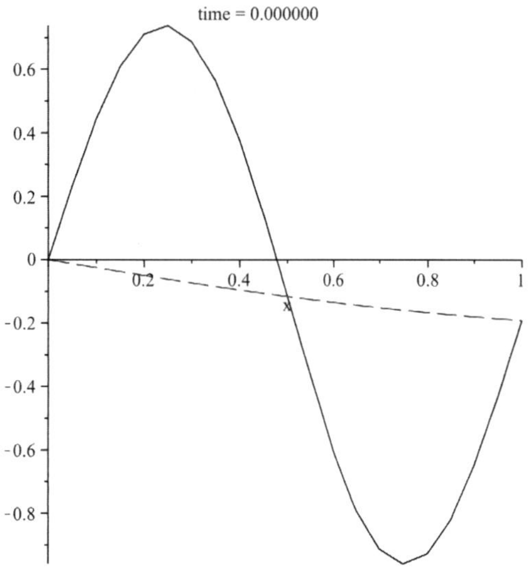

Figure 1. Initial profile $-\varepsilon^2\tanh(x) + 1.6\varepsilon\sin(2\pi x)$, $u(0,t) = 0$, $u(1,t) = -\varepsilon^2\tanh(1)$. Asymptotic limit (dash line) is the invariant solution $-\varepsilon^2\tanh(x)$; n=1

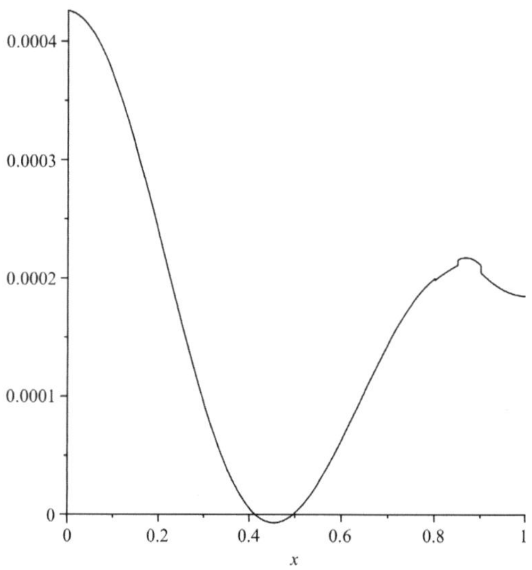

Figure 2. The graph of the integrand $\nu_x^2 + \mu_x\nu^2$ in equation (17), at $t = 2$.

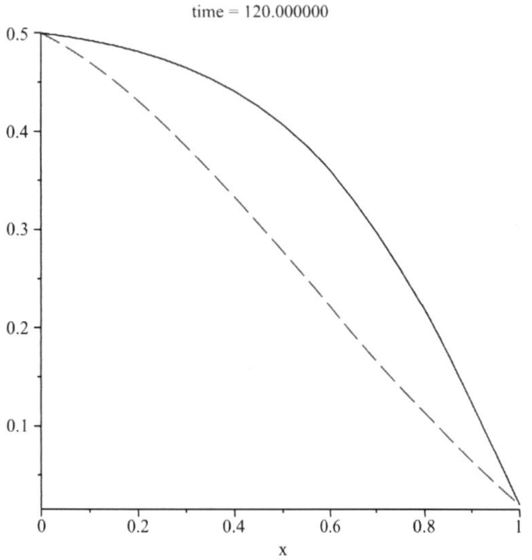

Figure 3. Initial profile $-\varepsilon^2\tanh(x) + \varepsilon((\mathrm{sech}^2(x))$ (dashed) and asymptotic limit $2.06\varepsilon^2\tanh(-2.06x + 2.1)$ (solid line).

3.1. Decay

Here are two examples of a decay towards a *decreasing* invariant solution. In both cases the initial profile is chosen in a vicinity of this solution and the right-hand side of (17) is negative.

Consider the equation $u_t = \varepsilon^2 u_{xx} - 2uu_x$.

(i) Choose IVBP: $u(x,0) = -\varepsilon^2\tanh(x) + 1.6\varepsilon\sin(2\pi x), u(0,t) = 0, u(1,t) = -\varepsilon^2\tanh(1)$; $\varepsilon = 0.05$. Here $\mu = -\varepsilon^2\tanh(x)$ is a decreasing invariant solution, $\nu = 1.6\varepsilon\sin(2\pi x)$ — the perturbation. Asymptotics at $t \to \infty$ coincides with μ, see fig. 1. The dissipation reigns in and no catastrophe develops. The explanation can be seen in fig. 2 where the typical graph of integrand $\nu_x^2 + \mu_x \nu^2$ in (17) is given at $t = 2$; clearly $\langle \nu_t \rangle|_{\mathcal{L}^2} < 0$.

(ii) For the same equation choose another IVBP: $u(1,t) = -\varepsilon^2 \tanh(1) + \varepsilon((\mathrm{sech}^2(1)),$ $u(0,t) = \varepsilon, u(x,0) = -\varepsilon^2 \tanh(x) + \varepsilon((\mathrm{sech}^2(x))$. The initial profile $u(x,0)$ gives an impression of being in vicinity of the invariant solution $-\varepsilon^2 \tanh(x)$ as it is modestly perturbed by $\varepsilon((\mathrm{sech}^2(x))$. In fact it tends to another (decreasing) invariant solution $2.06\varepsilon^2 \tanh(-2.06x + 2.1)$, see fig. 3.

3.2. Catastrophe

As it is known, for a general quasilinear transport equation ($x \in \mathbb{R}$)

$$w_t + f(w)w_x = 0 \tag{18}$$

the moment of gradient catastrophe can be defined as follows. Let $w = \varphi(x)$ be an initial profile. The solution of this problem may be given in a parametric form $w = \varphi(\xi)$, $x = \xi + \mathcal{F}(\xi)t$ where $\mathcal{F} = f(\varphi(\xi))$. The characteristics of the form $x = \xi + \mathcal{F}(\xi)t$ intersect in the case $\varphi'(\xi) < 0$ thus resulting in many-valued w (the tilting of a wave or a gradient catastrophe). If the inequality holds on a finite interval there exist a minimal value of time, t_c, when this problem arises. One may determine t_c by the formula $t_c = -1/\mathcal{F}'(\xi_c)$ where $|\mathcal{F}(\xi_c)| = \max |\mathcal{F}'(\xi)|$ on the interval $[\alpha, \beta]$ while $\mathcal{F}'(\xi) < 0$.

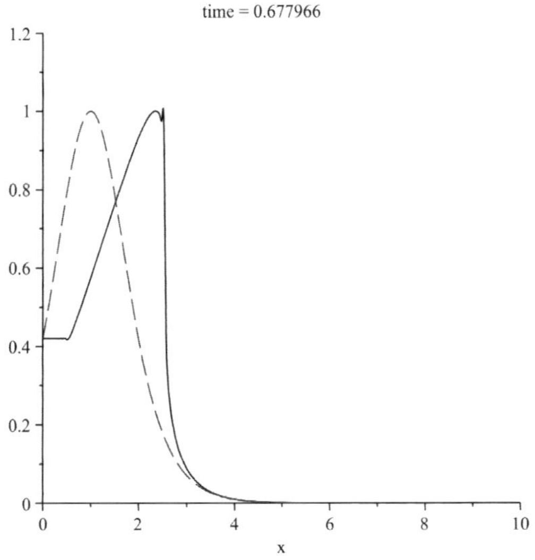

Figure 4. Start of gradient catastrophe at $t_c \approx 0.67$. Dash line is the initial profile $\mathrm{sech}^2(x-1)$. n=1

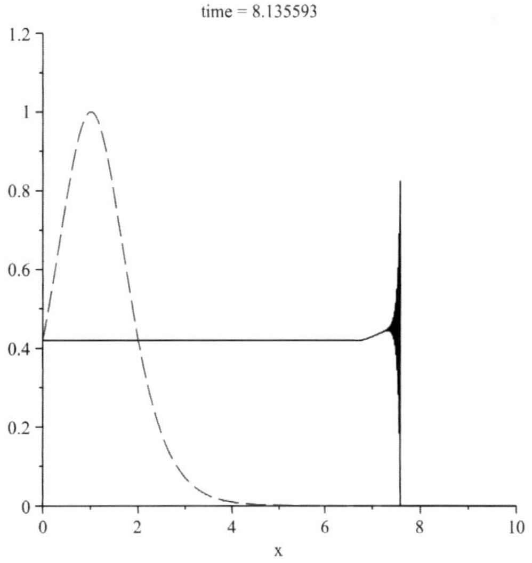

Figure 5. Multi-oscillations move to a Heaviside-type break $\tanh^2(1) - \tanh^2(9)$ at $x = 10$; $t \approx 8$.

We demonstrate this gradient catastrophe to be inherited by Burgers-like equations for some initial profiles, with modest dissipative effects added to a model (18); (cf [2, 1] dealing with a formation of dispersive shocks in a different class of extension of (18), namely Hamiltonian dispersive regularizations of (18) including KdV-likes and Kawahara equations.

In a complex environment of a finite interval combined with an added dissipation for the Burgers-like equation the catastrophe may be delayed or occur earlier, still the main features remain. We begin with the Burgers equation $u_t(x,t) = \varepsilon^2 u_{xx}(x,t) - 2u(x,t)u_x(x,t)$, with IVBP $\{u(x,0) = \mathrm{sech}^2(x-1), u(0,t) = \mathrm{sech}^2(1), u(10,t) = \mathrm{sech}^2(9)\}$ and $\varepsilon = 0.02$. The initial peak

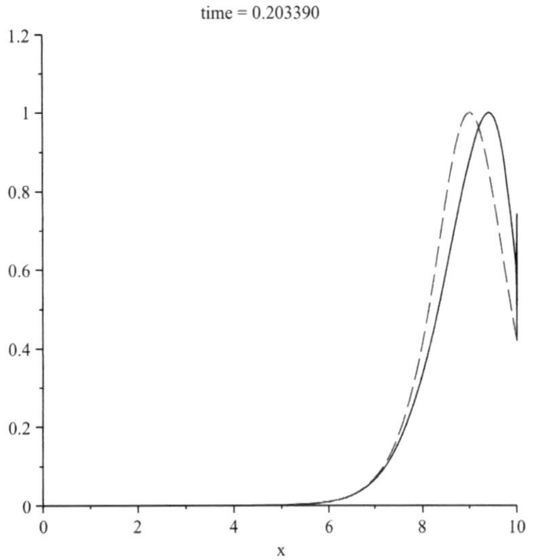

Figure 6. Earlier stage of multi-oscillations at $t_c \approx 0.2$. Dash line is the initial profile $\mathrm{sech}^2(x-9)$. n=1

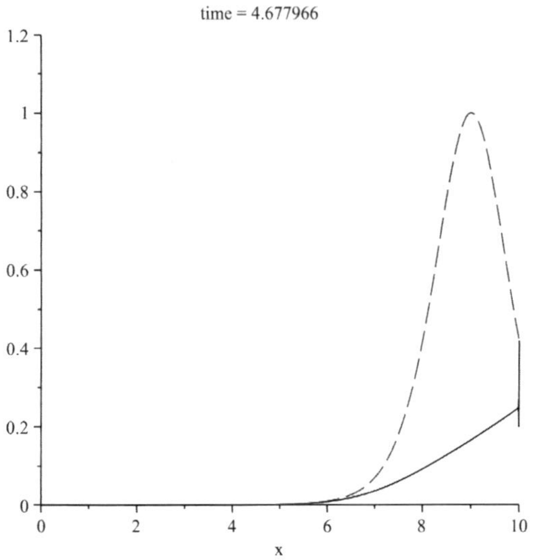

Figure 7. Multi-oscillations move to a Heaviside-type break $-\tanh^2(1)+\tanh^2(9)$ at $x=10$; $t \approx 4.7$.

moves to the right from the far left of the interval, so the moment t_c nearly coincides with that for a whole line. The ensuing multi-oscillating process results in a Heaviside-type break between boundary values at the right end of the interval, fig. 4, 5. Note that constants *are* invariant solutions.

In another example we change IVBP of the previous problem for $\{u(x,0) = \mathrm{sech}^2(x-9), u(0,t) = \mathrm{sech}^2(9), u(10,t) = \mathrm{sech}^2(1)\}$. Here, the right end of the interval being nearer, the catastrophe begins earlier, at $t \approx 0.1$. At $t = 0.2$ we see multi-oscillations fig. 6, gradually developing into a breakup, fig. 7.

Now take the generalized Burgers equation $u_t(x,t) = \varepsilon^2 u_{xx}(x,t) - 2u^2(x,t)u_x(x,t)$ and IVBP $\{u(x,0) = \mathrm{sech}^2(x-1), u(0,t) = \mathrm{sech}^2(1), u(10,t) = \mathrm{sech}^2(9)\}$ and $\varepsilon = 0.02$ as in the first example. The gradient catastrophe starts at $t = 0.45$ in agreement with characteristics method) and likewise develops into a breakup, fig. 8 and 9.

This is not a behavior specific for the sech^2-type initial data. In yet one more example change the IVBP of the previous example for $u(x,0) = -0.01x^2 + 0.9, u(0,t) = 0.9, u(10,t) = -0.1$. The overall picture changes only slightly, fig. 10, though the catastrophe starts at $t = 3.9$, much later than $t = 1,9$ predicted by the characteristics method.

3.3. Stabilization and/or frozen multi-oscillation

In some cases the evolution of the initial profile results early and clearly *not* in an invariant solution from the list (4), see for example fig. 11 with IVBP $\{u(x,0) = -k\varepsilon^2 \tanh(k)(2x^4 - x^2), u(0,t) = 0, u(1,t) = -k\varepsilon^2 \tanh(k)\}$, for $\varepsilon = 0.05$; $k = 50$.

The effect is stable, as the final profile (solid line) in this example does not change if the initial one is perturbated, provided boundary data is the same: identical asymptotics are obtained for $u(x,0) = -k\varepsilon^2 \tanh(k)x$ or $-k\varepsilon^2 \tanh(k)x^2$ (note that the invariant solution with the same boundary values is $\mu(x) = -k\varepsilon^2 \tanh(kx)$). The stabilization may be rather quick. The graph of $\mathcal{L}^2$-estimate for the difference ν, $\int_0^1 (u(s,t) - \mu(s))^2\, ds$ is presented in fig. 13.

The reason for his effect is not entirely clear. Calculus of variations suggests to seek such a

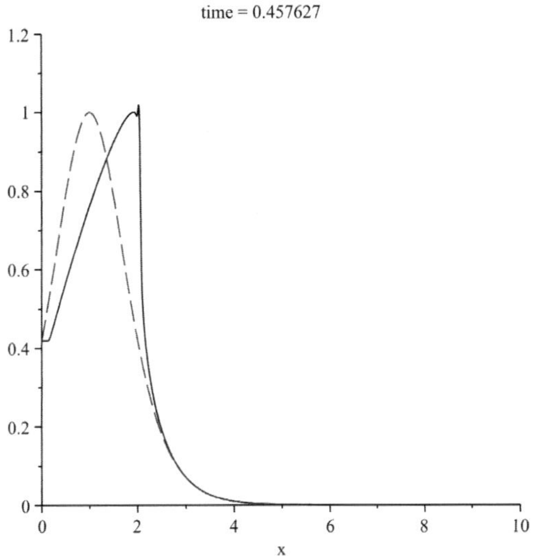

Figure 8. Start of gradient catastrophe at $t_c \approx 0.45$. Dash line is the initial profile $\mathrm{sech}^2(x-1)$. n=2

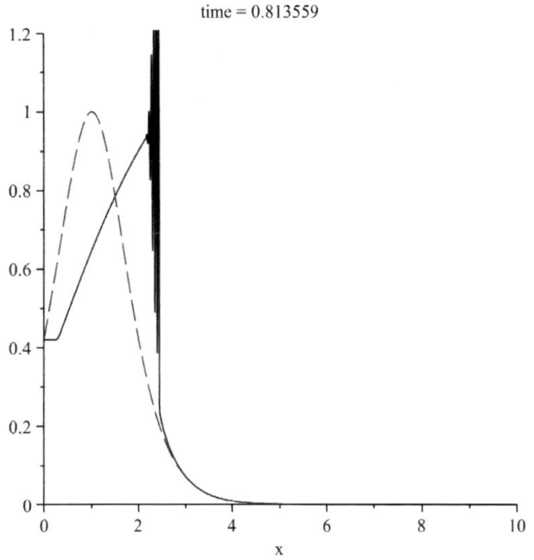

Figure 9. Multi-oscillations move to a Heaviside-type break $-\tanh^2(1)+\tanh^2(9)$ at $x=10$; $t \approx 0.8$.

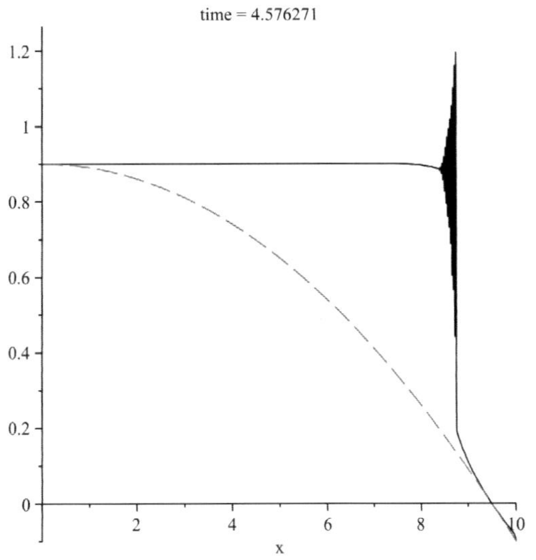

Figure 10. Multi-oscillations move to a Heaviside-type break at $x=10$; $t \approx 4.6$.Dash line is the initial profile $-0.01x^2+0.9$. n=2

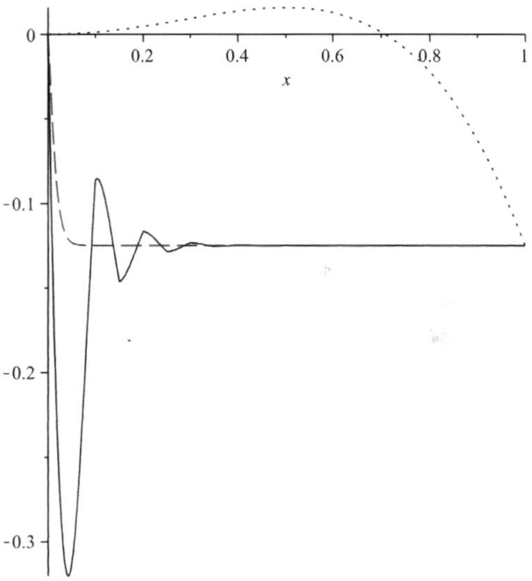

Figure 11. Initial profile (dots line) $-k\varepsilon^2\tanh(k)(2x^4-x^2)$, $u(0,t)=0$, $u(1,t)=-k\varepsilon^2\tanh(k)$. Asymptotic limit (solid) and the invariant solution (dash) $-k\varepsilon^2\tanh(kx)$; $\varepsilon=0.05$, $k=50$. n=1

stationary point as an extremal of the functional (17).

$$\frac{\partial}{\partial\tau}\bigg|_{\tau=0}\int_a^b \left((\nu+\tau h)_x^2-\mu_x(\nu+\tau h)^2\right)dx=0. \tag{19}$$

Journal of Physics: Conference Series **482** (2014) 012039 doi:10.1088/1742-6596/482/1/012039

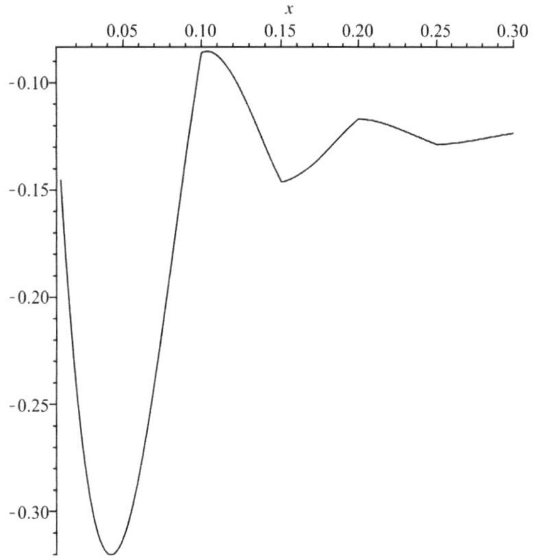

Figure 12. Enlarged part of fig. 11. Piecewise-smooth difference $\nu(x)$; $t = 20$.

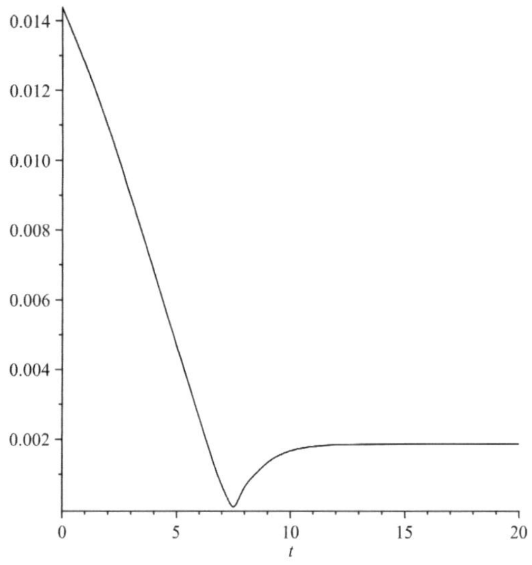

Figure 13. $\mathcal{L}^2$-estimate of difference between invariant $\mu(x) = -\alpha\varepsilon^2 \tanh(\alpha x)$ and stabilizing multi-oscillating solution of fig. 11, 12; n=1.

It follows

$$\nu_{xx} + \mu_x \nu = 0. \tag{20}$$

It is hard to compare the numeric extremal presented on figures 11 to solutions of (20) since these solutions are hard to obtain by numeric methods. The obstacle is that the decreasing solutions of the Burgers equation are of the form $\mu(x) = -k \tanh(kx + c)$ and the potential of the linear problem (20), μ_x, is numerically finite. Some of solutions of (20) are discontinuous (eg, the real part of complex solution of (20) is both discontinuous and multi-oscillating in some cases).

This discontinuity can generate a chaos on the numeric graph and may be a possible reason of a failed smoothness of $\nu(x)$, as presented on figure 12. The graph is composed with parts of different invariant solutions.

The exact mechanism for formations of such exotic solutions is yet to be described in detail; it will be published elsewhere.

References

[1] Dubrovin B, Grava T and Clein C 2011 *Siam J. Appl. Math* **71** 983
[2] Dubrovin B 2006 *Comm. Math. Phys.* **267** 117
[3] Dubrovin B and Elaeva M 2012 *Russian J. Math. Phys.* **19** 449
[4] Samokhin A and Dementyev Y 2009 it Proc. of the Int. Geometry Center (Odessa: ONAFT) vol 2 p 55
[5] Samokhin A 2007 *Sci. Herald of MSTUCA* **140** 29

Journal of Physics: Conference Series **482** (2014) 012040 doi:10.1088/1742-6596/482/1/012040

On singular elliptic equations involving critical Sobolev exponent

K Tahri

Preparatory School in Economics, Business and Management Sciences
Department of Mathematics. B.P 1085 Bouhannak, Tlemcen. Algeria.

E-mail: tahri_kamel@yahoo.fr

Abstract. Given a $n-$dimensional compact Riemannian manifold (M, g) with $n \geq 5$, we consider the following semi-linear elliptic equation :

$$P_g(u) := \Delta_g^2 u + div_g\left(a(x)\nabla_g u\right) + b(x)u = f(x)\left|u\right|^{N-2}u + \lambda h(x)\left|u\right|^{q-2}u$$

where the functions a, b and h are in suitable Lebesgue spaces, $2 < q < N$ and $\lambda > 0$ a real parameter, f is a smooth positive function and the operator P_g is coercive. Under some additional conditions, we obtain results concering the existence of strong solutions of the above equation in $H_2^2(M)$.

1. Introduction

In 1983 Paneitz discovered a particular conformally fourth-order operator defined on $4-$dimensional smooth Riemannian manifolds [1]. In 1987, Branson generalized the definition to higher dimensions in [2] as follows. Let (M, g) be smooth, compact $n-$dimensional Riemannian manifold with $n \geq 5$, and $u \in C^4(M)$. The Paneitz-Branson operator P_g^n is then defined via [2]:

$$P_g^n(u) = \Delta_g^2 u - div_g\left(a_n(x)du\right) + Q_g^n u$$

where

$$a_n(x) = \frac{(n-2)^2+4}{2(n-2)(n-1)}S_g.g - \frac{4}{n-2}Ric_g,$$
$$Q_g^n = \frac{1}{(n-1)(n-4)}\Delta_g S_g + \frac{n^3-4n^2+16n-16}{4(n-1)^2(n-2)^2(n-4)}S_g^2 - \frac{4}{(n-4)(n-2)^2}\left|Ric_g\right|^2,$$

being Δ_g, S_g and Ric_g the Laplace-Beltrami operator, the scalar and the Ricci curvatures of g, respectively. The Paneitz-Branson operator enjoys interesting conformal properties that are very similar to those of the conformal Laplacian operator. Remark that if $\tilde{g} = \varphi^{\frac{4}{n-4}}g$, with φ a positive function of class $C^4(M)$, is a conformal metric to g, then for all $u \in C^4(M)$, $P_g^n(u\varphi) = \varphi^{\frac{n+4}{n-4}}P_{\tilde{g}}^n(u)$. In particular, if $u \equiv 1$ then $P_g^n(\varphi) = Q_{\tilde{g}}^n\varphi^{\frac{n+4}{n-4}}$.

Many interesting results on Paneitz-Branson operator and related topics have been recently obtained by several authors, we refer the reader to Refs. [3]- [10]. Here we recall a few of these results that are pertinent to our investigation, see the list P1)-P3) below.

Let (M, g) be an $n-$dimensional compact, smooth and oriented Riemannian manifold with $n \geq 5$, $H_2^2(M)$ be the standard Sobolev space consisting of function in $L^2(M)$ whose derivatives up to the second order are in $L^2(M)$, and let $N = \frac{2n}{n-4}$ be the associated Sobolev critical exponent. Now, we define the best constant $K_\circ$ of the embedding $H_2^2(\mathbb{R}^n) \subset L^N(\mathbb{R}^n)$ given by

$$\frac{1}{K_o} = \frac{n(n^2-4)(n-4)\omega_n^{\frac{4}{n}}}{16}$$

where ω_n is the volume of the unit Euclidean $n-$sphere (S^n, h).

P1) In 2002, F. Robert and P. Esposito in [10] considered the following equation

$$\Delta_g^2 u + div_g(a(x)\nabla_g u) + b(x)u = f(x)|u|^{N-2} u + h(x)|u|^{q-2} u$$

where: i) $a \in \Lambda_{(2,0)}^{+\infty}(M)$ is a smooth symmetric $(2,0)$-tensor field, ii) b, h, f are smooth functions in M, with f positive, and iii) $2 < q < N$. They established the following remarkable result:

Theorem 1 *Let (M, g) be an $n-$dimensional compact Einsteinian manifold with $n \geq 8$. Assume that P_g^n is coercive and let $f \in C^\infty(M)$, $f > 0$ such that there exists $x_\circ \in M$ with $f(x_\circ) = \max_{x\in M} f(x)$, $\Delta f(x_\circ) = 0$ and*

$$\frac{4(n^2-4n-4)}{3(n+2)}|Weyl_g(x_\circ)|^2 + (n-6)(n-8)\frac{\Delta_g^2 f(x_\circ)}{f(x_\circ)} + 2(n-6)(n-8)\frac{\langle\nabla_g f(x_\circ), Ric_g(x_\circ)\rangle}{f(x_\circ)} > 0.$$

Then, there exists $\tilde{g}$ conformal to g such that $Q_{\tilde{g}}^n(x) = f(x)$.

P2) In 2010, M. Benalili in [5] considered the equation:

$$\Delta_g^2 u + div_g(a(x)\nabla_g u) + b(x)u = f(x)|u|^{N-2} u \tag{1}$$

where f is a positive $C^\infty-$function on M, $a \in L^r(M)$ and $b \in L^s(M)$, with $r > \frac{n}{2}$, $s > \frac{n}{4}$. He established the following result:

Theorem 2 *Let (M, g) is an $n-$dimensional compact manifold with $n \geq 8$ and for $2 < p < 5$, $\frac{9}{4} < s < 11$ or $n = 7$, $\frac{7}{2} < p < 9$, $\frac{7}{4} < s < 9$ assume that there exists $x_\circ \in M$ such that $f(x_\circ) = \max_{x\in M} f(x)$ and*

$$\frac{n^2+4n-20}{6(n-6)(n^2-4)}S_g(x_\circ) + \frac{(n-4)}{2n(n-2)}\frac{\Delta_g f(x_\circ)}{f(x_\circ)} > 0.$$

For $n = 6$, $\frac{3}{2} < p < 2$, $3 < s < 4$, assume that $S_g(x_\circ) > 0$. Then, the equation (1) has a weak solution in $H_2^2(M)$.

P3) Recently, M. Benalili and the author proved in [7], the following result:

Theorem 3 *Let (M, g) be a compact manifold of dimension $n \geq 6$, $a \in L^r(M)$, $b \in L^s(M)$, with $r > \frac{n}{2}$, $s > \frac{n}{4}$, $0 < q < 2$ and f a positive $C^\infty-$function on M. We suppose that P_g is coercive and the existence of a point $x_\circ \in M$ such $f(x_\circ) = \max_{x\in M} f(x)$ and*

$$\begin{cases} \frac{\Delta_g f(x_\circ)}{f(x_\circ)} < \frac{1}{3}\left(\frac{(n-1)n(n^2+4n-20)}{(n-6)(n-4)(n^2-4)}\left(1+\|a\|_r+\|b\|_s\right)^{-\frac{4}{n}} - \right) S_g(x_\circ) & \text{if } n > 6 \\ S_g(x_\circ) > 0 & \text{if } n = 6 \end{cases}$$

Then, there exists $\lambda^* > 0$ such that for any $\lambda \in (0, \lambda^*)$, the equation

$$\Delta_g^2 u + div_g\left(a(x)\nabla_g u\right) + b(x)u = f(x)\left|u\right|^{N-2} u + \lambda \left|u\right|^{q-2} u$$

has a weak solution in $H_2^2(M)$.

In this paper, we look for solutions to the following semi-linear elliptic equation:

$$P_g(u) := \Delta_g^2 u + div_g\left(a(x)\nabla_g u\right) + b(x)u = f(x)\left|u\right|^{N-2} u + \lambda h(x)\left|u\right|^{q-2} u \qquad (2)$$

where $a \in L^r(M)$, $b \in L^s(M)$ and $h \in L^d(M)$, with $r > \frac{n}{2}$, $s > \frac{n}{4}$ and $d > \frac{N}{N-q} := \alpha$, $2 < q < N$ and $\lambda > 0$ a real parameter. In doing so, we assume the following conditions:

(h^1) The operator P_g is coercive, that is: $\exists \Lambda > 0 : \langle P_g(u); u\rangle \geq \Lambda \left\|u\right\|^2_{H_2^2(M)}, \forall u \in H_2^2(M)$;
(h^2) The function h doesn't vanish almost everywhere on M.

(h^3) The parameter λ fulfills $0 < \lambda < \lambda_1$ with

$$\lambda_1 := \frac{q\,(N-2)}{2\,(N-q)}\Lambda^{\frac{q}{2}}(\max((1+\varepsilon)K_\circ, A_\varepsilon))^{-\frac{q}{2}}\left\|h\right\|_\alpha^{-1}.$$

Our main results state as follows:

Theorem 4 *Let (M, g) be an $n-$dimensional compact, smooth and oriented Riemannian manifold with $n > 6$ and f a smooth positive function on M. Let $a \in L^r(M)$, $b \in L^s(M)$ and $h \in L^d(M)$, with $r > \frac{n}{2}$, $s > \frac{n}{4}$, $d > \frac{N}{N-q}$ and $2 < q < N$. We assume that the conditions (h^1), (h^2) and (h^3) are satisfied and that there exists $x_\circ \in M$ such that $f(x_o) = \max_{x\in M} f(x)$ and*

$$\left(\frac{n(n-2\sqrt{6}+2)(n+2\sqrt{6}+2)-(n-6)(n-4)^3(n+2)}{3(n+2)(n-4)^2(n-6)\left(1+\left\|a\right\|_r+\left\|b\right\|_s\right)^{\frac{4}{n}}}S_g(x_\circ) - \frac{(n-4)\Delta f(x_\circ)}{2f(x_\circ)}\right) > 0.$$

Then, the equation (2) possesses a nontrivial solution in $H_2^2(M)$.

Theorem 5 *Let (M, g) be a compact, smooth and oriented Riemannian manifold of dimension $n = 6$ under the same conditions of theorem 4 with*

$$S_g(x_\circ) > 0$$

Then, the equation (2) possesses a nontrivial solution in $H_2^2(M)$.

2. Generic existence result

Throughout this section, we consider the energy functional J_λ, for each $u \in H_2^2(M)$,

$$J_\lambda(u) = \frac{1}{2}\int_M \left((\Delta_g u)^2 - a(x)\left|\nabla_g u\right|^2 + b(x)u^2\right)dv(g) - \frac{\lambda}{q}\int_M h(x)\left|u\right|^q dv(g) - \frac{1}{N}\int_M f(x)\left|u\right|^N dv(g)$$

First, we have the following lemma, whose proof is easy and can be found in [7].

Lemma 6 $\|u\| = (\int_M \left((\Delta_g u)^2 - a(x)\left|\nabla_g u\right|^2 + b(x)u^2\right)dv(g))^{\frac{1}{2}}$ *is an equivalent norm of the usual one of $H_2^2(M)$ if only if the operator P_g is coercive.*

The main tool to prove our result is the Mountain-Pass lemma of Ambrossetti-Rabinowitz given by the following lemma:

Lemma 7 *Let $J \in C^1(E, \mathbb{R})$ where $(E, \|.\|)$ is a Banach space. We assume that:*
(i) $J(0) = 0$.
(ii) $\exists r, R > 0$ such that $J(u) \geq R > 0$ for all $u \in E$ such that $\|u\| = r$.
(iii) $\exists v \in E$ such that $\limsup_{t \longrightarrow +\infty} J(tv) < 0$.

If

$$c = \min_{\eta \in \Gamma} \max_{t \in [0,1]} (J(\eta(t))) \qquad \text{where} \quad \Gamma = \left\{ \eta \in C^1([0;1]; E) : \eta(0) = 0, \eta(1) = v \right\}$$

then there exists a sequence $(u_n)_n$ in E such that:

$$J(u_n) \longrightarrow c \qquad \text{and} \qquad \nabla J(u_n) \longrightarrow 0 \qquad \text{in } E^*$$

where E^* is the dual space of E. Moreover, we have that: $c \leq \sup_{t \geq 0} J(tv)$.

It is easily seen that J_λ is a C^1 functional and its Fréchet derivative is given by:

$$\begin{aligned} \langle \nabla J_\lambda(u), v \rangle = & \int_M (\Delta_g u.\Delta_g v - a(x) g(\nabla_g u, \nabla_g v) + b(x) uv)\, dv(g) \quad + \\ & -\lambda \int_M h(x) |u|^{q-2} uv dv(g) - \int_M f(x) |u|^{N-2} uv dv(g). \end{aligned}$$

Moreover, the functional J_λ verifies the Mountain-Pass conditions, namely:

Lemma 8 *Suppose that the conditions of (h^1), (h^2) and (h^3) of section 1 are satisfied. Then J_λ fulfills the following properties*
1-There exist constants $r, R > 0$ such that $J_\lambda(u) \geq R > 0$, $\|u\| = r$.
2-There exists $v \in H_2^2(M)$, with $\|v\| > r$, such that $J_\lambda(v) < 0$.

Lemma 9 *Let (M, g) be a $n-$dimensional compact, smooth and oriented Riemannian manifold with $n \geq 5$ and suppose that conditions (h^1)- (h^2) are satisfied. Then each Palais-Smale sequence at level c_λ is bounded in $H_2^2(M)$.*

Proof. The proof follows from the coerciveness of the operator P_g, the Sobolev's inequality and the condition (h^2). ∎

Theorem 10 *Let (M, g) is an $n-$dimensional compact, smooth and oriented Riemannian manifold with $n \geq 5$. Let $(u_m)_m$ be a Palais-Smale sequence at level c_λ. Assume that conditions (h^1)-(h^2) and (h^3) are satisfied and that*

$$c_\lambda < \frac{1}{(1+\varepsilon)^{\frac{n}{n-4}} K_\circ^{\frac{n}{n-4}} \max_{x \in M} f(x)}.$$

Then, there is a subsequence of $(u_m)_m$ converging strongly in $H_2^2(M)$.

Proof. We follows closely the method used in [7]. ∎

3. The sharp case

Let $P \in M$, we define the distance function ρ on M by

$$\rho_P(Q) = \begin{cases} d(P,Q) & \text{if} \quad d(P,Q) < i_g(M) \\ \delta(M) & \text{if} \quad d(P,Q) \geq i_g(M) \end{cases}$$

and $i_g(M)$ is the injectivity radius of M. Furthermore, we define the space $L^p(M, \rho^\gamma)$ as follows.

Definition 11 *Let (M,g) be a compact $5 \leq n$-dimensional Riemannian manifold. We consider the space $L^p(M,\rho^\gamma)$ where $1 \leq p \leq +\infty$ of measurable functions u on M such that $\rho^\gamma |u|^p$ is integrable, i.e.*

$$\|u\|^p_{p,\rho^\gamma} := \int_M \rho^\gamma |u|^p \, dv(g) < +\infty$$

Now, we use the following Hardy-Sobolev inequalities proven in [5] (the Hardy-Sobolev inequalities for the singular Yamabe equation was proven in [9]).

Theorem 12 *[5] Let (M,g) be a compact $5 \leq n$-dimensional Riemannian manifold and p, q and γ three real numbers satisfying $\frac{\gamma}{p} = \frac{n}{q} - \frac{n}{p} - 2$ and $2 \leq p \leq \frac{2n}{n-4}$.*
For any $\epsilon > 0$, there is a constant $A(\epsilon, q, \gamma)$ such that

$$\forall u \in H_2^2(M) : \|u\|^2_{p,\rho^\gamma} \leq (1+\epsilon) K(n,2,\gamma)^2 \|\Delta_g u\|_2^2 + A(\epsilon,q,\gamma)\|u\|_2^2$$

In particular: $K(n,2,0)^2 = K_\circ$ is the optimal constant of Sobolev inequality.

Theorem 13 *[5] Let (M,g) be a compact $5 \leq n$dimensional Riemannian manifold and p, q and γ three real numbers satisfying: $1 \leq q \leq p \leq \frac{nq}{n-2q}$ and $\gamma < 0$.*
•.If $\frac{\gamma}{p} = n(\frac{1}{q} - \frac{1}{p}) - 2$, then the imbedding $H_2^q(M) \subset L^p(M,\rho^\gamma)$ is continuous.
•.If $\frac{\gamma}{p} > n(\frac{1}{q} - \frac{1}{p}) - 2$, then the imbedding $H_2^q(M) \subset L^p(M,\rho^\gamma)$ is compact.

We consider the following equation:

$$\Delta_g^2 u + div_g\left(\frac{a(x)}{\rho^\sigma}\nabla_g u\right) + \frac{b(x)}{\rho^\mu}u = f(x)|u|^{N-2}u + \lambda\frac{h(x)}{\rho^\beta}|u|^{q-2}u \tag{3}$$

where a, b and h are three smooth functions and the distance function defined before in section 1, $2 < q < N$ and $\lambda > 0$ a real parameter. The energy functional J_λ: $H_2^2(M) \longrightarrow \mathbb{R}$ associated to equation (3) is defined as:

$$J_\lambda(u) = \frac{1}{2}\int_M \left((\Delta_g u)^2 - \frac{a(x)}{\rho^\sigma}|\nabla_g u|^2 + \frac{b(x)}{\rho^\mu}u^2\right) dv(g) + \\ -\frac{\lambda}{q}\int_M \frac{h(x)}{\rho^\beta}|u|^q dv(g) - \frac{1}{N}\int_M f(x)|u|^N dv(g),$$

where $u \in H_2^2(M)$ and it is well-known that the critical points of J_λ are the weak solutions of (3).

Theorem 14 *Let $0 < \sigma < \frac{n}{r} < 2$, $0 < \mu < \frac{n}{s} < 4$ and $0 < \beta < \frac{N}{d} < N - q$. We suppose that the conditions (h^1), (h^2) and (h^3) are satisfied and*

$$\sup_{u \in H_2^2(M)} J_\lambda^{\sigma,\mu,\beta}(u) < \frac{2}{nK_\circ^{\frac{n}{4}}(f(x_\circ))^{\frac{n-4}{4}}}$$

Then, the equation (3) has a non trivial solution $u_{\sigma,\mu,\beta} \in H_2^2(M)$.

Proof. The result follows in that if we put $\tilde{a} = \frac{a(x)}{\rho^\sigma}$, $\tilde{b}(x) = \frac{b(x)}{\rho^\mu}$ and $\tilde{h}(x) = \frac{h(x)}{\rho^\beta}$, then $\tilde{a} \in L^r(M)$, $\tilde{b} \in L^s(M)$ and $\tilde{h} \in L^d(M)$, with $r > \frac{n}{2}$, $s > \frac{n}{4}$ and $d > \frac{N}{N-q}$. ■

4. Critical cases

Strategies developed in [7] and [8] enable us to derive another result, that refers to the critical cases when $\sigma = 2$, $\mu = 4$, and $\beta = \frac{n(q-2)}{2} - 2q$.

Theorem 15 *Let (M, g) be an $n-$dimensional compact, smooth and oriented Riemannian manifold with $n \geq 5$ and suppose that the conditions (h^1), (h^2) and (h^3) are satisfied. In addition, let $(u_m)_m := (u_{\sigma_m,\mu_m,\beta_m})_m$ be a sequence in $H_2^2(M)$ such that:*

$$\begin{cases} J_\lambda^{\sigma,\mu,\beta}(u_m) \to c_\lambda^{\sigma,\mu,\beta} & \text{for all} \quad n \in \mathbb{N} \\ \nabla J_\lambda^{\sigma,\mu,\beta}(u_m) \to 0 & \text{weakly in} \quad H_2^2(M) \end{cases} \quad \text{with } c_\lambda^{\sigma,\mu,\beta} < \frac{2}{nK_\circ^{\frac{n}{4}}(f(x_\circ))^{\frac{n-4}{4}}} \tag{4}$$

and

$$1 + a^- \max(K(n,2,\sigma); \quad A(\epsilon,\sigma)) + b^- \max(K(n,2,\mu); \quad A(\epsilon,\mu)) > 0. \tag{5}$$

Then, the equation

$$\Delta_g^2 u + div_g\left(\frac{a(x)}{\rho^\sigma}\nabla_g u\right) + \frac{b(x)}{\rho^\mu}u = f(x)\,|u|^{N-2}\,u + \lambda\frac{h(x)}{\rho^\beta}\,|u|^{q-2}\,u$$

has a nontrivial solution $u_{\sigma,\mu,\beta} \in H_2^2(M)$.

Proof. We follow closely the method used in [7] and [8]. First by using the condition (5) we obtain, as in [7], that the sequence $(\Lambda_{\alpha,\mu})_{\alpha,\mu}$ of constants of coerciveness of the operator $u \to \Delta_g^2 u + div_g\left(\frac{a(x)}{\rho^\sigma}\nabla_g u\right) + \frac{b(x)}{\rho^\mu}u$ is bounded below by a constant $\Lambda > 0$ as $(\alpha,\mu) \to (2^-, 4^-)$. Let $(u_m)_m \subset H_2^2(M)$, such that :

$$J_\lambda^{\sigma,\mu,\beta}(u_m) = c_\lambda^{\sigma,\mu,\beta} + o(1) \qquad \text{and} \qquad \nabla J_\lambda^{\sigma,\mu,\beta}(u_m) = o(1) \qquad \text{in } \left(H_2^2(M)\right)^*$$

Then we have:

$$J_\lambda^{\sigma,\mu,\beta}(u_m) - \frac{1}{N}\left\langle J_\lambda^{\sigma,\mu,\beta}(u_m), u_m\right\rangle = \left(\frac{1}{2} - \frac{1}{N}\right)\|u_m\|^2 - \lambda\left(\frac{1}{q} - \frac{1}{N}\right)\int_M h(x)\,|u_m|^q\,dv(g)$$

By Hölder and Sobolev inequalities, we get that

$$J_\lambda^{\sigma,\mu,\beta}(u_m) - \frac{1}{N}\left\langle \nabla J_\lambda^{\sigma,\mu,\beta}(u_m), u_m\right\rangle = c_\lambda^{\sigma,\mu,\beta} + o(1)$$

and

$$c_\lambda^{\sigma,\mu,\beta} + o(1) \geq \left(\frac{1}{2} - \frac{1}{N}\right)\|u_m\|^2 - \left(\frac{1}{q} - \frac{1}{N}\right)(\max((1+\varepsilon)K_\circ, A_\varepsilon))^{\frac{q}{2}}\,\|h\|_\alpha\,\|u_m\|^q_{H_2^2(M)}$$

In addition the hypothesis (h^1) and (h^2) are satisfied and if we have $\|u_n\| \geq 1$, then we obtain

$$\|u_m\| \leq \left[\left(\frac{N-2}{2} - \lambda\frac{N-q}{q}\Lambda^{-\frac{q}{2}}(\max((1+\varepsilon)K_\circ, A_\varepsilon))^{\frac{q}{2}}\,\|h\|_\alpha\right)^{-1} N c_\lambda^{\sigma,\mu,\beta}\right]^{\frac{1}{q}} + o(1)$$

Then $(u_m)_m$ is bounded in $H_2^2(M)$. The rest of the proof is the same as in Theorem 10. ∎

Concluding remark. To prove main Theorems given in the Introduction, let $\delta \in \left(0, \frac{i_g(M)}{2}\right)$ and $\eta \in C^\infty(M)$ such that:

$$\eta(x) = \begin{cases} 1 & \text{if} \quad x \in B(x_\circ, \delta) \\ 0 & \text{if} \quad x \in M - B(x_\circ, 2\delta) \end{cases}$$

For $\epsilon > 0$, we define the radial function u_ϵ by:

$$u_\epsilon(x) := \frac{\eta(x)}{\left(\epsilon^2 + (\xi\rho)^2\right)^{\frac{n-4}{2}}} \qquad \text{with} \quad \xi = \left(1 + \|a\|_r + \|b\|_s\right)^{\frac{1}{n}}. \tag{6}$$

We next point out that, by resorting to the strategy outlined in [7, 8], the function given by (6) can be proved to verify condition (4) of the generic theorem. This step completes our discussion on the solutions of Equation (2).

References

[1] Paneitz S 2008 A quartic conformally covariant differential operator for arbitrary peudo Riemannian manifolds *SIGMA* **4**
[2] Branson T P 1987 Group representation arising from Lorentz conformal geometry *J. Funct. Anal.* **74** 199
[3] Benalili M 2009 Existence and multiplicity of solutions to elliptic equations of fourth order on compact manifolds *Dynamics of PDE*, **3** 203
[4] Benalili M 2010 Existence and multiplicity of solutions to fourth order elliptic equations with critical exponent on compact manifolds *Bull. Belg. Math. Soc. Simon Stevin* **17**
[5] Benalili M 2013 On singular Q-curvature type equations *J. Diff. Eq.* **254** 547
[6] Benalili M and Tahri K 2011 Nonlinear elliptic fourth order equations existence and multiplicity results *Nonlin. Differ. Equ. Appl.* **18** 539
[7] Benalili M and Tahri K 2013 Existence of solutions to singular fourth-order elliptic equations *Electron. J. Diff. Equ.* **2013** No. 63, pp. 1
[8] Benalili M and Tahri K 2012 Multiple solutions to singular fourth order elliptic equations, (preprint: arXiv:1209.3764v1 [math.DG] 16 Sep 2012)
[9] Madani F 2008 Le problème de Yamabé avec singularités, (preprint: ArXix: 1717v1 [mathAP]).
[10] Robert F and Esposito P 2002 Mountain-Pass critical points for Paneitz-Branson operators *Calc. of Variations and Partial Diff. Eq.* **15** 493

Modified melting crystal model and Ablowitz-Ladik hierarchy

Kanehisa Takasaki

Human and Environmental Studies, Kyoto University, Kyoto 606-8501, Japan

E-mail: `takasaki@math.h.kyoto-u.ac.jp`

Abstract. This is a review of recent results on the integrable structure of the ordinary and modified melting crystal models. When deformed by special external potentials, the partition function of the ordinary melting crystal model is known to become essentially a tau function of the 1D Toda hierarchy. In the same sense, the modified model turns out to be related to the Ablowitz-Ladik hierarchy. These facts are explained with the aid of a free fermion system, fermionic expressions of the partition functions, algebraic relations among fermion bilinears and vertex operators, and infinite matrix representations of those operators.

1. Introduction

Recently, we extended our previous work [1, 2] on the integrable structure of the melting crystal model — a model of *random 3D Young diagrams* — to a modified model [3, 4]. In the context of string theory [5], the modified model is related to topological string theory on the so called *resolved conifold*. Our study was motivated by Brini's conjecture [6] that all genus Gromov-Witten invariants of this non-compact complex Calabi-Yau 3-fold can be captured by the Ablowitz-Ladik hierarchy [7]. Brini confirmed his conjecture to the first order (i.e., genus ≤ 1) of genus expansion. Proving the conjecture in the full genus case remained open. We believe that our results show an affirmative answer (or, at least, a decisive piece of evidence) to this question.

In our previous work [1, 2], the partition function of the melting crystal model is shown to be related to a tau function of the 1D Toda hierarchy. This fact is proven with the aid of a free fermion system, a fermionic realization of the *quantum torus algebra*, and algebraic relations called *shift symmetries* in this algebra. We could use these tools for the modified model as well to show that the partition function is related to a tau function of the 2D Toda hierarchy [3]. Although this was an important step, a new technique was wanted to reach the goal.

A final clue was found in the work of Brini et al. [8]. They presented a characterization of the Ablowitz-Ladik hierarchy (or, rather, the relativistic Toda hierarchy [1]) in the Lax formalism of the 2D Toda hierarchy. According to the characterization by Brini et al., the Ablowitz-Ladik hierarchy amounts to the case where the Lax operators of the 2D Toda hierarchy have a special *quotient* (or *rational*) form. We examined the Lax operators of the aforementioned tau function, and confirmed that they do have the special form [4].

[1] These two integrable hierarchies are known to be equivalent [9, 10], but it is the relativistic Toda hierarchy that can be embedded into the 2D Toda hierarchy in a natural manner.

2. Melting crystal model

2.1. Crystal corner, 3D Young diagrams and plane partitions

The melting crystal model is a statistical model of a crystal corner (Figure 1). The crystal consists of unit cubes and occupies an octant of the xyz space. The complement of the crystal in the octant is a *3D Young diagram*, which conversely determines the shape of the crystal (Figure 2).

Just as ordinary Young diagrams are identified with ordinary partitions

$$\lambda = (\lambda_i)_{i=1}^{\infty}, \quad \lambda_1 \geq \lambda_2 \geq \ldots \geq 0, \quad |\lambda| = \sum_{i=1}^{\infty} \lambda_i < \infty,$$

of integers [11], 3D Young diagrams are in one-to-one correspondence with *plane partitions*. Plane partitions are 2D arrays of non-negative integers that are decreasing in two directions:

$$\pi = (\pi_{ij})_{i,j=1}^{\infty} = \begin{pmatrix} \pi_{11} & \pi_{12} & \cdots \\ \pi_{21} & \pi_{22} & \cdots \\ \vdots & \vdots & \ddots \end{pmatrix}, \quad \begin{matrix} \pi_{ij} & \geq & \pi_{i,j+1} \\ \text{IV} & & \\ \pi_{i+1,j} & & \end{matrix}, \quad |\pi| = \sum_{i,j=1}^{\infty} \pi_{ij} < \infty.$$

Such a plane partition determines a 3D Young diagram that consists of stacks of unit cubes of height π_{ij} on the unit squares $[i-1, i] \times [j-1, j]$, $i, j = 1, 2, \ldots$, of the xy plane. For example, the 3D Young diagram of Figure 2 corresponds to

$$\pi = \begin{pmatrix} 3 & 2 & 2 \\ 3 & 2 & 1 \\ 1 & 1 & 1 \end{pmatrix}.$$

The partition function of this model is the sum

$$Z = \sum_{\pi \in \mathcal{PP}} q^{|\pi|} \tag{1}$$

of the Boltzmann weight $q^{|\pi|}$ $(0 < q < 1)$ over the set $\mathcal{PP}$ of all plane partitions.

2.2. Method of diagonal slicing

The partition function (1) can be calculated by the method of *diagonal slicing* [12]. Let us define the m-th diagonal slice $\pi(m)$, $m \in \mathbf{Z}$, of π as

$$\pi(m) = \begin{cases} (\pi_{i,i+m})_{i=1}^{\infty} & \text{if } m \geq 0, \\ (\pi_{j-m,j})_{j=1}^{\infty} & \text{if } m < 0. \end{cases}$$

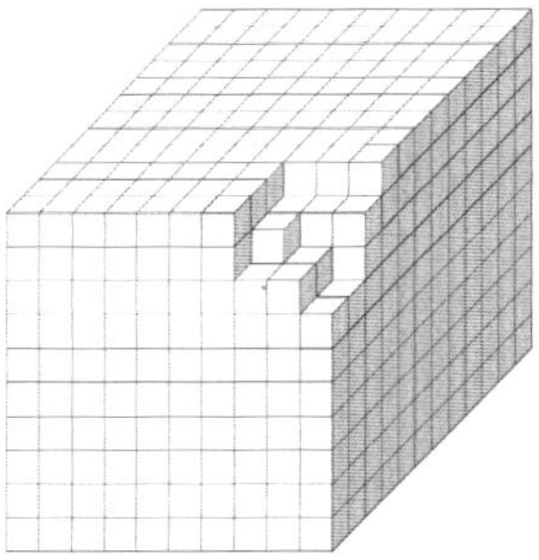

Figure 1. Melting crystal corner

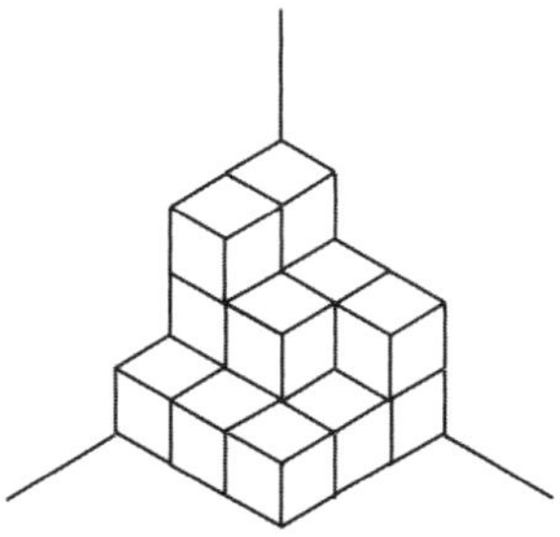

Figure 2. 3D Young diagram

$\pi(m)$ represents the slice of the 3D Young diagram along the vertical plane $y = x + m$ (Figure 3). The two sets $\{\pi(-m)\}_{m=0}^{\infty}$ and $\{\pi(m)\}_{m=0}^{\infty}$ of these slices give two increasing sequences of Young diagrams that fill up the Young diagram of shape $\lambda = \pi(0)$.

These increasing sequences of Young diagrams can be encoded to *semi-standard tableaux* T, T' of shape λ [11]. These tableaux are obtained by entering the numbers $m+1$, $m = 0, 1, 2, \ldots$, to the boxes of the skew Young diagrams $\pi(-m)/\pi(-m-1)$ and $\pi(m)/\pi(m+1)$ (Figure 4). It is easy to see that the entries $T(i,j)$ and $T'(i,j)$ of T, T' satisfy the conditions

$$\begin{array}{ccc} T(i,j) & \geq & T(i,j+1) \\ \vee & & \\ T(i+1,j) & & \end{array}, \qquad \begin{array}{ccc} T'(i,j) & \geq & T'(i,j+1) \\ \vee & & \\ T'(i+1,j) & & \end{array} \tag{2}$$

that characterize semi-standard tableaux [2]. Thus any plane partition π determines a triple (λ, T, T') of a partition λ and semi-standard tableaux T, T' of shape λ. It is also easy to see that any such triple determine a plane partition. The mapping $\pi \to (\lambda, T, T')$ thus turns out to be a bijection between $\mathcal{PP}$ and the set of all triples.

The sum in (1) can be thus converted to a sum over (λ, T, T')'s. Moreover, the weight $q^{|\pi|}$ can be factorized as

$$q^{|\pi|} = q^T q^{T'}, \quad q^T = \prod_{m=0}^{\infty} q^{(m+1/2)|\pi(-m)/\pi(-m-1)|}, \quad q^{T'} = \prod_{m=0}^{\infty} q^{(m+1/2)|\pi(m)/\pi(m+1)|}. \tag{3}$$

Because of this factorization, the sum in (1) can be separated to partial sums with respect to T, T' and a sum with respect to λ as

$$Z = \sum_{\lambda \in \mathcal{P}} \left(\sum_{T\text{: shape } \lambda} q^T \right) \left(\sum_{T'\text{: shape } \lambda} q^{T'} \right),$$

where $\mathcal{P}$ denotes the set of all partitions. According to the combinatorial definition

$$s_\lambda(\boldsymbol{x}) = \sum_{T\text{: shape } \lambda} \boldsymbol{x}^T, \quad \boldsymbol{x}^T = \prod_{(i,j)} x_{T(i,j)}, \tag{4}$$

of Schur functions of $\boldsymbol{x} = (x_1, x_2, \ldots)$ [11], the partial sums over T, T' become a special value of $s_\lambda(\boldsymbol{x})$:

$$\sum_{T\text{: shape } \lambda} q^T = \sum_{T'\text{: shape } \lambda} q^{T'} = s_\lambda(q^{-\rho}), \quad q^{-\rho} = \left(q^{1/2}, q^{3/2}, \ldots, q^{k+1/2}, \ldots\right). \tag{5}$$

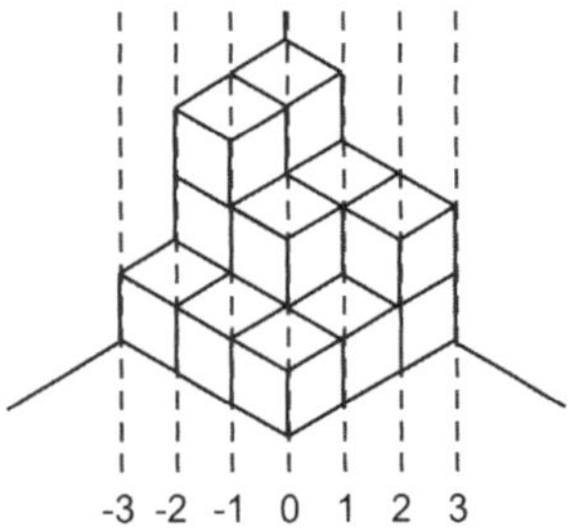

Figure 3. Diagonal slices

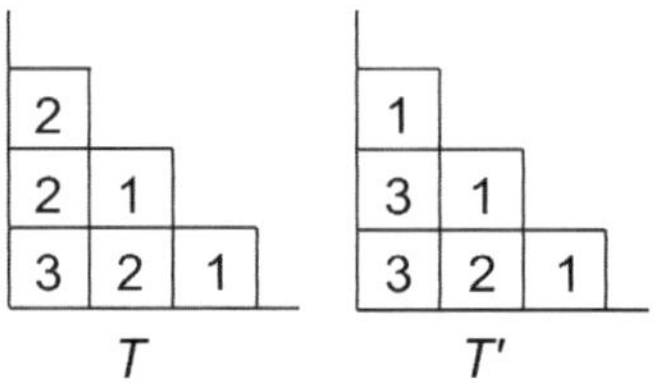

Figure 4. Semi-standard tableaux

[2] This is different from the common definition of semi-standard tableaux [11] in which all inequalities of (2) are reversed. Since Schur functions are symmetric functions, this difference does not affect the combinatorial definition (4) of Schur functions.

The partition function can be thus reduced to a sum over $\mathcal{P}$:

$$Z = \sum_{\lambda\in\mathcal{P}} s_\lambda(q^{-\rho})^2. \tag{6}$$

One can now use the Cauchy identity [11]

$$\sum_{\lambda\in\mathcal{P}} s_\lambda(x_1,x_2,\ldots)s_\lambda(y_1,y_2,\ldots) = \prod_{i,j=1}^{\infty}(1-x_iy_j)^{-1} \tag{7}$$

to calculate (6) explicitly:

$$Z = \sum_{\lambda\in\mathcal{P}} s_\lambda(q^{-\rho})^2 = \prod_{i,j=1}^{\infty}(1-q^{i+j-1})^{-1} = \prod_{n=1}^{\infty}(1-q^n)^{-n}. \tag{8}$$

This function is known as the *MacMahon function*.

2.3. Deformations by external potentials

A simplest deformation of this model is obtained by inserting the extra weight $Q^{|\pi(0)|}$, where Q is a new positive constant. The foregoing partition function (1) is thereby deformed as

$$Z = \sum_{\pi\in\mathcal{PP}} q^{|\pi|}Q^{|\pi(0)|}. \tag{9}$$

This is a deformation induced by the external potential $|\pi(0)|\log Q$. By the mapping $\pi \mapsto (\lambda, T, T')$, this sum, too, can be reduced to a sum of the form

$$Z = \sum_{\lambda\in\mathcal{P}} s_\lambda(q^{-\rho})^2 Q^{|\lambda|}, \tag{10}$$

and eventually boils down to a generalization of the MacMahon function:

$$Z = \prod_{i,j=1}^{\infty}(1-Qq^{i+j-1})^{-1} = \sum_{n=1}^{\infty}(1-Qq^n)^{-n}. \tag{11}$$

An integrable hierarchy emerges in a deformed partition function $Z(s,\boldsymbol{t})$ that depends on a discrete variable $s\in\mathbf{Z}$ and a set of continuous variables $\boldsymbol{t}=(t_1,t_2,\ldots)$. The deformations are induced by the external potential

$$\Phi(\lambda,s,\boldsymbol{t}) = \sum_{k=1}^{\infty} t_k\Phi_k(\lambda,s), \quad \Phi_k(\lambda,s) = \sum_{i=1}^{\infty}(q^{k(\lambda_i+s-i+1)} - q^{k(s-i+1)}) + \frac{1-q^{ks}}{1-q^k}q^k. \tag{12}$$

Note that the sum in the definition of $\Phi_k(\lambda,s)$ is actually a finite sum. This expression is an analytic continuation of the expression

$$\Phi_k(\lambda,s) = \sum_{i=1}^{\infty} q^{k(\lambda_i+s-i+1)} - \sum_{i=1}^{\infty} q^{k(-i+1)} \tag{13}$$

in the domain $|q|>1$ of the q-plane. The term $Q^{|\lambda|}$ in (10), too, has to be replaced by the s-dependent form $Q^{|\lambda|+s(s+1)/2}$. With the aid of 2D complex free fermion system (which will be briefly reviewed in the next section), we obtained the following result [1, 2]:

Theorem 1. *The deformed partition function*

$$Z(s,\boldsymbol{t}) = \sum_{\lambda\in\mathcal{P}} s_\lambda(q^{-\rho})^2 Q^{|\lambda|+s(s+1)/2} e^{\Phi(\lambda,s,\boldsymbol{t})} \tag{14}$$

is related to a special tau function $\tau(s,\boldsymbol{t})$ of the 1D Toda hierarchy as

$$Z(s,\boldsymbol{t}) = \exp\left(\sum_{k=1}^{\infty} \frac{t_k q^k}{1-q^k}\right) q^{-s(s+1)(2s+1)/6}\tau(s,\iota(\boldsymbol{t})), \tag{15}$$

where $\iota(\boldsymbol{t})$ denotes the alternating inversion

$$\iota(\boldsymbol{t}) = (-t_1, t_2, -t_3, \ldots, (-1)^k t_k, \ldots)$$

of $\boldsymbol{t} = (t_1, t_2, \ldots)$.

3. Modified melting crystal model

3.1. Partition function

The modified model is obtained by replacing

$$s_\lambda(q^{-\rho})^2 \ \to\ s_\lambda(q^{-\rho}) s_{{}^t\lambda}(q^{-\rho}),$$

where ${}^t\lambda$ denotes the conjugate partition that represents the transpose of the associated Young diagram. The Q-dependent partition function reads

$$Z' = \sum_{\lambda\in\mathcal{P}} s_\lambda(q^{-\rho}) s_{{}^t\lambda}(q^{-\rho}) Q^{|\lambda|}. \tag{16}$$

This is no longer a statistical model of 3D Young diagrams. $s_\lambda(q^{-\rho})$ and $s_{{}^t\lambda}(q^{-\rho})$ are partial sums of contributions from "half" pieces of 3D Young diagrams, but these pieces cannot be glued along the plane $x = y$. By the Cauchy identity

$$\sum_{\lambda\in\mathcal{P}} s_\lambda(x_1, x_2, \ldots) s_{{}^t\lambda}(y_1, y_2, \ldots) = \prod_{i,j=1}^{\infty} (1 + x_i y_j) \tag{17}$$

of another type [11], one can convert (16) to an infinite product of the form

$$Z' = \sum_{i,j=1}^{\infty} (1 + Qq^{i+j-1}) = \prod_{n=1}^{\infty} (1 + Qq^n)^n. \tag{18}$$

We deform (16) by the external potential

$$\Phi(\lambda, s, \boldsymbol{t}, \bar{\boldsymbol{t}}) = \sum_{k=1}^{\infty} t_k \Phi_k(\lambda, s) + \sum_{k=1}^{\infty} \bar{t}_k \Phi_{-k}(\lambda, s)$$

that depends on a discrete variable s and two sets of continuous variables $\boldsymbol{t} = (t_1, t_2, \ldots)$ and $\bar{\boldsymbol{t}} = (\bar{t}_1, \bar{t}_2, \ldots)$. The deformed partition function reads

$$Z'(s, \boldsymbol{t}, \bar{\boldsymbol{t}}) = \sum_{\lambda\in\mathcal{P}} s_\lambda(q^{-\rho}) s_{{}^t\lambda}(q^{-\rho}) Q^{|\lambda|+s(s+1)/2} e^{\Phi(\lambda,s,\boldsymbol{t},\bar{\boldsymbol{t}})}. \tag{19}$$

To identify an integrable hierarchy hidden therein, we translate the partition function to the language of the same 2D complex free fermion system as used in our previous work [1, 2].

3.2. Perspectives from fermions

Let ψ_n and ψ_n^*, $n \in \mathbf{Z}$, be the Fourier modes of 2D complex free fermion fields [14]

$$\psi(z) = \sum_{n\in\mathbf{Z}} \psi_n z^{-n-1}, \quad \psi^*(z) = \sum_{n\in\mathbf{Z}} \psi_n^* z^{-n}.$$

They satisfy the anti-commutation relations

$$\psi_m\psi_n^* + \psi_n^*\psi_n = \delta_{m+n,0}, \quad \psi_m\psi_n + \psi_n\psi_m = \psi_m^*\psi_n^* + \psi_n^*\psi_m^* = 0.$$

The associated Fock and dual Fock space can be decomposed to charge-s ($s \in \mathbf{Z}$) sectors. Let $\langle s|$ and $|s\rangle$ denote the normalized ground states in the charge-s sector:

$$\langle s| = \langle -\infty| \cdots \psi_{s-1}^*\psi_s^*, \quad |s\rangle = \psi_{-s}\psi_{-s+1}\cdots|-\infty\rangle$$

Excited states are labelled by partitions $\lambda \in \mathcal{P}$ as $\langle\lambda, s|$ and $|\lambda, s\rangle$:

$$\langle\lambda, s| = \langle -\infty| \cdots \psi_{\lambda_2+s-1}^*\psi_{\lambda_1+s}^*, \quad |\lambda, s\rangle = \psi_{-\lambda_1-s}\psi_{-\lambda_2-s+1}\cdots|-\infty\rangle.$$

Let us introduce the fermion bilinears

$$L_0 = \sum_{n\in\mathbf{Z}} n{:}\psi_{-n}\psi_n^*{:}, \quad W_0 = \sum_{n\in\mathbf{Z}} n^2{:}\psi_{-n}\psi_n^*{:},$$
$$H_k = \sum_{n\in\mathbf{Z}} q^{kn}{:}\psi_{-n}\psi_n^*{:}, \quad J_k = \sum_{n\in\mathbf{Z}} {:}\psi_{-n}\psi_{n+k}^*{:},$$

the vertex operators [12, 13]

$$\Gamma_\pm(z) = \exp\left(\sum_{k=1}^\infty \frac{z^k}{k} J_{\pm k}\right), \quad \Gamma'_\pm(z) = \exp\left(-\sum_{k=1}^\infty \frac{(-z)^k}{k} J_{\pm k}\right),$$

and the infinite products

$$\Gamma_\pm(q^{-\rho}) = \prod_{i=1}^\infty \Gamma_\pm(q^{i-1/2}), \quad \Gamma'_\pm(q^{-\rho}) = \prod_{i=1}^\infty \Gamma'_\pm(q^{i-1/2})$$

of the vertex operators specialized to $z = q^{i-1/2}$, $i = 1, 2, \ldots$. Building blocks of (19) can be expressed with these operators as

$$\begin{aligned} s_\lambda(q^{-\rho}) &= \langle s|\Gamma_+(q^{-\rho})|\lambda, s\rangle, \quad s_{{}^t\lambda}(q^{-\rho}) = \langle\lambda, s|\Gamma'(q^{-\rho})|s\rangle, \\ Q^{|\lambda|+s(s+1)/2} &= \langle\lambda, s|Q^{L_0}|\lambda, s\rangle, \quad \Phi_k(\lambda, s) = \langle\lambda, s|H_k|\lambda, s\rangle. \end{aligned} \tag{20}$$

Note that L_0 and H_k are diagonal with respect to the orthonormal bases $\langle\lambda, s|$ and $|\lambda, s\rangle$ ($\lambda \in \mathcal{P}$, $s \in \mathbf{Z}$) of the Fock and dual Fock spaces. Thus (19) turns out to be expressed as

$$Z'(s, \boldsymbol{t}, \bar{\boldsymbol{t}}) = \langle s|\Gamma_+(q^{-\rho})Q^{L_0}e^{H(\boldsymbol{t},\bar{\boldsymbol{t}})}\Gamma'_-(q^{-\rho})|s\rangle, \tag{21}$$

where

$$H(\boldsymbol{t}, \bar{\boldsymbol{t}}) = \sum_{k=1}^\infty t_k H_k + \sum_{k=1}^\infty \bar{t}_k H_{-k}.$$

For comparison, let us recall that the deformed partition function (14) of the previous model has a similar fermionic expression

$$Z(s,\boldsymbol{t}) = \langle s|\Gamma_+(q^{-\rho})Q^{L_0}e^{H(\boldsymbol{t})}\Gamma_-(q^{-\rho})|s\rangle, \quad H(\boldsymbol{t}) = \sum_{k=1}^{\infty} t_k H_k. \tag{22}$$

In our previous work [1, 2], we derived (15) from this fermionic expression of $Z(s,\boldsymbol{t})$ by manipulation of operators on the fermionic Fock space. The tau function $\tau(s,\boldsymbol{t})$ is defined by the fermionic formula

$$\tau(s,\boldsymbol{t}) = \langle s|\exp\left(\sum_{k=1}^{\infty} t_k J_k\right) g|s\rangle = \langle s|g\exp\left(\sum_{k=1}^{\infty} t_k J_{-k}\right)|s\rangle, \tag{23}$$

where g is an operator of the form

$$g = q^{W_0/2}\Gamma_-(q^{-\rho})\Gamma_+(q^{-\rho})Q^{L_0}\Gamma_-(q^{-\rho})\Gamma_+(q^{-\rho})q^{W_0/2}. \tag{24}$$

The existence of two apparently different expressions in (23) is a consequence of the algebraic relations

$$J_k g = g J_{-k}, \quad k = 1, 2, \ldots. \tag{25}$$

Extending this work, we have been able to show the following result [3].

Theorem 2. *The partition function $Z'(s,\boldsymbol{t},\bar{\boldsymbol{t}})$ of the modified model is related to a tau function $\tau'(s,\boldsymbol{t},\bar{\boldsymbol{t}})$ of the 2D Toda hierarchy as*

$$Z'(s,\boldsymbol{t},\bar{\boldsymbol{t}}) = \exp\left(\sum_{k=1}^{\infty}\frac{q^k t_k - \bar{t}_k}{1-q^k}\right)\tau'(s,\iota(\boldsymbol{t}),-\bar{\boldsymbol{t}}). \tag{26}$$

$\tau'(s,\boldsymbol{t},\bar{\boldsymbol{t}})$ is defined by the fermionic formula

$$\tau'(s,\boldsymbol{t},\bar{\boldsymbol{t}}) = \langle s|\exp\left(\sum_{k=1}^{\infty} t_k J_k\right) g' \exp\left(-\sum_{k=1}^{\infty}\bar{t}_k J_{-k}\right)|s\rangle, \tag{27}$$

where g' is an operator of the form

$$g' = q^{W_0/2}\Gamma_-(q^{-\rho})\Gamma_+(q^{-\rho})Q^{L_0}\Gamma'_-(q^{-\rho})\Gamma'_+(q^{-\rho})q^{-W_0/2}. \tag{28}$$

Derivation of this result is parallel to the case of $Z(s,\boldsymbol{t})$ [1, 2]. "Shift symmetries" in the quantum torus algebra of fermion bilinears imply the following algebraic relations ($k > 0$):

$$\begin{aligned}\Gamma_+(q^{-\rho})H_k\Gamma_+(q^{-\rho})^{-1} &= (-1)^k\Gamma_-(q^{-\rho})^{-1}q^{-W_0/2}J_k q^{W_0/2}\Gamma_-(q^{-\rho}) + \frac{q^k}{1-q^k},\\ \Gamma'_-(q^{-\rho})^{-1}H_{-k}\Gamma'_-(q^{-\rho}) &= \Gamma'_+(q^{-\rho})q^{-W_0/2}J_{-k}q^{W_0/2}\Gamma'_+(q^{-\rho})^{-1} - \frac{1}{1-q^k}.\end{aligned} \tag{29}$$

With the aid of these algebraic relations, one can derive (15) and (26) by rather straightforward calculations.

This is not the end of the story. To uncover hidden properties of $\tau'(s,\boldsymbol{t},\bar{\boldsymbol{t}})$, we now translate the fermionic machinery to the language of infinite matrices.

3.3. From fermions to infinite matrices

It is well known [14] that fermion bilinears are in one-to-one correspondence with $\mathbf{Z}\times\mathbf{Z}$ matrices as

$$\hat{X} = \sum_{i,j\in\mathbf{Z}} x_{ij}{:}\psi_{-i}\psi_j^*{:} \longleftrightarrow X = \sum_{i,j\in\mathbf{Z}} x_{ij}E_{ij}.$$

Apart from c-number corrections, they obey the same commutation relations, namely,

$$[\hat{X},\hat{Y}] = \widehat{[X,Y]} + c(X,Y), \tag{30}$$

where $c(X,Y)$ is a c-number term.

Actually, the set of infinite matrices is equipped with multiplication as well as Lie brackets. This provides us with more freedom. For example, matrix representation of our fundamental fermion bilinears L_0, M_0, H_k, J_k (let us use the same notations as fermion bilinears) can be written as

$$L_0 = \Delta, \quad W_0 = \Delta^2, \quad H_k = q^{k\Delta}, \quad J_k = \Lambda^k, \tag{31}$$

where

$$\Delta = \sum_{n\in\mathbf{Z}} nE_{nn}, \quad \Lambda = \sum_{n\in\mathbf{Z}} E_{n,n+1}, \quad E_{ij} = (\delta_{in}\delta_{jn})_{i,j\in\mathbf{Z}}.$$

Matrix representation of the vertex operators reveals an even more significant feature. Since

$$\sum_{k=1}^{\infty}\frac{z^k}{k}\Lambda^{\pm 1} = -\log(1-z\Lambda^{\pm}), \quad \sum_{k=1}^{\infty}\frac{(-z)^k}{k}\Lambda^{\pm k} = -\log(1+z\Lambda^{\pm}), \tag{32}$$

$\Gamma_{\pm}(z)$ and $\Gamma'_{\pm}(z)$ can be expressed as

$$\Gamma_{\pm}(z) = (1-z\Lambda^{\pm 1})^{-1}, \quad \Gamma'_{\pm}(z) = 1 + z\Lambda^{\pm 1}, \tag{33}$$

hence

$$\Gamma_{\pm}(q^{-\rho}) = \prod_{i=1}^{\infty}(1-q^{i-1/2}\Lambda^{\pm 1})^{-1}, \quad \Gamma'_{\pm}(q^{-\rho}) = \prod_{i=1}^{\infty}(1+q^{i-1/2}\Lambda^{\pm 1}). \tag{34}$$

The latter may be thought of as matrix-valued *quantum dilogarithm* [15, 16]. Moreover, these vertex operators show up in g and g' in such a form as $\Gamma_-(q^{-\rho})\Gamma_+(q^{-\rho})$ and $\Gamma'_-(q^{-\rho})\Gamma'_+(q^{-\rho})$. As one can see from Jacobi's triple product formula

$$\vartheta(z) = \prod_{n=1}^{\infty}(1-q^n)\prod_{n=1}^{\infty}(1+q^{n-1/2}z)\prod_{n=1}^{\infty}(1+q^{n-1/2}z^{-1}), \tag{35}$$

this indicates a possible relation with the theta function. This issue deserves to be studied in more detail [3].

[3] We learned from John Harnad that the same theta function emerges in his calculations on a class of tau functions. Those tau functions were studied by Sasha Orlov [17] in a different context.

3.4. Perspectives from infinite matrices

The set of $\mathbf{Z} \times \mathbf{Z}$ matrices is the place where the Lax formalism of the 2D Toda hierarchy can be reformulated [18]. Note that Δ and Λ amount to the scalar operator s and the shift operator e^{∂_s} that are fundamental in the Lax formalism based on difference operators [19]. In this setting, one can consider a *factorization problem* that captures all solutions of the 2D Toda hierarchy [20]:

Factorization problem Given a constant invertible $\mathbf{Z} \times \mathbf{Z}$ matrix U, find two $\mathbf{Z} \times \mathbf{Z}$ matrices $W = W(\boldsymbol{t}, \bar{\boldsymbol{t}})$ and $\bar{W} = \bar{W}(\boldsymbol{t}, \bar{\boldsymbol{t}})$ that satisfy the following conditions:

- W is lower triangular, and all diagonal elements are equal to 1.
- $\bar{W}$ is upper triangular, and all diagonal elements are nonzero.
- They satisfy the factorization relation

$$\exp\left(\sum_{k=1}^{\infty} t_k \Lambda^k\right) U \exp\left(-\sum_{k=1}^{\infty} \bar{t}_k \Lambda^{-k}\right) = W^{-1}\bar{W}. \tag{36}$$

If one can find such a pair $W, \bar{W}$, the Lax matrices

$$L = W\Lambda W^{-1}, \quad \bar{L} = \bar{W}\Lambda\bar{W}^{-1} \tag{37}$$

satisfy the Lax equations (hence become a solution) of the 2D Toda hierarchy.

To derive a solution that corresponds to $\tau'(s, \boldsymbol{t}, \bar{\boldsymbol{t}})$, we choose the matrix U to be matrix representation of g':

$$U = q^{\Delta^2/2}\Gamma_-(q^{-\rho})\Gamma_+(q^{-\rho})Q^{\Delta}\Gamma'_-(q^{-\rho})\Gamma'_+(q^{-\rho})q^{-\Delta^2/2}. \tag{38}$$

Remarkably, this matrix can be factorized as follows:

$$U = W^{-1}\bar{W}, \quad \begin{aligned} W &= q^{\Delta^2/2}\Gamma'_-(Qq^{-\rho})^{-1}\Gamma_-(q^{-\rho})^{-1}q^{-\Delta^2/2}, \\ \bar{W} &= q^{\Delta^2/2}Q^{\Delta}\Gamma_+(Qq^{-\rho})\Gamma'_+(q^{-\rho})q^{-\Delta^2/2}. \end{aligned} \tag{39}$$

This means that the factors $W, \bar{W}$ give a solution of the factorization problem (36) at the "initial time" $\boldsymbol{t} = \bar{\boldsymbol{t}} = \boldsymbol{0}$. This allows us to find an explicit form of the associated Lax matrices (37) as well:

$$\begin{aligned} L &= (\Lambda - q^{\Delta})(1 + Qq^{\Delta-1}\Lambda^{-1})^{-1}, \\ \bar{L}^{-1} &= (1 + Qq^{\Delta-1}\Lambda^{-1})(q^{\Delta} - \Lambda)^{-1}. \end{aligned} \tag{40}$$

Thus the "initial values" of the Lax matrices at $\boldsymbol{t} = \bar{\boldsymbol{t}} = \boldsymbol{0}$ take a very special form (so to speak, "quotients" of first order difference operators).

As pointed out by Brini et al. [8], the "quotient" ansatz

$$\begin{aligned} L &= BC^{-1}, \quad \bar{L}^{-1} = C(-B)^{-1}, \\ B &= e^{\partial_s} - b, \quad C = I - ce^{-\partial_s}, \end{aligned} \tag{41}$$

of the Lax operators, where b and c are scalar-valued functions of s, $\boldsymbol{t}$ and $\bar{\boldsymbol{t}}$, is preserved under time evolutions of the 2D Toda hierarchy. In other words, this is a reduction condition, and the reduced system is the Ablowitz-Ladik hierarchy [7] (or the relativistic Toda hierarchy [9, 10]). (40) amount to Lax operators of this type. We are thus led to the following conclusion:

Theorem 3. *$\tau'(s, \boldsymbol{t}, \bar{\boldsymbol{t}})$ is the tau function of a solution of the Ablowitz-Ladik hierarchy embedded in the 2D Toda hierarchy.*

Acknowledgements
We are very grateful to John Harnad for fruitful discussion. The remark on the theta function in Section 3.3 is inspired by his comments. This work is partly supported by JSPS Grants-in-Aid for Scientific Research No. 24540223 and No. 25400111 from the Japan Society for the Promotion of Science.

References
[1] Nakatsu T and Takasaki K 2009 Melting crystal, quantum torus and Toda hierarchy *Comm. Math. Phys.* **285** 445–468 (preprint arXiv:0710.5339 [hep-th])
[2] Nakatsu T and Takasaki K 2010 Integrable structure of melting crystal model with external potentials *Adv. Stud. Pure Math.* vol. 59 (Tokyo: Mathematical Society of Japan) pp. 201–223 (preprint arXiv:0807.4970 [math-ph])
[3] Takasaki K 2012 Integrable structure of modified melting crystal model *preprint arXiv:1208.4497 [math-ph]*
[4] Takasaki K 2013 Modified melting crystal model and Ablowitz-Ladik hierarchy *J. Phys. A: Math. Theor.* **46** 245202 (preprint arXiv:1302.6129 [math-ph])
[5] Mariño M 2005 *Chern-Simons theory, matrix models, and topological strings* (Oxford: Oxford University Press)
[6] Brini A 2012 The local Gromov-Witten theory of $\mathbb{CP}^1$ and integrable hierarchies *Comm. Math. Phys.* **313** 571–605 (preprint arXiv:1002.0582 [math-ph])
[7] Ablowitz M J and Ladik J F 1975 Nonlinear differential-difference equations *J. Math. Phys.* **16** 598–603
[8] Brini A Carlet G and Rossi P 2012 Integrable hierarchies and the mirror model of local $\mathbb{CP}^1$ *Physica D* **241** 2156–2167 (preprint arXiv:1105.4508 [math.AG])
[9] Kharchev S, Mironov A and Zhedanov A 1997 Faces of relativistic Toda chain *Int. J. Mod. Phys.* **A12** 2675–2724 (preprint arXiv:hep-th/9606144)
[10] Suris Yu B 1997 A note on the integrable discretization of the nonlinear Schrödinger equation *Inverse Problem* **13** 1121–1136 (preprint arXiv:solv-int/9606144)
[11] Macdonald I G 1995 *Symmetric functions and Hall polynomials*, (Oxford: Oxford University Press)
[12] Okounkov A and Reshetikhin N 2003 Correlation function of Schur process with application to local geometry of a random 3-dimensional young diagram *J. Amer. Math. Soc.* **16** 581–603 (preprint arXiv:math.CO/0107056)
[13] Bryan J and Young B 2012 Generating functions for coloured 3D Young diagrams and the Donaldson-Thomas invariants of orbifolds *Duke Math. J.* **152** 115–153 (preprint arXiv:0802.3948 [math.CO])
[14] Miwa T, Jimbo M and Date E 2000 *Solitons: Differential equations, symmetries, and infinite-dimensional algebras* (Cambridge: Cambridge University Press)
[15] Faddeev L D and Volkov A Yu 1993 Abelian current algebra and the Virasoro algebra on the lattice *Phys. Lett.* **B315** 311–318 (preprint arXiv:hep-th/9307048)
[16] Faddeev L D and Kashaev R M 1994 Quantum dilogarithm *Mod. Phys. Lett.* **A9** 427–434 (preprint arXiv:hep-th/9310070)
[17] Orlov A Yu 2003 Hypergeometric tau functions $\tau(\boldsymbol{t}, T, \boldsymbol{t}^*)$ as ∞-soliton tau function in T variables *preprint arXv:hep-th/0305001*
[18] Ueno K and Takasaki K 1984 Toda lattice hierarchy *Adv. Stud. Pure Math.* vol. 4 (Amsterdam: North-Holland) pp. 1–95
[19] Takasaki K and Takebe T 1995 Integrable hierarchies and dispersionless limit *Rev. Math. Phys.* **7** 743–808 (preprint arXiv:hep-th/9405096)
[20] Takasaki K 1984 Initial value problem for the Toda lattice hierarchy *Adv. Stud. Pure Math.* vol. 4 (Amsterdam: North-Holland) pp. 136–163

Generalized Lenard chains and multi-separability of the Smorodinsky–Winternitz system

Giorgio Tondo

E-mail: tondo@units.it
Dipartimento di Matematica e Geoscienze, Università degli Studi di Trieste, Piaz.le Europa 1, I–34127 Trieste, Italy

Abstract. We show that the notion of generalized Lenard chains allows to formulate in a natural way the theory of multi–separable systems in the context of bi–Hamiltonian geometry. We prove that the existence of generalized Lenard chains generated by a Hamiltonian function and by a Nijenhuis tensor defined on a symplectic manifold guarantees the separation of variables. As an application, we construct such a chain for the case I of the classical Smorodinsky–Winternitz model.

1. Introduction: superintegrability and separation of variables

This paper is part of a joint research program with P. Tempesta started in [1], where we have established a new connection between the theory of integrable and superintegrable systems on one side, and that of bi–Hamiltonian separation of variables on the other side. In fact, we have provided a theoretical framework for studying separation of variables for classical systems, by means of the notion of generalized Lenard (GL) chains. These chains, jointly with a couple of compatible Poisson tensors, are the main geometrical objects for our bi–Hamiltonian description of classical mechanics. These structures guarantee the separation of variables in a suitable bi–structured manifold.

In classical mechanics, superintegrable systems are Hamiltonian systems that possess more than N integrals of motion functionally independent, globally defined in a $2N$–dimensional phase space (see e.g. [2] for a monograph on the topic). These systems are also called noncommutatively integrable [3], [4]. Especially important are the maximally superintegrable ones, i.e. those having $2N-1$ integrals. It turns out that for these systems all bounded orbits are closed and the motion is periodic [5]. Among the physically most relevant superintegrable potentials are the harmonic oscillator and the Kepler potential, the Calogero–Moser potential, the Smorodinsky–Winternitz systems, the Euler top, etc. [2]–[7].

In quantum mechanics, superintegrable systems are also particularly interesting: they possess accidental degeneracy of the energy levels. This degeneracy can be removed by considering the quantum numbers associated to the additional integrals of motion. A paradigmatic example is offered by the Coulomb atom [8]. Recently, new examples of superintegrable systems have been discovered [9]–[12], both classical and quantum.

One of the most effective methods to solve Hamiltonian systems is to find a complete integral of the corresponding Hamilton–Jacobi equation through the technique of separation of variables. For the sake of clarity, we will recall briefly the geometric setting of Hamiltonian dynamics [13].

Let (M, ω) be a symplectic manifold, i.e. a $2n$–dimensional manifold endowed with a non degenerate closed 2–form ω, said to be a symplectic form. Such a geometrical structure selects a privileged dynamics on M, the one given by Hamiltonian vector fields defined by

$$i_{X_H}\omega = -\mathrm{d}H$$

(i_{X_H} denotes the contraction operator w.r.t. the vector field X_H and d denotes the exterior derivative operator) or, equivalently

$$X_H = (\omega^\flat)^{-1}\mathrm{d}H \ ,$$

where $\omega^\flat : TM \to T^*M$ denotes the fiber bundles isomorphism induced by ω. The function H is said to be the Hamiltonian function of the vector field X_H. A symplectic form acting on vector fields is equivalent to a non degenerate Poisson bracket defined as

$$\{F, G\} := \omega(X_F, X_G) =< \mathrm{d}F, X_G > \ , \qquad (1)$$

($<,>$ denotes the natural pairing between 1–forms and vector fields), i.e. as a skew–symmetric derivation on the ring $C^\infty(M)$, fulfilling the Jacobi identity. As we wish to study separable Hamiltonian systems that are Liouville–integrable, in principle we can start with a set of n independent Hamiltonian functions in involution w.r.t. the Poisson brackets (1). In this framework, in the tradition of the *Italian school* [14], an important result has been obtained by Benenti in [15]. It gives a characterization of separated coordinates in terms of Poisson bracket.

The Hamiltonian functions $\{H_i\}_{1\leq i\leq n}$ are separable in a set of canonical coordinates $(\boldsymbol{q}, \boldsymbol{p})$ if and only if they are in separable involution, i.e. if and only if they satisfy

$$\{H_i, H_j\}_{|k} = \frac{\partial H_i}{\partial q_k}\frac{\partial H_j}{\partial p_k} - \frac{\partial H_i}{\partial p_k}\frac{\partial H_j}{\partial q_k} = 0 \ , \quad 1 \leq k \leq n, \qquad (2)$$

where no summation over k is understood. However, such a condition is not intrinsic as it requires to know *a priori* the coordinates $(\boldsymbol{q}, \boldsymbol{p})$ in order to be applied.

Recently, a new geometric approach to SoV has been developed, based on the bi–Hamiltonian theory ([16], [17]) and on GL chains ([18]–[21]). It has succeeded in giving intrinsic and constructive criteria of separability and has connected the classical theory of SoV with the modern theory by Sklyanin [22]. The bi–Hamiltonian theory of SoV is formulated in phase spaces represented by manifolds endowed with two geometric structures satisfying two suitable compatibility conditions. Such structures are a symplectic form ω and a Nijenhuis (or hereditary) operator N acting on the tangent bundle of M. For this reason such manifolds have been called ωN manifolds. Whilst the symplectic form defines the algebra of Hamiltonian vector fields, the Nijenhuis operator defines sets of distinguished coordinates that are separated coordinates for a special class of Hamiltonian vector fields, those belonging to GL chains, so called as they are extensions of classical Lenard chains, widely known in soliton literature [23], [24].

In this paper we extend a theorem given in [1] proving that, given a generic integrable system on the cotangent bundle of the Euclidean plane, the existence of a GL chain ensures separation of variables on a ωN manifold, in a set of coordinates more general than those of [1]. Moreover, we will complete the study of an important physical model, namely the case I of the Smorodinsky–Winternitz (SWI) system, constructing explicitly a new bi–Hamiltonian structure related to elliptic separated coordinates in the plane.

The paper is organized as follows. In Section 2, the theory of bi–Hamiltonian manifolds is briefly reviewed. In Section 3, the main geometrical object of our theory, i.e. the GL chains, are introduced. The above mentioned general theorem on separation of variables for systems in T^*E^2 is proposed. In Section 4, the previous theory is applied to the study of the general class of integrable mechanical systems in the Euclidean plane, separable in elliptic coordinates. In Section 5, a GL chain is constructed for the classical SWI system. Some discussions are drawn in the final Section 6.

2. Bi–Hamiltonian manifolds and ωN manifolds

Generally, a Poisson bracket (see, e.g., [25]) can be defined by a Poisson bi–vector field, i.e. a skew–symmetric linear map $P : T^*M \mapsto TM$, eventually not invertible, such that

$$\{F, G\}_P :=< \mathrm{d}F, P\mathrm{d}G >$$

with vanishing Schouten bracket

$$0 = [P, P]\,(\alpha, \beta) := \mathcal{L}_{P\beta}(P)\alpha + P(i_{P\alpha}\mathrm{d}\beta) \qquad \forall\, \alpha, \beta \in T^*M, \tag{3}$$

($\mathcal{L}$ denotes the Lie derivative). In the special case of symplectic manifolds, $P := (\omega^\flat)^{-1}$ is a Poisson bi-vector. Generalizing (1) the vector field $X_G := P\,\mathrm{d}G$ is said to be the Hamiltonian vector field with Hamiltonian function G.

Bi–Hamiltonian manifolds were introduced by Magri [26] as models of phase space for soliton equations.

Definition 1 *A bi–Hamiltonian manifold (M, P_0, P_1) is a manifold M endowed with* two *Poisson bi-vectors fields such that*

$$0 = 2\,[P_0, P_1]\,(\alpha, \beta) := \mathcal{L}_{P_0\beta}(P_1)\alpha + P_1(i_{P_0\alpha}\mathrm{d}\beta) + \mathcal{L}_{P_1\beta}(P_0)\alpha + P_0(i_{P_1\alpha}\mathrm{d}\beta) \quad \forall\, \alpha, \beta \in T^*M\ . \tag{4}$$

Such a condition assures that the linear combination $P_1 - \lambda P_0$ is a Poisson *pencil*, i.e. it is a Poisson bi-vector for each $\lambda \in \mathbb{C}$, and therefore the corresponding bracket $\{,\}_{P_1} - \lambda\{,\}_{P_0}$ is a pencil of Poisson bracket. Condition (4) is known as the compatibility condition between P_0 and P_1.

If one of the Poisson tensor, say P_0, is invertible, and therefore its inverse is a symplectic operator $\omega^\flat := P_0^{-1}$, the bi–Hamiltonian manifold M turns out to be an ωN manifold (see [27]). Indeed, the composed operator $N := P_1P_0^{-1}$, thanks to the compatibility condition between P_0 and P_1, is a Nijenhuis (or hereditary) operator compatible with the symplectic form ω, induced by $\omega^\flat = P_0^{-1}$.

Definition 2 *A ωN manifold (M, ω, N) is a symplectic manifold endowed with an endomorphism of the tangent bundle of M, $N : TM \mapsto TM$ which satisfies the following conditions:*

- *its Nijenhuis torsion vanishes identically, i.e. $\forall X, Y \in TM$*

$$[NX, NY] - N([X, NY] + [NX, Y]) + N^2[X, Y]) = 0; \tag{5}$$

- *it is compatible with ω, i.e. the tensor $P_1 = N(\omega^\flat)^{-1}$ is again a Poisson tensor and is compatible with $P_0 := (\omega^\flat)^{-1}$, according to Definition 1.*

In short, the condition (5) can be rephrased by saying that the endomorphism N is a Nijenhuis (or hereditary) operator. The adjoint linear map w.r.t. the natural pairing will be denoted by $N^T : T^*M \mapsto T^*M$ and will be defined by

$$< N^T\alpha, X >=< \alpha, NX >\,.$$

The condition (5) on the operator N, introduced by Nijenhuis [28], has a relevant geometrical meaning: it implies that the distributions of its eigenvectors are integrable according to the Frobenius theorem. Consequently, under suitable completeness assumption to be introduced below, one can select local coordinate charts, in which N takes a diagonal form. We suppose that at each point x (or in a dense open subset) of M, the Nijenhuis tensor field N admits n distinct eigenvalues $\lambda_i(x)$ $(i = 1\ldots, n)$ (maximally distinct). Since in a generic ωN manifold the eigenspaces of N are even–dimensional, belonging to the kernel of the skewsimmetric tensor field $P_1 - \lambda P_0$, from the above assumption it follows that N (and the adjoint tensor N^T) can be put in diagonal form. Then we have the following result, proved in [29]. Let (M, ω, N) be a maximally distinct ωN manifold. In a suitable open neighborhood of each point, there exist a Darboux chart $(\boldsymbol{q}, \boldsymbol{p})$ for ω that, simultaneously, diagonalize N^T

$$N^T \mathrm{d}q_i = \lambda_i \mathrm{d}q_i \qquad N^T \mathrm{d}p_i = \lambda_i \mathrm{d}p_i \ . \tag{6}$$

As, in such a Darboux chart the Nijenhuis tensor N takes a diagonal form, the coordinates $(\boldsymbol{q}, \boldsymbol{p})$ are said to be Darboux–Nijenhuis (DN) coordinates, and are just separation coordinates in the bi–Hamiltonian theory of SoV. Hereafter, with an abuse of notation, we will identify an operator P with its matrix in a suitable basis.

Remark 1 *If the eigenvalues $\lambda_i(x)(i = 1, \dots, n)$ of N are functionally independent, they can be chosen as n coordinates of a DN chart. In [30], has been proved that the sets of conjugate momenta, denoted as $\mu_i(x)(i = 1, \dots, n)$, can be computed by quadratures. Such special DN coordinates $(\boldsymbol{\lambda}, \boldsymbol{\mu})$ are referred to as sDN coordinates [17].*

Remark 2 *The condition* (5) *is sufficient but not necessary for the integrability of the distribution of the eigenvalues of a generic operator N. In fact, Haantjes proved that the necessary and sufficient condition for such an integrability is the vanishing of the Haantjes tensor of N [31].*

Remark 3 *It can be easily verified that, given a DN chart $(\boldsymbol{q}, \boldsymbol{p})$, the* separated *canonical transformations*

$$\tilde{q}_i = f_i(q_i) \ , \quad \tilde{p}_i = \frac{p_i}{f_i'} \ , \tag{7}$$

with f_i a generic invertible smooth function of a single coordinate q_i, preserve the property (6), *i.e.,*

$$N^T \mathrm{d}\tilde{q}_i = \lambda_i \mathrm{d}\tilde{q}_i \qquad N^T \mathrm{d}\tilde{p}_i = \lambda_i \mathrm{d}\tilde{p}_i \ . \tag{8}$$

From a geometrical point of view, we can say that coordinates $(\tilde{\boldsymbol{q}}, \tilde{\boldsymbol{p}})$ and $(\boldsymbol{q}, \boldsymbol{p})$, related by transformations (7) *are DN coordinates adapted to the same coordinate web [32].*

3. Generalized Lenard chains

After having introduced the geometrical structures which define separation coordinates in the bi–Hamiltonian theory of SoV, let us characterize the class of Hamiltonian functions which are separable in DN coordinates. For the sake of concreteness, we will do this in the case of a 4–dimensional manifold, extending the theorem proved in [1] in the special case of sDN coordinates to generic DN coordinates.

Theorem 3 *Let (M, ω, N) be a 4-dimensional maximally distinct ωN manifold and (q_1, q_2, p_1, p_2) a DN local chart. Let H be a smooth function in M. The DN coordinates (q_1, q_2, p_1, p_2) are separated variables for H if and only if there exist two smooth functions f and g such that the one form*

$$\alpha = f \,\mathrm{d}H + g \, N^T \mathrm{d}H \tag{9}$$

is an exact one form, i.e., α is the differential of a function, say H_2

$$\alpha = \mathrm{d}H_2. \tag{10}$$

In this case, the function H_2 is an integral of motion in involution with $H_1 := H$ and the same DN coordinates are separated variables for H_2 as well.

Proof. In the above–mentioned chart, N takes the diagonal form

$$N = \lambda_1(\frac{\partial}{\partial q_1} \otimes \mathrm{d}q_1 + \frac{\partial}{\partial p_1} \otimes \mathrm{d}p_1) + \lambda_2(\frac{\partial}{\partial q_2} \otimes \mathrm{d}q_2 + \frac{\partial}{\partial p_2} \otimes \mathrm{d}p_2) \ , \tag{11}$$

being λ_1 and λ_2 the two double eigenvalues. Let us suppose that eqs. (9) and (10) are fulfilled. Then, it follows that

$$\begin{aligned} \frac{\partial H_2}{\partial q_k} &= f\frac{\partial H_1}{\partial q_k} + g\,\lambda_k\frac{\partial H_1}{\partial q_k} \ , \\ \frac{\partial H_2}{\partial p_k} &= f\frac{\partial H_1}{\partial p_k} + g\,\lambda_k\frac{\partial H_1}{\partial q_k} \ . \end{aligned} \qquad k = 1,2 \tag{12}$$

Therefore

$$\{H_1, H_2\}_{|k} = \frac{\partial H_1}{\partial q_k}\frac{\partial H_2}{\partial p_k} - \frac{\partial H_1}{\partial p_k}\frac{\partial H_2}{\partial q_k} \overset{(12)}{=} 0 \ , \tag{13}$$

i.e. H_1 and H_2 are in separable involution according to Benenti's theorem (see formulas (2)), w.r.t. a DN chart. Consequently, H_2 is an integral of motion for X_H.

Viceversa, let us suppose that (q_1, q_2, p_1, p_2) are separated variables for H_1 and H_2, i.e. conditions (13) hold, and let us consider the equation

$$f\mathrm{d}H_1 + gN^T\mathrm{d}H_1 = \mathrm{d}H_2 \tag{14}$$

in the unknown functions f and g. In the local chart (q_1, q_2, p_1, p_2), eq. (14) takes the form

$$f + g\lambda_1 = \frac{\frac{\partial H_2}{\partial q_1}}{\frac{\partial H_1}{\partial q_1}}, \qquad f + g\lambda_2 = \frac{\frac{\partial H_2}{\partial q_2}}{\frac{\partial H_1}{\partial q_2}}, \tag{15}$$

$$f + g\lambda_1 = \frac{\frac{\partial H_2}{\partial p_1}}{\frac{\partial H_1}{\partial p_1}}, \qquad f + g\lambda_2 = \frac{\frac{\partial H_2}{\partial p_2}}{\frac{\partial H_1}{\partial p_2}} \ . \tag{16}$$

We observe that the first equations (15) and (16) coincide in virtue of the conditions (13), so do the second equations (15) and (16). Thus the above system of four equations reduces to two equations that admit the unique solution

$$f = \frac{1}{\lambda_2 - \lambda_1}(\lambda_2\frac{\frac{\partial H_2}{\partial q_1}}{\frac{\partial H_1}{\partial q_1}} - \lambda_1\frac{\frac{\partial H_2}{\partial q_2}}{\frac{\partial H_1}{\partial q_2}}), \tag{17}$$

$$g = \frac{1}{\lambda_2 - \lambda_1}(-\frac{\frac{\partial H_2}{\partial q_1}}{\frac{\partial H_1}{\partial q_1}} + \frac{\frac{\partial H_2}{\partial q_2}}{\frac{\partial H_1}{\partial q_2}}). \tag{18}$$

Then we will say that Hamiltonian functions related by eqs. (9) and (10) belong to a GL chain generated by (ω, N, H) since, for $(f = 0, g = 1)$, a GL chain reduces to a classical Lenard chain.

Remark 4 *We note that, if $f = -(\lambda_1 + \lambda_2), g = 1$, we get a Quasi–Bi–Hamiltonian (QBH) chain of Pfaffian type generated by the function H [16, 33, 34, 35].*

Remark 5 *Let us observe that, in the proof of theorem 3, only the diagonal form of N has been exploited, without any reference to its null torsion. This implies that theorem 3 holds under the weaker assumption on N that its Haantjes tensor be null (see Rem. 2). Some results along this novel line of research are in preparation.*

Theorem 3 suggests the following procedure in order to classify Hamiltonian systems separable in DN coordinates:

(i) choose a Darboux chart (q_1, q_2, p_1, p_2) in a 4–dimensional symplectic manifold M;
(ii) construct a ωN structure which has (q_1, q_2, p_1, p_2) as DN coordinates;
(iii) search for Hamiltonian function H and for functions f and g such that make the one form (9) locally exact.

This procedure can be considered as an *inverse problem*, with respect to the *direct approach* that starts from a given Hamiltonian and aims to find separation coordinates.

Let us observe that the above method provides the integral of motion H_2 together with a set of separated variables both for H and H_2.

4. Bi–Hamiltonian geometry in T^*E^2: construction of GL chains

In [1], we have applied the procedure previously discussed, to the study of the bi–Hamiltonian properties of systems defined in the cotangent bundle of the Euclidean plane, $M = T^*E^2$ Precisely, we have studied mechanical Hamiltonian functions

$$H = \text{kinetic energy } + \text{potential energy}$$

and we have recovered the most general form of the potential in the Euclidean plane, which makes H separable in cartesian, polar and parabolic coordinates, considered as sDN coordinates of the (ω, N) structure given by the standard symplectic form and by the linear Nijenhuis tensor (11). Here, we recall the cartesian case for the sake of clarity, and complete the analysis of [1] adding the case of elliptic coordinates.

Classical separation of variables in cartesian and elliptic coordinates
Let us consider the natural Hamiltonian function

$$H = \frac{1}{2}\left(p_x^2 + p_y^2\right) + V(x, y), \tag{19}$$

(x, y, p_x, p_y) being cartesian coordinates and conjugate momenta. According to the requirement (11), we choose the linear Nijenhuis tensor $N_{car} : T(T^*E^2) \to T(T^*E^2)$

$$N_{car} = \text{diag}(x, y, x, y).$$

The one–form (9) reads

$$\alpha = f\,(x, y, p_x, p_y)\, dH + g\,(x, y, p_x, p_y)\, N^T dH, \tag{20}$$

and the closure condition $d\alpha = 0$ provides a system of nonlinear PDEs for f, g and V reported in formula (A1) of the Appendix of [1]. From such a system, we have deduced the differential consequence:

$$V_{xy} = 0\ , \tag{21}$$

therefore recovering the known result that

$$V(x, y) = V_1(x) + V_2(y) \tag{22}$$

is the most general potential that makes the Hamiltonian function (19) separable in cartesian coordinates. Furthermore, observing that the functions

$$f = \frac{x}{x - y}\ , \qquad g = -\frac{1}{x - y} \qquad x \neq y\ ,$$

fulfills the above mentioned system (A1), we get a GL chain generated by (ω, N, H). The second integral of motion is given by a primitive function of the 1–form (20) which is just the energy associated with the coordinate y

$$H_2 = \frac{p_y^2}{2} + V_2(y). \tag{23}$$

Let us introduce elliptic coordinates (ξ, η) in the plane

$$x = \frac{\xi\eta}{4c}, \qquad y = \pm\frac{1}{4c}\sqrt{(\xi^2 - 4c^2)(4c^2 - \eta^2)}, \quad \xi \geq 0 \quad \eta \in \mathbb{R}\ ,$$

where $(\pm c, 0)$ are the foci of the confocal ellipses and hyperbolas given by $\xi = const$, $\eta = const$, respectively. In such coordinates, the generic Hamiltonian function (19) reads

$$H = 2\frac{(\xi^2 - 4c^2)p_\xi^2 + (4c^2 - \eta^2)p_\eta^2}{\xi^2 - \eta^2} + V(\xi, \eta)\,, \tag{24}$$

being (p_ξ, p_η) conjugate momenta to (ξ, η). In order to get a QBH chain, thanks to the Rem. 3, we can choose the quadratic Nijenhuis tensor

$$N_{ell} = \mathrm{diag}(\xi^2, \eta^2, \xi^2, \eta^2), \tag{25}$$

with

$$\alpha = f\,(\xi, \eta, p_\xi, p_\eta)\, dH + g\,(\xi, \eta, p_\xi, p_\eta)\, N_{ell}^T\, dH. \tag{26}$$

The closure condition for α provides the system (40) of the Appendix. By combining such Eqs. we get the consequence

$$V_{\xi\eta} + 2\frac{\xi V_\eta - \eta V_\xi}{\xi^2 - \eta^2} = 0. \tag{27}$$

The general solution of (27) is

$$V(\xi, \eta) = \frac{V_1(\xi) + V_2(\eta)}{\xi^2 - \eta^2} \qquad \xi \neq \eta \tag{28}$$

which is the most general potential on E^2 that makes (24) separable in elliptic coordinates. Moreover, we deduce a particular solution of the system (40)

$$f = -(\xi^2 + \eta^2)\ , \qquad g = 1\ , \tag{29}$$

which enables us to construct, by means of the same procedure as in the cartesian case, the following GL chain related to the elliptic coordinates

$$H_1 := \frac{2}{\xi^2 - \eta^2}\Big((\xi^2 - 4c^2)p_\xi^2 + (4c^2 - \eta^2)p_\eta^2 + V_1(\xi) + V_2(\eta)\Big) \tag{30}$$

$$H_2 := \frac{2}{\xi^2 - \eta^2}\Big(\eta^2(4c^2 - \xi^2)p_\xi^2 + \xi^2(\eta^2 - 4c^2)p_\eta^2 - \eta^2 V_1(\xi) + \xi^2 V_2(\eta)\Big) \tag{31}$$

Remark 6 *Let us note that, thanks to the choice (25), the above GL chain for elliptic coordinates is indeed an example of the QBH chain defined in Rem. 4. In contrast, if one choose a Nijenhuis tensor linear in the elliptic coordinates, one can verify that a genuine GL chain generated by such a Nijenhuis tensor and H exists, but a QBH chain does not exist.*

Remark 7 *Choosing* $V_1(\xi) = (b_1 + b_2)\xi$ *and* $V_2(\eta) = (b_1 - b_2)\eta$ *in (28), one recovers the classical Euler problem for the potential of two Coulomb centers, fixed at the foci. Thus, also such a problem admits a QBH formulation in a* (ω, N) *manifold [36].*

5. Multi–separation of variables and superintegrable systems

In this Section, we use the bi–Hamiltonian structures constructed in the previous discussion to construct potentials admitting more than a system of separation coordinates. In fact, it can achieve that a Hamiltonian function belongs to GL chains generated by different and incompatible bi–Hamiltonian structures. In this case, we get a Hamiltonian system separable in different coordinates system or a *multi–separable* system together with additional integrals of motion that, if they are independent, assures superintegrability of the model. Thus, we recover in a natural way one of the Smorodinsky–Winternitz potentials in the plane, first discovered in a quantum–mechanical context in [37], [38] and studied again in [39] and [40] from a group theoretical point of view. These are the only potentials that are multi–separable in terms of orthogonal coordinates in E^2.

5.1. Cartesian and Elliptic coordinates

Let us search for the most general potential $V(x, y)$ that admits SoV both in cartesian and in elliptic coordinates. To this end, let us write down every equation in cartesian coordinates. Eq. (27) is equivalent to

$$xy(V_{xx} - V_{yy}) - (x^2 - y^2 - c^2)V_{xy} + 3yV_x - 3xV_y = 0. \tag{32}$$

Thus, the potential has to satisfy the system of the two PDEs (21) and (32). By substituting the solution (19) of the eq. (21) into eq. (32) we get the separated equations

$$V_1'' + \frac{3}{x}V_1' = V_2'' + \frac{3}{y}V_2' = 4a, \tag{33}$$

where a is an arbitrary constant. Their general solution is

$$V_1(x) = \frac{1}{2}ax^2 + \frac{c_1}{x^2}, \tag{34}$$

$$V_2(y) = \frac{1}{2}ay^2 + \frac{c_2}{y^2}. \tag{35}$$

Thus, the general solution of the system (21) and (32) is

$$V(x, y) = \frac{1}{2}a(x^2 + y^2) + \frac{c_1}{x^2} + \frac{c_2}{y^2}, \tag{36}$$

which is nothing but the SWI potential [38], sum of an isotropic elastic potential and an anisotropic Rosochatius potential. The Hamiltonian system with the SWI potential inherits the integral of motion (23) from SoV in cartesian coordinates

$$H_2^{(car)} = \frac{p_y^2}{2} + \frac{a}{2}y^2 + \frac{c_2}{y^2}, \tag{37}$$

together with the integral (31) from SoV in elliptic coordinates. The latter integral written down in cartesian coordinates, reads

$$H_2^{(ell)} = -2\left(c^2p_x^2 + (xp_y - yp_x)^2 + ac^2x^2 + 2c_1\frac{y^2 + c^2}{x^2} + 2c_2\left(\frac{x}{y}\right)^2\right). \tag{38}$$

A simple check shows that the Hamiltonian SWI, $H_2^{(car)}$ and $H_2^{(ell)}$ are independent. Consequently, the potential (36) is maximally superintegrable. Finally, we can state that the SWI Hamiltonian function generates two GL chains, starting from the two incompatible structures (ω, N_{car}) and (ω, N_{ell}), with the operator N_{ell} (25) represented in cartesian coordinates by the simple quadratic matrix

$$N_{ell} \equiv 4 \begin{bmatrix} x^2 & xy & 0 & 0 \\ xy & y^2+c^2 & 0 & 0 \\ 0 & xp_y - yp_x & x^2 & xy \\ -(xp_y - yp_x) & 0 & xy & y^2+c^2 \end{bmatrix} . \tag{39}$$

Indeed, it can be checked that the two Poisson tensors fields

$$P_{car} := N_{car}(\omega^\flat)^{-1} , \quad P_{ell} := N_{ell}(\omega^\flat)^{-1}$$

have non vanishing Schouten brackets, i.e.,

$$[P_{car}, P_{ell}] \neq 0 .$$

6. Future perspectives

In this work, we have illustrated the general bi–Hamiltonian setting for treating the geometry of both integrable and superintegrable systems on the relevant physical example of the SWI system. The present approach for the sake of concreteness has been formulated in the Euclidean plane. However, there is no theoretical restriction in extending it to higher–dimensional cases. Also, it seems interesting to include in the present analysis integrable and superintegrable systems defined in curved spaces. It would be very interesting to derive a quantum formulation of the present theory. Furthermore, its generalization to recursion operators with vanishing Haantjes tensor is under investigation.

Appendix

We report the explicit expressions of the systems of differential equations quoted in Section 4.

$$\begin{aligned}
&4\left(\eta^2-4c^2\right)\left(f_{p_\xi}+g_{p_\xi}\eta^2\right)p_\eta+4\left(\xi^2-4c^2\right)\left(f_{p_\eta}+g_{p_\eta}\xi^2\right)p_\xi=0, \\
&4\left(\eta^2-4c^2\right)\left(\xi\left(f_{p_\eta}+g_{p_\eta}\xi^2\right)\left(p_\xi^2-p_\eta{}^2\right)-\left(\xi^2-\eta^2\right)\left(g_\xi\eta^2+2g\xi+f_\xi\right)p_\eta\right)+ \\
&-\left(\xi^2-\eta^2\right)^2\left(f_{p_\eta}+g_{p_\eta}\xi^2\right)V_\xi=0 , \\
&4\left(\xi^2-4c^2\right)\left(\eta\left(f_{p_\xi}+g_{p_\xi}\eta^2\right)\left(p_\eta{}^2-p_\xi{}^2\right)+\left(\xi^2-\eta^2\right)\left(g_\eta\xi^2+2g\eta+f_\eta\right)p_\xi\right)+ \\
&-\left(\xi^2-\eta^2\right)^2\left(f_{p_\xi}+g_{p_\xi}\eta^2\right)V_\eta=0 , \\
&4\xi\left(\eta^2-4c^2\right)\left(f_{p_\xi}+g_{p_\xi}\xi^2\right)\left(p_\xi{}^2-p_\eta{}^2\right)+4\left(\xi^2-4c^2\right)\left(\xi^2-\eta^2\right)\left(g_\xi\xi^2+2g\xi+f_\xi\right)p_\xi+ \\
&-\left(\xi^2-\eta^2\right)^2\left(f_{p_\xi}+g_{p_\xi}\xi^2\right)V_\xi=0 , \\
&4\eta\left(\xi^2-4c^2\right)\left(f_{p_\eta}+g_{p_\eta}\eta^2\right)\left(p_\eta{}^2-p_\xi^2\right)-4\left(\eta^2-4c^2\right)\left(\xi-\eta^2\right)\left(g_\eta\eta^2+2g\eta+f_\eta\right)p_\eta+ \\
&-\left(\xi^2-\eta^2\right)^2\left(f_{p_\eta}+g_{p_\eta}\eta^2\right)V_\eta=0 , \\
&4\Big(2\xi\eta\left(8c^2-\eta^2-\xi^2\right)g+\xi\left(4c^2-\eta^2\right)\left(f_\eta+\xi^2g_\eta\right)+\eta\left(4c^2-\xi^2\right)\left(f_\xi+\eta^2g_\xi\right)\Big)\left(p_\eta{}^2-p_\xi{}^2\right)+ \\
&-\left(\xi^2-\eta^2\right)^2\left(f_\eta+\xi^2g_\eta\right)\left(V_\xi-V_\eta\right)+g\left(\xi^2-\eta^2\right)^3V_{\eta,\xi}=0 .
\end{aligned} \tag{40}$$

ACKNOWLEDGEMENTS

Its a pleasure to thank P. Tempesta, with whom this research program has been carried on, and the kind organizers of the PMNP2013 in Gallipoli, where these results have been presented. This work has been supported by the grant PRIN–2010 of the Italian MIUR.

References

[1] Tempesta P and Tondo G 2012 *Phys. Rev.* E **85** 46602

[2] Tempesta P, Winternitz P, Harnad J, Miller W Jr, Pogosyan G and Rodríguez M A eds 2004 *Superintegrability in Classical and Quantum Systems* (CRM Proceedings and Lecture Notes vol 37) (Montréal: AMS)

[3] Mischenko A S and Fomenko A T 1978 *Funct. Anal. Appl.* **12** 133

[4] Sparano G G and Vilasi G 2000 *J. Geom. Phys.* **36** 270–84

[5] Nekhoroshev N N 1972 *Trans. Moscow Math. Soc.* **26** 180

[6] Bertrand J 1873 *C. R. Acad. Sci. Paris* **77** 849

[7] Evans N W 1990 *Phys. Rev.* A **41** 5666

[8] Jauch J and Hill E H 1940 *Phys. Rev.* **57** 641

[9] Rodríguez M A, Tempesta P and Winternitz P 2008 *Phys. Rev.* E **78** 046608

[10] Rodríguez M A, Tempesta P and Winternitz P 2009 *J. Phys.: Conf. Series* **175** 012013

[11] Verrier P E and Evans N W 2008 *J. Math. Phys.* **49** 022902

[12] Post S and Winternitz P 2010 *J. Phys.* A **43** 222001

[13] Abraham R and Marsden J E 1978 *Foundations of Mechanics, 2nd ed.* (Reading: The Benjamin/Cummings Publishing Company)

[14] Levi–Civita T 1904 *Math. Ann.* **59** 383–97

[15] Benenti S 1980 *Lect. Notes Math.* vol 836 (Berlin: Springer) pp 512–38

[16] Morosi C and Tondo G 1997 *J. Phys.* A **30** 2799–806

[17] Falqui G and Pedroni M 2003 *Math. Phys., Anal. and Geom.* **6** 139–79

[18] Falqui G, Magri F and Tondo G 2000 *Theor. Math. Phys.* **122** 176–92

[19] Ibort A, Magri F and Marmo G 2000 *J. Geom. Phys.* **33** 210–28

[20] Magri F 2003 *Theoret. Math. Phys.* **137** 1716–22

[21] Tondo G 2006 *Generalized Lenard chains and separation of Variables*, (Quaderni Matematici vol 573) (Trieste: Università degli Studi) pp 1–27

[22] Sklyanin E K 1995 *Prog. Theor. Phys. Suppl.* **118** 35–60

[23] Magri F 1980 *Lect. Notes Phys* vol 120, ed M Boiti et al (Berlin: Springer–Verlag) pp 233–63

[24] Praught J and Smirnov R G 2005 *SIGMA* **1** 005

[25] Vaisman I 1994 *Lectures on the geometry of Poisson Manifolds* (Progress in Mathematics vol 118) (Basel: Birkhauser Verlag)

[26] Magri F 1978 *J. Math. Phys.* **19** 1156–62

[27] Magri F and Morosi C 1984 *Characterization of Integrable Systems through the Theory of Poisson–Nijenhuis Manifolds* (Quaderno S vol 19) (Milan: Università degli Studi)

[28] Nijenhuis A 1955 *Indag. Math.* **17** 390-403

[29] Turiel F J 1989 *C. R. Acad. Scien. Sér. I* **308** 575–78

[30] Magri F and Marsico T 1996 *Gravitation, Electromagnetism and Geometric Structures* ed G Ferrarese (Bologna: Pitagora editrice) pp 207–22

[31] Haantjes 1955 *Indag. Math.* **58** 158-62

[32] Benenti S 1997 *J. Math. Phys.* **38** 6578–602

[33] Caboz R, Ravoson V and Gavrilov L 1991 *J. Phys.* A **24** L523–25

[34] Brouzet R, Caboz R, Rabenivo J and Ravoson V 1996 *J. Phys.* A **29** 2069–76

[35] Morosi C and Tondo G 1998 *Phys. Lett.* A **247** 59-64

[36] Tondo G and Morosi C 1999 *Rep. Math. Phys.* **44** 255-66

[37] Fris I, Mandrosov V, Smorodinsky Ya A, Uhlr M and Winternitz P 1965 *Phys. Lett.* **16** 354–56

[38] Makarov A A, Smorodinsky Ya A, Valiev Kh and Winternitz P 1967 *Il Nuovo Cimento* A **52** 1061

[39] Sheftel M B, Tempesta P and Winternitz P 2001 *J. Math. Phys.* **42** 659–73

[40] Tempesta P, Turbiner A V and Winternitz P 2001 *J. Math. Phys.* **42** 4248–57

On integrability of $D0-$brane equations on $AdS_4 \times \mathbb{CP}^3$ superbackground

D V Uvarov

NSC Kharkov Institute of Physics and Technology, 61108 Kharkov, Ukraine

E-mail: d_uvarov@hotmail.com

Abstract. Equations of motion for the $D0-$brane on $AdS_4 \times \mathbb{CP}^3$ superbackground are shown to be classically integrable by extending the argument previously elaborated for the massless superparticle model.

1. Introduction

Integrable structures have become nowadays the basic objects of study both in AdS_5/CFT_4 [1] and Aharony-Bergman-Jafferis-Maldacena (ABJM) [2] dualities. In the former case, classical integrability of the $AdS_5 \times S^5$ superstring equations was proved in the seminal work [3] based on the $AdS_5 \times S^5$ superstring description as a $2d$ sigma-model on the $PSU(2,2|4)/(SO(1,4)\times SO(5))$ supercoset manifold [4, 5]. In the ABJM case, theories conjectured to be dual share lower space-time (super)symmetry so that integrable structure is more difficult to unveil and explore. Gravity dual of the ABJM gauge theory just in the special sublimit of the 't Hooft limit reduces to the IIA superstring theory on $AdS_4 \times \mathbb{CP}^3$ superspace, that is not isomorphic to a supercoset manifold. Only its subspace of dimension $(10|24)$ can be described as the $OSp(4|6)/(SO(1,3) \times U(3))$ supermanifold. The $OSp(4|6)/(SO(1,3)\times U(3))$ sigma-model [6, 7][1] corresponds to gauging away 8 fermionic coordinates for the supersymmetries broken by the $AdS_4 \times \mathbb{CP}^3$ superbackground in the complete superstring action constructed in [9]. In analogy with the $AdS_5 \times S^5$ superstring case the $OSp(4|6)/(SO(1,3)\times U(3))$ sigma-model equations are classically integrable [6, 7]. But integrability of the full set of $AdS_4 \times \mathbb{CP}^3$ superstring equations, which depend non-trivially on those 8 fermions, is by no means obvious. However, in [10] and [11] it was verified perturbatively up to the second order in such fermionic coordinates that this is indeed the case.

In the present and companion papers [12, 13] we explore integrability of equations for the point-like dynamical objects in IIA superstring theory on $AdS_4 \times \mathbb{CP}^3$ superbackground. In [13] it was proved integrability of the equations of massless superparticle describing dynamics of the superstring zero modes. Here we extend the proof to the case of $D0-$brane, for which various formulations of the action and $\kappa-$symmetry gauge conditions were considered in [14].

2. Aspects of supergeometry of $AdS_4 \times \mathbb{CP}^3$ superspace

In this Section, to make the presentation self-contained, we sketch the construction of supervielbein and Ramond-Ramond (RR) 1-form on the $OSp(4|6)/(SO(1,3) \times U(3))$ supercoset

[1] Alternative approach to construction of the $OSp(4|6)/(SO(1,3) \times U(3))$ sigma-model action relies on the introduction of the pure spinor variables [8].

manifold and the $AdS_4 \times \mathbb{CP}^3$ superspace, that are 'building blocks' of the $D0-$brane action, with the emphasis on the relation to extended superconformal symmetry of the ABJM gauge theory [15]. Thorough treatment of the supergeometry of $AdS_4 \times \mathbb{CP}^3$ superspace is given in [9].

Using the isomorphism between $osp(4|6)$ superalgebra and $D = 3$ $\mathcal{N} = 6$ superconformal algebra, left-invariant $osp(4|6)$ Cartan forms can be expanded over the $D = 3$ $\mathcal{N} = 6$ superconformal generators [16]

$$\begin{aligned}\mathscr{C}(d) = \mathscr{G}^{-1}d\mathscr{G} = \ & \Delta(d)D + \omega^m(d)P_m + c^m(d)K_m + G^{mn}(d)M_{mn} \\ + \ & \Omega_a(d)T^a + \Omega^a(d)T_a + \widetilde{\Omega}_a{}^b(d)\widetilde{V}_b{}^a + \widetilde{\Omega}_b{}^b(d)\widetilde{V}_a{}^a \\ + \ & \omega_a^\mu(d)Q_\mu^a + \bar{\omega}^{\mu a}(d)\bar{Q}_{\mu a} + \chi_{\mu a}(d)S^{\mu a} + \bar{\chi}_\mu^a(d)\bar{S}_a^\mu.\end{aligned} \tag{1}$$

Bosonic subalgebra is spanned by $D = 3$ conformal generators (D, P_m, K_m, M_{mn}) and $su(4) \sim so(6)$ $R-$symmetry generators, that we divided into the $U(3)$ ones $\widetilde{V}_b{}^a$ forming the stability algebra of $\mathbb{CP}^3 = SU(4)/U(3)$ manifold and the coset generators (T_a, T^a). 24 fermionic generators correspond to $D = 3$ $\mathcal{N} = 6$ Poincare $(Q_\mu^a, \bar{Q}_{\mu a})$ and conformal $(S^{\mu a}, \bar{S}_a^\mu)$ supersymmetries. They are equipped with the $SL(2,\mathbb{R})$ spinor index $\mu = 1, 2$ and $SU(3)$ (anti)fundamental representation index $a = 1, 2, 3$ in accordance with the decomposition $\mathbf{6} = \mathbf{3} \oplus \bar{\mathbf{3}}$ of the $SO(6)$ vector on $SU(3)$ representations.

Geometric constituents of the $(10|24)-$dimensional $OSp(4|6)/(SO(1,3) \times U(3))$ supermanifold that enter the $2d$ sigma-model equations and Lax connection [6, 7] are identified with the Cartan forms associated with the generators $g_{(\mathrm{k})}$, with $\mathrm{k} = 0, ..., 3$, having definite eigenvalues w.r.t. the $\mathbb{Z}_4$ automorphism Υ of the $osp(4|6)$ superalgebra: $\Upsilon(g_{(\mathrm{k})}) = i^{\mathrm{k}} g_{(\mathrm{k})}$. In particular, invariant eigenspace $g_{(0)}$ is spanned by the $so(1,3) \oplus u(3)$ stability algebra generators. So (1) can be presented in the manifestly $\mathbb{Z}_4-$graded form

$$\mathscr{C}(d) = \mathscr{C}_{(0)}(d) + \mathscr{C}_{(2)}(d) + \mathscr{C}_{(1)}(d) + \mathscr{C}_{(3)}(d), \tag{2}$$

where

$$\begin{aligned}\mathscr{C}_{(0)}(d) = \ & 2G^{3m}(d)M_{3m} + G^{mn}(d)M_{mn} + \widetilde{\Omega}_a{}^b(d)\widetilde{V}_b{}^a + \widetilde{\Omega}_a{}^a(d)\widetilde{V}_b{}^b \in g_{(0)}, \\ \mathscr{C}_{(2)}(d) = \ & 2G^{0'm}(d)M_{0'm} + \Delta(d)D + \Omega_a(d)T^a + \Omega^a(d)T_a \in g_{(2)}, \\ \mathscr{C}_{(1)}(d) = \ & \omega_{(1)}{}_a^\mu(d)Q_{(1)}{}_\mu^a + \bar{\omega}_{(1)}{}^{\mu a}(d)\bar{Q}_{(1)\mu a} \in g_{(1)}, \\ \mathscr{C}_{(3)}(d) = \ & \omega_{(3)}{}_a^\mu(d)Q_{(3)}{}_\mu^a + \bar{\omega}_{(3)}{}^{\mu a}(d)\bar{Q}_{(3)\mu a} \in g_{(3)}.\end{aligned} \tag{3}$$

The generators

$$M_{0'm} = \frac{1}{2}(P_m + K_m), \quad M_{0'3} = -\frac{1}{2}D, \quad M_{3m} = \frac{1}{2}(K_m - P_m), \quad M_{mn} \tag{4}$$

span the $so(2,3)$ algebra and one can write

$$g_{(0)} = (M_{3m}, \ M_{mn}, \widetilde{V}_a{}^b), \quad g_{(2)} = (M_{0'm}, \ D, \ T^a, \ T_a). \tag{5}$$

Fermionic generators from $g_{(1)}$ and $g_{(3)}$ eigenspaces are defined by the linear combinations of those of Poincare and conformal supersymmetries

$$Q_{(1,3)}{}_\mu^a = Q_\mu^a \pm iS_\mu^a, \quad \bar{Q}_{(1,3)\mu a} = \bar{Q}_{\mu a} \mp i\bar{S}_{\mu a}. \tag{6}$$

As a result, bosonic and fermionic Cartan forms from $\mathscr{C}_{(1,2,3)}$ eigenspaces

$$G^{0'm}(d) = \frac{1}{2}(\omega^m(d) + c^m(d)), \quad \Delta(d), \quad \Omega_a(d), \quad \Omega^a(d) \tag{7}$$

and

$$\omega_{(1,3)}{}^{\mu}_{a}(d) = \frac{1}{2}(\omega^{\mu}_{a}(d) \pm i\chi^{\mu}_{a}(d)), \quad \bar{\omega}_{(1,3)}{}^{\mu a}(d) = \frac{1}{2}(\bar{\omega}^{\mu a}(d) \mp i\bar{\chi}^{\mu a}(d)) \tag{8}$$

are identified with the $OSp(4|6)/(SO(1,3) \times U(3))$ supervielbein components, while remaining Cartan forms

$$G^{3m}(d) = \frac{1}{2}(c^m(d) - \omega^m(d)), \quad G^{mn}(d), \quad \widetilde{\Omega}_a{}^b(d) \tag{9}$$

describe the $SO(1,3) \times U(3)$ connection.

Since the $AdS_4 \times \mathbb{CP}^3$ superspace cannot be realized as a supercoset manifold, the approach described above is not applicable directly to the construction of its geometric constituents, that depend not only on the coordinates of the $OSp(4|6)/(SO(1,3) \times U(3))$ subsuperspace but also on 8 fermionic coordinates for the broken supersymmetries. However, the $AdS_4 \times \mathbb{CP}^3$ background of IIA supergravity can be promoted to the maximally supersymmetric $AdS_4 \times S^7$ background of $D = 11$ supergravity thanks to the Hopf fibration realization of the 7-sphere $S^7 = \mathbb{CP}^3 \times S^1$ [17], [18]. The $AdS_4 \times S^7$ superspace is isomorphic to the $OSp(4|8)/(SO(1,3) \times SO(7))$ supercoset manifold and one can construct $D = 11$ supervielbein, connection and 4-form field strength out of the $osp(4|8)$ Cartan forms similarly to the above discussion, albeit the $\mathbb{Z}_2$ automorphism of $so(2,3) \oplus so(8)$ subalgebra consistent with the coset-space description of $AdS_4 \times S^7$ space-time cannot be extended to the $\mathbb{Z}_4$ automorphism of the full $osp(4|8)$ superalgebra. Then performing dimensional reduction [19, 20] yields supervielbein and other constituents of the $AdS_4 \times \mathbb{CP}^3$ supergeometry [9].

Using the isomorphism between $osp(4|8)$ superalgebra and $D = 3$ $\mathcal{N} = 8$ superconformal algebra, left-invariant $osp(4|8)$ Cartan forms admit decomposition over the $D = 3$ $\mathcal{N} = 8$ superconformal generators [21, 22]

$$\hat{\mathscr{G}}^{-1}d\hat{\mathscr{G}} = \hat{\mathscr{C}}_{\mathrm{so}(2,3)} + \hat{\mathscr{C}}_{\mathrm{so}(8)} + \hat{\mathscr{C}}_{32\mathrm{susy}}, \tag{10}$$

where

$$\hat{\mathscr{C}}_{\mathrm{so}(2,3)} = \underline{\Delta}(d)D + \underline{\omega}^m(d)P_m + \underline{c}^m(d)K_m + \underline{G}^{mn}(d)M_{mn}, \tag{11}$$

$$\begin{aligned} \hat{\mathscr{C}}_{\mathrm{so}(8)} = {} & \Omega_a(d)T^a + \Omega^a(d)T_a + \widetilde{\Omega}_a{}^b(d)\widetilde{V}_b{}^a + \widetilde{\Omega}_b{}^b(d)\widetilde{V}_a{}^a + h(d)H \\ + {} & \widetilde{\Omega}_a(d)\widetilde{T}^a + \widetilde{\Omega}^a(d)\widetilde{T}_a + \Omega_a{}^4(d)V_4{}^a + \Omega_4{}^a(d)V_a{}^4 \end{aligned} \tag{12}$$

and

$$\begin{aligned} \hat{\mathscr{C}}_{32\mathrm{susy}} = {} & \underline{\omega}^{\mu}_{a}(d)Q^{a}_{\mu} + \underline{\bar{\omega}}^{\mu a}(d)\bar{Q}_{\mu a} + \underline{\chi}_{\mu a}(d)S^{\mu a} + \underline{\bar{\chi}}^{a}_{\mu}(d)\bar{S}^{\mu}_{a} \\ + {} & \omega^{\mu}_{4}(d)Q^{4}_{\mu} + \bar{\omega}^{\mu 4}(d)\bar{Q}_{\mu 4} + \chi_{\mu 4}(d)S^{\mu 4} + \bar{\chi}^{4}_{\mu}(d)\bar{S}^{\mu}_{4}. \end{aligned} \tag{13}$$

Eq. (11) contains $so(2,3)$ Cartan forms in conformal basis and Eq. (12) introduces Cartan forms for the $so(8)$ generators in a basis corresponding to the Hopf fibration realization of the 7-sphere $S^7 = \mathbb{CP}^3 \times S^1$. In this basis one explicitly singles out generators of the $su(4) \oplus u(1)$ isometry algebra of $\mathbb{CP}^3 \times S^1$. The generators $(\widetilde{V}_a{}^b, T_a, T^a)$ span the $su(4)$ algebra and commute with the $u(1)$ generator H. Remaining 12 generators $(\widetilde{T}_a, \widetilde{T}^a, V_a{}^4, V_4{}^a)$ belong to the $so(8)/(su(4) \times u(1))$ coset. Commutation relations and transformation to conventional form of the $so(8)$ generators are discussed in [21, 22] and rely on a particular convenient realization [23] for the Kähler 2-form on $\mathbb{CP}^3$ manifold.[2] The advantage of that choice of the Kähler 2-form consists in diagonalization of two projectors [17, 9] used to divide 32 fermionic generators of the $osp(4|8)$ superalgebra (and associated coordinates) into 24 generators of the $osp(4|6)$ superalgebra and 8 generators corresponding to the supersymmetries broken by the $AdS_4 \times \mathbb{CP}^3$ superbackground. Thus, the

[2] For general consideration of the embedding of $su(4)$ algebra into $so(8)$ compatible with the Hopf fibration of the 7-sphere see [9].

first line in (13) contains fermionic generators of $D = 3$ $\mathcal{N} = 6$ superconformal algebra while the second the generators of the broken supersymmetries.

$AdS_4 \times S^7$ supervielbein components tangent to Anti-de Sitter part of the background are equal to the $so(2,3)/so(1,3)$ Cartan forms

$$\hat{E}^{m'}(d) = \left(\frac{1}{2}(\underline{\omega}^m(d) + \underline{c}^m(d)),\ -\underline{\Delta}(d)\right), \quad m' = (m, 3), \tag{14}$$

and those tangent to S^7 in the basis corresponding to the Hopf fibration realization of the 7-sphere are given by

$$E_a(d) = i(\Omega_a(d) + \widetilde{\Omega}_a(d)), \quad E^a(d) = i(\Omega^a(d) + \widetilde{\Omega}^a(d)), \quad \hat{E}^{11}(d) = h(d) + \widetilde{\Omega}_a{}^a(d), \tag{15}$$

where the 11-th space-time dimension was identified with the S^1 fiber. Supervielbein fermionic components are identified with the odd Cartan forms in (13). Explicit form of the supervielbein depends on the choice of $OSp(4|8)/(SO(1,3) \times SO(7))$ representative $\hat{\mathscr{G}}$. To fulfil the reduction to 10 dimensions supervielbein components should match the Kaluza-Klein (KK) ansatz form [19, 20]. In the present case the reduction is performed along the S^1 parametrized by $y \in [0, 2\pi)$ so that considering

$$\hat{\mathscr{G}} = \mathscr{G} e^{yH} \mathscr{G}_{\rm br}(\upsilon), \tag{16}$$

where $\mathscr{G} \in OSp(4|6)/(SO(1,3) \times U(3))$ and $\mathscr{G}_{\rm br}$ is a function of 8 Grassmann coordinates for the broken supersymmetries $\upsilon_\mu = (\theta_\mu, \bar{\theta}_\mu, \eta_\mu, \bar{\eta}_\mu)$, ensures that the $osp(4|8)$ Cartan forms do not depend on y. Underlining in (11), (13) and (14) was used to indicate those of the $osp(4|6)$ Cartan forms that acquire dependence on dy, as well as υ and $d\upsilon$ in addition to that on the coordinates of the $OSp(4|6)/(SO(1,3) \times U(3))$ supermanifold. In particular, the supervielbein components tangent to AdS_4 and S^1 acquire additive contributions proportional to dy

$$\hat{E}^{m'}(d) = G^{m'}(d) + G_y^{m'} dy, \quad \hat{E}^{11}(d) = \Phi dy + a(d), \tag{17}$$

where $G_y^{m'}$ and Φ are functions of υ. Expression (17) deviates from the Kaluza-Klein ansatz [19], [20] so that the $SO(1,4)$ tangent-space Lorentz transformation L should be applied to remove the term proportional to dy in $\hat{E}^{m'}(d)$

$$\mathsf{L}\begin{pmatrix} \hat{E}^{m'}(d) \\ \hat{E}^{11}(d) \end{pmatrix} = \begin{pmatrix} E^{m'}(d) \\ \Phi_L(dy + A_L(d)) \end{pmatrix}. \tag{18}$$

Then $E^{m'}(d)$ is identified with the tangent to AdS_4 components of the $D = 10$ supervielbein in KK frame, $\Phi_L = \sqrt{\Phi^2 + G_y^2}$ ($G_y^2 = G_{ym'} G_y^{m'} \equiv G_y \cdot G_y$) is identified with $\exp(2\phi/3)$, where $\phi(\upsilon)$ is the $D = 10$ dilaton superfield, and A_L is the Ramond-Ramond (RR) 1-form potential. Explicit form of the entries of the Lorentz rotation matrix is[3]

$$\mathsf{L} = \begin{pmatrix} \delta_{n'}^{m'} + \frac{\Phi - \Phi_L}{\Phi_L G_y^2} G_y^{m'} G_{yn'} & -\Phi_L^{-1} G_y^{m'} \\ \Phi_L^{-1} G_{yn'} & \Phi_L^{-1}\Phi \end{pmatrix} \in SO(1,4). \tag{19}$$

$D = 11$ supervielbein bosonic components $E_a(d)$ and $E^a(d)$ in (15) do not depend on dy and thus can be directly identified with the tangent to the $\mathbb{CP}^3$ components of $AdS_4 \times \mathbb{CP}^3$ supervielbein in the KK frame, hence they were not endowed with 'hats'. This concludes characterization of

[3] Associated Lorentz rotation acting on the supervielbein fermionic components is discussed in [9].

the $AdS_4 \times \mathbb{CP}^3$ supervielbein bosonic components and RR 1-form potential, i.e. the constituents of the $D0-$brane action.

For the analysis of $D0-$brane equations in analogy with those of the massless superparticle, it turns helpful to expand $G^{m'}(d)$, $E_a(d)$, $E^a(d)$ and $a(d)$ on the $\mathbb{Z}_4-$graded $osp(4|6)$ Cartan forms (2) and dv [13]

$$\begin{aligned} G^{m'}(d) &= G^{0'n}(d)M_n{}^{m'} + G^{3n}(d)N_n{}^{m'} + \Delta(d)L^{m'} + G^{kl}(d)K_{kl}{}^{m'} \\ &+ q^{m'\mu}d\theta_\mu + \bar{q}^{m'\mu}d\bar{\theta}_\mu + s^{m'\mu}d\eta_\mu + \bar{s}^{m'\mu}d\bar{\eta}_\mu, \\ E_a(d) &= i\Omega_a(d) + u_{(1)}{}^\mu\omega_{(1)\mu a}(d) + u_{(3)}{}^\mu\omega_{(3)\mu a}(d), \\ E^a(d) &= i\Omega^a(d) + \bar{u}_{(1)}{}^\mu\bar{\omega}_{(1)\mu}^{\ a}(d) + \bar{u}_{(3)}{}^\mu\bar{\omega}_{(3)\mu}^{\ a}(d) \end{aligned} \tag{20}$$

and

$$\begin{aligned} a(d) = &\ \widetilde{\Omega}_a{}^a(d) + G^{0'm}(d)m_m + G^{3m}(d)n_m + \Delta(d)l + G^{mn}(d)k_{mn} \\ + &\ h^\mu d\theta_\mu + \bar{h}^\mu d\bar{\theta}_\mu + p^\mu d\eta_\mu + \bar{p}^\mu d\bar{\eta}_\mu. \end{aligned} \tag{21}$$

Coefficients at the differentials of the 'broken' fermions $dv_\mu = (d\theta_\mu, d\bar{\theta}_\mu, d\eta_\mu, d\bar{\eta}_\mu)$ represent corresponding $AdS_4 \times S^7$ supervielbein components, while those at the $osp(4|6)$ Cartan forms can be named 'previelbeins'. Due to the choice of the supercoset element (16) they are functions of v only. To verify integrability of the $D0-$brane equations it is useful to have explicit expressions for them, as well as for $G_y^{m'}$ and Φ, that can be derived by specifying $\mathscr{G}_{\rm br}$ (see [13] for further details).

3. $D0-$brane equations and their integrability

The $D0-$brane action

$$S = -m\int d\tau \Phi_L^{-1}\sqrt{-\left(E_{\tau m'}E_\tau^{m'} - E_{\tau a}E_\tau{}^a\right)} + m\int A_L \tag{22}$$

is the sum of kinetic and Wess-Zumino (WZ) terms defined by the world-line pull-back of the RR 1-form potential A_L with m measuring its flux. Introducing auxiliary $1d$ metric, the action can be brought to the form without the square root

$$S = \frac{1}{2}\int d\tau\left[e^{-1}\Phi_L^{-2}\left(E_{\tau m'}E_\tau^{m'} - E_{\tau a}E_\tau{}^a\right) - em^2\right] + m\int A_L. \tag{23}$$

Let us note arbitrariness in the definition of the Lagrange multiplier $e(\tau)$. The choice made in (23) gives conventional mass term em^2 independent of the superspace coordinates. It is also possible to 'absorb' Φ_L^{-2} factor into the definition of e resulting in the action reducing in the $m \to 0$ limit to that of the massless superparticle. The variation of (23) on e produces mass-shell constraint that is irrelevant for establishing integrability of other (dynamical) equations of motion and hence in the subsequent discussion we set $e = 1/2$, i.e. concentrate on the '$1d$ sigma-model' case.

Considering the $osp(4|6)/(so(1,3) \times u(3))$ Cartan forms (2) as independent variation parameters allows to derive the set of 10 bosonic and 24 fermionic equations of motion. They can be written as a system of the first order ordinary differential equations with the coefficients given by the world-line pullbacks of the $osp(4|6)$ Cartan forms to facilitate derivation of the Lax representation

$$
\begin{aligned}
-\frac{\delta S}{\delta G^{0'}{}_{m}(\delta)} &= \dot{a}^{0'm} + 2G_{\tau}{}^{m}{}_{n}a^{0'n} - 4l^{mn}G_{\tau}{}^{0'}{}_{n} + 4fG_{\tau}{}^{3m} - 2\Delta_{\tau}b^{3m} \\
&- 4i\left(\omega_{(1)\tau a}\sigma^{m}\bar{\varepsilon}_{(1)}{}^{a} - \varepsilon_{(1)a}\sigma^{m}\bar{\omega}_{(1)\tau}{}^{a} + \omega_{(3)\tau a}\sigma^{m}\bar{\varepsilon}_{(3)}{}^{a} - \varepsilon_{(3)a}\sigma^{m}\bar{\omega}_{(3)\tau}{}^{a}\right) = 0, \\
-\frac{1}{2}\frac{\delta S}{\delta\Delta(\delta)} &= \dot{f} + G_{\tau}{}^{0'}{}_{m}b^{3m} - G_{\tau 3m}a^{0'm} \\
&- 2\left(\omega_{(1)\tau}{}^{\mu}_{a}\bar{\varepsilon}_{(1)}{}^{a}_{\mu} - \varepsilon_{(1)}{}^{\mu}_{a}\bar{\omega}_{(1)\tau}{}^{a}_{\mu} - \omega_{(3)\tau}{}^{\mu}_{a}\bar{\varepsilon}_{(3)}{}^{a}_{\mu} + \varepsilon_{(3)}{}^{\mu}_{a}\bar{\omega}_{(3)\tau}{}^{a}_{\mu}\right) = 0, \\
-\frac{\delta S}{\delta\Omega_{a}(\delta)} &= \dot{y}^{a} + iy^{b}\left(\widetilde{\Omega}_{\tau b}{}^{a} + \delta^{a}_{b}\widetilde{\Omega}_{\tau c}{}^{c}\right) - 4iw\Omega_{\tau}{}^{a} \\
&+ 4i\varepsilon^{abc}\left(\omega_{(1)\tau b}{}^{\mu}\varepsilon_{(1)\mu c} - \omega_{(3)\tau b}{}^{\mu}\varepsilon_{(3)\mu c}\right) = 0
\end{aligned}
\tag{24}
$$

and

$$
\begin{aligned}
\frac{1}{4}\frac{\delta S}{\delta\bar{\omega}_{(1)}{}^{a}_{\mu}(\delta)} &= \dot{\varepsilon}_{(3)}{}^{\mu}_{a} + \tfrac{1}{2}\left(G_{\tau}{}^{mn}\varepsilon_{(3)}{}^{\nu}_{a} - l^{mn}\omega_{(3)\tau}{}^{\nu}_{a}\right)\sigma_{mn\nu}{}^{\mu} + i\tilde{\sigma}^{\mu\nu}_{m}\left(G_{\tau}{}^{3m}\varepsilon_{(3)\nu a} - \tfrac{1}{2}b^{3m}\omega_{(3)\tau\nu a}\right) \\
&+ i\tilde{\sigma}^{\mu\nu}_{m}\left(G_{\tau}{}^{0'm}\varepsilon_{(1)\nu a} - \tfrac{1}{2}a^{0'm}\omega_{(1)\tau\nu a}\right) + \Delta_{\tau}\varepsilon_{(1)}{}^{\mu}_{a} - f\omega_{(1)\tau}{}^{\mu}_{a} \\
&- i\left(\widetilde{\Omega}_{\tau a}{}^{b} - \delta^{b}_{a}\widetilde{\Omega}_{\tau c}{}^{c}\right)\varepsilon_{(3)}{}^{\mu}_{b} - 2iw\omega_{(3)\tau}{}^{\mu}_{a} - i\varepsilon_{abc}\left(\Omega_{\tau}{}^{b}\bar{\varepsilon}_{(1)}{}^{\mu c} - y^{b}\bar{\omega}_{(1)\tau}{}^{\mu c}\right) = 0, \\
-\frac{1}{4}\frac{\delta S}{\delta\bar{\omega}_{(3)}{}^{a}_{\mu}(\delta)} &= \dot{\varepsilon}_{(1)}{}^{\mu}_{a} + \tfrac{1}{2}\left(G_{\tau}{}^{mn}\varepsilon_{(1)}{}^{\nu}_{a} - l^{mn}\omega_{(1)\tau}{}^{\nu}_{a}\right)\sigma_{mn\nu}{}^{\mu} - i\tilde{\sigma}^{\mu\nu}_{m}\left(G_{\tau}{}^{3m}\varepsilon_{(1)\nu a} - \tfrac{1}{2}b^{3m}\omega_{(1)\tau\nu a}\right) \\
&- i\tilde{\sigma}^{\mu\nu}_{m}\left(G_{\tau}{}^{0'm}\varepsilon_{(3)\nu a} - \tfrac{1}{2}a^{0'm}\omega_{(3)\tau\nu a}\right) + \Delta_{\tau}\varepsilon_{(3)}{}^{\mu}_{a} - f\omega_{(3)\tau}{}^{\mu}_{a} \\
&- i\left(\widetilde{\Omega}_{\tau a}{}^{b} - \delta^{b}_{a}\widetilde{\Omega}_{\tau c}{}^{c}\right)\varepsilon_{(1)}{}^{\mu}_{b} - 2iw\omega_{(1)\tau}{}^{\mu}_{a} - i\varepsilon_{abc}\left(\Omega_{\tau}{}^{b}\bar{\varepsilon}_{(3)}{}^{\mu c} - y^{b}\bar{\omega}_{(3)\tau}{}^{\mu c}\right) = 0
\end{aligned}
\tag{25}
$$

and c.c. equations. In Eqs. (24), (25) the following quantities have been introduced

$$
\begin{aligned}
\begin{pmatrix} a^{0'}{}_{m} \\ 2f \\ -b_{3m} \\ -l_{kl} \end{pmatrix} &= 2\Phi_{L}^{-2}\begin{pmatrix} M_{m}{}^{n'} \\ L^{n'} \\ N_{m}{}^{n'} \\ K_{kl}{}^{n'} \end{pmatrix}\{G_{\tau n'} - [\Phi_{L}^{-2}(G_{\tau}\cdot G_{y} + \Phi a_{\tau}) + \tfrac{m}{2}]G_{yn'}\} \\
&+ \Phi_{L}^{-2}\begin{pmatrix} m_{m} \\ l \\ n_{m} \\ k_{kl} \end{pmatrix}[2\Phi_{L}^{-2}(G_{y}^{2}a_{\tau} - \Phi G_{\tau}\cdot G_{y}) + m\Phi],
\end{aligned}
\tag{26}
$$

$$
y^{a} = -i\Phi_{L}^{-2}E_{\tau}{}^{a}, \quad \bar{y}_{a} = -i\Phi_{L}^{-2}E_{\tau a}, \quad w = \frac{1}{2}\Phi_{L}^{-4}(G_{y}^{2}a_{\tau} - \Phi G_{\tau}\cdot G_{y}) + \frac{1}{4}m\Phi_{L}^{-2}\Phi \tag{27}
$$

and

$$
\begin{aligned}
\varepsilon_{(1)}{}^{\mu}_{a} = -\tfrac{1}{4}\Phi_{L}^{-2}E_{\tau a}\bar{u}_{(3)}{}^{\mu}, &\quad \varepsilon_{(3)}{}^{\mu}_{a} = \tfrac{1}{4}\Phi_{L}^{-2}E_{\tau a}\bar{u}_{(1)}{}^{\mu}, \\
\bar{\varepsilon}_{(1)}{}^{\mu a} = \tfrac{1}{4}\Phi_{L}^{-2}E_{\tau}{}^{a}u_{(3)}{}^{\mu}, &\quad \bar{\varepsilon}_{(3)}{}^{\mu a} = -\tfrac{1}{4}\Phi_{L}^{-2}E_{\tau}{}^{a}u_{(1)}{}^{\mu}.
\end{aligned}
\tag{28}
$$

Compared to the massless superparticle case, the above introduced quantities take into account the WZ term contribution and the overall factor of Φ_{L}^{-2} in the kinetic term. Finally there are 8 equations for the 'broken' fermions υ that can be brought to the following form

$$
\begin{aligned}
&- 2\tfrac{d}{d\tau}\Big\{\Phi_{L}^{-2}\left[G_{\tau m'} + (\tfrac{m}{2} - \Phi_{L}^{-2}(G_{\tau}\cdot G_{y} + \Phi a_{\tau}))G_{ym'}\right]\tfrac{\partial G_{\tau}{}^{m'}}{\partial\dot{\upsilon}_{\mu}} + \Phi_{L}^{-2}[\tfrac{m}{2}\Phi \\
&+ \Phi_{L}^{-2}(G_{y}^{2}a_{\tau} - \Phi G_{\tau}\cdot G_{y})]\tfrac{\partial a_{\tau}}{\partial\dot{\upsilon}_{\mu}}\Big\} + \tfrac{\partial L}{\partial\Phi_{L}}\tfrac{\partial\Phi_{L}}{\partial\upsilon_{\mu}} + \Phi_{L}^{-4}a_{\tau}\left[a_{\tau}\tfrac{\partial G_{y}^{2}}{\partial\upsilon_{\mu}} + (m\Phi_{L}^{2} - 2G_{\tau}\cdot G_{y})\tfrac{\partial\Phi}{\partial\upsilon_{\mu}}\right] \\
&+ \Phi_{L}^{-2}\Big\{2G_{\tau m'}\tfrac{\partial G_{\tau}{}^{m'}}{\partial\upsilon_{\mu}} - E_{\tau a}\tfrac{\partial E_{\tau}{}^{a}}{\partial\upsilon_{\mu}} - E_{\tau}{}^{a}\tfrac{\partial E_{\tau a}}{\partial\upsilon_{\mu}} + [m - 2\Phi_{L}^{-2}(G_{\tau}\cdot G_{y} + \Phi a_{\tau})]\tfrac{\partial G_{\tau}\cdot G_{y}}{\partial\upsilon_{\mu}} \\
&+ [m\Phi + 2\Phi_{L}^{-2}(G_{y}^{2}a_{\tau} - \Phi G_{\tau}\cdot G_{y})]\tfrac{\partial a_{\tau}}{\partial\upsilon_{\mu}}\Big\} = 0,
\end{aligned}
\tag{29}
$$

where the following shorthand notation was introduced: $\frac{\partial G_{\tau}{}^{m'}}{\partial \dot{\upsilon}_{\mu}} = (q^{m'\mu}, \bar{q}^{m'\mu}, s^{m'\mu}, \bar{s}^{m'\mu})$,[4] $\frac{\partial a_{\tau}}{\partial \dot{\upsilon}_{\mu}} = (h^{\mu}, \bar{h}^{\mu}, p^{\mu}, \bar{p}^{\mu})$ and

$$\begin{aligned} \frac{\partial G_{\tau}{}^{m'}}{\partial \upsilon_{\mu}} &= G_{\tau}{}^{0'n} \frac{\partial M_{n}{}^{m'}}{\partial \upsilon_{\mu}} + G_{\tau}{}^{3n} \frac{\partial N_{n}{}^{m'}}{\partial \upsilon_{\mu}} + \Delta_{\tau} \frac{\partial L^{m'}}{\partial \upsilon_{\mu}} + G_{\tau}{}^{kl} \frac{\partial K_{kl}{}^{m'}}{\partial \upsilon_{\mu}} \\ &- \dot{\theta}_{\nu} \frac{\partial q^{m'\nu}}{\partial \upsilon_{\mu}} - \dot{\bar{\theta}}_{\nu} \frac{\partial \bar{q}^{m'\nu}}{\partial \upsilon_{\mu}} - \dot{\eta}_{\nu} \frac{\partial s^{m'\nu}}{\partial \upsilon_{\mu}} - \dot{\bar{\eta}}_{\nu} \frac{\partial \bar{s}^{m'\nu}}{\partial \upsilon_{\mu}}, \end{aligned} \tag{30}$$

$$-\frac{\partial E_{\tau a}}{\partial \upsilon_{\mu}} = \omega_{(1)\tau\nu a} \frac{\partial u_{(1)}{}^{\nu}}{\partial \upsilon_{\mu}} + \omega_{(3)\tau\nu a} \frac{\partial u_{(3)}{}^{\nu}}{\partial \upsilon_{\mu}}, \quad -\frac{\partial E_{\tau}{}^{a}}{\partial \upsilon_{\mu}} = \bar{\omega}_{(1)\tau\nu}{}^{a} \frac{\partial \bar{u}_{(1)}{}^{\nu}}{\partial \upsilon_{\mu}} + \bar{\omega}_{(3)\tau\nu}{}^{a} \frac{\partial \bar{u}_{(3)}{}^{\nu}}{\partial \upsilon_{\mu}}. \tag{31}$$

Analogously are defined $\frac{\partial a_{\tau}}{\partial \upsilon_{\mu}}$ and $\frac{\partial G_{\tau \cdot} G_{y}}{\partial \upsilon_{\mu}}$.

Equations of motion (24), (25) and (29) are equivalent to the Lax equation

$$\frac{d\mathscr{L}}{d\tau} + [\mathscr{M}, \mathscr{L}] = 0 \tag{32}$$

with the Lax component $\mathscr{M}$ given by the world-line projection of the $osp(4|6)$ Cartan forms (2) and the component $\mathscr{L}$ that can be presented as a sum

$$\mathscr{L} = \mathscr{L}_{\mathrm{so}(2,3)} + \mathscr{L}_{\mathrm{su}(4)} + \mathscr{L}_{\mathrm{24susys}} \in osp(4|6), \tag{33}$$

where each summand is given by the linear combination of the quantities introduced in (26)-(28)

$$\begin{aligned} \mathscr{L}_{\mathrm{so}(2,3)} &= a^{0'm} M_{0'm} + fD + b^{3m} M_{3m} + l^{mn} M_{mn}, \\ \mathscr{L}_{\mathrm{su}(4)} &= y^{a} T_{a} + \bar{y}_{a} T^{a} + 4w \widetilde{V}_{a}{}^{a}, \\ \mathscr{L}_{\mathrm{24susys}} &= \varepsilon_{(1)}{}^{\mu}_{a} Q_{(1)}{}^{a}_{\mu} + \bar{\varepsilon}_{(1)}{}^{\mu a} \bar{Q}_{(1)\mu a} + \varepsilon_{(3)}{}^{\mu}_{a} Q_{(3)}{}^{a}_{\mu} + \bar{\varepsilon}_{(3)}{}^{\mu a} \bar{Q}_{(3)\mu a}. \end{aligned} \tag{34}$$

$\mathscr{L}$ takes the same form as for the massless superparticle modulo the definition of coefficients (26)-(28). Analogously to the superparticle case the Lax component $\mathscr{L}$ can be presented in the form of $osp(4|6)$−valued differential operator acting on the $D0$−brane action (23)

$$\begin{aligned} \mathscr{L} = & \Big(M_{0'm} \tfrac{\partial}{\partial G_{\tau}{}^{0'}{}_{m}} + \tfrac{1}{2} D \tfrac{\partial}{\partial \Delta_{\tau}} - M_{mn} \tfrac{\partial}{\partial G_{\tau mn}} - M_{3m} \tfrac{\partial}{\partial G_{\tau 3m}} \\ + & T_{a} \tfrac{\partial}{\partial \Omega_{\tau a}} + T^{a} \tfrac{\partial}{\partial \Omega_{\tau}{}^{a}} + \widetilde{V}_{a}{}^{a} \tfrac{\partial}{\partial \widetilde{\Omega}_{\tau b}{}^{b}} \\ - & \tfrac{1}{4} Q_{(1)}{}^{a}_{\mu} \tfrac{\partial}{\partial \bar{\omega}_{(3)\tau}{}^{a}_{\mu}} + \tfrac{1}{4} \bar{Q}_{(1)\mu a} \tfrac{\partial}{\partial \omega_{(3)\tau\mu a}} + \tfrac{1}{4} Q_{(3)}{}^{a}_{\mu} \tfrac{\partial}{\partial \bar{\omega}_{(1)\tau}{}^{a}_{\mu}} - \tfrac{1}{4} \bar{Q}_{(3)\mu a} \tfrac{\partial}{\partial \omega_{(1)\tau\mu a}} \Big) S. \end{aligned} \tag{35}$$

4. Conclusion

In this note we focused on the proof of the classical integrability of equations of the $D0$−brane on $AdS_4 \times \mathbb{CP}^3$ superbackground. They can be written as the system of the first order ordinary differential equations that admit Lax representation with the Lax pair components taking value in the $osp(4|6)$ isometry superalgebra of $AdS_4 \times \mathbb{CP}^3$ superbackground. This generalizes the proof of integrability of the massless superparticle equations given in [13]. However, the most important problem is to find a generalization to the superstring case, i.e. to find a zero curvature representation for the $AdS_4 \times \mathbb{CP}^3$ superstring equations. As follows from the discussion in [12] on the relation between the superparticle's Lax pair and the Lax connection of the superstring, the component $\mathscr{L}$ of the Lax pair captures the form of the contribution linear in ℓ_2 to the $2d$ Hodge dualized part of the Lax connection up to insertions of the functions ℓ_1, ℓ_3, ℓ_4 of the spectral parameter that is convenient to identify with ℓ_3. It is thus interesting to find out whether the form of the remaining part of the Lax connection is also determined by the constituents of $AdS_4 \times S^7$ supervielbein bosonic components (17) and 'previelbeins' analogously to $\mathscr{L}$.

[4] We assume that the fermionic derivative acts from the right.

Acknowledgments
The author is grateful to A A Zheltukhin for stimulating discussions.

References

[1] Maldacena J M 1998 *Adv. Theor. Math. Phys.* **2** 231 (*Preprint* hep-th/9711200)
Gubser S S, Klebanov I R and Polyakov A M 1998 *Phys. Lett.* B **428** 105 (*Preprint* hep-th/9802109)
Witten E 1998 *Adv. Theor. Math. Phys.* **2** 253 (*Preprint* hep-th/9802150)
[2] Aharony O, Bergman O, Jafferis D L and Maldacena J 2008 *J. High Energy Phys.* JHEP10(2008)091 (*Preprint* 0806.1218 [hep-th])
[3] Bena I, Polchinski J and Roiban R 2004 *Phys. Rev.* D **69** 046002 (*Preprint* hep-th/0305116)
[4] Metsaev R R and Tseytlin A A 1998 *Nucl. Phys.* B **533** 109 (*Preprint* hep-th/9805028)
[5] Kallosh R, Rahmfeld J and Rajaraman A 1998 *J. High Energy Phys.* JHEP09(1998)002 (*Preprint* hep-th/9805217)
[6] Arutyunov G and Frolov S 2008 *J. High Energy Phys.* JHEP09(2008)129 (*Preprint* 0806.4940 [hep-th])
[7] Stefanski B J 2009 *Nucl. Phys.* B **808** 80 (*Preprint* 0806.4948 [hep-th])
[8] Fre P and Grassi P A *Preprint* 0807.0044 [hep-th]
Bonelli G, Grassi P A and Safaai H 2008 *J. High Energy Phys.* JHEP10(2008)085 (*Preprint* 0808.1051 [hep-th])
D'Auria R, Fre P, Grassi P A and Trigiante M 2009 *Phys. Rev.* D **79** 086001 (*Preprint* 0808.1282 [hep-th])
[9] Gomis J, Sorokin D and Wulff L 2009 *J. High Energy Phys.* JHEP03(2009)015 (*Preprint* 0811.1566 [hep-th])
[10] Sorokin D and Wulff L 2010 *J. High Energy Phys.* JHEP11(2010)143 (*Preprint* 1009.3498 [hep-th])
[11] Cagnazzo A, Sorokin D and Wulff L 2012 *J. High Energy Phys.* JHEP01(2012)004 (*Preprint* 1111.4197 [hep-th])
[12] Uvarov D V 2013 *Nucl. Phys.* B **867** 354 (*Preprint* 1205.5388 [hep-th])
[13] Uvarov D V *Preprint* 1212.3157 [hep-th]
[14] Grassi P A, Sorokin D and Wulff L 2009 *J. High Energy Phys.* JHEP08(2009)060 (*Preprint* 0903.5407 [hep-th])
[15] Bandres M, Lipstein A and Schwarz J H 2008 *J. High Energy Phys.* JHEP09(2008)027 (*Preprint* 0807.0880 [hep-th])
[16] Uvarov D V 2009 *Phys. Rev.* D **79** 106007 (*Preprint* 0811.2813 [hep-th])
[17] Nilsson B E W and Pope C 1984 *Class. Quantum Grav.* **1** 499
[18] Sorokin D P, Tkach V I and Volkov D V 1985 *Proc. Quantum Gravity (Moscow)* (Singapore: World Scientific) p 376
Sorokin D P, Tkach V I and Volkov D V 1985 *Phys. Lett.* B **161** 301
[19] Duff M J, Howe P S, Inami T and Stelle K S 1987 *Phys. Lett.* B **191** 70
[20] Howe P S and Sezgin E 2005 *Class. Quantum Grav.* **22** 2167 (*Preprint* hep-th/0412245)
[21] Uvarov D V 2010 *Nucl. Phys.* B **826** 294 (*Preprint* 0906.4699 [hep-th])
Uvarov D V 2010 *Mod. Phys. Lett.* A **25** 1251 (*Preprint* 0912.1044 [hep-th])
[22] Uvarov D V 2012 *Int. J. Mod. Phys.* A **27** 1250118 (*Preprint* 1203.3041 [hep-th])
[23] Uvarov D V 2011 *Class. Quantum Grav.* **28** 235010 (*Preprint* 1011.5457 [hep-th])

Remarks on quadratic bundles related to Hermitian symmetric spaces

Tihomir Valchev

School of Mathematical Sciences, Dublin Institute of Technology,
Kevin Street, Dublin 8, Ireland
Institute for Nuclear Research and Nuclear Energy, Bulgarian Academy of Sciences,
72 Tsarigradsko chaussèe, 1784 Sofia, Bulgaria

E-mail: Tihomir.Valchev@dit.ie

Abstract. We consider quadratic bundles related to Hermitian symmetric spaces of the type $SU(m+n)/S(U(m)\times U(n))$. We discuss the spectral properties of scattering operator, develop the direct scattering problem associated with it and stress on the effect of reduction on these. By applying a modification of Zakharov-Shabat's dressing procedure we demonstrate how one can obtain reflectionless potentials. That way one is able to generate soliton solutions to the nonlinear evolution equations belonging to the integrable hierarchy associated with quadratic bundles under study.

1. Introduction

Derivative nonlinear Schrödinger equation (DNLS)

$$\mathrm{i}q_t + q_{xx} + \mathrm{i}(|q|^2 q)_x = 0 \tag{1}$$

is one of classical S-integrable nonlinear evolution equations (NLEE). It occurs in plasma physics to describe the propagation of nonlinear Alfvén waves with circular polarization [18, 19].

The zero curvature representation $[L, A] = 0$ of DNLS was discovered by Kaup and Newell [12] who picked up L and A in the form:

$$L(\lambda) = \mathrm{i}\partial_x + \lambda Q(x,t) - \lambda^2\sigma_3, \tag{2}$$

$$A(\lambda) = \mathrm{i}\partial_t + \sum_{k=1}^{3} A_k(x,t)\lambda^k - 2\lambda^4\sigma_3, \tag{3}$$

where $\lambda \in \mathbb{C}$ is a spectral parameter and

$$Q(x,t) = \begin{pmatrix} 0 & q(x,t) \\ q^*(x,t) & 0 \end{pmatrix}, \qquad \sigma_3 = \begin{pmatrix} 1 & 0 \\ 0 & -1 \end{pmatrix}.$$

DNLS is deeply connected to 2-dimensional Thirring model [13, 15] and the Gerdjikov-Ivanov equation [5, 6], both related to certain reductions of quadratic bundle L operator of generic form:

$$L(\lambda) = \mathrm{i}\partial_x + U_0(x,t) + \lambda U_1(x,t) - \lambda^2\sigma_3 \tag{4}$$

for $U_1(x,t)$ being an off-diagonal 2×2 matrix and $U_0(x,t)$ being a traceless 2×2 matrix otherwise arbitrary.

Derivation and study of multicomponent generalizations of classical scalar integrable equations is a trend of current interest [7, 8, 11] in theory of integrable systems. It was pioneered by Manakov [14] who studied 2-component counterpart of nonlinear Schrödinger equation to later become known as Manakov system. A. Fordy et al. developed that idea and established a geometric relation between linear and quadratic bundles and Hermitian symmetric spaces [1, 2, 3]. In [2] multicomponent versions of DNLS like the following one:

$$\mathrm{i}\mathbf{q}_t + \mathbf{q}_{xx} + \frac{2m\mathrm{i}}{n+m}\left(\mathbf{q}\mathbf{q}^\dagger\mathbf{q}\right)_x = 0 \tag{5}$$

for $\mathbf{q}$ being a smooth $n\times m$ matrix-valued function, were derived. Equation (5) is related to symmetric space of the type $SU(m+n)/S(U(m)\times U(n))$ which is to say the operator L has the same form as (2) but now Q is a block $(m+n)\times(m+n)$ matrix of the form

$$Q(x,t) = \begin{pmatrix} 0 & \mathbf{q}^T(x,t) \\ \mathbf{q}^*(x,t) & 0 \end{pmatrix}$$

and σ_3 is replaced by matrix $J = \mathrm{diag}\,(\frac{n}{m}\mathbb{1}_m, -\mathbb{1}_n)$ where $\mathbb{1}_m$ is the $m\times m$ unit matrix. Like the scalar DNLS (5) is a Hamiltonian system [2, 20] with Hamiltonian given by

$$H = \int_{-\infty}^{\infty} \mathrm{d}\,x\,\mathrm{tr}\left(\mathrm{i}\mathbf{q}^\dagger\mathbf{q}_x - \frac{m}{n+m}(\mathbf{q}^\dagger\mathbf{q})^2\right),$$

provided the Poisson bracket is defined as

$$\{F,G\} = \int_{-\infty}^{\infty} \mathrm{d}\,y\,\mathrm{tr}\left(\frac{\delta F}{\delta Q}\partial_y\frac{\delta G}{\delta Q}\right)$$

for F and G being functionals of the potential Q.

All afore-mentioned facts make a deeper study of quadratic bundles associated with symmetric spaces very important. Our purpose here is to shed some light on certain basic properties of equation (5) and the corresponding Lax pair. In doing so we shall partially extend some results already published in [20]. The report is structured as follows. Second section is preliminary in its nature. We discuss some basic properties of the scattering operator L and the linear problem $L\psi = 0$ related to the nonlinear evolution equation (5) to be used further in text. In next section we show how one can modify Zakharov-Shabat's dressing method for the case of quadratic bundles. This allows one to generate special types of solutions in an algebraic manner. Last section contains summary of our results and some additional remarks.

2. Quadratic bundles and Hermitian symmetric spaces

Let us consider the Lax pair:

$$L(\lambda) = \mathrm{i}\partial_x + \lambda Q(x,t) - \lambda^2 J, \tag{6}$$

$$A(\lambda) = \mathrm{i}\partial_t + \sum_{k=1}^{2N}\lambda^k A_k(x,t), \tag{7}$$

where $Q(x,t)$, J and $A_k(x,t)$, $k = 1,\ldots 2N$ are $(m+n)\times(m+n)$ traceless Hermitian matrices. Moreover, L and A are subject to additional $\mathbb{Z}_2$ reduction [16, 17]

$$\mathbf{C}L(-\lambda)\mathbf{C} = L(\lambda), \qquad \mathbf{C}A(-\lambda)\mathbf{C} = A(\lambda), \qquad \mathbf{C} = \mathrm{diag}\,(\mathbb{1}_m, -\mathbb{1}_n). \tag{8}$$

The constant matrix $\mathbf{C}$ defines an action of Cartan's involution connected with symmetric space $SU(m+n)/S(U(m)\times U(n))$ in the Lie algebra $\mathfrak{sl}(m+n)$, see [10] for more details. That way the Lax pair (6), (7) is related toa space of the type **A.III**. Cartan's involution induces a $\mathbb{Z}_2$ grading in $\mathfrak{sl}(m+n)$ as follows:

$$\mathfrak{sl}(m+n) = \mathfrak{sl}^0(m+n) + \mathfrak{sl}^1(m+n), \tag{9}$$

where

$$\mathfrak{sl}^\sigma(m+n) = \{X \in \mathfrak{sl}(m+n) |\, \mathbf{C}X\mathbf{C} = (-1)^\sigma X\}, \qquad \sigma = 0, 1$$

are eigensubspaces of the adjoint action of $\mathbf{C}$. Due to reduction condition(8) the potential Q has the block structure:

$$Q(x,t) = \left(\begin{array}{cc} 0 & \mathbf{q}^T(x,t) \\ \mathbf{q}^*(x,t) & 0 \end{array}\right),$$

where $\mathbf{q}(x,t)$ is a $n\times m$ rectangular matrix and J is picked up in the form

$$J = \mathrm{diag}\,(n_1, n_2, \ldots, n_m, -\mathbb{1}_n), \qquad \sum_{k=1}^{m} n_k = n. \tag{10}$$

From now on we shall assume that $\mathbf{q}$ is an infinitely smooth function to obey boundary conditions

$$\lim_{x\to\pm\infty} \mathbf{q}(x,t) = \mathbf{0}.$$

Let us now consider auxiliary linear problem

$$L(\lambda)\psi(x,t,\lambda) = \mathrm{i}\partial_x\psi + \lambda(Q - \lambda J)\psi = 0. \tag{11}$$

The function ψ is viewed as a fundamental set of solutions hence $\psi(x,t,\lambda)$ is a unimodular matrix. Since L and A commute we have as well

$$A\psi = \mathrm{i}\partial_t\psi + \sum_{k=1}^{2N} \lambda^k A_k \psi = \psi f. \tag{12}$$

Quantity

$$f(\lambda) = \lim_{x\to\pm\infty} \sum_{k=1}^{2N} \lambda^k A_k(x,t)$$

is called dispersion law of NLEE and it carries all essential characteristics of NLEE.

A special case of solutions to (11) are Jost solutions defined as follows:

$$\lim_{x\to\pm\infty} \psi_\pm(x,t,\lambda)\mathrm{e}^{\mathrm{i}\lambda^2 Jx} = \mathbb{1}, \qquad \lambda^2 \in \mathbb{R}. \tag{13}$$

Due to the special choice of the right handside of (12) the definition of the Jost solutions is correct. The transition matrix

$$T(t,\lambda) = \hat{\psi}_+(x,t,\lambda)\psi_-(x,t,\lambda), \qquad \hat{\psi} \equiv \psi^{-1}$$

between the Jost solutions is called scattering matrix. It can be proven that the scattering matrix evolves with time according to:

$$\mathrm{i}\partial_t T + [f(\lambda), T] = 0 \quad \Rightarrow \quad T(t,\lambda) = \mathrm{e}^{\mathrm{i}f(\lambda)t}T(0,\lambda)\mathrm{e}^{-\mathrm{i}f(\lambda)t}.$$

The Jost solutions are defined for λ lying on real and imaginary axis. To see this one introduces functions $\xi_\pm = \psi_\pm \exp(\mathrm{i}\lambda^2 Jx)$ to satisfy modified linear equation

$$\mathrm{i}\partial_x \xi_\pm + \lambda Q \xi_\pm - \lambda^2 [J, \xi_\pm] = 0. \tag{14}$$

Equivalently $\xi_\pm$ can be viewed as solutions to the following integral equations of Voltera type

$$\xi_\pm(x,\lambda) = \mathbb{1} + \mathrm{i}\lambda \int_{\pm\infty}^{x} \mathrm{e}^{-\mathrm{i}\lambda^2 J(x-y)} Q(y) \xi_\pm(y,\lambda) \mathrm{e}^{\mathrm{i}\lambda^2 J(x-y)} \mathrm{d}\, y. \tag{15}$$

By analysing it one can see that analytic continuation outside the real and imaginary axis is possible for the first and the last columns of $\xi_\pm$ only. Quite similarly to $\mathfrak{sl}(2)$ case [5] starting from the Jost solutions one is able to construct another set of solutions to have such analytic properties. Namely, the following theorem holds true:

Theorem 1 *There exists a pair of solutions χ^+ and χ^- analytic for $\operatorname{Im}\lambda^2 > 0$ (i.e. in the first and third quadrant in λ-plane and $\operatorname{Im}\lambda^2 < 0$ (in the second and the forth quadrants). They can be constructed from the Jost solutions as follows:*

$$\chi^\pm(x,\lambda) = \psi_-(x,\lambda) S^\pm(\lambda) = \psi_+(x,\lambda) T^\mp(\lambda) D^\pm(\lambda).$$

where the matrices $S^\pm(\lambda)$, $T^\pm(\lambda)$ and $D^\pm(\lambda)$ are

$$\begin{aligned} S^+(\lambda) &= \begin{pmatrix} \mathbb{1}_m & \mathbf{s}_+^T(\lambda) \\ \mathbf{0} & \mathbb{1}_n \end{pmatrix}, & T^+(\lambda) &= \begin{pmatrix} \mathbb{1}_m & \mathbf{t}_+^T(\lambda) \\ \mathbf{0} & \mathbb{1}_n \end{pmatrix}, \\ S^-(\lambda) &= \begin{pmatrix} \mathbb{1}_m & \mathbf{0}^T \\ \mathbf{s}_-(\lambda) & \mathbb{1}_n \end{pmatrix}, & T^-(\lambda) &= \begin{pmatrix} \mathbb{1}_m & \mathbf{0}^T \\ \mathbf{t}_-(\lambda) & \mathbb{1}_n \end{pmatrix}, \\ D^+(\lambda) &= \begin{pmatrix} d_m^+(\lambda) & \mathbf{0}^T \\ \mathbf{0} & d_n^+(\lambda) \end{pmatrix}, & D^-(\lambda) &= \begin{pmatrix} d_m^-(\lambda) & \mathbf{0}^T \\ \mathbf{0} & d_n^-(\lambda) \end{pmatrix}. \end{aligned}$$

The latter are involved in generalised LDU decomposition

$$T(\lambda) = T^\mp(\lambda) D^\pm(\lambda) \hat{S}^\pm(\lambda)) \tag{16}$$

of the scattering matrix[1].

The reductions imposed on our Lax operators give rise to certain symmetry conditions on the Jost solutions, the scattering matrix and fundamental analytic solutions [16, 17], as follows:

$$\begin{aligned} \hat{\psi}_\pm^\dagger(x,\lambda^*) &= \psi_\pm(x,\lambda), & \hat{T}^\dagger(\lambda^*) &= T(\lambda), && (17) \\ \mathbf{C}\psi_\pm(x,-\lambda)\mathbf{C} &= \psi_\pm(x,\lambda), & \mathbf{C}T(-\lambda)\mathbf{C} &= T(\lambda), && (18) \\ \left[\chi^+(x,\lambda^*)\right]^\dagger &= \hat{\chi}^-(x,\lambda), & \mathbf{C}\chi^\pm(x,-\lambda)\mathbf{C} &= \chi^\pm(x,\lambda). && (19) \end{aligned}$$

As a simple consequence of their construction the fundamental analytic solutions χ^+ and χ^- are interrelated through:

$$\chi^+(x,\lambda) = \chi^-(x,\lambda) G(\lambda) \tag{20}$$

for some function $G(\lambda) = \hat{S}^-(\lambda) S^+(\lambda)$. This means that they can be viewed as solutions to a local Riemann-Hilbert problem [9, 21] with boundary given by real and imaginary axis

[1] The decomposition (16) is called generalised since all factors have a block structure which respects the splitting (9), that is $d_n^\pm(\lambda)$ are $n \times n$ matrices for example while $\mathbf{s}_\pm(\lambda)$ are $n \times m$ matrices.

in λ-plane. To be more precise solutions to a local Riemann-Hilbert problem are functions $\eta^{\pm} = \chi^{\pm}\exp(\mathrm{i}\lambda^2 Jx)$ which satisfy (14). The latter pair of solutions can be expanded [9]

$$\eta^{\pm}(x,\lambda) = \eta_0^{\pm}(x) + \frac{\eta_1^{\pm}(x)}{\lambda} + \frac{\eta_2^{\pm}(x)}{\lambda^2} + \dots \tag{21}$$

Substituting the asymptotic expansion (21) into (14) leads to the following sequence of recurrence relations

$$\lambda^2 \qquad [J, \eta_0^{\pm}(x)] = 0, \tag{22}$$
$$\lambda \qquad Q(x)\eta_0^{\pm}(x) - [J, \eta_1^{\pm}(x)] = 0, \tag{23}$$
$$\lambda^0 \qquad \mathrm{i}\partial_x \eta_0^{\pm}(x) + Q(x)\eta_1^{\pm}(x) - [J, \eta_2^{\pm}(x)] = 0, \tag{24}$$
$$\dots \qquad \dots$$

It is seen from (22) that $\eta_0^{\pm} \in \mathfrak{sl}^0(m+n)$ and it depends on x due to the second term in (24). All this tells us that the Riemann-Hilbert problem is not canonically normalized.

One application of the fundamental analytic solutions is in spectral theory of the scattering operator $L(\lambda)$ [20]. The resolvent of $L(\lambda)$ is defined by

$$L(\lambda) \circ R(\lambda) = \mathbb{1},$$

where $\circ$ means composition of linear operators. $R(\lambda)$ is an integral operator of the form

$$\left(R(\lambda)\, F\right)(x,t) = \int_{-\infty}^{\infty} \mathcal{R}(x,y,t,\lambda) F(y)\,\mathrm{d}\,y$$

for $F: \mathbb{R} \to \mathbb{C}^n$ being a continuous function. The following theorem holds true:

Theorem 2 *The kernel $\mathcal{R}(x,y,t,\lambda)$ is expressed through the fundamental analytic solutions as follows:*

$$\mathcal{R}(x,y,\lambda) = \left\{ \begin{array}{ll} \mathrm{i}\chi^{+}(x,\lambda)\Theta^{+}(x-y)\hat{\chi}^{+}(y,\lambda), & \mathrm{Im}\,\lambda^2 > 0 \\ -\mathrm{i}\chi^{-}(x,\lambda)\Theta^{-}(x-y)\hat{\chi}^{-}(y,\lambda), & \mathrm{Im}\,\lambda^2 < 0 \end{array} \right. \tag{25}$$

where $\Theta^{\pm}$ are matrix-valued functions given by

$$\Theta^{\pm}(x-y) = \theta(\pm(y-x))P - \theta(\pm(x-y))\,(\mathbb{1} - P)$$

and

$$P = \left(\begin{array}{cc} \mathbb{1}_m & \mathbf{0} \\ \mathbf{0} & \mathbf{0} \end{array} \right)$$

being a constant projector. The kernel $\mathcal{R}$ is a meromorphic function in $\mathbb{C}$ with a finite number of poles to form the discrete spectrum of the scattering operator L.

The proof of theorem 2 is quite similar to that one in the case of linear bundles this is why we shall skip it. The reader can find a detailed exposition of the proof in [4, 9]. As a direct consequence of theorem 2 one derives:

Corollary 1 *The spectrum of the scattering operator L comprises a continuous part and a discrete part. The continuous part of the spectrum is determined by requirement*

$$\mathrm{Im}\,\lambda^2 J = 0, \tag{26}$$

i.e. it coincides with the real and the imaginary axis in the spectral λ-plane. The discrete spectrum belongs to orbits of the reduction group $\mathbb{Z}_2 \times \mathbb{Z}_2$, i.e. all discrete eigenvalues go together in quadruplets $\{\pm\mu_k, \pm\mu_k^*\}_{k=1}^r$.

As in linear bundle case the fundamental analytic solutions allow one to construct nonlinear Fourier transform to map the potential Q onto spectral data. For that purpose one introduces the so-called adjoint solutions (or squared solutions)

$$\mathcal{E}_{ij}^{\pm}(x,\lambda) = \left(\chi^{\pm}(x,\lambda)E_{ij}\hat{\chi}^{\pm}(x,\lambda)\right)^{\perp},$$

which satisfy the adjoint representation of the linear problem (11):

$$\mathrm{i}\partial_x \mathcal{E}_{ij}^{\pm} + \lambda\left[Q - \lambda J, \mathcal{E}_{ij}^{\pm}\right] = 0, \tag{27}$$

where $(E_{ij})_{rs} = \delta_{ir}\delta_{js}$ are Weyl generators of the Lie algebra $\mathfrak{sl}(m+n)$ and $\perp$ stands for taking the block off-diagonal part of matrix.

3. Dressing Method and Special Solutions

This section is dedicated to an effective way of finding particular solutions of NLEE whose scattering problem is (11). The method we shall discuss here is dressing Zakharov-Shabat's method [21, 22]. Conceptually the dressing method is an indirect way of integration, i.e. it generates a solution to a given NLEE starting from a known one by substantially using auxiliary problem associated with it. Let ψ_0 be a fundamental solution to problem

$$L_0\psi_0 = \mathrm{i}\partial_x\psi_0 + \lambda(Q_0 - \lambda J)\psi_0 = 0, \tag{28}$$

where

$$Q_0(x) = \begin{pmatrix} 0 & \mathbf{q}_0^T(x) \\ \mathbf{q}_0^*(x) & 0 \end{pmatrix} \tag{29}$$

for some known $n\times m$ matrix $\mathbf{q}_0$ satisfying (5). Next we apply a gauge transform $\psi_0 \to \psi_1 = g\psi_0$ such that the auxiliary linear system remains covariant, i.e. we have

$$L_1\psi_1 = \mathrm{i}\partial_x\psi_1 + \lambda(Q_1 - \lambda J)\psi_1 = 0, \tag{30}$$

where Q_1 has the form (29) but for some other $n\times m$ matrix $\mathbf{q}_1$ to be found. Then the dressing factor g satisfies:

$$\mathrm{i}\partial_x g + \lambda Q_1 g - \lambda g Q_0 - \lambda^2[J,g] = 0. \tag{31}$$

Similarly, by comparing linear problems:

$$A_0(\lambda)\psi_0 = \mathrm{i}\partial_t\psi_0 + \sum_{k=1}^{2N}\lambda^k A_k^{(0)}\psi_0 = \psi_0 f(\lambda), \tag{32}$$

$$A_1(\lambda)\psi_1 = \mathrm{i}\partial_t\psi_1 + \sum_{k=1}^{2N}\lambda^k A_k^{(1)}\psi_1 = \psi_1 f(\lambda) \tag{33}$$

one derives another p.d.e. for the dressing factor

$$\mathrm{i}\partial_t g + \sum_{k=1}^{2N}\lambda^k A_k^{(1)} g - g\sum_{k=1}^{2N}\lambda^k A_k^{(0)} = 0. \tag{34}$$

Due to constraints (17)–(19) the dressing factor must obey the following symmetry conditions:

$$\hat{g}^{\dagger}(x,t,\lambda^*) = g(x,t,\lambda), \tag{35}$$

$$\mathbf{C}g(x,t,-\lambda)\mathbf{C}^{-1} = g(x,t,\lambda). \tag{36}$$

Journal of Physics: Conference Series **482** (2014) 012044 doi:10.1088/1742-6596/482/1/012044

The gauge transform acts on all fundamental solution including the Jost solutions. To ensure that the dressed Jost solutions have proper asymptotics (30) should be modified into:

$$\psi_{0,\pm} \to \psi_{1,\pm} = g\psi_{0,\pm}\,\hat{g}_{\pm}\,, \qquad g_{\pm} := \lim_{x\to\pm\infty} g\,. \tag{37}$$

Hence the scattering matrix and the fundamental analytic solutions transform into:

$$T_0 \quad\to\quad T_1 = g_+ T_0\,\hat{g}_-, \tag{38}$$

$$\chi_0^{\pm} \quad\to\quad \chi_1^{\pm} = g\chi_0^{\pm}\hat{g}_-. \tag{39}$$

Relations (39) and (25) lead to the conclusion that the resolvent kernel transform as:

$$\mathcal{R}_1(x,y,t,\lambda) = g(x,t,\lambda)\mathcal{R}_0(x,y,t,\lambda)\hat{g}(y,t,\lambda). \tag{40}$$

If g does not depend on λ then it follows straight from (31) and (34) that it is constant. To have non-trivial dressing we assume that g depends on the spectral parameter. At this point we recall that $\chi_0^{\pm}$ and $\chi_1^{\pm}$ satisfy Riemann-Hilbert problem with no canonical normalization. This means that the asymptotic value of the dressing factor as $|\lambda| \to \infty$ should be a nonvanishing function of x and t. The simplest possible type is a dressing factor with simple poles only. Due to constraint (36) we pick up g in the form:

$$g(x,t,\lambda) = \mathbb{1} + \sum_{j=1}^{r} \frac{\lambda}{\mu_j}\left(\frac{B_j(x,t)}{\lambda-\mu_j} + \frac{\mathbf{C}B_j(x,t)\mathbf{C}}{\lambda+\mu_j}\right), \qquad \operatorname{Re}\mu_j \neq 0,\ \operatorname{Im}\mu_j \neq 0. \tag{41}$$

while according to (35) its inverse reads:

$$\hat{g}(x,t,\lambda) = \mathbb{1} + \sum_{j=1}^{r} \frac{\lambda}{\mu_j^*}\left(\frac{B_j^\dagger(x,t)}{\lambda-\mu_j^*} + \frac{\mathbf{C}B_j^\dagger(x,t)\mathbf{C}}{\lambda+\mu_j^*}\right). \tag{42}$$

Setting $|\lambda| \to \infty$ in (31) and taking into account (41) and (42) we get the following interrelation

$$Q_1 = Q_0 + \sum_{j=1}^{r} [J, B_j - \mathbf{C}B_j\mathbf{C}]A^\dagger \tag{43}$$

between the seed solution Q_0 and dressed one. The matrix $A(x,t)$ is the asymptotic value of g given by:

$$A = \mathbb{1} + \sum_{j=1}^{r} \frac{1}{\mu_j}\left(B_j + \mathbf{C}B_j\mathbf{C}\right).$$

Thus to obtain a new solution one needs to know the residues of g. The latter are found from identity $g\hat{g} = \mathbb{1}$. Indeed, after evaluating the residue of $g\hat{g}$ at $\lambda = \mu_k$ we obtain:

$$B_k\left\{\mathbb{1} + \sum_j \frac{\mu_k}{\mu_j^*}\left(\frac{B_j^\dagger}{\mu_k-\mu_j^*} + \frac{\mathbf{C}B_j^\dagger\mathbf{C}}{\mu_k+\mu_j^*}\right)\right\} = 0, \qquad k = 1,\ldots,r. \tag{44}$$

To ensure that we shall obtain non-trivial result we assume the residues are degenerate matrices, i.e. we have $B_k = X_k F_k^T$ for some rectangular matrices $X_k(x,t)$ and $F_k(x,t)$. The factors X_k can be found by solving linear system:

$$F_k^* = \sum_{j=1}^{r} \frac{\mu_k^*}{\mu_j}\left(X_j\frac{F_j^T F_k^*}{\mu_j-\mu_k^*} - \mathbf{C}X_j\frac{F_j^T\mathbf{C}F_k^*}{\mu_j+\mu_k^*}\right). \tag{45}$$

On the other hand the matrix factors $F_k(x,t)$ are expressed in terms of fundamental solutions of the bare linear problem, namely:

$$F_k^T(x,t) = F_{0,k}^T(t)\hat{\psi}_0(x,t,\mu_k), \tag{46}$$

where $F_{0,k}^T$ are constants of integration, i.e. they depend on t only, see [20] for more detailed explanations. $F_{0,k}$ turn out to be exponential functions on time and one can propose the following simple rule

$$F_{k,0}^T \to F_{k,0}^T \, \mathrm{e}^{-\mathrm{i}f(\mu_k)t}, \tag{47}$$

where $f(\lambda)$ is the dispersion law of NLEE.

In order to illustrate those general constructions let us consider in more detail the case when the dressing factor has a single pair of simple poles only ($r = 1$), i.e. it looks like:

$$g(x,\lambda) = \mathbb{1} + \frac{\lambda B(x)}{\mu(\lambda-\mu)} + \frac{\lambda \mathbf{C}B(x)\mathbf{C}}{\mu(\lambda+\mu)}, \qquad \operatorname{Re}\mu \neq 0, \ \operatorname{Im}\mu \neq 0. \tag{48}$$

Assume now that $Q_0 = 0$. Therefore the bare solution ψ_0 is a plane wave:

$$\psi_0(x,t,\lambda) = \mathrm{e}^{-\mathrm{i}\lambda^2 Jx} \tag{49}$$

and the T matrix is simply equal to the unit $(m+n)\times(m+n)$ matrix. Taking into account (33) and the fact that the asymptotic values of g as $x \to \pm\infty$ are block-diagonal matrices we see that dressed scattering matrix T_1 is a block diagonal matrix. This situation generalizes in a natural way the concept of a reflectionless potential for the $\mathfrak{sl}(2)$ case.

Now linear system (45) is easily solved for X to give

$$X = \frac{\mu}{\mu^*}\left(\frac{F^TF^*}{\mu-\mu^*} - \frac{F^T\mathbf{C}F^*}{\mu+\mu^*}\mathbf{C}\right)^{-1} F^*, \tag{50}$$

where F is expressed in terms the seed solution. We shall restrict ourselves here with the case when $\operatorname{rank}B = 1$, that is X and F are $m+n$-vectors. Then after collecting all information from (50) and (47) and plug it into (43) the reflectionless potential acquires the form:

$$\begin{aligned} q_{1,ji}(x) &= Q_{1,ij+m}(x) = \sum_{k=m+1}^{m+n} \frac{2\mathrm{i}(n_i+1)\rho\sin(2\varphi)\mathrm{e}^{-\mathrm{i}\sigma_{ik}(x)}\mathrm{e}^{-\theta_{ik}(x)}}{\mathrm{e}^{-2\mathrm{i}\varphi}\sum_{p=1}^{m}\mathrm{e}^{-2\theta_{pk}(x)} + \sum_{p=m+1}^{m+n}\mathrm{e}^{2\xi_{0,pk}}} \times \\ & \left(\delta_{kj+m} - \frac{2\mathrm{i}\sin(2\varphi)\mathrm{e}^{\mathrm{i}(\delta_{j+m}-\delta_k-2\varphi)}\mathrm{e}^{2\xi_{0,j+mk}}}{\mathrm{e}^{-2\mathrm{i}\varphi}\sum_{p=1}^{m}\mathrm{e}^{-2\theta_{pk}(x)} + \sum_{p=m+1}^{m+n}\mathrm{e}^{2\xi_{0,pk}}}\right), \end{aligned} \tag{51}$$

where

$$\begin{aligned} \theta_{pk}(x) &= (n_p+1)\rho^2\sin(2\varphi)x - \xi_{0,pk}, \qquad \xi_{0,pk} = \ln|F_{0,p}/F_{0,k}|, \qquad \delta_p = \arg F_{0,p} \\ \sigma_{pk}(x) &= (n_p+1)\cos(2\varphi)x + \delta_p - \delta_k - \varphi, \qquad \mu = \rho\exp(\mathrm{i}\varphi). \end{aligned}$$

In order to obtain the soliton solution for matrix DNLS one needs to recover the t-dependence. This can be done by using (47) and taking into account that

$$f_{\mathrm{MDNLS}}(\lambda) = -\frac{n+m}{m}\lambda^4 J,$$

which leads to the following rule:

$$\begin{aligned}\delta_p \quad &\rightarrow \quad \delta_p + \begin{cases} \frac{(n+m)n}{m^2}\rho^4\cos(4\varphi)t, & p = 1,\ldots,m, \\ -\frac{n+m}{m}\rho^4\cos(4\varphi)t, & p = m+1,\ldots,m+n, \end{cases} \\ \xi_{0,\,pk} \quad &\rightarrow \quad \begin{cases} \xi_{0,\,pk} - \left(\frac{n+m}{m}\right)^2\rho^4\sin(4\varphi)t, & p = 1,\ldots,m, \\ \xi_{0,\,pk}, & p = m+1,\ldots,m+n. \end{cases}\end{aligned}$$

The result we have just obtained represents a generalization of Kaup-Newell's soliton derived in the case of the scalar DNLS equation [12].

It is clear that one can recursively apply the dressing procedure we demonstrate here, to build a whole sequence of exact solutions to a NLEE.

4. Conclusions

We have formulated the direct scattering problem for quadratic bundles related to Hermitian symmetric spaces of the type **A.III** in terms Jost solutions, scattering matrix, fundamental analytic solutions etc. We have discussed the spectral properties of the scattering operator L. In this sense our results generalize those obtained by Gerdjikov et al. in [5, 6] treating the scalar case as well as author's in [20] treating spaces of the type $SU(n)/S(U(1)\times U(n-1))$.

Zakharov-Shabat's dressing method has been adapted to quadratic bundles of aforementioned type. This allowed us to establish an algebraic procedure for construction of reflectionless potentials. For this to be done it suffices to use a dressing factor with simple poles. As a special case we have considered in more detail the simplest case when the dressing factor has a couple of poles. All reflectionless potentials give rise to multisoliton solutions to certain NLEEs. For instance, applying simple time recovery procedure we have easily produced one soliton solutions to the multicomponent DNLS (5). This result naturally generalizes the classical result by Kaup and Newell [12] for the soliton solution to scalar DNLS. The latter can be derived by using a dressing factor chosen in the form:

$$g(x,t,\lambda) = \mathbb{1} + \frac{\lambda B(x,t)}{\mu(\lambda-\mu)} + \frac{\lambda\sigma_3 B(x,t)\sigma_3}{\mu(\lambda+\mu)}\,.$$

The results presented here could be extended in several directions. Firstly, one can search for solutions of different type, say rational type solutions. One possible way to find rational solutions consists in using a factor g whose poles lie in the continuous spectrum of L, i.e. $\mu\in\mathrm{i}\mathbb{R}$. In this "degenerate" case one can obtain the following result for the scalar DNLS:

$$q(x,t) = \frac{4\mathrm{i}\mu^3(x+4\mu^2 t)}{\left[2\mathrm{i}\mu^2(x+4\mu^2 t)-1\right]^2}\,\mathrm{e}^{-2\mathrm{i}\mu^2(x+2\mu^2 t)}. \tag{52}$$

It is seen that it obeys zero boundary conditions as $x\to\pm\infty$. As in the case of linear bundles it is also possible to derive rational solutions but obeying more complicated non-trivial background conditions.

Another meaningful direction of further developments is to study quadratic bundles associated with other types Hermitian symmetric spaces or to put it even in a more general setting, bundles related to homogeneous spaces like the Lax operator given below:

$$L(\lambda) = \mathrm{i}\partial_x + U_0 + \lambda U_1 - \lambda^2 J\,, \tag{53}$$

where U_0 splits into a diagonal and off-diagonal part, U_1 is strictly off-diagonal and J is a diagonal matrix. It is evident that the theory of complete quadratic bundles like this one is more complicated than in the case we have considered in that report.

Acknowledgments
The author would like to acknowledge financial support from the Government of Ireland Postdoctoral Fellowship in Science, Engineering and Technology.

References

[1] Athorne C and Fordy A 1987 Generalised KdV and MKdV equations associated with symmetric spaces *J. Phys.* A **20** 1377–86
[2] Fordy A 1984 Derivative nonlinear Schrödinger equations and Hermitian symmetric spaces *J. Phys.* A: Math. Gen. **17** 1235–45
[3] Fordy A and Kulish P 1983 Nonlinear Schrödinger equations and simple Lie algebras *Commun. Math. Phys.* **89** 427–43
[4] Gerdjikov V 2009 On spectral theory of Lax operators on symmetric spaces: vanishing versus constant boundary conditions *J. Geom. & Symm. in Phys.* **15** 1–41
[5] Gerdjikov V and Ivanov M 1983 The quadratic bundle of general form and the nonlinear evolution equations. I. Expansions over the "squared" solutions — generalized Fourier transform *Bulg. J. Phys.* **10** 13–26
[6] Gerdjikov V and Ivanov M 1983 The quadratic bundle of general form and the nonlinear evolution equations. II. Hierarchies of Hamiltonian structures *Bulg. J. Physics* **10** 130–43
[7] Gerdjikov V, Grahovski G, Mikhailov A and Valchev T 2011 Polynomial bundles and generalised Fourier transforms for integrable equations on **A.III**-type symmetric spaces *SIGMA* **7** 096–143
[8] Gerdjikov V, Kostov N and Valchev T 2009 Solutions of multi-component NLS models and spinor Bose-Einstein condensates *Physica* D **238** 1306–10
[9] Gerdjikov V, Vilasi G and Yanovski A 2008 *Integrable Hamiltonian Hierarchies. Spectral and Geometric Methods* (Berlin, Heidelberg, New York: Springer Verlag)
[10] Helgasson S 1978 *Differential Geometry, Lie Groups and Symmetric Spaces* (New York, Academic Press)
[11] Ieda J, Miyakawa T and Wadati M 2004 Exact analysis of soliton dynamics in spinor Bose-Einstein condensates *Phys. Rev Lett.* **93** 194102
[12] Kaup D and Newell A 1978 An exact solution for a derivative nonlinear Schrödinger equation *J. Math. Phys.* **19** 798–801
[13] Kuznetsov E and Mikhailov A 1977 On the complete integrability of the two-dimensional classical Thirring model *Theor. & Math. Phys.* **30** 303–14
[14] Manakov S 1974 On the theory of two-dimensional stationary self-focusing of electromagnetic waves *Soviet Physics-JETP* **38** 248–53
[15] Mikhailov A 1976 Integrability of the two-dimensional Thirring model *JETP Letts.* **23** 356–58
[16] Mikhailov A 1980 Reductions in integrable systems. The reduction groups *Lett. in J. of Exper. & Theor. Phys.* **32** 187–92
[17] Mikhailov A 1981 The reduction problem and the inverse scattering method *Physica* D **3** 73–117
[18] Mio K, Ogino T, Minami K and Takeda S 1976 Modified nonlinear Schrödinger equation for Alfvén waves propagating along the magnetic field in cold plasmas *J. Phys. Soc. Japan* **41** 265–71
[19] Mjølhus E 1976 On the modulational instability of hydromagnetic waves parallel to the magnetic field *J. Plasma Phys.* **16** 321–34
[20] Valchev T 2013 On the quadratic bundles related to Hermitian symmetric spaces *J. Geom. & Symm. in Phys.* **29** 83–110
[21] Zakharov V, Manakov S, Novikov S and Pitaevskii L 1984 *Theory of Solitons: The Inverse Scattering Method* (New York: Plenum)
[22] Zakharov V and Shabat A 1979 Integration of nonlinear equations of Mathematical Physics by the method of the inverse scattering transform II *Funct. Anal. Appl.* **13** 13–22 (in Russian)

Physics and Mathematics of Nonlinear Phenomena 2013 (PMNP2013) IOP Publishing
Journal of Physics: Conference Series **482** (2014) 012045 doi:10.1088/1742-6596/482/1/012045

Kink–Antikink Scattering in φ^4 and ϕ^6 Models

H Weigel

Physics Department, Stellenbosch University, Matiland 7602, South Africa

E-mail: weigel@sun.ac.za

Abstract. For kink–antikink scattering within the φ^4 non–linear field theory in one space and one time dimension resonance type configurations emerge when the relative velocity between kink and antikink falls below a critical value. It has been conjectured that the vibrational excitation of the kink would be the source for these resonances because (simplified) collective coordinate calculations, that emphasized on this excitation, qualitatively reproduced those resonances. Surprisingly a numerical study in the ϕ^6 field theory also exhibited such resonances even though it does not contain the vibrational excitation. To explore this contradiction we start from the working hypothesis that in either model any collective coordinate ansatz which includes a degree of freedom similar to the vibrational excitation leads to resonances, regardless of whether or not this mode emerges as a solution to the (linearized) field equations. To this end we compare numerical results in the φ^4 and ϕ^6 models that arise from the full set of partial differential equations to those from the ordinary differential equations for a collective coordinate ansatz. An inaccuracy in literature formulas for the collective coordinate approach in the φ^4 model requires to revisit those calculations.

1. Motivation

The kink soliton in the φ^4 model is often considered as the prototype [1] configuration for more elaborate soliton systems that occur in field theory. The range of physics disciplines in which solitons are relevant is huge. Solitons appear in cosmology [2,3], particle and nuclear physics [4,5], as well as condensed matter physics [6]. Kink–antikink configurations are of particular interest because they can mimic particle–antiparticle interactions and might eventually provide more inside into fundamental concepts like crossing–symmetry in particle scattering [7]. Using modern desktop computers, numerical solutions to the field equations that initially represent a widely separated but approaching kink–antikink pair are feasible. This amounts to numerically integrating partial differential equations (PDE) for time and space dependent fields. Nevertheless it is interesting to simplify those equations by assuming collective coordinate ansätze. This reduces the PDE to coupled ordinary differential equations (ODE) for a limited number of time dependent functions. Equally interesting, collective coordinates may also answer the question of which are important modes of the system. In this context, the so–called *shape mode* in the φ^4 model has attracted particular attention. This mode is a bound state in the vibration spectrum about the kink [8]. It has been assumed to be responsible for the *bounce* type solutions to the PDE in kink–antikink interactions [9]. These bounce solutions are resonating kink–antikink configurations. The perception is that the shape mode can absorb sufficient energy from the kink–antikink system to prevent it from falling apart after collision. This role of the shape mode has been doubted as the ϕ^6 model also contains bounce type solutions in the kink–antikink sector(s) [10]. However, apart from the translational zero mode there is no bound state in the vibration spectrum of the kink in the ϕ^6 model. In this model thus a thorough investigation of the collective coordinate analysis is required to better understand this contradiction. We will do so in sections four and five and adopt the working hypothesis that a collective coordinate ansatz with a shape mode type component will be a suitable approximation in either case. Furthermore it has turned out that the collective coordinate calculations in the φ^4 model have inherited (typographical) errors in the formula for the source term of the shape mode from ref. [11]. We will therefore revisit those calculations in section three. In all cases we will compare results from the corresponding ODE to solutions of the PDE, the full field equations. The latter are obtained with programming code adapted from that used in ref. [12]. We conclude and comment on the relevance of collective coordinate calculations in section six. We will, however, commence with a brief review of kink–antikink approaches.

2. Kink–Antikink Concept

By now there have been many detailed numerical studies [9, 13–16] of the PDE for the field in the φ^4 model in one time and one space dimension[1]. The model is defined by the Lagrangian

$$\mathcal{L}_4 = \frac{1}{2}\partial_\mu\varphi\partial^\mu\varphi - \frac{1}{2}\left(\varphi^2 - 1\right)^2 , \tag{1}$$

where all model parameters have been absorbed by appropriate redefinitions of the coordinates (t, x) and the field φ. The field equation reads

$$\ddot{\varphi} - \varphi'' = 2\varphi\left(1 - \varphi^2\right) , \tag{2}$$

where dots and primes denote time and coordinate derivatives, respectively. The (anti)kink $\varphi_{K,\overline{K}} = \pm\tanh(x)$ is a static solution to this equation. It connects the two vacuum solutions at $\varphi_{\rm vac} = \pm 1$. Kink–antikink configurations

$$\varphi_{K\overline{K}}(x, X(t)) = \varphi_K(\xi_+) + \varphi_{\overline{K}}(\xi_-) - 1 = \tanh(\xi_+) - \tanh(\xi_-) - 1 \tag{3}$$

with

$$\xi_\pm = \frac{x}{\sqrt{1 - v_{\rm in}^2}} \pm X(t) \qquad \text{and} \qquad \dot{X}(0) = \frac{-v_{\rm in}}{\sqrt{1 - v_{\rm in}^2}} \tag{4}$$

are solutions for wide separation $X(t) \gg 0$. For these conventions $X(t)$ essentially measures the position of the antikink. More interestingly these configurations may serve as initial conditions for the equation of motion by choosing $X(0)$ large enough to avoid interference and the constant velocity $v_{\rm in}$ such that kink and antikink approach each other. A typical solution is shown in figure 1. The most interesting feature is the appearance of bounces for initial velocity $v_{\rm in}$ below a critical value $v_{\rm cr}$. The example in figure 1 has two bounces but solutions with many bounces (and traps) exist too [9]. The occurrence of such bounces has frequently been linked to the existence of the so–called shape mode

$$\chi(x, t) = {\rm e}^{i\omega_1 t}\chi_1(x) \qquad \text{with} \qquad \chi_1(x) = \frac{\sinh(x)}{\cosh^2(x)} \tag{5}$$

with eigen–frequency $\omega_1^2 = 3$ in the vibration spectrum of the kink [8]. (The continuum spectrum starts at $\omega^2 = 4$ for the present units.) The common argument is that the collective coordinate parameterization

$$\varphi_{\rm cc}(x, t) = \varphi_{K\overline{K}}(x, X(t)) + \sqrt{\tfrac{3}{2}}\,A(t)\left[\chi_1(x + X(t)) - \chi_1(x - X(t))\right] , \tag{6}$$

[1] See ref. [9] for a comprehensive summary of these computations.

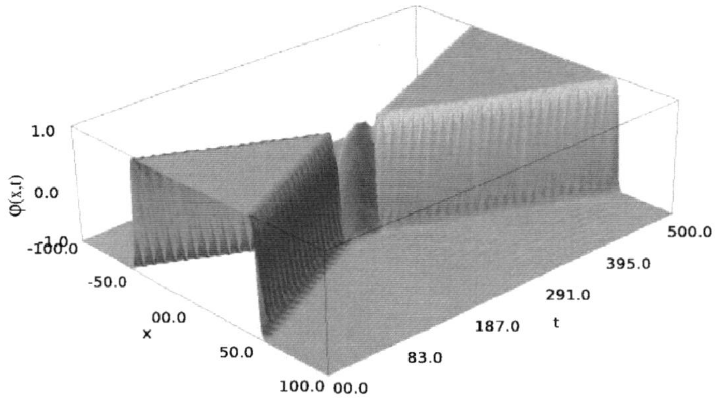

Figure 1. A typical solution to the PDE with kink–antikink initial conditions. Figure adopted from ref. [7].

that reduces the PDE (2) to the (simpler) ODE for the time dependent collective coordinates $A(t)$ and $X(t)$, not only approximates the full solution reasonably well, but also reproduces $v_{\rm cr}$ within 10% [9][2]. In the above ansatz the difference of the shape modes at $\xi_\pm$ is assumed because the (anti)kinks do not directly excite the sum.

Similar bounces and traps have been recently observed [10] in the ϕ^6 model which is defined by the Lagrangian

$$\mathcal{L}_6 = \frac{1}{2}\partial_\mu\phi\partial^\mu\phi - \frac{1}{2}\phi^2\left(\phi^2-1\right)^2 . \tag{7}$$

Although the corresponding field equation

$$\ddot{\phi} - \phi'' = -\phi\left(3\phi^4 - 4\phi^2 + 1\right) , \tag{8}$$

also allows for (anti)kink solutions $\phi_{K,\overline{K}} = [1+\exp(\pm 2x)]^{-\frac{1}{2}}$, there is no bound state with non–zero energy in the vibration spectrum about the kink. This suggests that the existence of the shape mode does not serve as a rigorous criterion for the occurrence of bounces and traps.

In view of the above observed puzzle we reverse that line of argument into a working hypothesis. We want to test the assumption that the solutions of the ODE for the collective coordinate ansätze reasonably well approximate the solutions to the full PDE; regardless of whether or not $\chi(x,t)$ solves the field equation for vibrations in the kink background. For the φ^4 model the ansatz is given in eq. (6). There are two analogs in the ϕ^6 model

$$\begin{aligned}\phi_{\rm cc}(x,t) &= \phi_K(\xi_+) + \phi_{\overline{K}}(\xi_-) + \sqrt{\tfrac{3}{2}}\,A(t)\left[\chi_1(x+X(t)) - \chi_1(x-X(t))\right] , & (9)\\ \overline{\phi}_{\rm cc}(x,t) &= \phi_K(\xi_-) + \phi_{\overline{K}}(\xi_+) - 1 + \sqrt{\tfrac{3}{2}}\,A(t)\left[\chi_1(x+X(t)) - \chi_1(x-X(t))\right] , & (10)\end{aligned}$$

since there are three possible vacuum solutions, $\phi_{\rm vac} = 0, \pm 1$. We substitute these ansätze into the Lagrangian and integrate over the spatial degree of freedom. Formally the resulting Lagrangian for the collective coordinates reads

$$L_6(A,\dot{A},X,\dot{X}) = \int dx\, \mathcal{L}_6(\varphi_{\rm cc})$$

[2] Another argument is based on the perturbed sine–Gordon model [17].

$$= a_1\dot{X}^2 - a_2 + a_3\dot{A}^2 - a_4A^2 + a_5A + a_6\dot{X}^2A + a_7\dot{X}\dot{A}$$
$$+a_8\dot{X}^2A^2 + a_9A\dot{X}\dot{A} - a_{10}A^3 - a_{11}A^4 - a_{12}A^5 - a_{13}A^6 . \qquad (11)$$

The coefficient functions depend on the (relative) distance parameter, *i.e.* $a_i = a_i(X)$ for $i = 1, \ldots, 13$. They also have a parametrical dependence on the initial velocity $v_{\rm in}$ via the construction in eq. (4). The φ^4 model has a similar collective coordinate Lagrangian, just that a_{12} and a_{13} are absent. The actual form of the a_i, of course, depends on whether eq. (9), eq. (10) or the φ^4 model is considered. As an example, we list

$$a_1(X) = \frac{1}{2}\int_{-\infty}^{\infty} dx \left[{\rm e}^{2\xi_+}\left(1+{\rm e}^{2\xi_+}\right)^{-\frac{3}{2}} + {\rm e}^{-2\xi_-}\left(1+{\rm e}^{-2\xi_-}\right)^{-\frac{3}{2}}\right] \qquad (12)$$

for the system of eq. (9) in the ϕ^6 model. It is straightforward to compute these coefficients numerically for any prescribed value of X. This is completely sufficient for the subsequent numerical integration of the ODE for $X(t)$ and $A(t)$ that follow from the variational principle for the collective coordinate Lagrangians. For the φ^4 model analytic expressions for some of the coefficients have been derived some time ago [11, 14]. Unfortunately, the formulas presented in ref. [11] contain some misprints which propagated through the literature and make the quoted numerical results for the collective coordinate approach in the φ^4 model unreliable. We therefore revisit those calculations in the next section. The subsequent sections contain novel studies on the collective coordinate approach in the ϕ^6 model.

3. Revisiting the φ^4 model

The most impressive result from the collective coordinate approach in the φ^4 model is the prediction for the critical velocity $v_{\rm cr} \approx 0.289$ [9] above which bounces and traps cease to exist. This is only about 10% off from the exact result from the PDE, 0.26 [14]. However, a number of (uncontrollable) approximations to the system were used: All non–harmonic terms were omitted ($a_{10}, \ldots, a_{13} = 0$), the ODE were diagonalized ($a_6, \ldots, a_9 = 0$) and the direct interactions between the shape modes at $\pm X(t)$ were discarded via $a_i \mapsto \lim_{X\to\infty} a_i(X)$ for $i = 3, 4$ [16]. The latter approximation also avoids the null–vector problem discussed in ref. [18]

The source term for the shape mode, a_5 is interesting because it represents a small amplitude variation about the kink–antikink configuration. It should hence vanish only when $\varphi_{K\overline{K}}$ is a static solution to the field equation. To gain some insight into the form of a_5 before computing it explicitly it is instructive to consider

$$\varphi_{K\overline{K}}(0, X) = 2\tanh(X) - 1 \longrightarrow \begin{cases} 1 & \text{for} \quad X \to +\infty \\ -3 & \text{for} \quad X \to -\infty . \end{cases} \qquad (13)$$

The second case suggests that $\varphi_{K\overline{K}}(x, -\infty)$ is not a solution to the field equations because then the center between widely separated kink and antikink profiles is not a vacuum configuration. But $\varphi(x, +\infty)$ is a valid solution. Hence, a_5 cannot be symmetric under $X \to -X$. However, that symmetry is erroneously reflected by the expression in the appendix of ref. [11], wherein it is called $F(X)$. Unfortunately, this incorrect expression has been used in all subsequent studies of the ODE. Redoing the calculation shows that the correct result is (for $v_{\rm in} = 0$)

$$a_5(X) = 3\sqrt{6}\pi\left[2 - 2\tanh^3(X) - \frac{3}{\cosh^2(X)} + \frac{1}{\cosh^4(X)}\right], \qquad (14)$$

which obviously is not symmetric under $X \to -X$. Changing the power of the tanh(X) term from 3 to 2 and doubling the arguments of all hypergeometric functions reproduces the expression

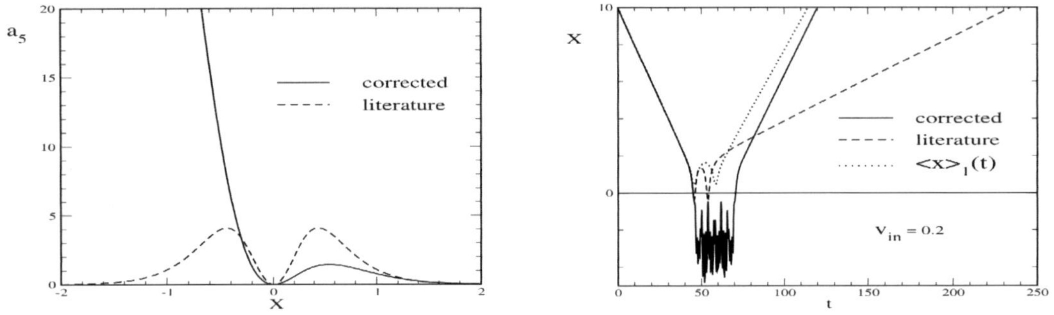

Figure 2. Left panel: Correction of the coefficient function $a_5(X)$ for $v_{\rm in} = 0$ (*corrected* refers to eq. (14), *literature* denotes the expression derived in ref. [11] and frequently adopted thereafter). Right panel: comparison of the resulting time dependence of the collective coordinate $X(t)$; the dotted line stems from the PDE with $n = 1$ in eq. (16).

in ref. [11]. In figure 2 the consequences of correcting this coefficient function are studied. As discussed above, the invariance of a_5 under reflection is lost. More importantly the solutions to the Euler–Lagrange equations for $X(t)$ and $A(t)$ that we solve for the initial conditions[3]

$$X(0) \to \infty\,, \quad \dot{X}(0) = \frac{-v_{\rm in}}{\sqrt{1-v_{\rm in}^2}}\,, \qquad A(0) = 0 \quad \text{and} \quad \dot{A}(0) = 0\,, \tag{15}$$

change drastically when corresponding initial conditions and all of the above listed approximations are imposed. In figure 2 this is exemplified for the kink–antikink distance for the initial relative velocity $v_{\rm in} = 0.2$. Most notably, once the correction is installed, multiple bounces occur at any initial velocity.

A main intention is to quantitatively compare the solutions to the ODE to those of the PDE for the full field equation. We therefore estimate the time–dependent position of the the antikink as the expectation value

$$\langle x\rangle_n(t) = \frac{\int_0^\infty dx\, x\, \epsilon_4^n(t,x)}{\int_0^\infty dx\, \epsilon_4^n(t,x)}\,. \tag{16}$$

Here

$$\epsilon_4(t,x) = \frac{1}{2}\left[\ddot{\varphi} + \varphi'' + \left(\varphi^2 - 1\right)^2\right] \tag{17}$$

is the energy density for the solution $\varphi = \varphi(x,t)$ to the PDE (2) for initial configurations described by equation (6) at $t = 0$. The integer n can be freely chosen to eventually emphasize effects. Since the energy density is typically strongly localized, increasing n turns the distribution against which x is tested, into a δ–function. From the right panel of figure 2 we also see that the corrected ODE result shows almost no resemblance with the PDE solution, while the literature (though incorrect) one correctly reproduces the number of bounces. However, with the correction for a_5 installed the out–going velocity is well reproduced.

In view of these results we must question the above listed approximations. This is even more the case as the consistency conditions emerging from $\varphi_{K\overline{K}}(x,\infty)$ being a solution for $v_{\rm in} \neq 0$ read[4]

$$\lim_{X\to\infty}\left[a_5(X) + a_6(X)\,\dot{X}^2(0)\right] = 0 \qquad \text{and} \qquad \lim_{X\to\infty} a_7(X) = 0 \tag{18}$$

[3] Numerically we find that $X(0) \sim 10$ is an adequate representation of infinity as no overlap between the initial structures occurs.

[4] We have verified them both analytically and numerically.

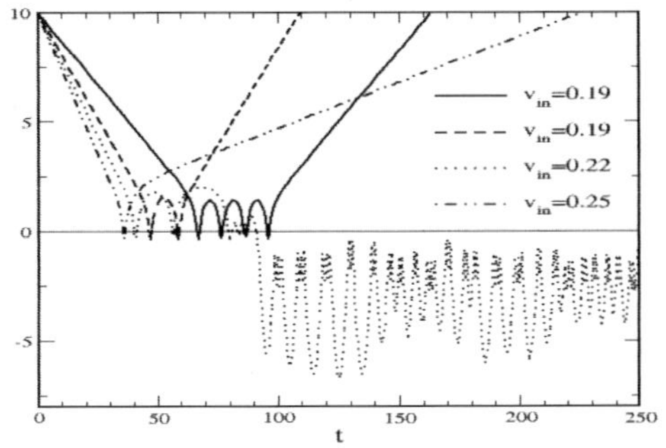

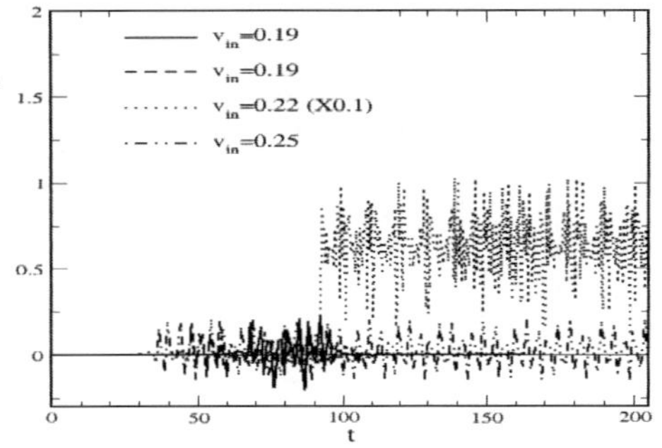

Figure 3. Solution to the ODE in the φ^4 model. Left panel: kink–antikink separation, right panel: amplitude of shape mode (note the change of scale in the $v_{\rm in} = 0.22$ entry).

and thus prohibit the omission of a_6. We have numerically solved the ODE for $X(t)$ and $A(t)$ with the sole approximation being that on $a_{3,4}$ to avoid the null–vector problem. (We comment on possible improvements in the summary section.). The results of these calculations are shown in figure 3. It turns out that with the correction on a_5, the corresponding contribution to the energy, $E_5 = -a_5(X)A$ may absorb much energy when X becomes negative. Hence trapping type solutions with large amplitudes of the shape mode emerge. A posteriori, this again disqualifies the harmonic approximation. The entry with $v_{\rm in} = 0.22$ in figure 3 is a typical example thereof. Most notably, however, we find that without the many approximations, the ODE indeed predict a critical velocity of $v_{\rm cr} = 0.247$ above which trapping or bounce type solution cease to exist. This compares favorably to the PDE result of 0.26 [14].

4. Antikink–Kink in ϕ^6

In the ϕ^6 model there are two types of initial structures that contain kink and antikink configurations. We first consider the so–called *antikink–kink* ansatz [10] of eq. (9) that, as in the φ^4 model, eq. (3), parameterizes a legitimate solution only for $X \to \infty$ but not for $X \to -\infty$. Hence it cannot describe penetration either. To compare the solutions to the ODE to those of the PDE for the full field equation in the ϕ^6 model we need to replace the energy density in eq. (16) by

$$\epsilon_6(t,x) = \frac{1}{2}\left[\ddot{\phi} + \phi'' + \phi^2\left(\phi^2 - 1\right)^2\right] . \tag{19}$$

Here $\phi = \phi(x,t)$ are the solutions to the PDE (8) with initial conditions taken as the $t = 0$ configuration in eq. (9).

The PDE yields a critical velocity of $v_{\rm cr,PDE} = 0.05$ above which no traps or bounces occur. As shown in figure 4 the solutions to the ODE and PDE differ substantially at low initial velocities. Typically the bounce frequency from the ODE is much larger. So is the predicted critical velocity $v_{\rm cr,ODE} = 0.357$. It is thus suggestive that the collective coordinate approach over–estimates the attraction between kink and antikink. For velocities above $v_{\rm cr,ODE}$ this can also be observed as the ODE prediction for the final velocity is lower than that from the PDE. In the close vicinity of that critical velocity the technical parameters that enter the numerical treatment of the ODE seem to matter, thereby indicating chaotic behavior. In particular, the results are sensitive to the coordinate value that is supposed to represent infinity; as can be seen from the $v_{\rm in} = 0.35$ entry.

For large initial velocities the PDE results actually exhibit some pseudo–bounces shown in figure 5. The kink and antikink linger on top of each other for a moderate time interval. If the

Journal of Physics: Conference Series **482** (2014) 012045 doi:10.1088/1742-6596/482/1/012045

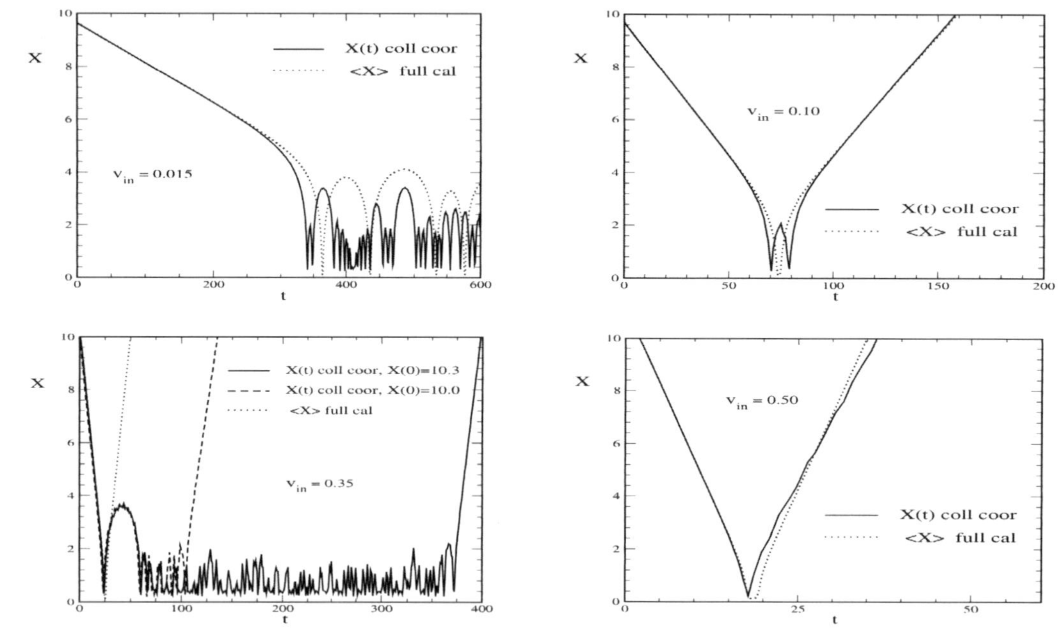

Figure 4. Solutions to the ODE for antikink–kink scattering in the ϕ^6 model and comparison to PDE results (*full cal*) with $n = 1$ in eqs. (16) and (19).

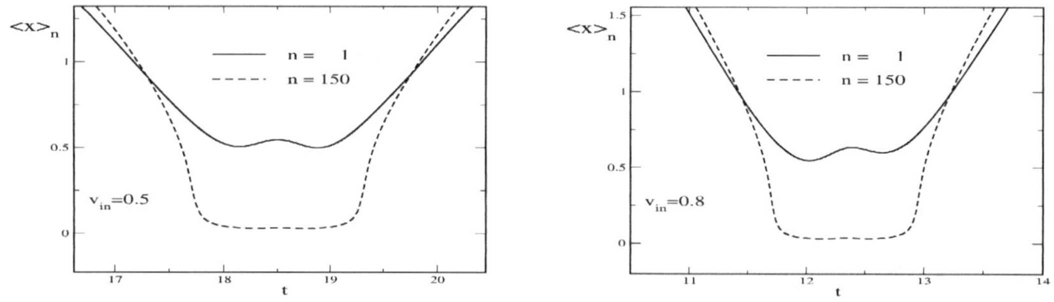

Figure 5. Pseudo–bounces in the antikink–kink system of the ϕ^6 model from the PDE . The parameter n enters the computation the expectation value of the position of the antikink via eq. (16).

distribution in eq. (16) is taken to be broad, the trajectory feigns mini–bounces.

5. Kink–Antikink in ϕ^6

In contrast to the previously discussed configurations the so–called *kink–antikink* structure from eq. (10) has topological properties which would allow it to be a solution even in the case $X \to -\infty$. However, the potential a_2 in eq. (11) is not symmetric under $X \to -X$ and $a_2(-\infty) > a_2(\infty)$. Hence there is a velocity threshold for solutions that correspond to the kink penetrating the antikink. Simple kinematical considerations on the ODE coefficients a_i indicate this threshold to be $v_{\rm th} = 0.296$. The numerical solution to the ODE yields a slightly larger value, 0.302. The left panel in figure 6 shows that the configuration with initial velocity $v_{\rm in} = 0.30$ bounces while bigger values of $v_{\rm in}$ yield $X(t) \to -\infty$ at large times. The ODE yields a window $0.216 \leq v_{\rm in} \leq 0.302$ in which the system bounces exactly twice. As for the φ^4 model,

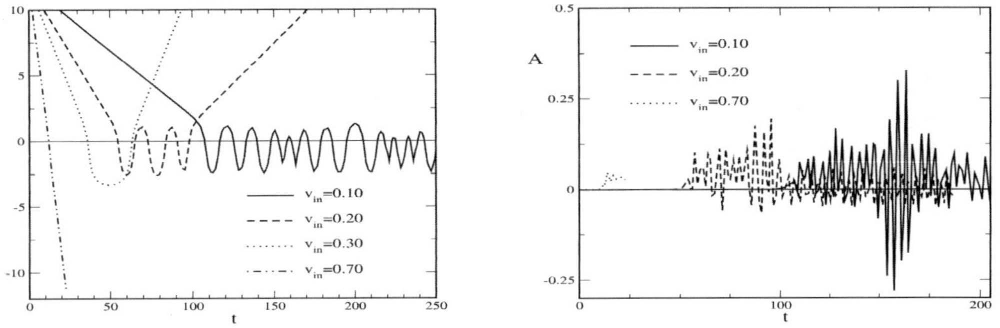

Figure 6. Solutions for kink–antikink interactions from the ODE system in the ϕ^6 model. Left panel: kink–antikink separation, right panel: amplitude of shape mode.

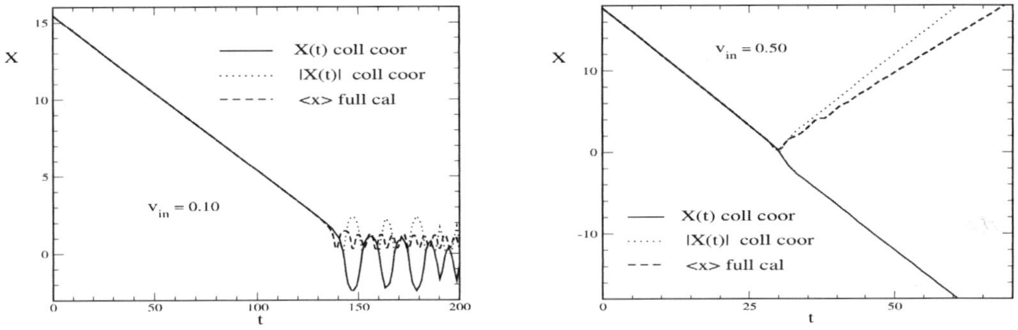

Figure 7. Comparison of PDE and ODE solutions for kink–antikink interactions in the ϕ^6 model. *Full cal* refers to $\langle x \rangle_1$ in eqs. (16) and (19).

multiple bounce solutions are characterized by a sizable amplitude of the shape mode. Even though this is not a valid vibrational solution it supports the working hypothesis that, once the collective coordinate ansatz allows for its excitation, it will do so. It should also be noted that the upper bound of the above mentioned window agrees reasonably well (but not perfectly) with the critical velocity (0.289) from the PDE calculation.

For $v_{\rm in} = 0.10$ and $v_{\rm in} = 0.50$ we compare the PDE and ODE results for the distances between kink and antikink figure 7. The PDE calculation for $\langle x \rangle$ is as in eqs. (8), (16) and (19). However, the initial condition at $t = 0$ is now taken from eq. (10). The structure within ODE and PDE solutions agree. However, if applicable, the frequency of bounces differs considerably.

6. Summary

Collective coordinate calculations are quite useful in non–linear field theory not only because they simplify the field equations considerably but also because they potentially provide insight into the underlying dynamics by focusing on particular modes. Within the φ^4 and ϕ^6 models in one time and one space dimension we have compared the ODE solutions from the collective coordinate approaches to those from the PDE for the space–time dependent fields. Our analysis, however, has revealed that care in needed when attempting to draw rigorous conclusions from collective coordinate calculations. Not only do they depend on the specific ansätze but also on the approximations made. For example, the harmonic approximation in the φ^4 model incorrectly yields solutions with repeated bounces for any relative initial velocity of the kink–antikink configuration (after correcting a literature error within the collective coordinate Lagrangian).

On the other hand the PDE yields quite a small critical velocity beyond which multiple bounces cease to exist.

Our studies support the working hypothesis that a shape mode in the vibration spectrum of the kink is not a necessity for bounces and traps to occur in kink–antikink scattering. We started from the conjecture that a collective coordinate ansatz including such a mode would represent the full field configuration reasonably well regardless of whether or not it is a true bound state of the kink. We have explored the (dis)agreements between the ODE and PDE results. These (dis)agreements are qualitatively equal in the φ^4 and ϕ^6 models. Since the shape mode only exists within the φ^4 model the collective coordinate approach can therefore not be used to establish that the shape mode is responsible for the existence of multiple bounces in kink–antikink scattering.

Though we have successfully waved a number of previous approximations within the ODE calculation, we have still omitted the interaction between the shape modes at $\pm X$. Otherwise the calculations run into the null–vector problem [18]. This problem arises because by pure construction of the ansatz, the coefficient of A vanishes for $X = 0$, *cf.* eq. (6). The particular combination of shape modes at $\pm X$ is commonly assumed because the more general parameterization with three collective coordinates (X, A, B)

$$\varphi_{\rm cc}(x,t) = \varphi_{K\overline{K}}(x, X(t)) + \sqrt{\tfrac{3}{2}}\left[A(t)\chi_1(x + X(t)) + B(t)\chi_1(x - X(t))\right] \tag{20}$$

only yields a linear source term for $A - B$. Of course, the combination $A + B$ will nevertheless be excited by non–linear and/or higher order effects. It will therefore be interesting to investigate the above parameterization. This will complicate the collective coordinate approach slightly as it turns it into a 3×3 problem. Beyond that, the ODE approach is probably no longer a sensible simplification to the PDE.

Acknowledgments

The author is grateful to A. M. H. H. Abdelhady for his collaboration at early stages of this study and his comments on the manuscript. The organizers of the conference are thanked for providing the opportunity to present these results. Support from the National Research Foundation of South Africa is acknowledged.

References

[1] Rajaraman R 1982 *Solitons and Instantons* (Amsterdam: North Holland).
[2] Vilenkin A and Shellard E P S 1994 *Cosmic Strings and other Topological Defects* (Cambridge: Cambridge University Press).
[3] Vachaspati T 2006 *Kinks and domain walls: An introduction to classical and quantum solitons* (Cambridge: Cambridge University Press).
[4] Manton N and Sutcliffe P 2004 *Topological Solitons* (Cambridge: Cambridge University Press).
[5] Weigel H, *Chiral 2008 Soliton Models for Baryons* Lect. Notes Phys. **743** 1.
[6] Bishop A R. and Schneider T (Eds.) 1978 *Solitons in Condensed Matter Physics* (Berlin: Springer Verlag).
[7] Abdelhady A M H H 2012 *Scattering in Soliton Models and Crossing Symmetry*, MSc Thesis, Stellenbosch University, unpublished.
[8] See, for example, chapter 5 in Ref. [1].
[9] Goodman R H and Haberman R 2005 *SIAM J. App. Dyn. Sys.* **4** 1195.
[10] Dorey P, Mersh K, Romanczukiewicz T and Shnir Y 2011 *Phys. Rev. Lett.* **107** 091602.
[11] Sugiyama T 1979 *Prog. Theor. Phys.* **61** 1550.
[12] Abdelhady A M H H and Weigel H 2011 *Int. J. Mod. Phys.* A **26** 3625.
[13] Kudryavtsev A E 1975 *Pisma Zh. Eksp. Teor. Fiz.* **22** 178.
[14] Campbell D K, Schonfeld J F and Wingate C A 1983 *Physica* D **9** 1.
[15] Belova T I and Kudryavtsev A E 1988 *Physica* D **32** 18.
[16] Anninos P, Oliveira S and Matzner R A 1991 *Phys. Rev.* D **44** 1147.
[17] Goodman R H and Haberman R 2004 *Physica* D **195** 303.
[18] Caputo J G and Flytzanis N 1991 *Phys. Rev.* A **44** 6219.

Journal of Physics: Conference Series **482** (2014) 012046 doi:10.1088/1742-6596/482/1/012046

Laboratory realization of KP-solitons

Harry Yeh[1] and Wenwen Li[2]

[1] Oregon State University, Corvallis, Oregon 97331, USA
[2] URS Corporation, Los Angeles, California 90017, USA

E-mail: harry@engr.orst.edu; wenwen.li@urs.com

Abstract. Kodama and his colleagues presented a classification theorem for exact soliton solutions of the quasi-two-dimensional Kadomtsev-Petviashvili (KP) equation. The classification theorem is related to non-negative Grassmann manifold, $\mathrm{Gr}(N, M)$ that is parameterized by a unique chord diagram based on the derangement of the permutation group. The cord diagram can infer the asymptotic behavior of the solution with arbitrary number of line solitons. Here we present the realization of a variety of the KP soliton formations in the laboratory environment. The experiments are performed in a water tank designed and constructed for precision experiments for long waves. The tank is equipped with a directional-wave maker, capable of generating arbitrary-shaped multi-dimensional waves. Temporal and spatial variations of water-surface profiles are captured using the Laser Induces Fluorescent method – a nonintrusive optical measurement technique with sub-millimeter precision. The experiments yield accurate anatomy of the KP soliton formations and their evolution behaviors. Physical interpretations are discussed for a variety of KP soliton formations predicted by the classification theorem.

1. Introduction

Based on the assumption of weakly nonlinear, weakly dispersive water waves that have small deviation in the transverse direction y from the main wave propagation direction x, we take the following orders in the effects of wave nonlinearity, dispersion and directivity:

$$\frac{a_0}{h_0} = \mathcal{O}(\epsilon), \qquad \left(\frac{h_0}{\lambda_0}\right)^2 = \mathcal{O}(\epsilon), \qquad \tan^2 \psi_0 = \mathcal{O}(\epsilon),$$

respectiverly, where a_0 is the wave amplitude, h_0 is the constant water depth at the quiescent state, λ_0 is the wavelength scale, and ψ_0 is the small oblique angle from the x direction. Then, the Euler formulation is approximated by the Kadomtsev-Petviashvili (KP) equation of the form

$$\left(\eta_t + c_0\eta_x + \frac{3c_0}{2h_0}\eta\eta_x + \frac{c_0 h_0^2}{6}\eta_{xxx}\right)_x + \frac{c_0}{2}\eta_{yy} = 0, \qquad (1)$$

in which η is the water-surface elevation from h_0, and $c_0 = \sqrt{gh_0}$. The KP equation (1) can be expressed in the following canonical non-dimensionalized form,

$$\left(4u_T + 6uu_\xi + u_{\xi\xi\xi}\right)_\xi + 3u_{YY} = 0, \qquad (2)$$

where the non-dimensional variables (ξ, Y, T) and u are defined by

$$\eta = \frac{2h_0}{3}u, \quad x - c_0 t = h_0\,\xi, \quad y = h_0\,Y, \quad t = \frac{3h_0}{2c_0}T.$$

With the form of the KP equation (2), we consider the solution in the form

$$u(\xi, Y, T) = 2\partial_\xi^2\left(\ln \tau(\xi, Y, T)\right),$$

where τ is referred to as the τ-function defined in the Wronskian determinant. For a 2-soliton case, τ-function can be written as

$$\tau = Wronskian(f_1, f_2) = \begin{vmatrix} f_1 & \partial_\xi f_1 \\ f_2 & \partial_\xi f_2 \end{vmatrix}.$$

In this paper, the term '2-soliton' is used to describe a wave condition with the presence of two solitons each as $y \to \pm\infty$. The functions f_1 and f_2 satisfy the linear equations: $\partial_Y f_i = \partial_\xi^2 f_i$ and $\partial_T f_i = -\partial_\xi^3 f_i$ (see e.g. Hirota [1]). Soliton solutions can be written by exponential functions of the form:

$$E_j = \exp(\theta_j) = \exp(k_j\xi + k_j^2 Y - k_j^3 T).$$

Therefore,

$$\begin{pmatrix} f_1 \\ f_2 \end{pmatrix} = \begin{pmatrix} a_{11} & a_{12} & a_{13} & a_{14} \\ a_{21} & a_{22} & a_{23} & a_{24} \end{pmatrix} \begin{pmatrix} E_1 \\ E_2 \\ E_3 \\ E_4 \end{pmatrix} = A \begin{pmatrix} E_1 \\ E_2 \\ E_3 \\ E_4 \end{pmatrix}.$$

Here the coefficient matrix $A = (a_{ij})$ is a constant 2×4 matrix. Thus each solution $u(\xi, Y, T)$ is completely determined by the A-matrix and the k-parameters. The foregoing is generalized for arbitrary number N f-functions and M E-exponential functions. The τ-function leads to the notion of Grassmannian variety: $\mathrm{Gr}(N, M)$. When constraints are applied for the regular soliton solutions, the τ-function is identified as a point of the totally nonnegative Grassmannian cell. The asymptotic soliton solutions for $y \to \pm\infty$ can be parameterized by the permutations, which lead to the introduction of chord diagram to express the classification of soliton solutions as a chord joining a pair of k_j's following its permutation representation. Detailed mathematical descriptions are given by Chakravarty and Kodama [2] and Kodama [3]. Figure 1 shows seven possible types of $N = 2$ solutions.

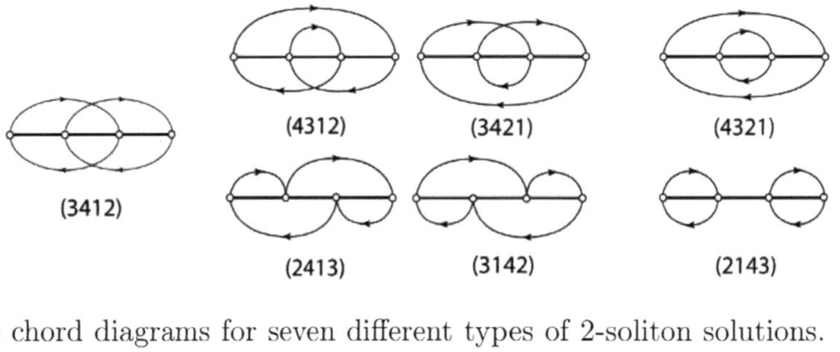

Figure 1. The chord diagrams for seven different types of 2-soliton solutions. The number in the parenthesis is the permutation.

As for the simplest case, consider the solution of single line-soliton that can be expressed as

$$\begin{aligned}\tau &= E_1 + aE_2 \\ &= 2\sqrt{a}\exp\frac{1}{2}(\theta_1 + \theta_2)\cosh\frac{1}{2}(\theta_1 - \theta_2 - \ln a).\end{aligned}$$

Therefore,

$$u = 2\partial_\xi^2(\ln\tau) = \frac{1}{2}(k_1 - k_2)^2 \mathrm{sech}^2 \frac{1}{2}(\theta_1 - \theta_2 - \ln a),$$
$$= A_{[i,j]}\mathrm{sech}^2 \sqrt{\frac{A_{[i,j]}}{2}}(\xi + Y\tan\Psi_{[i,j]} - C_{[i,j]}T - x^0_{[i,j]}),$$

where $A_{[i,j]} = \frac{1}{2}(k_j - k_i)^2$; $\tan\Psi_{[i,j]} = k_i + k_j$; $C_{[i,j]} = k_i^2 + k_i k_j + k_j^2 = \frac{1}{2}A_{[i,j]} + \frac{3}{4}\tan^2\Psi_{[i,j]}$, and $\Psi_{[i,j]}$ is the direction of wave propagation. The soliton formation and the corresponding code diagram is shown in Fig. 2.

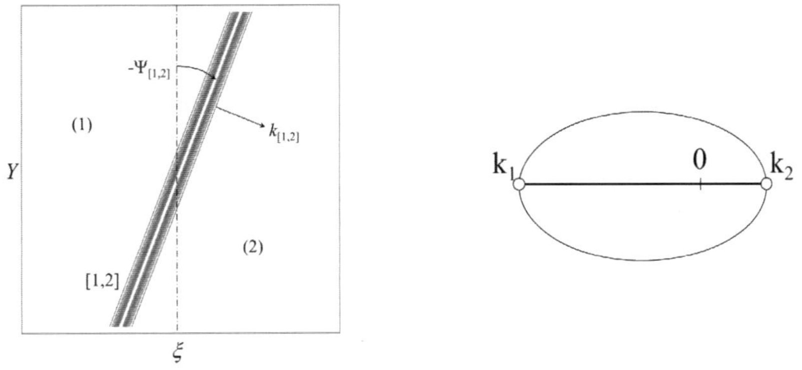

Figure 2. A line-soliton contour plot and the corresponding chord diagram.

2. Laboratory Experiments

Laboratory experiments are conducted in the wave tank, 7.3 m long, 3.6 m wide, and 0.30 m deep: see Fig. 3. The tank is elevated 1.2 m above the laboratory floor to enable us to make measurements with optical instruments. The bottom and sidewalls are made of 12.7 mm thick glass plates. The wave basin is equipped with a 16-axis directional-wavemaker system along the 3.6-m long headwall, capable of generating arbitrary-shaped, multi-directional waves. Each wave paddle is pushed through hinge connections by two adjacent linear-servo-motor motion devices. The paddles are made of PVDF (Polyvinylidene fluoride) plates that are driven horizontally in piston-like motions; they are optimally flexible to form a smooth curve between the driving axes. Each paddle has a maximum horizontal stroke of 55 cm – more than adequate to generate very long and multiple waves in a water depth. This wavemaker system, together with the precise wave tank, is ideal for investigation of nonlinear dynamics of long-wave motion. It is emphasized that the generated waves are precisely replicable: for example, the maximum error is less than 0.06 mm (or 0.1% of the depth) for a solitary wave with amplitude $a = 1.73$ cm in the water depth of $h_0 = 6.0$ cm. Note that, the horizontal dimensions of the tank scaled by the depth are $120h_0 \times 60h_0$, which is considered extremely large. Incidentally, the wave Reynolds number is on the order of 12,000 that we consider large enough to circumvent serious viscous scale effects.

The Laser Induced Fluorescent (LIF) method is used to examine temporal and spatial variations of water-surface profiles. Figure 4 (left) shows a setup for the LIF method. A laser beam (from a 5W diode-pumped solid-state laser) is converted to a thin laser sheet using a cylindrical lens. Two front-surface mirrors direct the laser sheet to illuminate the vertical plane. With the aid of fluorescein dye dissolved in water, the vertical laser-sheet illumination induces the dyed water to fluoresce and identifies the water-surface profile directly and non-intrusively.

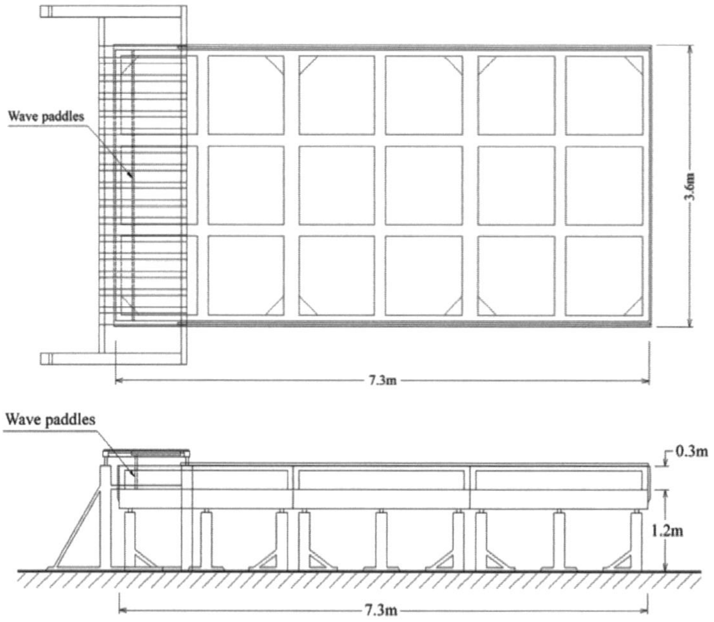

Figure 3. Schematic drawings of the laboratory apparatus: (upper) a plan view showing the waveguide that creates an oblique wave reflection and the tank frame that can be seen through the bed that is made of glass plates; (lower) an elevation view.

The illuminated profiles, as the wave passes through the light sheet, are recorded by a high-speed high-resolution video camera. The captured images are rectified and processed with the calibrated image so that the resulting images can be analyzed quantitatively. The transparent glass bed of the tank minimizes the reflection of the laser illumination that could have caused contamination in the image analysis for the wave profiles. The temporal and spatial variations of the water-surface profiles captured with the LIF technique are shown in Fig. 4 (right).

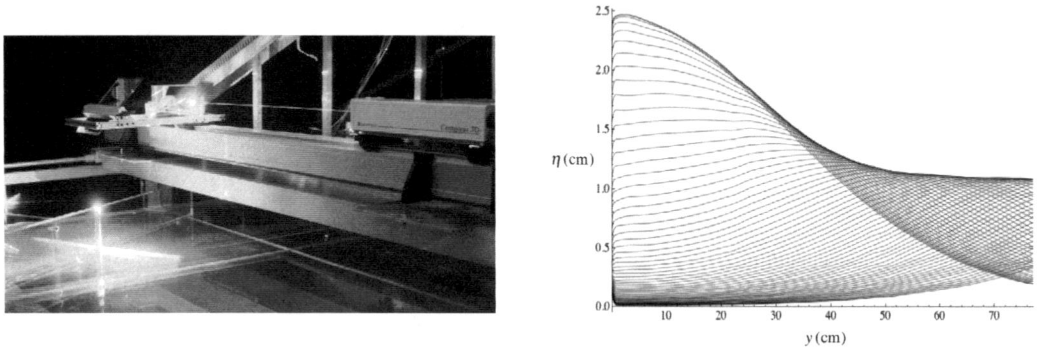

Figure 4. (Left) Setup for the Laser-Induced Fluorescent (LIF) method. (Right) Temporal variation of the water surface profile along the vertical wall at $y = 0$. The profiles were constructed by making a montage of three LIF images. The time interval of each profile is 1/100 sec.

One of the difficulties associated with the LIF technique for measuring long waves is the limitation in resolution. Unlike experiments for capillary waves or breaking waves, long waves have an inherently small vertical-to-horizontal scale ratio. This causes insufficient resolution in the vertical direction. Our laboratory experiments require measurements of small wave amplitudes (a few centimeters) in a large horizontal span (more than 100 cm). To circumvent this difficulty, LIF water-surface profiles are repeated on approximately 30 cm segments, creating a montage of the segment profiles.

3. Results

We first demonstrate the relation between a wave pattern of two solitons and the corresponding chord diagram. Let us consider, for example, two line solitons at $y \to \pm\infty$: namely the one has the wave amplitude $a_{[1,2]} = 0.35$ cm with the inclination angle $\psi_{[1,2]} = -20°$, and the other has the amplitude $a_{[3,4]} = 0.70$ cm with $\psi_{[3,4]} = 30°$ in the water depth $h_0 = 6.0$ cm: see Fig. 5 (left). Using the four relations $A_{[i,j]} = \frac{1}{2}(k_j - k_i)^2$ and $\tan\Psi_{[i,j]} = k_i + k_j$, we determine four k-parameters: $k_1 = -0.3785$, $k_2 = 0.01457$, $k_3 = 0.03250$, and $k_4 = 0.5448$. Because the asymptotic line-solitons are presented with [1,2] and [3,4] for $y \to \pm\infty$, the corresponding chord diagram is $\pi = (2143)$, which is also called the "O-type" 2-soliton because this solution was "originally" found to express the two-soliton solution [2]. Because the chords are closed independently, representing two individual line-solitons, the two solitons are not resonantly interacting and the waveform simply translates in time, maintaining the small phase shift at the intersection of the two solitons. It is noted that the chord diagram does not contain the initial phase information, which is represented in $A = (a_{ij})$ matrix.

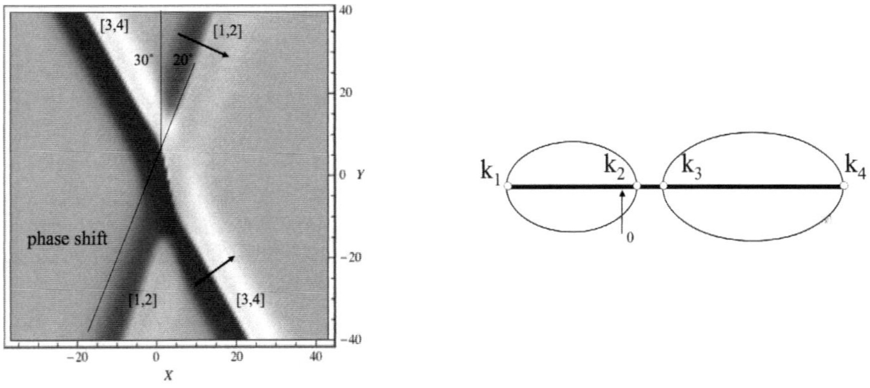

Figure 5. Theoretical water-surface profile of O-type 2-soliton solution of the KP equation, and the corresponding chord diagram with $\pi = (2143)$.

3.1. $\pi = (2143)$ and $\pi = (3142)$

In the laboratory, we create 2-soliton waveforms that are symmetric about the x-axis by utilizing the wave reflection at the vertical wall: in other words, only a half of the domain, $0 < y < \infty$, is realized in the laboratory taking the advantage of the symmetric wave pattern. Figure 6 shows the measured 2-soliton patterns of $\pi = (2143)$ (O-type) and $\pi = (3142)$, respectively. The tank sidewall is located along the top edge of the plot. The horizontal direction in the plot represents time that increases to the right. Negligibly small capillary-wave noise emanating from the wall can be detected in the figure. As shown in the figure, we observe non-resonant interaction in $\pi = (2143)$ and the stem-like formation along the wall remains unchanged, because this stem wave is a consequence of the phase shift resulted from the wave intersection.

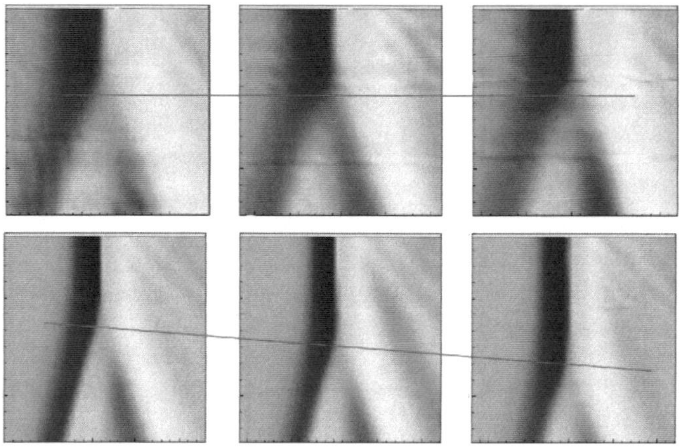

Figure 6. (Upper) O-type wave-profile evolution $a_i/h_0 = 0.076$ ($\kappa = 1.395$) and (lower) (3142)-type $a_i/h_0 = 0.277$ ($\kappa = 0.731$). $\psi_i = 30°$, $x/h_0 = 71.1$, 96.1, and 121.1 from the left panel to the right. Positive t-axis points to the right and positive y-axis points downward.

On the other hand, in the case of the (3142)-type, the stem-wave along the wall continually grows as it propagates. According to Miles [4], π = (2143) type and π = (3142) type are distinguished by the parameter $\psi_i/\sqrt{2A_0}$; recently, Li et al. [5] found that the parameter should be presented by $\kappa = \dfrac{\tan\psi_i}{\cos\psi_i\sqrt{3a_i/h_0}}$ for real-world solitons, in which a_i is the physical wave amplitude. When $\kappa > 1$, the wave pattern is represented by π = (2143) type, and when $\kappa < 1$, π = (3142). (Note that Kodama [6] formally derived the higher-order correction for the parameter κ, which includes the correction in terms of a_i.)

Figure 7 depicts the chord diagram of π = (2143) and π = (3142), respectively. The amplitude of [1,4]-soliton (the stem wave) can be inferred from the chord diagram of π = (3142). According to Miles [4], the stem-wave amplification is predicted by $(1+\sqrt{\alpha_r})^2$ where α_r is the reflected wave amplification of the Mach reflection or equivalently, the ratio of the smaller wave amplitude to the larger one: i.e. $\alpha_r = A_{[3,4]}/A_{[1,3]}$ at $y \to +\infty$ or $A_{[1,2]}/A_{[2,4]}$ at $y \to -\infty$. The theoretical amplitude prediction of [1,4]-soliton can be found in the chord diagram as

$$A_{[1,4]} = \frac{1}{2}(k_4-k_1)^2 = \frac{1}{2}\left[(k_3-k_1)+(k_4-k_3)\right]^2 = \frac{1}{2}\left[(k_3-k_1)\left(1+\sqrt{\frac{\frac{1}{2}(k_4-k_3)^2}{\frac{1}{2}(k_3-k_1)^2}}\right)\right]^2,$$

which represents the amplification of in fact $(1+\sqrt{\alpha_r})^2$. The transition between (2143)-type and (3142)-type can be depicted as the condition $(k_3-k_2) \to 0$. At this critical state, the

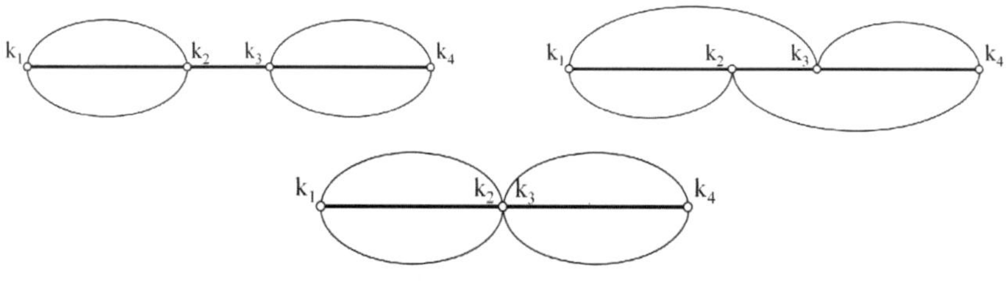

Figure 7. Chord diagrams of π = (2143) and π = (3142) types and their critical state.

amplification of the stem wave can be correctly interpreted from the chord diagram: namely the 4-fold amplification, $A_{[1,4]} = \frac{1}{2}(k_4 - k_1)^2 = 4 \times \frac{1}{2}(k_2 - k_1)^2$.

3.2. π = (3412)

According to Chakravarty and Kodama [2], π = (3412)-type solution is called the T-type solution, and the T-type solution has four asymptotic solitons that form an exact "X" shape near the origin. Nonetheless, the wave pattern deviates from the exact "X" shape near the soliton intersection. Figure 8 compares the KP solution with our laboratory measurements for the T-type solitons. At $t = 0$, the exact X-shaped wave pattern is created. The continual growth of [1,4] and [2,3] waves (see Fig. 1) is observed in both theoretical prediction and the laboratory observation, and the observed "box-shaped" wave pattern is consistent with the prediction.

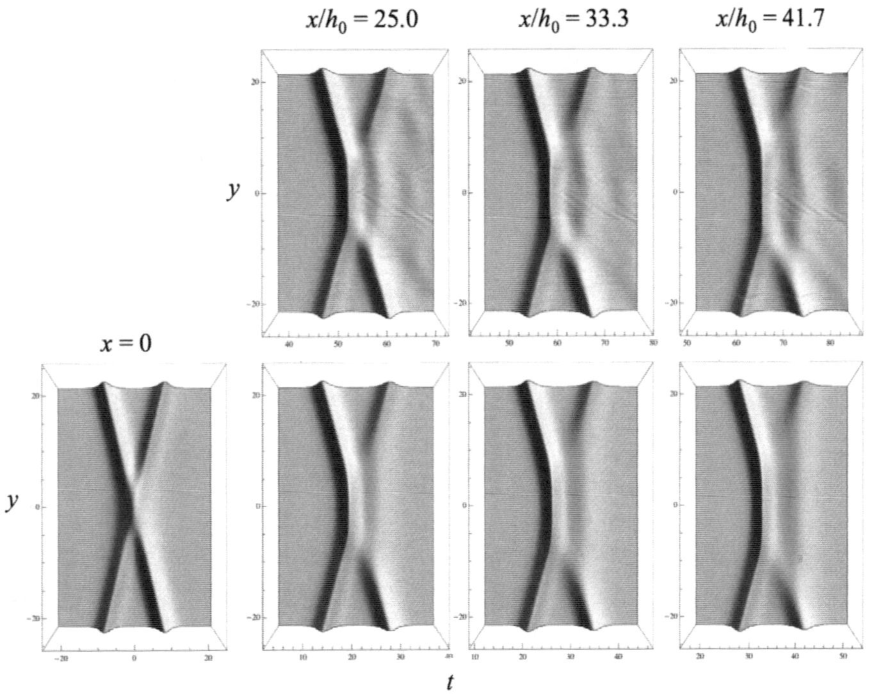

Figure 8. Laboratory (upper) and theoretical (lower) wave patterns for π = (3412)-type (T-type). $A_i = 0.280$; $\psi_i = 20°$.

The exact X-shaped wave pattern is made in the laboratory wave tank as shown in Fig. 9. Unlike (3412)-type, this is the O-type solution. Immediately after the formation of the exact X-shaped waveform at $t = 0$, the waves adjacent to the intersection are bent so as to accommodate the formation of the phase shift. Once it happens, the waveform remains steady.

3.3. V-Shaped Initial Waves

Initially V-shaped waves are generated along the sidewall in the laboratory with the aid of directional wavemaker system (see Fig. 10). This initial waveform creates the incomplete chord diagrams. Kao and Kodama [7] conducted the numerical experiments for the similar cases based on the KP theory. Here we verify their numerical findings in the real-fluid environment.

Figure 11 shows the case with $\kappa = \dfrac{\tan\psi}{\cos\psi\sqrt{3a_{[2,3]}/h_0}} = 1$, and $a_{[1,3]}/a_{[2,3]} = 4.0$. This is the

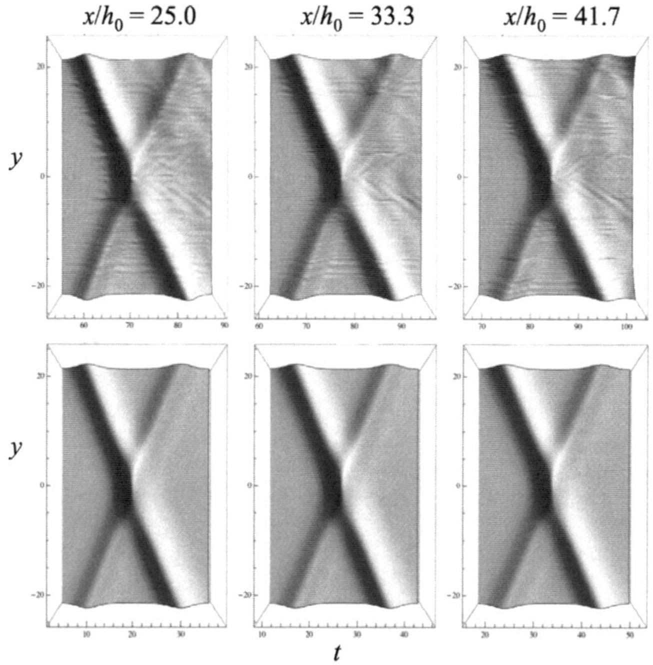

Figure 9. Laboratory (upper) and theoretical (lower) wave patterns for $\pi = (2143)$-type (O-type). $A_i = 0.090$; $\psi_i = 30°$.

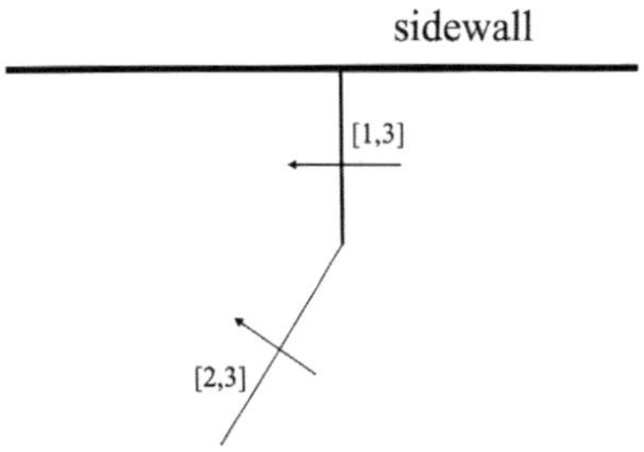

Figure 10. A sketch of the V-shaped wave generation in the laboratory.

critical condition of the Mach reflection [4]. The observation shows that the V-shaped wave is quickly transformed to the Y-shaped wave by completing the chord diagram by linking k_2 to k_1 (Fig. 11). Once the Y-shaped wave is formed by completing the chord diagram, the wave becomes stable maintaining the constant crest length of the stem wave in the laboratory setup. The actual wave amplitudes measured at $x/h_0 = 50$ are $a_{[2,3]}/h_0 = 0.133$, $a_{[1,2]}/h_0 = 0.130$, and $a_{[1,3]}/h_0 = 0.459$, consequently $\kappa = 1.05$. Note that $a_{[2,3]}/a_{[1,2]} \approx 1$ as expected. The discrepancy must be due to viscous dissipation that is unavoidable in the real-fluid experiments. Because of the amplitude attenuation, the value of κ continually but slowly increases in time.

Figure 12 shows the cases for $\kappa > 1$ and $\kappa < 1$, respectively. In the case of $\kappa > 1$, the crest length of the stem wave continually decreases as it propagates, which means that the origin of k in the chord diagram practically shifts to the right and $\psi_{[1,3]}$ is now slightly negative to make the Y-shaped wave formation steady; in other words, the observation coordinates should be rotated slightly in the clockwise direction. Also observed is $a_{[2,3]} < a_{[1,2]}$: the actual wave amplitudes measured at $x/h_0 = 50$ are $a_{[2,3]}/h_0 = 0.090$ and $a_{[1,2]}/h_0 = 0.124$, consequently $a_{[2,3]}/a_{[1,2]} < 1$.

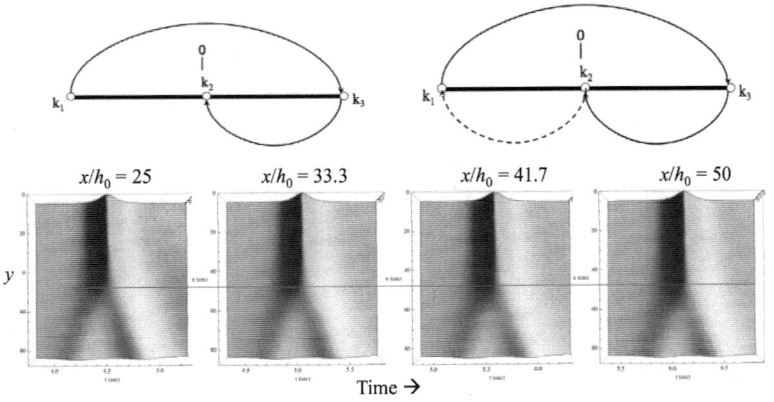

Figure 11. Incomplete Y-shaped soliton generated in the laboratory for the case of $\kappa = 1$; the generated wave amplitudes are $a_{[2,3]}/h_0 = 0.148$ and $a_{[1,3]}/h_0 = 0.592$ with the oblique angel $\tan\psi_{[2,3]} = 0.577$.

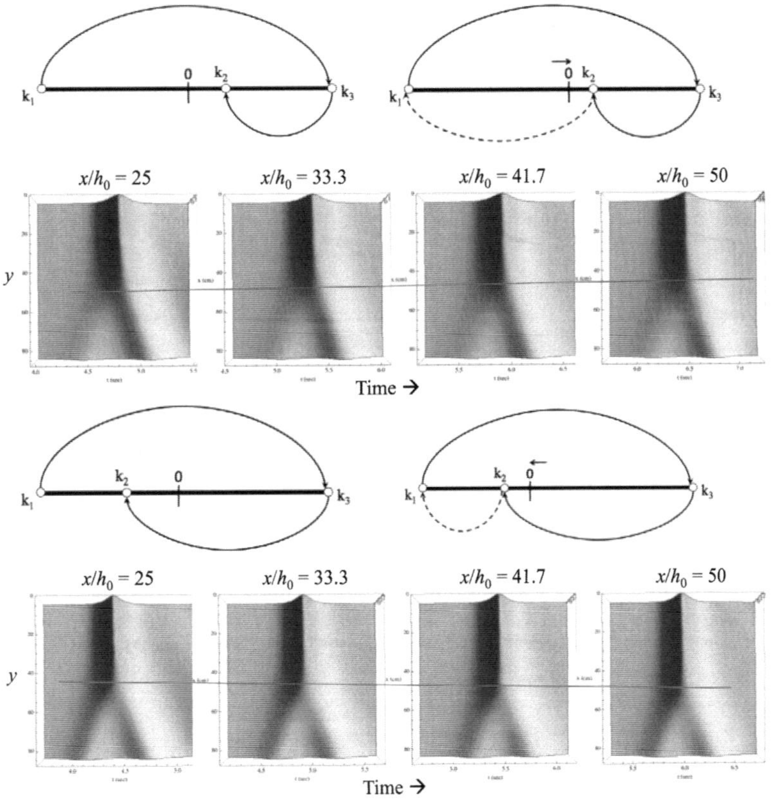

Figure 12. Incomplete Y-shaped soliton generated in the laboratory for the case of (top) $\kappa = 1.22 > 1$: the generated wave amplitudes are $a_{[2,3]}/h_0 = 0.10$ and $a_{[1,3]}/h_0 = 0.491$ with the oblique angel $\tan\psi_{[2,3]} = 0.577$; (bottom) $\kappa = 0.92 < 1$: the generated wave amplitudes are $a_{[2,3]}/h_0 = 0.175$ and $a_{[1,3]}/h_0 = 0.645$ with the oblique angel $\tan\psi_{[2,3]} = 0.577$.

In the case of $\kappa < 1$, the crest length of the stem wave continually increases as it propagates. To make the Y-shaped wave pattern steady, the origin $k = 0$ in the chord diagram should be shifted to the left and $\psi_{[1,3]}$ is now slightly positive, or the observation coordinate system is rotated slightly in the counterclockwise direction. Opposite to the case of $\kappa > 1$, this case yields $a_{[2,3]} > a_{[1,2]}$: the actual wave amplitudes measured at $x/h_0 = 50$ are $a_{[2,3]}/h_0 = 0.155$ and $a_{[1,2]}/h_0 = 0.129$, consequently $a_{[2,3]}/a_{[1,2]} > 1$. Those observations are in accord with the KP theory and well described in the chord diagram.

4. Summary

The classification theorem for 2-soliton solutions of the KP theory is validated by performing the experiments in the precisely controlled laboratory apparatus. The chord diagrams resulted from the classification theorem do not only represent the asymptotic solitons at $|y| \to \infty$, but also heuristically imply the interaction behaviors, even quantitatively in some cases.

We first demonstrated the difference between the (2143)-type (i.e. O-type) and the (3142)-type. The stem-wave-like formation is a consequence of the phase-shift associated with the (2143)-type and the intersecting wave formation remains constant and steady. On the other hand, the (3142)-type shows the resonant interaction behavior; consequently, the interacting stem wave continually grows its crest length. The chord diagram can provide the quantitative wave amplitude of the stem wave. Furthermore, the critical four-fold amplification at $\kappa = 1.0$ can be illustrated by the chord diagram.

The (3412)-type (i.e. T-type) soliton is realized in the laboratory, which creates the box-shaped wave formation at the intersection. The box-shaped wave formation grows in time in accordance with the KP theory. On the other hand, the exact X-shaped initial wave formation does not necessarily lead to the (3412)-type. When the soliton amplitude is sufficiently small, the X-shaped wave formation becomes the (2143)-type (i.e. O-type). In this case, the wave crests near the intersection are bent so as to accommodate the necessary phase shift associated with the (2143)-type.

Lastly, we studied the V-shaped KP-soliton formation at $t = 0$. This V-shaped soliton formation can be interpreted as an incomplete Y-shaped wave with missing one upper arm, or the corresponding chord diagram is not completely closed. It is observed that immediately after the generation, the V-shaped wave becomes the Y-shaped wave formation by completing the chord diagram. Three distinct cases of the incomplete Y-shaped wave formation are demonstrated in the laboratory, presenting the wave transformation behaviors in accordance with the three sets of the chord diagrams.

In short, we demonstrated the KP soliton formations in the real-fluid environment, and presented the physical implications of the chord diagram without solving the KP equation.

Acknowledgments

We are grateful to Yuji Kodama of Ohio State University for inspiring discussions. This work was supported by Oregon Sea Grant Program, and the Edwards Endowment through Oregon State University.

References

[1] Hirota R 2004 *The Direct Method in Soliton Theory* (New York: Cambridge University Press)
[2] Chakravarty S and Kodama Y 2009 *Studies in Applied Mathematics.* **87** 83–151
[3] Kodama Y 2010 *Journal of Physics A: Mathematical and Theoretical.* **43** 434004
[4] Miles J W 1977 *Journal of Fluid Mechanics.* **79** 157–69
Miles J W 1977 *Journal of Fluid Mechanics.* **79** 171–79
[5] Li W, Yeh H and Kodama Y 2011 *Journal of Fluid Mechanics.* **672** 326–57
[6] Kodama Y 2012 KP solitons and Mach reflection in shallow water *Preprint* 1210.0281
[7] Kao C and Kodama Y 2012 *Mathematics and Computers in Simulation.* **82,7** 1185–218

Quantum Gaudin model and classical KP hierarchy

A. Zabrodin

Institute of Biochemical Physics, 4 Kosygina, 119334, Moscow, Russia; ITEP, 25 B. Cheremushkinskaya, 117218, Moscow, Russia; National Research University Higher School of Economics, 20 Myasnitskaya Ulitsa, Moscow 101000, Russia

Abstract. This short note is a review of the intriguing connection between the quantum Gaudin model and the classical KP hierarchy recently established in [1]. We construct the generating function of integrals of motion for the quantum Gaudin model with twisted boundary conditions (the master T-operator) and show that it satisfies the bilinear identity and Hirota equations for the classical KP hierarchy. This implies that zeros of eigenvalues of the master T-operator in the spectral parameter have the same dynamics as the Calogero-Moser system of particles.

1. Introduction

In [1], a remarkable correspondence between the quantum Gaudin model and the classical Kadomtsev-Petviashvili (KP) hierarchy was established. It is a limiting case of the correspondence between quantum spin chains with the Yangian $Y(gl(N))$ symmetry algebra and the classical modified KP (mKP) hierarchy based on the construction of the master T-operator [2, 3]. The master T-operator was introduced in [2] (in a preliminary form, it previously appeared in [4]). It is a special generating function for commuting integrals of motion in the quantum model. In the Gaudin model case, any eigenvalue of the master T-operator appears to be the tau-function of the KP hierarchy, with polynomial dependence on the spectral parameter. Taking into account the well known story about dynamics of poles of rational solutions to soliton equations [5, 6, 7], this implies a link between the quantum Gaudin model [8] and the classical Calogero-Moser (CM) system of particles [9]. This link was also observed earlier in [10] using different arguments.

In this paper we present the results of [1] in a short compressed form. All proofs and technical details are omitted. Here we outline the results reviewed in the paper. Most of them can be obtained by a limiting procedure from the corresponding results for quantum spin chains proved in [2].

Using the matrix derivative operation, we construct commuting integrals of motion for the $gl(N)$ Gaudin model, with twisted boundary conditions and vector representations at the marked points in the quantum space, corresponding to arbitrary representations in the auxiliary space. The master T-operator is their generating function. It depends on an infinite number of auxiliary "time variables" $\mathbf{t} = \{t_1, t_2, t_3, \ldots\}$, where t_1 can be identified with the spectral parameter x. The master T-operator satisfies the bilinear identity for the classical KP hierarchy and hence any of its eigenvalues is a KP tau-function [13, 14]. This is a development of earlier

Journal of Physics: Conference Series **482** (2014) 012047 doi:10.1088/1742-6596/482/1/012047

studies [15, 16, 17] clarifying the role of the Hirota bilinear equations [18] in quantum integrable models.

Moreover, all eigenvalues of the master T-operator are *polynomial* tau-functions in $x = t_1$. Therefore, according to [6, 7], the dynamics of their roots in t_i with $i > 1$ is given by equations of motion of the classical CM system of particles. The marked points x_i in the Gaudin model (the inhomogeneities at the sites in the spin chain language) should be identified with initial coordinates of the CM particles while eigenvalues of the Gaudin Hamiltonians are proportional to their initial momenta. Eigenvalues of the Lax matrix for the CM model coincide with eigenvalues of the twist matrix (with certain multiplicities). Therefore, with fixed integrals of motion in the CM model determined by invariants of the twist matrix, there are finite number of solutions for the values of initial momenta which correspond to different eigenstates of the Gaudin model. In other words, the eigenstates of the Gaudin Hamiltonians are in one-to-one correspondence with (a finite number of) intersection points of two Lagrangian submanifolds in the phase space of the classical CM model. This "quantum-classical correspondence" was also discussed in [11, 12] in the context of supersymmetric gauge theories and branes.

In distinction to paper [1], here we consider the Gaudin model with a formal Planck's constant $\hbar$. In this case the master T-operator is the tau-function of the $\hbar$-dependent version of the KP hierarchy [19] and the coupling constant of the CM model becomes proportional to $\hbar^2$.

2. The quantum Gaudin model

Let $\mathsf{e}^{\hbar}_{ab}$ be generators of the "$\hbar$-dependent version" of the universal enveloping algebra $U(gl(N))$ with the commutation relations $[\mathsf{e}^{\hbar}_{ab}, \mathsf{e}^{\hbar}_{a'b'}] = \hbar(\delta_{a'b}\mathsf{e}^{\hbar}_{ab'} - \delta_{ab'}\mathsf{e}^{\hbar}_{a'b})$. The parameter $\hbar$ will play the role of the Planck's constant. Let π_λ be the finite dimensional irreducible representation of $U(gl(N))$ with the highest weight λ. We identify λ with the Young diagram $\lambda = (\lambda_1, \lambda_2, \ldots, \lambda_\ell)$ with $\ell = \ell(\lambda)$ non-zero rows, where $\lambda_i \in \mathbb{Z}_+$, $\lambda_1 \geq \lambda_2 \geq \ldots \geq \lambda_\ell > 0$. For example, $\pi_{(1)}$ is the N-dimensional vector representation corresponding to the 1-box diagram $\lambda = (1)$. We have $\pi_{(1)}(\mathsf{e}^{\hbar}_{ab}) = \hbar e_{ab}$, where e_{ab} is the standard basis in the space of $N \times N$ matrices: the matrix e_{ab} has only one non-zero element (equal to 1) at the place ab: $(e_{ab})_{a'b'} = \delta_{aa'}\delta_{bb'}$. Note that $I = \sum_a e_{aa}$ is the unity matrix and $P = \sum_{ab} e_{ab} \otimes e_{ba}$ is the permutation operator acting in the space $\mathbb{C}^N \otimes \mathbb{C}^N$.

In the tensor product $U(gl(N))^{\otimes n}$ the generators $\mathsf{e}^{\hbar}_{ab}$ can be realized as $\mathsf{e}^{\hbar\,(i)}_{ab} := \mathrm{id}^{\otimes(i-1)} \otimes \mathsf{e}^{\hbar}_{ab} \otimes \mathrm{id}^{\otimes(n-i)}$. Clearly, they commute for any $i \neq j$ and any a, b because act non-trivially in different spaces. Similarly, for any matrix $g \in \mathrm{End}(\mathbb{C}^N)$ we define $g^{(i)}$ acting in the tensor product $\mathcal{V} = (\mathbb{C}^N)^{\otimes n}$: $g^{(i)} = I^{\otimes(i-1)} \otimes g \otimes I^{\otimes(n-i)} \in \mathrm{End}(\mathcal{V})$. In this notation, $P_{ij} := \sum\limits_{a,b} e^{(i)}_{ab} e^{(j)}_{ba}$ is the permutation operator of the i-th and j-th tensor factors in $\mathcal{V} = \mathbb{C}^N \otimes \ldots \otimes \mathbb{C}^N$.

Fix n distinct numbers $x_i \in \mathbb{C}$ and a diagonal $N \times N$ matrix $g_0 = \mathrm{diag}\,(k_1, \ldots, k_N)$. (We assume that the k_i's are all distinct and non-zero.) We will call g_0 the twist matrix. The commuting Gaudin Hamiltonians are

$$\mathsf{H}_i = \sum_{a=1}^{N} k_a \mathsf{e}^{\hbar\,(i)}_{aa} + \sum_{j \neq i} \sum_{a,b=1}^{N} \frac{\mathsf{e}^{\hbar\,(i)}_{ab} \mathsf{e}^{\hbar\,(j)}_{ba}}{x_i - x_j}, \quad i = 1, \ldots, n. \tag{1}$$

The Hamiltonians of the quantum Gaudin model [8] with the Hilbert space $\mathcal{V} = (\mathbb{C}^N)^{\otimes n}$ are restrictions of the operators (1) to the N-dimensional vector representation:

$$H_i = \hbar \sum_{a=1}^{N} k_a e^{(i)}_{aa} + \hbar^2 \sum_{j \neq i} \sum_{a,b=1}^{N} \frac{e^{(i)}_{ab} e^{(j)}_{ba}}{x_i - x_j} = \hbar g_0^{(i)} + \hbar^2 \sum_{j \neq i} \frac{P_{ij}}{x_i - x_j}, \quad i = 1, \ldots, n. \tag{2}$$

It is easy to check that the operators

$$M_a = \sum_{l=1}^{n} e_{aa}^{(l)} \tag{3}$$

commute with the Gaudin Hamiltonians: $[H_i, M_a] = 0$. Therefore, common eigenstates of the Hamiltonians can be classified according to eigenvalues of the operators M_a. Let

$$\mathcal{V} = \bigotimes_{i=1}^{n} V_i \;=\; \bigoplus_{m_1,\ldots,m_N} \mathcal{V}(\{m_a\})$$

be the decomposition of the Hilbert space of the Gaudin model $\mathcal{V}$ into the direct sum of eigenspaces for the operators M_a with the eigenvalues $m_a \in \mathbb{Z}_{\geq 0}$, $a = 1, \ldots, N$. Then the eigenstates of H_i's are in the spaces $\mathcal{V}(\{m_a\})$. Clearly, $\sum_a M_a = nI^{\otimes n}$, and hence $\sum\limits_{a=1}^{N} m_a = n$. Note also that $\sum\limits_{i=1}^{n} H_i = \sum\limits_{a=1}^{N} k_a M_a$.

A more general family of commuting Gaudin Hamiltonians was discussed in [20, 21, 22]. In [1], it was shown that the $gl(N)$ Gaudin model with vector representations at the sites admits a very simple construction of the higher commuting Hamiltonians, which is a version of the one suggested in [23] for the spin chains of the XXX-type. The main technical tool is the matrix derivative.

Let g be an element of the Lie algebra $gl(N)$ and f be any function of g, with values in $\mathrm{End}(L)$, where L is the space of any representation of $gl(N)$. The matrix derivative is defined as follows:

$$\mathsf{d}f(g) = \frac{\partial}{\partial \varepsilon} \sum_{ab} e_{ab} \otimes f(g + \varepsilon \mathsf{e}_{ba})\Big|_{\varepsilon=0}. \tag{4}$$

The right hand side belongs to $\mathrm{End}\,(\mathbb{C}^N \otimes L)$. For example:

$$\mathsf{d}\,(\operatorname{tr} g)^k = k(\operatorname{tr} g)^{k-1} I, \quad \mathsf{d} \operatorname{tr} g^k = k g^{k-1}, \quad \mathsf{d} g^k = P \sum_{i=0}^{k-1} g^i \otimes g^{k-i-1} \quad \text{for} \quad k \in \mathbb{Z}_{\geq 0}$$

When the number of tensor factors is more than two another notation is more convenient. Let $V_i \cong \mathbb{C}^N$ be copies of $\mathbb{C}^N$ and $\mathcal{V} = V_1 \otimes \ldots \otimes V_n$ as before. Then, applying the matrix derivatives to a scalar function f several times, we can embed the result into $\mathrm{End}(\mathcal{V})$ according to the formulas

$$\mathsf{d}_i f(g) = \frac{\partial}{\partial \varepsilon} \sum_{ab} e_{ab}^{(i)} f(g + \varepsilon \mathsf{e}_{ba})\Big|_{\varepsilon=0},$$

$$\mathsf{d}_{i_2} \mathsf{d}_{i_1} f(g) = \frac{\partial}{\partial \varepsilon_2} \frac{\partial}{\partial \varepsilon_1} \sum_{a_2 b_2} \sum_{a_1 b_1} e_{a_2 b_2}^{(i_2)} e_{a_1 b_1}^{(i_1)} f\left(g + \varepsilon_1 \mathsf{e}_{b_1 a_1} + \varepsilon_2 \mathsf{e}_{b_2 a_2}\right)\Big|_{\varepsilon_1=\varepsilon_2=0}$$

and so on. The lower indices of d show in which tensor factors the resulting operator acts non-trivially.

Let $\chi_\lambda(g) = \operatorname{tr}_{\pi_\lambda} g$ be the character of the representation π_λ calculated for a matrix g. It is given in terms of the Schur polynomials $s_\lambda(\mathbf{y})$ of the variables $\mathbf{y} = \{y_1, y_2, \ldots\}$, $y_k = \frac{1}{k} \operatorname{tr} g^k$:

$$\chi_\lambda(g) = s_\lambda(\mathbf{y}) = \det_{i,j=1,\ldots,\ell(\lambda)} h_{\lambda_i - i + j}(\mathbf{y}) \tag{5}$$

(the Jacobi-Trudi formula), with the complete symmetric polynomials $h_k(\mathbf{y}) = s_{(k)}(\mathbf{y})$ defined by

$$\exp\Big(\xi(\mathbf{y}, z)\Big) = \sum_{k=0}^{\infty} h_k(\mathbf{y}) z^k, \tag{6}$$

where $\xi(\mathbf{y}, z) := \sum_{k\geq 1} y_k z^k$. For example, $\chi_{(1)}(g) = \mathrm{tr}\, g$. It is convenient to set $h_k = 0$ at $k < 0$. Let $p_1, \ldots, p_N$ be eigenvalues of g realized as an element of End $(\mathbb{C}^N)$. Then $y_k = \frac{1}{k}\,(p_1^k + \ldots + p_N^k)$ and $\chi_\lambda(g) = \dfrac{\det_{1\leq i,j\leq N}\big(p_j^{\lambda_i+N-i}\big)}{\det_{1\leq i,j\leq N}\big(p_j^{N-i}\big)}$ (see [24]). This formula implies that $\chi_\emptyset(g) = s_\emptyset(\mathbf{y}) = 1$. The characters form a special class of scalar functions on the space of $N\times N$ matrices which is of primary importance for us.

Now we are ready to construct the family of commuting operators for the Gaudin model:

$$\mathsf{T}_\lambda^G(x) = \Big(1 + \frac{\hbar \mathsf{d}_n}{x - x_n}\Big) \ldots \Big(1 + \frac{\hbar \mathsf{d}_1}{x - x_1}\Big)\chi_\lambda(g_0) \tag{7}$$

By the analogy with spin chains, we will call them Gaudin transfer-matrices. The first few are: $\mathsf{T}_\emptyset^G(x) = 1$, $\mathsf{T}_{(1)}^G(x) = \mathrm{tr}\, g_0 + \sum\limits_i \dfrac{\hbar}{x - x_i}$ and

$$\mathsf{T}_{(1^2)}^G(x) = \chi_{(1^2)}(g_0) + \mathrm{tr}\, g_0 \sum_i \frac{\hbar}{x - x_i} + \sum_{i<j} \frac{\hbar^2}{(x - x_i)(x - x_j)} - \sum_i \frac{H_i}{x - x_i}, \tag{8}$$

From the last formula we see that the Gaudin Hamiltonians H_i belong to this family. For one-column diagrams, this construction agrees with Talalaev's prescription [22]. At $n = 0$ the transfer matrix is just the character: $\mathsf{T}_\lambda^{G\,(n=0)}(x) = \chi_\lambda(g_0)$.

In what follows, the normalization such that any Gaudin transfer matrix is a polynomial in x will be more convenient. In the polynomial normalization, the Gaudin transfer matrices are introduced by the formula

$$T_\lambda^G(x) = (x - x_n + \hbar \mathsf{d}_n)\, \ldots\, (x - x_1 + \hbar \mathsf{d}_1)\chi_\lambda(g_0). \tag{9}$$

All these operators commute for any x and λ.

3. The master T-operator and the KP hierarchy

Let $\mathbf{t} = \{t_1, t_2, \ldots\}$ be an infinite set of "time parameters". The master T-operator for the Gaudin model is the following generating function for the transfer-matrices $T_\lambda^G(x)$:

$$T^G(x, \mathbf{t}) = \sum_\lambda T_\lambda^G(x) s_\lambda(\mathbf{t}/\hbar). \tag{10}$$

These operators commute for different values of the parameters: $[T^G(x, \mathbf{t}), T^G(x', \mathbf{t}')] = 0$. Since $\sum_\lambda \chi_\lambda(g_0) s_\lambda(\mathbf{t}/\hbar) = \exp\Big(\frac{1}{\hbar}\sum_{k\geq 1} t_k\, \mathrm{tr}\, g_0^k\Big)$ (the Cauchy-Littlewoodn identity, see, e.g., [24]), we can define the master T-operator more explicitly as

$$T^G(x, \mathbf{t}) = (x - x_n + \hbar \mathsf{d}_n)\, \ldots\, (x - x_1 + \hbar \mathsf{d}_1) \exp\Big(\frac{1}{\hbar}\sum_{k\geq 1} t_k\, \mathrm{tr}\, g_0^k\Big). \tag{11}$$

Note that because $e^{-t\,\mathrm{tr}\, g}\mathsf{d} e^{t\,\mathrm{tr}\, g} = t$, the role of the variable t_1 is to shift $x \to x + t_1$, so that $e^{\frac{1}{\hbar}x\,\mathrm{tr}\, g_0} T^G(x, \mathbf{t})$ depends on x, t_1 only through their sum $x + t_1$.

The master T-operator contains the complete information about the spectrum of all transfer-matrices. They can be restored from it according to the formula

$$T_\lambda^G(x) = s_\lambda(\hbar\tilde{\partial})T^G(x,\mathbf{t})\Big|_{\mathbf{t}=0}, \tag{12}$$

where $\tilde{\partial} = \{\partial_{t_1}, \frac{1}{2}\partial_{t_2}, \frac{1}{3}\partial_{t_3}, \ldots\}$. In particular,

$$T_{(1)}^G(x) = \hbar\partial_{t_1}T^G(x,\mathbf{t})\Big|_{\mathbf{t}=0}, \quad T_{(1^2)}^G(x) = \frac{1}{2}(\hbar^2\partial_{t_1}^2 - \hbar\partial_{t_2})T^G(x,\mathbf{t})\Big|_{\mathbf{t}=0}. \tag{13}$$

For any $z \in \mathbb{C}$ we put $\mathbf{t} \pm \hbar[z^{-1}] := \left\{t_1 \pm \hbar z^{-1},\, t_2 \pm \frac{\hbar}{2} z^{-2},\, t_3 \pm \frac{\hbar}{3} z^{-3}, \ldots\right\}$. As we shall see below, $T^G(x, \mathbf{t} \pm \hbar[z^{-1}])$ regarded as functions of z with fixed $\mathbf{t}$ play an important role. Here we only note that equation (12) implies that $T^G(x, 0 \pm \hbar[z^{-1}])$ is the generating series for T-operators corresponding to the one-row and one-column diagrams:

$$T^G(x, \hbar[z^{-1}]) = \sum_{s\geq 0} z^{-s}T_{(s)}^G(x), \qquad T^G(x, -\hbar[z^{-1}]) = \sum_{a=0}^{N}(-z)^{-a}T_{(1^a)}^G(x). \tag{14}$$

The following theorem is the main result of [1].

Theorem 3.1 *The master T-operator (11) satisfies the bilinear identity for the $\hbar$-dependent KP hierarchy [13, 14, 19]:*

$$\oint_\infty e^{\frac{1}{\hbar}\xi(\mathbf{t}-\mathbf{t}',z)}\, T^G\left(x, \mathbf{t} - \hbar[z^{-1}]\right) T^G\left(x, \mathbf{t}' + \hbar[z^{-1}]\right) dz = 0 \quad \text{for all } \mathbf{t}, \mathbf{t}'. \tag{15}$$

The integration contour is chosen in such a way that it encircles all singularities coming from the T^G's and none of those coming from $e^{\frac{1}{\hbar}\xi(\mathbf{t}-\mathbf{t}',z)}$.

This means that each eigenvalue of the master T-operator is a tau-function of the KP hierarchy. The general bilinear identity (15) implies many bilinear funtional relations for the master T-operator (the Hirota equations). Some of them are written explicitly in [1]. Equation (10) is the expansion of the tau-function in Schur polynomials [13, 25, 26].

As soon as the Gaudin model is linked to the KP hierarchy, it is tempting to ask about the role of the other standard ingredients of the KP theory. We will be mostly interested in the Baker-Akhiezer (BA) function. According to the general scheme, the BA function corresponding to the tau-function (11) is given by the formula

$$\psi(x,\mathbf{t};z) = e^{\frac{1}{\hbar}(xz+\xi(\mathbf{t},z))}(T^G(x,\mathbf{t}))^{-1}\, T^G\big(x, \mathbf{t} - \hbar[z^{-1}]\big). \tag{16}$$

and satisfies the differential equation

$$\hbar\partial_{t_2}\psi(x,\mathbf{t};z) = \hbar^2\partial_x^2\psi(x,\mathbf{t};z) + 2u(x,\mathbf{t})\psi(x,\mathbf{t};z), \tag{17}$$

where $u(x,\mathbf{t}) = \hbar^2\partial_x^2 \log T^G(x,\mathbf{t})$. Using the definition (11), we have:

$$\psi(x,\mathbf{t};z) = z^{-N}e^{\frac{1}{\hbar}(xz+\xi(\mathbf{t},z))}(T^G(x,\mathbf{t}))^{-1}\overleftarrow{\prod_{i=1}^{n}}(x - x_i + \hbar\mathsf{d}_i)\left[\det(zI - g_0)\exp\Big(\frac{1}{\hbar}\sum_{k\geq 1} t_k \operatorname{tr} h^k\Big)\right] \tag{18}$$

(here and below $\overleftarrow{\prod_{i=1}^{n}} A_i = A_n \dots A_2 A_1$). From this formula we see that $e^{-\frac{1}{\hbar}(xz+\xi(\mathbf{t},z))}\psi(x,\mathbf{t};z)$ is a polynomial in z^{-1} of degree N. Regarded as a function of x, it is a rational function with n zeros and n poles which are simple in general position. From (11), (18) it follows that

$$\lim_{x\to\infty}\left(e^{-\frac{1}{\hbar}(xz+\xi(\mathbf{t},z))}\psi(x,\mathbf{t};z)\right) = z^{-N}\det(zI-g_0). \tag{19}$$

We also note the formula for the *stationary* BA function $\psi(x,z) := \psi(x,0;z)$, which directly follows from (18):

$$\psi(x,z) = z^{-N}e^{\frac{1}{\hbar}xz}\overleftarrow{\prod_{i=1}^{n}}\Big(1+\frac{\hbar \mathsf{d}_i}{x-x_i}\Big)\det(zI-g_0). \tag{20}$$

4. The link to the classical Calogero-Moser model

Integrability of the Gaudin model implies that eigenvalues of the master T-operator in the polynomial normalization are polynomials in the spectral parameter x of degree n:

$$T^G(x,\mathbf{t}) = e^{\frac{1}{\hbar}t_1 \operatorname{tr} g_0 + \frac{1}{\hbar}t_2 \operatorname{tr} g_0^2+\dots}\prod_{k=1}^{n}(x+t_1-x_k(t_2,t_3,\dots)). \tag{21}$$

The roots of each eigenvalue have their own dynamics in the times t_k. This dynamics is known [6, 7] to be given by the rational CM model [9]. The parameters x_i of the Gaudin model play the role of coordinates of the CM particles at $t_i = 0$: $x_j = x_j(0)$. In particular, we have $T^G(x,0) = T^G_{\emptyset}(x) = \prod_{k=1}^{n}(x-x_k)$.

Using (13), we easily obtain the formula for $\mathsf{T}^G_{(1)}(x) = T^G_{(1)}(x)/T^G_{\emptyset}(x)$ given in (8). For $\mathsf{T}^G_{(1^2)}(x)$ the second equation in (13) yields

$$\mathsf{T}^G_{(1^2)}(x) = \chi_{(1^2)}(g_0) + \operatorname{tr} g_0 \sum_i \frac{\hbar}{x-x_i} + \sum_{i<j}\frac{\hbar^2}{(x-x_i)(x-x_j)} + \frac{1}{2}\sum_i \frac{\hbar \dot{x}_i}{x-x_i},$$

where $\dot{x}_i = \partial_{t_2}x_i(t_2)\Big|_{t_2=0}$. Comparing with the third equation in (8), we conclude that the initial velocities are proportional to eigenvalues of the Gaudin Hamiltonians:

$$\hbar\dot{x}_i = -2H_i. \tag{22}$$

This unexpected connection between the quantum Gaudin model and the classical CM model was observed in [10] using different arguments.

Following [6], one can derive equations of motion for the t_k-dynamics of the x_i's performing the pole expansion of the linear problem (17) for the BA function ψ. It is convenient to denote $t_2 = t$ and put all othet times equal to 0 because they are irrelevant in this derivation. The pole ansatz for the BA function is

$$\psi = e^{\frac{1}{\hbar}(xz+tz^2)}\left(c_0(z)+\sum_{i=1}^{n}\frac{c_i(z,t)}{x-x_i(t)}\right), \tag{23}$$

where $c_0(z) = \det(I-z^{-1}g_0)$. One should substitute it into the linear problem (17) with $u = -\sum_{i=1}^{n}\frac{\hbar^2}{(x-x_i)^2}$ and cancel all the poles at the points x_i. This yields an overdetermined

Journal of Physics: Conference Series **482** (2014) 012047 doi:10.1088/1742-6596/482/1/012047

system of equations for the coefficients c_i. Their compatibility implies the the Lax representation for the CM model:

$$\dot{Y} = [T, Y], \tag{24}$$

where the $n \times n$ matrices Y, T are given by

$$Y_{ik} = -p_i \delta_{ik} - \hbar \frac{1-\delta_{ik}}{x_i - x_k}, \qquad p_i := \frac{1}{2}\dot{x}_i, \tag{25}$$

$$T_{ik} = -\delta_{ik} \sum_{j \neq i} \frac{2\hbar}{(x_i - x_j)^2} + \frac{2\hbar(1-\delta_{ik})}{(x_i - x_k)^2}. \tag{26}$$

The equations of motion are:

$$\ddot{x}_i = -8 \sum_{j \neq i} \frac{\hbar^2}{(x_i - x_j)^3}. \tag{27}$$

Set $X = X(\mathbf{t}) = \mathrm{diag}(x_1(\mathbf{t}), \ldots, x_n(\mathbf{t}))$. For the function ψ itself we then have:

$$\psi = \det(I - z^{-1} g_0)\, e^{\frac{1}{\hbar}(xz + tz^2 + t_3 z^3 + \ldots)} \Big(1 - \hbar \mathbf{1}^{\mathrm{t}} (xI - X)^{-1} (zI - Y)^{-1} \mathbf{1}\Big), \tag{28}$$

where $\mathbf{1} = (1, 1, \ldots, 1)^{\mathrm{t}}$ is the n-component vector. As is well known (and easy to check), the matrices X, Y satisfy the commutation relation

$$[X, Y] = \hbar(I - \mathbf{1} \otimes \mathbf{1}^{\mathrm{t}}) \tag{29}$$

(here $\mathbf{1} \otimes \mathbf{1}^{\mathrm{t}}$ is the $n \times n$ matrix of rank 1 with all entries equal to 1).

The matrix Y is the Lax matrix for the CM model. As is seen from (24), the time evolution preserves its spectrum, i.e., the coefficients $\mathcal{J}_k$ of the characteristic polynomial

$$\det(zI - Y(t)) = \sum_{k=0}^{n} \mathcal{J}_k z^{n-k} \tag{30}$$

are integrals of motion. The highest integral, $\mathcal{J}_n$, was found explicitly in [27], where a recurrence procedure for finding all other integrals of motion was also suggested. In fact this procedure is equivalent to the following explicit expression for the characteristic polynomial:

$$\det\big(zI - Y(t)\big) = \exp\Big(\sum_{i<j} \frac{\hbar^2 \partial_{p_i} \partial_{p_j}}{(x_i - x_j)^2}\Big) \prod_{l=1}^{n} (z + p_l). \tag{31}$$

Note that this expression is well-defined because the sum obtained after expansion of the exponential function in the r.h.s. contains a finite number of non-zero terms.

One can see that eigenvalues of the Lax matrix Y are the same as eigenvalues of the twist matrix g_0 (with appropriate multiplicities). Indeed, let us compare expansions of (20) and (28) at large x. From (20) we have:

$$\psi(x, z) = \det(I - z^{-1} g_0)\, e^{\frac{1}{\hbar} xz} \left(1 - \frac{\hbar}{x} \sum_i \sum_a \frac{e_{aa}^{(i)}}{z - k_a} + O(x^{-2})\right).$$

Using the commutation relation (29) it is easy to check that for any $k \geq 0$ it holds $\mathbf{1}^{\mathrm{t}} Y^k \mathbf{1} = \mathrm{tr}\, Y^k$. Taking this into account, we can expand (28) at $\mathbf{t} = 0$:

$$\psi(x, z) = \det(I - z^{-1} g_0)\, e^{\frac{1}{\hbar} xz} \left(1 - \frac{\hbar}{x} \mathrm{tr} \frac{1}{z - Y_0} + O(x^{-2})\right),$$

Journal of Physics: Conference Series **482** (2014) 012047 doi:10.1088/1742-6596/482/1/012047

where $Y_0 := Y(0)$. Therefore, we conclude that $\mathrm{tr}\,\frac{1}{zI - Y_0} = \sum_i \sum_a \frac{e_{aa}^{(i)}}{z - k_a}$ and, since $\mathrm{tr}\,(zI - Y_0)^{-1} = \partial_z \log\det(zI - Y_0)$, we have

$$\det(zI - Y_0) = \prod_{a=1}^{N} (z - k_a)^{\sum_{i=1}^{n} e_{aa}^{(i)}} = \prod_{a=1}^{N} (z - k_a)^{M_a}, \tag{32}$$

where M_a is the operator (3). Hence we see that the M_a is the "operator multiplicity" of the eigenvalue k_a. In the sector $\mathcal{V}(\{m_a\})$ the multiplicity becomes equal to m_a. This argument allows one to prove the following important statement:

Theorem 4.1 *The eigenvalues of the Lax matrix Y of the CM model are numbers from the set $\{k_1, k_2, \ldots, k_N\}$ (the eigenvalues of the twist matrix) with multiplicities $m_a \geq 0$ such that $m_1 + \ldots + m_N = n$, with the m_a's being eigenvalues of the operators M_a.*

The hamiltonian form of equations of motion (27) is $\begin{pmatrix} \dot{x}_i \\ \dot{p}_i \end{pmatrix} = \begin{pmatrix} \partial_{p_i}\mathcal{H}_2 \\ -\partial_{x_i}\mathcal{H}_2 \end{pmatrix}$ with the Hamiltonian

$$\mathcal{H}_2 = \mathrm{tr}\, Y^2 = \sum_i p_i^2 - \sum_{i<j} \frac{2\hbar^2}{(x_i - x_j)^2}\,. \tag{33}$$

This result can be extended to the whole hierarchy [7]:

$$\begin{pmatrix} \partial_{t_k} x_i \\ \partial_{t_k} p_i \end{pmatrix} = \begin{pmatrix} \partial_{p_i}\mathcal{H}_k \\ -\partial_{x_i}\mathcal{H}_k \end{pmatrix}, \qquad \mathcal{H}_k = \mathrm{tr}\, Y^k. \tag{34}$$

The $\mathcal{H}_k$'s are higher integrals of motion for the CM model. They are known to be in involution [27, 28, 29]. This agrees with commutativity of the KP flows. The integrals $\mathcal{H}_k$ are connected with the integrals $\mathcal{J}_k$ introduces in (30) by Newton's formula [24] $\sum_{k=0}^{n} \mathcal{J}_{n-k}\mathcal{H}_k = 0$ (we have set $\mathcal{H}_0 = \mathrm{tr} Y^0 = n$).

The results of [7] imply an explicit determinant representation of the tau-function. It is easy to adopt it for the master T-operator $T^G(x, \mathbf{t})$ (21). Let $X_0 = X(0)$ be the diagonal matrix $X_0 = \mathrm{diag}(x_1, x_2, \ldots, x_n)$, where $x_i = x_i(0)$ and Y_0 be the Lax matrix (25) at $\mathbf{t} = 0$, with the diagonal elements being proportional to the Gaudin Hamiltonians $H_i = -\hbar p_i(0)$:

$$Y_0 = \begin{pmatrix} \frac{H_1}{\hbar} & \frac{\hbar}{x_2 - x_1} & \frac{\hbar}{x_3 - x_1} & \cdots & \frac{\hbar}{x_n - x_1} \\ \frac{\hbar}{x_1 - x_2} & \frac{H_2}{\hbar} & \frac{\hbar}{x_3 - x_2} & \cdots & \frac{\hbar}{x_n - x_2} \\ \vdots & \vdots & \vdots & \ddots & \vdots \\ \frac{\hbar}{x_1 - x_n} & \frac{\hbar}{x_2 - x_n} & \frac{\hbar}{x_3 - x_n} & \cdots & \frac{H_n}{\hbar} \end{pmatrix}. \tag{35}$$

Theorem 4.2 *The master T-operator for the Gaudin model is given by*

$$T^G(x, \mathbf{t}) = e^{\frac{1}{\hbar}\sum_{k\geq 1} t_k \mathrm{tr}\, h^k} \det\left(xI - X_0 + \sum_{k\geq 1} k t_k Y_0^{k-1} \right). \tag{36}$$

It follows from the above arguments that eigenvalues of the Gaudin Hamiltonians H_i, $i = 1, \dots, n$ (2), can be found in the framework of the classical CM system with n particles. Namely, the spectrum of H_i's in the space $\mathcal{V}(\{m_a\})$ is determined by the conditions

$$\operatorname{tr} Y_0^j = \sum_{a=1}^{N} m_a k_a^j \quad \text{for all} \quad j \geq 1, \tag{37}$$

i.e., given the initial coordinates x_i and the action variables $\mathcal{H}_j = \operatorname{tr} Y_0^j$ one has to find possible values of the initial momenta $p_i = -H_i/\hbar$. Taking into account equations (30) and (31), we can represent the equations for H_i in the form of the equality

$$\exp\Big(\hbar^4 \sum_{i<j} x_{ij}^{-2} \partial_{H_i} \partial_{H_j}\Big) \prod_{l=1}^{n} (z - \hbar^{-1} H_l) = \prod_{a=1}^{N} (z - k_a)^{m_a}, \qquad x_{ij} \equiv x_i - x_j \tag{38}$$

which has to be satisfied identically in z. This identity is equivalent to n algebraic equations for n quantities $H_1, \dots, H_n$.

We see that the eigenstates of the Gaudin Hamiltonians correspond to the intersection points of two Lagrangian submanifolds: one obtained by fixing the x_i's and the other obtained by fixing the $\mathcal{H}_i$'s, with values of the latter being determined by eigenvalues of the twist matrix. This purely classical prescription appears to be equivalent to the Bethe ansatz solution and solves the spectral problem for the quantum Gaudin Hamiltonians.

Example. Consider the vector $\mathsf{v}_a \in \mathbb{C}^N$ with components $(\mathsf{v}_a)_b = \delta_{ab}$. Since $P_{ij}(\mathsf{v}_a)^{\otimes n} = (\mathsf{v}_a)^{\otimes n}$, the vector $(\mathsf{v}_a)^{\otimes n}$ is an eigenstate for the Gaudin Hamiltonians H_i with the eigenvalues $k_a + \sum_{j \neq i} \frac{\hbar}{x_i - x_j}$. It is also an eigenvector for the operators M_b with eigenvalues $m_b = n\delta_{ab}$. The matrix (35) in this case is the $n{\times}n$ Jordan block with the only eigenvector 1 with eigenvalue k_a and $\operatorname{tr} Y_0^j = n k_a^j$.

Acknowledgments

Discussions with A.Alexandrov, A.Gorsky, V.Kazakov, S.Khoroshkin, I.Krichever, S.Leurent, M.Olshanetsky, A.Orlov, T.Takebe, Z.Tsuboi, and A.Zotov are gratefully acknowledged. Some of these results were reported at the International conference "Physics and Mathematics of Nonlinear Phenomena" (22-29 June 2013, Gallipoli, Italy). The author thanks the organizers and especially professor B.Konopelchenko for the invitation and support. This work was supported in part by RFBR grant 12-01-00525, by joint RFBR grants 12-02-91052-CNRS, 12-02-92108-JSPS and by Ministry of Science and Education of Russian Federation under contract 8207 and by grant NSh-3349.2012.2 for support of leading scientific schools.

References

[1] A. Alexandrov, S. Leurent, Z. Tsuboi and A. Zabrodin, *The master T-operator for the Gaudin model and the KP hierarchy* [arXiv:1306.1111].

[2] A. Alexandrov, V. Kazakov, S. Leurent, Z. Tsuboi and A. Zabrodin, *Classical tau-function for quantum spin chains*, JHEP **1309** (2013) 064 [arXiv:1112.3310].

[3] A. Zabrodin, *The master T-operator for vertex models with trigonometric R-matrices as classical tau-function*, Teor. Mat. Fys. **171:1** (2013) 59-76 (Theor. Math. Phys. **174** (2013) 52-67) [arXiv:1205.4152]; A. Zabrodin, *Hirota equation and Bethe ansatz in integrable models*, Suuri-kagaku Journal (in Japanese), Number 596 (2013) 7-12 [arXiv:1211.4428].

[4] V. Kazakov, S. Leurent and Z. Tsuboi, *Baxter's Q-operators and operatorial Bäcklund flow for quantum (super)-spin chains*, Commun. Math. Phys. **311** (2012) 787-814 [arXiv:1010.4022].

[5] H. Airault, H. P. McKean, and J. Moser, *Rational and elliptic solutions of the Korteweg-De Vries equation and a related many-body problem*, Commun. Pure Appl. Math., **30** (1977) 95-148.

[6] I.M. Krichever, *Rational solutions of the Kadomtsev-Petviashvili equation and integrable systems of N particles on a line*, Funct. Anal. Appl. **12:1** (1978) 59-61;
I.M. Krichever, *Rational solutions of the Zakharov-Shabat equations and completely integrable systems of N particles on a line*, J. Sov. Math., **21:3** (1983) 335-345.

[7] T. Shiota, *Calogero-Moser hierarchy and KP hierarchy*, J. Math. Phys. **35** (1994) 5844-5849.

[8] M. Gaudin, *Diagonalisation d'une classe d'hamiltoniens de spin*, J. de Phys. **37** (1976), no. 10 1087-1098.

[9] F. Calogero, *Solution of the one-dimensional N-body problems with quadratic and/or inversely quadratic pair potentials*, J. Math. Phys. **12** (1971) 419-436;
J. Moser, *Three integrable hamiltonian systems connected with isospectrum deformations*, Adv. Math. **16** (1976) 354-370.

[10] E. Mukhin, V. Tarasov and A. Varchenko, *Gaudin Hamiltonians generate the Bethe algebra of a tensor power of vector representation of gl_N*, St. Petersburg Math. J. **22** (2011) 463-472 [arXiv:0904.2131];
E. Mukhin, V. Tarasov and A. Varchenko, *KZ characteristic variety as the zero set of classical Calogero-Moser Hamiltonians*, SIGMA **8** (2012) 072 (11 pages) [arXiv:1201.3990].

[11] N. Nekrasov, A. Rosly and S. Shatashvili, "Darboux coordinates, Yang-Yang functional, and gauge theory", Nucl. Phys. Proc. Suppl. **216** (2011) 69-93 [arXiv:1103.3919].

[12] D. Gaiotto and P. Koroteev, *On three dimensional quiver gauge theories and integrability*, JHEP **05** (2013) 126 [arXiv:1304.0779].

[13] M. Sato and Y. Sato, *Soliton equations as dynamical systems on infinite dimensional Grassmann manifold*, Lect. Notes in Num. Appl. Anal. **5** (1982) 259-271.

[14] E. Date, M. Jimbo, M. Kashiwara and T. Miwa, *Transformation groups for soliton equations*, in "Nonlinear integrable systems – classical and quantum", eds. M. Jimbo and T. Miwa, World Scientific, pp. 39-120 (1983);
M. Jimbo and T. Miwa, *Solitons and infinite dimensional Lie algebras*, Publ. RIMS, Kyoto Univ. **19** (1983) 943-1001.

[15] I. Krichever, O. Lipan, P. Wiegmann and A. Zabrodin, *Quantum Integrable Models and Discrete Classical Hirota Equations*, Commun. Math. Phys. **188** (1997) 267-304 [arXiv:hep-th/9604080].

[16] A. Zabrodin, *Discrete Hirota's equation in quantum integrable models*, Int. J. Mod. Phys. **B11** (1997) 3125-3158;
A. Zabrodin, *Hirota equation and Bethe ansatz*, Teor. Mat. Fyz., **116** (1998) 54-100 (English translation: Theor. Math. Phys. **116** (1998) 782-819).

[17] V. Kazakov, A. S. Sorin and A. Zabrodin, *Supersymmetric Bethe ansatz and Baxter equations from discrete Hirota dynamics*, Nucl. Phys. B **790** (2008) 345-413 [arXiv:hep-th/0703147];
A. Zabrodin, *Bäcklund transformations for difference Hirota equation and supersymmetric Bethe ansatz*, Teor. Mat. Fyz. **155** (2008) 74-93 (English translation: Theor. Math. Phys. **155** (2008) 567-584) [arXiv:0705.4006].

[18] R. Hirota, *Discrete analogue of a generalized Toda equation*, J. Phys. Soc. Japan **50** (1981) 3785-3791.

[19] K. Takasaki and T. Takebe, *Integrable hierarchies and dispersionless limit*, Rev. Math. Phys. **7** (1995) 743-808;
K. Takasaki and T. Takebe, *$\hbar$-Dependent KP hierarchy*, Theor. Math. Phys. **171** (2012) 683-690 [arXiv:1105.0794].

[20] B. Feigin, E. Frenkel and N. Reshetikhin, *Gaudin model, Bethe Ansatz and critical level*, Commun. Math. Phys. **166** (1994) 27-62.

[21] B. Enriques and V. Rubtsov, *Hitchin systems, higher Gaudin operators and R-matrices*, Math. Res. Lett. **3** (1996) 343-357.

[22] D. Talalaev, *Quantization of the Gaudin system*, Funct. Anal. Appl. **40**(1) (2006) 86-91.

[23] V. Kazakov and P. Vieira, *From characters to quantum (super)spin chains via fusion*, JHEP **0810** (2008) 050 [arXiv:0711.2470 [hep-th]].

[24] I. Macdonald, *Symmetric functions and Hall polynomials*, 2nd ed., Oxford University Press, 1995.

[25] A. Orlov and T. Shiota, *Schur function expansion for normal matrix model and associated discrete matrix models*, Phys. Lett. **A343** (2005) 384-396.

[26] V. Enolski and J. Harnad, *Schur function expansions of KP tau functions associated to algebraic curves*, Uspekhi Mat. Nauk **66:4** (2011) 137-178 (Russian Math. Surveys **66:4** (2011) 767-807) [arXiv:1012.3152].

[27] K. Sawada and T. Kotera, *Integrability and a solution for the one-dimensional N-particle system with inversely quadratic pair potential*, J. Phys. Soc. Japan, **39** (1975) 1614-1618

[28] S. Wojciechowski, *New completely integrable Hamiltonian systems of N particles on the real line*, Phys. Lett. **59A** (1976) 84-86.

[29] M. Olshanetsky and A. Perelomov, *Classical integrable finite dimensional systems related to Lie algebras*, Phys. Reps. **71** (1981) 313-400.

AUTHOR INDEX

Institute of Physics
Dirac House, Temple Back
Bristol BS1 6BE UK

CURRAN ASSOCIATES INC
proceedings.com